T0192705

In order for biodiversity to be conserved, it is important to know how and where diverse assemblages of plants and animals exist, to understand the effects of human impacts on them and to find the means by which these impacts can be lessened and even reversed.

While tropical systems are known to be amongst the most diverse and most threatened globally, tropical freshwater systems have been neglected, and the tremendous variety of fish, amphibians, invertebrates and plants that live in them is poorly known yet seriously threatened.

This comprehensive book brings together a wealth of information on the fish of tropical African systems, and discusses how these systems evolved, what hold them together and what is tearing them apart. It will be an important reference work not only for those interested in fish, but for all those concerned with biodiversity conservation anywhere.

Biodiversity dynamics and conservation

the freshwater fish of tropical Africa

Biodiversity dynamics and conservation

the freshwater fish of tropical Africa

Christian Lévêque

Directeur de Recherche
Institut Français de Recherche Scientifique
pour le Développement en Coopération (ORSTOM)

CAMBRIDGE
UNIVERSITY PRESS

CAMBRIDGE UNIVERSITY PRESS
Cambridge, New York, Melbourne, Madrid, Cape Town, Singapore, São Paulo

Cambridge University Press
The Edinburgh Building, Cambridge CB2 2RU, UK

Published in the United States of America by Cambridge University Press, New York

www.cambridge.org
Information on this title: www.cambridge.org/9780521570336

First published 1997
This digitally printed first paperback version 2006

A catalogue record for this publication is available from the British Library

Library of Congress Cataloguing in Publication data

Lévêque, C.
    Biodiversity dynamics and conservation : the freshwater fish of
tropical Africa / Christian Lévêque.
        p. cm.
    Includes bibliographical references and index.
    ISBN 0 521 57033 6 (hardback)
    1. Freshwater fishes – Africa. 2. Freshwater fishes – Tropics.
3. Fish communities – Africa. 4. Fish communities – Tropics.
5. Biological diversity conservation – Africa. 6. Biological
diversity conservation – Tropics. I. Title.
QL635.A1L48   1997
597.092′96 – dc20   96-19436 CIP

ISBN-13 978-0-521-57033-6 hardback
ISBN-10 0-521-57033-6 hardback

ISBN-13 978-0-521-03197-4 paperback
ISBN-10 0-521-03197-4 paperback

Contents

This book is dedicated to

Alicia and Annie

for so much time stolen
from the private side of life.

Introduction

Conscious of the intrinsic value of biological diversity and of the ecological, genetic, social, economic, scientific, educational, cultural, recreational and aesthetic values of biological diversity and its components,

Conscious also of the importance of biological diversity for evolution and for maintaining life sustaining systems of the biosphere,

Affirming that the conservation of biological diversity is a common concern of humankind . . .

Determined to conserve and sustainably use biological diversity for the benefit of present and future generations,

Article 1. Objectives

The objectives of this Convention, to be pursued in accordance with its relevant provisions, are the conservation of biological diversity, the sustainable use of its components and the fair and equitable sharing of the benefits arising out of the utilisation of genetic resources, including by appropriate access to genetic resources and by appropriate transfer of relevant technologies, taking into account all rights over those resources and to technologies, and by appropriate funding.

Preamble of the Convention on Biological Diversity, Rio, 1992

This book takes a holistic view of biological diversity, from the molecular, through the organismal to the ecosystem levels of organisation. This is done using the freshwater fish of tropical Africa as a model. Some readers may find this a surprising approach, but I hope that this book will be read not just by fish enthusiasts but also by those with an interest in all aspects of biodiversity and its conservation, even if they are not fish specialists.

Biological diversity (shortened to biodiversity) was defined (Huntley, 1989) as 'the variety and variability of living organisms. This includes the genetic variability within species and their populations, the variety of species and their life forms, the diversity of the complexes of associated species and of their interactions, and of the ecological processes which they influence or perform'. But it also includes the variety of life-history styles, and the number of interactions between organisms and their environment (Bruton, 1990a). In the Convention on Biodiversity, biological diversity means 'the variability among living organisms from all sources including, inter alia, terrestrial, marine and other aquatic ecosystems and the ecological complexes of which they are part; this includes diversity within species, between species and of ecosystems'.

The topic of biological diversity receives significant attention because it can be considered as a global resource to be preserved. The concept arose because of the drastically accelerated transformation of natural landscapes that is taking place around the world, with associated loss of species and reduction in genetic diversity (Wilson & Peters, 1988; Solbrig, 1991*a*). As a consequence, the genetic, species and ecosystem diversity that was created by the long history of coevolution and adaptation to changes is now seriously threatened by the overall negative impact of many human activities associated with economic development. In the long term, man-induced climate changes may completely modify aquatic ecosystem dynamics through alteration of their water budgets or temperature, but the most serious immediate threats stem from management of aquatic systems, pollution, overexploitation of natural resources and species invasions or introductions.

The central question in biodiversity conservation is the assessment of human impacts on the different components of the biological hierarchy, and the search for means, both technical and socio–economic, to reduce these impacts and eventually restore damaged systems. The conservation of biodiversity is therefore a major challenge for the coming decades. It cannot be addressed through technological innovation alone and a wide array of approaches will be necessary, including research, education, *ex situ* collections, economic incentives and the establishment of protected areas. *In situ* protection is especially important for aquatic systems because their conditions are often difficult to replicate elsewhere, and theories already developed for terrestrial species conservation will be useful. It should also be recalled that the integrity of ecosystems must be maintained in order to preserve species *in situ*, in their habitats, and to ensure conservation of various ecological services. The conservation of biodiversity requires a respect for and conservation of the integrity of ecosystems and linkage between diverse ecosystems. There is an urgent need to provide the scientific rationale that will allow the proposal of effective strategies for comprehensive conservation.

In its broadest sense, from gene to species to ecosystems, biological diversity is an umbrella concept that covers all the ecological, sociological, ethical and economical aspects of relationships between man and the living biosphere, and provides a unifying approach to the different questions dealing with these relationships. As a result, it should be clear that biodiversity does not just mean species inventories for which we already have an established vocabulary, such as species number or species richness. Nor is biodiversity just a magic word that will reopen the doors (which have been closed for a long time) to funding and recognition for the importance of descriptive and taxonomic studies. Moreover, biodiversity should not be thought of as an alternative term for biological resources, because the debate about biological diversity is also part of the philosophical debate about the relationship between Man and nature (Lévêque, 1994*b*). For some, nature exists to serve human beings who must control it to use its resources. For others, people are part of nature and must obey the same ecological laws as other species, thus global ecological processes that guarantee survival of the human species must be protected. Abuses of the term biodiversity will certainly result in misunderstanding, and probably in a disaffection with the concept, and thus less concern for the conservation of biodiversity.

Why fish ?

While a great deal of attention has been given to the loss of biodiversity in tropical rain forests, or in coastal areas, the diversity of and within freshwaters has been widely neglected. Freshwaters too contain a tremendous diversity of fish, amphibians, invertebrates and aquatic plants that are among the most poorly known on Earth and now seriously threatened (Global Biodiversity Strategy, 1992). Teleost fish, which evolved during the Mesozoic (230–65 million years BP), outnumber all other vertebrates put together. They are now represented by approximately 22 000 living species (Nelson, 1994) of which about 9 to 10 000 are freshwater species. Some 3000 species inhabit the inland waters of Africa (Daget *et al.*, 1984, 1986*a*, *b*, 1991).

Why freshwater ?

Because fish cannot easily move from one aquatic system to another, they have to adapt to changes or die. This makes them very appropriate indicators of environmental conditions and trends in aquatic biodiversity as a whole, because they also have a major impact on the distribution and abundance of other aquatic organisms, both directly through predation or indirectly through nutrient releases or mechanical effects. There is little doubt that freshwater fishes represent the most threatened set of vertebrates. One-quarter of all vertebrate biodiversity is concentrated into less than 0.01% of the planet's water (Stiassny & Raminosoa, 1994), that is to say about 1% of the dry land. To this one should also add the other vertebrate groups or species that are dependent upon the existence of freshwater habitats, such as many amphibians, reptiles, birds, etc. This remarkable concentration of vertebrate biodiversity is also the most vulnerable, given that freshwater aquatic resources have undergone severe deterioration on a global scale and we are already losing freshwater species at an unprecedented rate. The many fish species now threatened with extinction (Moyle & Leidy, 1992) have so far received little attention compared to more charismatic fauna such as elephants, pandas or whales. Out of sight, out of mind, one-third of the native freshwater species of North America are extinct or endangered to some degree. For example, all the native fishes of the valley of Mexico are already extinct. A recent survey in Malaysia found fewer than half of the 266 fish species previously known from that country. On the island of Singapore, 18 out of 53 species of freshwater fish collected in 1934 could not be located in exhaustive searches only 30 years later (Global Biodiversity Strategy, 1992). The introduction of *Lates* into Lake Victoria, long-term eutrophication of the lake and the use of new fishing gear, has led, within ten years, to the virtual disappearance of some 200 species of endemic cichlids (see Chapter 14).

The distribution of biodiversity in freshwater systems is fundamentally different from that of most marine or terrestrial ecosystems. River drainages or lakes have been compared to islands. Most freshwater fish are only able to move naturally from one 'island' to another when climatic or geological events allow connections between drainages. This has a number of implications. First, freshwater biodiversity is usually highly localised and high levels of endemism are particularly evident in lakes isolated for millions of years, such as some of the Great Lakes of East Africa. There may also be physical barriers, such as waterfalls, within a system that contribute to the appearance and maintenance of locally evolved forms. Second, this patchy distribution results in great genetic variability between subpopulations. Third, freshwater species must survive *in situ* or they will disappear when exposed to climatic and ecological changes or when threatened by the impact of human activities.

Why Africa ?

In tropical Africa the human species has existed alongside other species for longer than anywhere else. It is a very old continent containing some of the oldest and largest lakes and rivers in the world, as well as many old fish lineages.

Only occasionally does this book stray beyond the tropical regions of Africa, because most aspects of biodiversity can be adequately illustrated within the tropical zone with which I am most familiar. Even within that zone there is an unfortunate division between the human communities. This is to some extent reflected in the different foci of their ichthyologists – lakes and cichlids in English speaking East Africa; rivers and floodplains in French speaking West Africa. This book is an attempt to link these two cultures and to make a synthesis that will facilitate co-operation, in what must be an all out effort, to conserve the amazing biodiversity of African freshwater fish so that it continues to be a rich resource for the people of Africa and their sustainable development.

Why this particular approach to the broader diversity issue ?

I have tried to focus attention on the role of past events in shaping the present fish species diversity and community structure. Fish will be viewed as a heritage of evolution in a changing environment, living in habitats whose characteristics and dynamics are the result of present day climate and geomorphology.

Genetic population structure and variability, life-history traits and physiological capacities, all shape and influence the way species interact with their abiotic and biotic environments. The ability of organisms to adapt to extrinsic constraints is regulated by intrinsic physiological or morphological constraints, and is controlled by genetic and phylogenetic histories. Different life-history strategies may well be developed by organisms that complete their life cycles (growth, survival, reproduction) under changing environmental constraints. Thus, the diversity contained within species is the ultimate source of biodiversity at higher levels, as well as the major determinant of how the species will be able to respond to environmental disturbances, either of natural or anthropogenic origin. It is not yet clear what are the actual processes really acting to maintain diversity (Signor, 1990).

Nowadays, our aquatic heritage (fish and their habitats) is subjected to different impacts, both on a large scale (global change) and at a regional or local level (e.g. pollution, habitat modifications, fishing, introduction of alien species). Fish conservation, either *in situ* or *ex situ*, raises many specific problems, and challenges water management as a whole. As an important source of food, fish are of particular interest and have a future in the world of biotechnology. They have exerted a strong influence on humans who developed not only a mystic relationship with this group of vertebrates, but also a very specific technological culture with an amazing variety of fishing equipment. Fish biological diversity in the broadest sense is, therefore, a multidisciplinary approach to a group of animals in their diversity of life and diversity of their relationships with environment and people.

It should be clear that the ultimate goal of research is to provide tools and methods that can be used to protect biodiversity. This book reviews the current state of our knowledge about African freshwater fish, as well as offering a conceptual framework of their role and function in ecosystems. It also provides some guidelines for better long-term use of biological resources within the framework of sustainable management of freshwater systems.

However, biodiversity will be conserved effectively only in a stable world in which major ecological and economic problems have been resolved. During the last few years scientists and non-governmental organisations have raised the biodiversity problem, defined the issue's scope and generated world-wide public interest that lead to an international treaty. Having achieved this remarkable political progress, they are now faced with an unforeseen situation where the causes of biodiversity loss, as well as solutions, are to be found largely in the realm of the social, economic and political sciences. Consequently, decline in biodiversity should be regarded as a social problem. A better scientific understanding of socio-economic forces is central to appreciating the nature and magnitude of the problems affecting biodiversity and to devising enduring solutions to them. Biological insights alone are not enough. In such a perspective, the future role of natural scientists in addressing the biodiversity problem is not clearly defined, but, in any case, is not nearly as straightforward as their role has been to date. However, social science research on environmental issues requires good biophysical science information, and the linkage between both is essential to assesss the impacts of biodiversity loss. The field of research now opened up in such a collaboration is certainly one of the most exciting challenges to science and society in the coming decades.

Acknowledgements

Without doubt this book was born from the fascination and unravelling pleasure of entering the world of biodiversity. The link between science and society, the fundamental interplay between environment and development, as well as the (re)discovery of life-history strategies developed by freshwater fishes, have been a constant stimulation. For me, coming back time after time to the 'fish world' was also particularly refreshing after days, sometimes weeks, of more or less boring administrative meetings, during which I often dreamed of the better times when I was out in the field, collecting, measuring, weighting and occasionally sexing baskets of more or less fresh fish.

Looking back on this experience and its formative role in developing my ideas, which led to this book, I would like to thank Professor Jacques Daget, who trained me in ichthyology and aquatic ecology. I am very much indebted to one of the pioneers of African freshwater ecology and ichthyology for his willingness to share his great experience and his readiness to pursue discussions and foster interest.

I would also like to thank all my colleagues from ORSTOM who participated in the extraordinary scientific adventure that was the multidisciplinary study of the Lake Chad basin during the late 1960s. My special thanks also to colleagues of the ichthyological team in Bouaké, Côte d'Ivoire (Jean-Jacques Albaret, Rémy Bigorne, Bernard de Mérona, Didier Paugy) with whom quite a lot of beers were downed between electrofishing 'parties' during the 1970s. I would like also to acknowledge my younger ORSTOM colleagues (J. F. Agnèse, J. F. Guégan, B. Hugueny) who participated in the PEDALO venture (Poissons des Eaux Douces d'Afrique de L'Ouest) during the 1980s, and colleagues from the Paris National Museum of Natural History where I spent many happy years in a friendly atmosphere.

But ichthyology also provided the opportunity to meet and to discuss with foreign colleagues, including close neighbours from Tervuren, colleagues from the Natural History Museum, London and from the J. L. B. Smith Institute, Grahamstown. I have particularly enjoyed collaboration with many African colleagues in the course of the Aquatic Monitoring Programme of the Onchocerciasis Control Programme in West Africa, especially K. Abban, K. Traoré and L. Yaméogo.

My particular thanks to colleagues who gave their time to revising preliminary drafts of different chapters: J. F. Agnèse, B. Hugueny, M. Legendre, R. Lowe McConnell, J. Maley, B. de Mérona, D. Paugy, D. Ponton, L. Pouyaud and G. Teugels. My special thanks to J. F. Guégan who read most of the chapters and made valuable comments on the manuscript, as well as to my old friend Bernhard Statzner, who encouraged me in the early stages of the project and made very useful comments on the preliminary manuscripts.

I am also grateful to ORSTOM (Institut Français de Recherches Scientifiques pour le Développement en Coopération), which gave me the opportunity to spend so many years working on African fishes, and provided facilities for writing and publishing this book.

Ultimately, this book would have never reached the final editorial stage without the expertise and encouragement of my old friends Mary and Pat Morris. I am very indebted to both for their advice in finalising the present version and their efforts to make the original manuscript readable by English speaking people.

I The diversity of African freshwater fish

1   The diversity and variability of freshwater ecosystems in tropical Africa

For many countries in tropical Africa, having in mind the abundance of their
lakes and rivers, limnology provides great opportunities for research and teaching
in some basic biological subjects, such as evolution of faunas and floras in the
light of geological and climatic history, biogeography, ecological problems in a
wide range of ecosystems, energy and nutrient cycles, photosynthesis, the seasonal
cycles of primary and secondary production under tropical conditions, and the
physiological and biochemical adaptation of organisms to peculiar conditions
(high salinity, periodic dessication, lack of oxygen).

Beadle, 1981

The principal difference between inland water and marine ecosystems is their patchy nature. As a result, each freshwater catchment may be considered as an island with its own environmental characteristics. This situation has allowed allopatric speciation to occur. Hundred of species are known only from one drainage or lake system, which may help to explain why tropical freshwater fish are so diverse but also explains why freshwater fish are particularly threatened by environmental change. Any serious threat to that ecosystem may result in the disappearance of endemic species.

Africa is an ancient continent. Her inland waters have a much longer history than those of the temperate zone, which were reformed at the end of the Pleistocene. They offer an enormous number and variety of habitats that have resulted from the transformations, by tectonic action during the past 20 million years, of less varied systems (see Chapter 6), and the present distribution of fish species across the continent has been determined largely by these events rather than by the differences in the conditions to be found in the various water systems. This has been demonstrated by the recent successful introduction of species of fish into water systems from which they

were previously absent. But it is also obvious that there are features of several of the water systems that make them favourable or unfavourable for certain species and that there are some systems that can support only a few species.

In this book, it is not possible to go into details of the structure and the functioning of all African freshwater systems. Nevertheless, biodiversity includes the diversity of ecosystems and for a better understanding of the species diversity, distribution and richness of African freshwater fish it is important to know what are the principal physical or chemical constraints encountered by fish species in the ecosystems of tropical Africa. The constraints may be perennial or seasonal and help to explain the population dynamics of the fish and the structure of fish assemblages.

Broad geographic and climatic patterns

In terms of relief, the African continent can broadly be divided into two: Low and High Africa (Fig. 1.1). Low Africa is composed largely of sedimentary basins and upland plains below 600 m a.s.l. (above sea level) (the

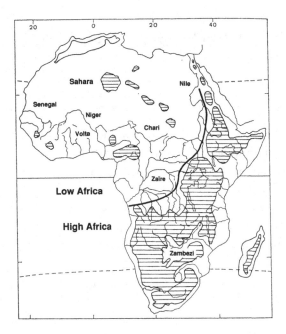

Fig. 1.1. The general topography of Africa showing the main watersheds, the areas over 1000 m and the approximate division between High and Low Africa (from Beadle, 1981).

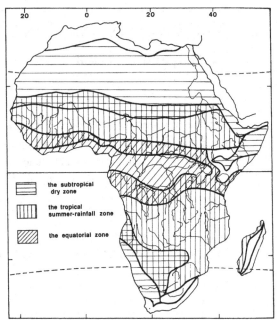

Fig. 1.2. The major climatic zones of tropical Africa (modified from Walter et al., 1975)

catchment areas of the Nile, Chari, Niger, Senegal, Volta and Zaïre rivers). Within this area, there are only a few mountainous regions such as, for example, Jebel Marra (3042 m) between the Chari and Nile basins, and the headwaters of the Niger (the Guinean highlands, 1853 m). In contrast, High Africa to the south and east is mainly above 1000 m. In East Africa, two major faultlines running approximately NE/SW from Ethiopia to Zimbabwe provide some of the highest mountains. This Rift Valley divides into two branches, the Western and the Eastern Rift Valleys, which are trenches lying some 1000 m below the top of steep escarpments and bearing signs of intense volcanic activity. They are ideal traps for freshwater, and most of the East African lakes are located in these depressions with the exception of Lake Victoria, which is situated on the uplifted plateau between the two Rift Valleys.

The climate is of utmost importance in determining the distribution of aquatic systems. Since the African continent stretches symmetrically across approximately 35° of latitude on either side of the Equator, it is subject to a very great range of climate, especially with respect to temperature, rainfall and wind, all of which have considerable biological influence. However, the pattern of climate is not regularly graded with latitude: the northern and southern dry desert belts are striking interruptions, and to some extent mark the limits of intertropical Africa. In the tropics, high mountains and plateaux have temperatures much below that of the surrounding country.

Within tropical Africa there are three major climatic types (Walter et al., 1975): (i) equatorial: hot, humid or with two rainy seasons; (ii) tropical: hot with summer rain; and (iii) subtropical: hot and arid (Fig. 1.2). In the north and south, Mediterranean types of climate also occur, with arid summers and winter rains. The amount of rain (Fig. 1.3) and its distribution throughout the year vary greatly within Africa. The highest rainfall occurs in equatorial Zaïre and along the humid littoral areas of the Gulf of Guinea, while the desert areas to north and south of the equatorial belt receive very small and unpredictable amounts of rain. Both the within and between year variations in rainfall are reflected in river flows and lake

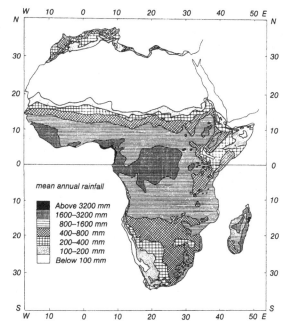

mean annual rainfall

Above 3200 mm
1600–3200 mm
800–1600 mm
400–800 mm
200–400 mm
100–200 mm
Below 100 mm

Fig. 1.3. Mean annual rainfall in Africa (from Balek, 1977).

levels in all areas and thus greatly influence the ecology of all aquatic systems.

Diversity of habitats

Traditionally, limnologists distinguish two main groups of freshwater environments: (i) running waters or lotic environments, including streams and rivers; and (ii) standing waters or lentic environments such as lakes, ponds, swamps and marshes. In fact, there is not always a clear cut boundary between them, and many freshwater systems are rather a mixture of biotas which, moreover, vary seasonally in relation to their flood dynamics. Since fish populations have evolved independently in different freshwater systems and those of Africa include a significant proportion of endemic species, some of them highly adapted to specific environments, the structural heterogeneity of habitats, both spatially and temporally, is a major ecological factor contributing to the existence and sustainability of fish diversity.

Wetland is often used as a generic term for a combination of shallow standing and running water. The one feature that distinguishes a wetland from other ecosystems is an abundance of aquatic vegetation. It could be defined loosely as a vegetation area that is flooded, either permanently or seasonally (Denny, 1985). African wetlands have been well investigated where their flora and fauna, including non-aquatic vertebrates, were attractive to scientists. There are many different types of wetland, from peat bogs to shallow lakes and they do not all have the same role in fish biology; only a few wetlands of specific interest for fish are listed here. Detailed information and references are available in Beadle (1981), Denny (1985), Burgis & Symoens (1987), Davies & Gasse (1988), Whigham et al. (1993). Several species use wetlands as temporary habitat at some time of their life, and the importance of floodplains in the functioning of riverine systems has been discussed by Welcomme (1979).

Running waters

The present courses of the main river systems of Africa are shown in Fig. 1.4. Their catchments are basically the ancient basins that existed before the Miocene when major earth movements, which produced many abrupt changes of level and altered the river courses, began to transform the drainage systems of eastern tropical Africa. Consequently, there are more rapids and waterfalls in African rivers than in any of the other great rivers of the world, and none has provided unimpeded access into the interior of the continent. This has had considerable influence on its history and development (Beadle, 1981).

The Zaïre drainage is situated entirely within the humid tropical belt. In the Northern Hemisphere, the Senegal, the Niger, the Chari and the Nile are bounded by very arid countries, but a significant part of their water is provided by rains in their equatorial headwater catchments. This is also the case for the Zambezi in the Southern Hemisphere.

Monographs have been published on the Nile (Rzoska, 1976), the Niger (Grove, 1985), the Jonglei Canal (Howell et al., 1988) and the Kafue Flats (Ellenbroek, 1987); and various papers have been published on the Bandama (Lévêque et al., 1983), the Volta (Petr, 1986) and the Zambezi (Davies, 1986).

Floodplains are typically associated with a river and are flooded regularly. They are also called flats, as in the Kafue Flats, or 'toiche' in the Nile system. They appear flat, but the microrelief can lead to great differences in the period of immersion. In fact, a floodplain consists of

Fig. 1.4. The distribution of river systems in Africa.

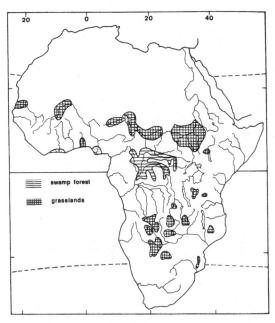

Fig. 1.5. Situation of the major African wetlands (modified from White, 1983; Denny, 1993).

a complex pattern of lagoons, pools, ox-bow lakes and seasonal marshes. Welcomme (1979), distinguished three main types.

1. Fringing floodplains that are relatively narrow strips of floodable land lying between the river valley walls.
2. Internal deltas occurring when river systems spread laterally over very large alluvial plains. The main stream is divided into a number of branches that join the main channel below the delta. Examples are the inner delta of the Niger (Mali) and the Sudd (Sudan) (Fig. 1.5).
3. Coastal deltaic floodplains develop where the main channel breaks down into smaller branches and produces the classic fan-shaped delta. Examples are the Nile and the Niger deltas.

Many inland floodplains are grasslands, that is to say land covered with grasses and other herbs, with no more than 10% of the ground covered with woody plants.

Lakes and swamps

In Africa the term swamp generally applies to wet areas flooded to a shallow depth, either permanently or for most of the year, that are densely covered with tall her-

baceous vegetation bottom rooted or floating, as in *Cyperus* swamps or *Phragmites* swamps that usually occur in the vicinity of permanent water bodies. The limiting factors for fish are generally deoxygenation or desiccation and only species that have structural or behavioural adaptations are able to survive difficult periods.

Tropical Africa has enormous areas of swampland (Beadle, 1981). The inner delta of the Niger in Mali, includes a complex of small shallow lakes. Much of Lake Chad is choked with swamp, and in the southern Sudan, the Nile spreads out into a wide area of papyrus and grass known as the Sudd. In the Zaïre basin, Lake Bangweulu, at the head of the Luapula River, has a great expanse of associated swamps, and in the ancient basin of the Cubango-Kalahari, lies Lake Ngami and the large Okavango swamps. Large swamps also occur in association with Lakes Victoria, George and Kyoga. Smaller swamps occur in numerous other places.

Papyrus (*Cyperus papyrus*) is the characteristic and most widely known swamp plant of central and eastern tropical Africa. In addition to papyrus, permanent swamps are populated by several types of emergent plants including other species of *Cyperus*. The grasses

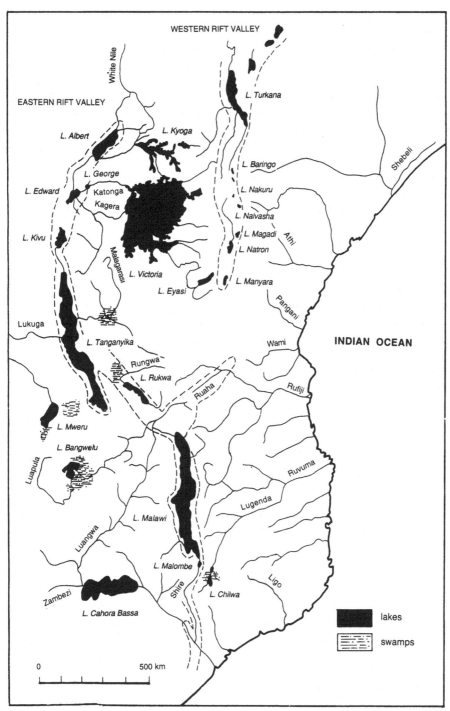

Fig. 1.6. The two Rift Valleys of eastern Africa and their associated lakes.

Vossia cuspidata, Loudetia phragmitoides, the reedmace *Typha*, the reed *Phragmites mauritianus* and *Echinocloa pyramidalis* are also common in the inner delta of the Niger, and form dense and extensive stands in some places (see Denny, 1985, for details).

Swamp forests may be either permanent or seasonal, and are widespread in the Guineo–Congolean region but extend into adjacent zones, such as that of Lake Victoria and the wetter parts of the Zambezian region (Denny, 1993). They contain dense stands of trees from 10–50 m in height that are specifically adapted to, or are tolerant of, flooding and waterlogged soils. We have very little information about fish assemblages and biology in such habitats.

In comparison with other standing water habitats, lakes are large bodies of water, the surfaces of which are mostly devoid of vegetation. They vary enormously in size, from mere ponds to the 70 000 km² of Lake Victoria, and in depth from a few metres to the 1470 metres of Lake Tanganyika. Although the concept is well known among limnologists, there is no simple and universally recognised definition of a 'shallow lake'. One possibility is to define them as lakes, less than 10 m deep, that do not have permanent stratification of the water column (Lévêque & Quensière, 1988). This definition applies to many natural African waterbodies, all over the continent. The largest and best-known shallow lakes in Africa are Lake Chad (Carmouze *et al.*, 1983), Lake Ngami in Botswana, Lake Turkana in Kenya (Hopson, 1982), Lake Bangweulu in Zambia (Toews, 1975). Smaller freshwater lakes have also been well studied: Lake George in Uganda (Burgis *et al.*, 1973; Ganf & Viner, 1973; Talling, 1992), Lake Chilwa in Malawi (Kalk *et al.*, 1979), Lake de Guiers in Senegal (Cogels, 1984), Lake Naivasha (Litterick *et al.*, 1979) and Lake Nakuru (Vareschi, 1979). See also Beadle (1981), Serruya & Pollingher (1983) and Burgis & Symoens (1987) for more detailed information.

Shallow soda lakes in closed basins, such as Lakes Elmenteita, Magadi, Nakuru and Natron in Kenya, include some extreme environments for fish. But the most striking feature of all shallow lakes is their sensitivity to climatic changes: they sometimes respond to variations in rainfall by large variations in size, as well as by changes in water salinity. Well-documented examples

are Lake Chad (Carmouze *et al.*, 1983) and Lake Chilwa (Kalk *et al.*, 1979).

In contrast, the very deep lakes generally have permanent stratification and are less sensitive to year-to-year changes in water volume. The Great Lakes of East and Central Africa have been much studied, partly because of their spectacular fauna of endemic cichlids. They include Lakes Victoria, Tanganyika (Coulter, 1991a), Malawi and Kivu, but some smaller deep lakes have also been studied (Fig. 1.6). Several hundred crater lakes are known, for example in volcanic areas, such as those of Cameroon or East Africa. They are mostly rather small, not exceeding about 2 km in diameter, as are their catchment areas, and they support a generally poor indigenous fauna.

Some natural lakes, mostly in volcanic regions, originated from blockage of the previous river system by lava flows during the Pleistocene. Lake Kivu, in the Western Rift, was formed when eruption of the Virunga Volcanoes formed a blockage and reversed the previously north-flowing river system. Lakes Bunyoni in Uganda and Lake Luhondo in Rwanda were formed in the same

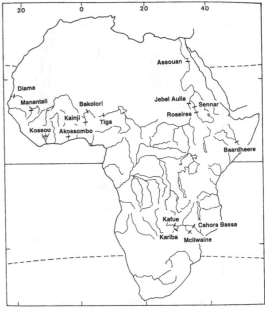

Fig. 1.7. Location of major man-made lakes in Africa.

manner, as well as Lake Tana in the highlands of Ethiopia.

Finally, during recent decades dams have been built on many major rivers (Fig. 1.7) forming man-made lakes. The Aswan Lake on the Nile, the Volta Lake on the Volta River, the Kariba and Cahora Bassa on the Zambezi, the Kossou Lake on the Bandama River are large, well-known reservoirs. Hundreds other smaller reservoirs have been built on headwater tributaries in arid zones, in order to store water for the dry season, for domestic purposes, cattle, irrigation or other uses including fish production. These standing water systems are new to the riverine fish fauna, which has to adapt to new environmental conditions.

Current spatial and temporal variabilities of abiotic environmental factors

Environmental conditions in continental waters are determined by an interacting array of external and internal influences. All contribute to the variability in time and space, yet a small number of components dominate, e.g. solar radiation and hence temperature regime, the magnitude and distribution of rainfall and hence the hydrological regime (cycle). Conditions are far from constant in the tropics and seasonal climatic changes do occur. The factors involved and their amplitude become increasingly seasonal with increasing latitude, both north and south of the Equator. A great range of seasonality is therefore encountered.

For shallow African lakes, Talling (1992) identified three types of environmental changes in time that involve concurrent changes in many variables.

1. The diel cycle that is impressed by the daily shortwave radiation but often modified by wind-stress. It involves a temperature rise of the superficial water and density stratification with its constraints on vertical distribution.
2. The annual or seasonal cycle that may be accompanied by extensive volume changes of a flooding–drying cycle. In shallow basins, large changes of water volume due to hydrological factors lead to correspondingly large changes in mean

depth and inundated area. A cycle of ionic content normally accompanies a flooding–drying cycle.
3. The interannual variability related to climatic variability. The trends are not predictable, but this variability may have dramatic direct or indirect effects on the ecology of aquatic ecosystems.

Rainfall and flood seasonality

Hydrology and water budget are the main driving forces in aquatic ecology. The very existence of the waterbodies and their long-term dynamics, strongly depend on the amount of rainfall, its fate and its distribution patterns.

Annual patterns

Rainfall is undoubtedly the most significant feature of the weather in Africa. The yearly pattern and amount of rain in different regions relates to complex migrations of air masses, which result in a seasonal distribution of rainfall. At the Equator, rain falls throughout the year, with two well-identified peaks, one in April and the other in November (Table 1.1). North and south of the Equator, the pattern becomes increasingly seasonal and the length of the rainy season becomes shorter, with a single rainy season culminating in July–August in the Northern Hemisphere, and January–February in the Southern Hemisphere (Table 1.1). Moving from the Equator to the tropics, the relative humidity also decreases while the evaporation rate increases.

Rivers show considerable variation in flow rate in relation to rainfall over the year. In equatorial areas, runoff tends to be spread evenly throughout the year, whereas towards the tropics, it tends to concentrate during a short period (July–September in the north, January–March in the south). Some streams or small tributaries of major rivers dry out during the drier months, and above 12°N, for example, the flood is highly seasonal and some water courses in the Sudanese zone remain completely dry for at least six months of the year. In contrast, many of the larger rivers spread out across floodplains over large areas during wet periods. These floodplain rivers are usually highly productive for fisheries (Welcomme, 1979).

Two major flood regime patterns have been identified that clearly reflect the seasonal distribution of rainfall

Table 1.1. *Mean monthly and annual rainfalls (in mm) according to latitude on the African continent*

Latitude	Jan.	Feb.	Mar.	Apr.	May.	Jun.	Jul.	Aug.	Sep.	Oct.	Nov.	Dec.	Mean annual rainfall (mm)
40–35° N	51	47	52	34	31	19	7	9	46	58	64	50	468
35–30° N	37	36	32	22	15	7	2	3	11	25	40	42	272
30–25° N	2	2	2	1	1			1	2	1	3	4	19
25–20° N	1	1	1	1	2	1	3	8	6	2	1	1	28
20–15° N	1	1	1	2	5	12	40	71	28	5	2	1	169
15–10° N	3	2	6	17	57	100	198	254	159	48	9	2	855
10–5° N	10	25	54	89	153	180	195	206	201	140	42	13	1308
5–0° N	32	51	102	153	147	120	114	141	155	166	107	49	1337
0–5° S	131	143	163	195	136	45	39	57	100	152	199	166	1526
5–10° S	155	153	169	147	30	3	3	10	39	82	146	180	1117
10–15° S	241	233	210	73	11	2	1	3	18	44	130	204	1170
15–20° S	163	176	119	42	9	5	3	3	4	20	69	130	743
20–25° S	124	121	95	40	19	15	11	9	10	24	57	90	615
25–30° S	53	57	56	29	16	10	9	10	14	27	43	47	371
30–35° S	32	41	46	38	41	36	35	33	37	35	37	33	444

From Jaeger, 1976.

1. Equatorial regime: while the flow is always sustained, the river discharges show two separate periods of high water as in the flood patterns of the Ogowe and the Zaïre (Fig. 1.8).
2. Tropical regime: there is a single high-water season. The dry season may be more or less pronounced and in extreme situations, the discharge may be very low. The flood patterns of the Niger and the Senegal are good examples (Fig. 1.8).

Of course there are many variations on these broad patterns. The tropical regime for instance may be further subdivided into a tropical transition type, with a longer period of high water and shorter period of low water than the classical tropical type. Such is the case for the flood pattern of the Sanaga river in Cameroon or the Ubangui in Central African Republic (Fig. 1.8). Conversely, in the Sahelian type, the flow is restricted to July to September and there is no flow for 6 to 9 months. The extreme situation is that of the desert type regime, where only one or two floods occur each year as a result of periodic storms.

The fate of the flow depends on the form of the river basin. Small water courses show more rapid changes in flood condition than larger courses, where the flood is smoothed as a result of cumulative discharge of many tributaries and so-called reservoir rivers have internal features that reduce the variability of flow. Actually, most of the large, long rivers have mixed regimes, which are further complicated when the river or its tributaries cross various climatic areas. The damming of several major rivers within the past 30 years has resulted in a reduced and more stable flow downriver of the reservoirs. As a result, the lower course of many rivers has been transformed from a flood type into a reservoir type river.

In many lakes, and particularly in a 'river lake', the river inflow provides the major water input. In small reservoirs and man-made lakes, the year to year river discharge variations may have a major influence in the filling of the lake. In the natural Lake George, some 80 % of the annual water input comes from inflows. In contrast, in a case like Lake Victoria direct rainfall accounts for some 83% of the total water input.

In many river-associated lakes and wetlands, the natural increases in water level occur during the rainy season, when temperature is high. Most fish take advantage of the favourable conditions and breed at this time. However, in areas with little local rainfall, e.g. Lake Chad or the Okavango Delta that depend on high rainfall in distant regions, the flood may only reach the lake in winter when conditions are less suitable for fish breeding.

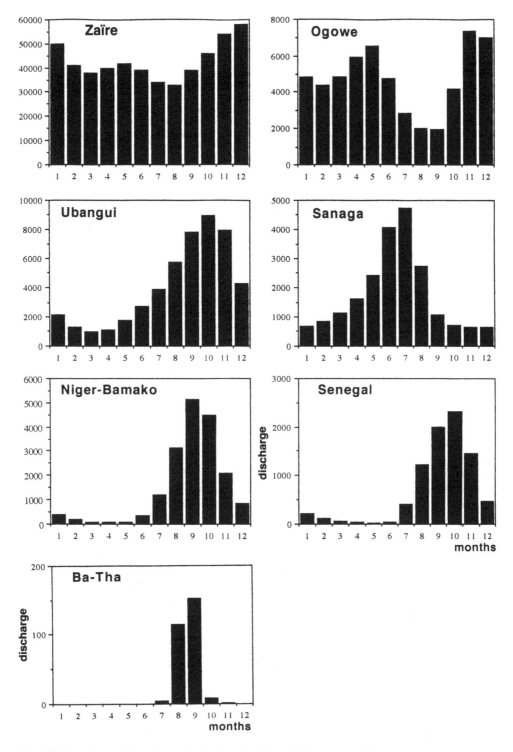

Fig. 1.8. Comparison of selected river flood regimes in Africa. Sahelian type, Ba-Tha at Ati (Chad) (data ORSTOM). Tropical type, Niger at Koulikoro (Mali) (data ORSTOM); Senegal (mean 1903–68, data ORSTOM). Tropical transitional type, Sanaga at Edéa, Cameroon (period 1970–80, from Olivry, 1986); Ubangui at Bangui, RCA (data ORSTOM). Equatorial type, Ogowe at Lambarene, Gabon (mean 1929–67, data ORSTOM); Zaïre at Brazzaville (data from Service Hydrologique–ORSTOM). Discharge in m^3 s^{-1} .

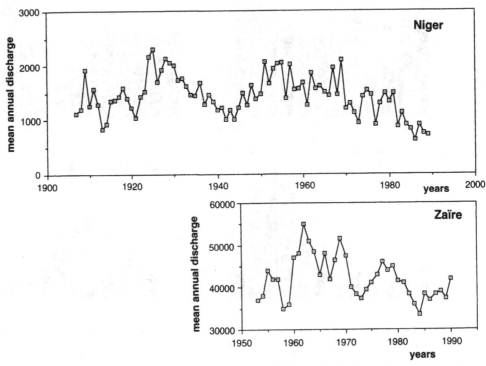

Fig. 1.9. Long-term year-to-year changes in the mean annual discharge ($m^3 s^{-1}$) of rivers Niger and Zaïre (data from Service Hydrologique-ORSTOM).

Year-to-year changes

The year-to-year changes in rainfall are the result of a complex array of interaction between the ocean, the atmosphere and the continent, which are not yet clearly understood. The great variability in rainfall and runoff observed in Africa, and the Sahel drought during the 1970s, initiated a great deal of research on long-term trends of the African climate. Mahé (1993), for example, studied the annual and interannual variations of both rainfall and runoff in West and Central Africa for the period 1950–90. The wettest period extended from 1951 to 1970, with a maximum in 1962–3. The driest years were nearly all recorded during the last decade or during the dry periods of the 1970s with a minimum in 1983. The rainfall not only decreased in the Sahelian areas, but also during the monsoon flow in the equatorial regions (Mahé, 1993; Mahé et al., 1990).

Records from the Senegal and Niger Rivers and from Lake Chad (Fig. 1.9) show that other periods of low discharge have occurred earlier this century, in the 1910s

and the 1940s. Nevertheless, the Sahel drought apparently did not significantly modify the long-term discharge of the Sanaga River (Fig. 1.9), for which the range of annual discharge is relatively low. For the period under study, it also appears that the long-term trend in the discharge of the Zaïre River (Fig. 1.9) was different to the others, which is an indication that regional factors are also important in understanding the climatic patterns observed on the African continent.

The amount of rain in a year has a direct effect on the river runoff in that year, but there are also delayed consequences of long-term climatic changes. For instance, the retention and release of groundwater contributes to sustaining the duration of a river flood during the dry season in tropical areas. In the Sahelian zone, after 20 years of drought, the groundwaters are now (1995) severely depleted and do not sustain the post flood discharge as efficiently as they did before the drought. The so-called depletion coefficient is estimated from the pattern of the drop in flow of the rivers considered, and

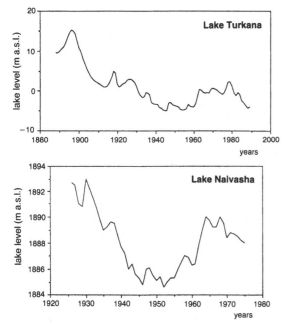

Fig. 1.10. Changes in lake level of Lake Turkana (Kolding, 1989) and Lake Naivasha (Litterick *et al.*, 1979) during the twentieth century.

is a measure (expressed in days^{-1}) of the rate of emptying of the aquifers after the annual flood. The depletion coefficient has been relatively stable since the beginning of the century, but increased dramatically after 1975 in many rivers (Senegal, Niger, Chari) (Olivry, personal communication). This situation reflects an overall degradation in the hydrology of Sahelian rivers, resulting in more severe and longer low-water periods, occurring sooner after the flood and for longer.

The long-term changes in rainfall and water budgets have dramatic consequences for the dynamics of the freshwater biotas. Changes in river discharge or lake level are often associated with marked changes in ecological conditions and consequently in the richness and structure of fish assemblages (see Chapters 11 and 12). The Sahel drought, for instance, resulted in a prolonged low-water period in tropical rivers and is probably responsible for changes in the ecological conditions for fish species, even if it is not well documented. Very spectacular changes in the aquatic environment were studied in the case of the endorheic Lake Chad (Carmouze *et al.*, 1983) for the period 1965–78. A comparatively small

change in water input resulted in a significant change in the size of the lake due to its shallowness: it decreased from 25 000 km^2 in the mid 1960s to some 5000 km^2 in the mid 1970s, and the whole system broadly shifted from lacustrine to marshy conditions. The ecological consequences for fish populations were severe (see Chapter 10). In the case of Lake Chilwa (Kalk *et al.*, 1979) (see Chapter 10), the drying up in 1968 and the recovery phase from 1969 to 1972 was also associated with marked changes in water chemistry and biological characteristics.

Nevertheless, the Sahelian drought of the 1970s was not a unique event, and since the beginning of the century (and the beginning of recorded measurements) there have been occasional wide fluctuations from the 'normal' with dramatic effects on the level of other lakes, including those of the Rift Valleys. The very high levels of the lakes in the Western Rift Valley, especially Lake Victoria during the mid 1960s are an example. Other large fluctuations in the level of shallow lakes, such as Lake Turkana and Lake Naivasha, have also been observed during this century (Fig. 1.10).

Water temperature

Seasonal patterns
Seasonal water temperature variations are restricted in equatorial aquatic systems and at Lake George (0°), Lake Ihema (1° 50′ S) and Lake Tanganyika (Kwetuenda, 1988) differences do not exceed 2 °C (Fig. 1.11). However, when moving north or south from the Equator, insolation does vary with season and the seasonal pattern becomes more apparent, with a pronounced winter minimum related to the latitudinal-geometrical factor of solar radiation (Talling, 1992). At Lake Chad, for instance, the seasonal differences in temperature exceed 10 °C (Fig. 1.11). Falls in surface water temperature with altitude have also been demonstrated (see Talling, 1992).

The pattern and the range of seasonal variation in temperature encountered in tropical waters strongly influence the biology and physiology of fish, such as the timing of their reproduction for example. Moreover, the low temperatures that occur seasonally in subtropical waters may also limit the distribution of thermophilic species. Most tropical African fish are restricted to a water temperature range of 20–30 °C, but a few species are adapted to much higher temperatures, e.g.

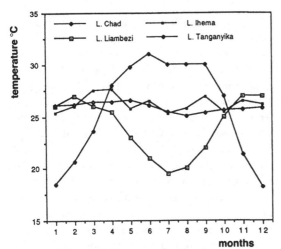

Fig. 1.11. Annual variations of surface temperature in some
African lakes: Lake Chad (from Lemoalle, 1974);
Lake Ihema (from Plisnier, 1990); Lake Liambezi
(from Seaman *et al.*, 1978); Lake Tanganyika
(average 1977–86, from Kwetuenda, 1988).

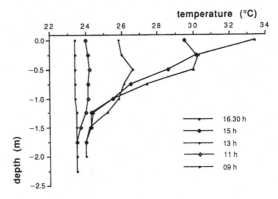

Fig. 1.12. Typical changes in water temperature profiles
during the day in the shallow Lake George
(Uganda). The profiles represent one half cycle of
the diurnal stratification and destratification (from
Viner & Smith, 1973).

Oreochromis alcalicus grahami, which occurs at temperatures up to 40 °C in the hot springs of Lake Magadi in the Eastern Rift.

Temperature stratification in lakes

The warming of the upper layer of water is not transferred equally down through the whole water column. Daily or seasonal changes in surface water temperature in lakes are usually associated with vertical gradients of temperature. A distinct diurnal pattern of thermal stratification followed by nocturnal destratification has been reported in several lakes (Beadle, 1981). For instance, in Lake George (Viner & Smith, 1973) the temperature of the bottom layers are about equal to the mean air temperature (23–25 °C), but the water surface heats up progressively during the day so that the maximum can reach 35 °C by mid-afternoon. A pronounced diel stratification may therefore be observed in waters deeper than a few metres (Fig. 1.12), and the vertical temperature gradient can be up to 10 °C within 2.4 m. Similar observations have been made in Lake Opi (Nigeria) and Lake Chad.

Seasonal lacustrine thermal stratification has been well studied all over the world and Beadle (1981) made a review of the information available for African waters. Mixing of the water column by wind is a major factor

preventing or delaying the onset of stratification. As a result, thermal stratification is never permanent in shallow West African lakes (John, 1986), and it only develops during drier spells of warm and relatively calm weather. These lakes having a more or less constant circulation of water are called polymictic, whereas those where seasonal circulation only occurs in the upper part of the lake are called meromictic. These lakes have a deep hypolimnion that does not circulate as in deep lakes, such as Kivu and Malawi, where there is frequently a permanent stratification and the thermal discontinuity, called the thermocline, separates the warm and lighter upper water (epilimnion) from the cooler denser water below the thermocline (hypolimnion). The difference in temperature may be quite small. In Lake Malawi for instance, there is a stable thermocline at between 200 and 250 m depth, but the difference in temperature across this boundary is only 1 °C (Eccles, 1974). In Lake Tanganyika, the thermal stratification is well marked and varies seasonally above an apparently permanent anoxic hypolimnion. The thermocline occurs at a depth of 50–100 m in the north basin and at about 250 m at the south end (Coulter & Spigel, 1991). The difference in water temperature is usually low from 0.5 °C to 2.0 °C.

The stratification may be seasonal, with a mixing of the water column occurring at some periods of the year. In Lake Victoria for instance, the thermocline is usually

Table 1.2. *Selected data on conductivity, salinity and pH of tropical African lakes*

Lake	Country	Conductivity µS/cm	Salinity g/l	pH range
Bangweulu	Zambia	20–32	0.04–0.07	7.0–8.3
Tumba	Zaïre	24–32		4.0–4.9
Victoria	Kenya	96	0.1	7.1–8.5
Mweru	Zambia	49–125	0.04–0.07	7.0–9.3
Itasy	Madagascar	65–105		6.8–7.5
Tana	Ethiopia	137–240	0.15	7.5–8.4
Alaotra	Madagascar	80–250		6.8–7.3
George	Uganda	200	0.14	8.5–9.8
Malawi	Malawi	210	0.2	8.2–8.9
Chad	Chad/Niger/Nigeria	60–800	0.07–0.7	8.0–9.0
Kyoga	Uganda	210–300		7.6–9.0
Naivasha	Kenya	311–353	0.23	8.4–9.1
Ziway	Ethiopia	370–427	0.33	8.7
Tanganyika	Zaïre/Tanzanie	610	0.53	8.0–9.0
Albert	Zaïre/Uganda	735	0.6	8.5–9.5
Abaya	Ethiopia	670–900	0.51	8.0
Kivu	Zaïre/Rwanda	1240	1.12	9.1–9.5
Turkana	Kenya	3300	2.5	9.5–9.7
Chala	Ethiopia	20 000–30 000	15.7–16.7	9.7–10.1
Kanem Lakes	Chad			
Mombolo		6100	4.5	9.4–10.2
Rombou		16–19 000	16	10.4
Latir		98 000	175	
Nakuru	Kenya	9500–165 000		10.5
Magadi	Kenya	28 000–160 000		

From Beadle, 1981; Burgis & Symoens, 1987; Talling, 1992.

observed at 30–40 m depth, but complete mixing occurs once a year and partial mixing occurs at other times. In Lake Turkana, periods of temperature stratification were observed during the first half of the year, in spite of the frequent strong winds. The temperature difference may reach 5 °C, and the thermocline in the deeper part of the lake occurs between 30 and 50 m (Källqvist *et al.*, 1988).

Oxygen stratification in lakes

Aquatic organisms need oxygen for their respiration, and the distribution of dissolved oxygen within the aquatic ecosystem is one of the main factors influencing the distribution of fish. It is also very dependent on water temperature.

River waters are usually well oxygenated due to the turbulent flow. However, in rivers with marked low water periods, deoxygenation may occur in the remaining pools. In lakes, vertical temperature gradients create

physical stability of the water column and reduce vertical mixing. Stratification caused by density gradients influences the vertical distribution of dissolved oxygen, which is also dependent on the processes that produce and consume oxygen, i.e. photosynthesis and respiration. Below the upper water layer, where oxygen is produced by phytoplankton and exchanged with the atmosphere, it is continuously consumed by the respiration of living organisms and in the decomposition of organic matter, sinking from the surface layer, through bacterial respiration and chemical oxidation. As a result, if mixing of the water is restricted, oxygen depletion may occur in the deep water. In stratified lakes, the thermocline acts as a physical barrier to the diffusion of dissolved oxygen, and if stratification is permanent the hypolimnion is usually anaerobic. Several lake fish species can spend some time in deoxygenated waters, but they cannot occupy permanently anaerobic situations. As a result, part of the volume and bottom area in stratified lakes is not suitable for fish life. Moreover, in some processes of destratification, the mixing of the water column results in the upwelling of anoxic waters, usually also rich in ammonia. Fish that are trapped in these waters may suffer hypoxia, and mass mortalities have been recorded from many waterbodies following lake turnover (see Chapter 9).

In shallow waters with a high organic productivity, the oxygen demand for the decomposition of decaying material may be high. This is the case in swamps, where the high production of aquatic macrophytes results in a great deal of detritus. Floating mats of vegetation are also common in tropical waters, and the density of plant cover may influence the extent to which water becomes deoxygenated. Moreover, local and temporary deoxygenation often occurs in recently inundated areas, such as floodplains during flooding, as well as in newly created aquatic environments such as man-made lakes. Only species that are physiologically or behaviourally adapted to extreme oxygen conditions are able to colonise, or survive in, these environments.

Salinity and water chemistry

The ionic concentration of waters results largely from the interaction of rainwater with the rocks, soil and vegetation in the catchment area. Freshwaters, compared to seawater, show considerable variability in both concentration and composition from place to place (Table 1.2). In tropical regions, including Africa, ionic concentration of river waters is usually low (Meybeck, 1979) compared to temperate rivers. It reaches higher values in lakes, where it ranges from very dilute waters on hard mountain rocks (conductivity 10–20 μS cm^{-1}), to evaporation pans and shallow lakes where the saturation limit depends upon the predominant anions (conductivity up to 160 000 μS cm^{-1} in Lake Magadi) (Talling, 1992). The chemical composition of the water is greatly influenced by variation in flow and varies seasonally: there is usually an increase in the salt concentration during the dry period, as a result of water evaporation. Only some species of fish are able to live in extremely saline waters (see Chapter 9) and the chemically heterogeneous environments, resulting from the patchy nature of freshwaters, have considerable influence on the distribution of organisms in these environments.

Conclusion

The diversity of the inland waters of tropical Africa is probably greater than on any other continent. The flowing waters range from snowfed high mountain streams to some of the longest and largest (in terms of discharge) rivers in the world; but Africa also contains some of the largest (both in depth and surface area) and oldest lakes in the world with habitats absent from other tropical regions. The smaller lakes, which are found on every continent, range from riverine expansions, where the water is dilute and constantly renewed, to closed basins whose high concentrations of salts fluctuate seasonally but are steadily increasing from year to year. The enormous variety of conditions both within and between these aquatic ecosystems, evolving as they have over many millions of years, has encouraged the evolution of an equally varied fish fauna, which is the principal subject of the following chapters.

2 The fish fauna of Africa

In Africa there is a comparatively greater variety of distinct freshwater types (than in India), imparting to the study of its fauna an unflagging pleasure.

Günther, 1880

A hundred and three years later, Günther's remarks still hold true but the problems associated with that study are now far more complex.

Greenwood, 1983a

The African ichthyofauna is fairly well diversified, and the present day composition and distribution are the result of a long and complex history, during which speciation and extinction processes interacted with climatic and geological events that isolated fish populations, or provided opportunities for migration and colonisation of new habitats. Most of the African continent has remained above sea-level since the Precambrian, more than 600 million years ago, though large areas, such as in the Sahara, Somalia and Ethiopia, were at some time inundated by the sea. This long period of exondation may explain why Africa has a far more diverse fish fauna than South America, and an unparalleled assemblage of archaic families, mostly endemic, that has evidently been there for a long time. The families Denticipitidae, Distichodontidae, Pantodontidae, Phractolemidae, Kneriidae, Mormyridae and Gymnarchidae are all endemic to Africa and probably date back to the Early Mesozoic (Briggs, 1979). The denticipitids, known from the Tertiary of Tanzania (Greenwood, 1974b), may have branched from an Upper Jurassic or Lower Cretaceous clupeoid stock (Goody, 1969). The mormyrids and gymnarchids are possibly related to *Lycoptera*, a fossil from the freshwater deposits of the Lower Cretaceous of north-east Asia (Patterson, 1975).

Composition of the ichthyofauna

As elsewhere, ostariophysian fish dominate the freshwater fish fauna of Africa, but the families are unequally represented. Over 2000 non-cichlid species, belonging to 340 genera and 75 families (fresh and brackish waters), have been recorded in Africa according to the *Check-List of the Freshwater Fishes of Africa* (CLOFFA) (Daget et al., 1984, 1986a, b, 1991). Cyprinidae (Skelton et al., 1991), Characidae and a few Siluriformes families constitute the bulk of the riverine fish fauna (Tables. 2.1 and 2.2), with Cyprinodontidae (Acanthopterygii) and Mormyridae (Osteoglossomorpha). Many other families including most of those that are endemic, are represented by only one or a few species, while the Cichlidae (Lowe-McConnell, 1991; Ribbink, 1991) are by far the most abundant with some 870 species and 143 genera recorded, most of them endemic to East African lakes (Daget *et al.*, 1991) (Table 2.1). Many lacustrine species still remain to be described.

Ichthyologists, following Myers (1951), used to distinguish three major groups of fish according to their tolerance of saltwater and their ability to disperse across the sea (Table 2.1).

1. The primary division comprises the fish that are 'strictly intolerant to salt water' and are therefore limited exclusively to freshwaters. They are usually much longer established in continental waters than the other groups and their ancestors entered freshwaters by continental routes.

2. The secondary division includes fish 'rather strictly

17

Table 2.1. *African fish families with representatives in fresh and brackish waters*

Hierarchy		Division	No. genera in Africa	No. species in Africa	Distribution beyond Africa
Super Class	Gnathostomata				
Class	Chondrichthyes				
Subclass	Elasmobranchii				
Order	**Carcharhiniformes**				
Family	Carcharhinidae	peripheral	1	1	widespread, marine
Order	**Rajiformes**				
Families	Pristidae	peripheral	1	1	widespread, tropical
	Dasyatidae	peripheral	1	4	widespread, tropical
Class	Sarcopterygii				
Subclass	Dipnoi				
Order	**Lepidosireniformes**				
Family	Protopteridae	primary	1	4	South America
Subclass	Neopterygii				
Order	**Polypteriformes**				
Family	Polyteridae	primary	2	10	fossils in South America
Subclass	Actinopterygii				
Division	**TELEOSTEI**				
Subdivision	Osteoglossomorpha				
Order	**Osteoglossiformes**				
Suborder	Osteoglossoidei				
Families	Osteoglossidae	primary?	1	1	Asia, Australia, South America
	Pantodontidae	primary	1	1	endemic to Africa
Suborder	Notopteroidei				
Families	Notopteridae	primary	2	2	Asia
	Mormyridae	primary	18	198	endemic to Africa
	Gymnarchidae	primary	1	1	endemic to Africa
Subdivision	Elopomorpha				
Order	**Elopiformes**				
Families	Elopidae	peripheral	1	3	widespread
	Megalopidae	peripheral	2	2	
Order	**Anguilliformes**				
Families	Anguillidae	peripheral	1	6	widespread
	Ophichthyidae	peripheral	2	4	widespread
Subdivision	Clupeomorpha				
Order	**Clupeiformes**				
Suborder	Denticipitoidei				
Family	Denticipitidae	peripheral	1	1	endemic to West Africa
Suborder	Clupeoidei				
Families	Clupeidae	peripheral	20	38	widespread
	Pristigasteridae	peripheral	1	1	widespread, tropical
	Congothrissidae	peripheral	1	1	endemic to Zaïre
Subdivision	Euteleostei				
Superorder	Ostariophysi				
Order	**Gonorhynchiformes**				
Suborder	Chanoidei				
Family	Chanidae	peripheral	1	1	Indian Ocean, W. Pacific
Suborder	Knerioidei				
Families	Kneriidae	primary	2	24	endemic to Africa

Table 2.1. (*cont.*)

Hierarchy		Division	No. genera in Africa	No. species in Africa	Distribution beyond Africa
	Cromeridae	primary	1	1	endemic to Africa
	Grasseichthyidae	primary	1	1	endemic to Africa
	Phractolemidae	primary	1	1	endemic to Africa
Order	**Characiformes**				
Suborder	Characoidei				
Families	Hepsetidae	primary	1	1	endemic to Africa
	Characidae	primary	18	109	Central and South America
	Distichodontidae	primary	17	90	endemic to Africa
	Citharinidae	primary	3	8	endemic to Africa
Order	**Cypriniformes**				
Suborder	Cyprinoidei				
Families	Cyprinidae	primary	23	475	Eurasia
	Cobitidae	primary	2	2	Eurasia
Order	**Siluriformes**				
Suborder	Siluroidei				
Families	Bagridae	primary	19	108	Asia
	Schilbeidae	primary	5	34	Asia
	Amphilidae	primary	7	45	endemic to Africa
	Clariidae	primary	12	74	Asia, Syria
	Malapteruridae	primary	1	2	endemic to Africa
	Mochokidae	primary	10	167	endemic to Africa
	Ariidae	peripheral	4	13	tropical
	Plotosidae	peripheral	1	3	Indian Ocean, W. Pacific
Superorder	Protacanthopterygii				
Order	**Salmoniformes**				
Families	Salmonidae	peripheral	1	1	N. temperate, Arctic
	Galaxiidae	peripheral	1	1	temperate
Superorder	Paracanthopterygii				
Order	**Lophiiformes**				
Family	Antennariidae	peripheral			
Superorder	Acanthopterygii				
Series	**ATHERINOMORPHA**				
Order	Atheriniformes				
Family	Atherinidae	peripheral	5	8	Pacific and Mediterranean
Order	**Beloniformes**				
Families	Belonidae	peripheral			widespread
	Hemiramphidae	peripheral	2	3	widespread
Order	**Cyprinodontiformes**				
Family	Cyprinodontidae	secondary	21	243	Africa
Series	**PERCOMORPHA**				
Order	**Gasterosteiformes**				
Families	Gasterosteidae	peripheral	1	1	North Africa, Europe
	Syngnathidae	peripheral	4	12	widespread
Order	**Synbranchiformes**				
Families	Synbranchidae	peripheral	2	2	widespread
	Mastacembelidae	primary	2	46	Asia
Order	**Perciformes**				
Suborder	Percoidei				
Families	Centropomidae	peripheral	1	7	widespread

Table 2.1. (*cont.*)

Hierarchy		Division	No. genera in Africa	No. species in Africa	Distribution beyond Africa
	Ambassidae	peripheral	1	5	east coast of Africa
	Serranidae	peripheral	2	4	widespread
	Moronidae	peripheral	1	2	
	Teraponidae	peripheral	2	2	Indo-Pacific, east coast of Africa
	Kuhliidae	peripheral	1	3	Indo-Pacific, east coast of Africa
	Carangidae	peripheral	10	24	circumtropical
	Lutjanidae	peripheral	1	8	widespread
	Gerreidae	peripheral	2	6	widespread
	Pomadasydidae	peripheral	3	12	circumtropical
	Sparidae	peripheral	2	2	
	Haemulidae	peripheral			
	Sciaenidae	peripheral	4	6	widespread
	Ephippidae	peripheral	2	3	
	Scatophagidae	peripheral	1	1	east coast
	Monodactylidae	peripheral	1	3	Indian Ocean, W. Pacific
	Nandidae	primary	2	2	Asia, South America
Suborder	Labroidei				
Family	Cichlidae	secondary	143	870	Asia, Central and South America
Order	**Mugiliformes**				
Family	Mugilidae	peripheral	5	14	widespread
Suborder	Sphyraenoidei				
Family	Sphyraenidae	peripheral	1	4	
Suborder	Polynemoidei				
Family	Polynemidae	peripheral	3	5	widespread
Order	**Beryciformes**				
Suborder	Blennioidei				
Family	Blenniidae	peripheral	4	4	widespread
Suborder	Gobioidei				
Families	Gobiidae	peripheral	41	90	world-wide
	Eleotridae	peripheral	11	25	world-wide
Suborder	Anabantoidei				
Family	Anabantidae	primary	2	28	Asia
Suborder	Channoidei				
Family	Channidae	primary	1	3	Asia
Order	**Pleuronectiformes**				
Suborder	Pleuronectoidei				
Families	Bothidae	peripheral	2	2	cosmopolitan
	Soleidae	peripheral	5	6	world-wide
	Cynoglossidae	peripheral	2	4	Indo-Pacific
Order	**Tetraodontiformes**				
Suborder	Tetraodontoidei				
Family	Tetraodontidae	peripheral	1	6	world-wide

Sources: families and number of species from CLOFFA I, II and IV (Daget *et al.*, 1984, 1986*a*, 1991); higher phylogenetic levels from Nelson (1994) and Lecointre (1994).

Table 2.2. *Composition of the fish fauna in representative river systems*

Fish families	Nile	Chad	Niger	Volta	Konk.	Jong	Sass.	Band.	Sana.	Ogowe	Ruaha	Zaïre	Zamb.	Oran.
	NS	NS	NS	NS	UG	UG	EG	EG	LG	LG	EC	ZA	ZZ	Cape
Dasyatidae			1									1		
Protopteridae	2	1	1	1			1		1			3	2	
Polypteridae	3	3	4	3	1		1	1	1	1		9		
Anguillidae	1										1		4	1
Denticipitidae			1											
Clupeidae	1		5	3			1	1	1			13	3	
Osteoglossidae	1	1	1	1					int			1		
Pantodontidae			1							1		1		
Notopteridae	1	1	2		1	2		1	1	1		2		
Mormyridae	15	14	27	16	10	13	8	10	15	22	6	109	9	
Gymnarchidae	1	1	1	1										
Cromeriidae	1		1	1										
Kneriidae											1	14	3	
Phractolemidae			1							1		1		
Grasseichthyidae										1		1		
Hepsetidae		1	1	1	1	1	1	1	1	1		1		1
Characidae	8	11	16	15	4	7	7	8	12	14	7	55	8	
Distichodontidae	7	10	14	8	1	4	3	4	3	14	3	48	2	
Citharinidae	2	3	4	3			1	1			1	3	4	
Cyprinidae	25	23	35	24	16	13	17	18	26	22	16	128	45	8
Cobitidae	1													
Bagridae	6	5	10	7	4	5	3	3	11	8	2	45	3	1
Schilbeidae	5	5	5	6	2	2	2	3	5	5	3	13	3	
Amphilidae	1	1	5	3	4	2	1	1	1	8	1	25	1	
Clariidae	7	8	14	7	4	5	3	7	9	8	3	28	8	
Malapteruridae	1	1	2	1	1	1	1	1	1	1		2	1	
Mochokidae	15	12	26	13	8	5	5	3	5	6	6	82	10	
Ariidae			3	1										
Cyprinodontidae	7	8	23	9	8	8	6	10	15	41	5	59	8	
Channidae	1	1	2	1		1	1	1	1	1		2		
Centropomidae	2	1	1	1	1	1	1	1	1			1		
Synbranchidae														
Nandidae			1							1				
Gobiidae			3	?		1	1	3			1	3	2	
Eleotridae	1	1	5	1	3	3	1	1		1	1	4		
Cichlidae	10	10	17	9	15	16	8	9	7	17	3	90	28	
Anabantidae	2	4	4	1	1	1	1	1	3	4		15	2	
Mastacembelidae		1	3	1	1	3	1	1	6	6		23	2	
Tetraodontidae	1	1	1	1								4	?	
Cynoglossidae			1									1		
Soleidae			1											
No. families	27	25	36	27	19	20	22	24	21	23	16	31	21	3
No. species	127	128	243	139	85	94	76	90	126	185	60	787	149	10

Sources: Nile, Chad, Niger and Volta from Lévêque *et al.* (1991); Konkoure (Konk.) from Lévêque *et al.* (1989); Jong from Paugy *et al.* (1989); Sassandra (Sass.) and Bandama (Band.) from Teugels *et al.* (1988); estimations for Sanaga (Sana.), Ogowe and Zaïre from CLOFFA (Daget *et al.*, 1984, 1986a, b, 1991); Ruaha from Skelton (1994); Zambezi (Zamb.) from Skelton (1994); Orange-Vaal (Oran.) native species only, from Skelton (1994). Abbreviations for the provinces: NS (Nilo–Sudan), UG (Upper Guinea), LG (Lower Guinea), EG (Eburneo–Ghanean), ZA (Zaïre), ZZ (Zambezi), Cape (Cape of Good Hope). int, introduced (see also Tables 2.3 and 2.4).

confined to freshwater but evidently capable of occasionally crossing narrow sea barriers'. Many of them are salt-tolerant for short periods. Some species of Cichlidae and Cyprinodontidae are able to live in hypersaline waters.

3. The peripheral division includes representatives of extant marine families that have colonised inland waters from the periphery, i.e. from the sea. Myers (1951) distinguished four divisions:

   (i) Diadromous: 'fishes that regularly migrate between fresh and salt water at a defined period of their life-cycle'. A few species, such as *Anguilla* spp., occurring in the Nile, in the Zambezi and in Madagascar belong to this group.

   (ii) Vicarious: 'presumably non-diadromous freshwater representatives of partially or primarily marine groups'. Banarescu (1990) suggested that the term 'vicarious freshwater fishes' should be restricted to the exclusively freshwater genera for which there is evidence that they dispersed by continental routes, and that the species resident in freshwaters that reached their present range by marine routes should be ascribed to the complementary division (see below). As described here, vicarious fishes therefore have continental ranges and may have evolved from complementary or diadromous fishes. An example is the genus *Lates* (Centropomidae), of marine origin but now widespread in Africa where it has colonised most freshwater habitats. The pufferfish, *Tetraodon lineatus* (Tetraodontidae), is also well adapted to freshwaters and six tetraodont species live in the Zaïre basin. The Clupeidae, a family of marine origin, succeeded in colonising freshwaters, and many representatives are found in rivers and in the landlocked open waters of many lakes, including man-made lakes.

   (iii) Complementary: 'freshwater fishes, often or usually diadromous and belonging to marine groups that become dominant in fresh water only in the paucity or absence of primary, secondary and probably also vicarious freshwater faunas'. Complementary fishes that dispersed by marine routes have a peripheral distribution pattern. The freshwater stingray, *Dasyatis garouaensis* (Dasyatidae), is, for instance, a notable inhabitant of the Benue River where a breeding population has been observed as far as 1200 km from the coast. *Dagetichthys lakdoensis* (Soleidae), a pleuronecti-

form species, is found in the Upper Benue River at a distance of 1300 km from the sea. The gobiid, *Chonophorus lateristriga*, occurs far upstream in most West African rivers. In the Lower Zambezi, peripheral species were found as far inland as Zimbabwe or the Lower Shire River in Malawi (Bell-Cross, 1976). In islands such as Madagascar, where the original primary freshwater fish fauna was probably poor or became extinct, most of the native endemic fish species occurring strictly in freshwaters are of marine origin.

   (iv) Sporadic: 'fishes that live and breed indifferently in salt and fresh water, or enter freshwater only sporadically and not as part of a true migration'. This is the case for some Sciaenidae, Lutjanidae, Mugilidae, etc.

In Africa, 27 out of 76 families are primary division fishes and represent 50% of the recorded species. Another 38% belong to the 2 families of secondary division fishes, of which the cichlids alone represent around 30%. The remaining 12% belong to 47 families of the peripheral division.

Looking at the African fish fauna particularly, Lowe-McConnell (1988) suggested that it currently consists of five elements.

1. Families endemic to the African continent that includes 14 of the 27 primary freshwater families (Table 2.1). They are probably among the most ancient families of fish.

2. Remnants of archaic elements of wide distribution, extending beyond Africa in South America and Australia. The Protopteridae and Osteoglossidae are representatives of this group.

3. Elements shared with South America and South-East Asia, indicating their Gondwanaland origin, as is the case for the Notopteridae, Characidae, Bagridae, Clariidae, Schilbeidae, Cyprinodontidae and Channidae.

4. Elements shared with South-East Asia that could result from much more recent land connections and faunal exchanges. Cyprinids for instance are considered to be of Asian origin, having colonised Africa during the Miocene.

5. Elements of marine origin that either evolved in freshwaters, or move seasonally or sporadically between the sea and freshwaters and are classified

as secondary or peripheral fishes. Perciformes, Pleuronectiformes and Tetraodontiformes provide examples.

Sources of taxonomic information

Daget (1988) distinguished three major periods in the history and the development of the systematics of the African fish fauna: 1758 to 1850 (or since Linneaus to the issue of *Histoire naturelle des poissons* by Cuvier & Valenciennes) was a period of early discoveries, and description of different species, mainly from the Nile. About 100 species were described before 1850. From 1850 to 1940 many species were described, based on numerous collections made by travellers or military expeditions and stored in European Museums. Faunas or Catalogues were published by prominent scientists such as Günther, Boulenger and Pellegrin. Since 1940, the improvement in transport and communication has opened up much of the continent, which was previously poorly explored, and large new collections were made available. With the establishment of field laboratories much more attention has also been given to regional faunas, as well as to fish biology and ecology. Our knowledge of the taxonomy and distribution of fish in Africa has improved considerably during the last two decades.

A major step forward was the publication of the *Check List of the Freshwater Fishes of Africa* (CLOFFA) that provides a current list of brackish and freshwater fish species, with full synonymies as well as a full bibliography (Daget *et al.*, 1984, 1986*a*, *b*, 1991). The CLOFFA is particularly useful for taxonomists as a major source of information for further taxonomic revisions.

The only general fish Fauna available on a pan-African scale is the classical Catalogue of Boulenger (1909–16) which now has only historical value. At the regional level, a fish Fauna is presently available for West Africa (Lévêque *et al.*, 1990, 1992), and for South Africa (Jubb, 1967). Similar work has still to be done for the other main regions such as the Zaïre system and eastern Africa.

Besides the production of regional accounts, many taxonomic revisions of fish families or genera at the continental level have recently been published sometimes introducing important taxonomic changes. These include :

Characidae: *Alestes* and *Brycinus* (Paugy, 1986), *Hydrocynus* (Brewster, 1986; Paugy & Guégan, 1989);
Cyprinidae: neobolines (Howes, 1984);
Bagridae: *Chrysichthys* (Risch, 1986), *Auchenoglanis* and *Parauchenoglanis* (Teugels *et al.*, 1991);
Schilbeidae: de Vos (1984);
Clariidae including *Clarias* (Teugels, 1986) and *Heterobranchus* (Teugels *et al.*, 1990);
Cichlidae :*Oreochromis* and *Sarotherodon* (Trewavas, 1983), Haplochromines (Greenwood, 1979), Pelmatochromines (Greenwood, 1987), *Tylochromis* (Stiassny, 1989); and
Channidae: *Parachanna* (Bonou & Teugels, 1985).

The following taxonomic revisions, at the regional level, published since the issue of the CLOFFAs should also be mentioned. Most of them concern West Africa :

Clupeidae: Gourène & Teugels (1994), *Limnothrissa* (Gourène & Teugels, 1993), *Microthrissa* (Gourène & Teugels, 1989), *Odaxothrissa* (Gourène & Teugels, 1991*a*) and *Pellonula* (Gourène & Teugels, 1991*b*);
Mormyridae: *Hippopotamyrus* (Lévêque & Bigorne, 1985*a*), *Marcusenius* (Jégu & Lévêque, 1984*b*), *Mormyrus* (Lévêque & Bigorne, 1985*b*), *Mormyrops* (Bigorne, 1987), *Brienomyrus* and *Isichthys* (Bigorne, 1989), *Pollimyrus* (Bigorne, 1990*a*), *Gnathonemus* (Bigorne, 1990*b*), *Petrocephalus* (Bigorne & Paugy, 1991);
Characidae: *Micralestes* and *Rhabdalestes* (Paugy, 1990);
Distichodontidae: *Phago* and *Ichthyborus* (Lévêque & Bigorne, 1987);
Cyprinidae: *Barbus* (Lévêque, 1989*a*, *b*; Lévêque *et al.*, 1988*c*; Lévêque & Guégan, 1990), *Raiamas* and *Leptocypris* (Lévêque & Bigorne, 1983; Howes & Teugels, 1989), *Labeo* (Jégu & Lévêque, 1984*a*; Reid, 1985); and
Anabantidae: *Ctenopoma* (Norris & Teugels, 1990; Norris & Douglas, 1992).

The discovery and description of new species is still a very active business for African fish taxonomists. The rich endemic cichlid faunas of the East African Great Lakes, for example, are assumed to include many undescribed species (Ribbink & Eccles, 1988). Poorly investigated areas, such as Angola, will probably also provide new material and new species as a result of further investigations and research. Even in comparatively 'well-known' regions, the introduction of new tools, such as molecular techniques, may lead to reconsideration of previous taxonomic works based on morphological and anatomical criteria.

Patterns of fish distribution in intertropical Africa

A simple comparison of the occurrence of fish families in representative tropical African river systems (Table 2.2) shows an overall similarity between them. The largest rivers (Niger and Zaïre) have the most diverse fauna, but faunal composition does not exhibit dramatic changes across the intertropical area.

The present day distribution of fish in the tropical lakes and river catchments of Africa is the result of several, very different, interacting factors.

1. The historical climatic or geological events that made possible connections between or isolation of river catchments. There have been continual climatic changes, but their number and nature is reasonably well established for only the past 20 000–30 000 years. Study of lake sediments provides a good source of information on the past hydrology of Africa, but there are comparatively few data on the paleogeography of rivers.
2. The ability of fish to disperse (Hugueny, 1990a) and the existence of refuge zones during drought periods, or of relict zones usually isolated by physical barriers such as falls impassable to fish.
3. Speciation processes that depend on the duration of isolation and the evolutionary potential of fish families. Lack of knowledge of the phylogenetic relationships among species and genera, as well as the sometimes still confused taxonomy and poor distribution data, is, in spite of great improvements during recent decades, a serious problem for biogeographic studies in Africa.
4. The size of the river basin and diversity of aquatic habitats available to fish. Relationships between the number of fish species present and catchment area and/or river discharge were established for Africa by Daget & Iltis (1965), Livingstone et al. (1982) and Hugueny (1989). There are undoubtedly extinction processes that can explain the lower number of species recorded in smaller river catchments.
5. The consequences of diseases (viruses, parasites, bacteria, fungi) that may have affected fish populations, even if this type of impact is still poorly documented in wild populations.

Main ichthyological provinces

The search for and recognition of areas characterised by a specific ichthyofauna assemblage has long been a favourite game for many African ichthyologists. In an historical sense, the number and extent of ichthyofaunal provinces in many ways reflect the state of knowledge of African freshwater fishes. As pointed out by Greenwood (1983a): 'provincial boundaries have been modified somewhat over the years: new provinces were defined as more taxa were described or the status of others was reassessed, and as more information was obtained on the distribution of various species. Overall, however, the changes have not been profound.'

Nevertheless, the concept of ichthyofaunal provinces proved to be a useful tool for descriptive and comparative purposes (Skelton, 1988), even if it suffers limitations in its application. They are in some ways subjective constructions, often based upon insufficient data on the distribution and phylogeny of fish groups. Moreover, different approaches to defining such provinces in Africa have mainly used the distributions of restricted taxa as criteria in determining their boundaries.

Boulenger (1905) was the first to be tempted to divide the so-called 'Ethiopian Region' into faunistic provinces. He and Pellegrin (1911, 1921) identified some large-scale patterns that, in most cases, are the foundations of the present day ichthyofaunal provinces identified by Roberts (1975), who synthesised the attempts of Blanc (1954), Matthes (1964) and Poll (1973) to divide Africa into homogeneous faunistic zones. Roberts (1975) finally recognised ten ichthyofaunal provinces. They are briefly discussed here, and some modifications are proposed as a result of new information now available.

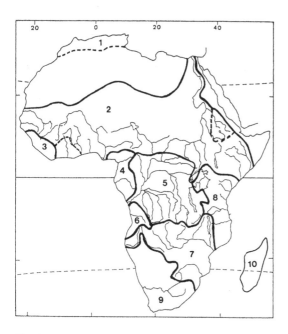

Fig. 2.1. Main ichthyological provinces in Africa. Modified from Roberts (1975). For numbers see text.

1. The **Maghreb** has an extremely poor fish fauna, only five families have been recorded (Doadrio, 1994) (Table 2.3).

2. The **Nilo–Sudan** that extends from the Atlantic Coast to the Indian Ocean and includes the major drainage basins of the Nile, Chad, Niger, Volta and Senegal. The northern limit of this province varies according to climatic conditions, but one can suggest that it also includes the relict rivers of the present day Sahara (see Fig. 2.1). Pellegrin (1923) provided a useful regional account of West African fishes, which has recently been updated by the publication of the 'Fauna of fresh- and brackish-water fishes of West Africa' (Lévêque *et al.*, 1990, 1992). This work builds on previous local contributions, the most important being for Gambia (Svensson, 1933; Johnels, 1954), Ghana (Irvine, 1947), Niger (Daget, 1954), Chad (Blache *et al.*, 1964), Côte d'Ivoire (Daget & Iltis, 1965), Volta (Roman, 1966). Boulenger's (1907) detailed account is still the most comprehensive work for the Nile.

The Nilo–Sudanian fish fauna is currently well known (Lévêque *et al.*, 1991; Paugy *et al.*, 1994),

uniformly distributed through most of its range and fairly diversified (Table 2.3). Nevertheless, it is possible to identify two small subprovinces: the Eburneo–Ghanean subprovince on the western side, characterised by a few endemic fish species mixed with a typical Nilo–Sudan fauna (Teugels *et al.*, 1988), and the Abyssinian subprovince, considered as a full province by Roberts (1975), with an impoverished nilotic fish fauna and few endemics. The Cross River is the south-western limit of that province (Teugels *et al.*, 1992).

3. The **Upper Guinea** province includes coastal rivers from south of the Kogon River in Guinea to Liberia, and exhibits faunistic affinities with the lower Guinea province and the Zaïre. The fauna, diversified and fairly well investigated, includes many taxa endemic to that area (Schültz, 1942; Daget, 1962*a*, *b*; Lévêque *et al.*, 1989, 1990, 1992; Paugy *et al.*, 1989).

4. The **Lower Guinea** covers the coastal rivers from Cameroon to the mouth of the Zaïre (which is not included). There is a scattered literature on the fish fauna that is still poorly investigated despite some recent surveys (Thys van den Audenaerde, 1966, 1967; Roman, 1971; Daget, 1978, 1979; Daget & Depierre, 1980; Amiet *et al.*, 1987; Teugels *et al.*, 1992*c*; Mamonekene & Teugels, 1993).

5. The **Zaïre** province, according to Roberts, includes Lakes Kivu and Tanganyika. The Zaïre drainage basin is the largest in Africa and the ichthyofauna is much richer than in any other river on the continent. This fauna has been investigated by a number of different scientists and large collections are available. However, the available information on taxonomy and distribution needs to be sorted and synthesised, and further field and taxonomic investigations will certainly be necessary while part of the basin remains poorly investigated as at present. Among local contributions, one should mentioned Gosse (1963, 1966, 1968), Poll & Gosse (1963), de Kimpe (1964), Matthes (1964), Poll (1967, 1976) and Banister & Bailey (1979).

6. The **Quanza**, which covers the Angolan coastal drainages, is certainly still more poorly known than all the other provinces. The inventory and affinities of the ichthyofauna need further investigation

Table 2.3. *Indigenous primary and secondary division fish families* (sensu *Myers, 1951*) *occurring in the different ichthyological provinces (see Fig. 2.1 and Table 2.1)*

Fish families	Mag	Nilo–Sudan			UG	LG	ZA	ZZ	Qua	E.Co.	Cape	Mad.
		EG	Nilo	Aby								
Protopteridae		*	*		*	*	*	*		*		
Polypteridae		*	*		*	*	*					
Denticipitidae						*						
Osteoglossidae		int	*			*						
Pantodontidae						*	*					
Notopteridae		*	*		*	*	*					
Mormyridae		*	*		*	*	*	*		*		
Gymnarchidae			*									
Kneriidae			*				*	*	*	*		
Phractolemidae						*	*					
Hepsetidae		*	*		*	*	*	*	*			
Characidae		*	*		*	*	*	*	*	*		
Distichodontidae		*	*		*	*	*	*	*	*		
Citharinidae		*	*			*	*			*		
Ichthyboridae			*		*	*	*					
Cyprinidae	*	*	*	*	*	*	*	*	*	*	*	
Cobitidae	*			*								
Bagridae		*	*		*	*	*	*	*	*	*	
Schilbeidae		*	*		*	*	*	*	*	*		
Amphilidae		*	*	*	*	*	*	*	*	*		
Clariidae	*	*	*	*	*	*	*	*	*	*		
Malapteruridae		*	*		*	*	*	*				
Mochokidae		*	*		*	*	*	*	*	*		
Cyprinodontidae	*	*	*		*	*	*	*	*	*		*
Channidae		*	*		*	*	*					
Synbranchidae					?							
Nandidae		*			?	*						
Cichlidae	*	*	*	*	*	*	*	*	*	*		*
Anabantidae		*	*		*	*	*	*		*	*	
Mastacembelidae		*	*		*	*	*	*	*	*		
No. families	5	21	24	5	20	27	24	17	14	16	3	2

Abbreviations: Mag (Maghreb), EG (Eburneo–Ghanean), Nilo (Nilotic), Aby (Absyssian), UG (Upper Guinea), LG (Lower Guinea), ZA (Zaïre), ZZ (Zambezi), Qua (Quanza), E.Co. (East Coast), Cape (Southern Africa), Mad. (Malagasy).

before any further speculation can be made about the faunal characteristics of that province. The paper from Poll (1967) that deals mainly with tributaries of the Zaïre is the most recent major contribution on the ichthyofauna of that area.

7. The **Zambezi**, including the river basins of the Cunene, Ovambo, Okavango, Zambezi and Limpopo, as well as Lake Malawi, has a moderately rich fauna (Table 2.3) that has been fairly well documented (Jubb, 1967; Bell-Cross, 1972, 1976;

Gaigher & Pott, 1972; Bruton & Kok, 1980; Skelton *et al.*, 1985; Jackson, 1986; Skelton, 1994).

8. The **East Coast** province covers the coastal drainages from the Juba in the north to the Zambezi in the south. There is surprisingly little recent literature on Kenyan rivers, which were investigated by Copley (1958), Whitehead (1959) and Bailey (1969). Eccles (1992) and Skelton (1994) provided a synthesis of the current knowledge.

9. The **Southern** province includes the basins of the

Orange–Vaal and all the systems to the south. The fauna is well documented, but rather poor and dominated by cyprinids (Table 2.3). Papers from Barnard (1943), Jubb (1965, 1967), Gaigher *et al.* (1980), Skelton & Cambray (1981), Cambray (1984) and Skelton (1986) provide most of the information on the ichthyofauna of that province.

10. **Malagasy** is not an ichthyofaunal province recognised by Roberts, but is usually considered as part of the African continent. The history of the island is an enigma to biogeographers, and in spite of its large area only 66 fish species are regarded as indigenous to its freshwaters (Arnoult, 1959; Kiener, 1963; Stiassny & Raminosoa, 1994). All of its freshwater fish belong to widely distributed secondary or peripheral division families, most of which are faunistically part of the Indian Ocean–Western Pacific marine province. All but two of these provinces (the Maghreb and the Cape of Good Hope) are within the tropical boundaries of this book.

Fish zoogeography in north tropical Africa

Most of the attempts to recognise different ichthyofaunal regions or provinces in north tropical Africa have been based on superficial, and sometimes poor knowledge of the occurrence and distribution of fish species in the dense coastal drainage network, which extends from Senegal to the Niger Delta.

Daget & Iltis (1965), in their major contribution to the West African fish fauna, distinguished three categories of freshwater fishes: Sudanian fishes living in the waters of the extensive savannah-covered floodplains, Guinean fishes inhabiting forest streams, and indifferent. Such a division enabled them to calculate a Sudano–Guinean index (which is the ratio of the number of Sudanian fish species to the number of Guinean ones) in order to characterise different river basins or parts of these basins.

Roberts (1975) identified three main ichthyofaunal provinces in northern tropical Africa: the Nilo–Sudan, the Upper Guinea and the Lower Guinea. He pointed out the faunistic relationships between the Upper and Lower Guinean regions, demonstrated by the existence of shared species or closely related taxa. All the above geographical delimitations were more or less empirical

and not supported by quantitative data. Nevertheless, they reflected the state of the art at the time they were published. Since Roberts' paper, a few works have been published that provide new information on the regional fish fauna and suggested improvements to his division.

The recognition of ichthyological provinces based on comparisons of faunistic lists for the different river basins (Teugels *et al.*, 1988; Lévêque *et al.*, 1989, 1990; Paugy *et al.*, 1989), partially confirms the empirical division proposed by Roberts (1975), but the boundaries must be slightly modified (Hugueny & Lévêque, 1994; Paugy *et al.*, 1994):

1. The Nilo–Sudan province remains almost unchanged with the Nile, Chari, Niger, Senegal, Volta and the Gambia (included in Upper Guinea by Roberts), but it appears that some differences in fish fauna occur in basins, or lower reaches of rivers close to the Lower Guinea province (*sensu* Roberts), such as the Cross (included in Lower Guinea by Roberts), Ogun, Oueme, Mono and Lower Niger. The Geba, the Corubal and the Kogon rivers basically belong to the Nilo–Sudan group, but also exhibit some affinities with the Upper Guinea fauna.

2. The Upper Guinea province (*sensu* Roberts) is not homogeneous. It should be divided into the Upper Guinea province *sensu stricto*, including coastal basins from the Konkoure in Guinea to the Saint John River in Liberia, and the Eburneo–Ghanean province, from the Cess (Nipoue) in Côte d'Ivoire to the Pra in Ghana, which is closer to the Nilo–Sudan province than to the Upper Guinea one. Faunistically, the Eburneo–Ghanean province is not completely homogeneous, the fish fauna of the Comoe, Sassandra and Bandama rivers for instance, being closer to the Nilo–Sudan fauna, whereas the Nipoue and Cavally have more obvious affinities with the Upper Guinean rivers. The distinction of the Eburneo–Ghanean province results from the existence of some fish species with a geographical distribution restricted to that area: *Marcusenius furcidens*, *Marcusenius ussheri*, *Citharinus eburneensis*, *Barbus trispilos*, *Barbus bynni waldroni*, *Amphilius atesuensis*, *Schilbe mandibularis*, *Synodontis bastiani* and *Aethiomastacembelus nigromarginatus. Brycinus imberi*,

a widely distributed species, is only recorded from the Eburneo–Ghanean area and not from the Nilo–Sudan river basins, but the fish fauna as a whole is very similar to that of the Nilo–Sudan, and the Eburneo–Ghanean province should be considered as a subregion of that province.

3. The Lower Guinea group includes the Mungo, Wouri, Sanaga, Nyong, Lokundjé, Lobé, Kribi, Ntem, Rio Ekudo, Ogowe and Niari-Kouilou.

The major difference from Roberts' zoogeographical division lies therefore in the larger Nilo–Sudan province, including parts of what he placed in his Guinean provinces (the Cross River and the Eburneo–Ghanean area). It is not clear why Roberts included the small coastal basins of Côte d'Ivoire in the Upper Guinean province, but the reason could be their location in the forest area. However, from the results above, it is clear that there is no direct correlation between fish species distribution and the present extent of the rain forest.

There is no quantitative biogeographic study dealing with other aquatic groups in West Africa. As a result, it is not possible to check if the provinces recognised for fishes are shared with other animals. A study of aquatic birds (Guillet & Crowe, 1985) gave different results, but that is not really surprising given that birds are independent of connections between watersheds for migration. For aquatic molluscs, Van Damme (1984) recognised different biogeographic areas in Africa that agree only in part with those for fish. At the generic level, Van Damme identified the equivalent of the Eburneo–Ghanean region, where there is high endemicity. The same is true for the Lower Guinea, whereas the Upper Guinea does not appear to be different from the Nilo–Sudan area. The zoogeography of some Monogeneans (branchial parasites of freshwater fishes) is fairly similar to that observed for fish (Guégan & Lambert, 1990, 1991). The *Dactylogyrus afer* and *D. brevicirrus* groups are mainly distributed in Nilo–Sudan rivers and not recorded from Upper Guinea. Conversely, the *Dactylogyrus ruahae* and *D. archaeopenis* groups are characteristic of the Guinean zones. Moreover, as a result of an enzymatic study of different populations of the siluriform *Chrysichthys maurus*, three main patterns were distinguished: Western Guinean province, Eastern Guinean province, and Côte d'Ivoire (Agnèse, 1989). This study confirms that the

Cess River is a boundary, separating the Guinean and Eburneo–Ghanean provinces.

Endemism and fish species flocks in the Great East African Lakes

The most striking feature of the fish of the East African Great Lakes (Victoria, Tanganyika, Malawi) is that each has its own highly endemic lacustrine cichlid fauna that apparently evolved from a riverine ancestral stock (Table 2.4). The term 'species flock' is used to refer to those monophyletic groups of closely related species coexisting in the same area (Greenwood, 1974b, 1984). The Great Lakes are therefore exceptional natural sites for the study of speciation and ecology. Why are there so many species and how do they manage to coexist in each lake? For some authors, these lakes are natural laboratories for the study of speciation. But this endemic fauna is particularly threatened by trawl fisheries and the introduction of alien species. The documented disappearance of many cichlid species from Lake Victoria resulted in international concern for the knowledge and preservation of this unique fish fauna, and stimulated taxonomic and biological research. An extensive literature has been published in the last ten years (van Oijen et al., 1981; Ribbink et al., 1983a, b; Eccles & Trewavas, 1989; Coulter, 1991; Ribbink, 1991; Witte et al., 1992a, b). Cichlid speciation has also occurred, to a lesser extent, in other smaller lakes of the Rift Valley, such as Albert, Turkana, Edward, George and Kivu. Mechanisms responsible for this speciation will be discussed in later chapters. The non-cichlid fauna of the Great Lakes is also highly endemic and reflects the fauna inhabiting their present and former drainage systems.

Roberts' (1975) inclusion of Lakes Kyoga and Victoria in the East Coast province was not really convincing. He argued that their ichthyofauna may be derived from a depauperate riverine fauna more or less similar to that of the eastern coastal rivers, but it is known that during the Miocene the fauna of Lake Victoria was quite similar to that of the Nile. Moreover, fluvial connections exist between the Nile and Lake Victoria, and the latter probably overflowed through the Victoria Nile into the Nile itself during the last pluvial. Lake Victoria, as well as Lakes Edward and Albert, could perhaps be considered as latecomers to the Nile system (Greenwood, 1976a). The present or past existence of a Nile-type

Table 2.4. *Composition of the fish fauna in the East African Great Lakes*

Fish families	Tanganyika		Malawi		Victoria		Kivu		Turkana	
	no. sp.	no. gen.	no. sp.	no. gen.	no. sp.	no. gen.	no. sp.	no. gen.	no. sp.	no. gen.
Protopteridae	1	1			1	1				
Polypteridae	2	1							2	1
Anguillidae			1	1						
Clupeidae	2	2					1 int			
Osteoglossidae									1	1
Mormyridae	6	6	6		7				2	2
Gymnarchidae									1	1
Kneriidae	1	1								
Characidae	7	5	2	1	2	1			9	4
Distichodontidae	3	1							1	1
Citharinidae	1	1							1	1
Cyprinidae	35	8	16		12		3	2	10	6
Bagridae	17	7	1	1	2				4	3
Schilbeidae	2	1			1	1			1	1
Amphilidae	2	1	1						1	1
Clariidae	6	4	14		6		2	1	2	2
Malapteruridae	1	1							1	1
Mochokidae	9	2	2		2				3	2
Cyprinodontidae	2	2	1		3				2	1
Centropomidae	4	1							2	1
Cichlidae	185	50	600+	53+ (48E)	200–250	21 (5E)	14	3	7	5
Anabantidae	1	1			1					
Mastacembelidae	12	2	1		1	1				
Tetraodontidae	1	1							1	1
Total	300+	103	645+	56+	238–288	25	20	6	51	35

Sources: for cichlids, data from CLOFFA IV (Daget *et al.*, 1991). For non-cichlids: Lake Tanganyika, data from Coulter (1991*b*, 1994*b*); Lakes Malawi and Victoria, data from Ribbink & Eccles (1988), Ribbink (1994), Greenwood (1994); Lake Turkana, data from Lévêque *et al.* (1991). E, endemics.

ichthyofauna in these lakes could be explained by the existence of an archaic fauna widely distributed over Africa before the formation of the lakes. Whatever the reasons, it seems more correct to include Lake Victoria in the Nilo–Sudan province, taking into account the affinities of the non-cichlid fauna with the Nile fauna. The currently endorheic Lake Turkana, which had connections with the Nile a few thousand years ago, also belongs to the Nilo–Sudan province.

Lake Tanganyika belongs to the Zaïre province and the presently diverse fauna of the Tanganyika basin must owe much of its character and diversity to the original Zaïrean stocks. All the fish families now present in the lake tributaries were probably represented in a swamp or river-dwelling palaeofauna of the Zaïrean region (Coulter, 1991*a*). Evolution in Lake Tanganyika itself has led to the occurrence of species flocks within a few families: 7 mastacembelid species, 6 species of the bagrid *Chrysichthys*, 7 species of *Synodontis* and 4 species of the centropomid *Lates*. The fauna of Lake Kivu, which formerly overflowed to the Lake Edward basin, is greatly impoverished compared to Lake Tanganyika with which it is now linked. This situation is probably the result of volcanic events that caused overturn of the lake, introduction of toxic gases into the surface waters and faunal extinctions as a consequence (Haberyan & Hecky, 1987).

Lake Malawi is connected with the Zambezi system via the Shire River, and the entire fish fauna could conceivably have been derived from the Lower Zambezi (Roberts, 1975). The diverse cichlid fauna (Table 2.4) of this lake is probably underestimated and according to some authors (Ribbink, 1988) it could be more than 500 species. Species flocks are also reported for the clariid catfish *Dinotopterus* (10 species).

Fish zoogeography in southern Africa

An overview of the zoogeography of freshwater fish in southern Africa has been recently provided by Skelton (1994). The primary freshwater fish of East and Southern Africa comprise a similar assemblages of relatively low diversity.

The East Coast fauna lacks many distinctive freshwater fish families including polypterids, clupeids, denticipitids, osteoglossids, pantodontids, notopterids, gymnarchids, phractolaemids, malapterurids, nandids and the genus *Hepsetus*. However, at least four species of *Anguilla* are reliably recorded from this region. The most notable features are the radiation of the aplocheilid genus *Nothobranchius* and the number of endemic species of the cichlid genus *Oreochromis*. Elements of the freshwater fish fauna of the East Coast province range as far south as northern Natal in South Africa and there is an overlap with the Zambezi ichthyofaunal province.

The Zambezian fauna is dominated by cyprinids (31% of species), cichlids (21%) and silurids (21%). As in the East Coast province, many fish families are absent. The greatest degree of endemic speciation in the Zambezian fauna has occurred within the mochokid catfish (genera *Chiloglanis* and *Synodontis*) and the serranochromine cichlids. Subunits have been recognised in the Zambezian province: western-sector subunits include the Kunene, Okavango, Upper Zambezi and Kafue Rivers; and the eastern-sector subunits are the Middle and Lower Zambezi, Limpopo, Incomati and Pongola. Lake Malawi and the Shire River are also part of the Zambezian province. The diversity of the fish fauna is consistently higher in the western than in the eastern sector of this province, and the fauna of the Upper Zambezi is the most diverse in terms of species and genera.

Characteristic 'East Coast' species, such as *Marcusenius macrolepidotus*, *Petrocephalus catostoma*, *Barbus annectens*, *Barbus marequensis*, *Barbus radiatus*, *Labeo congoro*, *Synodontis zambezensis*, *Oreochromis mossambicus* and *Ore-*

ochromis placidus, form essential components of the eastern-sector diversity. The absence of several prominent large mainstream dwelling species from the western sector, such as *Mormyrops anguilloides*, *Mormyrus longirostris*, *Heterobranchus longifilis*, *Malapterurus electricus*, *Distichodus mossambicus* and *Distichodus shenga*, which occur in the Middle and Lower Zambezi, profoundly underlines the differences between eastern- and western-sector drainages.

The Madagascan ichthyofauna

Prior to understanding the mechanisms by which Gondwana fragmented, early workers proposed various scenarios of ancient land connections (Sauvage, 1891), or massive extinctions due to competition from colonising marine fauna (Pellegrin, 1934) to explain the peculiarities of the Madagascan ichthyofauna: the total absence of socalled primary freshwater fish groups and the predominance of peripheral or marine groups in freshwaters. Only two secondary families are native to Madagascar: Cichlidae and Cyprinodontidae. However, Stiassny (1991) demonstrated that the cichlids are probably archaic and form the basal sister groups of the clade. That is also the case for other groups of marine origin such as the Ancharidae, Mugilidae, Bedotiidae. Madagascar therefore harbours a concentration of basal taxa particularly important for phylogenetic studies.

Madagascar separated from India long after they parted from Africa. The absence from Madagascan waters of several fish clades shared by India and Africa (channid, notopterid, anabantid, mastacembelid, clariid, schilbeid and bagrid) is therefore particularly noteworthy (Stiassny & Raminosoa, 1994). The notopterids, for instance, were established in the eastern portion of Gondawana from the Late Jurassic, and their absence from Madagascar may therefore be secondary as a result of pre- or post-drift extinction. That is also the case for polypterids as well as for Dipnoi, which were present on Madagascar at least until the Lower Triassic.

Lundberg (1993) questioned the pre-drift origin of various fish clades. He considered that for the majority of fish groups with so-called Gondwanan distributions, there is little evidence that their current distributions can be explained by recourse to a single drift-related vicariance model. A pre-drift origin of the Cichlidae or cyprinodontiforms is therefore highly unlikely according to this theory that would explain the absence of many clades from Madagascar (see Stiassny & Raminosoa, 1994). The presence of Cichlidae in Madagascar, would for example, be understood in terms of some type of post-drift dispersal. This hypothesis certainly needs more well-established evidence before it can be considered as an explanation of the relative poverty of freshwater fish on Madagascar.

The fossil record

A comprehensive review of fossil freshwater fishes of the African Cenozoic was given by Greenwood (1974a). Only a few studies have been published since (Patterson, 1975; Van Couvering, 1977, 1982; Mahboudi et al., 1984; Van Neer & Gayet, 1988; Gauthier & Van Neer, 1989; Van Neer, 1989).

The geographical and temporal distribution of fish remains, based on published records, is very uneven. For instance, there are very few or no remains from the Upper Guinean province, or the Great Rift Valley lakes where the contemporary fauna is of outstanding interest. The Lower Nile provides the most continuous record.

The information to be derived from fossil records is generally relatively slender (Greenwood, 1974a). Using osteological characters it is possible to identify the remains to the family or generic level but, with few exceptions, not to species level for which anatomical or morphometric characters are needed. At the generic level, there is little difference between the present day faunas of African freshwater systems, at least north of 10°S, and the situation was apparently the same throughout the Cenozoic. As a result, we have very few data on which to discuss the early evolution of African fish faunas and their differentiation. There is no indication from fossils dating from before the end of the Miocene of any large endemic fauna that might have evolved in an isolated part of the continent.

However, some observations are particularly relevant to understanding some aspects of the past history of the African fish fauna.

1. The discovery in fish remains from late Miocene deposits of Tunisia of a diversified freshwater ich-thyofauna, showing strong affinities with the contemporary fauna of Egypt and tropical Africa (Priem, 1914, 1920; Greenwood, 1951; Greenwood & Howes, 1975; Arambourg, 1963). It included Nilotic genera such as *Lates, Clarias, Heterobranchus, Polypterus*, etc. (Greenwood, 1973a), and contrasts strongly with the fish fauna of the present day. *Protopterus* sp. was also identified in Lower Eocene fossils from El Kohol, near Brezina on the southern flank of the Saharan Atlas, Algeria (Mahboudi et al., 1984). The extinction of this Nilotic fauna, and its replacement by a depauperate fauna showing affinities with Europe and Asia Minor, cannot be precisely dated.

2. Later Van Couvering (1982) reported that *Palaeochromis*, a Late Miocene cichlid discovered in the Seybouse Valley near Guelma (Algeria), is closely affiliated with the Central and West African pelmatochromines.

3. Fish remains from a Miocene lake that lay in the basin now occupied by Lake Victoria were more similar to the Nile fauna than is the fauna of Lake Victoria today. It included two genera no longer present in that area (*Lates* and *Polypterus*) (Greenwood, 1951, 1974a). A similar assemblage occurred from the Pliocene to the Holocene in the surroundings of Lake Edward (Uganda), where *Lates, Hydrocynus* and *Polypterus*, etc. are no longer present (Greenwood, 1959; Greenwood & Howes, 1975). According to Greenwood (1973a): 'the faunal similarity between different Miocene deposits gives the impression that during the Miocene there was a fairly uniform freshwater fauna (at least at the generic level) widely distributed in Africa north of the equator. Due to the relative scarcity of materials, it is difficult to determine just how widespread this fauna was or when it made its first appearance. But Eocene deposits in Libya (Arambourg & Magnier, 1961) have a very similar fish fauna (*Polypterus, Lates* and unidentified Siluriformes)'. The occurrence of specimens referred to as Pelmatochromines in the Early Miocene at the foot of Mount Elgon (Uganda) and of *Paleaofulu kuluensis* in the Early Miocene of Rusinga Island (Kenya) suggests zoogeographic ties between East Africa and either Central

Table 2.5. *Fish fossil records from different sites in Africa*

Fish fauna	Eocene 54–36 Myr			Oligocene 35–23 Myr	Miocene 23–6 Myr																Pliocene 6–1.8 Myr										Pleistocene 1.8–0.01 Myr									Holocene 0.01–today		
References	1	2	3	4	5	6	7	8	9	10	11	12	13	14	15	16	17	18	19	20	21	22	23	24	25	26	27	28	29	30	31	32	33	34	35	36	37	38	39	40	41	42
Polypterus	•																																								•	
Protopterus				•		•	•					•	•									•					•			•	•					•	•			•	•	•
Lates				•		•	•	•		•		•	•	•	•	•	•		•		•	•	•	•		•	•	•	•	•	•			•		•	•		•	•	•	•
Characidae					•				•							•				•			•	•																•		
Alestes																								•																		
Hydrocynus											•	•																														
Sindarax lep											•	•											•					?														
Auchenoglanis																																						•				
Arius								•																																		
Bagrus													•		•	•	•		•			•	•				•		•	•	•	•	•	•	?	•		•	•	•	•	•
Clarias				•		•							?				•					•	?	?			•		•			•	?	•		•	?	•	•	•	•	•
Clarotes																								?									?				?					
Chrysichthys									•				•		•		•					•	•	•			•		•	•		•	•	•		•	•		•	•	•	•
Heterobranchus																															?											
Synodontis													•		•		?					•	•	?	?				?	?	?	•	•	•	?	•		•	•	•	•	•
Barbus																						•	•								?											
Labeo								•																																		
Hyperopisus																						?	?																			
Cichlidae								•	•	•																														•		
Tilapia									•	•	?						•	•	•				•				•					•		•		•	•	•		•	•	•

Based mainly on reviews by Greenwood (1974a) and Van Couvering (1977). **Middle Cretaceous–Lower Eocene:** 1, Batarije (Egypt) (Greenwood, 1974a). **Eocene:** 2, Eocene ? Tamaguilelt (Mali) (Greenwood, 1974a); 3, El Kholol near Brezina (Algeria) (Mahboudi et al., 1984). **Oligocene:** 4, North of Fayum (Egypt) (Greenwood, 1974a). **Miocene:** 5, Daban formation, Biyo Gora (Somalia) ? 23–36 Myr (Van Couvering, 1977); 6, Mount Elgon (Uganda) , 23 Myr (Van Couvering, 1977); 7, Karunga, south Nyanza (Kenya), ? 23 Myr (Van Couvering, 1977); 8, Uyoma Peninsula, central Nyanza (Kenya), ? 23 Myr (Van Couvering, 1977); 9, Turkana basalt, Loperot (Kenya), ? 17 Myr (Van Couvering, 1977); 10, Rusinga Island, Nyanza Gulf (Kenya) 17–19 Myr (Van Couvering, 1977); 11, Lower Miocene, Sinda–Mohari (Greenwood & Howes, 1975); 12, Lower Miocene, Lake Albert (Greenwood & Howes, 1975); 13, Lower Miocene, Rusinga Island, Lake Victoria (Kenya) (Greenwood, 1974a); 14, Moghara, 150 km south-west of Alexandria (Egypt) (Greenwood, 1974a); 15, Upper Miocene, Chalouf, 25 km from Suez (Egypt) (Greenwood, 1974a); 16, Burdigalian, Cyrenaica, 150 km south of Syrte Major (Libya) (Greenwood, 1974a); 17, Vindobonian, Bled ed Douarah (Tunisia) (Greenwood, 1974a); 18, Malembe (Cabinda enclave), Zaïre basin (Greenwood, 1974a); 19, Ngrorora formation, Tugen Hills (Kenya), 10–12 Myr (Van Couvering, 1977); 20, Mpesida beds, Tigen Hills (Kenya), 7 Myr (Van Couvering, 1977); 23, Middle Pliocene, Wadi el Natrun (Egypt) (Greenwood, 1972); 24, Pliocene or earlier Pleistocene, Sinda Mohari (Greenwood & Howes, 1975); 25, Tessa formation, Gouv. Beja (Tunisia), 2–3 Myr (Van Couvering, 1977). **Pleistocene:** 26, Lower Pleistocene, Olduvai Gorge (Tanzania) (Greenwood, 1974a); 27, Lower Pleistocene, Omo, north of Lake Turkana (Kenya) (Greenwood, 1974a); 28, Lower Pleistocene, Lake Malawi (Greenwood, 1974a); 29, Lower Pleistocene, Sinda Mohari (Greenwood & Howes, 1975); 30, Lower Pleistocene, Lake Albert (Greenwood & Howes, 1975); 31, Lower Pleistocene, Lake Edward (Greenwood & Howes, 1975); 32, Middle Pleistocene, Rawe (Kenya) (Greenwood, 1974a); 33, Middle Pleistocene, Ishango, Lake Edward (Zaïre) (Greenwood, 1974a); 34, Upper Pleistocene, Ishango, Lake Edward (Zaïre) (Greenwood, 1974a); 35, Upper Pleistocene, Lake Eyasi (Tanzania) (Greenwood, 1974a); 36, Upper Pleistocene, Kangatotha, west of Lake Turkana (Kenya) (Greenwood, 1974a); 37, Upper Pleistocene to Holocene, Nubian Sudan near Wadi Halfa (Greenwood, 1974a); 38, Upper or Neo Pleistocene, Sahara (Greenwood, 1974a); 39, Upper or Neo Pleistocene, Chad and Tibesti (Greenwood, 1974a). **Holocene:** 40, Fayum (Egypt) (Greenwood, 1974a); 41, Ishango, Lake Edward (Greenwood, 1974a); 42, Taoudeni basin (Gayet, 1983).

and/or West African forest areas, the modern centre of Pelmatochromine distribution (van Couvering, 1982). Such results support the hypothesis suggested above of a widespread ichthyofauna during the Miocene.

4. The earliest fossil records of African fishes (Cretaceous–Eocene) are from Egypt, and of one of the most primitive bony fish represented by the living genus *Polypterus*. Pliocene deposits at Wadi el Natrun, showed a typical Nilotic fauna (Stromer, 1916; Weiler, 1926; Greenwood, 1972), which is fairly similar to the present day fauna, at least at the generic level. It includes the earliest record for characids in Africa, *Alestes deserti* Greenwood, 1972, possibly related to *Brycinus macrolepidotus*. Other material from Wadi el Natrun include the earliest record of mormyrids (genera *Hyperopisus*), an endemic African family, from the Middle Pliocene, as well as the first definite record for the genera *Labeo* and *Barbus* in Africa.

The fossil fish record from western Africa as a whole is poor, but the now desert Sahara region, during the Pleistocene and Early Holocene, had a widespread fish fauna that was associated with large lakes and a dense drainage (see Talbot, 1980; Riser & Petit-Maire, 1986). A diversified fish fauna was collected (Daget, 1959b, 1961b; Gayet, 1983; Van Neer & Gayet, 1988). *Lates maliensis* Gayet, 1983 was described from the Taoudeni depression in the Malian Sahara, but it is now considered as a synonym of *Lates niloticus* (Van Neer, 1987). At present, 15 species inhabit small waterbodies in the Sahara, and are relict populations of widespread species occurring in surrounding basins (Lévêque, 1990b). The isolation of these populations probably happened 5000–6000 years ago, at the end of the last Holocene humid period.

A fossil denticipitid, *Palaeodenticeps tanganikae*, was described from the Oligocene of East Africa (Greenwood, 1960), and it should be noticed that the only living species of this family is at present known from coastal rivers from Benin to the Niger Delta.

A rich fossil fish fauna has been reported from Zaïre and includes the presence, in Middle Cretaceous deposits of marine coastal origin, of an osteoglossid, *Paradercetis kipalaensis*, close to the present *Heterotis* (Taverne, 1975); three species of Salmoniformes in Lower Cretaceous deposits of north-east Zaïre (Taverne, 1976) and, in the Lower Cretaceous from Gaboon and Equatorial Guinea, *Parachanos aethiopicus* which is probably the oldest known teleost Gonorhynchiforme (Taverne, 1974).

Table 2.5 is a tentative summary of fossil freshwater fish records in Africa, the interpretation of which needs caution. The absence of fossils does not mean that the genus or the family was not present at that time. It simply means that there are no fossil records for the species at that geological period. Obviously, strong bones of large species, such as *Polypterus* or *Lates*, are more likely to be preserved and found in sediments than fragile bones of tiny species, but that is not always the case. The oldest fossils (Eocene) recorded belong to *Protopterus* and *Polypterus*. Many genera are already recorded during the Miocene and most of them were found in Pliocene deposits.

Intercontinental affinities of the African ichthyofauna

Despite the special features of the African ichthyofauna mentioned at the beginning of this chapter, many other African freshwater fish families exhibit intercontinental affinities mainly with the fish of Asia and South America. Such affinities have been used by different authors to propose dispersal patterns through land connections between continents.

Gondwana and continental drift

Two million years ago, during the Late Triassic, all the separate continents seen today were amalgamated together as one supercontinent, Wegener's Pangaea whose southern portion is referred to as Gondwana or Gondwanaland, and the northern part as Laurasia (Fig. 2.2). By the Mid Jurassic, the separation between West Gondwana (Africa and South America) and East Gondwana (Madagascar, India, Australia) had begun, and about this time the modern groups of freshwater fish began to appear. By the Late Jurassic (c. 140–150 Myr BP), Madagascar–India had broken away from Africa and Euamerica still had a narrow connection to Africa. The Atlantic Ocean had developed extensively.

The separation of Africa from South America probably started in the Lower Cretaceous (c. 125 Myr ago)

180–200 Myr

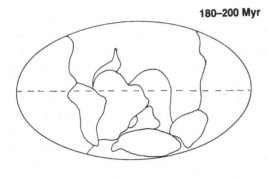

120 Myr

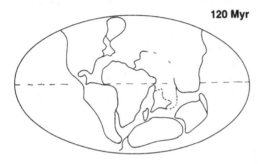

56 Myr

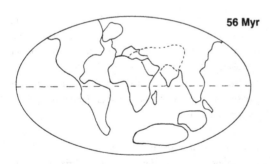

Fig. 2.2. Reconstruction of continental displacement, assuming an expanding Earth during the last 200 million of years (diameter respectively 94%, 87% and 80%, at 56, 120 and 180–200 Myr ago) (simplified from Owen, 1983).

Table 2.6. *The different geological periods and their approximate limits in time*

Era	Period	Limits in Myr
Cenozoic including		0–65
Quaternary		0.0–1.9
	Holocene	
	Pleistocene	
Tertiary		1.9–65.0
	Pliocene	1.9–5.3
	Miocene	5.3–23.0
	Oligocene	23–35
	Eocene	35–54
	Palaeocene	54–65
Secondary or Mesozoic		65–235
	Cretaceous	65–135
	Jurassic	135–195
	Triassic	195–235
Primary or Palaeozoic		235–570
Precambrian		<570

From Pomerol & Renard, 1989.

Africa, still joined to Arabia, had become an island and it remained so for some 25 million years.

The separation of each of the constituent continental segments of the former Gondwanaland was associated with a slow clockwise rotation and northwards movement of the African plate, which took place during the ensuing 65 million of years of Cenozoic time. During the Cretaceous, Africa was positioned such that the Equator ran from SW Nigeria, through central Chad into NW Sudan. In the course of the Middle Cenozoic, the African island was moving slowly northwards from a latitudinal position about 15° south of the present day.

During the Eocene, India was encroaching northwards on Eurasia, elevating the Himalayan ranges. In the Miocene, 17 million years BP, Arabia and the north-east corner of Africa came into contact with the Iranian and Turkish portion of south-west Asia, allowing faunal exchanges between Africa and Asia. The Bering land bridge was well developed and South America was still isolated.

but was not completed until the Late Cretaceous (90–80 Myr BP). By the Mid Cretaceous, the northern continents were completely separated from the southern ones and India had separated from Madagascar. Westamerica became joined to Asia across the Bering Strait. By the Late Eocene about 40 million years ago (Table 2.6),

Continental drift and the intercontinental affinities of the fish fauna

The modern African freshwater fish fauna was already differentiated by the Mid Jurassic, when Gondwanaland broke up. One might therefore expect that representatives of the different fish families were present on the main drifting subcontinents, and that they evolved by vicariance after the isolation. Indeed, some families with few species have a world-wide distribution. This is true of the Osteoglossidae (Fig. 2.3), for which we have one of the most complete biogeographical histories of any freshwater fish group in Africa. Today it includes different genera living in Africa (*Heterotis*), South America (*Arapaima* and *Osteoglossum*) and Australia (*Sclerophages*), but fossil osteoglossids have also been found in Asia, Australia and North America (Greenwood, 1973*c*). A fairly complete phylogeny is available for the Osteoglossomorpha (Greenwood, 1973*c*; Patterson, 1981), and the biogeographical history of the group was reconstructed from Pangaea (up to 145 Myr BP) onwards. There is a real need of similar studies for other groups.

The Dipnoi (Fig. 2.3), which are the most archaic group of living bony fishes, originated in the Lower Devonian and have a fairly well-documented history (Rosen *et al.*, 1981). Among the lungfishes, the family Protopteridae includes the four African species of *Protopterus* and the single species of *Lepidosiren* from South

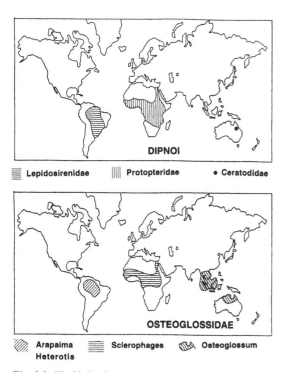

DIPNOI

≡ Lepidosirenidae   ⦀ Protopteridae   • Ceratodidae

OSTEOGLOSSIDAE

Arapaima   Sclerophages   Osteoglossum
Heterotis

Fig. 2.3. World distribution of some fish taxa. Dipnoi, Osteoglossidae, Notopteridae, Nandidae from Banarescu (1990). Characidae, Siluriformes, Cypriniformes, Cyprinodontidae, Bagridae, Clariidae, Schilbeidae from Berra (1981). Cichlidae from Stiassny (1991). Bariliine from Howes (1980). Mastacembeloidei from Travers (1984).

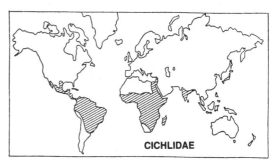

CICHLIDAE

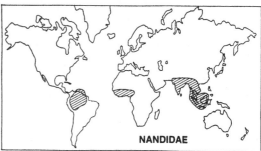

NANDIDAE

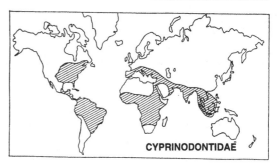

CYPRINODONTIDAE

Fig. 2.3. (cont.)

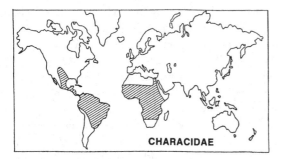

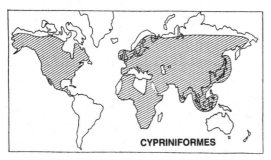

Fig. 2.3. (cont.)

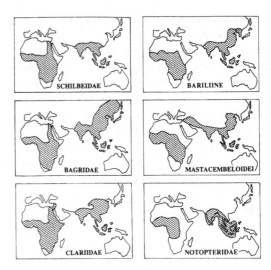

Fig. 2.3. (cont.)

America (however, some authors consider that *Protopterus* and *Lepidosiren* belong to a common family, the Lepidosirenidae). The related family Ceratodidae has a living representative in Australia (genus *Neoceratodus*), but many fossil lungfishes could be referred to this family that once had a much wider distribution. The present intercontinental distribution of Protopteridae can probably be ascribed to the dispersal of their common ancestor during a Late Mesozoic interconnection between Africa and South America (Patterson, 1975). The problem of relationships between Afro-American and Australian lungfishes is more complex. According to Patterson (1975), their common ancestor could have been marine.

The family Notopteridae, closely related to the Mormyridae, is represented in Africa by two species with widespread distributions: *Xenomystus nigri* and *Papyrocranus afer*. The latter superficially resembles the Asian *Notopterus* (4 species throughout southern Asia) (Fig. 2.3).

For a long time, the family Polypteridae was considered as endemic to Africa. Polypterid remains have been recorded in deposits from Middle Cretaceous–Lower Eocene in Egypt (Greenwood, 1974a) but later, Gayet *et al.* (1988) questioned the identification of the Eocene remains and reported the discovery of the oldest fossil polypterid remains in Lower Senonian beds at In Becetem (Niger). Recently however, remains of Polypteridae have also been found in Maastrichtian beds and Palaeocene levels of Central Bolivia (Gayet & Meunier, 1992). This first record of Polypteriformes in South America, and their absence from India (at least they have not yet been recorded from that continent), may suggest an age between 150 Myr (the probable date for the separation of India from Africa) and 110 Myr (the separation of America from Africa) for the origin of this group.

The ostariophysan fish comprise a group of related orders and families. With few exceptions, they have been confined to freshwater throughout their history. It is therefore unlikely that their historical dispersal could have been accomplished by other means than freshwater pathways, and they are of particular interest for studying past continental relationships (Briggs, 1979).

One major group, the characoids, is generally considered to be the most primitive, and living families are now restricted to Africa (4 families) and South America (15 families). Only one family, the Characidae, is common to both continents (Fig. 2.3). Fossil characoids have been identified from Palaeocene and Eocene deposits in Europe (Patterson, 1975, 1981). Why they disappeared and did not spread into North America is an open question. In Africa, fossil characoids are known from the Pliocene and Pleistocene (Greenwood, 1972), but remains were also found in Palaeocene beds at Ouarzazate (Maroco) (Cappetta et al., 1978). It has been suggested (Briggs, 1979) that the early characoids originated in tropical Asia in the Upper Jurassic, spread to Europe and then to Africa through the west end of the still incomplete Tethys Sea. They probably entered South America from Africa by means of a connection that was in existence in the Upper Jurassic. Characiformes, and probably Cypriniformes, are as old as the late Cretaceous or Palaeocene, and at least Cyprinidae were in evidence by the Middle Eocene (Cavender, 1991).

A similar biogeographical pattern was also presented for siluriform fishes that developed shortly after the characoids and dispersed from the same centre of origin: three families (Bagridae, Schilbeidae and Clariidae) occur only in Africa and Asia (Fig. 2.3), but there are no families common to South America and Afro-Asia. Another hypothesis was suggested by Gayet (1982a): an ancestral pre-characoid inhabiting the western Mesogea may have spread into South America, Africa and Europe. The discovery of the pre-characid Lusitanichthys, a euryhaline form from the Upper Cenomanian of Portugal (Gayet, 1980, 1981), may support this hypothesis.

The third major group (and the most modern) of ostariophysan fishes are the Cypriniforme (Fig. 2.3). The Catostomidae have a highly distinct distribution (North America and China), and few other families are present in Eurasia. Two families, the Cyprinidae and the Cobitidae, occur in African freshwaters. The family Cobitidae (loaches) occurs with greatest diversity in South-East Asia, but also has representatives in Africa (Morocco and Ethiopia). The Cyprinidae, a huge family of about 1600 species, is spread over North America, Eurasia and Africa, but is completely absent from South America. It

has been suggested (Briggs, 1979) that the cyprinoids also developed in tropical Asia in the Cretaceous, later than the characoids and siluriforms, and spread over Europe. They were unable to enter Africa before a connection was established in the Miocene (18–16 Myr according to Van Couvering, 1977). But the Asian–Africa dispersion of the genus Barbus has been the subject of debate. Géry (1969) and Roberts (1969) consider that the cyprinids are of African origin and that the dispersion was towards Asia among other areas. According to Géry (1969): 'Ostariophysans originated somewhere in Africa where two lines (Siluroids and Characoids) diverged during the Cretaceous or earlier, and quite early passed into South America. The origin of the Cyprinoids (from a Characoid-like ancestor) must have taken place later on in the Cretaceous or in the early Eocene after the isolation of South America (where Siluroids and Characoids, then Gymnotoids, radiated without competition, the most primitive ones being pushed into Patagonia) and in all probability somewhere in Africa according to the monophyly principle and the restricted Characoid fossils. Then together with catfish, the Cyprinoids would have spread northwards and eastwards radiating into the Oriental Region and finally reaching North America'. On the other hand, Novacek & Marshall (1976) set the origin in South America: 'if the origin of Ostariophysans was a post Jurassic event, then South America is the most likely centre of origin for the group. We postulate that Ostariophysans originated in South America in the earliest Cretaceous and there split into two sister groups, the Cypriniformes and Siluriformes. These groups dispersed throughout South America and the adjoining west shield area of Africa by middle Cretaceous. After regression of the west epicontinental sea and rifting apart of west Africa and south America in the late Cretaceous, characoids and siluroids dispersed throughout Africa. In South America and Africa these groups underwent independent adaptative radiation'. Briggs (1979) returned to the idea of a South-East Asian origin and according to hypotheses proposed by various authors (Darlington, 1957; Banarescu, 1973; Almaça, 1976), the genus Barbus originated from Asia and dispersed in two ways: (i) an Asian–Europe dispersion, the 'Siberian branch' that would have occurred between the Oligocene and Pliocene; and (ii) an Asia–Africa dispersion called the 'Ethiopian branch'. But the absence from Europe and Africa of fossil *Barbus* until the Upper Miocene is the only fact that seems to confirm the theory of Asian origin (Doadrio, 1990). However, no earlier record has been found in Asia either and the European fossils are the oldest known. Moreover, ancestral ostariophysarian remains were discovered in the Cretaceous of Portugal (Gayet, 1981) and from the Middle East (Gayet, 1982*b*, *c*).

Fink & Fink (1981) presented new evidence for relationships among ostariophysans, and their new phylogeny may stimulate competing hypotheses. The group is probably older than previously supposed, and it may well be that the details of the earliest distributions have been obscured by the long history of subsequent distribution/geological events. There is possible evidence that the characiform evolution took place before the Gondwana separation, and that most of the major characiform lineages had originated before the Africa–South America split. If this is the case, the more primitive sister-group, the cypriniforms, would have also evolved before the split. For Howes (1980), the bariliine (Fig. 2.3) and neoboline lineages could have a pre-drift, widely distributed, common ancestor of a Late Mesozoic age. But for Banarescu (1990), cyprinids as a whole are a modern group and some African genera (*Garra, Barbus*) are identical or closely related to southern Asia ones. As a result, speciation occurred in Asia and the distribution range extended to Africa from western Asia, probably in Miocene times. An old pre-drift occurrence of Cyprinidae in Africa cannot be accepted (Banarescu, 1990).

As stressed by Howes (1991), if cyprinids were differentiated very early, the question is why is not one group represented in South America? If competitive exclusion by characiforms is advocated, the more likely explanation could be that the West African part of Gondwanaland was not inhabited by cypriniforms during the time of continental drifting. Another possibility according to Howes (1991) would be that Fink & Fink (1981) are wrong, and that gymnotiforms are the sister-group of cypriniforms rather than siluriforms. Gayet (1982*a*, *b*) reported Cypriniform-like remains from the Upper Cretaceous of Bolivia, but until now little emphasis has been given to these findings. *Molinichthys inopinatus*, is the only specimen of a cyprinid recorded from South America, and the oldest known in the world. It is assumed to be a plesiomorph form, and that may lead to a funda-

mental reconsideration of the phylogeny of this family. No doubt the question of the origin of cyprinids will initiate other impassioned debates in the future.

The family Cyprinodontidae is widespread all over the world (Fig. 2.3). The order Cyprinodontiforme may have arisen in the New World, and a primitive species could have reached Africa in the Early Cretaceous when that continent was still close to South America.

Special mention has to be made of the cichlids, whose natural geographical distribution conforms to an essentially Gondwana pattern: Africa, Madagascar, Central and South America, India (Fig. 2.3). According to estimates, 80% of all cichlid species are to be found in African freshwaters, the great majority being represented by lacustrine taxa. There are no comparable lacustrine biotopes and radiations in the Neotropics (Stiassny, 1991). The African cichlids (excluding *Heterochromis*) and Neotropical cichlids are sister-groups (Stiassny, 1991). Remains of cichlids were found in Early (*c.* 17–18 million years ago) to Late Miocene (*c.* 11–13.5 Myr ago) brackish to freshwater deposits of the Paratethys in Europe (Gaemers, 1989). The fossil tilapiine genus *Eurotilapia* seems to be close to *Tristramella*, which is endemic to Israel. At least one large regression in the Early Miocene precedes the first occurrence of *Eurotilapia* in central Europe. The immigration of cichlids from Africa may have been possible via the Arabian Peninsula and south-western Europe (Gaemers, 1989). Later Miocene transgressions must have isolated the European cichlids from the African ones, and they later became extinct owing to the cooling of the European climate.

Some other percomorph families are shared only by Africa and Asia: the Channidae, Anabantidae and Mastacembelidae (Fig. 2.3). The family Channidae (= Ophicephalidae) occurs in Africa (*Parachanna*), and east and south Asia (*Channa, Micropletes, Ophicephalus*). The sister-group of African Anabantidae (*Ctenopoma, Sandelia*) is in the Belontiidae, represented by many species in southern and east Asia. The Oriental region could have been their centre of origin before they made their way to Africa, as well as the cyprinoids (Briggs, 1979; Banarescu, 1990), but the question remains open.

The distribution of freshwater representatives of primarily marine families (also classified as peripheral and vicarious freshwater fishes) reflects a rather marine zoogeography (Banarescu, 1990) and dispersal by marine routes probably played the major role. There are rather important differences between the peripheral freshwater fish faunas of the western and eastern coasts of Africa. Those of the eastern side have marked Indo–West Pacific affinities and the higher diversity. According to Banarescu (1990), the peripheral freshwater fishes from eastern Africa, Madagascar, south and east Asia, and northern Australia–Tasmania are, in most cases, related to each other, belonging to Indo–West Pacific lineages. The peripheral fish faunas of western Africa are poorer than those of the eastern coast.

One major question is why so many living primitive fish groups are only represented in Africa? The Lepidosireniformes had probably evolved during the Cretaceous and their world-wide distribution could result from the occurrence of ancestral forms in Gondwanaland. But the Denticipitidae were probably already present in African inland waters before the break up of Gondwanaland. In that case, why did these families not occur in South America and Asia? In the absence of fossil records, there is no proof that they dispersed from other continents, and there are examples of limited continental distributions at present. Conversely, it could also be possible that they became extinct early after separation of the continents. For Banarescu (1990), low interspecific competition could explain the occurrence in Africa of many archaic fish lineages and the abundance of others. To support his hypothesis, he pointed out that two of the most competitive groups of fish in the world are the Characiformes, which include only a few unspecialized genera in Africa, and the Cypriniformes that entered Africa recently and had no time to undergo an active specialisation. If a lower competitive pressure favoured the maintenance of archaic groups in Africa, it seems unlikely that it is the only reason. Past climatic and geological events are undoubtedly also responsible for this situation.

3 Genetic diversity and mechanisms of speciation

Ever since Darwin, the theory of evolution has been the main unifying idea in biology. It is natural selection that has made biological systems different from physical or chemical ones. Today, there is an increasing tendency for biology students to specialise either in molecular and cellular biology, or in the biology of whole organisms and populations ... A course in evolution, however, should unite both streams. Much of molecular biology makes sense only in the light of evolution: the techniques of molecular genetics are essential to a population biologist.

Maynard Smith 1989

Genetic diversity is the basis of the fundamental processes of biological change, adaptation and evolution. An understanding of the organisation of genetic diversity is therefore essential, both for knowledge of the processes at work and to ensure the existence of biological diversity. Genetic variability can be thought of as existing at two levels: genetic differences between individuals within local populations, and genetic differences between local populations within the same species. The study of how individual variability becomes transformed into differences between populations is fundamental to the study of evolution. Highly variable characters are useful in assessing geographic patterns of gene flows among populations, whereas less variable characters allow the study of family or generic relationships and evolutionary patterns.

Molecular and cytogenetic studies of African freshwater fish have been emphasised during the last few years. They have been conducted to describe the attributes of a group of taxa, to solve systematic or phylogenetic questions, or to assess the genetic variability of a species which could be used in aquaculture.

The basis of genetic diversity

Genetics has been defined as the study of differences among individuals. The study of inheritance depends upon finding individual differences so that the similarity of parents and their offspring can be compared. A genotype of an individual is the sequence of bases in a DNA or RNA molecule that carries the genetic information. It may vary following recombination and mutation.

Life originates at the molecular level, and is intimately connected to the properties of nucleic acids (deoxyribonucleic acid or DNA and ribonucleic acid or RNA) that are large and complex polymers of repeating subunits called nucleotides. The DNA, which is the hereditary material, is a giant, double-stranded molecule that is the informative chemical substance of the gene. The genetic information is contained in different sequences of four nitrogenous bases (adenine and guanine, thymine and cytosine) attached to a deoxyribose sugar. The order of the nitrogenous bases on the DNA molecule constitutes a code and determines (via RNA) the structure of proteins, which are similar to DNA but made up of different components called amino acids. The genotype of any individual is the set of instructions carried in the transcribable portion of the DNA.

The basic unit of inheritance is the gene, which contains the blueprint of the biological code for production of the phenotype. Most genes in fish are contained in chromosomes that occur in pairs, one set inherited from each parent via their gametes (sperm and egg cells). Two broad conceptual categories of genes are usually recognised: structural genes coding for a protein that is produced by the series of molecular events described above, and the regulatory genes that control the timing and location of expression of those structural genes.

The location of a gene on a chromosome is called a *locus* (plural loci). In diploid forms, the components of a paired set of genes may be different within a single individual, and different forms of a gene are called *alleles*. In a species, there may be many alleles for a particular locus, but within one diploid individual there can be no more than two alleles at one locus. When one allele only is known the locus is monomorphic, whereas when two or more are known it is polymorphic. If the paired genes are identical at a particular locus, an individual is called homozygous. In contrast, heterozygous individuals have two different forms (or alleles) of the gene.

The genetic diversity observed in nature is the accumulation of the regular appearance of mutations that are changes in the genetic message. Mutations that influence protein function can produce new kinds of enzymes, thus permitting the evolution of complex structures. Mispairings of bases in DNA, which are occasional 'mistakes' occurring regularly during replication, are one of the major sources of mutations that could result in amino acid alterations in proteins. On the present scientific evidence, the appearance of new mutations is a purely random process.

Many mutations that have no visible effect (or are assumed so) on the function of cell or organism are called neutral mutations. Other kinds of mutation (lethal mutations) may impede the proper function of cells and organisms, in which case the organisms are usually not able to survive and reproduce, and such mutations are eventually eliminated by natural selection. The remaining mutations result in slight changes in the shaping of the proteins. The genetic variants are integrated into the physiological and biochemical functions of the organism, as well as into the ecological framework of the species. The great diversity of living creatures is, therefore, the result of the spread of favourable mutations that have taken place since the origin of life (Solbrig, 1991*a*).

Mutations therefore generate the genetic diversity within a species, which determines its potential for subsequent evolutionary change and its ability to respond successfully to natural or anthropogenic disturbances.

One of the key questions is the identification of the sources of genetic variation and their variability among species in time and space.

Karyological diversity

The study of chromosome numbers and morphology long predates modern molecular methods of analysing their component genes. In fish species, chromosome and/or chromosome arm numbers exhibit great variability, and one can assume that the karyotype is specific. Although molecular methods have become an essential tool for the evolutionary biologist, cytogenetic and cytotaxonomic studies provide additional information to the biochemical markers resolved by electrophoretic techniques and are still a promising way to investigate fish genetic diversity.

Chromosome types and numbers

The study of fish chromosomes is a promising research area in terms of taxonomy, evolutionary studies, variation within and between populations, genetic control in fish breeding, etc., and techniques for harvesting and preparing fish chromosomes have improved a great deal during recent years (see Klinkhardt, 1991; Ozouf-Costaz & Foresti, 1992, for review). A detailed description of the karyotype provides a first approach to characterising a species genetically.

The number of chromosomes per cell, as well as the karyotypic configurations, vary from species to species but is characteristic within a species. In most fish, chromosomes occur in pairs (2n) one of maternal and one of paternal origin, and the double state is termed diploid. A single set of chromosomes, as found in a mature egg or in a spermatozoan, is described as a haploid set, and when more than two sets occur per cell, the condition is termed polyploid. There are also two basic types of chromosomes: autosomes that are morphologically similar in males and females, and sex chromosomes that carry the sex-determining genes. In a pair, each of the sex chromosomes may be morphologically different from the other, or may be similar.

Chromosome classification depends on their size and on their morphology, and is based on the centromere position according to the nomenclature proposed by Levan *et al.* (1964). Different types are identified as

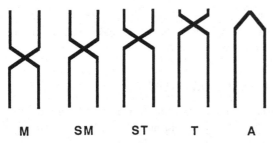

M   SM   ST   T   A

Fig. 3.1. Different types of chromosomes based upon their centromeric index CI, ratio of the length of the longer arm to the length of the shorter arm. M, metacentric (CI < 1.7); SM, submetacentric (1.7 < CI < 3); ST, subtelocentric (3 < CI < 7); T, telocentric (7 < CI); A, acrocentric with only one arm (from Agnèse, 1989).

meta-, submeta-, subtelo-, acro- or telocentrics (Fig. 3.1). However, the distinction between these types is partly subjective and highly dependent on the chromosome size and quality. Some authors consider subtelocentric chromosomes as having just one arm, while others allocate two arms. This could explain differences observed for the same species and by different authors for the calculation of the Fundamental Number (FN).

Discrimination and pairing of chromosomes is not easy, and chromosome marking techniques have improved the accuracy of karyotyping. Cytological and molecular techniques produce a specific banding pattern on each chromosome that can be used for comparative analysis (Ozouf-Costaz & Foresti, 1992).

Until now only a few studies have adapted the chromosome banding methods used in higher vertebrates to the study of fishes. Chromosomal banding is usually limited primarily to visualisations of the chromosomal Nucleolus Organiser Regions (NORs), which are chromosomal sites responsible for the synthesis of ribosomal RNA. The silver staining of NORs is a simple means of localising the genes coding for 18s and 28s ribosomal RNA in chromosomes. This method is not very powerful compared to the banding method, but much easier to apply. In fish, different studies have revealed only one NOR-bearing chromosome pair for the species studied, except for several cyprinids and a few other species. Those chromosomes and the position of their NORs usually differ between species and could be used for their

differentiation. For six species of West African Mochokidae, Oberdorff *et al.* (1990) reported the presence of only one nucleolar organiser region (NOR) bearing chromosome pair. These homologues showed the same morphology and their sizes were relatively similar to each other among the species. In that case, the NOR-bearing chromosomes could not therefore be used as a criterion for species identification.

C-banding is another technique that has been successfully applied to fish. It allows one to visualise the constituent heterochromatin corresponding to highly repeated DNA sequences, which are frequent in the centromere region and sometimes close to the NOR locations in fish chromosomes. C-bands may be effective for identifying homologous chromosomes, and thus characterise and differentiate species (Ozouf-Costaz & Foresti, 1992). Other different banding techniques have been used such as the physical localisation of DNA sequences (so-called markers) on the chromosomes, which has been realised with fluorescence labelled probes by *in situ* hybridisation techniques.

C-banding studies in several tilapiines (*Tilapia zillii, Oreochromis mossambicus, O. niloticus, O. aureus, O. macrochir, O. spilurus, Sarotherodon galilaeus*) show that the C-heterochromatin distribution is different according to species (Majumdar & McAndrew, 1986). Heterochromatin is localised in and around the centromere of all chromosomes in all species of *Oreochromis* and *Sarotherodon*, but *T. zillii* has more heterochromatin, with six chromosomes having completely C-positive short arms, and 10 to 12 chromosomes that show no banding.

Although we have quite a lot of descriptive information about the karyotypes, and we know something about the mechanisms of their changes, the general significance of karyotype evolution is obscure (Maynard Smith, 1989). The karyotype of many African fish is not well known despite the existence of a great deal of data on chromosome numbers (Table 3.1).

Siluriformes show a great diversity in the number and shape of chromosomes. Among the Bagridae, *Chrysichthys maurus, Chrysichthys auratus, Clarotes laticeps* have a higher number of chromosomes than other catfish species. An analysis of 37 specimens from three strains of *Clarias gariepinus*, originating from different localities in Africa and Asia Minor (the distribution limits of the species), revealed a stable karyotype (2n = 56) with a pair

Table 3.1. *Details of karyotypes for different African species*

Species	2n	FN	Authors
Protopteridae			
Protopterus annectens	34		Wickbom, 1945
Protopterus dolloi	68	104	Vervoort, 1980*a*
Polypteridae			
Calamoichthys calabaricus	36	72	Denton & Howell, 1973
Polypterus delhezi	36	72	Cataudella *et al.*, 1978
Polypterus endlicheri congicus	36	72	Cataudella *et al.*, 1978
Polypterus ornatipinnis	36	72	Urushido *et al.*, 1977
Polypterus palmas	36	72	Denton & Howell, 1973
Polypterus senegalensis	36	72	Urushido *et al.*, 1977
Notopteridae			
Papyrocranus afer	34	38	Uyeno, 1973
Cyprinidae			
Barbus aeneus	148	196	Oellermann & Skelton, 1990
Barbus bynni	150	22	Golubtsov & Krysanov, 1993
Barbus bynni occidentalis	148		Guégan *et al.*, 1995
Barbus capensis	150	208	Oellermann & Skelton, 1990
Barbus aethiopicus	150	190	Golubtsov & Krysanov, 1993
Barbus intermedius	150	216	Golubtsov & Krysanov, 1993
Barbus intermedius	150	240	Golubtsov & Krysanov, 1993
Barbus kimberleyensis	148	204	Oellermann & Skelton, 1990
Barbus natalensis	150	200	Oellermann & Skelton, 1990
Barbus petitjeani	150	186	Guégan *et al.*, 1995
Barbus polylepis	150	206	Oellermann & Skelton, 1990
Barbus wurtzi	148		Guégan *et al.*, 1995
Barbus ablabes	48	96	Rab *et al.*, 1996
Barbus anema	50	92	Golubtsov & Krysanov, 1993
Barbus bigornei	50	98	Rab *et al.*, 1996
Barbus holotaenia	50	100	Rab, 1981
Barbus kerstenii	50	84	Golubtsov & Krysanov, 1993
Barbus macrops	50	92	Rab *et al.*, 1996
Barbus paludinosus	50	96	Golubtsov & Krysanov, 1993
Barbus viviparus	48	96	Post, 1965
Garra dembeensis	50	82	Krysanov & Golubtsov, 1993
Garra makiensis	50	84	Krysanov & Golubtsov, 1993
Garra quadrimacukata	50	88	Krysanov & Golubtsov, 1993
Labeo senegalensis	50		Paugy *et al.*, 1990
Labeo coubie	50		Paugy *et al.*, 1990
Labeo roseopunctatus	50		Paugy *et al.*, 1990
Varicorhinus beso	150	216	Golubtsov & Krysanov, 1993
Cyprinodontidae			
Aphyoplatys duboisi	48	52	Scheel, 1972
Aphyosemion ahli	36		Scheel, 1968
Aphyosemion arnoldi	38	72	Scheel, 1968
Aphyosemion bivittatum	40		Scheel, 1966
Aphyosemion, bualanum	40	68	Scheel, 1968
Aphyosemion calliurum	26		Scheel, 1972
Aphyosemion cameronense	34	46	Scheel, 1972

Table 3.1. (*cont.*)

Species	2n	FN	Authors
Aphyosemion christyi	18	36	Scheel, 1968
Aphyosemion exiguum	36	72	Scheel, 1968
Aphyosemion filamentosum	36	48	Scheel, 1968
Aphyosemion franzwerneri	22	44	Scheel, 1972
Aphyosemion gardneri	40	54	Scheel, 1968
Aphyosemion guineense	38		Scheel, 1968
Aphyosemion gulare	32	32	Scheel, 1968
Aphyosemion labarrei	28	50	Scheel, 1972
Aphyosemion louessense	20	40	Scheel, 1972
Aphyosemion lujae	40		Scheel, 1968
Aphyosemion mirabile	32	60	Scheel, 1972
Aphyosemion obscurum	34		Scheel, 1968
Aphyosemion roloffi	42		Scheel, 1968
Aphyosemion scheeli	40	70	Scheel, 1972
Aphyosemion sjoestedti	40	40	Scheel, 1968
Aphyosemion walkeri	36	48	Scheel, 1968
Aplocheilichthys macrophthalmus	48	48	Scheel, 1972
Aplocheilichthys normani	48	48	Scheel, 1972
Epiplatys annulatus	50	86	Scheel, 1972
Epiplatys barmoiensis	34	50	Scheel, 1972
Epiplatys bifasciatus	40		Scheel, 1968
Epiplatys chaperi	50	52	Scheel, 1972
Epiplatys dageti	50		Scheel, 1968
Epiplatys fasciolatus	38		Scheel, 1968
Epiplatys sexfasciatus	48		Scheel, 1968
Epiplatys spilargyreius	34		Scheel, 1968
Nothobranchius guentheri	36		Ewulonu *et al.*, 1985
Nothobranchius kirki	36	58	Scheel, 1972
Nothobranchius melanospilus	36		Ewulonu *et al.*, 1985
Nothobranchius palmqvisti	34	46	Scheel, 1968
Nothobranchius palmqvisti	36		Ewulonu *et al.*, 1985
Nothobranchius patrizii	36		Ewulonu *et al.*, 1985
Nothobranchius rachovii	16		Ewulonu *et al.*, 1985
Nothobranchius rachovii	18		Scheel, 1981
Cichlidae			
Astatotilapia burtoni	40		Thompson, 1981
Aulonocara baenschi	44		Foerster & Schartl, 1987
Aulonocara korneliae	44		Foerster & Schartl, 1987
Aulonocara hueseri	44		Foerster & Schartl, 1987
Aulonocara stuartgranti	44		Foerster & Schartl, 1987
Hemichromis bimaculatus	44	88	Zahner, 1977
Heterotilapia multispinosa	48	96	Zahner, 1977
Melanochromis auratus	46	56–58	Thompson, 1981
Oreochromis alcalicus	48		Park, 1974
Oreochromis andersonii	44	48	Vervoort, 1980*a*
Oreochromis aureus	44	54	Kornfield *et al.*, 1979
	44	44–50	Thompson, 1981
	44	58	Majumdar & McAndrew, 1986
Oreochromis macrochir	44		Jalabert *et al.*, 1971
	44	48	Vervoort, 1980*c*

Table 3.1. (*cont.*)

Species	2n	FN	Authors
	44	54	Majumdar & McAndrew, 1986
Oreochromis mossambicus	44	62	Fukuoka & Muramoto, 1975
	44	44–50	Thompson, 1981
	44	50	Majumdar & McAndrew, 1986
Oreochromis niloticus	44	62	Jalabert *et al.*, 1974;
			Arai & Koike, 1980
	44	64	Majumdar & McAndrew, 1986
Oreochromis spilurus	44	50	Majumdar & McAndrew, 1986
Sarotherodon galilaeus	44	54	Kornfield *et al.*, 1979
	44	50	Vervoort, 1980*c*
	44	48	Majumdar & McAndrew, 1986
Tilapia busumana	44		Nijjhar *et al.*, 1983
Tilapia congica	44	54	Vervoort, 1980*c*
Tilapia guineensis	44	52	Vervoort, 1980*c*
Tilapia mariae	40	44	Vervoort, 1980*c*
	40	44–48	Thompson, 1981
Tilapia rendalli	44	52	Michele & Takahashi, 1977
Tilapia sparrmanii	42	50	Vervoort, 1980*c*
	42	46–50	Thompson, 1981
Tilapia zillii	44	54	Kornfield *et al.*, 1979
	44	66	Majumdar & McAndrew, 1986
Bagridae			
Auchenoglanis occidentalis	56	104	Agnèse, 1989
Bagrus docmak	54	98	Agnèse, 1989
Chrysichthys auratus	72	109	Agnèse, 1989
Chrysichthys maurus	70	104	Agnèse, 1989
Clarotes laticeps	70	102	Agnèse, 1989
Clariidae			
Clarias anguillaris	56	100	Agnèse, 1989
Clarias gariepinus	56	88	Teugels *et al.*, 1992*a*
Heterobranchus longifilis	52	82	Teugels *et al.*, 1992*a*
hybrid *C. gariepinus*	54		Teugels *et al.*, 1992*a*
× *H. longifilis*			
Mochokidae			
Hemisynodontis membranaceus	54	98	Agnèse *et al.*, 1990*b*
Synodontis bastiani	54	95	Agnèse *et al.*, 1990*b*
Synodontis budgetti	54	102	Agnèse *et al.*, 1990*b*
Synodontis courteti	54	101	Agnèse *et al.*, 1990*b*
Synodontis filamentosus	56	102	Agnèse *et al.*, 1990*b*
Synodontis ocellifer	54	96	Agnèse *et al.*, 1990*b*
Synodontis schall	54	95	Agnèse *et al.*, 1990*b*
Synodontis sorex	54	96	Agnèse *et al.*, 1990*b*
Synodontis violaceus	54	100	Agnèse *et al.*, 1990*b*
Phractolaemidae			
Phractolaemus ansorgii	28	54	Vervoort, 1979

Table 3.1. (*cont.*)

Species	2n	FN	Authors
Pantodontidae			
Pantodon buchholzi	48		Uyeno, 1973
Channidae			
Parachanna obscura	34	50	Nayyar, 1966

2n, diploid number of chromosomes; FN, fundamental number (total number of arms).

of heteromorphic sex chromosomes (ZW) (Ozouf-Costaz *et al.*, 1990). The representative karyotype of the Siluriformes has a diploid number of 2n = 54 ± 6 (Volckaert & Agnèse, 1996).

Among cyprinids two groups are recognised in Africa: those that have *c.* 2n = 50 (presumably plesiomorphic diploids) and those that have *c.* 2n = 148–150, indicating that these species are evidently polyploid on the basis of chromosome number (see below).

The karyotype is highly variable among the cyprinodonts (2n from 16 to 50), while it seems relatively stable among cichlids. It has been claimed (Majumdar & McAndrew, 1986) that karyotype evolution in the tribe Tilapiini is of a highly conservative nature. The diploid chromosome number is around 44 for these Old World species and most tilapia examined have a very large pair of acrocentric chromosomes. The diploid value for *Oreochromis alcalicus* (2n = 48) is seen to be plesiomorphic and is characteristic of the majority of cichlids, possibly predating the overall reduction to 44 in the other tilapiine species. The two *Tilapia* species with a different diploid number, *Tilapia sparrmanii* (2n = 42) and *Tilapia mariae* (2n = 40) can be explained by chromosome fusion within a 2n = 44 chromosome complement (Vervoort, 1980c). Conflicting reports of chromosome numbers have been found in the literature for some tilapiine species, which have not been substantiated by other work. On the other hand, the description of the karyotype relies heavily on the accuracy of the chromosome classification used, and discrepancies in the fundamental number are frequently observed between different reports.

Only a few karyotypes of haplochromines species are known from East African Lakes, but there appears to be little relationship between karyotypic diversification and phylogeny within major cichlids groups.

Changes in chromosome numbers and structures and speciation

Although we have quite a lot of descriptive information about karyotypes, and we know something about the mechanisms of their changes, the general significance of karyotype evolution is obscure (Maynard Smith, 1989).

The highly variable chromosome number, and the probably even more variable karyotypes of cyprinodonts, appear to be unique among fish and even among animals. According to Scheel (1968), the basic chromosome number for Cyprinodontidae is n = 24. This basic karyotype seems to be characterised by a smoothly graded series of rather small chromosomes that are generally acrocentric. Processes of reduction in the chromosome number have been discussed by Scheel (1975a, b). They can sometimes be guessed from the chromosome morphology alone, but they could also be confirmed by using some chromosome markers.

A change in karyotype requires at least two chromosome breaks, followed by a joining of the broken ends (Fig. 3.2), and chromosome evolution is often associated with Robertsonian translocations (fusions and fissions at the chromosome centromere level) and inversions. Reduction of the chromosome number usually takes place when two acrocentrics break during cell division and the broken parts fuse in a new way. One large metacentric with long arms and one small metacentric with small arms may be the result of fusions, but the small metacentric element is usually lost in African Cyprinodontidae. One reason is that the short arms of the chromosomes do not contain any active genetic material (genes), and this element is therefore not able to pair with the corresponding element during meiotic divisions. Together with loss of the small metacentric element, the corresponding centromere is also lost. As a result, if a

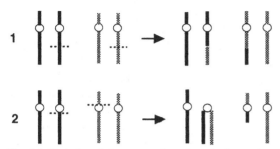

Fig. 3.2. Evolution of chromosome form through changes in chromosome structures. 1, translocation; 2, centric fusion.

large metacentric element breaks, the original acrocentrics cannot be reconstructed. Thus an increase in the haploid number is more unlikely than a decrease. On this assumption, in a series of closely related forms, the form with the lowest chromosome number is probably also the youngest, being derived from a form that had the highest number of chromosomes.

In *Epiplatys*, the haploid chromosome number has been reduced to 17 or 18. The reduction of the chromosome number appears to take place through production of large metacentric elements, the small ones being lost. Within *Aphyosemion*, the haploid chromosome number (n) may vary from 24 to 9 as in *Aphyosemion rectogoense* and *Aphyosemion christyi*. Among *Aphyosemion* and *Roloffia* species, the reduction in chromosome number also results in production of large metacentric elements, mainly within species exhibiting very low chromosome numbers. But the basic type of reduction appears to be different. There are two acrocentric breaks, one near the centromere as when metacentrics are produced, the other near the far end of the long arm. The large broken parts fuse producing a 'double-sized acrocentric element' and the small broken parts produce a small metacentric element which is usually soon lost.

Within *Aphyosemion* species there is an interesting example of evolution and speciation with reduction of n while the fundamental number remains constant (18): n decreases from 15 to 13,11,10 and 9 respectively in the series *Aphyosemion congicum–cognatum–schoutedeni–elegans–christyi*. The reduction in n from one species to another, could occur through Robertsonian type fusion of two telocentrics (or acrocentrics) into one metacentric. The species are sexually isolated but their phenotypes

are very similar, and cannot be distinguished either with meristic characters or morphometry. They differ only in details of the colour pattern of adult males. These differences would have been attributed to intraspecific polymorphism, but studies of chromosome numbers, as well as experimental hybridisation, had demonstrated that they were genetically isolated populations.

Two groups of species have also been identified among West African bagrids using chromosome numbers: the first group includes the genera *Chrysichthys* and *Clarotes* having 2n = 70 or 72 and the second group the genera *Auchenoglanis* and *Bagrus* with 2n = 54 or 56 (Agnèse, 1989). The first group is monophyletic and probably derived from the second. It is assumed that the higher number of chromosomes in *Chrysichthys* and *Clarotes*, results from chromosome fissions. The existence of small-sized acrocentric chromosomes, as well as the fairly similar number of arms in both groups, support this hypothesis.

Polyploidy

Polyploidisation is another mechanism of speciation. Natural polyploidisation is not a rare event in fish, compared with other vertebrates, and it plays an important role in the evolution of fish karyotypes. Some lineages of the subfamily Cyprininae (*sensu* Howes, 1987), for example, appear to have undergone a progressive polyploidisation in the course of their evolution. Certain varieties of *Carassius auratus* and species of the Asian genus *Schizothorax* are known to be polyploids (Zan *et al.*, 1986). In Africa, large southern African *Barbus* species (Table 3.1) are known to have modal numbers of 148–150 chromosomes (Oellermann & Skelton, 1990), whereas the majority of cyprinid species have 2n = 50. Those species that have in the order of three times the basic chromosome number of a diploid cyprinid are considered to be of hexaploidic origin. That is also the case for a few other cyprinids, including *Barbus marequensis* and *Varicorhinus nelspruitensis* (see Oellermann & Skelton, 1990). Large *Barbus* from Ethiopia (*B. bynni, B. intermedius, B. aethiopicus*) and *Varicorhinus beso* also have high numbers of chromosomes (2n = 150) (Golubtsov & Krysanov, 1993). The latter questioned the supposed tetraploid status of the large *Barbus* from West Africa as suggested by Berrebi *et al.* (1990), and predicted that future investigations will reveal hexaploidy in all forms related to the large *Barbus* species group in Africa. Indeed, recent data have confirmed that large West African *Barbus*, such as *B. bynni occidentalis*, *B. petitjeani* and *B. wurtzi*, are also hexaploids (Guégan *et al.*, 1995).

Polyploidisation must have occurred long ago in large African *Barbus*. For Oellermann & Skelton (1990), hexaploidy in large *Barbus* may be due to a combination of both autopolyploidy and alloploidy. 'An independent origin of the hexaploid forms of the large *Barbus* group from southern Africa and from Ethiopia seems to be unlikely unless the existence of ancestral tetraploid forms of the large *Barbus* in Africa is shown, because the most probable pathway of the origin of hexaploidy is via a tetraploid stage' (Golubtsov & Krysanov, 1993). They postulated that the large African *Barbus* originated from a tetraploid form close to *Barbus sensu stricto* or to the Asian genus *Tor*. They believe that the formation of the hexaploid happened before the wide dispersal of the large *Barbus* over the African continent and that the small African *Barbus* may be phylogenetically close to the diploid forms of the Asian genus *Puntius*, in which case one would assume an independent origin for the large and small *Barbus*, the large *Barbus* being a monophyletic group without a sister-group relationship to any small African *Barbus*.

The advantages of polyploidy for fish are considered to be larger size, longer life, faster growth and greater ecological adaptability. That explains why aquaculturists have encouraged the production of polyploid strains of selected species. Induced polyploidy involves the production of individuals with extra sets of chromosomes. Much experimental work has focused on production of polyploids in tilapiines. Triploid fish which appear to be sterile are of special interest for fish culture, given that the development of their gonads is reduced, or completely inhibited, allowing for increased somatic growth. Triploids of *Oreochromis niloticus*, *Oreochromis mossambicus* and *Oreochromis aureus* were experimentally produced by cold and heat shocks (Don & Avatlion, 1988a). Tetraploids have been induced by the combination of thermal (cold) and pressure shocks in *O. niloticus*, *O. mossambicus* and their hybrid (Myers, 1986).

Sex chromosomes
The genetic sex of Teleostei varies from macroscopically heterogametic karyotypes (in about 10% of the species

studied) over molecular heterogametic to suspect homogametic karyotypes (Chourrout, 1991a). The occurrence of a pair of morphologically differentiated sex chromosomes is apparently rare or difficult to detect in bony fish. Sola *et al.* (1981) found only 3.6% of the fish species studied have sex chromosomes and, until recently, very few data were available for African fish. The most common system among fish is the XY sex determining system. Homogametic fishes (XX) are females, whereas males are heterogametic (XY). Sons receive their father's Y chromosome and one of their mother's X chromosomes; daughters receive X chromosomes from both parents. In the second system (WZ), males are homogametic (ZZ) and females heterogametic (ZW). Other systems investigated have involved multiple sex chromosome mechanisms.

The karyotypes of some African siluriforms have been investigated by Agnèse *et al.* (1990b), Ozouf-Costaz *et al.* (1990) and Teugels *et al.* (1992a), and the most striking result of their observations was the existence of heteromorphic chromosomes in all the females of the species studied. The presence of sex chromosomes (particularly the female heterogamy, of the ZZ/ZW type) is likely to be widespread among African Siluriformes. With the improvement of techniques, sex chromosomes will probably be observed more frequently.

Until now, distinct sex chromosomes and sex-linked genes have not been discovered in tilapiines. Heteromorphic sex chromosomes have been reported in tilapia by Nijjhar *et al.* (1983), but no karyotype was provided by these authors. Recent studies have shown that a chromosomal sex-determination system is very unlikely in this group (Majumdar & McAndrew, 1986).

A multiple sex chromosomes system has been discovered in the annual cyprinodont fish, *Nothobranchius guentheri* (Ewulonu *et al.*, 1985). In all males examined, the modal diploid chromosome number was 35, while in all females it was 36. In other related species, such as *Nothobranchius melanospilus*, *Nothobranchius palmqvisti*, *Nothobranchius patrizii* and *Nothobranchius rachovii*, both females and males have the same chromosome number (see Table 3.1). The sex difference in diploid chromosome number observed in *N. guentheri* indicates the presence of a 'multiple' or 'fusion' sex chromosome system of the $X_1X_2Y/X_1X_1X_2X_2$ type. This type of heterogamety has so far been observed in few other fish species.

An understanding of sex determination in fish could be central for the development of aquaculture. Sustained all-male cultures of cichlids for instance could significantly increase yield through greater somatic conversion and faster growth.

Genetic diversity at the molecular level

There are two reasons for measuring genetic variation. One is to understand the relationships among organisms, through the diversity within and the divergence between them. The other is to test theories about the forces acting on genetic variants. The two sets of problems are obviously connected, as long as the central debate in evolutionary genetics is about whether most of the genetic variation seen in natural populations is maintained by natural selection or is neutral, and therefore subject only to the laws of chance.

Molecular methods

For geneticists, the genetic variability of a population is the range of alleles observable for the genes revealed by the method used, that is to say the degree of polymorphism shown by the genes. Heterozygosity or mean genetic variability (Nei, 1978) is often used to measure the amount of genetic variation in a population. Heterozygosity (H) at an individual locus is defined as the proportion of individuals that are heterozygous at that locus: $H = \Sigma\, h/r$ where r is the number of loci and h is the index of genetic diversity at each locus, such that $h = 1 - \Sigma\, q_i^2$, where q_i is the frequency of the ith allele of the gene at this locus. The mean heterozygosity in a population is usually calculated for a number of loci. This parameter is, of course, sensitive to variations in the sample size. There is an expected loss of heterozygosity for neutral alleles due to random genetic drift when a small population experiences a 'bottleneck' (panmictic populations of low effective size).

The level of polymorphism (P) is the percentage of polymorphic loci in the sample studied. This parameter is calculated either for genes for which the frequency of the most common allele is less than 0.99, or 0.95. The mean number (A) of alleles at a locus is also used to evaluate the genetic diversity. There is a wide range of indices used to evaluate genetic divergence between taxa or populations. Each has specific mathematical and bio-logical properties and cannot be used without caution. Nei's (1972) index and Rogers' (1972) index are the most widely used. Given that a genome contains thousands of gene loci, it is not possible to measure the genetic variation of all loci, but only for a small fraction of them. It can be detected, at the molecular level, by the changes in the structure of proteins coded by specific genes or by the study of changes in the structure of DNA.

Allozyme electrophoresis, which has been used on a wide variety of organisms, is a powerful method for comparing populations. In the jargon of protein electrophoresis, 'allele' is a widely used synonym for allozyme. The detection of protein polymorphisms assumes that allelic variants at a given locus code for proteins are usually similar in structure but differ in the electrical charge they carry. This can be detected by studying the differences in electrophoretic mobility of proteins when exposed to an electric field (see Pasteur et al., 1988).

An electrophoretic sample of 100 loci still represents far less than 1% of the total number of genes of a diploid organism (Crow, 1976). Thus, detection of differences between populations is a positive indicator of genetic differences, while the absence of differences is not proof of genetic identity at the DNA level. The number of allozyme products required for a study depends upon the problems and taxa being investigated. Usually, many enzymatic systems must be screened in order to identify informative patterns of variation. Based on electrophoretic data, the genetic distance is an estimate of the number of amino acid substitutions in proteins that separate two species. The most popular is the estimate devised by Nei (1972).

Genetic drift refers to chance changes in allele frequencies as a result of random sampling among gametes from generation to generation. The magnitude and importance of genetic drift varies inversely with population size. In large breeding groups, the gene frequencies in a new generation are identical to the gene frequencies in the parental generation, whereas in small populations, gene frequencies may fluctuate dramatically from generation to generation. The three main effects of genetic drift associated with small populations are: (i) heterozygosity is expected to decline; (ii) rare alleles are expected to be lost; and (iii) variance among populations is expected to increase. The direction of gene-frequency changes due to genetic drift is completely random, but one can predict that genetic variation will be lost at a rate that is inversely proportional to population size.

The finest resolution of genetic variation occurs at the level of the coding and non-coding nucleotide sequences of the mitochondrial and nuclear DNA, and ribosomal RNA. Coding sequences show considerable variation at the DNA level (the gene) despite the highly conservative nature of functional proteins. Molecular variability includes deletions and mutations, transpositions, common base substitutions, and nucleotide sequence alterations due to intragenic recombination and gene conservation. One of the major advantages of DNA sequence analysis in molecular systematics is data standardisation. While results obtained from protein electrophoresis may depend upon experimental conditions, DNA sequences are universally and independently replicable and understood. This technique still needs improvements, particularly in laboratory procedures, but will probably be the most efficient in the near future.

The rich potential for understanding genetic population structures through analysis of mitochondrial DNA molecule (mtDNA) has only just begun to be realised (Meyer, 1994). Mitochondria are self-replicating organelles found in every eukaryotic cell. The vertebrate mtDNA is much less complex than the nuclear genome, and almost all the enzymes that are located in the mitochondria are encoded in the nucleus. Thus, the inheritance of most mitochondrial enzymes follows Mendelian principles. But the mtDNA is not heritated in a Mendelian fashion, because individuals inherit only their mother's mtDNA in a clonal form through the eggs. The advantage of mitochondrial DNA is that it evolves 5–10 times more rapidly than nuclear DNA, providing a magnified view of the differences between populations and closely related species. Phylogenies based on mtDNA can therefore provide the finer resolution necessary for examining evolutionary relationships among young species. On the other hand, the maternal inheritance allows the tracing of maternal lineages. Studies dealing with evolution or systematics have focused upon differences at the sequence level of specific DNA molecules through restriction endonuclease analysis of nucleotide sequencing.

Another class of genetic markers is satellite DNAs that are highly repeated sequences found in most eukaryotic genomes. They have attracted considerable interest as genetic markers for chromosome mapping and phylogenetic studies. The monomeric units of satellite DNAs are organised in long tandem arrays that constitute much of the heterochromatin of chromosomes, predominately that of centromeres and telomeres. Their nucleotide sequence and genomic organisation is often species as well as chromosome specific. Little is known about the organisation of satellite DNAs in teleost fish, but the highly-ordered architecture of satellite DNAs may offer a general source of genetic markers for DNA fingerprinting of fish (Franck et al., 1992; Queller et al., 1993).

Genetic variability within populations

A fish species population rarely consists of a single large randomly mating population, but is usually divided into several subpopulations by space and time (see metapopulation concept, Chapter 12). When genetic subpopulations occur in the same river basin, they are not completely isolated and some genetic exchange could be expected among them. The less genetic exchange there is among subpopulations, the less genetically similar they will be. An extreme case, frequent in freshwater fish, is the complete isolation of populations in different river basins physically isolated for a long time. In this case large divergences among subpopulations could be expected, depending on the length of time since the isolation occurred. The longer the isolation, the more time they have to accumulate genetic differences.

The genetic diversity of West African populations of the bagrid Chrysichthys, which is used in fish culture, was investigated. Thirteen populations of the Chrysichthys auratus complex were studied using enzymatic protein electrophoresis to estimate genetic differentiation and verify their taxonomic status (Agnèse, 1991). Twenty-seven alleles were observed at 19 loci, but only five loci were polymorphic. The maximum genetic distances between two populations (D = 0.112) is lower than the level of maximum divergence observed between conspecific populations of other related species: D = 0.289 in Chrysichthys maurus, and D = 0.304 in Chrysichthys nigrodigitatus. This result tends to support those of Risch (1986), who concluded after a morphological study that coastal lagoon populations, previously known as Chrysichthys filamentosus, were in fact synonyms of C. auratus. The average heterozygosity observed (H = 0.024) is low. Very low values (H = 0.003) observed in upper Senegal tributaries might be explained by the severe reduction of the habitat during the dry season, which results in a considerable reduction of the population size (bottleneck effect). Zero values of heterozygosity were also observed in the White Volta and the Mono River where samples were monomorphic at all the loci. The latter drainages also experience extremely low water levels, while samples from the Niger drainage, coming from habitats flowing throughout the year, had a greater diversity (H = 0.033–0.063).

The populations of *Chrysichthys nigrodigitatus* in West Africa show a clear genetic differentiation (based on 5 polymorphic loci) between the coastal plains (Côte d'Ivoire) and the inland Niger basin at Bamako (Mali): 0.233 < D < 0.266 (Agnèse *et al.*, 1989). Samples from Ebrié and Aby lagoons (Côte d'Ivoire) display a high polymorphism (H = 0.045 and 0.063 respectively) .

In an extensive study of the West African catfish, *Chrysichthys maurus*, it was demonstrated using 20 loci, 12 of them polymorphic (Agnèse, 1989), that there are three population groups exhibiting closer genetic affinities within each group, than between groups. It was possible to show that group 1 (rivers and lagoons from Côte d'Ivoire) exhibits the most ancestral characteristics, and to conclude that groups 2 (Liberia) and 3 (Guinea) are derived from group 1. *Chrysichthys maurus* probably colonised West Africa from the south, taking the opportunity of favourable climatic and hydrographic conditions to migrate along the coast and to colonise suitable biotopes. The colonisation of the Guinean rivers and of the Senegal River is probably recent. The average heterozygocity is low (0.028) but exhibits large variations: from 0.004 (Comoe, Cavally and Nipoue rivers) to 0.061–0.088 in the Bandama and Makona rivers and Aby lagoon.

Low heterozygosity (H = 0.001) was also reported for *Schilbe intermedius* in the Volta system (Abban & Skibinski, 1988), but the results also suggested that the species consists of different types belonging to different gene pools in the Volta basin. For the authors, the low variability may result partly from human activities, such as intensive fishing using poisons during low-water periods, or indirect effects of pesticides. But, as for *C. auratus*, it is difficult to determine exactly what caused the low variability observed. In contrast, it is remarkable that 15 polymorphic loci out of 51 protein loci were scored for

Synodontis leopardinus from the Upper Zambezi River (Van der Bank, 1993).

Results of enzymatic polymorphism obtained on a few species of Clariidae, pointed out that *Clarias anguillaris*, *Clarias gariepinus* (both belonging to the subgenus *Clarias*) and *Heterobranchus longifilis* seem closer to each other than to *Clarias* (*Anguilloclarias*) *ebriensis* (Teugels *et al.*, 1992b). These data also support the view that *C. anguillaris* and *C. gariepinus* form two distinct genetic entities (Teugels, 1986). The heterozygosity of natural populations (River Niger) of *C. gariepinus* and *C. anguillaris* was respectively 0.06 and 0.07 (Teugels *et al.*, 1992b). The average heterozygosity observed for African catfish populations is close to the average heterozygosity observed for 183 species of fishes (H = 0.051) (Nevo *et al.*, 1984).

The cichlid complexes of the East African Great Lakes have received considerable attention, and several electrophoretic studies have demonstrated small genetic distances among morphologically divergent lineages within Lake Malawi and Lake Victoria (Sage *et al.*, 1984; Verheyen *et al.*, 1985b). Most of these endemics, which have radiated recently, lack diagnostic alleles and can only be distinguished by difference in allele frequencies (Kornfield, 1991). It has been suggested for instance that endemic Mbuna species from Lake Malawi may have evolved less than 1000 years ago (Owen *et al.*, 1990). Studies of mtDNA of the 'Haplochromis' cichlid flock from Lake Victoria showed also that mtDNA variation among them was extremely small (Meyer *et al.*, 1990), indicating a very recent divergence. More variation has been found in the homologous portion of mtDNA in *Homo sapiens* than was found among 14 species of nine representative endemic genera of Lake Victoria cichlids (Meyer, 1993).

A few studies on cichlid populations have been conducted along the West African coast in order to investigate the genetic variability of species distributed in isolated river basins. For *Sarotherodon melanotheron*, genetic polymorphism is greater in Senegal and Guinea than in its southern range. Heterozygocity varies for example from 0.060 in Senegal, to 0.032 in Côte d'Ivoire and 0.026 in Benin–Congo (Pouyaud, 1994). In the West African *Tilapia guineensis*, there is a high variability in heterozygocity: from 0.115 in the Ebrié lagoon, to almost 0 in a population of Pointe Noire (Congo). The existence of various subpopulations of *Sarotherodon melanotheron*, as distinguished by morphological and colour characteristics (Trewavas, 1983), has been confirmed by genetic analysis. Three subgroups have been identified: a northern group including *S. melanotheron heudelotti* and *S. m. paludinosus*, that does not appear to be genetically different from the subspecies *heudelotti*; a southern group that includes *S. m. melanotheron* populations from Côte d'Ivoire and Benin; and a third group that includes *S. m. nigripinnis* from Congo (Pouyaud, 1994). However, *S. m. nigripinnis* is genetically more similar to *S. m. heudelotti*, and this result seems to support the hypothesis of the close relationship between the Congolese and the Guinean fish fauna.

For *Tilapia guineensis*, another brackish water tilapiine from West Africa, different subpopulations, which occupy more or less similar areas to *S. melanotheron* have also been recognised: one group for Senegal and Guinea, another for the Ivorian populations, and a third for Benin and Congo populations (Pouyaud, 1994). In this case too, the Congolese group is genetically more similar to the Guinean group than to the Ivorian one.

Micro- and macromutation theories

Evolution consists of an aggregate of processes of various sorts, affecting different taxa in different ways (Endler & McLellan, 1988). It is a two-step process, bearing in mind that the factors which cause the appearance or origin of new variants (molecular mutations such as changes in DNA) are different from those that affect replacement of older by newer variants. Gene substitution in a single organism is the result of mutation and constraints, while gene replacement in a population is an allele frequency change. Phenotypic mutations are strongly dependant on where the molecular mutations occur, and any one molecular mutation does not necessarily result in a phenotypic mutation.

Two main theories explaining the mechanisms of evolution have been prevalent. The **micromutation theory**, proposed by Lamarck and adopted by Darwin, states that evolution is adaptive to environmental changes and progresses in small, almost imperceptible steps. If new properties are required to cope with new environmental conditions, they will be developed whereas the old ones, no longer necessary, will be lost through natural selection (Løvtrup, 1988). In other words, the gradualistic model holds that most evolutionary change is the result of slow gradual transformation of existing species (phyletic evolution).

The opposite idea, the **macromutation theory**, which has arisen from palaeobiology and fossil records, suggests that evolutionary modifications arise independently of the environment and proceed in large steps as the outcome of the epigenetic changes in ontogenesis. Schematically this theory assumes that new taxa evolve essentially instantaneously via a major genetic mutation that establishes reproductive isolation and new adaptations all at once. Macroevolution, which is an evolutionary change greater than that usually characterising species, is sometimes associated with major adaptive innovation, followed by the radiation of a lineage in a new adaptive zone. Stanley (1989) supported the macroevolution theory and characterised as speciation those events that are extreme discontinuities of stasis. The great longevity of many species recognised in fossil records (millions of years for freshwater fishes) should be taken to indicate that approximate evolutionary stasis is fairly common in nature.

The micromutation theory leads to the conclusion that living organisms developed properties to carry out some particular function. But with the macromutation theory, it can be suggested that the appearance of a new feature independently of any 'purpose' can be used by the modified organisms to perform certain functions that promote their existence (Løvtrup, 1988). An alternative hypothesis in this impassioned controversy is the theory of 'punctuated equilibria' or 'punctuational model of evolution' (Eldredge, 1971; Stanley, 1979), which asserts that most evolutionary change in the history of life is restricted to brief periods of drastic environmental and evolutionary modifications, while between these periods both remain constant. Both proponents and opponents of punctuated equilibria reject any saltational theory of speciation and agree that if the punctuations of the fossil record represent speciation, they probably represent adequate animal generation times for normal allopatric speciation to occur.

Greenwood (1984) considered the haplochromine flocks of the African Great Lakes to be an outstanding illustration of an extant punctuational evolutionary phase: it is a good example of rapid speciation associated with marked phenotypic changes, which occurred in a restricted area. The existence of various stages of development in particular characters, represented by distinct and contemporary species, is a process that could be identified as cladistic gradualism (Greenwood, 1981). Such a pattern has not been recognised in fossil assemblages used to exemplify punctuational phases in evolution, probably as a result of the difficulties in identifying close species in fossil records. The haplochromine flocks therefore provide empirical evidence unavailable to palaeontologists, and constitute a unique chance for studying evolution in action (Greenwood, 1991). This author also stated that anagenic speciation events (gradualistic patterns of evolution through minor changes) must 'have occurred, together with essentially stasigenic ones, to account for the adaptive radiation in lacustrine flocks. This hypothesis is supported by the existence in each of the major lakes of groups of several sibling species that barely differ from one another, except in the details of the males' breeding colorations. Van Couvering (1982) suggested from fossil records that repeated events of adaptive radiation have been important throughout cichlid history.

In contrast to the haplochromines, the African lineage of tilapiines appears to be characterised by a predominance of stasigenic over anagenic speciation. It has given rise to only two species flocks: one of 5 species in Lake Malawi and another of 9 species in Lake Barombi-Mbo (Trewavas, 1983). The more specialised feeding mechanisms of tilapiines would not be suitable for the production of trophically multiradiate species flocks (Trewavas, 1983).

II The past as a key to understanding the present

4 Species diversity: evolution at work

There is a great diversity of evolutionary processes, and the parameters that are valid for one group of organisms may not necessarily be at all applicable to another group. There are few, if any, universal laws in evolutionary biology.

Mayr, 1988

There exists no law of evolution, only the historical fact that plants and animals change, or, more precisely, that they have changed. The idea of a law which determines the direction and character of evolution is a typical nineteenth-century mistake, arising out of the general tendency to ascribe to the 'Natural law' the functions traditionally ascribed to God.

Popper, 1962

Species are almost universally used as the units in which biological diversity is measured, but there is no generally agreed definition of, or criteria for the delimitation of, fish species, despite the fact that this problem has been debated for more than two centuries. The main difficulty lies in the dynamics of evolution, and the many problem areas are thought to represent examples of speciation in progress. Thus, understanding the causes of species diversity depends on the evidence of how the species have evolved.

Speciation is the process by which one species gives rise to two or more species, often called sister-species. It is the splitting of an originally uniform gene pool into new independent gene pools, which can then acquire a unique set of biological and morphological characters (autapomorphies), and the evolution of genetic barriers to gene exchange among populations. Different hypotheses have been suggested for the various ways in which genes causing reproductive isolation may become established and then spread throughout the population. Although the mechanisms that create variation at the population level have been the focal point for a lot of research, the overall pattern of speciation is still poorly understood, and little material is available, particularly for African fish, to support the theories proposed.

What is a species?

The term 'species' refers both to a taxonomic category and to biological concepts, which evolved in parallel with the improving knowledge of evolutionary biology. Usually, but not always, taxonomic species are equivalent to biological species.

Until about the first half of the nineteenth century every organism was thought to be created by God and to correspond to some idealised plan. There was a limited number of species and the goal of taxonomy was to draw up an inventory and to describe their specific characters. Modern taxonomy, as founded by Linnaeus, has therefore traditionally been based on forms and patterns, and the typological species concept involves a morphological definition. Each species is identified by a binomial and the task of systematics is to order and to classify the identified species. The obvious difficulty with this is to take into account the natural variability of characters, a problem which, partly for philosophical reasons, was not considered by the founders of typological systematics.

While this concept of fixed species is no longer supported, one should nevertheless remember that some of the traditional practices of taxonomy and species description are inherited from it (see Daget, 1988). For example, the designation of a type specimen (holotype), and the priority in nomenclature given to the first described specimen are sometimes responsible for taxonomic changes, or even amazing situations. A famous case is the stuffed type specimen of *Synodontis xiphias* (Mochokidae) that has a pointed, elongated snout. This species has never been recorded again, and it has been demonstrated that the pointed snout is due to a long iron wire used during stuffing (Poll, 1971). Thus the type of *S. xiphias* is an abnormal specimen of a species later called *Synodontis labeo*. But despite its abnormality, *S. xiphias* was the first name assigned to this species and the type of *Synodontis labeo* is considered a synonym of *S. xiphias*.

As long ago as 1798, Cuvier used a biological criterion to define species and Mayr's (1942) definition, which is now widely accepted, is not fundamentally different: 'species are groups of actually or potentially interbreeding natural populations which are reproductively isolated from other such groups'. One of the major criteria to differentiate species is the sterility or inviability of hybrids, but such post-mating isolation is rather unusual given that, in most cases, pre-mating isolation prevents interbreeding. Nevertheless, there are exceptions to the biological concept of a species as in the case of parapatric populations, which may be only partially isolated reproductively and can interbreed along a hybrid zone that can persist for long periods.

The basis of a biological species is of a pool of genes that can be recombined through sexual reproduction within the population. This pool is 'protected' from mixing with other gene pools by biological, behavioural or physiological mechanisms. A fundamental tenet of the biological concept, particularly in sympatric situations, is that each species possesses a set of isolating mechanisms, which keeps it distinct from other species. Paterson (1980, 1982) discussed this Isolation Concept, arguing that pre-mating isolating mechanisms should be regarded as incidental consequences of speciation rather than as features that evolved in order to ensure reproductive isolation (Ribbink, 1988). He (Paterson, 1985) therefore proposed to replace the Isolation Concept with the Recognition Concept, which emphasises the origin of those mechanisms that ensure syngamy amongst members of the nascent species. This assumes that each species of mobile animal has a specific-mate recognition system

whose components are the ecological and behavioural coadaptations of partners, including courtship and spawning repertoire, synchronised breeding periods, etc. To illustrate this point, colour differences between parent and daughter species in Lake Victoria cichlids could be viewed either as a means of keeping species within their own niches and thereby minimising the competition between them (Isolation Concept) or as a consequence of genetic reorganisations which occurred in isolation (Recognition Concept).

Thus a species can be defined as those individuals sharing a common fertilisation system, which is a set of adaptations whose function is to bring about fertilisation, including the specific-mate recognition system that is crucial for motile organisms. The characters of a specific-mate recognition system are considered to be adaptive and particularly efficient in the preferred habitat of the species, this 'normal' habitat being viewed as a basic aspect of the Recognition Concept (Ribbink, 1988).

Species have also been defined wholly in terms of the gene pool (Meglitsch, 1954), but there is no simple answer to the question how great must a genetic difference be to make a species? Rather, species owe their existence to specific characters, such as sexual behaviour, the timing of reproduction and specific developmental pathways, governed by specific genes.

The biological and genetic concept of species are still widely accepted, although criticised from various directions, including denial of the existence of species. For example, Løvtrup (1979) suggested that taxonomists would do better to be concerned with populations, and to use a strictly phenetic approach to classification at the species level, as well as at the higher or lower levels. For some authors, only demes (local breeding units) and populations are real, species being artificial constructions. But one of their most significant limitations is the absence of reference to the evolutionary process.

It is now largely recognised that species are not immutable entities, but that they evolve: 'a lineage (an ancestor–descendant sequence of populations) evolving separately from others and with its own unitary evolutionary role and tendencies' (Simpson, 1961). The evolutionary concept of species is based on the distinctiveness and coherence of a population, or populations, in time. Species can change their appearance during evolution without losing their identity, and those prevailing today are therefore considered as the latest descendant populations. This leads to the problem of delimiting species in both space and time.

One approach to this is that taken by Ridley (1989), who advocated a specifically cladistic species concept based on Hennig's original views on the delimitation of species. For the Hennigian school of phylogenetics, species originate when one lineage branches into two. This phylogenetic concept emphasises cladistic monophyly and coherence in time, but also the status of the species category as taxonomically irreducible. According to Nixon & Wheeler (1990), a phylogenetic species is 'the smallest aggregation of populations (sexual reproduction) or lineages (asexual reproduction) diagnosable by a unique combination of character states in comparable individuals'. While the concept of evolutionary species takes into account the evolutionary processes, and is therefore not contradictory to the biological species concept, so that one could consider the latter as a special case of the former (Blackman & Day, 1981; Brooks & McLennan, 1991), the risk of the phylogenetic approach is to recognise far more species than does the biological species concept.

Diversity within species

Most individuals in a population exhibit a more or less limited amount of variability in almost all of their measurable phenotypic characteristics (i.e. size, shape, colour, physiological tolerances, growth, etc.). The phenotype is generated by an array of inherited and environmental effects, and it is often difficult to work out if the variability is the result of either external or internal (ecophenotype) environmental constraints, or is genetically controlled.

Phenotypic variability
The concept of **polytypic species**, which emerged from population biology, has had a major impact on taxonomy. It refers to a species (usually widespread) which shows more than one form and is frequently composed of several subspecies that may differ either in morphological, physiological, ecological or behavioural characters. Terms such as variety, race, form, etc. have been used to refer to any deviation from the 'typical' species, but the

subspecies is the lowest rank permitted in formal nomenclature. It is recognised on morphological evidence, sometimes on geographical criteria, and only rarely on behavioural or genetic characters (Lane & Marshall, 1981). For many authors, a subspecies is often seen as virtually a 'species in the making'. The related term polyphenism describes a situation in which one genotype produces two or more discrete phenotypes in response to an environmental signal. It is the analogue among phenotypes of the term polymorphisms among genotypes.

The recognition of variability among populations led to two different approaches, often known as lumping or splitting. Either varieties were described as local adaptations of only one nominal species, or large numbers of allopatric species were described differing only in details of coloration or other morphological characters. There are some situations, for instance, where the opposite ends of a geographical cline have been recognised as distinct species. The use of criteria other than morphology alone (such as genetics, biological labelling by parasites, reproductive behaviour, etc.) may now aid our understanding of the status of these 'varieties'. Another reaction was to group allopatric races or subspecies, which are more or less slightly differentiated, into a single polytypic species, and to call this assemblage of populations a superspecies.

Brycinus nurse (Characidae) is an example of a polytypic species. Dwarf populations have been recorded in some lakes: *B. nurse dageti* in Lake Chad, which was originally recognised as a distinct species, and *B. nurse nana* in Lake Turkana (Paugy, 1986). Many examples of polytypic species can be found among the cyprinids. For instance, the widely distributed *Barbus radiatus* is a polytypic species that had two recognised subspecies (Stewart, 1977): *Barbus radiatus radiatus*, and *B. radiatus aurantiacus*. A third subspecies, *B. radiatus profundus*, is in fact a distinct species. The results of a study on meristic and morphometric variation in *B. radiatus*, indicated that there are not only two discrete allopatric morphs, but that there is a population mosaic of highly variable morphs. Body depth, caudal peduncle length, counts of scales along the longitudinal line, dorsal fin height and orbit diameter accounted for most of the geographic variations (Stewart, 1977). It is hypothesised that this population mosaic of morphology may reflect ecotypic responses to environmental factors, such as water velocity and temperature. A genetic component could be associated with what seems to be an environmental morphocline, but laboratory and field studies are needed in order to estimate the relative contribution of ecological and genetic factors to this geographic variation in morphology. A fairly similar situation occurs in *Barbus wurtzi* and *Barbus sacratus* inhabiting coastal rivers of West Africa (Lévêque & Guégan, 1990), where meristic and morphometric characters vary markedly from one river to the next. *Barbus sacratus* also exhibits morphological variations in the mental lobe that are either developed or absent, even in the same river. In the widespread *Barbus bynni*, which range from the Nile to Senegal and Côte d'Ivoire, three subspecies are still distinguished based mainly on differences in the number of scales: *B. bynni bynni*, *B. b. occidentalis* and *B. b. waldroni* (Lévêque & Guégan, 1990).

When species are distributed discontinuously, as is the case for fish inhabiting different isolated river catchments, each population could be considered as a panmictic unit (one in which members form a single, randomly mating population over its entire geographic range), providing physical barriers do not prevent exchanges between subpopulations within catchments. There are numerous examples of phenotypic variability in panmictic fish populations many of which refer to meristic variations, that is to say any countable character (number of vertebrae, fin rays, scales). The systematic revisions of African genera or families, such as the Mormyridae (Bigorne, 1987,1989,1990a; Lévêque & Bigorne, 1985b), Cyprinidae (Lévêque & Guégan, 1990), Characidae (Paugy, 1986), Clariidae (Teugels, 1982, 1986) provide many examples. Morphological variations, such as those among several large *Barbus* species, were also noticed. The so-called rubber lip specimens have elongated mental lobes and a fleshy upper lip, whereas other specimens have thin lips. The adaptative significance of rubber lips is unknown, but Groenewald (1958) reported that a rubber lip specimen of *Barbus aeneus* (ex *Barbus holubi*) caught in fast flowing waters and subsequently kept in a small pond, after a period of one year became the thin lip type. It has been suggested that the degree of lip development could be related to feeding habits. Lévêque & Guégan (1990) also observed morphological variability in the mental lobe morphology of West African populations of *Barbus sacratus* and rubber lips

were also observed on some specimens of *B. parawaldroni* (Lévêque *et al.*, 1987).

Geographical variation

Two populations of a given species living in different parts of its geographical range are usually not identical, and this geographical variation could result either from the response of the organism to different environments, or from the limitation on gene flow between spatially separated populations. When conditions such as climate or other ecological variables change gradually over a given area, it can result in gradual changes (termed clines) in certain attributes of a species. A cline is 'a geographic gradient in a measurable character or gradient in gene, genotype or phenotype frequency' (Endler, 1977). Geographical changes in meristic characters, such as the number of scales, the number of anal or dorsal fin rays and the number of vertebrae, have been described for many African fish species and examples are available for *Alestes baremoze* (Paugy, 1978), *Mormyrops* (Bigorne, 1987), *Mormyrus* (Lévêque & Bigorne, 1985b) and *Barbus bynni* (Lévêque & Guégan, 1990).

A marked morphological cline has been observed along an east–west gradient in the species *Schilbe mandibularis* inhabiting the coastal river basins of Côte d'Ivoire (Lévêque & Herbinet, 1982). There is a sharp decrease in the number of anal fin rays (Table 4.1) as well as in the number of vertebrae. Smaller, but still significant, variations were also noted in the number of pectoral fin rays and of branchiospines, and the relative length of barbels to head length. This cline is also obvious in the maximum standard length recorded in the river basins and in the size at first maturation (Fig. 4.1), which means that in this case morphometric and meristic changes are closely related to biological characteristics.

Another example of geographical variation is that of *Brycinus macrolepidotus*. Prior to a morphometric study of this species throughout its distribution area in West Africa, two species were identified: *Brycinus rutilus* in the forest area, and *B. macrolepidotus* in the savannah area. Paugy (1982) pointed out the existence of a morphological cline (body depth, position of the dorsal versus the pelvic fins) so that only *B. macrolepidotus* is now recognised as a valid species. In addition to obvious clines, however, many cases of irregular distributions of vari-

Table 4.1. Schilbe mandibularis: *clinal changes in meristic characters along an east–west gradient in river catchments of Côte d'Ivoire and Liberia*

Characters	River basins						
	Comoé	Bandama	Sassandra	San Pedro	Cavally	Nipoue	Diani
No. anal fin rays	59.4(55–66)	60.7(53–69)	59.5(54–66)	57.7(52–64)	54.0(50–57)	49.3(43–55)	43.9(39–47)
No. pectoral fin rays		11.2(10–12)	10.8(9–12)	11.1(10–12)	10.9(10–12)	10.0(9–11)	9.2(8–10)
No. vertebrae	48.3(47–50)	48.4(46–50)	48.6(48–50)	47.6(46–48)	45.6(44–48)	43.5(43–45)	
No. branchiospines	12.2(11–14)	11.7(10–14)	11.2(9–13)	11.6(10–14)	11.2(10–14)	11.4(10–13)	9.5(9–10)

From Lévêque & Herbinet, 1982.

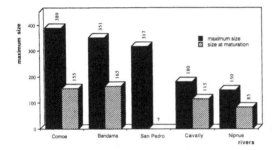

Fig. 4.1. Changes along an east–west gradient in the maximum observed standard length and size at first maturation (in mm) for *Schilbe mandibularis* from river catchments in Côte d'Ivoire and Liberia (data from Lévêque & Herbinet, 1982).

ation can be observed, which may be attributed to a variety of causes, sometimes simply a discontinuous distribution preventing gene flow between populations.

It is often questionable whether members of a vicariating series of populations should be classified as species or subspecies. For some populations, the distinction is not clear and rather subjective, as when Trewavas (1983) treated the brackish-water populations of *Sarotherodon* along the West African coast as subspecies of *Sarotherodon melanotheron* , although Thys van den Audenaerde (1971) had considered them all to be valid species.

Polymorphism without any evidence of a cline has been reported for cichlids of the East African Lakes. Lewis (1982a) suggested that there is polymorphism in the pharyngeal dentition of *Labidochromis caeruleus* in Lake Malawi, but no experimental demonstration is available. Witte (1984a) has demonstrated considerable plasticity in the oral dentition of *Haplochromis squamipinnis*, and Hoogerhoud (1986) did confirm experimentally the observations on pharyngeal dimorphism in *Astatoreochromis alluaudi* made by Greenwood (1965a).

Ontogenetic variability

Morphological changes occur during growth, and in most fish there is an allometric growth of organs or parts of body with size. For example, the size of the eye is often proportionally greater in juveniles than in adults. When using ratios for taxonomic purposes it is therefore important to know the size, or the range of sizes, of the specimens examined. Developmental changes in morphology can also occur and in a few cases juveniles have been described as different species from their adults. In the bagrid *Clarotes*, for example, a bony spine is present in front of the adult adipose fin but is not present in juvenile fish, which may be confused with *Chrysichthys*. This is also the case for *Synodontis filamentosus*, whose juveniles, lacking the long dorsal filament characteristic for the species, were described under the name *Synodontis augeriasi*.

One of the most obvious ontogenic changes concerns colour pattern. It is well known for instance in *Synodontis*, whose juveniles exhibit a colour pattern clearly distinct from that of the adults (see Daget, 1954; Poll, 1971). In *Hemisynodontis membranaceus*, individuals less than 80–90 mm SL have irregular brownish spots on the flanks and smaller round spots on the fins, while adults have uniform greyish flanks and fins. Similar ontogenetic changes in colour patterns have been observed in many cichlids and cyprinodonts, where males usually develop their bright colour after maturation. There are also a few species of Pleuronectoformes adapted to freshwater that undergo metamorphosis during their development.

Sexual dimorphism

Sexual dimorphism occurs in the morphology of some fish families. The long dorsal rays of male *Brycinus longipinnis* (see Paugy, 1986) are an example. Among characids, as well as mormyrids, the anal fin of males is often characteristically enlarged. More conspicuous sexual dimorphism is seen in the bagrid genus *Chrysichthys* whose mature males exhibit an enlarged head and are morphologically very different from the females (Risch, 1986).

Among cichlids and cyprinodonts, males are usually much more brightly coloured than females. In Lake Tanganyika, remarkable sexual dimorphism is observed in maternal mouthbrooders with male mating territories, while non-territorial mouthbrooders rarely exhibit sexual dichromatism (Kuwamura, 1986).

Recognition of new species

If the conceptual understanding of what is a species is an interesting issue in the philosophy of biology, it is, nevertheless, also of basic importance for biologists to be able to identify species. The bases upon which a new species is eventually recognised are complex. The biological species concept is of little use to museum taxonomists working almost entirely on dead specimens, and therefore one of the traditional methods of distinguishing fish species populations has been, and is still, the comparative examination of morphological characters, both meristic and metric. For experienced taxonomists the first diagnostic approach is the 'overall impression', which is a way of mentally integrating subtle differences, and it should not be underestimated although it has to be confirmed later by meristic counts or by measurements.

During recent decades new criteria have been developed to seriously aid identification of species. The use of genetics is difficult, given that it is not possible either to interbreed all expected species, or to relate genetic distances estimated by enzymatic polymorphism with the existence of true biological species. Other criteria used concern comparison of biological characteristics and

behaviour of expected species, and the use of biological markers such as parasites.

The use of biological criteria

The separation of species on morphological grounds becomes difficult where speciation has taken place unusually rapidly (explosive speciation *sensu* Greenwood, 1981). Morphologically similar species may be reproductively isolated, but how is it technically possible to check that this is the case? The biological species concept is often very difficult to apply. For fish species, such as cyprinodonts, that easily breed in captivity, crossbreeding experiments have been conducted to check the existence or not of genetic barriers between populations, and Scheel (1968) gave a few results for African species. In the case of African Great Lakes cichlids, many closely related species are morphologically very similar and their identification is very difficult when they are dead. The term 'sibling species' (also called cryptic species) applies to those groups of sympatric anatomically very similar species that can hardly be distinguished morphologically. In the classical example of the four *Oreochromis* endemic to Lake Malawi that exhibit marginal morphological differences, Lowe (1953) was able to demonstrate that males have distinctive breeding coloration, they breed at different periods of the year, in different depths of water and have colonised different habitats.

When working on living specimens, even quite subtle differences in coloration can be of considerable taxonomic value. For cichlid species that display complex courtship behaviour, males have distinctive breeding colorations. But coloration differences between allopatric populations may represent early stages in the speciation process, and splitting species solely on the basis of small colour differences would be confusing. Some authors may therefore attribute geographical colour differences to intraspecific variations (Lewis, 1982b) as is the case, for instance, for *Labeotropheus* in Lake Malawi (Ribbink et al., 1983a).

The use of biological criteria is, therefore, the only practical way to identify morphologically close species (Ribbink, 1988), but the greater problem is to decide what constitutes a species. For Lake Malawi cichlid species, Ribbink et al. (1983b) used, on a wide scale, the specific-mate recognition system, giving priority, when patterns were unclear, to differences rather than to similarities. The result was probably an overestimation of the number of species, judged preferable by the authors to 'lumping' different species (Ribbink, 1988). The difficulties faced in recognising different species could also be a source of misinterpretation concerning niche segregation in Great Lakes cichlids (van Oijen, 1982). For many other species, the task is too big, if not impossible, and we are not in a position to investigate their biological isolation. In practice, the phenetic approach remains the main way of identifying species. Phenotypic characters are nevertheless assumed to be highly sensitive to environmental changes.

Parasites as biological markers

The biology and evolution of parasites are connected to some extent with those of their hosts. The existence of parasites specific to a particular host, which is the final product of a long specialisation on a particular host, has led parasitologists to postulate the existence of co-evolution between them. This term encompasses both the degree of mutualistic phylogenetic association (cospeciation) and the degree of mutual modification through interactions (coadaptation) (Brooks & McLennan, 1991).

Euzet & Combes (1980) distinguished three types of specificity between a parasite and its host.

1. Strict or oioxenic, when a parasite species lives only on a single host species.
2. Close or stenoxenic, when it occurs on phylogenetically related species.
3. Croad or euryxenic, when the parasite is observed on different hosts whose similarity is more ecological than phylogenetic.

In the case of strict specificity, it is easy to imagine that the speciation of a host into two distinct populations might be followed by a concomitant speciation of the parasite into two different parasite infrapopulations. In such conditions, the Farenholz's rule postulates that 'parasite phylogeny mirrors host phylogeny', and this has been observed in a number of studies on fish parasites.

Monogenean gill parasites of fish have a direct life cycle and frequently display a strict host specificity. Parasitological studies of four sympatric West African cyprinids, *Labeo coubie*, *Labeo senegalensis*, *Labeo roseopunctatus*, and *Labeo parvus*, revealed 8 species of *Dogiel-*ius and 13 of *Dactylogyrus*. Five *Dogielius* and seven *Dactylogyrus* were found on *L. coubie*. Two species of *Dogielius* and five of *Dactylogyrus* occur in *L. senegalensis* and *L. roseopunctatus*, while only one species of each parasite taxon has been recorded on *L. parvus* (Table 4.2). None of these *Labeo* species harboured monogeneans found on other hosts (Paugy *et al.*, 1990). This strong host specificity of parasites is probably a reflection of the absence of gene exchange between reproductively isolated sympatric host taxa. This lack of gene exchange has been demonstrated by allozyme investigations.

Seeking the reasons for the different parasite species richness on these four species, Guégan & Agnèse (1991) showed that the parasites encountered on *L. coubie* and *L. senegalensis* seem to have evolved concomitantly with their host species, or by sequential colonisations between these two related hosts. The preferential presence of parasite forms on *L. coubie* and *L. senegalensis*, displays the close phylogenetic affinities between these two cyprinids, and thus partially obeys Fahrenholz's rule. However, this possible concomitant speciation (or cospeciation), even if it results in topologically congruent phylogenies, does not necessarily prove host–parasite coevolution. The parasitism observed on *L. roseopunctatus* and *L. parvus*, may result from switching of parasites present on *L. coubie*. In fact, the monogeneans of *L. parvus* and *L. roseopunctatus* are phylogenetically closer to some of the parasites of *L. coubie* than to any other West African cyprinid parasites. Switching from one host species to another (also called capture) could probably account for the observations in this case, and constitute non-phylogenetic evolution, or evolution that is horizontal in time and thus does not obey Fahrenholz's rule at all. This hypothesis of host switching is supported by parasitological geographical data, since parasites gained by capture are absent on host populations where *L. coubie* is absent (Guégan & Lambert, 1990).

Fish parasite captures appear to be frequent, and we can imagine that this change to a new host species, followed by speciation of the parasite, must undoubtedly be the source of reciprocal and intensive selection pressures, that is to say a true host–parasite coevolution. At present, we have very little knowledge about the role of parasites in fish evolution and diversity. What happens when a fish meets a new parasite species? What are the coevolutionary processes? What are the final products of evolution?

Table 4.2. *Parasitic specificity of Monogenea* Dactylogyrus *and* Dogielius *for* Labeo *from West Africa*

Parasites	Hosts			
	Labeo parvus	*Labeo roseopunctatus*	*Labeo senegalensis*	*Labeo coubie*
Genus Dactylogyrus				
D. digitalis				*
D. decaspirus				*
D. oligospirophallus				*
D. retroversus				*
D. titus				*
D. falcilocus				*
D. jaculus				
D. cyclocirrus			*	
D. senegalensis			*	
D. labeous			*	
D. rastellus			*	
D. tubarius			*	
D. nathaliae		*		
D. brevicirrus	*			
Genus Dogielius				
D. harpagatus				*
D. clavipenis				*
D. anthocolpos				*
D. flagellatus				*
D. complicitus				*
D. tropicus			*	
D. flosculus			*	
D. grandijugus		*		
D. parvus	*			

Source Guégan & Agnèse, 1991.

It would certainly be premature to reply to all these fascinating questions. However, what we can say is that parasites have certainly played a great role in evolution, diversification and biodiversity, and such a research perspective should promote further studies of fish-host parasite coevolution.

That the Monogenea should be considered as good biological markers of host species, and thus of value as taxonomic criteria, was demonstrated for African bagrids (*Chrysichthys*) in a double blind test, using genetic and parasitological (Monogenea) criteria. Comparison of the electrophoretic characteristics and the monogenean species assemblage of each fish revealed a very good relationship, enabling identification of 3 species among a series of 40 individuals. Each host species is characterised by two monogenean species (Euzet *et al.*, 1989). Platyhelminths parasiting *Chrysichthys* are therefore as good an indicator of their hosts as the allozymes. The existence of very strict host specificity has also been demonstrated for a variety of other African fish families, including *Dactylogyrus* parasiting *Labeo* (Guégan *et al.*, 1988*b*; Paugy *et al.*, 1990), or barbs (Guégan & Lambert, 1990) and *Annulotrema* parasiting *Hydrocynus* (Guégan *et al.*, 1988*a*). For the large West African species of *Barbus*, the characteristics of gill parasitism in *B. bynni*, *B. occidentalis* and *B. waldroni* (Guégan & Lambert, 1990) confirm the opinion that they are a single complex of isolated populations exhibiting clinal morphological variations along an west–east gradient (Lévêque & Guégan, 1990). The other large *Barbus* have specific parasites (Table 4.3). For cichlids, branchial Monogenea are also good biological markers (Table 4.4).

Of particular interest for methodological purposes is the example in which a new species of *Labeo* was charac-

Table 4.3. *Parasitic specificity of Monogenea Dactylogyridae for large* Barbus *from West Africa*

Parasites	Hosts					
	Barbus bynni occidentalis	Barbus bynni waldroni	Barbus petitjeani	Barbus sacratus	Barbus wurtzi	Barbus parawaldroni
Genus Dactylogyrus						
D. pseudanchoratus	*	*	*	*	*	*
D. ruahae			*	*	*	
D. archeopenis			*	*		*
D. aferoides	*	*	*			
D. sahelensis	*	*	*			
D. clani			*			
D. petitjeanii			*			
D. sacrati				*		
D. wurtzii					*	
D. parawaldroni						*
Genus Dogielius						
D. djolibaensis	*	*	*			
D. phrygieus				*		
D. vexillus					*	
D. pedaloe					*	*

From Guégan & Lambert, 1990; Lévêque & Guégan, 1990.

terised using three simultaneous and independent approaches: morphology, enzymology, parasitology (Paugy *et al.*, 1990). The existence of specimens morphologically distinct from the sympatric widespread species *L. coubie* and *L. senegalensis* was recognised in the Baoule River (Upper Senegal basin in Mali), but it was difficult, using morphological data alone, to conclude whether these specimens belonged to a new species or were hybrids of the two other species inhabiting the same river. The existence of a third species, *Labeo roseopunctatus*, was established since it could be distinguished allelically by four homozygous loci that are not present in *L. coubie* and *L. senegalensis*. Moreover, the Nei's genetic distances are shorter between *L. coubie* and *L. senegalensis* than between either of these two and the suspected new species. All three species can be further separated by the existence of specific monogenean gill parasites. There is thus no doubt about the validity of *L. roseopunctatus* as a sympatric species of *L. coubie* and *L. senegalensis* after the use of these complementary diagnostic tools. This simultaneous approach should be more widely used for species diagnosis in the future, but the limit is that parasite hosts must be sympatric.

Speciation

Speciation and morphological evolution

Speciation and morphological evolution are two independent processes that are usually synchronised but may proceed independently. The cichlid fish of Lake Victoria and Malawi for instance, originated extremely rapidly through adaptive radiation and are genetically very closely related yet they have undergone major morphological differentiations. In the older Lake Tanganyika, which harbours cichlid lineages with a long, independent evolutionary history, many geographically distinct populations have been described, only distinguishable by minor if any morphological variation, but by pronounced differences in coloration. That is the case for the six species of the *Tropheus* lineage, for which more than 50 distinctly coloured 'races' are currently known, whereas

Table 4.4. *Branchial parasites of the genus* Cichlidogyrus *encountered in West African* Tilapia *(subgenus* Coptodon*)*

Parasites	Hosts				
	Tilapia dageti	*Tilapia guineensis*	*Tilapia louka*	*Tilapia walteri*	*Tilapia zillii*
C. vexus		*			*
C. arthracanthus	*	*		*	*
C. digitatus	*	*	*	*	*
C. ergensi	*	*	*	*	
C. ergensoides	*			*	
C. retroversus	*		*	*	*
C. tiberianus	*	*		*	*
C. yanni	*		*		
C. aegypticus		*			*
C. agnesei		*			
C. anthemocolpos		*			*
C. bilongi		*			
C. cubitus		*	*	*	*
C. flexicolpos		*			
C. gallus		*		*	
C. louipaysani		*			
C. microscutus		*			
C. ornatus		*			*
C. amphoratus			*		
C. tilapiae					*

From Pariselle, unpublished data.

all the species are morphologically almost identical even where they live in sympatry (Brichard, 1989). These species are strictly confined to rocky habitats and have limited capacity for dispersal.

Mitochondrial DNA from several populations of the six described species of *Tropheus* was sequenced in order to investigate the distribution of genetic variation within and between populations and species in relation to their taxonomy, coloration and geographic distribution (Sturmbauer & Meyer, 1992). A surprisingly large genetic divergence was observed among populations of the endemic *Tropheus* lineage that might be about 1.25 million years old. It contains twice as much genetic variation as the entire, morphologically diverse cichlid assemblage of Lake Malawi, and six times more variation than the Lake Victoria species flock (Sturmbauer & Meyer, 1992). Conversely, the DNA also demonstrated that genetically closely related species can differ considerably in coloration.

Another example is given by the genetic relationships between several West African mormyrids (Agnèse &

Bigorne, 1992). Enzymatic studies have demonstrated that the genera *Petrocephalus* and *Marcusenius* are genetically very distinct, while the genera *Mormyrops*, *Hippopotamyrus* and *Pollimyrus* are closely related. The last three are morphologically well differentiated so this result emphasizes the lack of correlation between genetic distances and fish morphology.

Modes of speciation

Broadly speaking, speciation could occur suddenly or gradually. Theoretically, a viable mutation might be able to promote sudden speciation, and such deviant individuals, often called 'hopeful monster' (Goldschmidt, 1940), may form a population of descendants that are reproductively isolated from the parental stock. But most of the current thinking deals with gradual speciation, through isolating mechanisms causing reproductive barriers between species. Different pathways are recognised as illustrated in Fig. 4.2 and major pre-mating isolating mechanisms, preventing gene flow between sympatric

Table 4.5. *Factors that prevent gene flow between fish species*

1. **Geographical**: species live in different areas

2. **Pre-mating**: factors preventing the formation of hybrids
   ecological separation: species occupy different niches or
      habitats
   temporal separation: different spawning seasons
   ethological separation: differences in courtship behaviour
      and signals used in mate recognition

3. **Post-mating**: factors acting after the formation of hybrids
   hybrid inviability: F1 hybrids inviable
   hybrid sterility disrupts gametogenesis: F2 (or F1?) hybrids
      infertile
   hybrid breakdown: F1 viable and fertile but F2 and
      backcross hybrids inviables or infertiles

Modified from Maynard Smith, 1989.

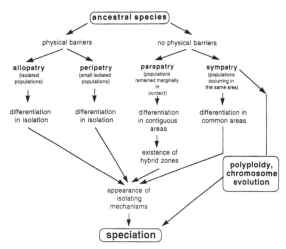

Fig. 4.2. Main pathways of speciation (modified from Endler, 1977).

species, have been identified (Table 4.5). All the evidence indicates that mate recognition, sterility and inviability of hybrids all increase gradually with time as estimated by genetic distances, which confirms the impression gained by a study of geographical variation that speciation is not a sudden event (Maynard Smith, 1989).

What are the causes and circumstances leading to the differentiation of parental populations into derived species is a major question in evolutionary biology, and has given rise to an extensive scientific literature. Different modes of speciation used to be recognised (Fig. 4.2). Phyletic speciation represents evolutionary change within a single species, also termed anagenesis (changes without speciation), in which case, the lineage has undergone evolutionary change without splitting so that the older and the younger members of the lineage are markedly dissimilar. The gradual progression of forms through this single lineage is assigned species status at different points in time. Additive speciation leads to an increase in the number of species by lineage splitting (cladogenesis). Reticulate evolution may, conversely, lead to a lower number of species. All speciation models (Fig. 4.2) are models of additive speciation that are based on several different mechanisms involving phylogenetic patterns, biogeographic information, population biological data, etc. (Wiley, 1981). Until now, no algorithm of phylogenetic reconstruction has been able to process data on reticulate evolution.

Classical models of speciation

Allopatric or geographical speciation is clearly the classical model of speciation. It postulates that ancestral populations become at some time spatially isolated and evolve independently, becoming genetically differentiated as a response to natural selection in the new, different restricted environments, or through genetic drift. Ultimately, they form new species, i.e. groups of organisms reproductively incompatible, but with a common ancestor. The implications of this model are that divergence is assumed to occur only when a physical barrier prevents gene flow, and that changes must occur gradually and evenly in each isolate.

The allopatric accumulation of changes to the genome in response to changed ecological conditions is the first step in speciation (West-Eberhard, 1983). It is called adaptative speciation, because the causal agent is adaptation to differing ecological situations, or 'divergent evolution'. Pre-mating isolation mechanisms are byproducts of this adaptative change. Alternatively, a nonadaptative speciation has been suggested, where premating isolation evolved in allopatry, but as a direct consequence of sexual or social selection.

The peripatric speciation model was proposed by

Mayr (1963), and assumes that if a small population becomes isolated from the bulk of the species range it might evolve more readily or more often. This was suggested by the rather frequently observed pattern of broad uniformity across continuous populations contrasting with striking divergence in peripheral isolates. The loss of genetic variability and lack of competition with other species in such isolates could support this theory, but the exact causes are difficult to identify, and it is not clear if peripatric speciation would give a significant supply of species to the main part of the taxon's range (Barton, 1988).

The term founder effect was used by Mayr (1963) to describe the genetic consequences of the colonisation of a new habitat by a few colonists. In such small populations the initial alteration in gene frequencies by genetic drift sets off a cascade of genetic change at other loci. New gene combinations may be selected, opening the way to speciation.

An example of possible peripatric speciation is given by Paugy (1986) for *Brycinus imberi*. Three endemic species are known on the margins of the main distribution area: *Brycinus abeli* from the Ubangui, *Brycinus carolinae* from the upper Niger and *Brycinus nigricauda* from the Nipoue River (Côte d'Ivoire and Liberia). All three species are very similar morphologically to *B. imberi* and differ mainly in their colour patterns. Another example might be found among the Cyprinodontidae (Scheel, 1968). The *Aphyosemion* species appear to be composed of a large number of very small populations consisting of a limited number of individuals inhabiting pools, and may or may not be eliminated during the dry season. The exchange of genes between these populations is certainly restricted and populations living at the extreme limit of the geographic range of the species are not so well adapted to the local conditions as are the populations living near the centre. Gene mutations may improve the adaptation of a border population, which can then succeed in building up a better-adapted genotype that enables it to eliminate or greatly reduce the influx of genes from other populations. Chromosome mutations, which generally result in a reduction of the chromosome number, may result in sterility of individuals, but if mutant individuals are able to produce offspring that survive and reproduce, the mutation may

become established in the border population. The new so-called biospecies is almost completely isolated from the populations of the species from which it arose, and will again establish a species border. In this way, a chain of very closely related species may develop, characterised by a steady reduction of the chromosome number outwards from the original centre of the superspecies.

There are many examples of pairs of taxa that each exhibit a constant form over a wide geographical area but meet along a narrow zone of overlapping distribution where they interbreed and intermediate forms are observed. In such a case, it is often difficult to decide whether the observed pattern represents a group of species, or a single species with distinct subspecies, that interbreed to a greater or lesser extent. The **parapatric speciation** model would state that they are two populations of an ancestral species differentiated into descendant 'forms', despite the maintenance of some gene flow and geographical overlap during the speciation process that is only partially completed. An alternative is proposed by the alloparapatric model which implies that allopatric populations of an ancestral species began to differentiate during a period of isolation, and that the resulting taxa have rejoined later becoming sympatric over an intermediate hybrid zone.

Speciation would be **sympatric** if it arose within a previous panmictic population in the same geographical area. This only happens if a biological barrier to interbreeding appears without spatial segregation. Most models of sympatric speciation proposed by different authors appear to be highly controversial (Futuyma, 1986; Barton, 1988), and for many authors there is little evidence that it occurs. The only exception seems to be speciation following change in the karyotype (fusion, fission, polyploidy), which confers relative reproductive isolation as soon as it appears. Nevertheless, according to population genetics theory, mating barriers can evolve, despite gene flow, under certain conditions of natural selection and population structure (Barton *et al.*, 1988). As a result, several studies of populations in nature could now be interpreted in terms of sympatric differentiation. Preference for new habitats may, for instance, evolve within a sympatric population, leading to behavioural, ecological and genetic changes that might be a precursor of speciation.

Trewavas (1983) suggested a few examples of sympatric speciation among tilapiines. The four species of the genus *Oreochromis* (*Nyasalapia*) in Lake Malawi have so much in common that they obviously had a common ancestor. Divergence occurred spatially (*Oreochromis saka* in the southern area and *Oreochromis karongae* in the northern area) and trophically, probably in sympatry, but also possibly aided by changing lake levels. In Lake Barombi-Mbo (Cameroon), the four endemic *Sarotherodon* species probably derived from *Sarotherodon galilaeus* or *Sarotherodon melanotheron*. It was suggested that the four species are the result of different waves of colonisation by the same ancestor from the riverine biota. They adapted both to new ecological conditions and to the presence of a resident population, and evolved independently. However, the data from a mitochondrial DNA analysis conducted by Schliewen *et al.* (1994) on cichlid species flocks endemic to the crater lakes Barombi-Mbo and Bermin (Cameroon) suggest that these species flocks (11 species in Barombi-Mbo and 9 in Bermin) are each monophyletic, and that they evolved within each lake after a single colonisation event. The size and shape of each lake is such that subsequent diversification would have been sympatric. There are no discernible microgeographical barriers within these lakes that would have allowed separation of microallopatric subpopulations. According to Schliewen *et al.* (1994), trophic diversification alone may have been a key factor responsible for speciation after colonisation of these small, ecologically monotonous crater lakes. For example, in Barombi-Mbo, the sequence analysis identified three ecological groups: one group containing the predatory genus *Stomatepia* (3 species); a second group with the genera *Sarotherodon* (4 species) and *Myaka myaka*, a fine-particle feeder; and a third group including the sponge feeder *Pungu maclareni* and two species of *Konia*. In Bermin, the two basal lineages of the flock separate the pelagic planktivorous species from the substrate orientated feeders.

The question of allopatric versus sympatric speciation has been widely discussed in relation to the African Great Lakes cichlids. The existence of numerous sympatric sibling species led some authors to postulate some process of sympatric speciation, but this hypothesis has been rejected (Mayr, 1988). Given that other, higher taxa in the same lakes did not evolve species flocks, as well as the sister-groups of lacustrine species in the nearby rivers, there must be some particular aspects of the life history and/or the genome that permit certain taxa to speciate so quickly in these lakes (Mayr, 1988).

To explain a possible allopatric speciation of cichlids in lakes, different but complementary mechanisms have been identified.

1. **Repeated colonisations** from rivers in connection with the lakes. A good example has been given by Trewavas (1983) in the four endemic species of *Sarotherodon* in Lake Barombi-Mbo (as described above). In a region subjected to alternate floods and recessions, successive colonisations of the lake by the same riverine species could have occurred, followed by periods of relative isolation. With enough time, successive invaders could have evolved in the new lacustrine environment in the presence of a resident population, giving rise to the present living species all sharing the common riverine ancestor. In this model, the existence of sympatric closely related species is therefore the result of successive waves of colonisation by a common ancestor. One should assume, nevertheless, that speciation occurred fairly rapidly, the adapted lacustrine populations of the invader being differentiated sufficiently not to interbreed with the following wave of colonisation.

2. **Fragmentation of the main waterbody** allowing allopatric speciation in smaller lakes. Following a rise in the water level, there is a fusion of the previously isolated waterbodies and mixing of the different species. There is geological evidence that the level of the Great Lakes fluctuated greatly in the past and for Lake Victoria marked changes in lake level have been recorded. According to Livingstone (1980), some 15 000 years ago the level was depressed by at least 75 m and the lake itself must have been virtually non-existent. Such fluctuations probably allowed the existence of various sized lakes, isolated or partly connected on a number of occasions. If there was sufficient time, speciation took place. This hypothesis is supported by what has been observed in Lake Nabugabo (30 km²), which was isolated from Lake Victoria about 4000 years ago by a narrow sand bar. There are now five endemic

Haplochromis species among the seven species enco-untered in the lake (Greenwood, 1965*b*). It should be noted, nevertheless, that there is apparently still some uncertainty over the age of Lake Victoria, which is usually given as 750 000 years but is thought to be much less by some authors who sug-. gested that it had almost entirely dried up during the Pleistocene.

In the Malagarazi swamps (Tanzania), two spec-ies of *Oreochromis* (*Nyasalapia*) occur (*Oreochromis malagarasi*, and *Oreochromis karomo*) but they have different feeding habits and breeding cycles. For instance, in August, *O. karomo* was observed breed-ing whereas no breeding activity was recorded among the *O. malagarasi*. The opportunity to segre-gate into subpopulations during dry periods and to evolve independently for a while is probably respon-sible for such divergences. *Oreochromis upembae*, an allopatric sister-species of *O. malagarasi*, also exists in the Upemba and the Lualaba region (Trewavas, 1983).

Lake Tanganyika, at an estimated 6–10 million years old, has experienced major fluctuations in water level. The shape of the lake basin developed gradually over many millions of years, from a system of rivers and more or less endorheic shallow struc-tural depressions (Early Pleistocene), to small lakes occupying elementary half-grabens that resulted from weak tectonic activity (Middle Miocene to Early Pleistocene). Some lakes were isolated, others possibly interconnected, as determined chiefly by climatic changes. The last evolutionary stage was marked by a relatively fast transition (at most 2 million years) to a large and deep basin, occupied by a single lake, as a result of the coincidence of major tectonic movements with wet climatic conditions (Tiercelin & Mondeguer, 1991). This long history and the lake fragmentation would partly explain why the Lake Tanganyika fish are, in general, strongly differentiated from one another. There is less mor-phological differentiation in endemic cichlids of the younger Lake Malawi and Lake Victoria, and many sibling species or 'intermediate species' between marked morphotypes have been reported.

3. **Allopatric intralacustrine speciation** by ecologi-cal segregation. This could possibly take place

within one lake, even if it is not divided into separate basins, as long as many species only occur in a por-tion of the lake or have specific and restricted eco-logical requirements. For instance, in Lake Malawi, Ribbink *et al.* (1983*b*) demonstrated that species belonging to the Mbuna group are so highly seden-tary that populations are readily fragmented by habi-tat discontinuities, depth and distance. Many species of the rock-dwelling *Labidochromis* are also known only from a single rock outcrop in Lake Malawi (Lewis, 1982*a*). Minor level fluctuation, which are common in all lakes, have probably influenced the evolution of the littoral fauna, creating dynamic con-ditions in the seemingly stable littoral communities where new ecological opportunities stimulated diversification (Martens *et al.*, 1994).

Witte (1984*b*) reported a similar trend in Lake Victoria where habitat partitioning is also important among piscivorous fish. The different modes of eco-logical segregation involved in the Lake Victoria haplochromines include: substrate types, distri-bution along the bottom profile, exposure to wind, vertical distribution in the water column, quantitat-ive and qualitative differences in food composition, differences in food collecting strategies and par-titioning of spawning areas. Various combinations of the above identified segregation parameters, make the understanding of niche differentiation very com-plex (Witte, 1984*b*). The rock-fishes in Lake Victoria are rarely caught more than a few metres from rocky substrate, and one of the presumed algal grazers feeds at night about 100 m offshore on surface phy-toplankton (Witte, 1984*b*). In Lake Tanganyika, algal grazers also exhibit strong substrate preferences (Hori *et al.*, 1983).

The habitat type preference of lake cichlids and their sedentary behaviour could be linked to their reproductive behaviour. Most lake cichlids practice parental care and are on the whole poor dispersers. In mouthbrooding species, few young are produced and are released when fully formed, eliminating larval dispersal.

Ribbink (1991), summarising the current knowl-edge of the ecology of cichlids in African Great Lakes, describes many species as geographically restricted to small areas within each lake. In Lake

Malawi almost every rocky outcrop and island has a unique Mbuna fauna with 'endemic' colour forms and species. He attributes such intralacustrine endemicity to philopatry, which is imposed by adult stenotopy and by parental care of offspring that ensures that the entire life history takes place within a single habitat. So strong is their philopatry and so narrow is their stenotopy that some species are endemic to localities that are no more than two or three thousand square metres in extent (Ribbink *et al.*, 1983*b*). The relatively small number of individuals within stenotopic populations 'predispose the cichlids to rapid speciation in the event of population subdivision consequent upon vicariant events'. Those haplochromine species flocks in Lake Malawi that are not strongly tied to specific habitats are less speciose as are some wide ranging forms, such as some predatory species that move between habitats and are fairly eurytopic (Ribbink, 1990).

If we assume that Lake Malawi experienced water level fluctuations of 250–300 m during the last 25 000 years and that lake level fell by at least 121 m for part of the time between AD 1390 and 1860 (Owen *et al.*, 1990), the localities now occupied by species endemic to restricted habitats have not existed for very long, only since the end of the last recession. Many of the islands and outcrops observed today were dry land within the last 200–300 years. As a consequence, one can imagine that the distinctive colour forms and even good species recorded in this restricted habitat have evolved within the space of 200 years. Owen *et al.* (1990) are of the opinion that periods of a few hundred years which were previously considered negligible by evolutionary standards may, on recent evidence, be sufficient for evolution to produce distinct species of cichlids recognisable both by colour and other morphological features. Thus, the numerous species or colour forms of Mbuna endemic to the recently inundated rocky habitats of southern Malawi (Ribbink *et al.*, 1983*b*) seem to confirm that intralacustrine speciation is more likely to have occurred in Lake Malawi than speciation in separate waterbodies, as suggested above for Lake Victoria.

The subsequent recessions must have severely affected the fish fauna, and provided several opportunities for splitting and recombination of populations and for the establishment of different founder populations. If the Lake was indeed reduced to 50% (Scholz & Rosendahl, 1988), numerous previously allopatric populations were forced into sympatry, and probably numerous species were forced to extinction following drastic changes in the abiotic environment. Comparative allozyme studies (Kornfield, 1978) and biological data (Ribbink *et al.*, 1983*b*) have demonstrated that sympatric forms of Mbuna that differ in colour really are distinct species (*sensu* Mayr, 1963). Comparative biochemical studies, including mitochondrial DNA differentiation, have also clearly established that the endemic haplochromine cichlid flock of Lake Malawi is of recent origin. In contrast, the endemic cichlid genera in Lake Tanganyika are all well separated suggesting much older divergence (Owen *et al.*, 1990).

During the long geological history of Lake Tanganyika (see above), there were important changes in habitats. A sedimentary environment prevailed in the previously existing rivers and shallow lakes. Faulting and subsidence created sharp bottom topography with, for instance, a zone of rocky substrates near the lake edges that offered more heterogeneous habitat, and opportunity for the appearance of lithophilous communities (Coulter, 1991*a*). Shallow littoral rocky niches were probably the first sites of opportunistic radiation in Tanganyika (Poll, 1950). There is also evidence for a lake level about 600 m below the present (Scholz & Rosendahl, 1988) that existed about 25 000 years ago, and is assumed to represent a prolonged Mid Pleistocene low stand (Tiercelin *et al.*, 1989).

Continuous changes in level of Lake Tanganyika have probably also influenced the evolution of its fish fauna. As pointed out by Coulter (1994*b*), episodes of accelerated speciation in littoral and benthic cichlids have coincided with major environmental changes involving large fluctuations in lake level. Conditions existing at these times are probably close to those invoked by the punctuated equilibrium model (Eldredge & Gould, 1972), involving multiple invasions of new environments by small populations, with frequent isolation of such populations on a stochastic basis during dispersal. The intervening period of stability, with only minor level fluctuations, might be regarded as equilibrium phases when low rates of diversification would be expected. Such a view is consistent with the usual concept of adaptative radiation as a macroevolutionary phenomenon, and the term 'explosive speciation', commonly applied to the radiation of cichlids in these lakes, should then be restricted to historic episodes when punctuation phases have occurred (Coulter, 1994b).

Hybridisation

It is known that species can be produced by hybridisation but the extent to which this occurs is unknown. Hybrid species are known amongst plants and Rosen (1978, 1979) has also suggested that the populations of the Central America fish genera *Heterandria* and *Xiphophorus* are intrageneric hybrids. Hybridisation does occur in natural environments, but rarely. It is difficult in the field to demonstrate possible hybridisation between species in the field, but the existence of suspected hybrids has been reported in a few cases, such as the three young specimens from the Benue River and from the Upper Niger, identified as hybrids between *Citharidium ansorgii* and *Citharinops distichodoides* by Daget (1963). Usually hybridisation is observed when changes in the environment suppress the reproductive isolation between sympatric species, or when a species is introduced into an aquatic habitat that is already occupied by a closely related species.

The recognition of hybrid species has a profound effect on the methodology of phylogenetic reconstruction. In a phylogenetic tree, 'intertaxa' must obviously be represented by lines joining the hybrid to its putative parents so that the pertinent part of the tree has a convergent as well as a divergent pattern. In other words, it is reticulate (Panchen, 1992). If speciation by hybridisation between sister-species were as common as speciation by cladogenesis producing sister-species, the trees (in the cladist sense) would appear as meshes or networks like fishing nets rather than representing a divergent hierarchical pattern.

Cichlid fish are extremely diverse in that hundreds of species have been described, but are also very similar genetically. Enzymological studies have shown that there is practically no enzyme differentiation among different species of the haplochromine fish in Lake Victoria: most

gene loci are allelically identical (Sage *et al.*, 1984; Verheyen *et al.*, 1985b). Analyses of general protein electrophoretic patterns did reveal differentiation at the supraspecific level, but the patterns are closely related to one another and probably differ in a very limited number of genes (possibly only two or three). Therefore, the usefulness of the concept of biological species for the taxonomy of cichlid fish has been questioned. Pre-mating barriers such as behavioural, morphological and ecological barriers between species prevent cichlids from spawning. Post-mating barriers prevent interspecific crosses from being completely fertile by means of genetic or cytoplasmic incompatibility (Mayr, 1982). The latter are confirmed by Crapon de Crapona & Fritzsch (1984), who have provided a review of the information available on hybridisation among cichlids (Table 4.6). Post-mating barriers of varying strength are found, which are reflected in the lack of viability of some of the crosses and the skewed sex ratio in others. When the F1 generation is fertile, the F2 or backcrosses may be either sterile or fertile.

Hybridisation between strains has been widely used to identify 'biological species' among cyprinodonts. Despite the great variability of their karyotypes, and the number of chromosome modifications observed, the general morphology and overall behaviour of cyprinodonts have not been very much modified. In fact, the differences are usually less important than those observed in other groups, such as the cichlids for instance, where the karyotype has remained constant. In some cases, it was possible to identify 'biospecies' differing in karyotype but morphologically so similar that preserved and live material cannot be differentiated. *Aphyosemion bualanum* for instance, a strain from Ndop on the eastern slopes of the Cameroon Mountain chain, has 19 haploid chromosomes, while *Aphyosemion rubrifasciatum*, the strain from Bandim (East Cameroon), has 20 haploid chromosomes. Adult males of both strains were very similar morphologically and in coloration patterns. Hybrids of these two strains are not completely sterile, but F1 × F1 = F2 individuals are too feeble to be raised to sexual maturity (Scheel, 1968). Moreover, *Aphyosemion exiguum*, a closely related species of *A. bualanum*, has 18 haploid chromosomes. When crossed with the Ndop strain, the *A. exiguum* produced hybrids that are viable and almost fertile in both sexes. With *A. rubrifasciatum*, the hybrids are

Table 4.6. *Hybrids in cichlid fish*

Species	Type	Result
Unsuccessfull crosses		
Oreochromis aureus × Tilapia zillii	FMB	
	SS	no fry obtained
Tilapia zillii × Tilapia sparrmanii	SS	
	SS	no fry obtained
♀ *Tilapia zillii ×♂ Sarotherodon galilaeus*	SS	
	FMMB	no fry obtained
Lethal crosses		
♀ *Oreochromis niloticus ×♂ Tilapia tholloni*	FMB	
	SS	high fry mortality
♀ *Sarotherodon melanotheron ×♂ Tilapia tholloni*	MMB	
	SS	high fry mortality
♀ *Haplochromis nubilus ×♂ Haplochromis burtoni*	FMB	
	FMB	high fry mortality
Viable crosses		
♀ *Oreochromis aureus × ♂ O. niloticus vulcani*	FMB	viable, more males
	FMB	F2 ?
♀ *Oreochromis mossambicus × ♂ O. niloticus*	FMB	viable, 75% males
	FMB	F2 ?
♀ *Oreochromis macrochir × ♂ O. andersonii*	FMB	viable, more males
	FMB	F2 ?
♀ *O. mossambicus × ♂ O. urolepis hornorum*	FMB	viable, more males
	FMB	F2 ?
♀ *Tilapia tholloni × ♂ Oreochromis niloticus*	SS	viable, 100% females
	FMB	no F2
♀ *Tilapia tholloni × ♂ Oreochromis mossambicus*	SS	viable, 100% females
	FMB	no F2
♀ *Oreochromis spilurus niger ×♂ O. leucostictus*	FMB	viable, 98% males
	FMB	F2: eggs, no fry
♀ *Oreochromis macrochir ×♂ O. niloticus*	FMB	viable, 60% males
	FMB	F2 viable, sterile
♀ *Oreochromis niloticus × ♂ Oreochromis leucostictus*	FMB	viable, 95% males
	FMB	F2 viable
♀ *Haplochromis nubilus ×♂ Haplochromis elegans*	FMB	viable, 95% females
	FMB	F2 viable
Successfull crosses		
♀ or ♂ *Tilapia guineensis × ♀ or ♂ Tilapia zillii*	SS	F1 viable
	SS	F2 viable and fertile
♀ *Haplochromis burtoni ×♂ Haplochromis nubilus*	FMB	F1 viable
	FMB	F2 viable and fertile

Selected data compiled by Crapon de Crapona & Fritzsch, 1984. SS, substrate spawners; FMB, female mouthbrooder; MMB, male mouthbrooder; FMMB, female and male mouthbrooder.

viable but almost completely sterile. Within the cyprinodonts, reproductive isolation is usually based on postmating factors and rarely on pre-mating ones (Scheel, 1968). The reason for the occurrence of strong post-mating mechanisms is not known, but it is likely that very often they are the results of reorganisation of the chromosome material during the evolution of these fishes.

Hybrids may occur more generally when related species are introduced into new habitats. An account of hybridisation between *Oreochromis macrochir* and *Oreochromis niloticus* in Lake Itasy (Madagascar), where both were introduced in 1958 and 1961 respectively, is given by Daget & Moreau (1981). In 1965 and 1966, a new fish that was partly similar to *O. macrochir* but easily distinguished from that species by fishermen was recognised and named '*Tilapia 3/4*'. The pharyngeal bone resembles that of *O. niloticus*, but the meristic numbers are nearer to those of *O. macrochir*. The fishery monitored from 1963 to 1975 showed an increase of the hybrid population from 5% by weight in the 1965 catch to 74% in 1969, followed by a decrease to 39% by 1975. Meanwhile, the catch of *O. niloticus* increased steadily whereas *O. macrochir*, which constituted 85% of the catch in 1963, had almost disappeared in 1971. The decline of the hybrid population was attributed to a decrease in fecundity of the females and an overall deterioration in their physiological condition. Their maturation size also became smaller from 1969 to 1975. The population of *O. niloticus* that finally dominated the catch was probably heterogeneous and not genetically the same as the original introduced stock. This was partly confirmed by the existence of two subpopulations with different growth rates. Hybridisation between *O. niloticus* and *O. macrochir* has also been demonstrated in experimental ponds by Jalabert *et al.* (1971). When *O. niloticus* was the female parent the F1 was 100% male. The reverse cross produced 75% males. Back-crossing with the parent species was possible and produced fertile offspring. These hybrids proved to resemble '*Tilapia 3/4*' and confirmed the hybrid nature of the latter (Trewavas, 1983).

A parallel situation to that in Lake Itasy has also been observed in Lake Naivasha, where hybrids of *Oreochromis spilurus niger* and *Oreochromis leucostictus* were recorded in 1960 (see Elder *et al.*, 1971 for review) both having been introduced into this lake. The hybrids were intermediate between the parent species in meristic characters, colour pattern and some anatomical features. They were abundant in the early 1960s, but they later decreased, and only a few specimens were observed in 1975 (Siddiqui, 1977, 1979). At this time, there was no trace of *O. spilurus niger*, and surviving hybrids were much closer structurally to *O. leucostictus*. The hybrids

have probably lost their intermediate characters either by back-crossing with other hybrids or with *O. leucostictus*. Proof that hybridisation can occur between these two species was obtained in experimental tanks, as well as confirmation that the hybrids were fertile and produce viable offspring (Elder *et al.*, 1971). A similar case has been reported from Lake Bunyoni, Uganda, where *Oreochromis niloticus* was introduced in 1927 from Lake Edward and *Oreochromis spilurus niger* from Lake Naivasha in 1932. Hybrids of these two species were recorded in 1937 with the parents, but by 1947 *O. s. niger* had disappeared as a pure species and only fish resembling *O. niloticus* were found (Lowe, 1958).

The hybridisation of *O. leucostictus* and *O. s. niger* could be explained by the spawning requirements and reproductive behaviour of the two species in Lake Naivasha (Siddiqui, 1979). When the water level is high, extensive lagoons and papyrus swamps occur all around the lake, and it was observed that *O. leucostictus* successfully colonised this extensive littoral zone. In contrast, *O. s. niger* remained beyond the edge of the papyrus. *Oreochromis s. niger* was first established in the lake where it apparently found good ecological conditions without competition, but it is a riverine fish that prefers clear water without mud or weeds for breeding so the breeding grounds of the two species are different. The introduction of *O. leucostictus* occurred in 1956, during a period of drought when the lake level was low and the papyrus lagoons either closed or inaccessible to the fish. Presumably at these times, *O. leucostictus* was forced to occupy the same breeding zone as *O. s. niger*, which could explain how hybridisation occurred. With the rise in the lake level after 1962, *O. leucostictus* was again able to colonise its favourite habitats and to develop successfully while the number of hybrids decreased.

Crapon de Crapona & Fritzsch (1984) suggested that hybridisation may be provoked in the field by situations resembling the 'forced conditions' prevailing in the laboratory. For example, the drops in the water level of Lake Victoria may create numerous pools along the shore in which different fish species are trapped for various periods of time. In such situations it is possible that accidental hybridisation could occur, particularly if one sex of a species is rare or absent. The accidental hybridisation of fish in Lake Victoria would also be favoured by the existence of turbid water all year round, limiting the use of species-specific optical cues. Until now, apart from *Tilapia* hybrids that resulted from the introduction of species not naturally found in the lake, no cichlid hybrids have been described from Lake Victoria, but the problem might be the difficulty of discriminating morphologically between hybrids and 'true' species. Moreover, many species in Lake Victoria represent graded morphological groups with no clear-cut morphological gaps between species, and are often identified only on the basis of their colour patterns. It is important to ask whether this gradation is caused by adaptation to graded biological niches, or whether it is caused by hybridisation (Crapon de Crapona & Fritzsch, 1984). If the latter hypothesis proved to be true, then hybridisation would be considered important for the explosive speciation of the haplochromines.

The potential contribution of hybridisation to speciation in fish, either directly or indirectly as the source of polymorphism, has not been given much attention, which may be due to the apparent rarity of naturally occurring hybrids. Experimental artificial hybridisations have been performed for aquaculture purposes, and also to check the biological species concept. These experiments should help to focus attention on the genetic mechanisms involved, and also on the question of specific-mate recognition cues. Crapon de Crapona & Fritzsch (1984) expressed the view that in a stable environment, hybrids may be at a disadvantage when competing with well-established species, while in an unstable environment, hybrids may be better able to compete for and adapt to changing environmental conditions, since they possess more 'flexibility' than the members of already established species.

Mosaic hybrid zones

The literature on mosaic hybrid zones is expanding, and promises to provide important insights into both the origin of barriers to gene exchange and the genetic architecture of speciation (Harrison & Rand, 1989). There is growing evidence that a large number of species across their range are often divided into patchworks of parapatric subspecies and races. Where two forms meet, mate and hybridise, a hybrid zone occurs and this phenomenon may represent a much more frequent species substructure than previously imagined (Hewitt, 1989). Most mosaic hybrid zones appear to represent secondary con-

tact between taxa (species or populations) that have differentiated in allopatry and are ecologically or behaviourally distinct, differing, for example, in habitat utilisation or in response to physical environmental factors. Endler (1977) defined hybrid zones as 'narrow belts with greatly increased variability in fitness and morphology . . . separating distinct groups of relatively uniform sets of populations'. It seems that the heterogeneity of the environment in which they occur and the complex internal structure of these hybrid zones promote the maintenance of diversity (species diversity or allelic diversity). In fact, quite a number of hybrid zones are broadly associated with an environmental transition and may include clines of alleles suited to different conditions (Hewitt, 1989).

Hybridisation between *Tilapia zillii* and *Tilapia guineensis* has been observed in the Ayame dam on the Bia River in Côte d'Ivoire (Pouyaud, 1994). These two species are usually parapatric, *T. zillii* being exclusively encountered in freshwater, while *T. guineensis* occurs in brackish waters. It is likely that after damming of the Bia River in 1958, populations of both species were trapped in the reservoir. Another example of hybridisation involving three tilapiine species (*T. zillii*, *T. guineensis*, *T. dageti*) has also been observed in the Comoe River in Côte d'Ivoire (Pouyaud, 1994). In that case hybridisation occurs preferentially between *Tilapia dageti* and *T. guineensis*, and a possible explanation is that the two species that were previously isolated came into contact only recently (8000 years ago).

Phenotypic plasticity, ecological processes and natural selection

The phenotype includes the morphological, physiological, biochemical, behavioural and other properties of an organism that develop through the interactions of genes and environment. Polymorphism is the existence of morphologically distinct alternatives in a population, and phenotypic plasticity is the ability of a single genotype to produce more than one alternative form in response to environmental conditions. Plasticity usually refers to variability induced by the environment, in a particular life stage. Compared to the role of genetics in evolution and selection, there have been relatively few studies that tried to understand the effects of the environment on the phenotype and their evolutionary consequences. However, plasticity, which could be considered as the production by a single genome of a diversity of potentially adaptative responses subjected to natural selection, is of special interest for evolutionists and a subject for experimental and theoretical studies. If different phenotypes can achieve different degrees of success in the environment, the latter can also strongly influence the ranges of phenotypes that could be expressed by a given genotype and therefore exposed to selection. In other words, only expressed phenotypes in a given environment could be genetically modified under selection (West-Eberhard, 1989).

Most studies on phenotypic plasticity in fishes have been conducted in the temperate zones, and concern the effects of temperature or other environmental factors during early ontogeny on the number of fin rays or vertebrae (Barlow, 1961; Lindsey & Arnason, 1981). Witte *et al.* (1990) presented a few cases of phenotypic plasticity in African cichlids that were presumed to be adaptative responses of the anatomical structures to environmental changes. This is the case, for example, in *Astatoreochromis alluaudi*, which exhibits considerable variation in the degree of hypertrophication of its pharyngeal jaw apparatus according to its diet. The lower pharyngeal jaw is stouter in individuals raised with a diet of thick-shelled molluscs, than in individuals raised on soft food (Hoogerhoud, 1986).

Attempts to understand the balance between genetic determinism and environmental influence in fish requires well-planned and often long-term breeding experiments. The genetics of colour pattern in tilapiines can illustrate this point. Adult colour is thought to be a stable attribute of a particular taxon, but transient alterations in body coloration and colour patterns are the rule in most cichlids (Kornfield, 1991). The red morph, for instance, has appeared independantly in a number of different *Oreochromis* species and is a common variant in many cichlid genera (Axelrod *et al.*, 1986). The appearance of a single mutant fish in a genetically 'pure' stock of *Oreochromis niloticus* made it possible to follow its inheritance under controlled experimental conditions (McAndrew *et al.*, 1988). Histology revealed that melanophore development rather than a biochemical pathway mutation was responsible for the colour in red fish. An almost complete absence of melanin granules in the skin was observed in red tilapias, whereas blotched individuals showed areas characteristic of red and normal skin. The histology of the skin of blond fish showed the incomplete development of melanophores and melanocytes. Given the ability to transfer the gene in a single cross in *O. niloticus*, red is an autosomal dominant, probably inhibiting melanophore development, while in red tilapias produced by hybridisation between *Oreochromis mossambicus* and *O. niloticus*, red body colour is inherited as a simple locus with incomplete dominance (Huang *et al.*, 1988). In *O. mossambicus*, gold body coloration is controlled by a single autosomal gene with incomplete dominance (Tave *et al.*, 1989).

Natural selection, the central principle of Darwin's theory, remains one of the major concepts in evolutionary biology. Broadly speaking, living organisms are able to adapt to the conditions prevailing inside and outside themselves. Adaptation occurs through small-step inheritable advances that are accumulated from generation to generation. Among the innumerable micromutations, natural selection will preserve those that improve the execution of a certain function presumed to give a competitive advantage. This implies that properties that permit further adaptation of the organisms in question are represented in the genotype and the phenotype of some individuals. For any given environment, some phenotypes will, on average, have a lower mortality or higher fecundity, or both, so that they are better represented in the next generation than are phenotypes with high mortality or low fecundity (Dobzhansky *et al.*, 1977). These phenotypes are said to have greater fitness, and the changes in their frequency over time is a measure of fitness. Fitness is, therefore, a comparison between the probabilities of survival and reproduction of different phenotypes that are a function of the environment in which they are located.

We know little about the ecological genesis of selective mortality or selective fertility (Endler, 1986), and most of our experience of selection is based on anthropogenic alteration of the environment in the face of catastrophic events (Travis & Mueller, 1989). A great number of selective pressures act simultaneously on a variety of traits in a natural population, and a single trait is usually subject to many selective forces that could act antagonistically and may vary spatially or temporally. In this changing ecological context, there is a pattern of fluctuating modes of selection (Travis & Mueller, 1989). Fluctuating selection pressures at the phenotypic level could be an important mechanism for reducing the rate of loss of genetic variation (Grant, 1986). One should keep in mind that selection is a two-step process: (i) the internal selection determines which mutant will be viable and what phenotypes will appear in a population; and (ii) external selection determines whether a mutant changes in frequency and will spread in a population. Natural selection, in any case, acts directly on phenotypes and only indirectly on genotypes.

Behaviour, which is a particularly plastic characteristic of phenotypes, has been considered by some authors as the first aspect of phenotype to evolve in a new direction (Mayr, 1963; Plotkin, 1988). Behavioural adaptability is important because the activity of individuals has the potential to diminish or exacerbate the influence of external environmental heterogeneity (Wcislo, 1989). The role of behaviour in evolution is not a new idea and was already considered an important speciation mechanism by Lamarck (1809). Mayr (1974, 1988) emphasised the role of behaviour in claiming that changes in behaviour are almost always the initial steps in evolutionary change. He stated that, in the endemic cichlids of the African Great Lakes: 'local fashions in mate preference can apparently change rather rapidly in small isolated populations, and there can develop into behavioural isolating mechanisms ... The first character to diverge should relate to courtship while characters relating to competition for shared resources will evolve later, particularly after sympatry is established'. Behaviour-induced divergence is particularly important in social behaviour, sexual behaviour, feeding behaviour and habitat selection (Wcislo, 1989).

Sexual selection theory
Sexual selection could be defined as the differential mating success due to competition within one sex for mates (usually between males), or choice by members of one sex of members of the other (usually female choice). Sexual selection could therefore occur among genetic variants which differ in mating characteristics (when, where and how mating takes place). These characteristics can be advantageous or disadvantageous in mate competition selection. If the latter, the variant will probably disappear soon, whereas sexually selected traits are likely to diverge rapidly facilitating reproductive isolation and speciation.

Darwin (1871) emphasised the role sexual selection has played in evolution, but the topic has apparently been rediscovered recently and the literature is expanding rapidly (West-Eberhard, 1983; Arnold, 1985). Divergent schools have developed (Kirkpatrick, 1987): 'the good genes' school which postulates that mate choice evolves under selection for females to mate with ecologically adaptive genotypes, and the 'non-adaptive' school which holds that preferences frequently cause male traits to evolve in ways that are maladaptive with respect to their ecological environment.

The importance of sexual selection to cichlid speciation has been discussed by Dominey (1984). This group provides an extraordinary opportunity to study the processes of mate choice, genetics and sexual selection by 'female choice' in particular, and their importance to evolutionary theory. An understanding of the role of sexual selection in the evolution of mating behaviour would certainly be useful in explaining the explosive speciation in the African Great Lakes, and the existence of so many closely related species (McKaye, 1991). A major element in this theory (Dominey, 1984) involves female choice among different variants in male coloration or other features, such as alternative mating tactics or competition for display sites, that could lead to a rapid divergence of those characteristics at intra- and interpopulation levels. There could be a consequent divergence of genotypes within the assorted mating individuals. The reproductively isolated, rock-dwelling cichlids of Lake Malawi include numerous colour variants (Ribbink et al., 1983b). They were first thought to be morphs of a single species but are now considered to be sibling species, after field behavioural observations of mate choice proved that female mate choice was correlated with male coloration (McKaye, 1991). Further observations also provided evidence of the existence of more than expected sand-dwelling cichlid species, which can be distinguished by the form of their display sites or 'bowers' (McKaye & Stauffer, 1988; McKaye, 1991). But it must be demonstrated, either by field observations or laboratory experiments, that female preference for bower size and form is heritable, in order to confirm the sexual selection hypothesis. An alternative hypothesis suggested by West-Eberhard (1983) is that social rather than sexual competition may be responsible for the rate of speciation, and the maintenance and divergence of the bright coloration of monogamous African cichlids. Competition among males and intrapopulational assertive mating certainly occur in cichlids (Greenwood, 1991), but the question is whether or not such a situation results in a stabilising selection for a particular mate recognition system rather than the evolution of new features. There is apparently no record among cichlids of females preferring atypical males, as in some other animals.

Sexually and socially selected traits might undergo more rapid diversification than traits undergoing survival selection, and might facilitate or fuel the explosive pace of speciation in cichlids. Coloration is important in intraspecific interactions and is a trait that might be more strongly shaped by sexual rather than survival selection; coloration might act as a reproductive barrier without concordant morphological diversification (Meyer, 1993).

Among the cyprinodonts, *Nothobranchius guentheri*, an inhabitant of seasonal waterbodies, possesses the extreme sexual dichromatism typical of the genus. There is no courtship display, indeed spawning often occurs within seconds after a pair make visual contact. This reproductive strategy is enhanced by the attractiveness to females of the brightly coloured males. Females approach males for spawning in response to the degree of intensity of red colour of the tail fin (Haas, 1976b). In the temporary habitats occupied by *N. guentheri*, the turbid water transmits longer wavelengths better than shorter, and red males are more visible to females, but also to predators such as birds. Sexual selection for males apparently overcomes the negative selective pressures by predators, and a high loss of males can be tolerated because they lack territoriality and because of the readiness of males to spawn immediately with any receptive female once she has approached and caught his attention.

An interesting observation made on North American fishes, which could certainly be extrapolated to African fishes, is that female *Gasterosteus aculeatus* prefer males with the brightest red on their bellies. But for this species, the intensity of the red breeding coloration correlates with the physical condition of the males: males infected by parasites are less intensively coloured, leading to an apparent reduction in their attractiveness to females (Milinski & Bakker, 1990).

Interpretations of the nature of colour are inevitably anthropocentric. Who knows how colour is perceived by fish? What is subtle to us may be obvious to them, but all kinds of visual cues must be more difficult to use underwater due to turbidity and also selective penetration of water by different wavelengths of light.

The role of predation

The influence of predation upon fish speciation in African freshwaters has received much attention and been extensively reviewed by Coulter (1991a). Fifty years ago Worthington (1937, 1940) had already reported that speciation among cichlids had been slight in Lakes Chad, Turkana and Albert where *Lates* and *Hydrocynus* are present, compared with that which had taken place in Lakes Victoria and Malawi where these predators are absent. He deduced that prey species, in the process of adaptive radiation into new ecological niches, are more vulnerable to predation than species already established and specialised in existing niches. Lake Tanganyika did not fit into this theory, since cichlid speciation must have occurred in the presence of *Lates* species that are evidently of ancient stock in the lake, given their degree of evolutionary diversification. They must have coevolved with prey species and, as Worthington suggested, the greater age and isolation of Lake Tanganyika compared to other lakes would have permitted this evolution. This theory overlooked the existence of other predator species.

In demonstrating the drastic impact of *Hydrocynus* on the ecology of small and juvenile fish, Jackson (1961) partially supported Worthington's view, and postulated that both the absolute number of prey fish and the number of their species were governed, in the presence of *Hydrocynus*, by the availability of vegetation cover. Jackson believed that the piscivores of Lake Malawi were by comparison too 'mild' to have the same effect, but Fryer (1959) took the view that the many other piscivores present in Lake Malawi may have facilitated speciation by reinforcing the isolation of populations segregated in discontinuous habitats, and reducing the interspecific competition in preying on the more abundant species. Fryer & Iles (1972) elaborated the opinion that *Lates* and *Hydrocynus* had no overall effect upon speciation different from that of other predatory fishes.

Lowe-McConnell (1987) concluded that predation is of major importance in the maintenance of diversity in tropical fishes, and that the presence of cover is a key factor in allowing the evolution of diverse aquatic communities. The behaviour of predators in patrolling areas containing cover helps to split up populations into isolated patches, and therefore facilitate speciation. In contrast, in pelagic communities, predator pressure seems to lead to uniformity and to reduce species diversity. Thus, depending on the amount of cover, predation could produce opposite effects.

Coulter (1991a) discussed the role of predation in the speciation of the fish in Lake Tanganyika. He stressed that predation exerts a profound influence upon the ecology and behaviour of prey fishes, specifically their reproductive strategies, their distribution (particularly of small and juvenile fish), their behaviour and their mortality rates, and came to the conclusion that Lowe-McConnell's interpretation appears to be born out by the Tanganyika fish. The pelagic community consists of large populations of few species with much trophic overlap, while the littoral community has many species, often territorial and split into numerous disjunct populations, with a complex trophic partitioning. A few dominant predators, chiefly the *Lates* spp. and *Boulengerochromis microlepis*, exert their effect chiefly by intimidatory behaviour that forces prey to hide, remain close to cover or to adopt schooling, migratory or other evasive behaviour. This intimidation affects most aspects of prey life history, and reactions by prey fish to these predators possibly correspond to a phenomenon called 'resource depression' in which the mere presence of the predator lowers the capture rates of prey in its vicinity.

According to Coulter (1991a), the tendency towards dominance by a few predators, and the resulting prey reactions, seem more highly developed in Lake Tanganyika than in Lakes Malawi and Victoria. The numerous fish predators in the two latter lakes tend to be habitat restricted. The dramatic impact of the introduced *Lates niloticus* in Lake Victoria indicates that most cichlid species did not possess the behavioural mechanisms to escape and/or that adequate cover was not available to them. In Lake Malawi, on the other hand, predatory pressure upon adult cichlids appears to be low but may be intense upon their fry and eggs (Ribbink, 1990).

Convergent evolution

Forms may be found in taxonomically distant groups that closely resemble each other. The analogy of phenetic states is a specific case of a situation when forms of vari-ous taxa (phenons) are more similar to each other than to other forms of their own taxa. For example, various genera of characoids greatly resemble representatives of other fish families. The African *Hydrocynus* for instance, strongly resembles the European whitefish (*Coregonus*, family Salmonidae), in the shape of its body, fins and coloration, and there is also an analogy with species of the family Bryconidae from South America. There are close phenotypic similarities between the African genus *Hepsetus* and the South American *Acestrorhynchus*, as well as with the European pike (*Esox*). A few species of the African genus *Brycinus* strongly resemble bleak (*Alburnus*, family Cyprinidae) in the shape of their bodies, mouths, length of anal fins and coloration. *Citharinus citharus* (Citharinidae) from Nilo–Sudanian basins is also similar to the bream (*Abramis*) in their shape of body and fins. Géry (1977) also draws an analogy between the African family Citharinidae and the American families Hemiodidae and Curimatidae, and between the small African genus *Nannocharax* and the South American family Characidae. Armour catfishes of unrelated families occurring in stony streams, also exhibit remarkable convergences: loricariids in South America, certain amphiliid genera in Africa, and *Sisor rhabdophorus* in India (Lowe-McConnell, 1987).

Comparing fish faunas of the African Great Lakes, mainly Malawi and Tanganyika, there are intriguing indications of convergent evolution. Similarities in feeding morphology and behaviour were already described as 'ecological equivalent species' by Fryer & Iles (1972). Convergent or parallel evolution in specialisation towards the same goal has often occurred independently in different waterbodies, and sometimes been achieved by different means. In the course of phylogenetic reconstruction, the term homoplasy is often used to qualify this phenomenon which has been investigated among the cichlid communities of the Great Lakes that exhibit a wide range of parallelism (Eccles & Trewavas, 1989). The development of heavy pharyngeal mills for mollusc crushing has occurred in *Mylacochromis* in Lake Victoria and in a number of species such as *Maravichromis sphaerodon*, *Lethrinops mylodon* and *Protomelas macrodon*, in Lake Malawi. Several species exhibit hypertrophied lips for extracting invertebrates from crevices in Lake Malawi, as do *Haplochromis labiatus* Trewavas, 1933, in Lake Victoria and the genus *Lobochilotes* in Lake Tanganyika.

The development of a predatory form with elongated head and body, and sharp oral teeth, is also common and has been investigated by Stiassny (1981), who compared the cichlid genera *Bathybates* and *Rhamphochromis*, endemic to Tanganyika and Malawi respectively. Both genera are open-water piscivores, sharing similar morphological features. Cladistic analysis revealed that *Bathybates* is a monophyletic genus and, despite apparent similarities, convergence rather than close phyletic relationship is indicated between the two genera. Another example is given by Poll (1987), who pointed out that the occurrence of similar hypertrophic development of cephalic laterosensory organs in the genera *Trematocara* and *Aulonocranus* in Lake Tanganyika, and *Trematocranus* in Lake Malawi, could not be attributed to close phyletic relationships, but rather demonstrated convergent evolution through adaptation to deepwater conditions.

Paedophagy is a peculiar behaviour that has evolved in the Great Lakes. Several paedophagous species of the genus *Haplochromis* occur in Lake Victoria, such as *Haplochromis barbarae* and *Haplochromis cronus* (Greenwood, 1974b). *Haplochromis taurinus* has the same habit in Lakes Edward and George. *Caprichromis orthognathus* in Lake Malawi has a special form of mouth, with steeply inclined gape and thick lips, and a peculiar tooth form, with the anterior teeth of the outer row in the dentary curved outwards at the tip, similar to those of *Lipochromis*. These features also occur in *Caprichromis liemi* and in *Protomelas spilopterus* .

Scale eating has also evolved independently in Lakes Malawi, Tanganyika and Victoria, but the end points are all different. The seven lepidophagous species from Lake Tanganyika belong to the genus *Perissodus*. They evolved in the lake, and are characterised by a dentition in which the teeth of the upper and lower jaws are recurved as a specialisation for removing scales from other fish. *Haplochromis welcommei* is known to eat scales in Lake Victoria, and *Docimodus evelynae*, *Genyochromis mento*, *Melanochromis* sp. and two species of *Corematodus* in Lake Malawi.

5 Classification of diversity

The characters which naturalists consider as showing true affinity between any two or more species, are those which have been inherited from a common parent, all true classification being genealogical.

Darwin, 1872: 346

Classification of organisms involves two processes: (i) taxonomy, which is the application of rules for naming the categories; and (ii) systematics, which is the process of classification itself. These terms are often used loosely, but the two operations should be distinguished. An important function of classification is that it establishes groupings about which generalisations can be made. To the extent that classifications are explicitly based on the theory of common descent with modification, they postulate that members of a taxon share a common heritage and thus will have many characteristics in common. 'A biological classification, like any other, must serve as the basis of a convenient information storage and retrieval system' (Mayr, 1988).

The category of a taxon indicates its rank or place in the hierarchy of the classification. Broadly speaking, the evolution of classifications reflects the current thinking in systematics. In earlier classifications, for instance, workers sought characters to differentiate genera or subfamilies but generally ignored their interrelationships, and different classifications were sometimes proposed without any convincing justification. In the 1950s and 1960s, with the establishment of new schools of taxonomy, there was an attempt to develop a more Cartesian approach (Mayr, 1988). More recently, the development and improvement of molecular genetics techniques has provided powerful tools for investigations of phylogenetic relationships. What is now called biosystematics is a modern approach to taxonomy and phylogeny that requires morphological, genetical, biological, behavioural and ecological information.

Current phenotypic methods

A natural taxon is a group of organisms that is given a name resulting from analysis of evolutionary processes. It is the task of systematics to discover taxonomic patterns of diversity and to relate these patterns to underlying evolutionary processes. Whatever the classification system, the search for monophyletic units is one of the major goals. Closely related pairs of species are called sister-species, and pairs of equivalent monophyletic units are termed sister-groups. For any classification, it should be stressed that the cladogram obtained depends in some way on the selection of characters retained for analysis, leaving room for discussion on the interpretation of data analysis. There are two main schools of modern systematic theory and practice.

Numerical phenetics, also known simply as 'phenetics' or 'numerical taxonomy', assumes that classifications are based exclusively on quantitative measures of 'overall phenotypic similarity'. Degree of similarity is determined by quantifying the state of many different characters in a large number of specimens and seeking correlations between these characters. Equal weight is given to every character investigated, and various classification methods using algorithms allow clustering of species with the most similar set of character states, more closely than those that are less similar. Most phenetists would agree that organisms sharing common patterns of characteristics have a common evolutionary history although, strictly speaking, the phenetic approach does not involve genealogical hypotheses in the formation and ranking of taxa.

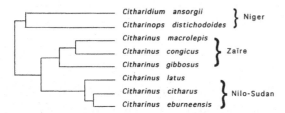

Fig. 5.1. Phenetic classification of the Citharinidae (from
    Daget, 1966).

Daget (1966) used numerical phenetics to obtain a
phenogram for the Citharinidae (Fig. 5.1). Results were
in concordance with the previous empirical classification,
showing the species *Citharinops distichodoides* to be rather
different from the other *Citharinus*. The three Zaïrian
and the three Nilo–Sudanic species probably evolved
independently and in parallel, but there are morphological convergences between species of the two groups.

Cladistics is a method of classification first introduced by Hennig (1950) that is exclusively based on gen-
ealogy. It therefore takes into account the branching pattern of phylogeny which assumes that an ancestral
species splits into two daughter species before disappearing, or not (during cladogenesis, one taxon may
remain close to the ancestral form while the other
strongly diverge from that ancestor). The 'holophyletic'
taxon includes the ancestral species and its descendants.
Characters evolve through different states during evolution and ancestral characters (or primitives, or
plesiomorphes) are distinguished from derived ones (or
advanced, or apomorphes). Hennig considered cladograms to be common ancestry trees in which monophyletic groups could be determined from the distribution of
shared derived characters (synapomorphies) inherited
from the most recent common ancestor. The characters
unique to a terminal taxon, called autapomorphies, are
informative with respect to the formation of groups. As
discussed by Nelson (1973), the derived or ancestral
nature of characters can be evaluated either by outgroup
comparisons involving the evaluation of the state of the
character in various groups in order to determine which
character states are possibly primitive, or from information available from ontogenic transitions.

Cladism is therefore a classification by means of
shared derived characters indicating **monophyletic**

groups, which are derived from a single ancestor. It is a
phylogenetic classification. One should also recognise the
existence of **polyphyletic** groups, based on resemblance
due to convergent, or independently derived, characters
not inferred to have occurred in the common ancestor,
and **paraphyletic** groups which are those that do not
contain all of the descendants of a common ancestor (see
Humphries *et al.*, 1988). The assumption of strict
dichotomies has been much discussed and the existence
of polyfurcations is now admitted.

One of the first important cladistic studies on African
freshwater fishes was the classification of the families
Citharinidae and Distichodontidae by Vari (1979). He
attempted to determine their phylogenetic relationships,
at the generic and suprageneric levels, using primarily
osteological characters and, to a lesser extent, myological
and soft anatomical characters. Vari used the Hennigian
methodology, which he considered to be more suitable
than numerical or evolutionary ones for the aims of his
study.

Vari's study also illustrates the changes in ideas about
the classification of this group of characoids (Table 5.1).
Since the beginning of the century, workers have recognised between two and five families or subfamilies. In
contrast to previous authors, Vari (1979) did not recognise the family Ichthyboridae, which he incorporated
into the family Distichodontidae to form a monophyletic
assemblage. Vari also apparently ignored the genus
Citharinops introduced by Daget (1962*b*, 1966). His conclusions were not widely accepted by ichthyologists.

Using 15 different morphological and anatomical
characters, Daget & Desoutter (1983) built up a tentative
scheme for the evolution of the living polypterids from a
common ancestral form (Fig. 5.2). The group *Polypterus
bichir–endlicheri* probably split very early, whereas the
genus *Calamoichthys* (= *Erpetoichthys*) is unexpectedly
close to other *Polypterus* suggesting that *Calamoichthys*
has only recently been isolated.

A phylogeny of bariliine and neoboline cyprinid fishes
(Fig. 5.3) was also established by Howes (1980, 1984)
using osteology, external anatomical features and gross
brain morphology, in which the distribution of certain
characters resulted in a reclassification of the groups previously considered as genera. The bariliine group is represented in Africa and Asia, and both plesiomorph and
derived taxa occur sympatrically in Africa and India. The

Table 5.1. *Changes in the classification of the families Citharinidae and Distichodontidae during the twentieth century according to various authors*

Author	Families and subfamilies	Genera
Boulenger, 1909	Citharininae	*Citharinus, Citharidium*
	Distichodontinae	*Nannaethiops, Neolebias, Distichodus, Nannocharax, Xenocharax*
	Ichthyborinae	*Ichthyborus, Neoborus, Mesoborus, Eugnatichthys, Paraphago, Phago*
Regan, 1911	Citharinidae	*Citharinus, Citharidium*
	Xenocharacinae	*Nannaethiops, Neolebias, Xenocharax,*
	Distichodontidae	*Distichodus, Nannocharax*
	Hemistichodidae	*Hemistichodus*
	Ichthyborinae	*Ichthyborus, Neoborus, Mesoborus, Eugnatichthys, Paraphago, Phago*
Greenwood *et al.*, 1966	Citharinidae	
	Distichodontidae	
	Ichthyboridae	
Poll, 1973	Citharinidae	*Citharinus, Citharidium, Nannaethiops, Neolebias, Xenocharax, Dundocharax, Distichodus, Paradistichodus, Nannocharax, Hemigrammocharax*
	Ichthyboridae	*Ichthyborus, Phagoborus, Gavialocharax, Hemistichodus, Neoborus, Microstomatichthyoborus, Eugnatichthys, Paraphago, Phago, Belonophago*
Vari, 1979	Citharinidae	*Citharinus, Citharidium*
	Distichodontidae	*Neolebias, Nannaethiops, Xenocharax, Distichodus, Paradistichodus, Nannocharax, Hemigrammocharax, Ichthyborus, Hemidistichodus, Microstomatichthyoborus, Mesoborus, Eugnatichthys, Paraphago, Phago, Belanophago*

Partly from Vari, 1979.

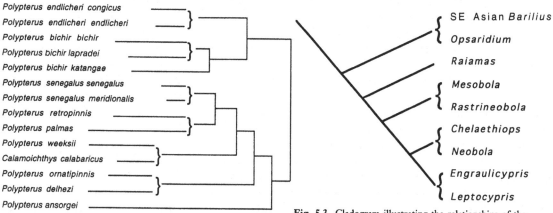

Fig. 5.2. Cladogram illustrating the relationships of the species of Polypteridae (from Daget & Desoutter, 1983).

Fig. 5.3. Cladogram illustrating the relationships of the African bariliine and neoboline genera (from Howes, 1984).

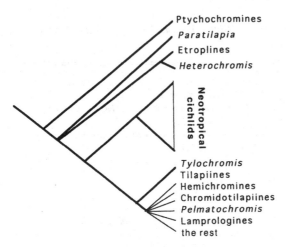

Ptychochromines
Paratilapia
Etroplines
Heterochromis

Neotropical cichlids

Tylochromis
Tilapiines
Hemichromines
Chromidotilapiines
Pelmatochromis
Lamprologines
the rest

Fig. 5.4. Summary cladogram of cichlid intrafamily relationships (from Stiassny, 1991).

relationships presented by Howes (1980, 1983, 1984) show both African genera, *Opsaridium* and *Raiamas*, as having Asiatic relatives and representatives. The South-East Asian *Barilius* is considered to be the sister-group of *Opsaridium*, and *Raiamas* is also represented in India. This indicates vicariance events occurring after the break-up of Gondwanaland. In Africa, the monophyletic neoboline group (*Neobola* and *Chelaethiops*) forms the sister-group of *Leptocypris* and *Engraulicypris*. Both neoboline and leptocyprine forms, in turn, are the sister-groups of *Mesobola* and *Rastrineobola*.

A similar distribution to that of the bariliines, i.e. including Africa and SE Asia, is observed for the Masta-cembeloidei, for which a phylogenetic hypothesis of their relationships has been proposed by Travers (1984).

In a phylogenetic study of African Pelmatochromines, Greenwood (1987) pointed out the existence of a mono-phyletic lineage called Chromidotilapiines (*Thysia, Chromidotilapia, Pelvicachromis, Nanochromis, Paranano-chromis, Limbochromis*), whose relationships cannot be satisfactorily resolved, and two other monophyletic taxa, *Pelmatochromis* and *Pterochromis*. They are not closely related to the genus *Tilapia*, nor to any of the fluviatile haplochromine lineages from eastern Africa. Stiassny (1991), following previous studies from Cichocki (1976) and Oliver (1984), proposed a cladogram of cichlid intrafamily relationships (Fig. 5.4) in which the Madaga-scan cichlids (Ptychochromines) form the sister-group of

the rest of the family. The Zaïrean genus *Heterochromis* is phylogenetically different from other African assem-blages and represents the sister-group of Asian and Indian cichlids (Etroplines). The remaining Cichlidae constitute a monophyletic lineage that includes two sister-groups: Neotropical cichlids and African cichlids. There is also some indication that the pan-African genus *Tylochromis* represents the sister-group of the remaining African lineage.

Gourène & Teugels (1994) established the phylogeny of the Pellonulinae from West and Central Africa. The genera *Limnothrissa* and *Stolothrissa* from Lake Tangan-yika form the sister-group of all other Pellonulini, in which three other clusters are recognised: *Microthrissa, Potamothrissa* and *Nannothrissa* from Zaïre, forming the sister-group of *Thrattidion* (Sanaga) and *Laeviscutella* (from Senegal to Zaïre); and *Pellonula* and *Odaxothrissa* form the sister-group of the above cluster.

Mo (1991) also used cladistics for establishing the systematic status and reconstructing the phylogenetic intra- and interrelationships of the family 'Bagridae', which, as previously described, contained many phylo-genetically heterogeneous taxa. As a result of an anatom-ical study, including some 200 characters, the taxa pre-viously assigned to Bagridae are now recognised as constituting three families: the Bagridae, the Clarotidae and the Austroglanididae that are all explicitly defined by synapomorphies.

The family Austroglanididae has been created for the single African genus *Austroglanis* with three species exclusively confined to Southern Africa. The family Bag-ridae, as proposed, is restricted to the '*Bagrus*-like group' and includes 15 Asiatic genera and a single African genus, *Bagrus*. The Clarotidae are recognised as a separ-ate family that includes two sister-groups: the *Clarotes*-like and the *Auchenoglanis*-like groups for which the subfamily names Claroteinae and Auchenoglanidinae, respectively, have been retained (Fig. 5.5). The Clarotei-nae includes only African genera: *Clarotes, Chrysichthys, Gephyroglanis, Bathybagrus, Lophiobagrus, Phyllonemus* and *Amarginops*. The subfamily Auchenoglanidinae also includes five African genera: *Auchenoglanis, Parauchenog-lanis, Platyglanis, Notoglanidium* and *Liauchenoglanis*. The West African sister-taxa *Leptoglanis* and *Zaïreichthys* have been assigned to the Amphilidae. In fact, the genera *Auchenoglanis* and *Parauchenoglanis* have been reviewed

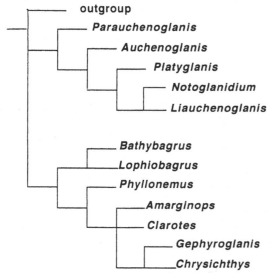

Fig. 5.5. Phylogenetic relationships among Claroteinae and Auchenoglanidinae (from Mo, 1991).

by Teugels *et al.* (1991) who discussed their status and described a new genus, *Anaspidoglanis*.

Mo (1991) also proposed a tentative phylogeny of the Siluroidei, keeping in mind that, while the monophyly of many families has not been ascertained, it could only be a preliminary testable hypothesis of relationships between families. The families Bagridae, Clarotidae and Austroglanididae are obviously phylogenetically distinct from each other. The family Mochokidae is recognised as a sister-group of a cluster comprising the Doradidae, Auchenipteridae, Ageniosidae and Centromochlidae. Those four families are in turn, the sister-group of the Ariidae. Mochokidae is a monophyletic taxon and the Madagascan genus *Ancharius*, which has so far been included in the Ariidae, is obviously a member of the Mochokidae. The family Schilbeidae is obviously a non-monophyletic assemblage.

If the cladistic approach is considered to be an improvement on more empirical approaches, the results obtained nevertheless need to be validated. For instance, the classification of genera of the family Cyprinodontidae is highly controversial and there is an intensive debate about the phylogeny established by Parenti (1981). The controversy is mainly concerned with the relationships between various subgenera and genera and their possible

links with the well-differentiated genus *Nothobranchius*. Results of biochemical investigations on different West African species (Agnèse *et al.*, 1987) tend to support the older classification proposed by Myers (1955) for the cyprinodonts, rather than Parenti's classification, and some authors have refuted the latter classification (Romand, 1992).

Each classification school has virtues and weakness, even if it believes that its approach is the best. For instance 'if Pheneticists and Cladists claim that their methods give non-arbitrary results, which is not really substantiated, both fail to reflect adequately the past evolutionary history of taxa' (Mayr, 1988). The tendency is now towards the incorporation of some of the criteria of the different schools in order to develop an eclectic methodology, containing a proper balance of phenetics and cladistics that will produce far more 'natural classifications' than any one-sided approach that relies exclusively on a single criterion. 'Evolutionary taxonomy has been characterised by the adoption of an eclectic approach that makes use of similarities, branching patterns, and degree of evolutionary divergence' (Mayr, 1988).

Evolutionary classification also includes all available attributes of the organisms, their affinities, ecological and distribution patterns. The major difference between the cladists and evolutionary systematists is the refutation of Hennig's statement that all taxa have to be holophyletic or strictly monophyletic (Ashlock, 1979). When cladists give the same rank to sister-groups, evolutionary taxonomy considers the relative weight of their unique characters (autapomorphies) as compared to their shared derived characters (synapomorphies). Autapomorphies are important in the way that they reflect the adaptation to new niches, and in the way that they allow the conversion of a cladogram into a phylogram which is a diagrammatic representation of the analysis, showing the branching points and the degree of divergence.

Molecular systematics and phylogeny

While numerical cladistics has been evolving since the 1960s, various techniques of phylogenetic reconstruction, involving the use of biochemical and molecular data, were developing and interacting with other schools of phylogenetic reconstruction.

Our understanding of the phylogenetic relationships among East African cichlid fish species flocks has increased rapidly since the recent invention of the polymerase chain reaction (PCR) which dramatically facilitated the collection of molecular data (Meyer, 1994). Mitochondrial DNA (mtDNA) serves as an excellent marker for the inference of hypotheses on phylogenetic relationships, at least among closely related species or groups of species and mtDNA restriction analysis has been extensively applied to the endemic haplochromines of Lakes Malawi and Victoria (Meyer *et al.*, 1994; Moran *et al.*, 1994). This technique apparently does not provide sufficient resolution to discriminate clearly between species of closely related Mbuna (Bowers *et al.*, 1994), and species assumed to be good biological species may have identical cleavage profiles. Conversely, the same technique applied to *Tropheus* lineages in Lake Tanganyika revealed a surprisingly high degree of genetic variation, and the evolutionary relationships among *Tropheus* species have been reconstructed based on DNA sequences (Sturmbauer & Meyer, 1993).

Mitochondrial DNA variation among fish of the Lake Victoria flock was found to be extremely small. This high degree of mtDNA similarity and the earlier allozyme data suggested a very young age for this flock, probably less than 200 000 years. The data also support the hypothesis of intra-lacustrine speciation: the adaptive radiation of this species flock is likely to have occurred in Lake Victoria itself rather than being due to several immigrations of different ancestral lineages (Meyer *et al.*, 1994). Moreover, it is likely that the Victoria super-flock originated from single ancestral species (Meyer *et al.*, 1990). The established monophyly of the Lake Victoria flock, which has already been suggested by electrophoretic data (Sage *et al.* 1984), does not support the previous tendency to believe that the Lake Victoria haplochromine cichlid assemblage had more than one ancestor (Fryer & Iles, 1972; Greenwood, 1983*b*). MtDNA sequences have tentatively identified the non-endemic *Haplochromis burtoni*, a generalist species found in Lake Tanganyika and surrounding waters, to be the closest living relative of the Lake Victoria flock (Meyer *et al.*, 1991). More recently it has been found that several other non-endemic East African riverine cichlids from the Malagarasi River, the Ruahu River, Lake Rukwa and Lake Kitangiri (i.e. *Haplochromis bloyeti*) are even more closely related to the

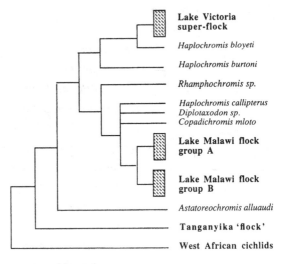

Fig. 5.6. Phylogenetic tree relating the endemic species flocks of Lake Victoria, Lake Malawi and some riverine species of haplochromine cichlids from East Africa to part of the Lake Tanganyika flock (from Meyer *et al.*, 1994). Presumed monophyletic assemblages are indicated with shaded boxes.

Victoria flock than *Haplochromis burtoni* (Meyer *et al.*, 1994).

Within Lake Malawi, two genetically monophyletic groups, each composed of about 200 species, were characterised based on mtDNA sequences. One group is largely confined to rocky habitats (the Mbuna), while the second lives over sandy habitats and is composed of species that were assigned to the genus *Haplochromis*. MtDNA data suggest that both groups can be traced back to a common ancestral species for probably the whole Lake Malawi flock, with the exception of the *Haplochromis callipterus* lineage (Meyer *et al.*, 1990; Moran *et al.*, 1994). More recent findings based on mitochondrial RFLP data, suggest that there are six independent lineages in Lake Malawi (Moran *et al.*, 1994). *Serranochromis robustus* (not included in studies of Meyer *et al.*, 1994) is a basal member lineage. Aside from the two major groups (Mbuna and non-Mbuna), the data also suggested the existence of other discrete endemic lineages: *Rhamphochromis*, *Diplotaxodon*, *Haplochromis callipterus* and *Copadichromis mloto* (see Fig. 5.6). *Haplochromis callipterus*, which is not strictly endemic to Lake

Malawi, might be representative of the ancestral stock, as had been previously suggested by morphological data.

Morphological and electrophoretic data both suggest that the lineages of cichlids from Lake Tanganyika are old and can be traced back to at least seven distinct ancestral lineages (Poll, 1986; Nishida, 1991). Comparisons of electrophoretic and mtDNA data demonstrated that several Tanganyikan lineages are much older than the lineages of Lakes Victoria and Malawi (Nishida, 1991; Sturmbauer & Meyer, 1993). The Ectodini, a large tribe of endemics, is probably about 3.5 to 4 million years old, and some lineages (Bathibatini, Lamprologini) might be ever older than 5 million years. It is not yet clear if the Lake Tanganyika species flock evolved within the lake basin from a single ancestral lineage or if its origin is polyphyletic.

Electrophoretic and mtDNA sequences also suggest that the Victoria and Malawi flocks are closely related to some Tanganyikan tribes: the Tropheini and Haplochromini (Nishida, 1991; Sturmbauer & Meyer, 1993). While considerable similarity between *Tropheus* (L. Tanganyika) and *Pseudotropheus* (L. Malawi) had been interpreted as favouring a polyphyletic origin of Lake Malawi cichlids, the molecular phylogeny strongly suggests that morphological similarities between taxa from these lakes are due to convergences, and they are not phylogenetically related despite recent origins and close affinities. Specialisation's probably evolved repeatedly and independently in each lake. This finding is contrary to the concept developed previously by Greenwood (1980) that cichlid lineages recognised using Hennigian principles have representatives in different lakes, cutting across the present-day lake boundaries. The Tanganyika flock can be viewed as a reservoir of old phylogenetic lineages that gave rise to the Victoria and Malawi flocks (Meyer *et al.*, 1990, 1994; Nishida, 1991).

A phylogenetic relationship has also been established for eight species of West African bagrids, belonging to the genera *Chrysichthys*, *Clarotes*, *Auchenoglanis* and *Bagrus* (Agnèse, 1989). Two main groups are distinguished according to their karyotypes (see Chapter 3): the group *Chrysichthys–Clarotes* with $2n = 70$ or $2n = 72$ chromosomes, and the other group *Auchenoglanis–Bagrus* with $2n = 54$ or $2n = 56$ chromosomes (Fig. 5.7). In comparison with Asian bagrids, it is assumed that the karyotype of the *Chrysichthys–Clarotes* group is derived from

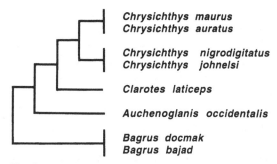

Fig. 5.7. Phylogenetic relationships between eight species of bagrids (from Agnèse, 1989).

the *Bagrus* type. This group may be considered as monophyletic. The analysis of genetic distances confirms the close relationships between *Chrysichthys* and *Clarotes*, and these two genera were later included in the same family Clarotidae by Mo (1991). But the analysis also confirms that the four *Chrysichthys* species under study belong to two of the subgenera proposed by Risch (1986): subgenus *Chrysichthys*, *C. maurus* (Valenciennes, 1839) and *C. auratus* (Geoffroy Saint-Hilaire, 1809)) on the one hand, and subgenus *Melanodactylus*, *C. nigrodigitatus* (Lacépède, 1803) and *C. johnelsi*, (Daget, 1959) on the other hand.

The genus concept

Quite a large number of papers have been published dealing with the species concept, while the genus concept has apparently been neglected by systematists. In many cases, the genus is a somewhat arbitrary taxonomic category that serves to group more or less similar species. It has been suggested (Dubois, 1988) that genera should be considered as discontinuous evolutionary units, which could be defined using a combination of genetic, phylogenetic and ecological criteria. In that way, interspecific hybridisation would be proof of a common phylogenetic origin.

The absence of standardisation in supraspecific systematics has led to numerous nomenclatural changes as knowledge of species phylogeny improves. That was the case for instance for the Mastacembelidae (Travers, 1984), bariliine (Howes, 1984), Schilbeidae (de Vos, 1984), etc. The splitting of the broad genus *Tilapia* (Tribe Tilapiini) into three by Trewavas (1983), which has been a matter of debate, is a good illustration of the difficulties that could be encountered in defining a genus. It is also one of the few examples of the introduction of spawning behaviours as a criterion in defining a genus. There have been numerous previous attempts by taxonomists to divide this genus into smaller, probably monophyletic groups. These divisions were based on morphology and meristics: number of gill rakers, number of scales, jaw structure, morphology of pharyngeal bones, etc. For instance, Thys van den Audenaerde (1968, 1971) recognised several subgenera of mouthbrooding tilapias *Sarotherodon*, *Oreochromis*, *Alcolapia*, *Neotilapia*, *Nyassalapia*, *Loruwiala*, *Danakilia* and *Nilotilapia*. Trewavas (1966) first included *Sarotherodon* in *Tilapia* as a subgenus, and then (1973) gave it the rank of genus, including all Thys's mouthbrooding subgenera within it. Later (1982, 1983), Trewavas proposed to divide the tilapias into three genera: *Tilapia*, which consists of substrate spawners; *Sarotherodon*, which includes the paternal and biparental mouthbrooders, and *Oreochromis*, which is restricted to maternal mouthbrooders. This classification is not yet widely accepted and Thys van den Audenaerde (1971) would prefer to leave the tilapias undivided (see also note in Teugels & Thys van den Audenaerde, 1992). Genetic information has therefore been collected to check the monophyly of the proposed classification.

Different hypotheses have been presented to explain the possible evolution of tilapiines, both agreeing that *Tilapia* has given rise to mouthbrooding branches. According to Trewavas, mouthbrooding could have arisen from one, or possibly two, splits from the ancestral line: one branch (*Sarotherodon*) remained conservative, whereas the other (*Oreochromis*) became more progressive. This is not the opinion of Peters & Berns (1978, 1982) who believe that any of these subdivisions are unjustified and that the various forms should all be called *Tilapia*, and at best given a subgeneric status. They believe that a number of splits from the ancestral substrate spawners may have occurred, possibly from different ancestors and at different times. The oldest branch represents maternal mouthbrooders (*Oreochromis*) compared with the younger paternal and biparental brooders.

This debate has excited some molecular taxonomists during the last decade. Kornfield *et al.* (1979) confirmed the broad taxonomic relationships derived from morpho-

logical data, but used only one species from each of the three genera. Later, McAndrew & Majumdar (1984) used electrophoresis on nine different tilapiine species, including at least one from each of the three genera proposed by Trewavas (1982, 1983). The results were inconclusive with regard to deciding between the different hypotheses, but the closer relationship of *Sarotherodon galilaeus* to the *Oreochromis* group of species, rather than to the *Tilapia* group, appeared not to favour the Peters & Berns (1982) hypothesis.

More recently, using more species and more loci, Sodsuk & McAndrew (1991) were able to reach rather clearer conclusions. All the analytical methods clearly separated the *Tilapia* clade from all *Sarotherodon* and *Oreochromis* species. But the relative position of the various species in the latter clade are more variable. A recent study by Kornfield (1991) based on mtDNA restriction, led to a clear split of the mouthbrooders *Sarotherodon* and *Oreochromis*. Within *Oreochromis*, subgroupings were observed. An unexpected one was between *Oreochromis spirulus spirulus* and *Oreochromis niloticus niloticus* in the studies of Sodsuk & McAndrew (1991) as well as in those of Kornfield (1991). *Oreochromis s. spirulus* was thought by Trewavas (1983) to be closer to the '*mossambicus* complex'. Three species of *Oreochromis* also consistently group together: *O. andersonii*, *O. mortimeri* and *O. mossambicus*, whereas they were placed in two separate subgenera by Trewavas (1983). Sodsuk & McAndrew (1991) concluded that 'despite the minor rearrangements within major clades, it is clear that the classification of Trewavas (1983) that divides the Tilapiini into three genera, not only describes the biological characteristics of the species, within each genus, but also reflects the evolution of this group'.

The most recent contribution to tilapiine phylogeny is a study of the enzymatic polymorphism and phylogeny of 21 tilapiine taxa, most of them from West Africa (Pouyaud, 1994). The results suggest that *Sarotherodon* may be a paraphyletic group. A subgroup, which includes *Sarotherodon occidentalis* and *Sarotherodon caudomarginatus*, is a monophyletic group intermediate between the group of species belonging to the genera *Oreochromis* and *Tilapia*. The other subgroup includes *Sarotherodon galilaeus* and various *Oreochromis* species. *Sarotherodon galilaeus* cannot be distinguished from *Oreochromis* species by diagnostic characters. The case of

Sarotherodon melanotheron is more difficult. Trewavas has already recognised that *S. melanotheron* belonged to a subgroup of *Sarotherodon* (with *S. mvogoi*). In the results obtained by Pouyaud (1994), this species is close to the *Oreochromis* group of species but differs from the closest species, *Oreochromis macrochir*, by five diagnostic characters. Seyoum (1989) came to a similar conclusion and suggested a specific taxonomic status for that species that is the only known male mouthbrooder in the tilapiines.

Pouyaud (1994) also concluded that the genus *Tilapia* is paraphyletic. One group, the subgenus *Coptodon* according to Thys van den Audenaerde (1968), is composed of genetically close species: *Tilapia guineensis*, *T. zillii*, *T. dageti*, *T. walteri*. The other group includes *Tilapia brevimanus*, *T. busumana*, *T. mariae*, as well as *T. cessiana* and *T. buttikoferi*, which were placed into the subgenus *Heterotilapia* by Thys van den Audenaerde (1968).

Higher classification and phylogeny of fishes

Evolution is an historical process concerned with the origin, ancestry and differentiation of organisms. There is a close relationship between the evolution of fish morphology, physiology or behaviour, and the various parameters of the environment in which they live. The phylogenetic and evolutionary patterns of fish trace the adaptation of these organisms to the changing environment, and the higher levels of systematics, derived from anatomical and morphological data, expresses their evolutionary relationships. The appearance of the revised classification of teleosts by Greenwood *et al.* (1966), in selecting not primitive but derived character states as the more useful for inferring monophyletic groups, was a decisive major step towards Hennig classification. Since this monograph, however, much progress has been made in the understanding of the relationships between the various major groups of ray-finned fishes. Here I shall give a general overview of the accepted ideas, focusing on taxa of particular interest for the African ichthyofauna. The synthesis of the current thinking about the classification of fish, proposed by Lecointre (1994), has been used to establish proposed cladograms.

There is apparently still no agreement on the interrelationships of the major groups of lower vertebrates, and the freshwater or marine origin of the ancestors of all

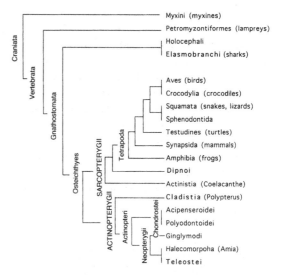

Fig. 5.8. Branching diagram representing the phylogeny of the major living groups of craniates. Established from different sources by Lecointre (1994).

fishes and fish-like vertebrates is still a matter of debate (see Banarescu, 1990). According to Greenwood *et al.* (1966) the earliest teleosts lived in freshwater, but this theory is not widely accepted by many palaeontologists and ichthyologists who consider that most primary freshwater fishes are offshoots of marine ancestors that recolonised freshwaters after a long evolution in the sea (Banarescu, 1990). Lauder & Liem (1983) suggested that the actinopterygians (Actinopterygii), coelacanths (Actinistia), lungfishes (Dipnoi) and tetrapods share a common ancestor (Fig. 5.8). Recent data based on the study of mitochondrial DNA sequences, show that the lungfish mtDNA is more closely related to that of a frog than is the mtDNA of the coelacanth. As a result, Dipnoi is the sister-group of tetrapods, and the coelacanths the sister-group of choanates including both Dipnoi and tetrapods. This result tends to support the theory that land vertebrates arose from an offshoot of the lineage leading to lungfishes (Meyer & Wilson, 1990). The Dipnoi is the more archaic group of living bony fish, and it originated in the Lower Devonian.

The subclass Actinopterygii (ray-finned fishes) forms by far the most diverse group of fishes and, with 23 000 species known, represents half of all extant vertebrate species. They have undergone an extensive radiation since their appearance in the Lower Devonian. The relationships of the primitive living actinopterygians have been relatively well established (Rosen *et al.*, 1981; Patterson, 1982; Janvier, 1986). The early actinopterygian fish were usually included in the infraclass Chondrostei, which comprised a diverse assemblage of fossil and living taxa, but following Patterson (1982), the Chondrostei will be restricted to a monophyletic clade including the two remaining groups of living survivors of actinopterygians: the families Acipenseridae (sturgeons) known from Upper Cretaceous fossils, and Polyodontidae (paddlefishes). Acipenseridae are found in Europe, Asia and North America. The Polyodontidae consists of two monospecific living genera, one in China, and the other in North America.

The relationships of the Cladistia, which contains the single family Polypteridae, was the subject of extensive debate. In recent years they were considered as a separate subclass of the Osteichthyes, the Brachiopterygii. But several authors, and especially Patterson (1982), provided evidence that they were highly specialised survivors of primitive actinopterygian fishes. The Polypteridae includes two genera (*Polypterus* and *Calamoichthys*), morphologically very similar and restricted to the freshwaters of tropical Africa. Young *Polypterus* exhibit a pair of relatively large external gills, and adults are able to breath air having a pair of highly vascularized lungs. These morphological characteristics may be a primitive feature of actinopterygian fishes that has be retained in *Polypterus*.

The Teleostei are the most diverse group of Actinopterygii and were first known from the Middle Triassic. Their phylogeny has been fairly well investigated (Greenwood *et al.*, 1973; Patterson & Rosen, 1977). Four major monophyletic groups are recognised among living teleosteans: Osteoglossomorpha, Elopomorpha, Clupeomorpha and Euteleostei (Fig. 5.9).

The osteoglossomorph ('bony-tongued') fishes comprise the most primitive group of living teleosts, and are known from the Upper Jurassic. The branching of the elopocephalan lineage, which includes the three other monophyletic groups of teleosteans, may be considerably older. The Osteoglossomorpha may be divided into two major clades: the Notopteroidei including the families Notopteridae (Africa and Asia), Mormyridae (Africa) and Hiodontidae (North America); and the Osteoglossoidei containing the Pantodontidae, the Osteoglossidae

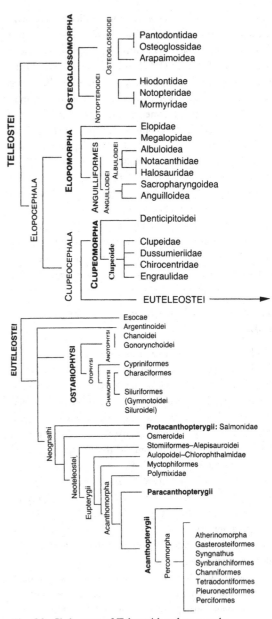

Fig. 5.9. Cladograms of Teleostei based on morpho-anatomical characters (from Lecointre, 1994).

and the Arapaimidae. A fairly complete phylogeny is available for the Osteoglossomorpha (Greenwood, 1973*c*; Patterson, 1981) and the biogeographical history of the group was reconstructed from the Pangaea (up to 145 Myr BP). For the Osteoglossoidei we have one of the most complete biogeographical histories of any fresh-water fish group. It includes different living genera in Africa (*Heterotis*) and South America (*Arapaima*), but fossil osteoglossids have been found in Asia, Australia and North America (Greenwood, 1973*c*). The Osteoglossidae includes three genera (each with one or a few species) living in Australia and South-East Asia (*Sclerophages*), Africa (*Pantodon*) and South America (*Osteoglossum*).

The Elopomorpha contains different families and species, almost exclusively marine but with a few species able to enter brackish waters. The most remarkable feature of this group is the occurrence of a leptocephalus larva in the family Anguillidae (eels), which is the only one containing species that spend part of their life in freshwaters.

The Clupeomorpha (herring-like fishes) first appear in the Lower Cretaceous. They are primarily marine, but a few species may move easily into, or are adapted to, brackish and freshwaters. The Denticipitoidei contains only the primitive species *Denticeps clupeoides* from western Africa. The Clupeoidei includes the remaining extant Clupeomorphs with freshwater representatives in the families Clupeidae and Congothrissidae.

The Euteleostei is a very large group that has been well investigated during the last two decades (Lauder & Liem, 1983; Rosen, 1985; Johnson, 1992). At present two main groups are recognised: the Ostariophysi and Neognathi, and a third one of primitive Euleostean fishes of uncertain relationships, including Esocoidei and Argent-inoidei.

There are four monophyletic groups within the superorder Ostariophysi (Fink & Fink, 1981): the Gonorynchiformes (milkfishes), Characiformes, Cypriniformes (minnows) and Siluriformes, including Siluroidei (catfishes) and Gymnotoidei (gymnotoids)

The order Gonorynchiformes contains five families, three with only one species: the Phractolaemidae (West Africa), Grasseichthyidae (Zaïre, Africa), Cromeriidae (Africa), Channidae (East Africa and South-East Asia) and Kneriidae (Africa). Their relationships with the other Ostariophysi have long been discussed by ichthyologists. Rosen & Greenwood (1970) after an extensive survey placed Gonorynchiformes within Anotophysi. Others were placed in series Otophysi. Fink & Fink (1981) provided additional evidence for this classifi-

cation. The Otophysi constitute a group of related orders and families characterised by the possession of a Weberian apparatus for the transmission of sound impulses from the swimbladder to the inner ear: Cypriniformes, Characiformes, Gymnotoidei and Siluroidei. The interrelationships of these groups are not yet well known despite their importance. One of the reasons is possibly the tremendous morphological and ecological diversity of each subgroup. For a long time, one of the major groups, the Characiformes, was considered to be the most primitive, ancestral to minnows and catfishes (Briggs, 1979). Characins and gymnotoids were considered to be closely related, and cyprinids more closely related to characids than to catfishes. Fink & Fink (1981) presented a new hypothesis of relationships among the Otophysi: the siluroids and gymnotoids form a monophyletic group (Siluriformes) that is the sister-group of the Characiformes; and together these taxa are called Characiphysi, a sister-group of the Cypriniformes, in a taxon Otophysi. A marine otophysan, *Chanoides*, from the Eocene of Monte Boca (Italy) could be the sister-group to all recent otophysan (Characiformes, Cypriniformes, Siluriformes) according to Patterson (1984).

Among living families of Characiformes only one family, the Characidae, is common to Africa and South America. In Africa the families Hepsetidae (one species), Citharinidae and Distichodontidae have representatives in freshwaters. The two latter families should be the primitive sister-group to all other Characiformes (Fink & Fink, 1981). The primitive sister-group of the Characiformes is the Cypriniformes (Lauder & Liem, 1983).

The Siluriformes (catfish) form a group of highly modified fishes, with a distinctive morphology: the body is naked or has thick bony plates and there are large spiny first rays to the pectoral and dorsal fins. The phylogenetic position of the catfish families is uncertain. Three families (including Ariidae and Plotosidae) are partly or completely marine, whereas the others (over 30 families) live in freshwaters. There is no common family between South America and Eurasia and Africa, but the Gymnotoidei, which are restricted to Central and South America, probably share a common ancestor with the catfishes (Fink & Fink, 1981).

The phylogeny of Neognathi is more complex. A group including Salmonidae (salmon) and Osmeroidei may constitute the sister-group of the Neoteleostei. They have no native representative in the tropical freshwaters of Africa, but the small family Galaxiidae (included in the Osmeroidei) is represented by one species in South Africa. The Neoteleostei covers a vast group of marine fishes (Stomiiformes, Alepisauroidei), and the group of Eurypterygii, itself including different groups, the most advanced of which (Acanthopterygii) are known from the Cretaceous.

Acanthopterygii can in turn be divided into two groups: the Atherinomorpha, and the poorly defined Percomorpha. Atherinomorpha were first known from the Eocene, have a nearly world-wide distribution and contain three lineages. The Cyprinodontiformes occurs in fresh and brackish waters in Africa (Cyprinodontidae), Central and South America (Rivulidae, Anablepidae, Poecilidae), North America (Fundulidae) and Europe (Aplocheilidae). The Beloniformes (including the family Hemiramphidae) have predominantly marine habits, as do the Atherionoidei (family Atherinidae).

The Percomorpha group is ill-defined and their classification is still confused and problematic despite much effort during recent years (Stiassny & Moore, 1992; Johnson, 1993; Johnson & Patterson, 1993). Many of the major taxa are probably polyphyletic groups. New relationships established by different authors are not always completely agreed by others, and quite a lot of work is still necessary to establish a more clear phylogeny of the Percomorpha, and more generally of the Acanthomorpha (Lecointre, 1994). Therefore the proposed classification (Fig. 5.9) is only tentative.

The Gasterosteiformes are widely distributed today in both marine and freshwaters, and include the families Gasterosteidae and Syngnathidae. The order Perciformes is not a monophyletic assemblage. It includes over 150 families, mainly marine, but some of them are well-adapted to freshwater and are of great importance in the African fish fauna: Cichlidae, Centropomidae (Greenwood, 1976b); Nandidae (Liem & Greenwood, 1981); and Anabantidae. The position of the Cichlidae in relation to other Percomorph taxa has been a point of contention for many years, but Stiassny & Jensen (1987) placed the Cichlidae, as the sister-group of other labroids, in the suborder Labroidei. The Cichlidae, as well as other Labroid families, probably arose early in the Cretaceous, and taxonomic differentiation was well under way prior to the separation of Gondwanaland (Stiassny, 1987, 1991).

The Mastacembeloidei, previously associated with the Perciformes, should be reallocated to the Synbranchiformes (Travers, 1984) and considered as a suborder. The family Mastacembelidae occurs in Africa (subfamily Afromastacembelinae) and from the Middle East to South-East Asia (subfamily Mastacembelinae). The Channiformes, traditionally aligned with the anabantoids, are also closely related to the Synbranchiformes (Lauder & Liem, 1983). The Tetraodontiformes date back to the Lower Eocene, and comprise mainly marine forms. Pleuronectiformes (flatfishes) also date back to the Eocene, and are a monophyletic assemblage based on the asymmetrical position of the eyes but their position in relation to other fish groups remains problematic.

Rapid progress in molecular systematics will certainly help to clarify the phylogeny of fish in the near future (see Lecointre, 1994, for review)

6 Chance and challenge in a changing environment

Since the eyes and minds of human observers are themselves diverse, and prone to a variety of biases, the results of their observations and deliberations inevitably will produce a variety of opinions. That side of diversity, too, is apparent in the individual approaches to a common problem: the analysis and interpretation of the processes which shaped and are even now changing the biosphere.

Introduction, The Evolving Biosphere, 1981

To understand the similarities and differences between the fish faunas of different aquatic systems it is necessary to consider the present distribution of fish as the result of past geological and climatic events. Indeed, living freshwater fish offer unique material for the study of past biogeographic events because most river and lake basins can be considered as isolated islands (Hugueny, 1989). Although some species are probably able to enter other basins by crossing the sea or occasionally during high floods, most fish species have been able to colonise those basins only during favourable situations in the past when physical connections existed between basins, or when one river was captured by another.

Biogeographical scenarios

Distribution patterns of animals (zoogeography) can be demonstrated to be non-random and must therefore be explained in terms of processes. Biogeography is the discipline that seeks to explain the observed distribution of organisms and the means by which the patterns of distribution have arisen. The raw material of biogeography is the distribution of species in space and time, and therefore lists of organisms in different areas have to be established before one can proceed to the reconstruction of the sequence of events that shaped the distribution patterns presently observed.

The existence and continuance of freshwater habitats

is strongly dependent upon two major environmental factors: the geomorphology and the water budget. On the one hand, geomorphology shapes the waterbody, and changes due to erosion, tectonics or volcanism will result in changes in its morphology. On the other hand, the water budget strongly depends on the climate, and stability of water level or discharge is controlled by the net balance between inputs (precipitation, runoff) and outputs (evaporation, infiltration, discharge). Slight changes in the balance may result in the expansion of many aquatic habitats, or conversely in their disappearance.

Morphological and climatic parameters interact at different time scales, from one year to millions of years, and the present state of African lakes and rivers is only a transitory one in the evolution of those aquatic landscapes. The pattern of interconnections between river systems and/or lakes has changed from time to time in response to geological or climatic changes, and will certainly change in the future too. When faced with these habitat changes, the fish fauna may be offered various possibilities: (i) to adapt to the changing environment by speciation and/or physiological adaptation; (ii) to move when possible to more suitable habitats or to perish if there is no other possibility. For instance, in the case of African freshwater fish, it has been emphasised that a major event in the biological history of African inland waters was the formation in East and Central Africa of a very great number and variety of habitats, as a result of the transformation of less varied systems by tectonic activity during the past 20 million years (Beadle, 1981). A large number of fish species, previously adapted to riverine conditions, have evolved in response to these events, which are more likely to have determined the present distribution of fish species over the continent than the present differences in conditions in the various water systems.

Various schools of biogeography have emerged during the last two decades. 'Traditional' or dispersal biogeography (see Nelson, 1978 for review) interprets present-day distribution as the result of dispersal of species from their native area (centre of origin), crossing barriers in their path, or taking the opportunity of temporary 'bridges' between two regions. Vicariance biogeography, as well as historical biogeography, rests on the assumption that taxa evolved where they live. When a widespread ancestral biota is split into two or more separate biotas, as a result of geological or climatic changes producing barriers within the ancestral area, species evolve separately and speciation occurs (Rosen, 1978). Patterson (1981) illustrated the differences in this way: 'Thus, in essence, finding two sister species of freshwater fish, one on each side of a mountain range, the dispersal biogeographer will ask how and when the fish crossed the mountains, and the vicariance biogeographer will ask when the mountains interrupted the original range of the fish. Dispersal biogeography sees the earth as relatively stable, and life as active and mobile, whereas vicariance biogeography sees life as relatively immobile, and the earth as active, imposing change on life as mountains rise and oceans open under the feet of animals . . . '.

Vicariance biogeography has grown out of three events. The first was the emergence of the major geological paradigm of plate tectonics, which completely modified the observations of dispersalist biogeographers, and was based on fixed continental positions. The second was the emergence of phylogenetic systematics that allowed reconstruction of evolutionary relationships among members of a group. Third was the idea that all hypotheses based on a priori centres of origin should be abandoned in favour of vicariance explanations (Wiley, 1988). There has been a growing interest in vicariance biogeography over the last two decades, but the tendency is now to take into account the different hypotheses in a synthetic approach.

Ecological biogeography interprets the distribution of organisms as being caused by environmental factors, including interspecific competition, and the pattern of occurrence is explained in terms of the ecological tolerance of individual species (see MacArthur, 1972).

Greenwood (1983a) noted that 'there have been few major zoogeographical studies on the African freshwater ichthyofauna during the last century, and our basic understanding of African freshwater fish biogeography is little different from that outlined by Günther in 1880'. This seems rather pessimistic and is only partly true, as we shall see.

Opportunities and barriers to dispersal

With a few exceptions, fish need water to survive and are therefore only able to migrate and colonise new habitats when physical connections between them are established. For instance, the great similarities between the ichthyofaunas of the Nilo–Sudan river basins can only be explained if faunal exchanges have occurred relatively recently. There is some geological evidence of past and present-day connections between these drainages, but it is difficult to date the time and the length of the contacts. Such 'bridges' allowing faunal exchanges and dispersion have appeared and disappeared in the past, as a result of environmental events.

Connections between river basins and river captures

River captures are assumed to be an important means of exchange between adjacent watersheds for primary division fish. There is a reasonable amount of field evidence for capture of some rivers in Africa, which would explain observed faunal similarities between basins, and no doubt a greater number are unknown. In flat areas, exchanges of fish probably also occurred between adjacent drainages through inundated zones resulting from overflow following heavy rains, or through more or less permanent swampy areas. Given the drastic ecological conditions prevailing in such aquatic habitats, fish physiologically adapted to survive in poor oxygen situations (see Chapter 9) had a greater chance of crossing such barriers.

Coastal connections

Apart from species particularly tolerant of saltwater, which are classified as peripheral, freshwater fish cannot disperse across the sea. Nevertheless, at present, there is a series of coastal lagoons along the West African coast, often in connection with two river basins. Many species that can tolerate brackish water, at least for a short period, are therefore able to migrate from one basin to another, and eventually to colonise a new basin. Such exchanges would be facilitated during flood periods, when input of freshwater reduces the salinity of the lagoons.

Some reputed primary freshwater groups may include species or populations more or less permanently inhabiting brackish waters, as in the case of *Chrysichthys auratus* in West Africa. This catfish, which is widespread in most Nilo–Sudan river basins, also inhabits brackish coastal waters of West Africa where it was known as *Chrysichthys filamentosus*. A morphological study (Risch, 1986) and the use of protein electrophoresis (Agnèse, 1991) has confirmed its synonymy with *C. auratus*. But it should be noted that, whereas *C. auratus* is able to colonise river basins through brackish water connections, this species has never been found upstream in the Atlantic coastal rivers from Côte d'Ivoire to Guinea, which are inhabited by another species, *Chrysichthys maurus*. The origins of such a distribution are unknown and some kind of competition between the two species could be involved. *Chrysichthys auratus* is currently a savannah species, and might be absent from the West African humid forest area as a result of adverse environmental parameters that have not yet been identified.

Bearing in mind the past climatic changes, there were probably pluvial periods during which high floods provided suitable conditions for fish to disperse coastwise between river mouths, either through very dilute brackish waters, or through freshwater marshes extending between the lower reaches of two adjacent rivers. Sydenham (1977) has noticed that today, 'the Ogun River was tenuously linked with the Niger via the lagoon creek network that extends from the Niger delta into the Republic of Benin. With this network, the Ogun is also directly allied to a whole series of lesser rivers from the Mono in the west to the Ohosu in the east'.

Changes in sea level

Changes in sea level have also resulted in modifications to the morphology of coastal areas. For instance, 18 000 years ago, with a sea level 110 m lower than nowadays,

there were obviously other possible coastal connections between rivers and possible faunal exchanges. In the Island of Fernando Po, now separated from the continent by a strait 35 km wide and up to 60 m deep, the depauperate native fish fauna (see Thys van den Audenaerde, 1967) is identical with that of the mainland and probably a relict fauna that was isolated during the late Pleistocene following subsequent rising of the sea level.

Falls as zoogeographical barriers

Waterfalls can act as physical barriers, isolating fish populations or preventing colonisation by upstream migrants. One can also imagine that watersheds are less easily crossed when the valleys are steeply embanked. According to the vicariant model of speciation, the long-term presence of such barriers to migration and dispersal can explain the existence of high endemicity areas on both sides.

The upthrust of mountains can cause the appearance of barriers. For example, the Guinean range, including the Fouta-Djalon, could have isolated the south-western Upper Guinea fauna from that of the north-western Nilo–Sudan. Such events are rather old (between the end of the Jurassic and the end of the Eocene for the Fouta-Djalon) and could explain the presence of endemic species in the Guinean provinces. In Guinea, the existence of only a few endemic species (*Barbus cadenati*, *Barbus guineensis*) in the upper reaches of rivers such as the Konkoure could be explained by the existence of falls, which keep this fauna safe from later new invaders (Daget, 1962c). Populations of this fauna were probably isolated during the upthrust of the mountains and have evolved independently since.

Another well-known example of such a barrier is the Gauthiot Falls between the Benue River (Niger basin) and the Mayo-Kebi (Chad basin). Species from the Niger basin (such as *Citharidium ansorgii*, *Arius gigas*, *Synodontis ocellifer*, *Cromeria nilotica*, etc.) occurring in the Benue River were not able to cross the falls and have never been recorded from the Chad basin (Daget, 1988). A similar situation occurs between Lake Malawi and the Zambezi River, the faunas of which are very different, with few species in common. The Murchison rapids, which are a series of falls and rapids in the Middle Shire River, are generally regarded as a barrier to upstream migration of Zambezian fish (Fryer & Iles, 1972), and inhibit Lower Zambezi fish from entering Lake Malawi, but are not a complete barrier to downstream migration (Banister & Clarke, 1980).

Other means of dispersal

Although apparently never demonstrated, it has been suggested that dispersal of some fish species could also occur through birds or other vertebrates. Drought resistant eggs of some cyprinodonts inhabiting temporary ponds could certainly be transported in mud for long distances on the feet of different animals, mammals or birds. Transportation of live fish by birds is certainly unlikely but should not be completely excluded. Nevertheless, one should recall a very simple fact: for most fish species, two mature fish, at least one male and one female, are necessary to establish a new population. The only exception would be a viviparous or a mouthbrooding fish, with its progeny, and in either case, such a chance is probably very low and the possibility largely anecdotal.

The full inventory of possible means of dispersal would also include the fish rains, which have been reported from different regions of the world, and result from the upward suction of a certain amount of water (including fish and other organisms) by local cyclones that then move and drop the transported material in other places.

Pre-Quaternary geological and climatic events in Africa and their consequences for aquatic biotas

At the beginning of the Miocene, the African landscape was apparently smoother than at present after a long period of tectonic stability. At that time, there were no very high mountain ranges and the watersheds between the drainage basins were generally much lower than at present. The surface of the continent is believed to have been moulded by differential uplift and subsidence into a gentle basin-and-swell pattern of large depressions separated by ridges. This structure is still apparent today in the Chad and Zaïre basins that have been little affected by subsequent deformation (Hamilton, 1982). It would seem that the faunal barriers between water systems were then easier to cross and that the fish fauna over the tropical and subtropical regions of the continent was there-

fore rather uniform in composition. This ancient pattern is still the basis of the hydrology of most of the continent.

The post-Miocene earth movements and volcanics, which have continued at intervals up to the present day, together with great fluctuations of climate, especially of rainfall during the Pleistocene, greatly altered the Miocene drainage systems. The Middle Tertiary (roughly 25 Myr ago) witnessed the beginnings of a general elevation of extensive areas of central and eastern Africa. Uplifting of several of the old watersheds has increased the isolation of some of the previous drainage basins and has been accompanied by a number of other disruptive activities, such as rifting, faulting and volcanism. The most dramatic event was the formation of the two Great Rift Valleys associated with the raising of the eastern highlands. It is generally agreed that the eastern part of the African continent is separating into two blocks: Africa to the west, Somalia to the east. (Tiercelin & Mondeguer, 1991). This long, slow process resulted in the formation of the East African Rift. During the Miocene (15 Myr ago), tectonic activity associated with rifting and the uplifting of domes along the rift margins resulted in the depression and downfaulting of the rift valleys. The rift system extends from Lake Malawi through the Western and Eastern (Gregory) Rifts, into Ethiopia, and further north as far as the Red Sea and the Jordan valley. One consequence of all this activity has been the disruption of drainage patterns in East Africa, with the reversal of previously west-flowing rivers and the creation of the Lake Victoria and Kyoga basins.

Volcanism associated with this faulting during the past 20 million years has resulted in falls and rapids that are partial or complete barriers to the movement of fish, and some lakes have been formed as a direct result of volcanic activity. Thus, in East and Central Africa, a new complex of catchment basins and drainage channels has been superimposed upon the old pattern and provided conditions for the evolution of a great variety of new fish species. In other parts of the continent, water courses have not been disrupted by earth movements to the same extent. The basic pattern has, therefore, been comparatively stable, but there have been considerable fluctuations in discharge (both quantity and seasonal pattern) and in the connections between them. This history is

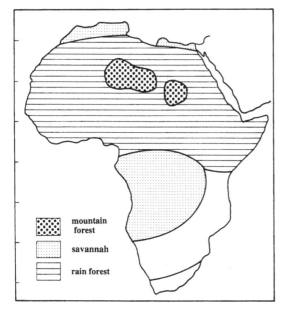

Fig. 6.1. Distribution of the main floras in Africa during the Upper Palaeocene/Lower Eocene, 60 million years ago. The Equator is further north than today, and rain forests developed along the northern coast (from Bonnefille, 1993).

mountain forest

savannah

rain forest

particularly well documented for the Late Pleistocene–Holocene period, and indicates extreme fluctuations that have on the whole been unfavourable to the long-term evolution of rich endemic faunas.

The geological events were also associated with dramatic changes in climatic conditions. During the Palaeocene–Eocene, as a result of the drifting of the African continent, the Equator was further north (10° to 15° in latitude) than its present position, and an equatorial climate prevailed in the Sahara (Fig. 6.1). The distribution of the flora was very different from the present (Bonnefille, 1993) and there is evidence that a dense forest, already present in the Palaeocene, developed along the northern coast in the Eocene, with savannah further south. A dense tropical forest was also present along the west coast from Senegal to Cameroon, and there were mangroves in Libya and along the Senegalese coast during the Eocene. It is therefore possible to understand why fossils of tropical species such as *Protopterus* spp., occur in North Africa (see Chapter 2).

By the end of the Eocene, there was a shift from an equatorial to a tropical climate in the Sahara, with a marked dry season (Maley, 1980). During the Oligocene, the drifting of the continent to the north, as well as the regression of epicontinental seas, resulted in a progressive drying of the northern coast, a regression of the dense forest, and progression of the savannah. In the Upper Miocene, the biotope was dry in northern Africa, and hygrophilous vegetation appeared to the south, at the level of the Tibesti. But, at the same time, dense forest was well developed along the western coast, particularly in Cameroon (Salard-Cheboldaeff, 1981). The present-day distribution of the main phytogeographic units was reached during the Middle Miocene (14 Myr), when the Equator was close to its present position.

Analysis of cores collected off the West African coast has revealed the long-term trends of the climate over the last 8 million years (Hamilton, 1988). Between 8.8 and 6.4 million years ago it was relatively warm and stable, and West Africa was humid. During the period 6.4 to 5.4 million years ago, which was associated with a major expansion of savannah, the Mediterranean sea became isolated and evaporated. Marked climatic fluctuations occurred between warm, humid periods and colder, drier periods. This fluctuating climate also prevailed between 4.6 and 2.4 million years ago, with a shift towards being drier and cooler after 3.5 million years. A major glaciation in the Northern Hemisphere started around 2.4 million years ago and the world became generally colder and drier, but with pronounced oscillations. Available data show that desert conditions appeared in the central and southern Sahara during the Early Pleistocene, while at the same time, rainfall increased southward (Maley, 1980).

A major result of research on deep-sea sediments was to establish that there have been many glacial periods during the Quaternary, which was a period of climatic instability. Up to 21 glacials or near glacials are estimated to have occurred during the last 2.3 million years (Hamilton, 1988). Most of the glacial maxima were probably associated with aridity in tropical Africa, whereas wet tropical conditions prevailed during interglacials. The result was cycles of major spread and retreat of aquatic habitats during the last 2.3 million years, probably with many opportunities for the isolation and divergence of fish populations.

Late Quaternary climatic changes and the refuge zones theory

The distribution pattern shown by African forest organisms is not only the result of the present-day environment (Hamilton, 1982) but can be partly explained by climatic changes that occurred throughout the Quaternary. Much attention has been given during the last two decades to these climatic changes, with their alternating dry and wet periods. A consequence of these changes in Africa was the extension and recession of forest zones, and some recent studies have tried to interpret present-day species distributions and endemism in the light of the forest refuge zones theory (Haffer, 1982) which states that during arid periods, forest was confined to core areas when the climate was unfavourable.

According to this theory of forest refugia, centres of speciation for forest organisms should occur at the sites of the last forest refugia during the last dry period, when the uniform dense forest divided into smaller patches. Forest species were able to survive only in those refugia during unfavourable climatic conditions. They therefore evolved separately for long periods of time, allowing speciation. When climatic conditions allowed the recovery of the extensive dense forest, some organisms extended their ranges, repopulating formerly depopulated areas. The rates of dispersal from the refugia varied with the ability of species to move and cross physical barriers. If the period of time since the last regression phase was not too long, some of the species would have remained in the refuge zones or their surroundings, such areas being consequently characterised by a high degree of endemicity.

By analogy with forest species, the refuge zones theory could apply to other organisms and biotopes that fragmented when unfavourable conditions occurred. There is possibly no strong correlation between forest and fish, and we have, as yet, no clear indication that some fish species are forest-dependant but ecological parameters, such as shade, temperature, water chemistry and water regime, may differ in forest streams compared to savannah streams. What is really important is that forest refuge zones were humid enough to maintain aquatic biotopes where fish could survive. Actually, in so far as forest refuges were very probably dependent on heavy rain, there should be some geographical similarity between forest and aquatic refuges. The reverse is not true, and aquatic refuges may have existed independently of any forest.

Late Quaternary climatic changes in Africa

It is now well known that during the Quaternary the African climate changed considerably with humid periods alternating with droughts. The last 30 thousand years are fairly well documented. Perrot & Street-Perrot (1982) presented evidence of a wet period between 25 000 and 22 000 BP in many tropical closed-basin lakes north of the Equator. They concluded that the Upper Pleistocene wet phase in northern intertropical Africa resulted from a combination of cooler temperatures and slightly greater precipitation than today. Environmental conditions in many areas most closely resembled the period 12 000–11 000 BP that was a time of climatic transition. This wet phase was more marked between the Equator and about 22° N, and is clearly evident in deep sea cores that record long-term fluctuations in the discharge of the Niger and the Nile. Results from Servant & Servant-Vildary (1980) and Maley (1989) also provide evidence that there was a wet climatic phase between 30 000 and 20 000 BP, which resulted in the formation of numerous lakes in the Chad basin. Between 26 000 and 20 000 BP, these lakes were shallow, with an extensive development of aquatic plants. In Ghana, the level of Lake Bosumtwi was also relatively high between 28 000 and 20 000 BP (Maley, 1989).

From 20 000 to 15 000 BP, aridity became general over West Africa, and dune fields covered a major part of the Sahel (Fig. 6.2). Van Zidderen Bakker (1982) gave an extensive survey of present knowledge of the period 18 000 BP, which is generally accepted to have been the coldest period coinciding with the last glacial maximum in the northern regions. During this period, the North Atlantic Ocean had changed into a polar sea, and the temperature of the surface water off the West African coast dropped substantially. Estimates for the decrease in temperature of the cold branch of the Canaries current vary from 3.5 to 10 °C according to different studies. As a consequence, evaporation was reduced and the winds brought much less humidity to Africa, which was one of the main causes of the glacial aridity. For the central Sahara, there is evidence that temperatures dropped as much as 10–14 °C in January and 6–8 °C in July and in the southern Sahara, the climate was hyperarid and lakes, such as Lake Chad, dried up. In western Africa, the lowering of temperature has been estimated to be 4–5 °C or 8–9 °C depending on the author. The climate was arid and rivers did not flow in Senegal and Mauritania. Temperature depression in East Africa was estimated to have been between 7 and 9 °C (see Hamilton, 1988), and would have caused major extinctions of terrestrial and aquatic organisms that had no possibility of moving toward more suitable conditions. Bonnefille *et al.* (1990) however, think that the last glacial cooling may have been overestimated in tropical regions and, using pollen data, they estimate the temperature decrease to have been 4 ± 2 °C in Burundi. At the same time, they also estimate that the mean annual rainfall decreased by 30% during the last glacial period.

The ice age aridity, having reduced the area of the rain forests and humid montane forests, savannahs, grassland and semideserts, spread over vast regions more or less replacing the lowland forests (Maley, 1987). Grassland became dominant in the Niger Delta while forest decreased significantly. The longitudinal vegetation belts of the western Africa coast were compressed southward and tropical rainforest could only survive in certain refugia that were exposed to the onshore westerly winds. Few lowland forest refugia have been identified (Diamond & Hamilton, 1980; Maley, 1987), the main ones being the Upper Guinea, the Cameroon–Gabon and the eastern Zaïre.

After 12 000 BP, humidity increased. Abundant rainfall permitted the existence of permanent lakes and ponds as well as large rivers crossing the Sahara. By 9000 BP, wet conditions entirely covered north tropical West Africa, and the wetter period is dated at 8500 BP. Wetter conditions than at present then extended up to the Tropic of Cancer (Hoggar Massif) (Fig. 6.2). For instance, at 9000 BP, the position of the 400 mm isohyet was near to 19°10′N in the eastern Sahara and 21°N in the western Sahara where the mean annual precipitation today is 5 and 60 mm per year respectively (Lézine, 1987; Ritchie *et al.*, 1985). From 12 000 to 8000 BP, the major Sahelian river systems probably passed through their maximum development with permanent high water (Talbot, 1980). During the Late Pleistocene and Upper Holocene, the rivers Niger and Senegal, and Lake Chad had a much greater effective catchment. A number of

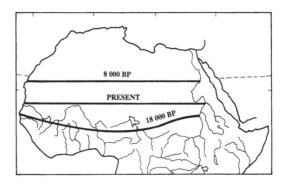

Fig. 6.2. Recent changes in Sahara–Sahel limits (from Petit-Maire, 1989).

prominent watercourses in the central Sahara, such as the Tilemsi, Azaouak, Dallol Bosso, Dallol Maouri and Tarka, flowed directly into the River Niger (Fig. 6.4). Under present climatic conditions, they are essentially relict features. The largest lacustrine expansion of Lake Chad 6000 yrs BP, the so-called Mega Chad, covered 300 000–400 000 km² (Servant, 1983). In the equatorial west African zone, a major change occurred around 9000–8500 BP: the sudden reappearance of the humid dense forest in the area it now covers (Maley, 1987; Lézine & Casanova, 1989).

Immediately after the wet phase (8000 BP), hydrological conditions became drier in the Sahelian zone and by about 7500 BP, rainfall had declined over the Tibesti Massif. Between 7000 and 5000 BP, the climate remained humid (Maley, 1981, 1989), but after 4500 BP the spatial pattern of moisture conditions changed over West Africa. The Sahara was arid and drier in the west than in the east (Sudan). Groundwater levels remained high in the Sahara until 8000 BP but decreased after 7000 BP. Another phase of increased moisture occurred between 4000 and 2500 BP, but of lower magnitude than during the Early Holocene period. North of 14° N latitude, hydrological conditions became drier as early as 2500 BP, and even drier around 2000 BP (Lézine & Casanova, 1989), and vegetation changed rapidly to its modern distribution.

Recognition of some major refuge zones

The humid periods allowed connections between river basins, mainly in flat areas, and favoured the dispersal

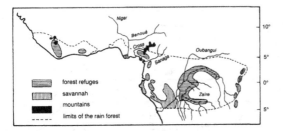

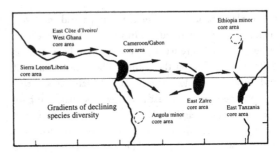

Fig. 6.3. Distribution of the main lowland forest refugia in equatorial Africa during maximum of the last arid phase (about 18 000 years BP) according to Maley (1991). The modern conditions (forest boundary, included savannahs) are adapted from White (1983).

Fig. 6.4. Distribution patterns of forest organisms in Central Africa: core areas and gradients of declining species richness (from Hamilton, 1982).

and extension of fish species' geographic ranges. Droughts, in contrast, resulted in the recession of waterbodies and sometimes complete drying. In the latter situation, it is assumed that fish which recolonised rivers had been able to survive in refuge zones.

The location and the extent of these refuge zones is controversial. One hypothesis claims that large river basins, such as the Nile and the Niger, were not totally dry and served as refuges, as was suggested for aquatic birds by Guillet & Crowe (1985). Another hypothesis is that fish took refuge in the more humid mountainous areas, such as the upper Nile, upper Benue (Roberts, 1975), upper Chari and Logone Rivers, as well as Guinean dorsal. Taking into consideration the number of endemics, the Niger and the Nile are much better candidates (with about 10% endemics each) as refugia than the Chari where almost no endemic species are known. This may have resulted from the drying up of most of the Chad basin, following severe droughts, and the possible extinction of the fish fauna. In comparison, the relatively large number of endemics in the Nile and the Niger suggests the existence of aquatic refuges, but their locations have still to be identified.

During dry periods, the rain forest occupied only limited areas in West Africa, as indicated by Maley (1991) (Fig. 6.3). It is assumed that rainfall in those areas not occupied by the rain forest were sufficient to maintain permanent aquatic habitats where fish survived. They later recolonised the continent when rains became sufficient to restore the hydrographic network.

The general distribution pattern of African forest organisms established by Hamilton (1982) for plants and different animal groups is of great interest to understanding the present fish distribution. Other studies (Diamond & Hamilton, 1980; Endler, 1982; Grubb, 1982; Mayr & O'Hara, 1986) have pointed out the existence of centres of endemism for birds, mammals and butterflies. Figure 6.3 shows the basic elements of the pattern shared by all groups: the core areas and gradients of declining species diversity. Core areas are rich in numbers of species and in endemics, and are the centres of isolated populations of disjunct species. Two important core areas are east Zaïre and Cameroon/Gabon. Three others are somewhat poorer in species: two in West Africa, in Sierra Leone/Liberia and east Côte d'Ivoire/western Ghana, one in eastern Tanzania (Fig. 6.4). For West African fish, the Hamilton pattern fits fairly well with the ichthyoprovinces of Upper Guinea and Lower Guinea, and with the Eburneo–Ghanean area of the Nilo–Sudan province.

The refuge zone theory is not widely accepted and Endler (1982), who refuted the forest refuge hypothesis, considered that centres of endemism can be explained by current environmental characteristics. However, pollen data obtained in Ghana and west Cameroon provide strong arguments in favour of the Pleistocene forest refuges (Maley et al., 1991). For fish, the observed endemicity in the Guinean provinces, for instance, is possibly the result of adaptation of fish species to prevailing highland ecological conditions (cooler waters, torrential hydrological regimes, etc.) over millions of years. It is true that many fish species adapted to the turbulent conditions encountered in rapids and swift flowing reaches have distributions more or less restricted to the Guinean rivers (Welcomme & de Merona, 1988). This is particularly the case for various Amphiliidae (genera *Amphilius*, *Paramphilius*, *Phractura*, *Doumea*), some Mochokidae (genera *Chiloglanis*, *Mochokus*) and Cyprinidae (genera *Garra*, *Labeo*), as well as for various anguilliform fish belonging to different families (see Lévêque *et al.*, 1989, 1991; Paugy *et al.*, 1989). Nevertheless, even if the refuge zone theory needs to be adapted to explain the distribution of freshwater fish, it provides a useful conceptual framework for biogeographical studies.

The relict tropical fish fauna in central Sahara

Since the beginning of this century, numerous Saharan expeditions have collected fish species in the widely scattered and small isolated patches of water in this area (Pellegrin, 1914, 1919*a*, *b*, 1931, 1934, 1936; Estève, 1949, 1952; Fowler, 1949; Monod, 1951, 1954; Daget, 1959*a*, 1968*a*; Dumont, 1979, 1987; Le Berre, 1989; Van Neer, 1989). If the occurrence of species known from Sudanese rivers far to the south, was quite surprising for the first discoverers, it is nowadays obvious that their occurrence results from the dynamics of changing climates across the area (Dumont, 1982; Maley, 1983).

As a result of taxonomic revisions (see Lévêque, 1990*b*), it appears that the relict fauna of the Borkou–Ennedi–Tibesti area, which exhibits the highest diversity (Table 6.1), is clearly related to that fauna occurring in the Chari and Nile basins. There is good evidence that during the last Holocene humid period, the Chad basin extended to the foot of the Ennedi–Tibesti whose drainages flowed into the so-called Mega Chad (see Talbot, 1980; Servant, 1983). The isolation of these populations occurred after 5000 BP and this apparently explains the relative richness of the ichthyofauna. It is also clear that in spite of the present extreme isolation of the area, this time period was apparently insufficient to allow speciation and the appearance of endemics, even if some morphological variability can be observed.

The record of *Barbus apleurogramma* in the Ennedi is nevertheless more surprising, as this species is now only known from Lake Victoria and the associated river systems. The importance of this record and its significance

have so far been rather neglected. Knowing that the species no longer occurs in the Chad basin and the lower Nile, it could well be the relict of a more ancient fish fauna extending northwards, and whose representatives later disappeared from Sudanese river basins.

The phylogenetic affinities of *Barbus deserti*, known from the northern slopes of the Tassili n'Ajjer, are not clear, but morphologically it appears to be related to the small tropical *Barbus* with which it was confused. There are superficially close similarities with Sudanese species and comparing morphometric characteristics and colour patterns, *B. deserti* resembles *Barbus callipterus*, a species presently widespread in the Niger and Chad basins. Some other tropical species were also collected in the Tassili: *Clarias gariepinus*, *Tilapia zillii* and *Hemichromis bimaculatus*. It is possible that they are relicts of the Chad basin fauna which were able to survive severe droughts and to colonise the northern slopes of the Tassili. To support this idea, it should be recalled that the Tassili represents the northern limit of the maximum potential catchment of Chad (see Talbot, 1980 and Fig. 6.4). During periods of maximum humidity, including the earlier Holocene, the whole catchment was probably active and relict drainages from Tassili to Chad are still apparent.

The relict fish fauna of the Ahaggar is poor compared to that of the Tibesti–Ennedi and there is no record from Aïr or Adrar Ifora. A depauperate vegetation has also been recorded in that area (Quézel, 1965). This is surprising if we consider the dense relict drainage of the central Sahara, including prominent watercourses such as the Tilemsi, Dallol Bosso, Dallol Maouri and Tarka that were active during the Late Pleistocene and Upper Holocene (Talbot, 1980). During the Early Holocene, there were large tropical lakes (Riser & Petit-Maire, 1986) and a diversified fossil fish fauna was collected all over this area (Daget, 1959*b*, 1961*b*; Gayet, 1983). The paucity of the present fish fauna could be explained by the scarcity in the area of suitable deep and perennial freshwater habitats, such as the 'gueltas' observed in the Ennedi, but this hypothesis can hardly be tested.

The Saharan relict tropical fish fauna has survived for thousands of years in isolated small waterbodies (named 'gueltas'), but it is likely that some of these records will become of historical interest only as man's impact

Table 6.1. *Occurrence of afrotropical fish species in different areas of the Sahara*

Species	1	2	3	4	5	6	7	8	9
Cyprinidae									
Barbus apleurogramma								*	
Barbus occidentalis						*			
Barbus deserti					*	*			
Barbus macrops	*			*			*	*	
Barbus pobeguini	*								
Labeo niloticus							*		
Labeo parvus							*	*	
Raiamas senegalensis							*		
Clariidae									
Clarias anguillaris	*								
Clarias gariepinus					*		*	*	
Cyprinodontidae									
Epiplatys spilargyreius									*
Cichlidae									
Hemichromis bimaculatus		*			*			*	*
Sarotherodon g. galilaeus	*								
Sarotherodon g. borkuanus							*	*	*
Tilapia zillii		*		*	*		*	*	

From Lévêque, 1990*b*. 1, Adrar (Mauritania); 2, Tunisia and South Algeria; 3, Aïr; 4, Ahaggar; 5, Tassili n'Ajjer; 6, Ghat; 7, Tibesti; 8, Ennedi; 9, Borkou.

(mainly water pollution) continues to progressively modify the present distribution, eradicating fish species in some of the waterbodies.

The history of river basins and the role of past connections in the dispersal of species

The present-day distribution of fish species has been shaped by the past climatic and geological events that eventually allowed the fish to disperse over a range of aquatic systems. A knowledge of the history of river systems is therefore useful in order to understand what the opportunities were for dispersal, and how it can explain the current zoogeographic patterns. We briefly review here some of the known or suspected interconnections that have occurred between major African drainages.

Relations between the Nile and Chad basins

The great similarity between the fish fauna of the Chad and Nile basins obviously suggests past connections, but the desert region now separating the two is largely devoid of fish. Gayet (1983) suggested that a connection could have occurred, during wet periods, in the Plio–Pleistocene when Lake Chad extended into a large hydrographic system. This would explain the colonisation of the Niger and Senegal Rivers by the Centropomidae *Lates niloticus*, first recorded during the Pliocene in the Nile region (Sorbini, 1973).

Livingstone (1980) claimed that the reduced River Nile of the Late Pleistocene could not be expected to have supported so many species of fish as the larger Nile today. From 20 000 to 12 500 years BP, the White Nile was a seasonal river (Adamson *et al.*, 1980) and the annual discharge from the Ethiopian headwaters was low (Street & Grove, 1976; Gasse, 1977; Gasse *et al.*, 1980). But despite the fact that flow along the main Nile was severely curtailed, it did reach the delta. No doubt it contained the 26 species that are now endemic (Greenwood, 1976*a*) but perhaps not many more. From about 12 500 years BP there was overflow from Lake Victoria and higher rainfall in Ethiopia, and the flow of the Nile became permanent and more regular. There were a few centuries of extremely high floods in Egypt (Butzer,

1980). The main Nile and its tributaries had established more stable channels, similar to the modern situation, by the Mid Holocene. With the resumption of heavy flow an opportunity existed for additional species to live in the river and this was apparently exploited largely by fish that came from well-watered West Africa. Immigration from such a distant source was facilitated by the Early and Mid Holocene wetness along the southern fringe of the Sahara and also by the general flatness of the terrain, where sand dunes dominate the local relief. Shifting dunes could generate the river captures that are a major mechanism of fish dispersal. Zaïrean fish, although much closer, depended on the much rarer accidents of tectonic, volcanic and erosion river capture, and were unable to colonise the expanded White Nile during the Holocene.

This hypothesis may be correct, but for the moment is not supported by any field evidence. Even if, during the pluvials, numerous river systems were probably perennial and inhabited by fish, we have, so far, no geological or palaeontological evidence as to when the last exchange of fish occurred across the intervening watershed (Roberts, 1975). According to Beadle (1981), the nearly identical Sudanian faunas of Lake Chad and the Nile must imply a recent (probably Late Pleistocene) connection between the two basins, but our geological knowledge of the intervening desert regions is very slight. There were two apparent paths for such fish exchanges: the first to the north-east of Lake Chad, through the gap between the Erdi Plateau and Ennedi, via the Mourdi depression; the second by the Upper Chari valley and the Bahr el Arab (Beadle, 1981). Unfortunately, little information is available on Jebel Mara, a volcanic mountain complex located almost midway between the Nile valley and the Chad basin, at the intersection of two major volcano–tectonic zones of Tertiary age. Williams *et al.* (1980) and Adamson & Williams (1980) have established that a major change in the western margin of the Nile basin occurred as a result of Tertiary and Quaternary volcanism. The eruption of the Jebel Mara volcanic complex, which commenced in the Miocene, resulted, in their view, in a westward diversion to Chad of previously eastward flowing drainage. More than 60 000 km^2 were thus deleted from the catchment area of the Nile. Adamson & Williams (1980) suggested that prior to the eruption of Jebel Mara, the Nile–Chad watershed lay almost 300 km west of its present position.

Recently, data obtained from Shuttle Imaging Radar and flight field investigations have been used to investigate paleodrainages in eastern Sahara. It suggests that during the Late Tertiary, a trans-African master stream system may have flowed from headwaters in the Red Sea Hills south-westward across North Africa, crossing the Chad basin (the Bahr el Ghazal may represent part of this master stream or one of the trunks) and flowing into the Atlantic by way of the Mayo Kebi River, the Benue and the Niger (McCauley *et al.*, 1986). Such a drainage would have extended some 4500 km from the crest line of the Red Sea Hills to the early Niger Delta. It lasted about 20 million years, from the Lower–Middle Eocene, with the uplifting of the Red Sea Hills (30–40 Myr) that were the highest mountains in north-east and central Africa during the Oligocene and Early Miocene, to the middle of the Miocene (*c.* 15 million years) when widespread domal uplift and volcanism complexes developed in the Sahara. The present-day topographic obstacles (Ennedi, Tibesti, Darfur) therefore did not exist at the time of the trans-African system, but appeared *c.* 15 million years ago. One might expect that this trans-Africa connection would be a highway for faunal exchanges between East and West Africa. The system probably did not disappear abruptly, and still worked partly and occasionally during high humid periods, allowing further exchanges. The hypothesis of a trans-African connection, however, remains controversial.

Relations between the Chad and Niger basins

At present, part of the floodwater of the Logone, a tributary of the Chari River, overflows to the west into the Mayo Kebi and over the Gauthiot Falls to join the Benue. There is good geological evidence that during the pluvial periods, this connection was already functioning. Comparison of the fish fauna above and below the Gauthiot Falls (45 m high) indicates that species are not able to pass upstream beyond the falls and as a result the connection functions only one way: from Chad to the Niger, which may help to explain the absence of any endemics in the Chad basin.

Past history of the Niger

During the Late Pleistocene and Upper Holocene (12 000–8000 years BP), at the height of the humid phases, the major river systems of western Africa had a

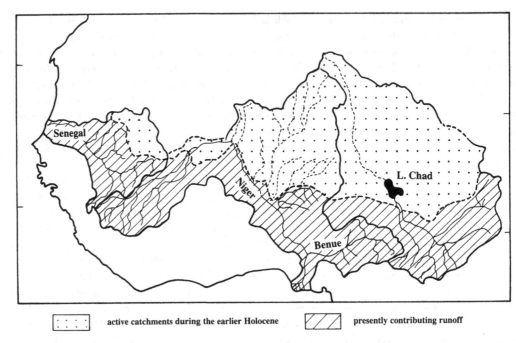

active catchments during the earlier Holocene   presently contributing runoff

Fig. 6.5. Maximum potential catchments of the Senegal and Niger–Benue river systems, and of the Chari and Chad basin. During periods of maximum humidity, such as the early Holocene, the whole of the catchments were probably active. Today, only part of the catchment (the shaded areas) contributes to the runoff (modified from Talbot, 1980).

much greater effective catchment (Talbot, 1980) than they have now. A number of prominent watercourses in the central Sahara, such as the Tilemsi, Dallol Bosso, Dallol Maouri and Tarka, flowed directly into the River Niger or joined seasonally active tributaries of those rivers (Fig. 6.5). Under present climatic conditions they are essentially relict features, but during the humid phases these rivers were active and operated as fully integrated drainage networks, maintained by catchments that stretched far into the present arid zone. According to Talbot (1980), the main valleys are probably old. The southern-most sections of Dallol Bosso and Dallol Maouri are quite deeply incised into the continental terminal table-lands, forming valleys that are comparable to the present valley of the Middle Niger. It appears from Landsat images that the Dallol Bosso may once have been the dominant stream, the present Niger valley being quite clearly a downstream continuation of the Dallol Bosso.

There is stratigraphic and sedimentological evidence of changes in the nature of these rivers from 12 000 BP

to the present time (Sombroek & Zonneveld, 1971; Talbot, 1980). From being strongly seasonal and ephem-eral, they changed to become seasonal, with broader stre-ams, during the Late Pleistocene. They became pro-gressively regular with meandering streams, from 10 000 to 8000 BP, as a result of a more regular rainfall regime, more evenly distributed through the year (Maley, 1977, 1981; Servant, 1983). At this time, the Niger–Benue system probably received a large amount of water from the overflowing Chad basin. Results obtained by Pastouret *et al.* (1978) in the Niger Delta confirmed that most of the rainfall over the drainage basin of the Niger occurred between 8500 and 7000 BP. During this early Holocene period, all the tropical lakes were high (Street & Grove, 1976; Talbot & Delibrias, 1977) and the discharge of the Senegal River was probably at a maximum (Michel, 1973). Water tables were high in the dune fields of the Sudan Qoz (Williams *et al.*, 1975) and probably also in the Sahel.

The Early Holocene fluvial activity in the Sahel was terminated by the onset of a brief period of aridity

(7500 BP). When humid conditions returned, rainfall was seasonal and braided streams of seasonal character prevailed during the Mid Holocene period (7500–4500 years BP). The climate was relatively wet and the annual rainfall may have been twice as great as at present in south central Mauritania. The drainage networks of the Sahel may have functioned seasonally, the larger rivers probably flowing throughout the year. In the Late Holocene, the decline in rainfall (4500–3500 BP) resulted in highly seasonal river regimes, and the present state of ephemeral rivers in the central Sahara has existed since about 2000 BP. Pastouret *et al.* (1978) reported that the large freshwater discharge of the Niger ended abruptly around 4000 years BP, but there was a humid recurrence in the Sahara between 3500 and 3000 BP, with lakes in the Tenere for example (Maley, 1981; Servant, 1983).

Riser & Petit Maire (1986) provided evidence for the existence of a broad interior delta of the Niger in Mali, up to 100 km into the Azawad, to the north-north-east of Timbuktu, between 8500 and 3500 BP. This delta was fed by effluents from the Niger as well as by local rainfall. A palaeolacustrine area was also irregularly linked with the Niger effluents as far as 70 km north of Arawan. It is suggested that the Arawan basin was already functioning during the humid phases of the Upper and Middle Pleistocene. *Lates maliensis* Gayet, 1983 was described from that area but is now considered to be a synonym of *Lates niloticus* (Van Neer, 1987).

The origin of the 'boucle du Niger' between Timbuktu and Bourem has been much discussed. It has been suggested that a river flowing south-west–north-east was captured by a river flowing north-west–south-east entering the Guinean gulf. According to Tricart (1965) and Blanck (1968), the capture was a Holocene event and contemporary with the humid phase of 10 000–7000 BP. Beaudet *et al.* (1977), on the contrary, claimed that the course of the Niger is older, at least from the Pleistocene, and data from Riser & Petit Maire (1986) tend to confirm this latter hypothesis.

Relations between the Niger and the Senegal–Gambia

Michel (1973) provided an extensive geomorphological study of the Senegal and Gambia river basins. During the Late Jurassic, an epeirogenic uplift gave rise to the central Fouta-Djalon while the western region subsided.

As a result, the Fouta-Djalon became a major watershed, the Upper Gambia and the Bafing flowing north-east, and the Koumba, Kakrima and Konkoure flowing west or south-west. The Bafing, Bakoy and Baoule possibly disappeared into a large endorheic basin (the Hodh) located north-east of Fouta-Djalon.

During the Cenomanian and Upper Eocene, new epeirogenic uplifts occurred in the upper parts of these basins. The Bafing and Bakoy joined, and gave rise to the Senegal flowing west–north-west, probably into a marine gulf in the region of Bakel. Tectonic movements took place probably during the Miocene and the occidental region emerged during the Pliocene. At that time, the Senegal River did not enter the sea near Saint Louis, but was probably flowing towards the north-west in the direction of Nouakchott. Other tectonic movements (probably after the Aïoujien transgression *c.* 120–100 000 BP) subsequently diverted the river westward into the deltaic region downstream of Bogue.

The last phases of this evolution were marked by the climatic changes that occurred in the second half of the Upper Pleistocene and in the Holocene. During the humid phase (40 000 BP), rivers deeply incised their beds and a highly hierarchic network was established in the headwaters. The following arid phase culminated between 20 000 and 15 000 BP, and large dune belts progressively obstructed the lower Senegal valley as well as the lower Gorgol valley. Thus, the Senegal basin was endorheic and possibly the Gorgol and Ferlo dried up. When precipitation increased (14 000–12 000 BP), the Senegal River cut its course progressively through the belt of dunes and again reached the ocean, while the Gorgol and Ferlo rejoined the Senegal. The large Hodh depression, covered by dunes, had many river systems, all flowing into the Senegal. It is assumed that the Baoule boucle is a capture of the Baoule (previously flowing to the Niger) by a small tributary of the Bakoy that intervened at the beginning of the humid phase. According to Michel (1973), the Bakoy, an upper tributary of the Senegal, is progressively eroding its basin, which is close to the Tinkisso an upper tributary of the Niger. In the future, the capture of the Tinkisso by the Bakoy is a possibility.

According to Roberts (1975), 'Nilo Sudan fishes presumably reached the Gambia from the Senegal by crossing the low-lying country in between their lower courses.

During the last interpluvial, the Senegal and Gambia rivers may have been greatly reduced or even ceased to flow, and the present relatively full complement of Nilo sudanic fishes may be largely or entirely the result of colonisation during the last pluvial'.

Relations between the Niger and the Volta

Two of the headwaters of the Volta basin captured north-east-flowing tributaries of the Niger, but the date of capture is unknown. One appears to be the headwater of the Black Volta in Burkina Faso, where the Sourou depression could be the relict valley of a river once flowing to the Niger now captured by a tributary of the Volta. This would explain the peculiar shape of the river at present. A similar phenomenon could have occurred with the Pendjari, a tributary of the Oti River. According to Roberts (1975), if the relict population of fishes described by Daget (1961b) from the eastern flank of the Bandiagara escarpment was stanched when the Black Volta captured its present north-east-flowing headwaters, then the capture probably occurred since the end of the last interpluvial (no more than 12 000 years ago). More probably however, the Bandiagara population was isolated directly from the present Niger system.

Relations between the Zaïre, Nile and Chad basins

The Zaïre system as a whole seems to have had a continuous existence since well before the Pleistocene (Beadle, 1981). The wide range of habitats, the environmental stability over a long period and the prolonged isolation of parts of the river system are conditions that may have favoured the evolution of its large endemic fish fauna.

The Zaïre basin covers over 4 000 000 km² and includes a flat central region, lying in one of the ancient cratons that was more than once invaded by the sea during the Mesozoic. The subsequent uplifting of the peripheral borders obstructed its drainage to the coast and a large lake formed during the Pliocene. In the Late Miocene–Early Pliocene, the Zaïre basin was an internal drainage system within which a large lake covered the 'cuvette centrale' (Cahen, 1954). Before the beginning of the Pleistocene, it was captured by a coastal stream that then became the Lower Zaïre River and drained east-

west to the Atlantic. The swampy area in the western half of the basin, including Lakes Tumba and Maindombe, could be regarded as remnants of this great lake (Beadle, 1981). The point of capture was probably just below Stanley Pool between Brazzaville and Kinshasa. Many of the rivers descending from the surrounding highlands are interrupted by falls and rapids which are barriers that have contributed to the relative isolation of specific fish assemblages, in preventing colonisation by other potentially competing species.

The fauna of the Upper Lualaba River is largely isolated from the rest of the Zaïre system by rapids. The fish include many endemics, some Zambezian species and several widespread species also reported from the Nilo–Sudan basins. Poll (1963), considering the presence of Sudanese relict fish species in the Lualaba River, suggested that the Upper Lualaba may once have been a tributary of the Nile system that was later blocked by upwarding and captured by the Zaïre River at the Portes d'Enfer, where it takes a turn to the north-west. But an alternative hypothesis is that these fish entered the Lualaba via Lake Tanganyika (Roberts, 1975). A reassessment of the supposed relict fauna by Banister & Bailey (1979) resulted in the reidentification of some species, and in new information about species distribution in the Zaïre basin. The conclusion is, therefore, that the Upper Lualaba fauna is not as distinct from that of the central Zaïre basin as previously thought. Geological investigations indicate that from well before the Miocene the drainage into the Zaïre basin from the east was via rivers rising east of the future Victoria basin, the Nile basin being confined to the north. These rivers were then disrupted by the rifting and consequent formation of the Rift Valley lakes.

Exchanges of fish fauna probably occurred between the Zaïre system and some coastal rivers of the Lower Guinea. A headwater of the Nyong, for instance, may have been captured by the Dja, a major tributary of the Sangha River (Zaïre basin). A recent influx of species from the Zaïre system into the Ogowe is reported by Roberts (1975). The exchange of fish species across the humid and moderately elevated watershed between the northern Zaïre basin and the Chad basin, via the upper reaches of the Oubangui River, has been also discussed. Examples may be *Tilapia zillii* and *Sarotherodon galilaeus* (Thys van den Audenaerde, 1963) and *Clarias albopunctatus* (Teugels, 1986). Exchanges might be possible during periods of heavy rains, but a reassessment of the fish species mentioned by Blache *et al.* (1964), and supposed to be recent immigrants from the Zaïre into the upper reaches of the Chari River, is certainly necessary before any further speculation (Lévêque *et al.*, 1991) is justified.

Rivers from the Tanzanian shield

The hydrology of the Tanzanian land (between the Athi and the Zambezi Rivers, to the east of the Rift Valley) has been drastically affected by geological events throughout the Tertiary and Quaternary: rivers have been captured or decapitated, and directions of flow reversed. Changes in sea level allowed confluences between rivers now isolated, shallow lakes developed and overflowed at various time in different directions, and there have been tremendous opportunities for fish to move from one river system to another (Banister & Clarke, 1980).

The Malagarasi River (now a tributary of Lake Tanganyika), and the Rungwa River (tributary of the endorheic Lake Rukwa) are assumed to be relict headwaters of the extended pre-rift Zaïre system (McConnell, 1972). The headwaters of the present Great Ruaha probably also drained into the Zaïre system (see Banister & Clarke, 1980, for a review). The history of the Rufiji and Pangani systems are fairly complex and strongly related to deformations of the land surface during Pleistocene rifting.

The Zambezi system

The past history of the Zambezian fish fauna is exceptionally difficult to unravel, and there have been speculative suggestions for routes by which the various elements of the fauna may have reached the Zambezi (Bell-Cross, 1972). It is likely that two major drainage basins occurred in Central Africa south of the Zaïre River (Bell-Cross, 1968; Bowmaker *et al.*, 1978). The first, the western basin, included the Upper Zambezi faunal region, that is to say the Upper Zambezi, the Upper Cunene, the Okavango complex, the Upper Kafue and the Bangweulu–Upper Lualaba subsystem. Prior to the Pleistocene, it is supposed that this drainage may have flowed westwards to join the Zaïre basin and was later reversed into the Zambezi. The Bangweulu–Upper Lualapa subsystem is now part of the Zaïre and

its capture probably occurred at the south end of the Bangweulu swamps. The Okavango River rises from watersheds shared with the west-flowing Angolan rivers Cuanza and Cunene, and flows into the arid Okavango basin. During the higher rainfall period in the Late Pleistocene, the Okavango River was a tributary of the Zambezi. Even now, the Okavango swamps overflow into the Zambezi via the Selida Spillway in seasons of heavy rains, which is a situation fairly similar to that observed between the Chad and Niger systems. Consequently, the fish fauna of the Okavango is very close to that of the Upper Zambezi and to that of the Kafue River which, though independently flowing into the Middle Zambezi, is partially isolated from it by the rapids in the Kafue Gorge (Beadle, 1981). Part of the discharge of the Okavango reaches Lake Ngami and, via the Botletle River, the Makgadikgadi depression. There is evidence that this depression at times experienced high lake levels. An earlier lake probably occupied some 60 000 km^2 between about 30 000 and 19 000 years ago and could have overflowed to the Chobe and Zambezi (Grove, 1983). During the last pluvial period, a lake with an area of about 35 000 km^2 may have also come into existence about 12 000 years BP.

The second, eastern basin, included the Middle and Lower Zambezi, the Kafue River, the Luangwa and the Lower Shire River. A direct link between the Zaïre and the Middle and Lower Zambezi occurred before the formation of the rift lakes. Following uplift in the late Cenozoic, the Lower Zambezi cut down strongly and the Victoria Falls, which mark the lower limit of the Upper Zambezi, came into existence. Victoria Falls constitute a major barrier to upstream movements of fish and the fish fauna of the Middle Zambezi is poorer than that of the upper basin. This was attributed by Jackson (1961) to the absence of calm waters in this section where fast-flowing waters dominate.

The drainage pattern of Southern Africa today can be traced back to the early post-Gondwana drainage of the continent. It has two main elements: large internal catchment basins with few outlets to the sea, surrounded by a perimeter of short coastal rivers rising on the escarpment divide (Skelton, 1994). The modern configuration of the Zambezi system and adjacent watersheds is the result of different development phases that can be summarised briefly as follows (Thomas & Shaw, 1988; Skelton, 1994).

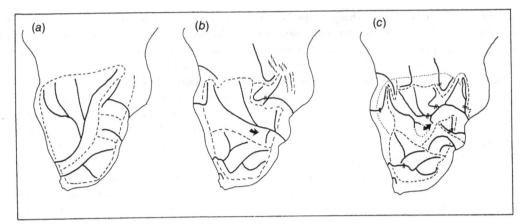

Fig. 6.6. Outline of drainage evolution in southern Africa: (*a*) Early Tertiary; (*b*) Mid Tertiary; (*c*) Late Tertiary. Heavy lines show main drainage lines, dashed lines show catchment divides, dotted line shows former extent of west Zambezi catchment. Arrows indicate major diversion of Okavango–Upper Zambezi drainage (from Skelton, 1994).

In the earliest phase, the western sector units were tributaries of a large southerly drainage, the so-called trans-Tswana system, reaching the sea via the proto-Lower Orange (Fig. 6.6*a*). The second phase involved the diversion of the western drainage from the south-west to the south-east, i.e. from the Orange to the Limpopo valley (Fig. 6.6*b*). The Upper Zambezi–Limpopo connections were operating by Mid Tertiary (Miocene) times and ended in the Pliocene as a result of tectonic uplift along the Kalahari–Zimbabwe axis. This tectonic movement may have caused some endorheism and damming of the Okavango and the Upper Zambezi to form a large shallow lake (Lake Magadigadi) before the Upper Zambezi was diverted to the Middle Zambezi.

The diversion and capture of the Upper Zambezi by the Middle Zambezi to form the modern configuration of the system started during the Miocene, when the proto-Middle Zambezi extended its course headwards (Fig. 6.6*c*). Renewed uplift in the Pliocene rejuvenated erosion and precipitated the capture of the Upper Zambezi by a headwater of the Middle Zambezi. Large physical barriers, such as the Victoria Falls on the Zambezi and the Chasunta Falls on the Kafue, have preserved the integrity of both upstream and downstream faunas of particular river sections during the process of drainage recombination. This would explain the present distribution of several species.

Southern headwaters of the Zaïre River system have tapped Zambezian headwaters along the divide which at present is remarkably flat in places so that, at times, extensive swamps flow between both the Zaïre and Zambezi, facilitating the two-way transfer of certain fish species. Capture of a headwater of the Zambezi by the Kasai is well documented (Bell-Cross, 1965–66*a*, *b*). A relatively early capture of the Kunene River by an advancing coastal tributary to form part of the west Zambezian drainage is also well documented.

Great Rift Valley lakes: the myth of stability?

The large isolated water masses of the East African Great Lakes, and their highly endemic fauna, may have contributed to the feeling that this fauna must have evolved over a long period of relative environmental stability. In fact these lakes have a complex history of falling and rising water levels, including dramatic shrinkage at some periods, even during the Holocene.

Lakes Malawi, Tanganyika and Victoria are nearly closed basins, with evaporation accounting for well over 90% of water loss; Lake Turkana is presently closed. The water budgets of these lakes are essentially balances between evaporation and rainfall on the catchment and lake surface, and levels reflect slight variations in this balance.

Evidence for fluctuations in the level of Lake Malawi (present maximum depth 770 m) has been presented by Scholz & Rosendahl (1988) and Owen et al. (1990). The lake level fluctuated greatly in the Late Pleistocene, and during the Late Quaternary there were several major low stands. Before about 25 000 years ago, the level was 250–300 m lower than today. Another major recession has been dated at 10 740 years BP, with further large falls between AD 1150 and 1250 and within the period 1500–1850. During this last, climatically controlled recession the lake fell by at least 121 m for part of the time. Such a fall implies the loss of the Shire River outflow, Lake Malombe, and the south-east and south-west arms of Lake Malawi.

Changes in level of Lake Tanganyika (present maximum depth 1470 m) were reviewed by Tiercelin & Mondeguer (1991). Seismic investigations provided evidence of a low lake stand above the present 600 m depth contour, which occurred at a minimum of 200 000 years BP. The lacustrine domain was therefore split into three discrete basins separated by interbasin ridges, and these isolated lakes were probably chemically, physically and biologically distinct. This situation could have resulted in the evolution of different fish populations. A second low lake level, above the present 400 m depth contour, is possibly dated to the Middle–Upper Pleistocene. During this time interval, the three paleolakes were connected. The general transgression that followed the −400 m lake level seems to have continued until 40 000 years BP, when the lake probably reached its present level, and would have been favoured by the onset of a cool and humid climate at this time in equatorial Africa. Then followed a progressive fall in the lake level down to the −150 m depth contour, estimated at around 35 000 years BP. The lake continued to fall until c. 22 000 years BP (Tiercelin et al., 1989), but certainly not below about the −300 m contour (Mondeguer et al., 1989). This environment apparently persisted until 15 000 years BP and levels from −200 to −300 m were recorded between 16 000 and 14 000 BP (Haberyan & Hecky, 1987). There are indications of a rising lake level after this time and from 12 000 years BP the climate became more humid and warmer.

Two stages have been recognised in the history of Lake Kivu. A stage of shallow waters that lasted between one and five million years (Degens et al., 1973) and resulted in the deposition of 400 to 500 m of sediment. Lake Edward, which formed in the Early Pleistocene, is assumed to have been in contact with Lake Kivu via the Rushuru River until the volcanic events of the Late Pleistocene (Beadle, 1981). The history of Lake Kivu, which resulted in its present form, started at about 13 700 years ago (Hecky & Degens, 1973). About 12 600 years ago, the water level started to rise and the lake was about 100 m deep. At 12 000 years BP the Virunga volcanic centre was active, and lava flows blocked the upper Rushuru River connecting Lake Kivu to Lake Edward (Degens et al., 1973). Maximum level of Lake Kivu was reached c. 10 000–9500 BP and the lake overflowed via the Ruzizi River into Lake Tanganyika. Many species probably disappeared from both Lakes Kivu and Edward (Beadle, 1981).

More catastrophic events occurred between 5500 and 5000 years BP due to increasing volcanic activity. Lake Kivu became strongly stratified (Haberyan & Hecky, 1987) and sudden overturn events, which introduced high concentrations of CH_4, H_2S and H^+ into the surface waters, probably resulted in sudden faunal extinctions. Under the influence of a drier climatic phase in East Africa, which commenced about 5000 years ago, Lake Kivu was closed between 3500 and 1400 years BP and Lake Tanganyika's level fell again to −75 m (Haberyan & Hecky, 1987). The surface waters became cooler and more saline, hydrothermal gases and heavy metals from the lower, anoxic depths gradually mixed with the surface water. A wetter and warmer climate followed this cool period, and about 1200 years BP the lake was reopened towards the south (Hecky & Degens, 1973). The absence of truly limnetic zooplankton in the pelagic zone of Lake Kivu (Dumont, 1986) supports the hypothesis of the destruction of the biotic system of the ancestral lake basin and its complete isolation since then. According to Haberyan & Hecky (1987), 'under modern conditions closure of Lake Kivu would also close Lake Tanganyika, as the Ruzizi discharge from Kivu is greater than the Lukuga discharge from Lake Tanganyika. The Lukuga outlet from Lake Tanganyika was blocked in 1854 when it was first seen by Europeans, and may have been closed (or nearly so) since the outflow from Kivu stopped at 3500 years BP. Loss of the Ruzizi discharge alone would lower the level of Tanganyika by 75 m'. In 1878 Lake Tanganyika reached +10 m above the present level, after which it overflowed. Flow have been continuous ever since.

The present-day ichthyofauna of Lake Kivu must, therefore, be considered as the remnant of a previously older lake fauna, with, theoretically, no possible recent immigration from any other major waterbody. The only chance of survival for part of the endemic lake fauna would have been rapid dispersion into surrounding stable riverine environments (Thys van den Audenaerde et al., 1982). But rivers do not offer suitable habitats for highly specialised lacustrine species, and probably only generalised lacustrine species could have survived a transition period in surrounding rivers and have constituted the inoculum for new recolonisations during stable phases of the lake.

If we consider the possibility of complete extinction of lacustrine species during the volcanic events, the present-day endemic fauna must have evolved within the lake from generalised riverine forms since the stabilisation of the lake some 1200 years ago. Is that realistic? (Losseau-Hoebeke, 1992). A more likely explanation is that the present fauna are remnants of a pre-existing fauna. Some generalised members of the ancestral stock may have been able to avoid the unfavourable lacustrine conditions by entering the surrounding rivers and the hypothesis of the survival of a relict fauna in an isolated bay has also been suggested (Losseau-Hoebeke, 1992). This bay would have been closed during the volcanic events and reopened when lake conditions returned to normal, a situation similar to that presently observed for Lake Nabugabo near Lake Victoria. The Lake Kivu flock would, therefore, be a relict and would not have evolved within the confines of the present phase of the lake.

The present-day Lake Victoria originated near the end of the Middle Pleistocene, and is believed to have been in contact with the Edward basin until the end of the Late Pleistocene (20 000–30 000 years BP). In the Early Miocene there is good evidence for the existence of an earlier lake named 'Lake Karunga' that occupied the Lake Victoria basin and whose fauna is represented by fossils in sediments of Miocene age from Rusinga Island. During the Pleistocene, the gradual uplift of land along the eastern shoulder of the Western Rift Valley (western side of the present lake basin) resulted, in the Mid Pleis-

tocene, in the disruption of drainage patterns and the reversal of previously west-flowing rivers such as the Kagera and Katonga. These rivers formed swampy lakes similar to the present Lake Kioga, but further flooding and continuous uplift caused these lakes to coalesce and this is the presumed origin of Lake Victoria. This lake probably overflowed westward into the Lake Edward basin through the Katonga valley well into the Late Pleistocene (less than 60 000 years ago). The connection was interrupted 30 000 years ago by the continued uplift and Lake Victoria, which was at a very high level (over 30 m above the present level), then overflowed to the north toward Lake Kioga and the Kabalega Falls, as it does at present. Sediment cores from Lake Victoria (Kendall, 1969; Livingstone, 1980) have suggested that this lake also exhibited large fluctuations in its water level. About 15 000 years ago, the lake was clearly much lower than now, and probably depressed by at least 75 m. It may have been reduced to a rather saline remnant with no outlet. According to Scholz et al. (1991), a hiatus in the sedimentation reveals a lake-wide desiccation at or about 14 000 years BP. It is more likely, however, that small ponds around the basin margin persisted during this lowstand and provided refuge for some fish species. The fragmentation of the major lake into smaller waterbodies may have favoured speciation, and 14 000 years is considered by McCune et al. (1984) as an adequate time for the speciation of Lake Victoria's endemic species. Between 12 000 and 10 000 BP the lake possibly rose to a much higher level (18 m above the present level).

Hypothetical connections between Tanganyika and Malawi

Lakes Malawi and Tanganyika are long, deep, ancient rift basins containing numerous endemic species. Evidence for possible former aquatic routes between these 'sister lakes' has been investigated by Banister & Clarke (1980) and widely discussed by Coulter (1991a). If any hydrological connections occurred between these lakes, it must have been via Lake Rukwa. There are indications of an overflow level of Lake Rukwa into Lake Tanganyika, and apparent zoogeographical links between the two lakes were described by Grove (1983), who suggested that lacustrine deposits about 150 m above the present level could be the mark of the last overflow to Tanganyika, some 33 000 years BP. Lower shell terraces 60 m and 50 m above the present lake level are dated at 10 000–8000 years BP and 6340 years BP respectively, but these results, as well as supposed fossil evidence, have been questioned given the tectonic instability of the region. As a result, while it is readily conceivable that Lake Rukwa at its maximum level has overflowed into Lake Tanganyika, perhaps many times, there is no definitive proof of biological links between them. As to a former connection between Lakes Rukwa and Malawi, the possibility of a single protolake as suggested by Banister & Clarke (1980) has also been debated in the light of new results. 'One must conclude there is still insufficient evidence as to whether or not faunal exchange took place within a united palaeo Tanganyika–Rukwa–Malawi. Exchange between Tanganyika and Rukwa was not unlikely, especially in the Pleistocene during large climatic fluctuations, depending on whether or not high lake stands occurred concurrently in both lakes' (Coulter, 1991a).

Hypothetical connections between Tanganyika and the Nile

Hypothetical connections with the ancient Nile system have also been discussed in detail and a review is given by Coulter (1991a). There are in fact analogies between the Tanganyika and Nile ichthyofaunas: the majority of fish families are common to the two provinces, as well as many genera. Different routes have been postulated by which Nilotic species may have reached Lake Tanganyika, but 'there is no reason to believe from the evidence available that the Tanganyika fish fauna has been influenced by direct connections between the lake and the Nile. A few species may have immigrated by passing over the intervening divide with Lake Victoria. It is more likely rather, that affinities at generic and higher taxonomic levels originate from a time before the formation of the lake when widespread ancient stocks were shared by both regions. Evidence from the palaeontology and modern distributions of some molluscs supports this conclusion' (Coulter, 1991a).

Faunal similarities between Tanganyika and Zaïre

There are strong resemblances in the taxonomic composition of the ichthyofauna of Lake Tanganyika and the Zaïre River basin, even though species common to both are not numerous (Roberts, 1975). Seven of the ten fam-

ilies that have contributed endemics to Tanganyika are also represented by endemics in the lower Zaïre River (Roberts & Stewart, 1976). There is now good evidence that when Lake Tanganyika was formed, the protolake region crossed the headwaters of the Zaïre transversally, and that the Malagarasi River, on the eastern side of Lake Tanganyika, is a vestige of the pre-rift Zaïre drainage. All the fish families now present in the tributaries of Tanganyika were probably represented in a swamp or river-dwelling paleofauna of the Zaïrean region. Once the lake was formed, probably only a few of these families adopted habitats in the lake itself where they would be expected to have diverged in isolation.

Biogeographical implications of past geological and climatic events

The biogeography of African freshwater fish is still poorly known compared with that of European or North American fish, which may be explained by the considerable taxonomic and ecological complexity of the ichthyofauna, but also by the poor fossil record. Moreover, little attention has been given to the study of the African fish fauna, and the phylogeny of many fish families has still to be worked out. However, the improvement in knowledge during recent decades allows some tentative biogeographical scenarios to be proposed.

At the continental level

Some fish taxa are widely distributed over the whole of tropical Africa. Good examples are the catfish *Malapterurus electricus*, *Schilbe intermedius*, *Clarias gariepinus* and *Heterobranchus longifilis* or the characins *Hydrocynus vittatus*, *Brycinus imberi*, *Brycinus macrolepidotus* and *Hepsetus odoe*, and the cichlids *Tilapia zillii* and *Sarotherodon galilaeus*. Many genera that seem to be holophyletic taxa, also have a similarly wide distribution. 'This almost pan-African distribution of certain taxa seems strongly to suggest that at some stage in the past the waterways of Africa were from the fishes point of view, accessibly interconnected. Such a situation would permit the wide dispersal of the fauna ancestral to that which we study today. The break-up of that early hydrographic pattern, the consequent isolation of biotas, and their differentiation (taxonomically and phylogenetically) in that isolation, are all elements of a classical vicariance pattern in historical biogeography (Greenwood, 1983a).

The investigations of the neoboline group by Howes (1984) is one of the well-documented studies on phylogeny and distribution of freshwater fish in Africa. The most notable feature of the present-day neoboline distribution is the geographical division of the genera by the Eastern Rift system. There is a broadly branching sequence in the cladogram of neoboline taxa, demonstrating a sister-group relationship between Nilotic and Zaïrean areas on the one hand (genus *Chelaethiops*), and East Coast and Zambezian on the other hand (genera *Neobola, Mesobola, Rastrineobola*). A similar pattern is observed for the sister-group of the neobolines, *Leptocypris–Engraulicypris*. At the species level, sister-groups are observed for *Chelaethiops* and *Leptocypris*, one being Zaïrean and Guinean, the other Nilo–Sudanian. This latter pattern was also observed for Distichodontidae by Vari (1979).

Biogeographical scenario in northern tropical Africa

The current distribution of freshwater fish species in western Africa is obviously the result of different process acting at different time scales: many millions of years for geological events to a few thousand years for recent climatic changes, centuries and even shorter periods when human impact (introductions, destruction) is taken into consideration. At present, most of West Africa still consists of flat sedimentary basins and upland plains ranging from 150 to 600 m above sea level (a.s.l.), with disjunct patches of highlands above 1000 m. Among the most important of these, the Fouta-Djalon and the Guinea range, along the western coast, separate those northward-flowing rivers that are actually the sources of the Niger, Senegal and Gambia Rivers, from the rivers of the Atlantic side (Guinea, Sierra Leone and Liberia) that constitute the Upper Guinea province, and are relatively short and partly torrential in their upper courses.

Typology of north tropical African fish zoogeography
Fish distribution in northern tropical Africa has now been well investigated (for review see Teugels *et al.*, 1988, 1992c; Lévêque *et al.*, 1989, 1990, 1991, 1992; Paugy *et al.*, 1989). From the list of fish species recorded

from the different river basins, it is possible to recognise some general patterns of distribution that are illustrated in Figs. 6.7 and 6.8. Such a typology, which is based on the current knowledge of fish distribution (which may of course improve in the future), is the result of past relationships between river basins, as well as of extinctions that have occurred within river systems.

Type 1 Species recorded from the Nile to West Africa, and, at least, to the Zaïre southwards. This includes many species with wide distribution in the continent, such as *Polypterus bichir, Polypterus lapradei, Xenomystus nigri, Brycinus macrolepidotus, Hydrocynus forskalii, Hydrocynus vittatus, Ichthyborus besse, Clarias gariepinus, Heterobranchus longifilis, Schilbe intermedius, Auchenoglanis occidentalis, Lates niloticus, Sarotherodon galilaeus* and *Tilapia zillii*, which are examples of species whose original distributions fit with type 1. Many of these species have a distribution range that extends far beyond the Zaïre basin into the southern hemisphere.

Type 2 Species recorded from the Nile to West Africa and not recorded southwards. They includes a number of species, such as *Gymnarchus niloticus, Brienomyrus niger, Hyperopisus bebe, Mormyrus hasselquistii, Alestes baremoze, Alestes dentex, Brycinus nurse, Distichodus engycephalus, Distichodus brevipinnis, Citharinus citharus, Citharinus latus, Barbus bynni, Chelaethiops bibie, Labeo coubie, Raiamas senegalensis, Chrysichthys auratus, Schilbe intermedius, Brachysynodontis batensoda, Hemisynodontis membranaceus, Synodontis clarias, Synodontis filamentosus, Hemichromis bimaculatus, Oreochromis niloticus, Oreochromis aureus* and *Tetraodon lineatus.*

Type 3 Species recorded from the Chad basin to West Africa and recorded also in the Zaïre, but not in the Nile. This group includes *Campylomormyrus tamandua, Gnathonemus petersii, Mormyrus rume, Hepsetus odoe, Labeo parvus, Clarias albopunctatus, Clarias camerunensis, Ctenopoma kingsleyae,* etc.

Type 4 Species recorded from the Chad basin to

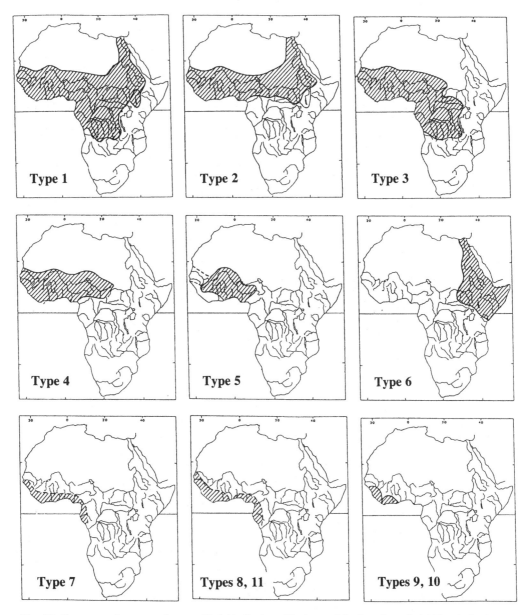

Fig. 6.7. Some general patterns of geographical distribution of freshwater fish of north tropical Africa. 1, Species recorded from the Nile to West Africa, and to the Zaïre southwards; 2, species recorded only from the Nile to West Africa and not recorded southwards; 3, species recorded from the Chad basin to West Africa and recorded also in the Zaïre; 4, species recorded from Chad basin to West Africa and not in the Zaïre; 5, species recorded in western Nilo–Sudan basins, but not from Chad and the Nile; 6, species only recorded from north-eastern Africa (the Nile and Omo basins and coastal rivers); 7, species recorded from Senegal to Zaïre but absent from Chad and the Nile; 8, species recorded from the Zaïre estuary or Lower Guinea to the Volta; 9, species only recorded from the Eburneo–Ghanean province; 10, species only recorded from the Upper Guinea province; 11, species recorded from Senegal to the Volta.

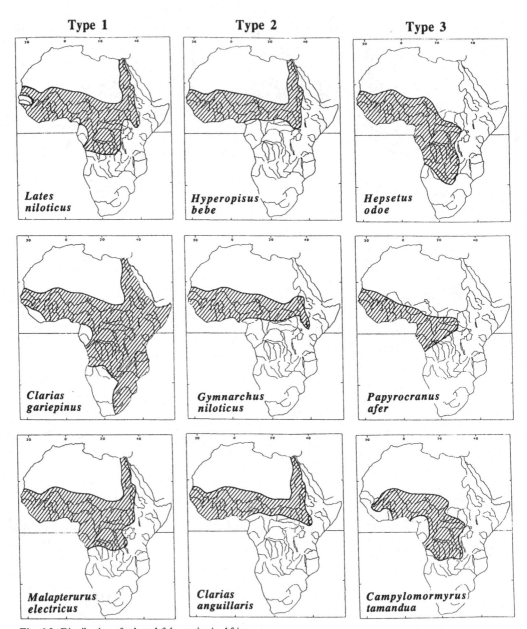

Fig. 6.8. Distribution of selected fish species in Africa.

West Africa but not in the Zaïre or in the Nile. This group includes a number of species, such as *Marcusenius senegalensis, Nannocharax ansorgii, Nannocharax fasciatus, Paradistichodus dimidiatus, Citharinops distichodoides, Barbus baudoni, Barbus callipterus, Barbus leonensis, Barbus macinensis, Barbus macrops,* *Labeo senegalensis, Heterobranchus isopterus, Synodontis courteti, Synodontis violaceus, Tylochromis sudanensis, Tilapia dageti,* etc.

Type 5 Species recorded in western Nilo–Sudan drainages, but not from Chad and the Nile. This group includes a small number of species including *Mormyrus macrophthalmus, Marcusen-*

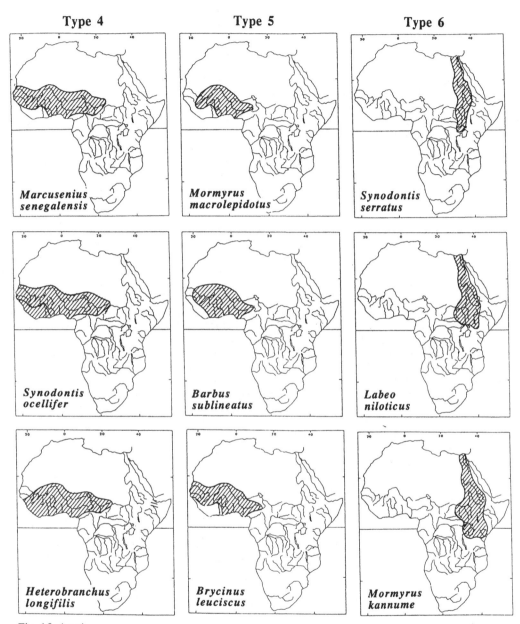

Type 4

Type 5

Type 6

Marcusenius senegalensis

Mormyrus macrolepidotus

Synodontis serratus

Synodontis ocellifer

Barbus sublineatus

Labeo niloticus

Heterobranchus longifilis

Brycinus leuciscus

Mormyrus kannume

Fig. 6.8. (*cont.*)

ius abadii, Brycinus leuciscus, Barbus sublineatus, Barbus pobeguini, Synodontis ocellifer and *Chiloglanis voltae.*

Type 6 Species only recorded from north-eastern Africa (the Nile and Omo basins and coastal rivers). Ten species are endemic to this area (Lévêque *et al.*, 1991).

Type 7 Species whose distribution is restricted to the western coast: species recorded in coastal river basins, both in the Upper and the Lower Guinean provinces, with some records in the Eburneo–Guinean area and in the Niger Delta, as is the case for *Isichthys henryi, Brienomyrus brachyistius, Brienomyrus longianalis,*

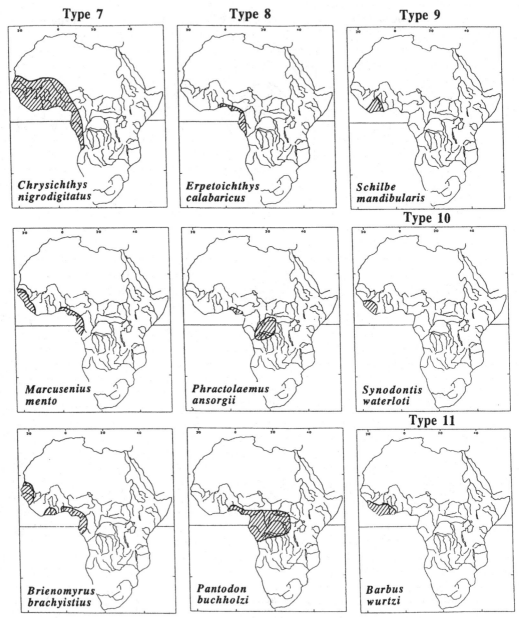

Fig. 6.8. (*cont.*)

Marcusenius mento, Mormyrus tapirus, Mormyrops caballus and *Brycinus longipinnis*; species or groups of species recorded from the Zaïre basin to the Senegal. Brackish-water species are restricted to the coast, but others may be encountered in the river systems far inland.

Typical representatives are *Papyrocranus afer, Chrysichthys nigrodigitatus, Pellonula leonensis* and *Sarotherodon melanotheron.*

Type 8 Species recorded along the coast from the Zaïre basin or Lower Guinean province, up to the Dahomey Gap, east to the Volta River, as is the case for *Calamoichthys calabaricus*, *Phractolaemus ansorgii*, *Clarias buthupogon* and *Pantodon buchholzi*. Some species (e.g. *Phago loricatus*) are only recorded between the Cross and Volta Rivers.

Type 9 Species only recorded from the Eburneo–Ghanean province: *Marcusenius ussheri*, *Marcusenius furcidens*, *Citharinus eburneensis*, *Barbus trispilos* and *Schilbe mandibularis*.

Type 10 Species endemic to the Upper Guinea province. Some of them may occur in the upper course of some Nilo–Sudan rivers but did not spread all over the basin: *Barbus petitjeani*, *Barbus salessei*, *Barbus foutensis*, *Amphilius rheophilus* and various Cyprinodontids.

Type 11 Species recorded from Senegal to the Volta: *Chrysichthys maurus*, *Barbus wurtzi*, *Barbus* species related to the *B. trispilos* group.

Tentative biogeographical scenario

The above typology shows clearly that two groups of species occur in West Africa: a group of Nilo–Sudan fish colonising most of the Nilo–Sudan rivers up to the headwaters; and a group of species with relatives in Lower Guinea or Zaïre that are mainly confined to the coastal areas in West Africa. This typology can be used to understand the biogeography of north tropical African fish and the following scenario can be tentatively suggested to explain the existence of ichthyoprovinces, and the presently observed faunistic relationships.

A large proportion of the fish species inhabiting the major Nilo–Sudan basins belongs to types 2 and 4 (Table 6.2), that is to say fish present all over the Nilo–Sudan province but absent from the Zaïre drainage. A significant number of type 1 species (widespread species) are also present in all these rivers. However, for the Niger and Oueme Rivers, and to a lesser extent for the Volta River, the fish fauna with Zaïrean–Lower Guinean affinities (types 7 and 8) make a significant contribution to the total species richness of the drainage. For the northern Nilo–Sudan basins, there is also a small proportion of type 10 species whose distribution is restricted to the Guinean range.

During the Miocene, 25 million years ago, faunal barriers between the water systems of Africa were easier to cross than today, and there was a rather uniform widespread fish fauna all over tropical and subtropical Africa.

The climate was probably humid and, during the early Miocene, forests occurred from the Gulf of Guinea to the south of Tethys. There is no fossil evidence from before the end of the Miocene for the existence of any water system that was isolated for a long enough time to produce a large endemic fauna (Beadle, 1981). Types 1 and 3 fish species are probably the survivors of this widespread ancestral fauna that already existed before the isolation (during the Pliocene) of the Zaïre basin (and possibly the Nile basin) from the Chad basin. Types 2 and 4 should conversely be considered as a typical Nilo–Sudan fish fauna which evolved since that province was isolated from southern river basins. Types 3 and 4 may result from the extinction in the Nile drainage of species belonging to the Nilo–Sudan pool, possibly following severe drought periods that eliminated part of the indigenous fish fauna. An alternative hypothesis would be that type 3 species are more recent colonisers from the Zaïre province that entered the Chad basin and then spread across West Africa via the Gauthiot Falls and the Niger. In that case, their absence from the Nile may be explained by the absence of connections between the Nile and the Chad basins after the colonisation of the latter. Types 5 and 6 are a pool of species that evolved independently, either in the Nile or in the western Nilo–Sudan basins, and are now endemic to these areas. For type 5 it may also be possible that the species disappeared from the Chad basin, while the Gauthiot Falls are presently a barrier preventing fish from the Niger colonising, or recolonising, the Chad. The absence of endemism in the Chad basin could also be partly explained by the overflow from Chad to the Niger through the Mayo Kebi and Benue Rivers.

If the Niger, and eventually the River Nile, were possibly refuge zones for the Nilo–Sudan fauna on account of their high number of endemic species (Table 6.2), it is difficult to know if there were one or more aquatic refuges in each basin, and where they were located. Nevertheless, in West Africa, the River Niger (originating close to the Upper Guinea and ending not far from the Lower Guinea) may have played a major role as a centre of origin for many Nilo–Sudan species that colonised the other coastal basins in more recent times. In fact, the Senegal, Gambia, Geba, Oueme, Mono, and Ogun have almost no endemic species. It is likely that these rivers dried up, at least partly, during

Table 6.2. *Relative composition of the fish fauna of the major Nilo–Sudan drainages in relation to the different types of distribution identified for the north tropical African fauna*

River	End.	Type 1	Type 2	Type 3	Type 4	Type 5	Type 6	Type 7	Type 8	Type 10	Type 11
Nile	27	24	63	0	0		10				
Chad	1?	24	61	12	23						
Niger	23	22	56	10	23	14		25	35	10	2
Senegal	2	17	42	5	15	7		12		15	2
Gambia		13	29	5	12	4		12		12	4
Geba		9	16	4	7			10		10	5
Volta	10	20	49	7	17	11		8	5		3
Mono											
Oueme		20	23	7	10	2		15	26		
Cross											

See Fig. 6.7 for explanation. End., number of species strictly endemic to the drainage system.

the later dry periods and were colonised again by Nilo–Sudan fishes through direct or indirect connections with the Niger system. The colonisation of the Eburneo–Ghanean subprovince by Nilo–Sudan species is probably more complex and older, while species hypothetically belonging to that pool had time to evolve and to differentiate, i.e. *Barbus bynni waldroni*, *Citharinus eburneensis* and *Schilbe mandibularis*.

The highly endemic fish fauna of the Upper Guinea province (type 10 distribution) (Lévêque *et al.*, 1989, 1991, 1992) has probably evolved for a long time in this isolated zone where wet conditions have prevailed more or less permanently (refuge zone). Moreover, the Fouta-Djalon is a massif formed by a series of compact plateaux subdivided into blocks delimited by abrupt cliffs. The existence of numerous waterfalls delimiting endemic areas and preventing recolonisation by migrant fish has also favoured vicariant speciation and may explain this high endemicity. The central Fouta-Djalon is an old massif that already existed during the Late Jurassic but was elevated during Miocene tectonic movements. The Guinean range acted as barrier to dispersal preventing exchanges between Sudanese rivers flowing northwards, and the coastal rivers of the Upper Guinea province flowing westwards.

A more or less similar situation probably prevailed in the Lower Guinea province. However, here the connections with the Zaïre system certainly occurred more frequently through river captures or heavy floods. The

Zaïre system has the greatest number of fish species and endemics of any African drainage (see Chapter 2). In addition to its great size, it is likely that the Zaïre River presents a greater number and variety of habitats for aquatic organisms than any of the other basins, and it has remained intact at least from the Miocene, in spite of climatic changes (Beadle, 1981). It is probable that a large Pliocene lake in the central Zaïre basin also contributed to the evolution of the present varied fauna.

Much more interesting is the distribution of fish species that appear to be of Zaïrean and/or Lower Guinean origin, along the West African coastal basins (types 7 and 8). The patchy distribution of many species raises the question of the timing and the dynamics of species migration northwards. At some time in the past, there was probably a more or less continuous distribution of Guinean fishes all along the Atlantic coast, from the Zaïre estuary to Guinea, corresponding to a continuous fringe of humid forest. We know, for example, that during the Early and Middle Holocene (from 9000 to 4000 BP), the present-day interruption of the rain forest in Togo and Benin known as the Dahomey Gap was probably absent (Maley, 1991), and the rain forest expanded greatly outside its present-day limits (Maley, 1989). At least there was northern invasion of a fauna, originating from the Zaïre or the Lower Guinean provinces, that may have occurred in successive waves. The observed similarities between the Upper Guinean ichthyofauna and the Lower Guinean and Zaïrean ichthyofaunas, as well as the presence of common taxa in Lower and Upper Guinea (type 7), help to support this idea of a colonisation from the south. The existence of closely related taxa, shared exclusively by the Upper and Lower Guinea, also supports the existence of faunistic relationships between these provinces. The genus *Afronandus* in Upper Guinea and the genus *Polycentropsis* in Lower Guinea are the only genera of the family Nandidae in Africa. The species *Ichthyborus quadrilineatus* from Upper Guinea is close to *Ichthyborus ornatus* from Zaïre as is the slightly modified *Ichthyborus monodi* from Lower Guinea (Lévêque & Bigorne, 1987). The colonisation of western African coastal basins by fish of supposed Zaïrean origin is also illustrated by type 7, which correspond to species widespread in the Zaïre basin but also occurring far inland in the river basins entering the Atlantic Ocean. Brackish-water species, such as *Sarotherodon melanotheron*, have a continuous distribution from the Zaïre to the Senegal. Migration and colonisation of primary freshwater fish along the west coast was probably made possible through connections between river basins at the time of heavy rains that allowed fish migration either through inundated areas, or through coastal lagoons during high floods that pushed away saline or brackish waters. Migration northwards also possibly occurred during periods of fluctuating sea levels.

To understand the causes of the patchy distribution of many species, it is necessary to know what has been the succession of climatic events in the coastal area of West Africa. During periods of drought, and certainly during the last dry period (18 000 years BP), the rainforest belt was divided and greatly reduced in western Africa. It disappeared in Ghana but persisted in West Cameroon, Gaboon and on the Guinean range (Maley, 1987, 1989). Fish and rain forests are both dependant upon a water supply, and where it is depleted the extensions of the rain forest regress, as do the freshwater habitats, and in an ultimate regression phase both disappear. One can expect, therefore, that what happened to the rain forest also happened to the freshwater fish fauna. During the dry periods, the current Upper and Lower Guinea ichthyological provinces were refuge zones where wet climatic conditions prevailed. There was also a possible, but less important, forest refuge zone in south-east Côte d'Ivoire–south-western Ghana (Hamilton, 1982). In between the Upper and Lower Guinea refuge zones, the

Zaïrean–Guinean fish fauna partly disappeared from areas where fish encountered harsh environmental conditions. However, local populations of a few species may have continued in a few spots where the environment permitted and that may explain the scattered distribution of species sometimes observed along the western coast (distribution type 8). The genus *Calamoichthys* for instance is confined to Lower Guinea. *Phractolaemus*, absent from the west side of the Dahomey Gap, occurs in Zaïre province.

The history of the colonisation of the western coastal habitats by the Zaïrean fauna was also probably complicated by the disruptions of the rain forest by southwards expansion of the savannah. The present-day Dahomey Gap is an illustration of a recent disruption (after *c.* 4000 BP according to Maley, 1987, 1989) of the forest belt by the savannah expanding far south to the coast in Benin, and also in Togo and Ghana. Its role as an ecological barrier for fish has probably been overemphasised (Howes & Teugels, 1989), but it marks a discontinuity in the distribution of flora and fauna and is often cited as an example in support of the forest refugia theory (see Maley, 1987, for a review). This situation is also encountered to a lesser extent in the Baoule-V Gap in Côte d'Ivoire, which is a disruption of the tropical rain forest in the Comoe–Bandama–Sassandra region, where it is replaced by savannah. On both sides of the Baoule-V Gap (rivers Cavally and Cess to the west and rivers Pra and Tano to the east) there are common shared taxa, such as *Limbochromis cavalliensis* and *Limbochromis robertsi* (see Greenwood, 1987), *Chrysichthys johnelsi* (see Risch, 1986) and *Clarias laeviceps* (see Teugels, 1986), that are absent from the area in between.

The close relationship of the Nilo–Sudan and the Eburneo–Ghanean faunas (type 9) probably results from the absence of physical barriers between the Eburneo–Ghanean area and the surrounding Niger or Volta drainages. The Nilo–Sudan fauna was therefore in a position to expand further south during wet periods, mixing with the relict fauna surviving in the neighbouring zones, including species moving from the Upper Guinea refuge area and expanding into the western part of Côte d'Ivoire. An alternative biogeographical hypothesis would be that the Eburneo–Ghanean area was a refuge zone for the Nilo–Sudan fauna and not for the Guinean one, but it is unlikely that this was the case because of

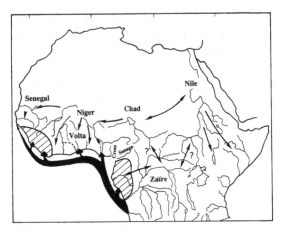

Fig. 6.9. Possible post-Miocene biogeographic scenario for fish assemblages of north tropical Africa.

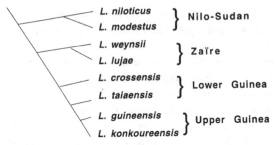

Fig. 6.10. Cladogram of *Leptocypris* relationships (from Howes & Teugels, 1989).

the high number of Nilo–Sudan endemics in the River Niger.

The biogeographical scenario suggested here (Fig. 6.9) is certainly the result of many recession–expansion cycles, each of different length and intensity. Successive invasions probably took place, and chance may also have played a role in the maintenance or disappearance of species. At the present stage of our knowledge, and in the absence of fish remains, it is not possible to propose a better chronology of events. Moreover, there were probably differences between species in relation with their ecology and their ability to survive or to disperse. New techniques and new approaches will probably provide useful information in the future. For example, a genetic study of the catfish, *Chrysichthys maurus*, across West Africa (Agnèse, 1989) supports the idea of an Eburneo–Ghanean refuge zone. This species evolved in that area and migrated later to the Upper Guinea, using the coastal connections.

The phylogenetic relationships of the cyprinids also support the hypothesis that the ichthyofaunas of Upper and Lower Guinea are closely related, and that both are derived from a primitive ichthyofauna shared also with the Zaïrean province. In *Leptocypris* for instance, the phylogeny indicates a basal dichotomy between Nilo-Sudan and Zaïrean–Guinean areas, followed by another dichotomy between Zaïrean and Guinean regions with

successive dichotomies within the Upper Guinean province (Howes & Teugels, 1989) (Fig. 6.10).

It is also be of interest to note that there are two major lineages of *Polypterus* (Daget & Desoutter, 1983), and within each lineage there are repeated dichotomies between Nilo–Sudan and Zaïrean–Guinean species. For instance, *P. palmas* from Upper Guinea has as sister-species *P. retropinnis* from Zaïre. The same relationship occurs between *P. ansorgei* and the Zaïrean species *P. delhezi* and *P. ornatipinnis* (see Chapter 5).

A biogeographic model for the Zambezian fish fauna

The traditional view considers that inhabitants of the freshwaters of Southern Africa are comparatively recent immigrants from the north. This dispersalist view states that groups such as cyprinids and catfish entered north-east Africa from Asia and 'worked their way southwards down the east side of the continent' (Lowe-McConnell, 1987).

Skelton (1994) presented a biogeographic model for the Zambezian fish fauna based on a reconstructed history of hydrographic development. This model proposes that the fauna developed in two main arenas (western and eastern) and that drainage development within these arenas resulted in the modern-day faunal composition.

As for northern tropical Africa (see above), the model is based on different categories of distribution patterns in the Zambezi province (Fig. 6.11)

1. Non-endemic species with widespread distributions throughout the province, such as *Marcusenius macro-lepidotus*, *Barbus paludinosus*, *Barbus radiatus*, *Labeo cylindricus*, *Schilbe intermedius*, *Clarias gariepinus*,

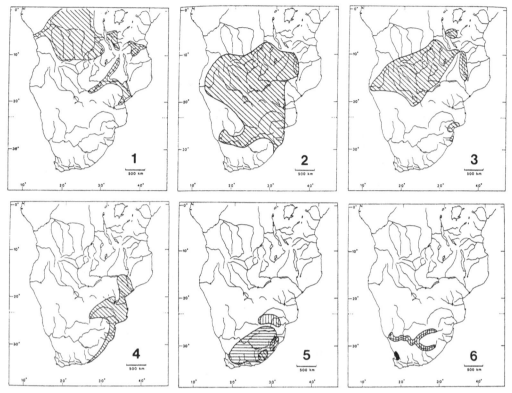

Fig. 6.11. Examples of distribution of Zambezi (1,2,3,4) and Southern (5,6) provinces primary freshwater species. 1, *Heterobranchus longifilis*; 2, *Tilapia sparrmanii*; 3, *Serranochromis* spp.; 4, *Oreochromis mossambicus*; 5, *Barbus anoplus* group; 6, *Austroglanis* spp. (from Skelton, 1994).

Tilapia sparrmanii, etc., are generally characteristic of the rivers of 'High Africa'.

2. Endemic species confined to western-sector drainages: *Kneria maydelli, Barbus barnardi, Labeo lunatus, Parauchenoglanis ngamensis, Clariallabes platyprosopos, Synodontis leopardinus, Nothobranchius kafuensis, Serranochromis carlottae, Serranochromis angusticeps, Haplochromis albolabris*, etc. The large number of species of different families clearly suggests that the region has been a strong evolutionary centre over time.

3. Non-endemic species confined to western-sector drainages: *Mormyrus lacerda, Hepsetus odoe, Kneria angolensis, Barbus afrovernayi, Barbus fasciolatus, Coptostomobarbus wittei, Schilbe yangambianus, Clarias stappersii, Serranochromis angusticeps, Tilapia ruweti*, etc. These species are confined to rivers or

parts of rivers that formerly were part of the 'greater western Zambezi' system and were diverted later by river capture.

4. Endemic species confined to eastern-sector drainages: *Parakneria mossambica, Barbus marequensis, Labeo molybdinus, Varicorhinus nelspruitensis, Distichodus mossambnicus, Chiloglanis emarginatus, Synodontis nebulosus, Chetia flaviventris, Serranochromis meridianus, Oreochromis mortimeri*, etc. The regional speciation of typical groups such as *Nothobranchius* provides support for a long period of isolation between the eastern and western sectors.

5. Non-endemic species confined to the eastern-sector drainages: *Protopterus annectens, Mormyrops anguilloides, Brycinus imberi, Malapterurus electricus, Synodontis zambezensis, Oreochromis mossambicus*, etc. These species are distributed further north in the

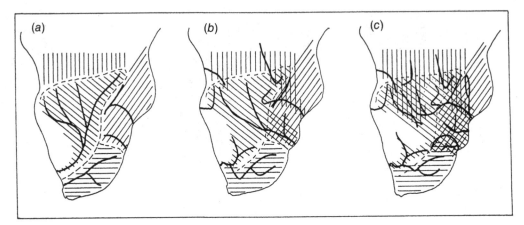

Fig. 6.12. Biogeographic model for Zambezian freshwater primary division fish. (a) Early Tertiary; (b) Mid Tertiary; (c) Late Tertiary. In the Early Tertiary the freshwater fish faunas of southern Africa became established in four evolutionary arenas: (1) the southern arena (horizontal hatching), (2) the western arena (left slanting hatching), (3) eastern arena (right slanting hatching) (4) Zaïrean arena (vertical hatching). By the Mid Tertiary there were infusions of western and Zaïrean fauna into the eastern arena, and marginal infusions of Zaïrean and southern fauna into the western arena. By the Late Tertiary there was a second infusion of western fauna into the eastern arena and further infusion of Zaïrean fauna into the western arena. Minor faunal infusions occur along contact zones of the faunal arenas, e.g. western/eastern elements into the southern arena (from Skelton, 1994).

East Coast province or through the Lualaba–Zaïre system.

6. Western-sector species with sporadic eastern-sector occurrence which are valuable indicators of historical drainage connections: *Hippopotamyrus ansorgii*, *Barbus bifrenatus*, *Barbus haasianus*, *Barbus kerstenii*, *Clarias ngamensis*, *Clarias theodorae*, *Ctenopoma intermedium*, etc.

This biogeographic model proposed by Skelton (1994) is that these ichthyofauna groupings have evolved in accordance with three main phases of drainage evolution in the Zambezi and Cape Provinces (Fig. 6.12). In Stage I (Fig. 6.12a), separate western (interior) and eastern (coastal) faunas evolved and became established. In Stage II (Fig. 6.12b), the western fauna underwent a par-

tial fragmentation and extended into the south-eastern sector via the Okavango–Upper Zambezi–Limpopo drainage gateway. A separate Lualaba–Middle/Lower Zambezi connection facilitate the infusion of Zaïrean elements into the proto Middle/Lower Zambezi. In Stage III (Fig. 6.12c), a new north-eastern extension was established through the capture of the Upper Zambezi by the Middle Zambezi in the Plio–Pleistocene, and there was a continued fragmentation of the original western arena through river captures and rifting. While there were several intrusions of the western fauna into the eastern arena via the Okavango–Upper Zambezi, the integrity of the western faunal was often shielded from intrusions from the east by means of major waterfalls acting as physical barriers. The northern section of the eastern fauna has also received an infusion of Zaïrean elements.

III The diverse lifestyles of African freshwater fish

7 Diversity of growth and feeding behaviours

In any study of evolutionary ecology, food relations appear as one of the most important aspects of the system of animate nature. There is quite obviously much more to living communities than the raw dictum 'eat or be eaten', but in order to understand the higher intricacies of any ecological system, it is most easy to start from this crudely simple point of view.

Hutchinson, 1959

From an evolutionary perspective 'growth constructs the framework and metabolic machinery necessary to synthesise and to protect the gametes until their release. The process of natural selection will lead to the evolution of patterns of growth that tend to maximise the lifetime production of offspring. Growth and reproduction are complementary processes but both depend on the limited resources of energy and nutrients made available by the foraging behaviour of the fish' (Wootton, 1990). Brett (1979) also noted that, although readily observed and easily measured, growth is one of the most complex processes of the organism. It represents the net outcome of the whole metabolic system that depends on the income of energy and nutrients generated by feeding activities. It is therefore not surprising that fish have developed a wide adaptive radiation in their feeding habits, including morphological, ecological and ethological specialisations. The more successful the search for food, the more likely it is that an individual will be able to contribute to the reproductive effort.

Ontogeny

The life history (ontogeny) of any fish, from beginning to end, consists of a maximum of five different periods (Balon, 1984, 1986, 1990).

1. The **embryonic** period that begins with fertilis-

ation and is characterised by exclusively endogenous nutrition from the yolk of the ovum.

2. The **larval** period starts with the transition to exogenous feeding and usually lasts until metamorphosis. The development of special larval temporary organs, as well as the persistence of temporary embryonic organs also characterise this period. At the beginning of this period, there could be an overlap of mixed endogenous and exogenous feeding that could be considered as a safety adaptation in case the appearance of prey is not quite synchronous. A larval period exists mostly in fish with distinct metamorphoses, but is absent in fish that produce large and well-developed young.

3. The **juvenile** period begins when fins are fully differentiated and all the temporary organs have been replaced by definitive organs, and it lasts until the maturation of the first gametes. It is usually a period of rapid growth and there maybe a specific juvenile coloration.

4. The **adult** period begins with the first maturation of the gametes. Somatic growth often decreases substantially.

5. A **senescent** period may also be distinguished when the multiplication rate of cells in most tissues has reached its limits.

If ontogeny consists of a sequence of stable states, the development of the phenotype cannot be gradual (Balon,

1986). The theory of saltatory development predicts that species should respond to their environmental cues by heterochronous rates of development, i.e. by taking different lengths of time for different intervals of development. According to this theory, structures align their rates of development to become complete simultaneously, and to initiate a new function via a rapid transition from one stabilised state to another. The organism passes through thresholds at much accelerated rates in order to achieve the relative security of the next stabilised state (Bruton, 1989). Heterochronic changes in developmental rates allow organisms to grow in different ways, either by increasing the growth rate, or by prolonging the length of a particular stage at the same growth rate. Some species also exhibit particular adaptations, such as diapause or dormancy that allow a particular development stage to pass through highly unfavourable periods during its life.

Not all fish have a larval period. Balon (1984) distinguished altricial fish, producing small incompletely developed young, and precocial fish producing large, well-developed young. The first (altricial or indirect ontogeny) is characterised by the production of a relatively large number of small eggs, small size and with a short life span (Noakes & Balon, 1982; Bruton, 1990a). The small amount of yolk in the eggs is insufficient in itself to produce the definitive phenotype (juvenile or adult), and the fish has to pass through an interval of external feeding in order to accumulate sufficient nutrients for its development (Balon, 1986). There is a well-identified larval stage and the larva as a 'feeding machine' is thus a key element in the altricial life style. Conversely, precocial fish are characterised by direct development from a large embryo to a juvenile that is comparatively well developed at the time it starts exogenous feeding. They produce relatively few large eggs with a large amount of yolk, the larval period is reduced or eliminated, the life span is longer and they reach a larger size and an advanced state of development at first exogenous feeding. The survival of young is enhanced by an increase in the endogenous food supply and parental care, and the vulnerable larval phase is eliminated, which is an important ecological and evolutionary phenomenon (Balon, 1990).

Altricial fish are best able to live in varying environments, such as floodplains, where the best reproductive strategy is to produce juveniles as soon as possible; whereas precocial fish are more adapted to relatively stable and predictable habitats where, as a result of density effects and competition, it is better to produce larger and more competitive young. However, little is known about the early life histories of African freshwater fish (Cambray & Teugels, 1988). African cichlids (especially tilapias) are considered to be phenotypically plastic and are able to alter the duration and timing of ontogenetic events so as to shift from an altricial to a precocial state (Balon, 1985).

Growth

Information on growth and age determination is considered to be essential for fishery managers to estimate of stock size and production. As a result, quite a number of growth curves have been established for several African freshwater fish to provide the parameters that could be included in population dynamics models. However, very few studies really took into account either the relationships between growth and other biological parameters, such as reproduction or food availability, or the influence of environmental factors on growth. As a result, they are of limited use for biological or ecological purposes.

Longevity, size and growth models

Fish species exhibit a wide range of longevity, from annual fish completing their life cycle in less than 1 year to fish that potentially may reach more than 20 years. Increase in fish mass or length slows with age but never ceases. This 'indeterminate' growth of fish differs from the determinate growth of many other vertebrates. Fish also display considerable intraspecific ranges of growth rates as the result of different conditions of food, space, competition, temperature and crowding. For this reason, again in contrast to mammals or birds, a particular species cannot be associated with a definite adult (final) size. The maximum observed size is often used as an indication of the maximum size that a species might expect to reach in wild conditions.

The year-to-year growth patterns of most fish species may be described by asymptotic growth curves. The von Bertalanffy model has gained the widest acceptance among fishery biologists, although other growth models have been used (see Wootton, 1990). This model is based on bioenergetics principles, with the assumption that the rate of growth is equal to the difference between the rate

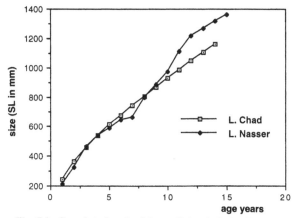

Fig. 7.1. Growth in length of *Lates niloticus* in Lake Chad (from Loubens, 1974) and in Lake Nasser (from Latif, 1984).

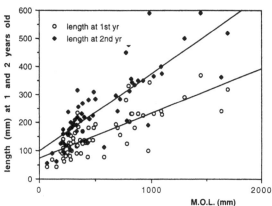

Fig. 7.2. Relationships between the maximum observed length (M. O. L. in mm) and the length at 1 and 2 years old (respectively L1 and L2) for various fresh- and brackish-water African fish species (data from Legendre & Albaret, 1991).

of anabolism and the rate of catabolism. Whereas the von Bertalanffy model describes a smooth year-to-year growth in length or weight, the within-the-year growth pattern is more complicated. Growth is not a continuous process and there are long periods in the year during which growth ceases completely or is considerably restricted, that is to say growth may occur during a relatively short period. For instance, Dudley (1972) recorded that 75% of the expected first year growth of *Oreochromis andersonii* and *Oreochromis macrochir* took place within six weeks of peak floods in the Kafue River. Because long periods of arrested growth interfere with the estimation of annual production, modified growth models have been proposed. Daget & Ecoutin (1976) produced such a model applicable to *Polypterus senegalus*, with the assumption that growth stops completely during the dry season. However, growth may also vary from year-to-year depending on environmental conditions, so more sophisticated models have been developed. Kapetsky (1974) observed in the Kafue flats that growth in weight of *Oreochromis* was considerably slowed, but does not stop completely in the dry season and proposed a growth equation adapted to his data. Welcomme & Hagborg (1977) adopted a model for growth within the year for floodplain fish which took into account a fast initial increase in length followed by a period of slower growth.

The von Bertalanffy model of growth does not work very well for a few species that normally reach a large size. For *Lates niloticus*, the rate of growth in length decreases very slowly with age (Fig. 7.1) and cannot be

described accurately by a von Bertalanffy model (Loubens, 1974). This is also true for *Clarias gariepinus* and *Heterobranchus longifilis* (de Mérona et al., 1988). It may be that this situation is no more than an artefact: populations of these species are decimated at an early age by predation and by exploitation, whereas they could potentially reach longer life and a size at which growth becomes asymptotic. But for some authors (Soriano et al., 1990), two phases of growth may be distinguished in long-lived fish, such as *Lates niloticus*: the first phase is that of the juveniles and young adults that feed on zooplankton and spend time and energy in searching for prey; the second phase consists of accelerated growth of large and piscivorous adults, with the assumption that their feeding behaviour is more efficient.

One would expect that the greater the length or weight that a fish species can reach, the greater its longevity. If so, there should be a relationship between a parameter expressing the maximum age or size of a species and its growth characteristics. Actually, for African freshwater and brackish-water fish, Legendre & Albaret (1991) found a highly significant relationship between the maximum observed lengths (MOL) in wild populations and the lengths recorded at the ages of 1 (L1) and 2 (L2) years (Fig. 7.2). The relationships were respectively L1 = 0.205 MOL + 45.57 and L2 = 0.333 MOL + 63.86, for 58 values. These results were confirmed for six species grown in culture conditions.

Table 7.1. *Maturation size for each sex, longevity, and maximum observed size for selected species of African fish*

Fish species	L	Sex	Locality	M	Long	TMO	Sex size	Authors
Petrocephalus bovei	SL	b	Côte d'Ivoire	65	2.5	110		de Mérona, 1980
Marcusenius macrolepidotus	FL	m	L. Liambezi	140		290		van de Waal, 1976
		f		150				
Hepsetus odoe	FL	m	L. Liambezi	200		370	f > m	van de Waal, 1976
		f		250	5	470		
Hydrocynus vittatus	SL	b	L. Kariba	300 m	9	610 m		Kenmuir, 1972
Alestes baremoze	SL	m	L. Chad	180	5	285	f > m	Durand, 1978
		f		205	6	326		
	SL	m	Côte d'Ivoire	165	4	266	f > m	Paugy, 1978
		f		175	5	284		
	SL	m	L. Turkana	300	5	390		Hopson, 1982
		f		315	5	425		
Ichthyborus besse		b	L. Chad		3	210		Lek & Lek, 1978*b*
Bagrus bajad	FL	b	L. Turkana	315	?	850 f		Lock, 1982
Barbus meridionalis	TL	m	L. Malawi	340	11	890	f > m	Tweddle, 1975
		f		360	17	970		
Schilbe uranoscopus	FL	m	L. Turkana	210	2 to 3	250	f > m	Lock, 1982
		f		240		340		
Schilbe intermedius	FL	m	L. Liambezi		5		f > m	van der Waal, 1976
		f			5	330		
Synodontis schall		m	L. Turkana	195	3		equal	Lock, 1982
		f		200	3	350		
Clarias gariepinus		m	L. Sibaya		8+	1088	m > f	Bruton & Allanson, 1980
		f			7+	1036		
Haplochromis 'Ihema'		m	L. Ihema	82.3			m > f	Plisnier, 1990
		f		76				
Oreochromis macrochir	TL	m	L. Ihema	182		402	f > m	Plisnier *et al.*, 1988
		f		185				
Oreochromis niloticus	TL	m	L. Ihema	180		530	f > m	Plisnier *et al.*, 1988
		f		187				

SL, standard length; FL, fork length; TL, total length; f, female; m: male; b, both sexes; M, size at maturity (mm); Long, longevity (years); TMO, maximum size observed (mm).

The ratios L1/MOL and L2/MOL show that on average African fish reach approximately one-third and one-half of their maximum length at the end of their first and second year respectively (Legendre & Albaret, 1991). This is in agreement with the results of de Mérona (1983), who observed a similar relationship between the asymptotic length ($L\infty$) and the growth coefficient (K) of the von Bertalanffy's equation: $\log K = 2.186 - 1.048 \log L\infty$ that could be simplified into $K = 153/L\infty$. In order to relate this equation to possible field data, de Mérona (1983) also established an empirical relationship between $L\infty$ and the mean maximum size observed in a population (MMS), which is the mean value (or the mode) of the group of individuals of largest size observed in a population: $L\infty = 1.248$ MMS. The combination of these two relationships allows a rapid estimate of the growth curve of any African fish species (de Mérona *et al.*, 1988). Therefore, the relative growth of a species, which is the rate at which an individual approaches its maximum size, decreases when maximum length (and longevity) increases.

Despite this demonstration that general trends do exist, one should keep in mind the existence of great interspecific variability, as well as large variations between populations of the same species inhabiting different river basins. In teleost fish, males are also frequently smaller than females (Table 7.1) and this is generally the case for cyprinids. However, there are exceptions, such as in the tilapias, where the male generally grows faster than the female. This appears to be genetically controlled and a practical consequence is that monosex culture of males gives faster growth in fish culture.

Despite the commonly held opinion that fish grow faster in tropical conditions than in temperate ones, the data available for African fish do not support this principle (de Mérona *et al.*, 1988).

Growth rates and environmental factors

Growth rates of the same species may differ widely in populations inhabiting different aquatic systems. For example, the growth of *Alestes baremoze* and its maximum size are much higher in Lake Turkana than in Lake Chad (Table 7.2), although the two lakes do not differ dramatically in their overall ecology. In this case it is difficult to explain the difference in growth. In other cases, the influence of environmental factors has been suspected.

For instance, the food quality of the diet is directly proportional to its ability to support growth. The most important dietary component in limiting the growth of herbivorous and detrivorous fish is protein (Bowen, 1982). Amino acids that are essential for growth must be obtained from the diet. Carnivorous fish consume prey that may be >80% protein by dry weight, but the diet of tilapias ranges from 50% to <1% protein and values below 15% are most common. In a study conducted at Lake Sibaya, Bowen (1979) observed that juveniles of *Oreochromis mossambicus* grew well and were in good condition, whereas the adults were stunted and showed signs of malnutrition. Although both juveniles and adults fed on benthic detritus, they collected their food in different parts of the lake: juveniles fed near shore, frequently at depths less than 30 cm, while adults fed in water 3 to 5 m deep. Analysis of the food samples from these different depths showed that food quality estimated by the protein–energy ratio was much higher near-shore and dropped rapidly below 1 m (Fig. 7.3). This result may explain why the juveniles have good growth, while adults lack adequate protein.

In floodplain rivers, there is considerable year-to-year variation in growth within the same species. The possible

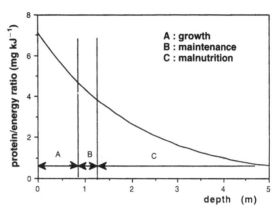

Fig. 7.3. Relationship between protein/energy ratios and depth for benthic detrital aggregate consumed by *Oreochromis mossambicus* in Lake Sibaya (South Africa), showing the expected nutritional significance of feeding (A,B,C) at different depths (simplified from Bowen, 1979).

causes of such variations were investigated in the Kafue River by Dudley (1972, 1974) and Kapetsky (1974) who found significant correlations between some physical variables and the mean growth increment. In particular, the intensity and duration of flooding could have accounted for much of the difference in growth rates of year class 1 and 2 of the cichlids *Tilapia rendalli*, *Oreochromis macrochir* and *Oreochromis andersonii*. Using different temperature and hydrological indices, it was possible to establish relationships between growth rates and environmental parameters, and these relationships were used to predict growth increments for certain year classes.

The Sahelian drought of the 1970s provided an opportunity to assess the effect of flooding on the growth of some fish species. In the River Niger where the floods of 1972 and 1973 were particularly low, Dansoko *et al.* (1976) reported poor growth rates for *Hydrocynus brevis* and *Hydrocynus forskalii*, and particularly for young of the year. In the Senegal River also, Reizer (1974) pointed out great differences in growth of *Citharinus citharus* between 1968, a year of extremely low flood, and other years. The first year class was totally missing for that year and the growth increment for year classes 2 and 3 was significantly lower than that observed during the previous year.

Temperature is another environmental factor that could explain differences in growth between habitats. Growth and metabolism are primarily enzymatic processes whose rates in cold-blood vertebrates are often exponential functions of temperature, often expressed by the quantity Q_{10}. This is the factor by which the rate of the process increases when temperature increases by 10 °C. In broad theoretical terms, metabolic energy demands are expected to approximately double with every 10 °C rise in body temperature, and therefore $Q_{10} = \sim 2$. Temperature affects the swimming activity and the frequency of encounter with prey, as well as the rates of digestion and assimilation of food.

It is generally accepted that tilapias cease growing significantly at temperatures below about 20 °C, but at the same time, a constant warm temperature is not optimal for growth. Tilapias are strongly thermophilic and experiments conducted in tanks have demonstrated that fish subjected to a thermal gradient swim actively toward water that is only marginally cooler than their upper lethal temperature tolerance (Caulton, 1982). The critical temperatures demonstrated for *Oreochromis mossambicus*, *O. niloticus*, *O. macrochir* and *Tilapia rendalli* varied between 30 and 36 °C, a range that is common in shallow marginal waters of tropical lakes during the day. But high temperatures in shallow waters are generally not stable and may decrease considerably at night. Tilapias respond to these diel oscillations by moving inshore during the day and offshore at night (see also Chapter 12). Such movements have been reported by a number of authors (Fryer & Iles, 1972; Bruton & Boltt, 1975; Caulton, 1975). High daytime temperatures coupled with lower night temperatures lead to an optimal use of energy resources and result in increased growth potential (Caulton, 1982).

Age and size at first maturation

The onset of sexual maturity represents a critical period in the life of any animal, and it has long been recognised that maturation is accompanied by a decrease in or cessation of growth. Maturation is a costly process which requires that stored resources, or resources previously available for growth or maintenance, be channelled to gonad growth and development, gamete production, production of secondary sexual structures and sexual behaviours. There are, therefore, potential conflicts in the util-

isation of the energy assimilated. The age at which this transition occurs is evolutionarily significant, because it affects fitness by weighting the value of offspring to future population growth against further growth of the present generation. There should be a trade-off between present reproduction and future growth.

Whether fish mature at a fixed size or at a fixed age is a question that has stimulated much discussion. According to Stearns & Crandall (1984) the question itself is not accurate, as long as fish mature along a trajectory of age and size that depends on demographic conditions and is determined both by genes and by environment. Whatever the answer, age and size at first maturation are intimately linked to growth rates, and therefore to the existence of suitable food conditions for the fish. There is a wide range of ages at first reproduction, from a few weeks in some cyprinodonts inhabiting temporary pools to a few years for large species such as *Lates niloticus*.

Size and age at maturation may, of course, be influenced by the prevailing environmental conditions when comparing different habitats. The adaptability of *Oreochromis niloticus* to different ecological conditions has been reviewed by Lowe-McConnell (1982). The species was stocked into many lakes where it grew and multiplied, and the size and age at which it matures vary greatly according to its habitat (Table 7.2). Fish from large lakes mature at a larger size (over 20 cm) than those from lagoons and ponds, and the two sexes do not differ significantly in maturation size. Most tilapias have the well-known capacity to reproduce at a smaller size when they are environmentally stressed, as is the case for *Sarotherodon melanotheron* (Legendre & Ecoutin, 1996; and Fig. 7.4*b*). In four wild populations of *Oreochromis mossambicus* inhabiting small waterbodies with different levels of environmental harshness, age at maturity varied between 1 and 3 years, and SL at 50% maturity between 118 and 263 mm for females (James & Bruton, 1992). This species demonstrated its ability to adopt a precocial life-history style in some habitats and to switch to an altricial style in more 'hostile' waterbodies.

For cichlids, change in size at maturation is also influenced by fishery pressure. In Lake George the maturation size of *O. niloticus* was smaller at the beginning of the 1970s compared to previous studies (Gwahaba, 1973): L50 that was 27.5 cm TL in 1960 had

Table 7.2. *Comparison of size at sexual maturity, maximum size observed and longevity in different stocks of tilapiines.*

Species	Locality	Maturation size (mm)	Maximum size (mm)	Longevity (years)
Tilapia zillii	L. Kinnereth	135	270	7
	Egypt, ponds	130	250	
	L. Naivasha	90		
	aquaria	70		
Oreochromis mossambicus	L. Kariba	300	390	8
	South Africa		390	11
	Sri Lanka	150	340	
	Egypt		300	7
	Hong Kong	165	310	4
	aquaria	45		
Oreochromis niloticus	L. Turkana	390	640	
	L. George	280	400	
	L. Albert	280	420	
	L. Edward	250	360	
	L. Baringo	180	360	
	Egypt	200	330	9
	Buhuku lagoon	140	260	

Data from Noakes & Balon (1982) for *Tilapia zillii* and *Oreochromis mossambicus*; from Trewavas (1983) for *Oreochromis niloticus*.

fallen to 20 cm TL in 1972. This was attributed to the effects of intensive fishing. A similar trend was observed with *O. mossambicus* whose L50 fell from 26 cm to 14 cm TL over about 20 years (Lowe-McConnell, 1987), but the reasons for this decline were unclear. In Lake Turkana, the length at maturation of *O. niloticus* was considerably higher in the 1950s (39 cm TL in Lowe, 1958) than in the 1970s (29.6 cm TL in Harbott & Ogari, 1982) and the 1980s (26 cm TL in Gjerstad, 1982)

There are, in fact, numerous examples of variations in age and size at first maturation in the same species, but the switch from a life-history tactic favouring growth to another favouring early reproduction at a smaller size has not, for the moment, received any convincing explanation. Causal factors are rarely identified but often related to food supply. Field studies have provided evidence that tilapias in poor condition (low weight for length) switch to reproduction at a smaller size than those in better condition (Fig. 7.4a) and such a pattern has also been observed for *S. melanotheron* in Côte d'Ivoire (Fig. 7.4b). It is quite obvious that other factors, including water oxygenation and space or habitats available, must also be involved.

In tilapiines, the ability to breed at a small size and/or early age, according to environmental circumstances and sometimes with increased frequency, has an adaptative value in unstable or at least cyclic environments, including those which dry up (Lowe-McConnell, 1982). They are able to survive in small ponds, and then to repopulate lakes and rivers when conditions permit (Lowe-McConnell, 1987).

The existence of dwarf populations that mature at a much lower size than usual, whether or not it can be attributed to a switch, is recorded for various species other than cichlids. In Lake Chad, *Brycinus nurse dageti* is a dwarf subpopulation of *Brycinus nurse nurse*, and the two subspecies coexist in the lacustrine environment (Bénech & Quensière, 1989). In Lake Turkana, *Brycinus nurse nana* is also a dwarf population of *B. nurse* (Paugy, 1986). Dwarf populations of *Schilbe mystus* in Lake Chad (Bénech & Quensière, 1985) or *Ethmalosa fimbriata* in specific parts of the Côte d'Ivoire lagoons, such as Bietri Bay (Albaret & Charles-Dominique, 1982), have also been recorded. The origin of this phenomenon is mainly speculative at the moment and no clear interpretation has been provided. The cline described for biological

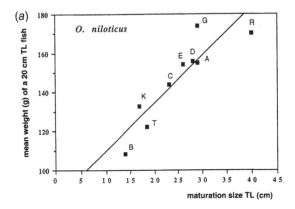

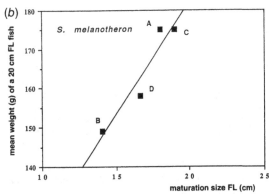

Fig. 7.4. The relationship between weight of 20 cm TL fish (condition) and maturation size of: (*a*) *Oreochromis niloticus* populations in various African waters. Data from Lowe (1958): A, Lake Albert; B, Buhuku lagoon of Lake Albert; C, Lake Chanagwora; D, Lake Katinda; E, Lake Edward; G, Lake George; K, Lake Kijanebatola; R, Lake Turkana; T, Tonya lagoon of Lake Albert. (*b*) *Sarotherodon melanotheron* in Côte d'Ivoire. A, Ebrié lagoon; B, enclosure; C, Acadja 1; D, Acadja 2 (from Legendre & Ecoutin, 1996). TL, total length; FL, fork length.

age, they are old for their size' (Noakes & Balon, 1982). Stunting has serious practical consequences in aquaculture. *Tilapia zillii* normally reach sexual maturity at an age of 2–3 years and a length of 200–300 mm, but when kept in small ponds, they reach sexual maturity within a few months at a length less than 100 mm (Fryer & Iles, 1972). *Oreochromis niloticus* also has a remarkable plasticity in maturation sizes. In Lake Turkana the median maturation size was 26–28 cm TL for wild fish during the 1980s, while juveniles from the lake kept in aquaria became sexually mature after 7 months and at 8 cm TL (Kolding, 1993). Populations in isolated springs around the lake were also found breeding at 8 cm TL (Harbott & Ogari, 1982). Iles (1973) suggested that stunting represents a unique adaptative mechanism that enables tilapia populations to withstand extremely high mortality rates under adverse conditions, such as predation by birds and crocodiles or physical factors such as desiccation.

In the tropical rain forest of Gabon (Brosset, 1982), eight small sympatric species of cyprinodonts (genera *Epiplatys*, *Aphyosemion*, *Raddaella*, *Hylopanchax*) live in the small creeks and pools, and provide a well-documented example of the influence of environmental factors on social and demographic structure. In the mainstream of the creeks (0.5–6 m wide, 1–50 cm deep), there is a patchy distribution of small groups of large adults (3 to 5 cm length) from 2 to 3 species, and their density is low. In the very small and shallow creeks less than 3 cm deep, groups of various species may be observed. They are small-sized adults (less than 3 cm length) belonging to all the species of the fish assemblage. They are brightly coloured and the females are mature. The social groupings of these sympatric cyprinodonts are extremely variable, ranging from monogamous pairs to unstructured promiscuous groupings, as well as harems and multimale groups. When moved to aquaria and well fed, these dwarf adults develop 'normally'. For each type of habitat, there is, therefore, a particular type of social and demographic structure that is developed by the species occupying that habitat.

Trophic guilds and trophic specialisations

Diet can rarely be studied by direct observation and it is difficult to know the relative importance of food items in the diet. It is therefore necessary to sample the stomach contents and several methods have been proposed to pro-

characteristics of *Schilbe mandibularis* in coastal West African rivers (Lévêque & Herbinet, 1982), including size and age at first maturity, is also unexplained (see Chapter 4).

An extreme of this phenomenon is known as stunting. In stunted populations, the fish reach sexual maturity at an unusually small size. The problem is not one of inhibited somatic growth, but one of accelerated ontogeny (sexual maturation) (Fryer & Iles, 1972; Noakes & Balon, 1982). 'The fish are not small for their

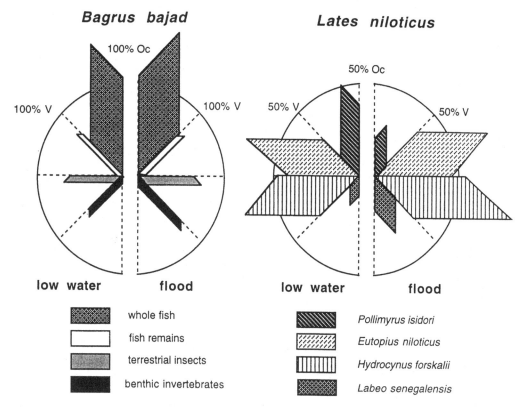

Fig. 7.5. Diet composition of *Bagrus bajad* and *Lates niloticus* in Lake Chad. The graphic method makes use of the occurrence index (%Oc) and of the volumetric method (%V) (from Lauzanne, 1976).

vide qualitative and quantitative descriptions of such samples, but none is entirely satisfactory (Wootton, 1990).

The simplest presence–absence method is to give the frequency of occurrence of food items in stomachs, expressed as a percentage of the total number of stomachs sampled. In the numerical method, the number of items in each food category is counted in all stomachs and expressed as a percentage of the total number of items counted in all stomachs. This method suffers severe limitations when plant material and detritus are involved and, therefore, the volumetric and gravimetric methods have been proposed. In both methods, the relative importance of a food item can be expressed as a percentage of the total volume or weight of each item present in the sample.

Lauzanne (1976, 1977), in his study of fish food habits in Lake Chad, proposed a feeding index (IA) that makes use of occurrence (%Oc) and volumetric (%V) methods: IA = %Oc × %V/100. This index, which can vary from 0 to 100, allows comparison of the relative importance of different food items in the diets. A graphical presentation has also been proposed, making easier intra- or interspecific comparisons. Figure 7.5 illustrates the results obtained for two species from Lake Chad.

The main trophic guilds
There have been various attempts to classify fish into broad trophic categories, which was certainly useful, but it should be remembered that many species show great flexibility in their trophic ecology. The diet of any one species can change during the life cycle, the seasons, the day (see Chapter 12), or the productivity of the habitat.

Following Lauzanne (1977, 1983), a few major trophic categories could be recognised in Lake Chad, and

Table 7.3. Selected data on the food habits of adult fish from north tropical Africa

	Localities	Author	Phytopl.	Superf. organic layer	Vegetat.	Zoopl.	Terrestr. insects	Aquatic insects	Molluscs	Other benthic invert.	Macroph.	Veget. détritus	Shrimps	Fish	Fish debr. scales
Predominantly phytoplanktivores															
Oreochromis niloticus	L. Turkana	Harbott, 1982	***												
Sarotherodon galilaeus	L. Chad	Lauzanne, 1976	***												
	Sudd	Hickley & Bayley, 1987b	***	*	*										
Browsers															
Citharinus citharus	L. Chad	Lauzanne, 1976		***											
Citharinus latus	Sudd	Hickley & Bayley, 1987b		***											
Labeo niloticus	Sudd	Hickley & Bayley, 1987b		***											
Labeo parvus	Bandama	de Mérona, unpub. data		***											
Labeo senegalensis	L. Chad	Lauzanne, 1976		**											
Herbivorous															
Distichodus brevipinnis	L. Kainji	Arawano, 1982	**		***										
Distichodus engycephalus	L. Kainji	Arawano, 1982	**		***										
Distichodus rostratus	L. Kainji	Arawano, 1982	**		***										
	Sudd	Hickley & Bayley, 1987b	**		***										
Tilapia zillii	Sudd	Hickley & Bayley, 1987b	**		***										
Predominantly zooplanktivores															
Alestes baremoze	L. Chad	Lauzanne, 1976				***									
	L. Turkana	Hopson, 1982				***									
Alestes dentex	L. Chad	Lauzanne, 1976				***	*	**			*			*	
	Sudd	Hickley & Bayley, 1987b	*			***		*	*					*	
Brachysynodontis batensoda	L. Chad	Lauzanne, 1976				***		*							
Hemisynodontis membranaceus	L. Chad	Lauzanne, 1976				***		*							
Micralestes acutidens	R. Chari	Lek & Lek, 1977				***		*							
	L. Chad	Robinson & Robinson, 1969				**									
Nothobranchius	Sudd	Hickley & Bayley, 1987b						*							
Predominantly molluscivores															
Hyperopisus bebe	L. Chad	Lauzanne, 1976					*	*	***	*	*				
Symodontis clarias	L. Chad	Lauzanne, 1976					*	*	***	*					
Symodontis frontosus	Sudd	Hickley & Bayley, 1987b		*					**						
Symodontis schall	L. Chad	Lauzanne, 1976						*	***	**					
S. schall	L. Turkana	Lock, 1982				**		*					*	*	
S. schall	R. Niger	Thiero Y., 1983						*	*	**					
S. schall	Bandama	de Mérona, unpub. data							**						
Tetraodon lineatus	L. Chad	Lauzanne, 1977							***						

Aerial and benthic invertebrates

Species	Location	Reference
Aphyosemio geryi	Guinea rivers	Pandare & Romand, 1989
Brycinus leuciscus		Ghaza et al., 1991
Brycinus longipinnis	R. Mono	Paugy & Bénech, 1989
Brycinus macrolepidotus	R. Mono	Paugy & Bénech, 1989
Brycinus macrolepidotus	L. Chad	Lauzanne, 1976
Schilbe mandibularis	Bandama	Vidy, 1976
Schilbe intermedius	Bandama	de Mérona, unpub. data
Schilbe intermedius	R. Mono	Paugy & Bénech, 1989

Benthic invertebrates and others

Species	Location	Reference
Amphilius atesuensis	Bandama	de Mérona, unpub. data
Aethiomastacembelus nigromarginatus	Bandama	de Mérona, unpub. data
Barbus ablabes	R. Mono	Paugy & Bénech, 1989
Barbus chlorotaenia	R. Mono	Paugy & Bénech, 1989
Barbus callipterus	R. Mono	Paugy & Bénech, 1989
Barbus bynni	L. Turkana	Mraja, 1982
Brienomyrus brachyistius	R. Mono	Paugy & Bénech, 1989
Brienomyrus niger	L. Chad	Lek, 1979
	R. Chari	Lek, 1979
	Nigeria	Hyslop, 1986
Brycinus nurse	R. Mono	Paugy & Bénech, 1989
Brycinus imberi	R. Mono	Paugy & Bénech, 1989
Chrysichthys auratus	R. Mono	Paugy & Bénech, 1989
Chrysichthys auratus	Nigeria	Nwadiaro & Okodie, 1987
Chrysichthys maurus	Bandama	de Mérona, unpub. data
Clarias agboyiensis	R. Mono	Paugy & Bénech, 1989
Clarotes laticeps	Sudd	Hickley & Bayley, 1987b
Ctenopoma petherici	Sudd	Hickley & Bayley, 1987b
Heterobranchus isopterus	R. Mono	Paugy & Bénech, 1989
Heterotis niloticus	L. Chad	Lauzanne, 1976
	Sudd	Hickley & Bayley, 1987b
Marcusenius brucii	R. Mono	Paugy & Bénech, 1989
Marcusenius brayeri	Bandama	de Mérona, unpub. data
Marcusenius senegalensis	Nigeria	Hyslop, 1986
Marcusenius furcidens	Bandama	de Mérona, unpub. data
Micraleste acutidens	Sudd	Hickley & Bayley, 1987b
Mormyrops anguilloides	R. Mono	Paugy & Bénech, 1989
Mormyrus rume	R. Mono	Paugy & Bénech, 1989
Pelmatochromis guentheri	Bandama	de Mérona, unpub. data
Petrocephalus bane	L. Chad	Lek, 1979

Table 7.3. (cont.)

	Localities	Author	Phytopl.	Superf. organic layer	Vegetat.	Zoopl.	Terrestr. insects	Aquatic insects	Molluscs	Other benthic invert.	Macroph.	Veget. détritus	Shrimps	Fish	Fish debr. scales
Predominantly phytoplanktivores															
Petrocephalus bovei	R. Mono	Paugy & Bénech, 1989				*		***							
	Bandama	de Mérona, unpub. data				*		***							
	R. Chari	Lek, 1979				**		**							
	L. Chad	Lek, 1979				***		***				*			
Pollimyrus adspersus	Nigeria	Hyslop, 1986				***						*			
Pollimyrus isidori	Nigeria	Hyslop, 1986				***		***							
	Lake Chad	Lek, 1979				**						*			
	R. Chari	Lek, 1979				*									
Polypterus senegalus	Sudd	Hickley & Bayley, 1987b					*					*			**
Synodontis frontosus	L. Chad	Im, 1977						**				***			
	L. Chad	Tobor, 1972							**			***			*
	R. Chari	Im, 1977						**				***			
Synodontis obesus	R. Mono	Paugy & Bénech, 1989						***							
Top consumers															
Bagrus bajad	L. Chad	Lauzanne, 1976						*					**	**	
	L. Turkana	Lock, 1982											***	***	
Bagrus docmak	L. Turkana	Lock, 1982						*					**	***	
Gymnarchus niloticus	L. Chad	Tobor, 1972												***	
Haplochromis macconneli	L. Turkana	Ogari, 1982											*	***	
Hemichromis fasciatus	R. Ogun	Adebisi, 1981						*						***	
Hepsetus odoe	R. Ogun	Adebisi, 1981												***	
Hydrocynus brevis	L. Chad	Lauzanne, 1976						*						***	
Hydrocynus forskalii	L. Chad	Lauzanne, 1976											**	***	
	L. Turkana	Hopson et al., 1982												***	
Ichthyborus besse besse	L. Chad	Lek & Lek, 1978a						*						***	fins
Lates niloticus	L. Chad	Lauzanne, 1976											***	***	
Lates longispinis	L. Turkana	McLeod, 1982											***	***	
Malapterurus electricus	L. Kainji	Sagua, 1979											*	***	
Mormyrops deliciosus	R. Ogun	Adebisi, 1981									*			***	
Parachanna obscura	Sudd	Hickley & Bayley, 1987b						*						**	
Schilbe mystus	L. Chad	Lauzanne, 1976					**			*				**	
	R. Mono	Paugy & Bénech, 1989					**	**				**		*	
Schilbe uranoscopus	L. Chad	Lauzanne, 1976					*	***		*			*	**	
	L. Turkana	Lock, 1982											**	**	

***, item almost exclusively consumed; **, item a major constituent of the diet; *, item occurring regularly in the diet but not dominant. The food of juveniles is often different from the food of adults and usually includes microinvertebrates and zooplankton.

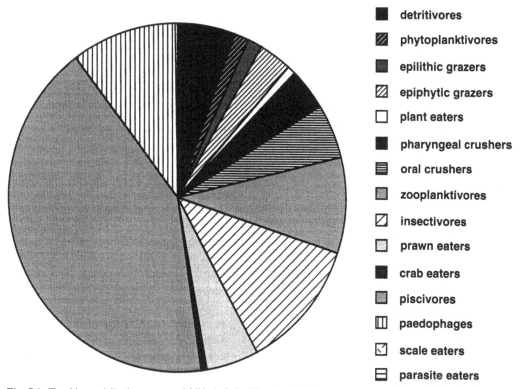

detritivores

phytoplanktivores

epilithic grazers

epiphytic grazers

plant eaters

pharyngeal crushers

oral crushers

zooplanktivores

insectivores

prawn eaters

crab eaters

piscivores

paedophages

scale eaters

parasite eaters

Fig. 7.6. Trophic specialisations among cichlids in Lake Victoria and relative number of species (data from Witte & van Oijen, 1990).

this classification can be extrapolated to most riverine tropical African fish communities.

1. **Dominant primary consumers** eating plant material or detritus. This group could be subdivided into:

   phytoplanktivorous filter feeders consuming mainly planktonic algae, such as *Sarotherodon galilaeus*;

   grazers or macrophyte consumers, such as *Brycinus macrolepidotus*;

   browsers eating the superficial layer of sediment usually composed of sedimented algae, or the periphyton growing on rocky substrates, e.g. *Labeo senegalensis, Labeo coubie, Citharinus citharus, Citharinops distichodoides, Distichodus rostratus*; and

   detritivores, such as some tilapiines.

2. **dominant secondary consumers** include :

   the zooplanktivores that should be mainly restricted to standing waters where zooplankton can develop abundantly throughout the year, but a few species living both in still and running waters can adapt their trophic regime to this mode of feeding, e.g. *Alestes baremoze, Alestes dentex, Brachysynodontis batensoda, Hemisynodontis membranaceus*;

   the benthivores, feeding mainly on benthic invertebrates e.g. *Synodontis schall, S. clarias, Hyperopisus bebe, Heterotis niloticus* and a few species that feed almost exclusively upon molluscs, such as *Tetraodon lineatus*; and

   surface feeders eating predominantly terrestrial invertebrates.

3. **top consumers** are either:

   strictly piscivorous fish feeding exclusively on living fish, such as *Lates niloticus* and *Hydrocynus brevis*; or

   less strictly piscivore species consuming both fish, shrimps and other invertebrates, e.g. *Schilbe mystus, Schilbe uranoscopus, Bagrus bajad* and *Hydrocynus forskalii*, which is a dominant piscivore but also eats large quantities of shrimps.

Some data on the food habits of the north tropical African fish fauna are summarised in Table 7.3.

Trophic specialisation and adaptive radiation
Adaptative radiation is commonly viewed as the evolutionary diversification of a lineage across a variety of resource types. Among the suspected causes of the process is competition between species for resources: divergence of new species in morphology and resource use is hypothesised as being driven by competition between them for food, and facilitated by an absence of competition from species in other taxa.

That cichlids show considerable adaptive radiation is demonstrated by the wide spectra of their trophic specialisations, correlated to anatomical and functional specialisations (Fryer & Iles, 1972; Greenwood, 1981, 1984) that have evolved in each of the African Great Lakes where they occupy virtually every available habitat and niche. It seems that every food source in each lake is used by cichlids while some resources are not utilised by any species of other families (Witte & van Oijen, 1990; Yamaoka, 1991, for review). In Lake Victoria for instance, Witte & van Oijen (1990) distinguished 15 trophic categories (Fig. 7.6). There are generalised detritivores bottom feeders, phytoplanktivores, epilithic and epiphytic algal grazers, plant eaters, benthic invertebrates feeders, zooplanktivores, molluscivores including oral shell crushers and pharyngeal crushers, prawn eaters, crab eaters, piscivores including parasite eaters removing leeches and crustacean parasites from other fish, scale eaters, and paedophages eating eggs and embryos from brooding female cichlids. Parallel adaptations have occurred in other lakes (Malawi, Tanganyika) where similar diversities of resources are used. Apparently, all the major potential food resources are utilised by cichlids in each lake, which is not the case for the fish species belonging to other families.

Paradoxically, species apparently specialised for certain food sources have the potential to exploit a wider range of food niches, while generalists cannot do the reverse and are unable to exploit exclusively specific trophic niches without the necessary anatomical specialisations (Greenwood, 1984; Mayr, 1988). In Lake Vic-

toria for instance, zooplanktivores, detritivores and insectivores all switch to *Melosira* spp. when there is a bloom (Witte, 1984*b*).

Morphological and physiological adaptations to feeding behaviour

It is assumed that the oldest teleost fish evolved as generalised predators feeding on comparatively large prey (Gosline, 1971). The evolution of feeding from large prey (macrophagy) to smaller prey (microphagy) involved the development of specialised structures, such as the modification of jaws from fixed to protrusible or the replacement of teeth by elaborate gill rakers on the branchial arches, and modification of some gill rakers into an epibranchial organ on the roof of the mouth. The more specialised microphages are the filter-feeding species.

Body form and feeding types
Behavioural strategies for foraging are constrained by morphology, including mouth shape and position as well as body form. Water is a dense, viscous medium that places a premium on the relationship between body form and effective propulsion through it. Fish have a remarkable variety of shapes and sizes, from the long narrow eel-shape to the flat fish and stone fish. Whatever the shape, the actinopterygians show enormous diversity of propulsive mechanisms and every part of the body or every appendage is employed for swimming by some fish or other, including the body itself, the tail, the paired and the median fins (Webb, 1982). Webb (1984*a*, *b*), looking for correlations between locomotor morphology and the feeding niche, distinguished locomotor specialists and locomotor generalists. For specialists, he recognised three basic functional locomotary mechanisms which may be linked to foraging.

1. To exploit food widely distributed in space and time, fish have to move at speeds that sample the greatest volume for the least expenditure of energy. This applies both to macrophage and filter-feeders. The body and caudal fin periodic propulsion (BCF periodic) is the optimal locomotor option. There are cyclically repeated patterns of wave generation providing thrust over periods from about a second

to several weeks; that is for sprinting or for cruising during chases, patrolling, searching, migration, etc. Narrow caudal peduncle, a streamlined anterior body and a high aspect ratio lunate tail, are among the morphological characteristics of this category.

2. When food is locally abundant and persistent, we may expect the specialisation to be related to specific food characteristics, such as size or evasive ability. For a predator, the objective is to minimise the duration of any interaction necessary to catch prey before they can escape and reach shelter. The pike (*Hepsetus odoe* is the African equivalent) is the archetype of the body and caudal fin transient propulsion (BCF transient). The wave generation is brief, non-cyclical, providing thrust for fast starts or powered turns, and used in prey capture or predator evasion. Fish in this category have deep, especially caudally, and flexible bodies.

3. When prey are concentrated in a bewildering diversity of locations, sometimes hidden in holes or lying on surfaces, it is necessary to take food in any plane and to move through restricted spaces. The median and paired fins type of propulsion (MPF) provides better manoeuvrability. The paired fins generate the thrust for locomotion, allowing slow swimming but great and precise manoeuvrability in searching, feeding, hiding, etc. A lateral insertion of the pelvic fins, extended anal and dorsal fins, and deep laterally flattened body are the major characteristics of this category. Many cichlids fit these characteristics as do the Mormyridae, Notopteridae, Tetraodontidae (Hoar & Randall, 1978).

It is not possible to combine all the optimum features for the different types of swimming in one fish: the optimum design for manoeuvring excludes the elements that favour high performance in accelerating or in cruising. The three basic designs are therefore mutually exclusive. The locomotor specialists represent only a small proportion of fish species and the majority of fish are locomotor generalists. However, locomotor adaptations for food capture are of diminished importance among locomotor generalists, and other adaptations (e.g. suction or protrusible jaws), which are vital to food selection and the extension of diet breadth, are more common in foraging. Suction appears to be necessarily correlated to the

exploitation of small-item resources that are too dispersed to filter, and no special fin/body plan is necessary for suction feeders (Rosen, 1982). However, for small locomotor generalists, fast-start capabilities have to be retained for predator evasion.

Peculiar adaptations have been developed among electric fish. The catfish, *Malapterurus electricus*, is a sluggish swimmer, but it feeds almost exclusively on fish and its electric organ discharge is the major prey capturing mechanism. A powerful volley of high frequency electric organ discharges emitted close to a pelagic school of prey fish can paralyse a number of fish simultaneously. This explains why, compared to *Hydrocynus forskalii*, *M. electricus* is a more effective predator of small clupeids and small fish in general (Sagua, 1979).

Head and mouth morphologies

It has been suggested that one key innovation, the pharyngeal mechanism, has been one of the major reasons for the extraordinary explosive speciation undergone by the cichlid fish in the African Great Lakes (Liem, 1973). Functional morphology has demonstrated that the trophic apparatus of these cichlids has undergone a spectacular functional diversification, for collecting a wide range of foods, unparalleled by any other family of fish (Liem, 1991 for review). It was noticed (Coulter *et al.*, 1986) that only slight morphological and anatomical changes, e.g. in the feeding apparatus (mouth, teeth, guts), has resulted in profound changes in the feeding habits of species and the great variety in shape of oral teeth reflect the adaptation to collecting many different types of food (van Oijen, 1989). Adaptative changes in the jaws are particularly striking, for example in the huge difference between the elongated mandibles of the piscivorous *Rhamphochromis*, and the squat, almost square-cornered jaws of the scraper *Labeotropheus* (Fryer, 1991).

From a functional perspective, cichlids feed in three basic ways: (i) suction feeding, in which the prey is sucked and move into the mouth; (ii) ram feeding, in which the predator opens its mouth and simply overtakes its prey; and (iii) biting, in which the teeth are applied to the prey (Liem, 1991). The extent to which various species are adapted to biting or sucking is clearly related to the type and size of food, and the habitat in which the species are found (van Oijen, 1982). Most haplochromine predators feeding on small items are suckers rather than

Table 7.4. *Selected information about the ratio of intestine length to fish standard length in various African species*

Family/Species	IL/SL	Regime	Source
Notopteridae			
Papyrocranus afer	0.18	omnivore	Teugels *et al.*, 1992*c*
Mormyridae			
Hyperopisus bebe	0.67	invertivore	Paugy, pers. com.
Marcusenius senegalensis	0.74	invertivore	Paugy, pers. com.
Mormyrus rume	0.78	invertivore	Paugy, pers. com.
Petrocephalus bovei	0.68	invertivore	Paugy, pers. com.
Characidae			
Alestes baremoze	0.89	omnivore	Paugy, pers. com.
Brycinus macrolepidotus	1.16	omnivore	Paugy, pers. com.
Brycinus nurse	1.31	omnivore	Paugy, pers. com.
Hydrocynus forskalii	0.85	piscivore	Paugy, pers. com.
Citharinidae			
Citharinus citharus	6.13	micro/macrophyto	Daget, 1962*b*
Citharinus congicus	4.40	micro/macrophyto	Daget, 1962*b*
Citharinus latus	6.78	micro/macrophyto	Daget, 1962*b*
Distichodontidae			
Distichodus engycephalus	4.71	micro/macrophyto	Paugy, pers. com.
Hemidistichodus vaillanti	0.37	invertivore	Daget, 1968*b*
Ichthyborus besse	1.10	piscivore	Daget, 1967
Nannocharax ansorgii	0.52	invertivore	Daget, 1961*a*
Cyprinidae			
Barbus anoplus	0.80	omnivore	Cambray, 1983
Barbus macrops	0.82	omnivore	Paugy, pers. com.
Barbus sacratus	1.63	omnivore	Paugy, pers. com.
Chelaethiops elongatus	0.75	zooplanktivore	Matthes, 1963
Garra congoensis	4.50	micro/macrophyto	Matthes, 1963
Labeo lineatus	16.10	limnivore	Matthes, 1963
Labeo niloticus	16.90	limnivore	Matthes, 1963
Labeo parvus	10.00	limnivore	Paugy, pers. com.
Labeo senegalensis	13.36	limnivore	Paugy, pers. com.
Opsaridium chrystyi	0.73	invertivore	Matthes, 1963
Raiamas senegalensis	0.38	invertivore	Paugy, pers. com.
Channidae			
Parachanna obscura	0.55	piscivore	Paugy, pers. com.
Amphilidae			
Amphilius atesuensis	0.51	invertivore	Paugy, pers. com.
Bagridae			
Auchenoglanis occidentalis	0.88	invertivore	Paugy, pers. com.
Chrysichthys auratus	1.98	invertivore	Paugy, pers. com.
Mochokidae			
Synodontis ocellifer	2.18	invertivore	Paugy, pers. com.
Synodontis schall	1.96	invertivore	Paugy, pers. com.
Schilbeidae			
Schilbe intermedius	0.86	omnivore	Paugy, pers. com.
Clariidae			
Clarias anguillaris	0.86	omnivore	Paugy, pers. com.
Heterobranchus isopterus	1.32	omnivore	Paugy, pers. com.
Centropomidae			
Lates niloticus	0.52	piscivore	Paugy, pers. com.
Cichlidae			
Sarotherodon galilaeus	6.54	micro/macrophyto	Lauzanne & Iltis, 1975
Tilapia zillii	5.03	micro/macrophyto	Paugy, pers. com

From Paugy, personal communication. IL, intestine length; SL, standard length.

biters. Barel (1983) also assigned oral jaws of piscivorous haplochromines to two basic types: suckers and biters, with a whole range of intermediates in between. He considered that morphological adaptations are more related to the feeding behaviour than to the type of food.

Many cichlids can combine the different ways of feeding, but Barel *et al.* (1989) argued that the number of resources which a species may exploit is limited by the compatibility of the anatomical requirements for coping with these resources. The size of the buccal cavity, for instance, is related to food but also to the locomotor apparatus, the gill apparatus, the eyes and the number and size of eggs produced in the ovary (Barel *et al.*, 1991). This implies that for haplochromines the various ways of feeding, swimming, breathing and breeding cannot be combined haphazardly and that there should be trade-offs between the above functions.

It has been claimed that the basic cichlid 'bauplan' that is the fundamental body plan and organisation of especially the neuro- and splanchnocrania and the associated musculature, combined with the nature of their oral and pharyngeal dentition, provides a uniquely plastic substrate for evolutionary change. Relatively slight changes in the shape or size of component parts can produce far-reaching effects with regard to the potential feeding capabilities of the individual so modified. This feature may explain the range of cichlid diversity, and the wide diversification in feeding behaviour observed (Greenwood, 1991). To understand the adaptive value of such morphological radiation, it is necessary to conduct field studies in order to assess the feeding habits and strategies as well as the food preferences of fish species. Experimental studies may contribute to our understanding.

In many species of cyprinids, there are feeding adaptations in the form of the mouth. In the so-called *Varicorhinus* mouth form, the horny (or keratinised) lips are adapted for scraping algae and aufwuchs from submerged rocks or other hard substrates. In some species, hypertrophied lips (called rubber lips) are most suited for grubbing between loose rocks and pebbles. The complex ventral mouths of labeines, with homodont pharyngeal teeth that are close together to form a single grinding surface, are specialised to feed on epibenthic algae.

Length of the gut

There is a correlation between the diet and the length of the gut relative to body length (Kapoor *et al.*, 1975). Carnivorous fish usually have a large stomach and a short gut (ratio less than 1). That is the case for *Lates* and *Hydrocynus*, but also for many invertebrate feeders (Table 7.4). Conversely, in sediment browser fish such as *Labeo*, the stomach is often absent or small, and the gut is usually more than ten times longer than the body length (Table 7.4). In herbivorous fish (*Tilapia, Citharinus, Distichodus*), the intestine length is 2 to 8 times the length of the fish. The long intestine of fish feeding on plant material suggests that one or more essential components of the diet such as amino acids are slow to be digested. A long residence time and extensive exposure to absorptive surfaces are therefore required (Bowen, 1988). pH values below 2.0 and sometimes as low as 1.2 allow the lysis of blue-green algae (Moriarty, 1973), diatoms (Bowen, 1976) and macrophytes (Caulton, 1976).

In haplochromines from Lake Kivu, Ulyel (1991) observed that the ratio of intestine length to fish standard length was 3.05 and 3.32 respectively for the phytophagous *Haplochromis kamiranzovu* and *Haplochromis olivaceaus*, while only 1.92 for the insectivore *Haplochromis graueri* and 1.06 for the piscivore *Hydrocynus vittatus* (Table 7.5). Similar observations were made on a sample of African Great Lakes cichlids (Fryer & Iles, 1972). There is also an allometric relationship between intestine length and fish length, and for some species, such as *H.*

Table 7.5. *Relationship between relative gut length and diet in a sample of haplochromine from Lake Kivu*

Species	Regime	IL/SL	Branchiospines
H. olivaceus	microphyto	3.32	10–2/11–4
H. kamiranzovu	microphyto	3.05	11–3/13–4
H. astatodon	detritivore	2.84	10–2/12–3
H. graueri	insectivore	1.92	8–2/10–2
H. nigroides	omnivore	1.88	12–3/13–4
H. scheffersi	omnivore	1.72	9–2/11–4
H. adolphifrederici	insectivore	1.68	8–2/9–3
H. gracilior	omnivore	1.56	9–3/11–4
H. paucidens	insectivore	1.42	9–2/10–3
H. vittatus	piscivore	1.06	9–2/11–3

From Ulyel, 1991. IL, intestine length; SL, standard length.

graueri and *H. gracilior*, the ratio is only 1.2–1.3 for planktophage juveniles.

Prey selection and ontogenic shifts of the feeding niche

Most fish species vary greatly in body size during their life and often undergo drastic changes in ecology as they grow. Their dietary requirements and their feeding behaviour also undergo changes that are usually related to the ability of growing animals to handle particular food types. According to Werner (1986), when ontogenic shifts occur, they almost always involve shifts to larger prey, as corroborated by theoretical and empirical studies. This ontogenic niche shift involves trade-offs of current growth, mortality and birth rates and is closely related to the problem of optimal life histories.

Meanwhile, resources are not evenly distributed, and prey usually occur in patches that are unlikely to contain items uniform in size, taxonomic category or nutritional value. According to the optimal foraging theory, the forager should always select the most profitable prey item, from the energetic point of view, when it is encountered. The most profitable prey for a predator will be that for which the net energetic gain is maximum, that is to say prey for which there is a maximum energy input for a minimum cost of capture.

The problem of prey choice, which is of major importance in the understanding of feeding behaviour, has been studied extensively. Field workers have used 'electivity indices' such as that devised by Ivlev, expressed as $E = (r_i - p_i)/(r_i + p_i)$ where r_i is the proportion of food item i in the diet, and p_i the proportion of item i in the environment. This index is useful when investigating whether a prey species is or is not being eaten in proportion to its abundance in the habitat, but does not distinguish mechanisms behind any selection revealed. Working on African fish, Gras & Saint-Jean (1982) also showed that the Ivlev index does not accurately represent the degree of prey selection as it is markedly influenced by the abundance of prey in the natural environment. They proposed another forage ratio, the collective efficiency, to relate the stock of planktivores to the fraction of plankton biomass useful for the predator.

Phytoplankton feeders

In Lake Chad, *Sarotherodon galilaeus* is a filter-feeder that feeds on bottom deposits composed of benthic algae. These fish positively select filamentous cyanobacteria (*Oscillatoria, Lyngbya, Raphidiopsis*) and negatively select diatoms (*Coscinodiscus, Tabellaria, Fragilaria*) and Chlorophycea (Lauzanne & Iltis, 1975). Similar results were obtained with *Oreochromis niloticus* by Moriarty (1973) in Lake George. In contrary, in Lake Turkana, *S. galilaeus* and *Oreochromis niloticus* apparently ingest all the major members of the phytoplankton in similar proportions to those observed in the plankton (Harbott, 1982).

In Lake Awassa (Ethiopia), juvenile *O. niloticus* less than 30 mm SL consume chironomid larvae and some adult insects in abundance, while above this size, the species becomes predominantly herbivorous. The change in diet with size may reflect a change in habitat (Tudorancea *et al.*, 1988).

Zooplankton feeders

Most fish feed on plankton during at least part of their life. Many fish larvae of non-planktivorous species consume zoo- or phytoplankton before switching to larger prey. Others feed on plankton over their entire life, exclusively (obligate planktivore) or facultatively (facultative planktivore). There is a good deal of evidence that zooplankton feeders are selective of their prey and various methods have been developed to compare the relative frequency of occurrence of prey types in a predator's gut and in its environment (review in Lazzaro, 1987).

Planktivorous fish exhibit two distinct feeding behaviours (Lazzaro, 1987). Particulate feeders attack and ingest single individual planktonic prey that are visually selected. For successful prey selection, a visual predator is dependent on its visual acuity, the optical characteristics of the water and the conspicuousness of the prey. On the other hand, pump filter feeders do not detect individual prey but engulf a volume of water that is then filtered to retain planktonic prey on gill rakers, microbranchiospines or branchial tooth plates.

Filter-feeding fish select their prey passively according to their size. Their retention efficiency depends on the structure and functioning of the branchial filtering

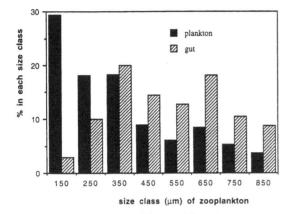

Fig. 7.7. Size frequency of crustaceans (copepods and cladocerans) in the zooplankton and in the gut of the filter feeder *Hemisynodontis membranaceus*, showing its prey selection as a function of size (from Im & Lauzanne, 1978).

apparatus, but many other factors are also involved, such as size and shape of the particles, mucus production, water flow, etc. (Lazzaro, 1987). *Brachysynodontis batensoda* is a microzooplanktivore in Lake Chad (Gras *et al.*, 1981). The smallest prey captured are about 80 μm in length. Nauplii and rotifers are progressively selected as a function of their size up to 260 μm. The larger microcrustaceans are mainly selected on the basis of their ability to avoid predation: the relatively big cladoceran *Moina micrura*, with low motility, is easily captured whereas the highly fragile diaptomids are not. On the other hand, large adults (230–250 mm SL) of *Alestes baremoze*, also in Lake Chad, do not retain small-sized nauplii and rotifers, and the branchial filter retains particles greater than 400 μm in length. All filtered items above a size of 880 μm are collected (Lauzanne, 1970). These two species, both feeding on zooplankton, have different behaviour patterns. *Alestes baremoze* feeds on bigger prey, and consumes more copepods than *B. batensoda*. The latter selects smaller prey, such as rotifers and nauplii, more efficiently.

Another filter-feeding species in Lake Chad, *Hemisynodontis membranaceus*, feeds preferentially on copepods but small-sized nauplii (<200 μm) are not collected by the branchial filter. The selection index increases with size of prey (Fig. 7.7), and above 460 μm most copepods

and cladocerans are retained (Im & Lauzanne, 1978). However, size is apparently not the only selection factor for *H. membranaceus*, since the cladoceran *Moina dubia* exhibits a strong negative electivity index, which has not been explained.

Benthic invertebrates feeders

Benthic fish are also able to select their food according to size. A striking example is given by the malacophagous fish from Lake Chad studied by Lauzanne (1975*a*). *Synodontis clarias*, *Synodontis schall* and *Hyperopisus bebe* feed almost exclusively on molluscs, but they select small-sized, chiefly immature molluscs. This was demonstrated for various species, and Fig. 7.8 illustrates the size selection for two Prosobranchia, *Cleopatra bulimoides* and *Melania tuberculata*. There is apparently no size

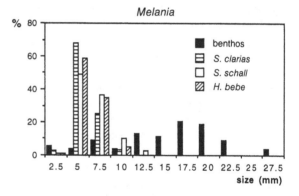

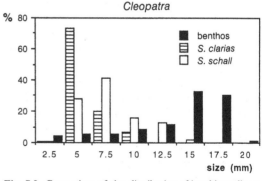

Fig. 7.8. Comparison of size distribution of benthic molluscs in the Lake Chad benthos and in the gut of malacophagous fish species (*Synodontis clarias*, *S. schall*, *Hyperopisus bebe*) (data from Lauzanne, 1975*a*).

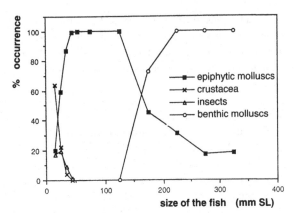

Fig. 7.9. Ontogenic change in diet of *Tetraodon lineatus* in Lake Chad (from Lauzanne, 1977).

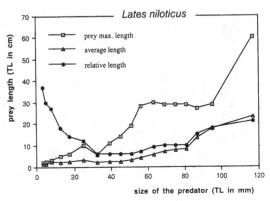

Fig. 7.10. The relationships between the length of *Lates niloticus* and the size of prey in Lake Chad (from Hopson, 1972).

selection for the small-sized mollusc, *Gabbia* sp. The impact of predation upon their young greatly influences the dynamics of the mollusc populations, and probably explains the scarcity of small individuals in their size frequency distributions despite more or less continuous reproduction (Lévêque, 1972).

The ontogenic shift in the diet of *Tetraodon lineatus* appears to be related to risk of predation: juveniles are confined to areas where risk of predation is lower, and they move to more open habitats as they grow, to exploit food there. In Lake Chad the juveniles up to 200 mm SL live in submerged vegetation banks where they feed mainly on microcrustacea and insect larvae (up to 50 mm SL) and then on molluscs inhabiting vegetation beds, e.g. *Gabbia* sp., *Anisus natalensis* and *Bulinus* spp. (Fig. 7.9). When larger, *Tetraodon lineatus* leave the shelter of the vegetation to move to open waters where they feed on benthic molluscs, such as *Cleopatra bulimoides*, *Bellamya unicolor*, *Melania tuberculata* and *Corbicula africana* (Lauzanne, 1977). A more or less similar shift in diet has been observed for the brackish-water fish *Trachinotus teraia*. Up to 125 mm it feeds on fry (mainly of the clupeid *Pellonula*) and young shrimps, whereas the diet of larger fish becomes exclusively malacophagous. In the Ebrie lagoon, the dominant prey is the bivalve *Corbula trigona* (Trebaol, 1991).

Fish feeders

In a number of fish species, the relationship between cost of capture and prey size, has been estimated for different predator sizes and is more or less U-shaped along the gradient of prey size. This means that for a predator there is an intermediate prey size, which minimises cost (Wootton, 1990). Larger predators have a bigger optimal prey size, but also a wider range of possible prey sizes. Another implication of the existence of a relationship between body size and prey size is that juveniles of large species have resource requirements similar to those of adults of smaller species. Such competition may result in a significant bottleneck to recruitment to the adult stages of larger species.

The relationship between the length of *Lates niloticus* and the size of prey was studied by Hopson (1972). *Lates* at all lengths are capable of eating fish up to approximately half their own length. The largest prey in relative size were eaten by the smallest *Lates*, 3 to 4 cm TL. Relative prey size fell steeply from 38% in the smallest *Lates* to 6% in the 30–36 cm TL group. Between 30 and 80 cm TL there was a slight increase in the relative size of prey from 6 to 10%. A marked change was noted in *Lates* over 84 cm TL with prawns and small characins virtually disappearing from the diet. The result is a steep rise in both relative size and average size (Fig. 7.10).

It has also been reported that such ontogenic changes could be related to morphological changes. For instance, in Lake Victoria, the percentage of *Lates niloticus* containing the prawn *Caridina nilotica* declines from 60–85% for fish between 5 to 40 cm TL to less than 10% for fish larger than 80 cm TL (Hughes, 1992). These percentages are similar to those recorded by Hopson (1972) for Nile perch in Lake Chad feeding on *Macrobrachium niloticum*, a prawn of similar size to *Caridina nilotica*. It has been demonstrated that gill raker spacing, which increases in direct proportion to fish length, may be the constraint that prevents fish larger than 70 cm in length from being effective prawn predators (Hughes, 1992). Above 70 cm TL, gill raker spacing exceeds mean body depth of the prey, and gill rakers are therefore inefficient at straining prawns from water expelled through the branchial basket during feeding.

The existence in some fish prey species of strong, locking, dorsal or pectoral spines, may reduce predation. In Lake Turkana, *Synodontis schall* and *Chrysichthys auratus* were rarely taken at lengths greater than 22% of the predator, whereas for other species, such as *Hydrocynus forskalii*, *Oreochromis niloticus* or *Bagrus bajad*, individuals of up to one-third the length of the predator were regularly eaten (Hopson, 1982). Moreover, the low incidence of *S. schall* in the diet of *Lates niloticus* suggests that this predator tends to avoid *S. schall* on account of their strong spines.

Size specific shifts in food or habitat type have been documented in many species. For *Hydrocynus forskalii* in the Chari River (Lauzanne, 1975*b*), juveniles up to 30 mm are almost strictly zooplanktophages. Between 30 and 45 mm they eat both zooplankton and insects. Above 50 mm length, they become strictly piscivorous (Fig. 7.11*a*). In the Lake Chad SE archipelago, a similar but less clear pattern has been observed, shrimps being another important food item in the diet of *H. forskalii* (Fig. 7.11*b*).

Nevertheless, the range of prey sizes and the mean prey size can also increase with the body size of *H. forskalii* (Fig. 7.12). It was demonstrated that between 10 and 20 cm, *H. forskalii* feeds mainly on the adults of small species (*Barbus, Micralestes, Pollimyrus, Raiamas*) throughout the year whereas, above 20 cm, young individuals of large species are a more important component of the diet during the flood (*Alestes, Labeo, Schilbe, Distichodus*).

Small tigerfish, *Hydrocynus vittatus*, also eat relatively larger prey than large tigerfish, with a mean ratio of 25% (Kenmuir, 1975). Lower values of the prey/predator length ratio have been published for other species: 20% for *Hydrocynus forskalii* and 10% for *Hydrocynus brevis*

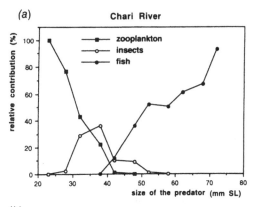

(a)

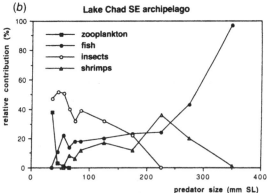

(b)

Fig. 7.11. Changes in the proportions of food items consumed by *Hydrocynus forskalii* in: (*a*) Chari River and (*b*) SE archipelago of Lake Chad.

(Lewis, 1974). In Lake Chad, the mean relative prey size of *H. forskalii* is about 10% for small fish (50–75 mm SL) eating mainly invertebrates, but varies between 20 and 33% for fish from 100 to 400 mm SL (Fig. 7.12). *Hydrocynus forskalii* only feeds on living prey that are ingested whole, whereas *H. brevis* is known to commonly bite pieces from larger prey.

In Lake Victoria, juveniles of *Bagrus docmak* up to 15–20 cm length feed mainly on invertebrates (insect larvae, shrimps), but fish larvae also occur in their stomachs. From 20 cm long, they show a preference for fish (Okach & Dadzie, 1988) and are almost strictly piscivorous above 50 cm. Ontogenic changes in food have also been reported for *Bagrus bajad* and *B. docmak* in the Nile (Khallaf & Authman, 1992).

A relationship between predator size and prey size has also been recorded for *Ichthyborus besse*, a less

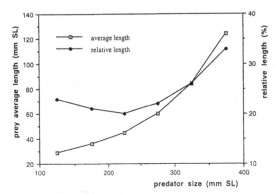

Fig. 7.12. *Hydrocynus forskalii:* relationship between the length of the predator and the size of the prey (from Lauzanne, 1975*b*).

common predator than those cited previously (Lek & Lek, 1978*a*). The mean relative size of prey varies from 20 to 30% for individual fish less than 100 mm SL and from 30 to 40% for *I. besse* above 100 mm SL. The highest relative prey length observed was 50% that of the predator. A peculiar behaviour in relation to this ontogenic shift is that *I. besse* were observed to feed on fish fins when juvenile and move progressively to entire fish with increasing size.

Diet also changes with individual growth in cichlids. A peculiar behavioural and morphological development has been reported for *Docimodus evelynae*, a cichlid species from Lake Malawi (Ribbink, 1984, 1990), in which small juveniles (less than 50 mm SL) are facultative cleaners removing fungi from other cichlids. From 50 to 70–80 mm SL, they are no longer cleaners, and feed upon plankton, insects and aufwuchs. Above 80 mm SL, their dentition changes (the juvenile tricuspid teeth are replaced by adult unicuspid teeth) and adult *D. evelynae* alter their diet to feed upon scales, fins and skin of their fish prey. Field data have also provided evidence that as *D. evelynae* grow they move into deeper water. As a result of the depth restricted distributions observed for many species in Lake Malawi, the cichlids cleaned by juvenile *D. evelynae* in shallow water are not the same as those preyed upon by adults in deeper water. Similar changes in feeding behaviour are reported for many other species that are simply too small as juveniles to utilise the same foods as the adults.

Feeding strategies

Foraging strategies and optimal foraging theory

The global problem for any fish is how to allocate its time to growth and reproduction in a pattern which maximises reproductive success. It is faced with a number of decisions with respect to food acquisition: it has to decide to feed and then to look for food on a search path that will maximise the energy gain per unit of search effort. It has to decide where and how long to stay at a feeding site, and what prey will be most appropriate to take (size or nutritional value). By minimising the time spent in feeding, a fish will have more time to allocate to other activities. Alternatively, by minimising the energy spent on capturing its prey, a fish will have more food energy left for metabolism and growth (Hart, 1986). As a result, time minimisation in searching for food, or energy maximisation, are the two ways in which a fish can be efficient as a forager.

The introduction of evolutionary concepts has helped to provide theories that predict decision rules for foraging animals. Foraging behaviour that is constrained by the physiology and the morphology of the fish, and the resulting strategies of food gathering, are a compromise between the benefits derived from food gathered and the costs associated with the strategy. Foraging strategies are thus adaptations that allow animals to deal efficiently with all kinds of environmental pressures, such as limited food supply, competition and unpredictable variation in food resources. Foraging strategies are also currently studied as decision-making processes in answer to questions such as where should an individual forage, or towards which prey should it direct searching? (Cézilly *et al.*, 1991).

The most developed and widely tested foraging models are concerned either with the choice among randomly dispersed prey in a homogeneous environment, or the choice among depletable patches of prey which are assumed to be small and locally abundant so that a forager might visit many patches during a single foraging bout. These models do not seek to describe mechanisms by which the forager achieves its maximisation, but only to predict outcomes such as what is eaten and which patch is searched for prey. Their application to fish has been reviewed by Townsend & Winfield (1985), Hart (1986) and Lazzaro (1987). Much of modern optimal foraging theory focuses on the choice of foods within a multiprey environment (Stephens & Krebs, 1987). It has been suggested that some foraging animals learn to detect a cryptic prey, and this may assist the predator towards greater efficiency at finding profitable items. Learning also makes it possible for the forager to adapt to systematic variation in its environment, and to survive for longer by using alternative behaviour. 'Fish are well able to learn, so there is every reason to expect significant changes in our understanding of the rules governing their behaviour as new results become available' (Hart, 1986).

At the moment, we have very few results in this area dealing with African fish, but we know that those used for fish culture, such as tilapiines or catfishes, are well able to learn how to feed from food distributors.

Variability and flexibility in diet composition

Different field studies have suggested that many fish broaden their diet to include less preferred prey as the availability of the preferred prey declines (and hunger consequently increases) (Dill, 1983). Examples of this flexibility are provided by the ontogenic and time-related changes observed. But fish in the same population can also have significantly different diets, as well as populations of the same species occupying different habitats. For example, *Brachysynodontis batensoda*, a detritivorous fish in the river systems of the Chari (Blache *et al.*, 1964), the Nile (Sandon & El Tayib, 1953) and the Niger (Daget, 1954), was exclusively a filter feeder in Lake Chad before the drought period (Lauzanne, 1972; Tobor, 1972). As demonstrated by Bishai & Abu Gideiri (1965) with fish maintained in aquaria, this species can modify its food habits and its feeding behaviour as environmental conditions change. When the food offered is composed of zooplankton and floating organisms, *B. batensoda* swim on their backs in order to filter the food at the water surface; if the food offered is composed of benthic detritus and mud, the fish swims in a ventral position and filters sediment. In 1974, during the drying phase of Lake Chad, when the zooplankton became less abundant, lake populations of *B. batensoda* exhibited both these feeding behaviours (Im, 1977).

Changes in diet may occur during spawning migrations as for example in the Chad basin, where *Alestes baremoze* is strictly zooplanktivore in the lake (Lauzanne, 1976), feeds very little during its upstream migration, and becomes partly phytophagous during the flood period when it is in the Chari River. A dominant zooplanktivore regime, but with a significant component of emerging insects, has also been reported for this species in Lake Turkana (Hopson, 1975) and Lake Albert (Verbeke, 1959). In the Niger River, *A. baremoze* feeds mainly on weeds and insects at high water, and on phytoplankton at low water (Daget, 1952). In Côte d'Ivoire, aquatic insects and terrestrial invertebrates constitute the bulk of the diet of *A. baremoze* in rivers (Paugy, 1978), whereas in the Kossou man-made lake, this species is zooplanktivore (Kouassi, 1978).

In lakes, feeding strategies and prey distribution may change with depth. In Lake Turkana for example, prawns are much more abundant in the diet of *Bagrus bajad* caught in deep water than in the shallows (Lock, 1982) (Fig. 7.13). The variation in the percentage of prawns at different depths probably reflects the prey distribution in the lake: densities of prawns are generally lower in shallow waters, and *B. bajad* inhabiting inshore areas depend to a greater extent on fish as a food item. Once *B. bajad* exceeds a length of 50 cm, fish are the preferred food at all depths, even when prawns are present. In water less than 30 m deep, *Alestes* spp., which are the main species of pelagic fish occurring in mid-water, accounted for more than half the total intake of fish by *B. bajad*. In water deeper than 30 m *Haplochromis macconneli*, which is restricted to water deeper than 20 m becomes the dominant fish eaten.

Fish may also switch from one type of prey to another as the relative abundance of the prey changes. *Lates niloticus* was introduced into Lake Victoria in the 1950s and appeared in the fishery at the beginning of the 1960s. The most important foods at that time were mainly the native haplochromines that represented the bulk of the catches (Gee, 1969; Hamblyn, 1966). At the beginning of the 1980s, the diet had completely changed. Native haplochromines were now virtually absent, whereas the prawn *Caridina*, juvenile *Lates*, and the pelagic cyprinid fish *Rastrineobola argentea*, which were absent or rare in the diet in the 1960s, are now the mainstay of this introduced predator (Hughes, 1986). This change in diet mirrors the drastic fall in the abundance of the native haplochromines. After depleting its major food sources, *Lates niloticus* switched to other available prey, including

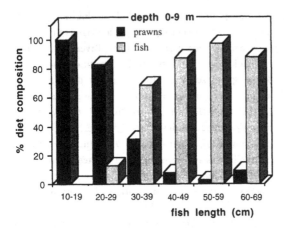

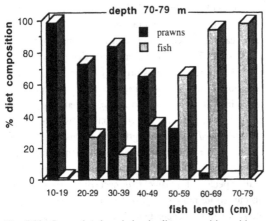

Fig. 7.13. *Bagrus bajad:* variation in diet composition with depths and fish size in Lake Turkana (from Lock, 1982).

Lake Turkana differs considerably from the diet of the same species observed in Lake Chad (Hopson, 1972, 1982), despite the fact that the two lakes have more or less similar food resources. Prawns and small characins were abundant in the diet of Lake Chad populations, particularly in offshore areas. These two food items also occur commonly in the open waters of Lake Turkana, but food resources were clearly partitioned between the two species of *Lates*. *Lates longispinis* exploited the prawns and the small characin *Brycinus minutus*. Small *Lates niloticus*, restricted to the inshore areas of Lake Turkana, feed chiefly on cichlids, whereas the larger fish, which extended their range offshore and overlapped in distribution with *L. longispinis*, exploited completely separate food resources such as larger species of open-water fish including *Bagrus bajad*, *Clarias gariepinus*, *Hydrocynus forskalii* and *Oreochromis niloticus*. Large *L. niloticus* also prey on *L. longispinis* (Hopson, 1982).

Switching from one food resource to another is common among Lake Malawi cichlids and appears to be facultative and opportunistic (Ribbink, 1990). Studies of functional anatomy (Liem, 1978, 1980) provided evidence that these cichlids could potentially feed more broadly upon the resources available than their specialised feeding apparatus would suggest. Individuals usually feeding on aufwuchs can move to a planktonic diet or opportunistically eat eggs or young of other fish. A striking example is given by the species belonging to the Mbuna species flock. Most of them are specialised to feed upon aufwuchs (Ribbink *et al.*, 1983*b*), but the main breeding peak appeared to be geared to a secondary food resource, the plankton, which is most readily available during the cold period of the year (Marsh *et al.*, 1986).

If so many species are able to feed on alternative resources, why do they develop trophic specialisation? One possible answer is that during periods when food resources are scarce, the individuals which are able to feed upon resources that are not available to the others could have an adaptative advantage. Evolution would favour individuals with attributes that give them a chance to succeed under conditions of severe intraspecific competition (Ribbink, 1990).

Social aspects of feeding behaviour
The feeding behaviour of an individual fish is influenced both by the presence of conspecifics and by predators. It can also involve the participation of many individuals

its own progeny. A similar situation occurred in Lake Kyoga (Ogutu-Ohwayo, 1990*a*, *b*), where *Lates* was also introduced but earlier than into Lake Victoria. After the late 1970s, the large Nile perch extended their prey selection to include the introduced tilapiine *Oreochromis niloticus*, whereas *R. argentea* remains an important food item for the intermediate sized *Lates* and *Caridina nilotica* is eaten by the juveniles. The situation observed for *Lates* could also occur in species that seem to be specialised for exploiting specific foods. Cichlids that are typically zooplanktivores or insectivores can switch to feeding on a diatom, *Melosira* spp., when it becomes superabundant (Witte, 1984*b*).

Changes in diet in the presence of related species have also been observed. The diet of *Lates niloticus* in belonging to the same species or to several species. The study of the feeding behaviour of African fish is still in its infancy, but several recent publications have demonstrated that it is a promising field of research.

In diverse communities such as the cichlid communities of the East African Great Lakes, symbiotic or commensal relationships, which may have evolved during their adaptive radiation, can exist between a large number of species (Takamura, 1984). Commensalism among fish of different feeding habit groups and the mutualism among species of the same feeding habit group are assumed to positively facilitate the coexistence of these species.

Hori (1987) and Takamura (1984) tried to find direct evidence of such facilitation in the coexistence of predatory and epilithic algal-feeding cichlids in Lake Tanganyika. Takamura (1983,1984), suggested a symbiotic relationship between *Petrochromis polyodon* and *Tropheus moorii*, in which they share intensive grazing sites. Both have the same temporal grazing pattern although their diets, of unicellular and filamentous algae respectively, were clearly different. The two species also have different feeding apparatus. A removal experiment suggested that *P. polyodon* appears to benefit *T. moorii*. *Petrochromis polyodon* combs algae from and at the same time removes sand and silt from the rock surface, in this way helping *T. moorii* to scrape the filamentous algae. In nature, *T. moorii* preferred to graze on rock surfaces where *P. polyodon* and other species had already grazed and left little sand. In other words, *P. polyodon* may benefit *T. moorii* by decreasing the number of obstacles for the latter's grazing.

Examples of mutualism were also found among scale eaters as well as three piscivorous *Lamprologus* species (Hori, 1987). By following large herbivorous fish, such as *Petrochromis polyodon*, *Lepidiolamprologus elongatus* and *Lepidiolamprologus profundicola* achieved greater hunting success than when patrolling alone. They use the large fish as cover to approach their prey and then dash suddenly from cover to attack them. When another piscivore is near, *L. profundicola* and *Neolamprologus fasciatus* showed a higher success ratio than when hunting alone.

A typical commensal relationship was seen in Lake Tanganyika among the benthic feeders *Lamprologus callipterus*, *Lobochilotes labiatus* and *Gnathochromis pfefferi* (Hori, 1987). They sometimes feed together, forming a mixed-species school and it was noticed that *L. callipterus*

and *G. pfefferi* catch more shrimps when following the caddis fly eater *L. labiatus*, than when they were solitary.

In Lake Malawi, three sibling species of *Petrotilapia* with similar dietary and microhabitat requirements feed selectively on rocky shores (Marsh & Ribbink, 1985) on the aufwuchs cover in shallow water. Dominant adult males of *Petrotilapia genalutea* are territorial, but females and adult males that have not yet established a territory, as well as subadults, are not territorial and normally feed alone, or in small groups, with other species such as *Labeotropheus fuelleborni*, usually in areas undefended by dominant males. Occasionally, however, members of these species band together to form feeding schools of 10 to 300 members composed of a nuclear species, *P. genalutea* (92% of the school members), and six occasionally associated species. These schools raid the rich feeding sites in the territories of highly aggressive territorial *Pseudotropheus* species. These sites support considerably greater standing crops of algae, the food of the schooling species, than undefended sites. After a period within a feeding school, their behaviour reverts to the normal solitary feeding pattern (Marsh & Ribbink, 1986). In this case, schooling is a peculiar behaviour that enables these species to gain access to prime feeding areas. They attempt to settle simultaneously in the territory thereby swamping the aggression of the tenant. They would probably never gain access to the defended rich feeding areas without joining schools.

Another example of temporary schooling is given by *Nimbochromis polystigma* (Ribbink, 1990). It is 'essentially an ambush predator that feeds upon young. While hunting, individuals approach a group of small fish and then lie on the substratum near the group, remaining very still until one of the prey is near enough to be caught by a short, rapid movement of the predator. These ambush predators usually hunt alone in this way, though several individuals may gather around a particularly vulnerable cloud of juveniles. Their alternative hunting tactic is to form large groups that move as a pack over a substratum of aquatic macrophytes, or in areas where sand and rocks are intermingled. In this way, they flush the young out of hiding, catch and devour them.'

Hunting by a group of animals, which potentially benefits all the individuals in the group, is practised by a wide variety of fish and this behaviour is well documented for the catfish, *Clarias gariepinus*, in Lake Sibaya (Bruton, 1979*b*). Social hunting in this species grades from mere scrambling to highly cooperative and synchronised behaviour. Formation feeding is performed in shallow waters by groups of *C. gariepinus*. 'A tightly-knit group in a rough sickle-shaped formation swim slowly inshore near the water surface with their mouths open, herding shoals of small (20–80 mm TL) cichlids. When a water depth of 50–100 mm is reached the catfish suddenly close their mouths with a loud noise, open them again and swim forwards with the mouth wide open and partly above the water surface. These loud, sudden actions cause the prey to panic and jump in all directions. The catfish swim steadily inshore, opening and closing their mouths more or less in unison. Their prey are eventually encircled and form a dense, panic-stricken mass, and are readily captured by the catfish. When all the prey have dispersed or been eaten, the catfish submerge and swim along shore before reforming a pack and swimming inshore as a group again. This cycle may continue every few minutes for over an hour, although it usually terminates after 20–30 min' (Bruton, 1979*b*). Groups of 15 to 40 individuals were observed in Lake Sibaya, but larger groups of about 400 individuals were observed in Botswana. This behaviour of *C. gariepinus* has several functions: to locate dispersed prey; to disorientate elusive prey, e.g. *Sarotherodon mossambicus*, and cause them to lose their normal defensive shoaling synchronisation; and to restrict the prey spatially or to herd prey away from shelter.

Pack-hunting has also been observed in the Okavango Delta, Botswana, in two species of catfish, *Clarias gariepinus* again and *Clarias ngamensis* (Merron, 1993). During the annual drawdown, large shoals of catfish migrate upstream in the main river channels, together with numerous small fish species. Both catfish feed predominantly on the mormyrids, *Marcusenius macrolepidotus* and *Petrocephalus catostoma*, and their organised hunting strategy allows a group to do what one individual cannot achieve, i.e. catch small, elusive mormyrids that are hidden in the edge of the papyrus mat.

Cannibalism occurs in many families of fish, particularly among piscivores and parental care-giving species. Its advantages have been discussed by Fox (1975) and Smith & Reay (1991). The latter defined it as the act of killing and consuming individuals of the same species irrespective of their stage of development. Egg cannibalism is common and may be passive, while consumption of larvae, juveniles and adults is active. It is often suggested that the high quality diet provided by conspecifics allows high growth rates particularly during early life, hence reducing the length of stages when they are at risk from other predators. That is obviously true in a situation of starvation or shortage of prey, but cannibalism has also been recorded among seemingly well-provisioned fish. Cannibalism has been considered by various authors to be an evolutionary competitive strategy, but its effect on the dynamics of wild populations, and its role in density regulation, is rather obscure.

While parental care can facilitate the occurrence of cannibalism it is often accompanied by some degree of inhibition. Limited filial cannibalism has been reported in cichlids (Coe, 1966), while in aquarium breeding the predation of eggs by parents is common and the destruction of broods has often been assumed to be a laboratory artefact. Schwanck (1986) demonstrated that cannibalism in *Tilapia mariae* occurs quite frequently at the egg-stage in aquaria, males being mostly responsible. From a series of experiments, he concluded that most of the observed cannibalism was caused by lowered motivation for parental care in the males, part of it being related to artificial conditions. Nevertheless, when paternal care occurs, the male receives several clutches from the females and his consuming part of them could be compensation for reduced foraging time as a result of parental duties (Rohwer, 1978). This permits continued care of the remaining brood, during the time when parents experience an increased risk of injury and predation and are excluded from other breeding attempts, and thus has adaptive value. Early destruction of the brood might be beneficial if the expected immediate success of reproduction is low.

In many cases, cannibalism occurs when large individuals prey on small conspecific ones. It has been recorded in *Clarias gariepinus* (Hecht & Appelbaum, 1988) and *Oreochromis niloticus* (Smith, 1989), while a spectacular case is that of the *Lates niloticus* introduced to Lake Victoria. In the 1980s, their diet included a large proportion of prawns, and juvenile *Lates* for fish less than 80 cm TL (Hughes, 1986).

Resource-sharing mechanisms

Niche theory predicts that the more the niches of different species overlap, the more likely it is that competition will occur. The coexistence of many related species, often sharing the same resources, raises the question of the existence of factors that allow sympatric species to avoid too intensive competition. This stimulated a number of studies on the patterns of resource use by species. Research on 'resource partitioning', which means how species differ in their resource use, has as its primary goal description of the limits that interspecific competition places on the number of species that can stably coexist (Roughgarden, 1976).

Fish are challenging subjects for studies of resource partitioning because, unlike most vertebrates, they exhibit indeterminate growth, and many fish assemblages are temporally structured (Ross, 1986). Analysis of 230 studies on the diets, spatial distribution, patterns of diel activity and use of other relevant dimensions by fish in an assemblage showed that temporal separation was much less important than trophic or habitat separation. The greater importance of trophic rather than habitat partitioning in some aquatic habitats may result from the often pronounced morphological specialisation of fish trophic mechanisms for capturing prey, as well as low habitat specialisation by fish living in habitats of limited duration (Ross, 1986).

If patterns in resource partitioning among fish represent effects of exploitative competition, one should be able to demonstrate that resource limitation is important in these communities. Although a large but diffuse literature indicates that resource limitation often occurs, different works on African lakes have also indicated that food is usually not an important factor regulating the number and distribution of fish. Fryer & Iles (1972) regarded food as superabundant on the rocky shores of Lake Malawi where numerous species often shared a similar food resource, and considered predation by larger fish to be a more important factor for population control, also reducing the effects of competition between the prey species of fish. Similar conclusions were reached in Lake Victoria (Corbet, 1961) and Lake Chad (Carmouze et al., 1972).

Many studies have pointed out that sympatric species sometimes exhibit very different spatial distributions.

Space-sharing mechanisms include diurnal/nocturnal activity cycles, separate feeding and hunting areas both for developmental stages of one species or for related species, shelter sites, or spawning sites. For example, the African pike, *Hepsetus odoe*, and the tigerfish, *Hydrocynus forskalii*, are sympatric piscivorous species in the Zambezi River. During the period of annual flooding, juveniles of both species coexist in flooded savannah regions. Adults of the two species have very similar diets but exhibit very little overlap in their use of river and aquatic floodplain habitats (Winemiller & Kelso-Winemiller, 1994). *Hydrocynus forskalii* occupies the open water of the main river channel almost exclusively, while *H. odoe* inhabits vegetated environments of river backwaters and lagoons. Differences in diet and habitat utilisation by these two sympatric species are best explained as resulting from the combined and interactive effects of predation threat, interspecific differences in foraging mode, and differential efficiencies in open-water versus structured environments.

When habitats and food habits are considered simultaneously, the seven Upper Zambezi *Serranochromis* species exhibit nearly complete ecological separation during the low-water season (Winemiller, 1991). Adults of the largest species (*S. robustus, S. altus, S. giardi*) inhabit the main channel and exhibit almost no dietary overlap. Adult *S. robustus* with massive jaws and pharyngeal plates specialise on small *Synodontis* catfish and are diurnal piscivores; *S. altus* specialise on nocturnal mormyrid fish and appear to be a crepuscular or nocturnal feeder; *S. giardi* feed primarily off the sand substrate on small bivalves and aquatic Trichoptera larvae. The four lagoon/backwater dwelling *Serranochromis* also show little diet overlap. *Serranochromis macrocephalus* consumes mostly crepuscular or nocturnal mormyrids, while *S. angusticeps* feed primarily on small diurnal *Barbus* species and characids. *Serranochromis codringtonii* has benthic foraging habits and feeds on snails and seeds, and *S. carlottae* feeds mostly on scales and diurnal aquatic insects.

Sharing food and space in Great African Lakes

It has been demonstrated (van Oijen et al., 1981; Witte, 1984b) that populations are spatially isolated in Lake Victoria, and intralacustrine allopatry of species is an important way of preventing direct competition between

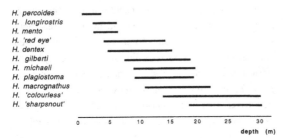

Fig. 7.14. Depth distribution of adults of some haplochromine piscivores species from the Mwanza Gulf (Lake Victoria) (from van Oijen, 1982).

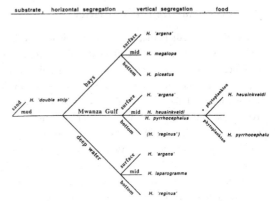

Fig. 7.15. Schematical representation of expected ecological segregation among zooplanktivore haplochromines in Lake Victoria (from Goldschmidt *et al.*, 1990).

species feeding on the same resources. Despite the fact that piscivorous haplochromine species are found in all major habitats in Lake Victoria, many piscivores, and most non-piscivorous species, are restricted to certain substrate types (van Oijen, 1982). Juveniles are generally found in shallower and less exposed areas than their parents. Comparably, in Mwanza Gulf, adults of many species have a restricted depth distribution, some species being only recorded in shallow waters whereas other species only occur in deeper zones (Fig. 7.14). A number of species are restricted to sandy bottoms that do not extend beyond 6 m depth in the area studied where muddy bottoms prevailed in shallow water. It is therefore difficult to determine whether the depth or the nature of the substrate was the limiting factor, but it seems that the depth preference range of benthic piscivores is much narrower in shallow water than in deeper habitats.

Thus, different rhythms of activity may reduce competition between species and allow the coexistence of species sharing the same resources. But species occurring together in the same water system may also have different feeding or spawning sites. Especially in a group of closely related species belonging apparently to the same trophic group, there should be mechanisms that allow spatial niche differentiation. This was investigated in a species flock of zooplanktivorous haplochromines from Lake Victoria (Goldschmidt *et al.*, 1990). Distinct segregation patterns (Fig. 7.15) were observed suggesting that these species were ecologically isolated to a great extent. Niche partitioning by habitat is the predominant means of reducing competition between zooplanktivorous species. Clear horizontal segregation patterns were observed in the Mwanza Gulf, both for adults and juveniles.

Bottom type, exposure to wind and depth of the water column are the most important abiotic factors that are related to horizontal distributions. Most juvenile zooplanktivores were restricted to shallow water (less than 9 m deep) where they find shelter from predators. Species with a relatively large overlap in horizontal distribution may segregate vertically, perhaps with a circadian rhythm of vertical migration. Finally, there may be slight differences in diet composition with some species eating more phytoplankton than others.

To summarise, zooplanktivore haplochromine species in Lake Victoria segregated along at least one of the niche dimensions: horizontal and/or vertical separation in space, and food. Niche segregation may be reinforced by reproductive patterns. Among the Lake Victoria zooplanktivores, the taxonomically most similar species differed most when breeding characteristics were taken into account (Goldschmidt & Witte, 1990).

Among the paedophage species in Lake Victoria, *Haplochromis barbarae* steal eggs from a spawning female in the short period during which eggs are laid and then taken into the female's mouth (Witte-Maas, 1981). In contrast, *H. 'rostrodon'* obtains its prey by ramming mouthbrooding females and engulfing their snouts (Wilhelm, 1980). The anatomical differences between the two paedophages reflect their behavioural differences: the oral features of *H. barbarae* are similar to those of bottom-feeding insectivores; whereas the suction apparatus of *H. 'rostrodon'* resembles that of an ambush-hunting piscivore. The anatomical differentiation of the other 20 species of paedophages suggest further differentiation of their feeding behaviour (Barel *et al.*, 1991).

The cichlid molluscivore species of Lake Victoria can be divided into the oral shellers and the pharyngeal crushers. The latter possess hypertrophied pharyngeal jaws and muscles, whereas the shellers have short, stocky oral jaws powered by thick adductor muscles (Barel, 1983). Shellers can feed on larger snails than crushers, but their feeding strategy requires a visual guide and accurate predation technique to grasp a large snail by its foot. Most oral-shelling species are restricted to sandy habitats, probably due to the need for visual cues. Conversely, crushers live over various bottom types, but feed on smaller prey, and swallowing is preceded by a time-consuming process of sorting out shell fragments (Barel *et al.*, 1991).

Most of the results obtained by different research teams on these cichlids suggest that when detailed studies are conducted, species that have been regarded as sharing the same trophic requirements exhibit slight but distinct interspecific differences in feeding ecology, feeding behaviour, feeding sites and habitat (Yamaoka, 1991). Hori (1987) insisted that differences in food preference or habitat use among coexisting related species are merely a condition that permits coexistence.

Resource partitioning in the pelagic fish assemblage of Lake Turkana

In Lake Turkana, where the tropho–dynamic relationships among 11 species of zooplanktivores were investigated (Hopson & Ferguson, 1982), resource partitioning was accomplished in several ways: (i) spatial separation of competitive species achieved by differences in vertical and horizontal distribution and by diel migration; (ii) interspecific differences in prey-size selection; (iii) interspecific differences in diel feeding patterns; and (iv) ontogenetic changes in diet.

During the day, *Alestes baremoze* fed principally on *Tropodiaptomus banforanus* in the surface layers while post-larval *Neobola stellae* ingested copepod nauplii and diatoms. At intermediate depths the two species of small *Brycinus* coexisted: *Brycinus minutus* preying chiefly on zooplankton; and *Brycinus ferox* preying on small fish and prawns in addition to zooplankton. Adult *Neobola stellae*, which fed entirely on zooplankton, were scattered over the lower part of the water column between mid-water and the bottom. In contrast, *Lates* spp. post-larvae ranged throughout the entire column, and thus overlapped with the preceding species as well as with *Synodontis schall*, which at times fed exclusively on zooplankton.

Considerable changes in the distribution of pelagic fish occurred towards night. *B. minutus* and *B. ferox*, which fed also at night, migrated upwards into the superficial layers, and their food changed considerably. Cladocera dominated in the food of *B. minutus* by day whereas *Tropodiaptomus banforanus* was eaten almost exclusively by night. *Tropodiaptomus banforanus* also became the predominant food of *B. ferox* at dawn. This shift from one prey to another may be important for survival in situations where particular prey species were likely to be periodically depleted. *Lates* spp. post-larvae and *A. baremoze* do not feed at night whereas *S. schall* is principally a nocturnal feeder.

As usual, selectivity of prey varied both within and between species. Ontogenetic changes were important in *Neobola stellae* where the smallest size groups fed entirely on diatoms, switching almost exclusively to adult copepods when they had grown above 20 mm length. *Lates* spp. post-larvae longer than 6 mm fed exclusively on adult copepods.

Interspecific competition among the inshore zooplanktophages was also decreased by variations in the diel feeding cycle and habitat segregation. *Aplocheilichthys rudolfianus*, small *Raiamas senegalensis* of under 22 mm FL and *Lates* spp. of 22–37 mm TL fed exclusively on zooplankton, the two latter species only in the daytime. Other species with a mixed diet, such as *Brycinus nurse*, large *Raiamas senegalensis* and inshore *Neobola stellae*, likewise fed on non-insect plankton during day, but switched to planktonic insect food after dark. At times of food shortage, the minute characid, *Micralestes acutidens*, that usually fed on zooplankton, largely nauplii, may ingest the blue-green algae *Microcystis* in large quantities.

Food webs

Food webs are diagrams depicting which species in a community interact in feeding. But these interactions change at least seasonally and not all interactions are equally strong. 'Food webs are thus caricatures of nature' (Pimm, 1982), but they give a picture of the processes at work in ecosystems. Food webs describe which kinds of organisms in a community eat which other kinds. A community food web describes the feeding habits of a set of organisms chosen on the basis of taxonomy, location, or other criteria without prior regard to the feeding habits among the organisms. 'Webs were invented in the natural-historical approach to community ecology as a descriptive summary of which species were observed to eat each other. If an ecological community is like a city, a web is like a street map of the city: it shows where road traffic can and does go.' (Cohen, 1989).

A trophic species is a collection of organisms that have the same diets and/or the same predators. That could be a biological species, or several species, or a stage in the life history of a biological species. A web is, therefore, a collection of trophic species, together with their feeding relations. Some species are basal (they eat no other species, such as phytoplankton), or top (no other species eats them) or intermediate (they eat at least one species and one species at least eats them). A food chain describes the energy transfer through different trophic levels, from producers to top consumers. A chain is a path of links from a basal species to a top species. The length of a chain is the number of links in it. The energy transfers are subject to the laws of thermodynamics: at each link, considerable energy is dissipated from the system in the form of heat.

The concept of a food web as a network of local trophic interactions is a fairly simple idea that dates back to the early development of ecology as a scientific discipline. For a long time food webs served principally as heuristic devices, useful in depicting complex ecosystems as diagrams composed of many interactive parts and enhancing our understanding of pathways of energy and material transfer in aquatic ecosystems. However, the recent surge of interest in food webs seems related to the question of the functional role of biodiversity, in relation with May's theoretical findings that defined conditions for mathematical stability within randomly constructed interactive networks (May, 1972, 1973).

Recognition of common features and properties of food webs is likely to reveal much about the structure and function of ecosystems. Ecology has been challenged to seek out and explain robust universal patterns among natural food webs despite the practical difficulties of dealing simultaneously with their spatial, temporal and intraspecific variation. Chain lengths are usually short and include four to five links (Cohen, 1989). Food webs in aquatic ecosystems differ primarily in the extent to which the main energy source is within the system (autochthonous) or is imported into the system (allochthonous). In lakes, the food webs of the pelagic zone are fundamentally based on the primary production by phytoplankton. In contrast, the food webs in the benthic zone are usually based on organic material, including living and dead organisms, imported from pelagic communities or from littoral zones. In streams and rivers, but also in some lakes, the food webs are often heavily dependent on allochthonous organic material falling into the water or washed in by rains.

Bowen (1988) pointed out that the majority of African fish do not feed directly on living plant material or detritus. Most species depend either on planktonic or benthic invertebrates for food. True primary consumers only belong to six genera (*Labeo*, *Oreochromis*, *Sarotherodon*, *Tilapia*, *Citharinus* and *Distichodus*) that is to say less than 7% of the total species. These observations raise the fundamental question: why are there so few primary consumers species in Africa? No clear answer can be given at the moment, and it is assumed that 'since there are few primary consumers, then the resource spectrum available to them must be narrow'. Possibly, the plant and detritus resources have not enough protein to support growth or are less digestible than diets utilised by secondary consumers. For instance, the macrophyte grazers *Tilapia rendalli* and *Tilapia zillii* assimilate respectively 55% and 30% of their diet, which is low compared to higher level consumers that often assimilate about 85% of what they ingest (Bowen, 1988). But that does not explain why the few primary consumer fish are so widely distributed and generally develop large populations.

In order to illustrate the existence of food webs, selected examples obtained mainly in African lakes are proposed below.

Phytoplanktivore communities: from producers to consumers

The shortest food chain could be illustrated by simple fish communities feeding on phytoplankton. The cichlid, *Oreochromis grahami*, is endemic to Lake Magadi where it feeds on benthic filamentous cyanophytes, insect larvae and copepods. It was introduced in the 1950s into Lake Nakuru to combat mosquito breeding. There it is the only fish species, and quickly became one of the major filter feeders exploiting the high-standing crops of the cyanobacterium, *Spirulina platensis*, which provide 95% of its diet. The other major filter feeder is the flamingo, which occurs in large flocks.

Phytoplankton, mostly blue–green algae, represented about 95% of the total biomass in Lake George, and varied only slightly throughout the year (Burgis *et al.*, 1973). It was the most abundant type of food and was consumed mainly by two species, *Oreochromis niloticus* and *Haplochromis nigripinnis*, which largely dominated the fish biomass (Gwahaba, 1975). These species are able to digest up to 70% of the cyanobacteria they ingest (Moriarty, 1973).

The pelagic fish community of Lake Tanganyika

The trophic hierarchy among the pelagic fish of Lake Tanganyika is relatively simple (Fig. 7.16), but their diets show considerable overlap and change during the life cycle of most species. This has been well investigated and most of the information available is synthesised by Coulter (1991*b*).

Stolothrissa tanganicae and *Limnothrissa miodon* are clupeid fish, which appear from their diet and life history to be narrowly specialised for pelagic life and are key members of the pelagic food chain, linking planktonic and piscivorous trophic levels. *Stolothrissa* depends on plankton for food in both the larval and adult stages. Juveniles living inshore feed mostly on phytoplankton, and switch to the adult diet of zooplankton (mainly copepods, and sporadically shrimps) as they start to move into the pelagic habitat at about 50 cm TL. *Stolothrissa* has a short life cycle (only one year) and a great fecundity: mature individuals occur throughout the year and may spawn several times in a year. *Limnothrissa* occupies a more inshore habitat than *Stolothrissa*. The diet of their young is similar to that of *Stolothrissa* and adults are zooplanktivores, but also feed regularly on juvenile and adult *Stolothrissa*. This predation is believed to be an important factor influencing the stock dynamics of *Stolothrissa*, especially in areas of moderate depth where *Limnothrissa* tend to be more numerous. Spawning occurs throughout the year, but with a peak during the rainy season. This

species usually lives two years. Maximum length is 175 mm for *Limnothrissa*, and 110–120 mm for *Stolothrissa*. Although they cohabit the same areas at times in their life cycle, for the most part *Stolothrissa* lives offshore and *Limnothrissa* inshore. But *Limnothrissa* can replace *Stolothrissa* successfully offshore, contributing to the maintenance of clupeid biomass at high levels.

These clupeids are the prime food for predators, which are essentially the entirely pelagic *Lates stappersii* and *Lates microlepis* feeding mainly on *Stolothrissa*. Other predators, such as *Lates mariae* and *Lates angustifrons*, also feed on *Stolothrissa*, mainly during its annual peak of abundance, as do other benthic predators, such as *Boulengerochromis* or *Dinotopterus*. Post-larvae and fry of *Lates* spp. may to some extent compete with *Stolothrissa* for planktonic food. This is particularly true for *Lates stappersii* and *Stolothrissa* in the same size range, given the local abundance of *L. stappersii*. Young of other *Lates* species also commonly feed on zooplankton until they reach about 30 mm TL. Young zooplanktivorous *L. stappersii* become almost exclusively piscivorous at 130 mm and their major prey is then *Stolothrissa*, but shrimps could also be heavily consumed and cannibalism is not rare. This ability to switch between foods gives *L. stappersii* considerable independence from the seasonal fluctuations that occur in *Stolothrissa*, provided that shrimps are abundant when *Stolothrissa* are not. *Lates stappersii* can eat fish up to 40% of its body length.

Lates microlepis is also a voracious predator of clupeids. After leaving the littoral weed cover, juveniles mainly feed on immature clupeids inshore and later upon adult clupeids in the pelagic zone. *Lates stappersii* are also eaten up to 40% of the predator's length.

Hecky (1991) pointed out that the pelagic food web of Lake Tanganyika is marine in nature: the primary grazer is a diaptomid copepod just as in many productive marine systems, and the primary planktivores (Clupeidae) are also of marine origin, as are the main predators that also belong to a marine family, the Centropomidae.

Fish communities in tropical waters are characterised by relatively large biomasses of predators (Lowe-McConnell, 1987), but in Lake Tanganyika, the proportion of predators (in biomass and number of species) in different biotopes seems exceptionally high compared with other African lakes (Coulter, 1991*a*). It could be

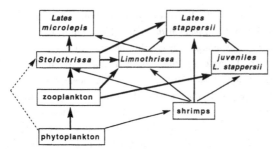

Fig. 7.16. Food web of the pelagic fish community of Lake Tanganyika. Heavy arrows for major food preferences (modified from Coulter, 1991*b*).

argued that one of the causes has been the evolution in the lakes of certain predators that dominate within their particular ecological communities, such as *Lates microlepis*, *L. mariae* and *Boulengerochromis microlepis*. Another cause of the high predator/prey ratio could stem from wide differences in the biological productivity between the small short-lived prey with high turnover rates, and the large long-lived predators (Coulter, 1991*a*). Whatever the cause, Lake Tanganyika seems to have an unusually high fish yield relative to its rate of algal production, and the pelagic community is made up of highly efficient species. The result is an extremely efficient ecosystem operating at high rates of trophic transfer (Hecky *et al.*, 1981).

Trophic interactions in Lake Kivu

Only 24 native and introduced fish species, including 16 cichlids, have been recorded from Lake Kivu (Ulyel, 1991). The open water, pelagic food webs are relatively simple including the introduced zooplanktophage, *Limnothrissa miodon*, the microphytophage, *Haplochromis kamiranzovu*, the micropredator, *Raiamas moorei* (feeding on insects but also on juvenile fish), and the piscivore, *Haplochromis vittatus* (Ulyel, 1991) that also feed on allochthonous terrestrial invertebrates.

In the benthic communities, endemic haplochromines demonstrate a wide spectrum of trophic adaptations despite their comparatively low number. Some feed on benthic and epilithic algae (*Haplochromis olivaceus)* or phytoplankton (*H. kamiranzovu*), or organic detritus (*H. astatodon*). Others are insectivores (*H. graueri*, *H. paucidens*) and *H. vittatus* is the only true piscivore of the

Lake (Ulyel, 1991). The clupeid *Limnothrissa miodon* is the main zooplanktivore, but the juveniles of many haplochromines feed at this trophic level. The tilapiines (*Tilapia rendalli*, *Oreochromis niloticus*, *O. macrochir*) mainly feed on benthic and epilithic algae and the two clariids (*Clarias gariepinus*, *C. liocephalus*) are omnivores.

Study of the feeding habits of the ten endemic species of *Haplochromis* demonstrated that each species occupies a sufficiently distinct ecological niche to avoid competition for important prey. This is assumed to be the consequence of adaptative radiation (Ulyel *et al.*, 1990).

Trophic interactions in the open waters of Lake Chad

In this shallow lake, the food chains are much more complex than in Lake Tanganyika, and Lauzanne (1976, 1983) distinguished a planktonic chain from a benthic and detritus chain (Fig. 7.17). The planktonic chain includes small characins, such as *Micralestes acutidens* and *Brycinus nurse dageti* (Robinson & Robinson, 1969), as well as *Pollimyrus isidori* feeding mainly on zooplankton. They also have a short life span and a high turnover rate. Larger zooplanktophages are *Alestes baremoze*, *Brachysynodontis batensoda* and *Hemisynodontis membranaceus*.

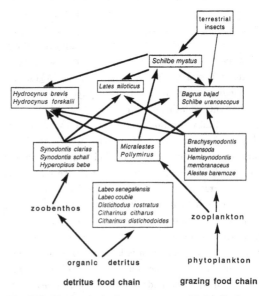

Fig. 7.17. Food webs in the open waters of Lake Chad (modified from Lauzanne, 1983).

Predators apparently did not feed upon the benthos feeders. Remains of the typical middle to large size fish species that feed on detritus or benthic items were scarcely ever found in predator's stomachs. The top predators therefore essentially feed on the secondary consumers of the grazing food chain. Less strict piscivores, such as *Schilbe uranoscopus*, *Schilbe mystus* and *Bagrus bajad*, consumed a lot of small pelagic fish species, such as *Micralestes* and *Pollimyrus*, that are abundant in the ecosystem, but they also ingest a considerable amount of fish debris and terrestrial insects, which were a significant food particularly for *Schilbe*.

Among the more strict piscivores, *Hydrocynus forskalii* also fed on small pelagic species, as well as on *Schilbe*. The latter was of considerable importance in the diet of most of the predator species. *Lates niloticus* mainly consumed *H. forskalii*, *S. intermedius* and *P. isidori*.

The presence of terrestrial allochthonous material in the diet of freshwater fish is not unusual. Worthington (1932) mentioned the abundance of terrestrial insects and fish in the diet of *S. mystus* in Lake Albert, and similar observations were made in the Nile where *Schilbe* feed on fish, terrestrial insects (Orthopteres) and insect larvae (Sandon & Al Tayib, 1953).

Thus, in summary, results from Lake Chad and Lake Tanganyika have suggested that zooplanktivorous fish are an essential link in open-water food chains and similar results were obtained in Lake Turkana (Hopson & Ferguson, 1982). In the Lake Chad archipelago, the food webs of the top consumers are more complex than in the open water, but the overall pathway is fairly similar. Nevertheless, the zooplanktivores dominate the fish biomass in the archipelago instead of the top consumers that play a major role in the open water (Lauzanne, 1983).

8 Diversity of reproductive strategies and life histories

Life histories lie at the heart of biology; no other field brings you closer to the underlying simplicities that unite and explain the diversity of living things and the complexities of their life cycles. Fascinating in themselves, life histories are also the keys to understanding related fields. Life history theory is needed to understand the action of natural selection, a central element of evolution, the only theory that makes sense of all biology. It also helps us understand how the other central element, genetic variation, will be expressed.

Stearns, 1992

The main goal of reproduction remains the transfer of the information stored in the parents' genomes to the next generation. A great deal of the variation among life histories relates to aspects of reproduction, while the reproductive success of a fish (as for other animals) depends on the amount of resources it allocates to reproduction as opposed to growth and maintenance, and on where and when spawning takes place. *The diversity of life histories that is the result of phylogeny and adaptations to changing environments also favours speciation in isolated populations.*

The reproductive cycle consists of two distinct parts: gametogenesis and spawning. Gametogenesis is the formation of gametes (oocytes or spermatozoa) the duration of which varies from species to species, but appears to depend on regular and long-term changes, and seems to be initiated by subtle changes in the environment in tropical zones (Munro, 1990). Spawning, which comprises the sequence of events (oocyte maturation, ovulation and the formation of sperm) leading to the liberation of gametes, appears to be the most critical period of the reproductive cycle and to require specific stimuli, even for fish with a continuous period of reproduction. In tropical and equatorial zones, these stimuli could be climatic events such as flooding or rainfall, or probably many other triggers (Billard & Breton, 1978). It is likely

that these environmental cues are adaptations which ensure that the larvae, hatched soon after reproduction, have adequate environmental conditions.

Modes of reproduction

Most fish are oviparous: eggs and sperm are expelled into the surrounding water where fertilisation occurs immediately. However, fish can either produce all their offspring simultaneously in a single reproductive event or they can produce them in a series of separate events. They may also provide parental care, and may have specific requirements (laying and incubation substrates) and specific behaviour for spawning. All these strategies have been developed to optimise reproductive success.

The reproductive strategy of a fish species is the overall pattern of reproduction typically shown by individuals in that species. It is a complex of reproductive traits including age at first reproduction, size- and age-specific fecundity, size of gametes, timing of the reproductive season, reproductive behaviour, etc. The term tactics focuses attention on the range of variation that these reproductive traits are able to exhibit when faced with environmental variations, as well as on the mechanisms of the interactions between environmental variables and reproductive responses. Each individual fish has a suite of reproductive traits that are genetically controlled, and are the inheritance of a long evolutionary and phylogenetic process. A major goal of population biologists is, therefore, to understand how and what particular reproductive strategies and tactics are adaptive to particular environmental conditions. One question for instance is whether similar reproductive strategies show similar tactical responses to fluctuations ?

Spawning strategies

Eggs and larvae are usually vulnerable to both unfavourable abiotic conditions and to predators while they cannot escape such hazards. Species also have to develop spawning strategies to prevent or avoid other potential hazards, such as lack of oxygen in the water, which is a major threat. So reproductive cycles are adapted to the seasons, and to the existence of favourable conditions for the survival of eggs and larvae. Many fish species exhibit an annual cycle of reproductive development with a restricted period for spawning, but, in tropical waters, many species also have extended spawning seasons, either with different individuals breeding at different times throughout the year, or multiple-spawning behaviour of individuals. Three main types of strategy have been recognised.

1. **Species with a short annual spawning period.** In 'total spawners' (*sensu* Lowe-McConnell, 1975), ova ripen all at the same time and are produced in one batch. They are usually fecund fish, and spawning generally occurs once a year. Many riverine fish undergo long-distance migrations and spawn just before the peak flood, at a time when the young are able to find food and refuges in the floodplains. Spawning may be stimulated either directly or indirectly by local rains or by floods coming downriver. Many lake fish such as *Alestes baremoze* retain the habit of moving seasonally into rivers to spawn (Paugy, 1978).

2. **Multiple spawners with a long annual reproductive period.** Their eggs ripen in batches and are laid either at intervals throughout a breeding season, or aseasonally. This is a strategy that may be considered to have an adaptive advantage if one of the batches may be endangered by unsuit-able environmental conditions. Many species make only local movements to spawn, an extreme case of which is observed in cyprinodonts that exhibit a daily spawning pattern. The daily fecundity of annual killifishes ranges from a few eggs to at least 50 (Simpson, 1979). This iteroparity is related to environments which are at best semipredictable. There are, of course, intermediate situations such as *Lates niloticus*, which is said to be a total spawner (Hopson, 1972) but has an extended breeding season.

3. **Small-brood spawners.** Fish with parental care tend to have smaller broods than do fish that leave their eggs unguarded and mouthbrooders have the smallest batches of eggs. The cichlids, for example, produce small batches of eggs at frequent intervals for most of the year. In concrete tanks, the time between two successive spawnings is about 2 weeks for *Sarotherodon melanotheron* and 3 weeks for *Tilapia guineensis* (Legendre & Ecoutin, 1989). In fish ponds, Moreau (1979) found a mean time of 29 days between two successive spawnings for *Tilapia rendalli*, 33 days for *Oreochromis niloticus* and 48 days for *Ooreochromis macrochir*. For some *Oreochromis* species, the time interval between two spawnings is estimated to be 4 to 6 weeks (Baroiller & Jalabert, 1989). In female mouthbrooders, incubation of eggs and larvae has an inhibitory effect on oocyte development (Smith & Haley, 1988).

The number of broods per year in the natural lakes of Madagascar was estimated to be 7 for *T. rendalli*, 3–4 for *O. niloticus* and 2–3 for *O. macrochir* (Moreau, 1979). In Lake Victoria, Fryer (1961) thought that the indigenous *Oreochromis variabilis* produced three or possibly five broods in eight months. For the cichlid, *Lepidiolamprologus elongatus*, in Lake Tanganyika, the duration of brood care by the parents is nearly 3–4 months, and females usually spawn at 3–5 month intervals (Nakai, 1988). River fish with parental care, often produce a batch of eggs in the low-water season before the floods. In aquaria with constant optimal conditions, *Labeotropheus* species produce from five to seven broods per year (Balon, 1977).

Selection, in the 'choice' by a fish species of a particular set of life-history tactics, is probably not towards a single 'big-bang' type reproduction (semelparity) involving a high reproductive effort, but towards iteroparity, with increased longevity and lower reproductive effort (Mann & Mills, 1979). Under natural conditions, such a strategy would be expected to protect populations against catastrophic crashes. Riverine populations that breed during the floods presumably try to spawn at a time and a place where those factors ultimately controlling reproductive success are optimal (Munro, 1990), but as a result of the year-to-year variability of floods this time is unpredictable. The reproductive strategy for long-lived ostariophysans (usually large species) may therefore be to spawn once each season over several years in an attempt to ensure that at least one season's spawning does occur at the right time. Conversely, for small and short-lived species, the breeding season tends to be more prolonged. In predictable environments, fish could become highly specialised in their reproductive tactics but the reduced fecundity associated with parental care, such as substrate brooding, may also be considered a disadvantage when unexpected changes in the environment occur.

Reproductive guilds

There have been a number of different attempts to develop an ecological classification by grouping animals that breed in similar ways. Balon (1975) proposed a comprehensive classification of reproductive styles based largely on the sites of spawning and the degree of parental care. He (Balon, 1981a, 1984) distinguished three main ethological sections (Table 8.1) irrespective of their phyletic origin: (i) non-guarders, with open substrate spawners and brood hiders; (ii) guarders, with substrate choosers and nest spawners, whose zygotes are cared for by one or both parents and may include oxygenation, removal of debris, or defence; and (iii) bearers, either external or internal, whose zygotes are carried for at least a portion of the embryonic period of development by one or both parents.

Each section includes different ecological groups and reproductive guilds (Table 8.2) (Balon, 1984; Bruton & Merron, 1990). Different reproductive styles will markedly differ in the duration, character and even presence of successive life-history periods in ontogeny. In such a hierarchical system, the succession of guilds in each eco-

Table 8.1. *The three ethological sections of reproductive guilds and their associated biological characteristics*

Non-guarder	Guarder	Bearer
open substrate spawner	substrate chooser	external
brood hider	nest spawner	internal
no parental care	some parental care	intensive parental care
high fecundity		low fecundity
indirect development	direct development	
high number investment at low cost		low number investment at high cost

From Balon, 1981a; Bruton, 1989.

Table 8.2. *Simplified classification of reproductive styles in fish*

NON-GUARDERS
Open substratum spawners
    Pelagic spawners: *Stolothrissa, Limnothrissa, Lates, Ctenopoma muriei, Alestes*
    Rock and gravel spawners: *Opsaridium microlepis*
Brood hiders
    Annual fishes: *Nothobranchius*

GUARDERS
Clutch tenders
    Plant tenders: *Polypterus*
    Rock tenders
Nesters
    Froth nesters: *Hepsetus odoe, Ctenopoma ansorgei*
    Hole nesters: *Protopterus*, many Tanganyika cichlids
    Sand nesters: *Tilapia zillii* and other cichlids
    Plant material nesters: *Pollimyrus, Gymnarchus, Heterotis*

BEARERS
External brooders
    Mouth brooders: *Oreochromis* and *Sarotherodon* spp., haplochromines, *Arius*
    Pouch brooders: *Syngnathus*
Internal live bearers
    ?

Sensu Balon 1990.

logical group and the succession of groups within the three ethological sections follows trends from less to more protective styles (Balon, 1981a).

Most fish are non-guarding, egg-scattering pelagic

Table 8.3. *Some broad characteristics of the eggs and of the reproductive behaviour for different African fish families*

Families	Mating	Breeding sites	Products	Parental care	Migration
Lepidosirenidae	?	swamps	adhesive eggs	male builds, guards and aerates nest	none known
Polypteridae	?	swamps	adhesive eggs	?	?
Clupeidae	school spawning	open water or surface of bottom	pelagic eggs or demersal adhesive eggs	?	some species anadromous
Phractolemidae	presumably all pair	?	?	?	?
Notopteridae	distinct pairing	swamps, lakes, rivers	adhesive eggs	eggs guarded by males	
Osteoglossidae	presumably all pair	swamps, lakes, rivers	non-adhesive eggs?	eggs guarded by males rivers or oral incubation	?
Gymnarchidae	distinct pairing	ponds and swamps	adhesive eggs	nest guarded	to overflow swamps
Mormyridae	pairing	rivers	?	nest guarded	some species
Characidae	pairing	on rocks, plants or out of water	usually demersal adhesive	male, seldom female guards eggs in some species	anadromous movement in some species
Citharinidae	pairing	among plants in swamps	usually demersal adhesive	male tends to remain near nesting site	*idem*
Cyprinidae	pairing or polyandry		usually demersal adhesive	in some species eggs in nests guarded or not	anadromous movement in stream species
Bagridae	distinct pairing	various	large adhesive eggs	none, or nests guarded by both parents	
Schilbeidae	distinct pairing	various	large adhesive eggs	eggs unguarded	
Clariidae	distinct pairing	various	large adhesive eggs	eggs in guarded nests for some species	?
Mochokidae	distinct pairing	various	large adhesive eggs		?
Ariidae			very large eggs	mouthbrooding by males	
Cyprinodontidae	distinct pairing		demersal eggs with adhesive threads	nest guarded by males for some species	none known
Eleotridae	distinct pairing	often under stones	demersal eggs with adhesive threads	males builds nest or selects nesting sites and guards and usually aerates eggs	none known
Gobiidae	distinct pairing		demersal eggs with adhesive threads	*idem*	none known
Cichlidae	pairing or polyandry		adhesive eggs or demersal non-adhesive	eggs guarded or orally incubated	none known
Nandidae				male care	
Mastacembelidae	distinct pairing		demersal non-adhesive eggs	none	not known

Modified from Breder & Rosen, 1966.

spawners, a reproductive style that seems to be the general, if not the ancestral, condition of reproduction in fish (Balon, 1990). More specialised guilds (guarders and bearers) have low fecundity but large-yolked ova, exercise extensive parental care and spawn in specially prepared nests. Embryos have an accelerated differentiation, and precocial forms produce large, well-developed young.

Balon's reproductive guild framework was based on the premise that environmental requirements and the adaptations of early life stages are likely to account for a large amount of the variance in densities and geographical distributions of fish populations. Reproductive guilds allow ichthyologists to identify common ecological features and problems in different areas involving different fish faunas. However, the reproductive guild concept that is qualitative, and emphasizes the physiological ecology of early life stages, is presently limited in its application due to the paucity of data about the early stages of many African fish.

Modes of reproduction in fish may also differ with the mating behaviour, the nature and location of breeding sites, the characteristics of the eggs, and the need for mature fish to undertake spawning migrations or not. Broad characteristics of the modes of reproduction prevailing in the African fish families are presented in Table 8.3. Within many families, deviating modes of reproduction may of course be encountered.

Timing of reproduction

According to Bye (1984), the timing of annual spawning has evolved to ensure that the young hatch and commence feeding at the season most conducive to their survival. This is an oversimplification of a complex situation, in which many environmental factors are probably involved in the timing of the reproductive cycle. The cycle of gametogenesis, for instance, needs to start many months in advance of the spawning season (and therefore to anticipate the spawning season), when environmental conditions are not particularly suitable for the survival of young. But it is obvious that production of young at the time when food is abundant and when cover is available to reduce predation is an adaptative advantage.

Gonad development and the timing of spawning are regulated by endocrine processes depending on environmental factors, such as temperature, photoperiod, rainy seasons and others. Reproductive cycles are therefore synchronised with environmental cycles by means of regularly occurring events that act as 'timing cues', and have become physiologically significant in that they stimulate or inhibit specific stages of gametogenesis or other reproductive processes. The existence of internal rhythms is also suspected. For example, adult *Clarias* transferred from outdoor stagnant ponds into an indoor hatchery maintained their annual reproductive cycle for one year, and then maintained a high gonadosomatic index during subsequent years (Janssen, 1985, cited in Huisman, 1986).

Gonad maturation

If the final stage of the reproductive cycle in many tropical species is often associated with seasonal flooding, it remains to identify the factors that are responsible for regulation of the earlier stages so that the fish are physiologically ready for spawning when the flood occurs. The exact stimulus is not clearly understood. It could be related to changes in water chemistry, flow rate, food supply, availability of spawning sites, etc. Although many environmental determinants have been invoked as regulators, there is little experimental confirmation.

Fish being poikilotherms, the role of temperature may be important in gonad maturation. There is evidence that the optimal temperature range for oocyte growth is much narrower and much more clearly defined than that of other physiological processes, and that this optimum range appears to be genetically fixed (Munro, 1990). In the tropics, water temperatures are usually highest at the end of the dry season, but at lower latitudes, the range of seasonal variation is generally lower than the range of fluctuation during any one day. The role of temperature as a potential proximate factor is documented for some cichlids. It has been reported that tilapias need a water temperature of at least 20 °C to breed (see Philipart & Ruwet, 1982), and experimental studies have confirmed that low temperatures inhibit ovarian development in *Oreochromis aureus* (Terkatin-Shimony *et al.*, 1980). In Lake Kariba, the introduced *Tilapia rendalli* breeds mainly when temperatures are high, despite the fact that the concurrent falling water levels expose many potential nursery areas and therefore decrease their potential reproductive success (Donnelly, 1969). Paugy (1978) also concluded that high dry-season temperatures were necessary for the maturation of *Alestes baremoze*, whereas the arrival of the flood triggered actual spawning. Temperature may also regulate reproductive processes by acting on metabolic activity (Terkatin-Shimony *et al.*, 1980) and growth (Lam, 1983; Yan, 1987). In males of a number of different teleosts (including *Oreochromis mossambicus*), the activity of different testicular enzymes may be affected by temperature in different ways (Kime, 1982; Kime & Hyder, 1983), and the result is that biologically active steroids are mainly secreted over a narrow temperature range.

Although photoperiodic control of reproduction is well documented in temperate regions, too little information is currently available to determine the importance of photoperiod changes for the reproductive biology of tropical fishes. If seasonal changes in temperature and daylength have smaller amplitudes at low compared with high latitudes, these factors may, nevertheless, play a role in spawning seasonality. For instance, Hyder (1970) suggested that light intensity changes could regulate the reproduction of fish in Lake Naivasha, but Siddiqui (1977) came to different conclusions (non-seasonal breeding) for the same fish population and Moreau (1979) did not find clear evidence for a correlation between photoperiod and the reproductive cycles of *Tilapia zillii* and *Oreochromis mossambicus* in a Malagasy lake. Data available for *Clarias gariepinus* (Richter *et al.*, 1987) also do not provide evidence for photoperiod as a predictive cue in populations from Cameroon or Israel.

In the River Niger, Bénech & Ouattara (1990) conducted both field observations, at two different stations (Mopti and Niamey), and experimental studies to investigate the role of environmental factors in gonad maturation of the characid *Brycinus leuciscus*. Changes in water temperature are similar at the two stations, with a maximum of 29 °C during the 19th week of the year. At the same time, the first rains occurred and the water conductivity ceased to increase. At Mopti (inner delta of the Niger), gametogenesis started in late May–early June (20th week of the year) with the seasonal climatic variations induced by the intertropical front moving northward. The same occurred at Niamey, far downstream, during the 22nd week. At Mopti, the gonadosomatic

index increased to reach a maximum during the 30th week, when spawning occurred. Conversely, at Niamey, ovarian growth appeared to be suspended at an early stage, and started again 6 to 9 weeks later, to reach a maximum during the 35th week. In both locations, gonad development coincided with a decrease of water conductivity, and the spawning period ended in week 39 at Mopti and in week 45 at Niamey. The 6 to 8 weeks timelag observed in gonad development between Mopti and Niamey is similar to that observed for the conductivity changes.

These field observations were tested in aquaria to check the relationship between conductivity change and gonad development. Fish maintained in aquaria with a constant water level, increasing conductivity and rain simulation exhibited earlier gonadal activity after ten weeks than river fish. As a consequence, this experiment did not demonstrate the existence of a causal relationship between decreasing conductivity and gonad maturation. The results are also different from those obtained by Kirshbaum (1984), who induced gonad maturation of mormyrids in aquarium experiments by decreasing the conductivity of the water, simulating rain and increasing the water level. At present it is difficult to say if the observed differences should be attributed to different physiological behaviour between species belonging to different families, or if the factors presumed responsible are not the relevant ones!

In tropical rivers that are highly seasonal, most feeding occurs at the high-water period, but gonad growth in many riverine species that spawn at flood occurs at a time when food availability is limited, and egg-guarding species stop feeding during the parental phase. The accumulation of energy reserves is therefore likely to be important for spawning migrations and reproduction, but experimental data are scarce. Small ration sizes or low dietary protein levels lead to female *Oreochromis niloticus* maturing at a smaller size, the combined low levels of both inhibiting reproduction (Santiago *et al.*, 1985). Underfed females of *O. mossambicus* in aquaria mature earlier at a smaller size and spawn more frequently (Mironowa, 1977). For some lacustrine species, there is evidence that reproduction tends to coincide with periods of high biological productivity and algal blooms. The Tanganyika clupeids, *Limnothrissa* and *Stolothrissa*, breed throughout the year in their native lake, but growth studies from spawnings at the time of seasonal plankton blooms suggest that recruitment is seasonal. (Coulter, 1970; Chapman & Well, 1978). Food availability in that case may be an important ultimate factor, but it is not clear, nevertheless, if the fish mature early and await the relevant synchronised cues, or spawn regardless.

Spawning

In large tropical rivers, the flood regime (or the assemblage of conditions that mark the beginning of the flood) seems to be the major regulator of breeding, and the onset of reproduction in many fish species tends to coincide with the flooding. Spawning may occur from low water to peak flood, but rarely during flood recession. Inundated floodplains provide large areas with abundant food and cover for newly hatched juveniles, and temperature changes are potential synchronised cues for species that spawn at the start of the floods.

Many species inhabiting rivers with a marked flood cycle spawn just before the peak, which can be illustrated (Fig. 8.1) by changes in the gonadosomatic index of different species in tributaries of the Senegal River (Paugy, unpublished data).

Observations on the reproductive biology of *Hepsetus odoe* in the Okavango system may help us to understand the relationships between spawning behaviour and external stimuli such as temperature and floods (Merron *et al.*, 1990). In most African floodplain systems, the flood is synchronous with the rainy season and the warmer summer months, but in the Okavango system the flood coincides with the dry winter months. Here, *H. odoe* spawns during summer with a peak in reproductive activity when water levels are lowest. In the drainage rivers, where the floodwaters arrive during the coldest time of the year, the species spawn in spring with a peak in summer, that is to say 3 to 4 months after the peak flood while water levels are receding. These observations seem to demonstrate that the floods are not the major ecological stimulus for the

Fig. 8.1. Hydrological cycle and changes in the gonadosomatic index (GSI) of different fish species from a tributary of the Senegal River. Spawning occurs before the flood peak (Paugy, unpublished data).

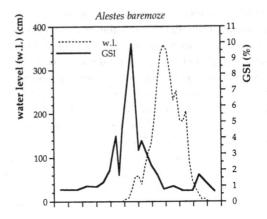

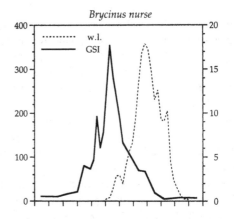

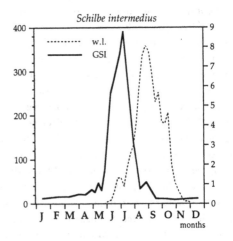

spawning periodicity of *H. odoe*, and similar observations on the marked influence of temperature changes on spawning periodicity have also been made for this species in Lake Liambezi (van der Waal, 1985). Similarly in the Lake Chad basin, the lake level rises during the winter period (December–January), several months after the peak river flood (September), but most species reproduce in the rainy season (July–August) (Bénech & Quensière, 1985).

For species with a long breeding period (partial or successive spawners), such as many cichlids, it remains difficult to assess which are the key factors that stimulate, inhibit or exert any regulatory influence on the various stages of the breeding cycles (Jalabert & Zohar, 1982). In some cases, it is possible to recognise peaks in breeding activity during the hot season, such as *Oreochromis macrochir* in Lake Sibaya and *Tilapia zillii* in Lake Quarun (see Philipart & Ruwet, 1982 for review), but in many cases, either continuous reproduction throughout the year, without a marked period of increased activity, or slight changes that were not easily related to specific climatic factors were reported (Johnson, 1974; Harbott & Ogari, 1982; Legendre & Ecoutin, 1989). However, there is a tendency to seasonality with increasing latitude. In Lake Malawi, Marsh *et al.* (1986) studied ten species of rock-frequenting cichlid, and observed that they were reproductively least active in the cold windy season (May–June), had a major peak of reproductive activity in spring (August–October) and a secondary one in late summer to autumn (February–March). The authors tentatively concluded that most Mbuna species follow this pattern of breeding activity. However, although Mbuna spawn throughout the year with fluctuations, the non-Mbuna haplochromines that come to inshore sandy areas to spawn tend to be more seasonal (e.g. McKaye, 1983, 1984; Tweddle & Turner, 1977); these include zooplanktivore Utaka species (originally called haplochromines, now *Copadichromis*).

The three endemic and very similar *Oreochromis* (*O. saka* (*karongae*), *O. lidole* and *O. squamipinnis*) from Lake Malawi have restricted spawning seasons (Table 8.4) and have sorted themselves out on the colours of breeding males and different breeding depths and seasons; *O. saka* and *O. lidole* mainly in hot weather before the rains (*saka* shallow water, *lidole* deeper), *O. squamipinnis* later, during the rains and deeper. Under natural conditions,

many species exhibit distinct fluctuations in their breeding activities.

A similar pattern exists in Lake Victoria where Witte (1981) found that most haplochromines have a peak of reproductive activity after the rains. These observations were largely confirmed by Goldschmidt & Witte (1990) who studied zooplanktivores, the peak of reproductive activity being observed during the dry season (June–October) for most species. A small proportion of ripe females was observed throughout the year, but the intensity and seasonality of reproductive activity varied widely from year to year suggesting that proximate factors are influential. In Lake Victoria, the endemic *Oreochromis esculentus* peak spawning was during the rains (twice a year at the north end of the lake, once at the south end) (Lowe-McConnell, 1979).

Synchronous breeding in Malawi haplochromines, which is clear in non-Mbuna species (McKaye, 1983, 1984), might be a predation response: the presence of a large number of young could reduce the chance of an individual being eaten by a predator and young may benefit from collective defence by parents in the surrounding area (Dominey, 1981).

It has been suggested that asynchronous breeding (occurring for example in Malawi tilapia species) may be an important mechanism for maximising resource use, particularly for closely related species with similar requirements. That is not true for Mbuna species that have similar resource requirements and exhibit considerable overlap in their breeding cycles. In mouthbrooding cichlids, females that spend approximately three weeks with developing fry in their mouths, in addition to egg production, need to build up considerable energy reserves prior to breeding. It is therefore likely that peaks of breeding activity would occur in response to seasonal fluctuations in food availability. Witte (1981) attributed the peak in reproductive activity during the dry season to nutrients washed into the lake during the rainy season, which stimulate plankton productivity and thereby increase food for fish. During the dry season the water column is completely mixed and peaks in primary and secondary production also occur (Goldschmidt & Witte, 1990). A similar situation occurs in Lake Malawi where major peaks in cichlid breeding activities (August–October) follow plankton blooms that occur in the cool windy season. The secondary peak in summer occurs at

Table 8.4. *Seasonality of spawning for Lake Malawi cichlids*

	J	F	M	A	M	J	J	A	S	O	N	D
Oreochromis squamipinnis	*	*	*	*								*
O. saka								*	*	*	*	
O. lidole									*	*	*	
O. shiranus chilwae								*	*	*	*	
Haplochromis virginalis			*	*	*							
H. quadrimaculatus				*	*	*	*					
H. pleurostigmoides				*	*	*	*	*				
H. borleyi	*	*	*	*	*	*	*	*	*	*	*	*
H. eucinostomus	*	*			*		*					
H. mloto								*	*	*		
H. intermedius						*	*	*				
Lethrinops parvidens									*	*		
L. longipinnis						*	*	*	*	*		
Mbuna	*	*	*	*	*	*	*	*	*	*	*	*

From Lowe-McConnell, 1979.

the time of the highest standing crop of aufwuchs. These results demonstrate that breeding activity of haplochromines is not directly related to temperature *per se*, the major peak being observed when water is warming up (Marsh *et al.*, 1986).

It is, therefore, still not possible to give a comprehensive answer to the question (Scott, 1979) 'How do teleost fish time their reproductive cycle? That they do, at any rate in habitats where timing confers a selective advantage, there is no doubt. That regularly recurring environmental events regulate the timing of the cycles, there is no doubt, but the nature of these cues is far from clear, especially during the complex of activities called spawning.'

Reproductive behaviour

Reproductive behaviour can include many components in relation to spawning activities. First is the choice of breeding site, which sometimes needs large-scale movements and migrations. Then, in some cases, fish spend time in preparing the spawning site as well as in courtship and mating. Finally, after spawning, many species defend the spawning site and exhibit some form of parental care. Such reproductive behaviour is costly in time and energy, both of which are not then available for other activities, e.g. growth.

Mating systems

The mating system is defined by the number of members of the opposite sex with which an individual mates (Turner, 1986), and is obviously intimately bound up with the type of parental care. Mating systems of African fish are particularly well documented for cichlids (Barlow, 1991) and three main types have been identified.

1. **Group sex.** The ancestral reproduction pattern in fish is likely to be a promiscuous mating system with no courtship or mate choice, and no parental care (Balon, 1981a). This occurs when a large number of fish congregate at regular breeding grounds and spawn simultaneously.

2. **Monogamy.** The system is said to be monogamous when each individual mates with only one member of the opposite sex, even if they do not remain together outside the breeding season. Among cichlids, the substrate guarders are monogamous, relatively isomorphic species with adhesive eggs and helpless larvae with large yolk sacs (Fryer & Iles, 1972; Barlow, 1991). Breeding starts with pair formation, and the parents share in parental care though males and females usually invest differently. The female is usually closely tied to the fry whereas the male repels intruders. A few cichlid mouthbrooder species are monogamous, as is the case for the genus

Chromidotilapia (Myrberg, 1965). In Lake Tanganyika, females of *Perissodus microlepis* brood the eggs and the pair then jointly defend the fry (Yanagisawa, 1986). Prolonged biparental care of fry, and hence monogamy, has been also observed in other species from Lake Tanganyika, such as *Xenotilapia longispinis* and *Haplotaxodon microlepis* (Kuwamura, 1986), and *Xenotilapia flavipinnis* (Yanagisawa, 1986).

Of special interest is the behaviour of *Sarotherodon galilaeus*. The male and female pair up briefly and dig a nest. After spawning, each parent takes eggs into its mouth and then they go their own way (Trewavas, 1983). This behaviour could give rise either to maternal or to paternal care. Whereas paternal care is the general rule among teleost fish, exclusive male paternal care is known for only two species of cichlid: *Sarotherodon melanotheron* (Trewavas, 1983) and *Sarotherodon occidentalis* (Loiselle, 1985). The male fertilises the eggs and broods them in his mouth. Monogamy is uncommon in African mouthbrooders, probably because in maternal mouthbrooding males remain on spawning sites while brooding females move away.

3. **Polygamy.** Theories about the evolution of territoriality and male care are often associated with polygamy and, in fact, a polygamous mating system is found in most teleosts. This may take the form of polygyny, with one male fertilising the ova of several females as frequently observed, for example, among cichlid substrate brooders in Lake Tanganyika (Kuwamura, 1986), but there can also be polyandry, when one female mates with several males, and 'promiscuity' when both sexes typically have several partners. In cichlids, the evolution of the mating system has evidently proceeded from monogamy with biparental care to polygamy with maternal care. Though *Tilapia sensu stricto* (substrate spawners) appear to be monogamous, in *Oreochromis* tilapias (mouthbrooders), females may mate with several males, and males with numerous females (Lowe-McConnell, personal communication).

Within the cichlid family most of the mouthbrooders are polygamous and usually dimorphic. In these, mating is brief so that no pair bond is formed. Instability in a pair bond can take the form of desertion of mate and brood by one partner, or establishment of additional pair bonds by one member, i.e. simultaneous polygamy. Usually, males persistently defend a site that is then visited by females, as happens in the cichlids of Lake Malawi (McKaye, 1983; Ribbink *et al.*, 1983*b*), and in mouthbrooders from Lake Tanganyika such as *Petrochromis*, *Simochromis* and *Cyphotilapia*. Originally, Tanganyikan mouthbrooders were thought to be strictly maternal. Most are, but there are exceptions. In some species, the female broods the eggs but then passes the newly-hatched young to the male who broods them in his mouth (e.g. *Xenotilapia boulengeri*). In *Limnochromis auritus*, the male and female sometimes brood simultaneously (Konings, 1988).

Polygamy also occurs among a few substrate-guarding cichlids, such as the lake-dwelling lamprologines that form harems (Konings, 1988; Kuwamura, 1986; Loiselle, 1985). Males of some *Lamprologus* have territories or domains encompassing the territories of two or more females. The domain of a *Neolamprologus furcifer* male reaches up to 100 m^2 and may contain the territories of up to 16 females (Yanagisawa, 1987).

Leks are a peculiar form of polygamous mating. A lek is an aggregation of males gathering in colonies, each of which defends a small territory and courts any arriving female (Fryer & Iles, 1972; Loiselle & Barlow, 1979; McKaye, 1983, 1984). Males do not provide parental care, and it is known that a few males always achieve the majority of matings. There were various suggestions to tentatively explain why males clump together in leks, but the question is still open. So far, lekking is only known for African mouthbrooding species that nest in open areas and is common among lake and riverine species of haplochromine and tilapiine species (Fryer & Iles, 1972). Konings (1988) provided many examples of lekking among Tanganyikan cichlids, such as *Ophthalmotilapia* and *Callochromis*.

Courtship and recognition

For the majority of fish, fertilisation occurs externally, and the fertilisability of the eggs and the fertilisation ability of sperm is time-limited. Thus, one of the functions of courtship behaviour is to synchronise spawning readiness in order that the gametes are extruded simultaneously.

To induce a response in a potential mate, it is necessary to follow a predetermined sequence of stereotyped behaviour patterns that do not appear to have many alternatives. In cichlids, courtship behaviour has two functions: (i) to stimulate the female so that she becomes receptive; and (ii) to contribute to a specific-mate recognition pattern (Ribbink, 1990). Any fish that does not perform appropriately is likely to be excluded from contributing to the gene pool. Recognition of the complexity of this behaviour led to the hypothesis that it could be one component of the speciation processes. Greenwood (1974*b*) stated that 'this complexity may be a pointer to the ease with which barriers to interspecific crosses are evolved . . . a slight deviation from an established pattern could provide an effective barrier to successful courtship and mating'. However, in contrast to the enormous diversity of trophic adaptations and variety of characteristics, which enable different species to use space and shelter, the components of courtship and spawning behaviour and their hierarchical sequencing are remarkably similar (Ribbink, 1990). The reproductive behaviour has apparently remained unmodified among the more than 200 Mbuna species, and the behavioural components are common to a wide variety of cichlids. This evolutionary conservatism of spawning and courtship behaviour contrasts with the evolutionary diversity shown by these fish in many other respects. The universality of these behavioural elements among mouthbrooders supports the theory that this behaviour evolved early in the evolutionary history of this group and has been subjected to stabilisation selection (Ribbink, 1990). Courtship displays are species-specific. In haplochromines, pheromones inform mates of the sexual status of their partner and may influence their courtship behaviour for several days. These effects have been studied within members of the same species and are considered to be one of the barriers against hybridisation (Crapon de Caprona, 1980).

It is usually acknowledged that the colours of breeding fish are an advantage for recognition, but it is not always very well understood how they manage to display in waters with very low visibility. Most mouthbrooding haplochromines bear one or a small number of ovoid

orange-yellow spots surrounded by a transparent ring on the anal fin which resemble haplochromine eggs. The spots are much more pronounced in males than females. The presence of these ocelli on the anal fin is regarded as an advanced character, their close mimicry of eggs deceiving the female into mouthing the sperm-laden area near the male's genital papilla, so ensuring in-the-mouth fertilisation of the eggs that she has taken into the comparative safety of her mouth before the male had a chance to fertilise them on the substratum (Eccles & Trewavas, 1989).

It has also been suggested that these spots play an essential role in courtship (Wickler, 1962). They mimic eggs and are therefore called 'egg dummies', and this was confirmed for Lake Victoria haplochromines (Goldschmidt, 1991). In Lake Tanganyika, males of *Simochromis pleurospilus* have a single egg marking on the rear margin of the anal fin. In *Callochromis macrops* and *Callochromis pleurospilus* the coloured tip of the anal fin rolls up during fertilisation, suggesting an egg in three dimensions (Konings, 1988). Egg spots are important as intraspecific sexual signals and may play a role in species recognition. Mimicking an egg to stimulate sexually active females is presumably an important demand on the form of the anal spots, whose size is negatively correlated with light intensities of the habitats of the species investigated in Lake Victoria. But conspicuousness may also be important in determining the form of the spots. Conspicuousness 'is likely to have a positive impact with regard to conspecific courting females but a negative one with regard to predators visually hunting eggs or adults' (Goldschmidt, 1991). The ultimate appearance of anal spots may be a compromise between the positive impact on the behaviour of conspecific ripe females (benefits) and predation risks (costs).

Courtship and spawning behaviour among cichlids are usually complex, but much of the behavioural diversity can be explained by the parental care strategy, and the duration of courtship as well as its level of complexity are similarly related to parental care. Malawi haplochromines (all mouthbrooders) for instance show a reduction in diversity of courtship processes and they lack many of the behavioural elements present among *Tilapia* (*sensu stricto*) substrate spawners. Courtship display by male Mbuna does not appear critical to species recognition and may represent an evolutionary relict (McElroy & Kornfield, 1990).

In recent years, interest has also focused on how electric fish may detect mates and on the nature of the signals that are exchanged during courtship and spawning (Kramer, 1990). Specific electric organ discharge interval patterns were recorded during the nocturnal courtship and spawning of the mormyrid *Pollymirus isidori*.

Territorial behaviour

The existence of territorial behaviour is fairly common among cichlid fish and one of the most conspicuous aspects of parental behaviour in fish is the defence of a breeding territory. A territorial fish defends an optimal breeding site from conspecifics of the same sex before and during spawning. Many authors, discussing the evolution of this parental behaviour, suggest that territory should be considered as a resource required for spawning. Moreover, Baylis (1981) suggested that pre-spawning territorial defence in a cichlid implies that suitable spawning sites are a resource of limited availability. That is strongly supported by studies of Tanganyika cichlids (Gashagaza, 1991).

Male territoriality obviously occurs in fish (Blumer, 1979). Barlow (1964) and many other authors have attributed the territorial role to the male sex because of its potential for multiple mating, but this idea has been questioned and Ridley (1978) suggested that the male territorial behaviour may be either a result rather than a cause of parental behaviour, or a result of female choice. The question yet remains as to why a male is territorial prior to spawning.

If we assume that being territorial gives an advantage in access to females, it is therefore in the interest of the male to stay within a territory as long as possible, bearing in mind that it also costs a lot in defence effort. There were very few attempts to test whether cichlids spend their whole lives within one area though, at least as adults, they are strongly philopatric. Hert (1992) demonstrates the existence of home site fidelity and homing ability in *Pseudotropheus aurora* and other rock-dwelling cichlid species from Lake Malawi. Some males stayed at least one and a half years in their territories, which is a surprisingly long time for a small fish. In most maternal mouthbrooding cichlids from Lake Tanganyika, large males established territories of 3–10 m in diameter and chased conspecifics and other fish. These territories were essential for mating with visiting females, but were also used as feeding sites by the males (Kuwamara, 1986). In

several species, within a male's territory an elaborate 'nest', such as a sand crater, was made for courtship and spawning.

Nest building

To protect their eggs and larvae from predation, several fish species construct more or less elaborate nests in which the eggs are spawned. A variety of nests are known among African fish.

Protopterus annectens practises a form of parental care by guarding its young in a nest-like depression, and details of the nest shapes are given by Greenwood (1986). The commonest form is a broadly U-shaped tunnel extending some 40 cm into the substrate. Between the two vertical arms, there is an enlarged chamber in which the eggs are deposited.

Nests of *Heterotis niloticus* are circular 'miniature lagoons' made in about 60 cm deep water in thick and high grass, 2 to 10 m from open water (Svensson, 1933). They measure 1.0 to 1.5 m in diameter, with compact walls made from the stems of the grass removed by the fish from the centre of the nest. The walls reach the surface of the water and are about 20 cm thick at the top. The floor of the nest which is made perfectly smooth and bare is the bottom of the swamp.

The nest of *Gymnarchus niloticus* is elliptical, made of plants, and floats in open water or swampy areas. Some nests are as large as 0.9 by 0.65 m and presumably built by males (Budgett, 1900). In mormyrids, the males of *Pollimyrus isidori* are territorial and build well-hidden nests constructed from plant material. During each spawning act, at repeated intervals of a few minutes, 2 to 4 eggs are laid. The male puts them into the nest and guards them for several weeks. The female is allowed access to the nest area only for the brief courtship and spawning bouts (Kirshbaum, 1987). *Pollimyrus petricolus*, endemic to the River Niger, also builds nests, using decaying roots and stems of *Echinocloa stagnina*, in floating mats of vegetation or attached near the bottom, on growing stems of *Echinocloa*. Males actively guard the eggs, while the decaying vegetation allows development of macroinvertebrates that serve as food for the young *Pollimyrus*, and also for juveniles of other species.

The African pike, *Hepsetus odoe*, is one of the few non-cichlid nest-guarding freshwater fish species in Africa, and belongs to the aphrophilic breeding guild

(Balon, 1981b) characterised by a moderate parental investment in relatively few eggs, the construction of a nest for guarding the young and the provision of an oxygen-rich environment. The nesting behaviour of *H. odoe* was first described by Svensson (1933) and thoroughly investigated by Merron *et al.* (1990). The foam nests are built among dense emergent reeds. Newly built nests have a roughly circular basis and appear as a high dome of tightly packed foam-bubbles, while older nests are more irregularly shaped and covered with leaf and stalk debris. Eggs are deposited in the nest and were found embedded in the foam, up to 25 mm above the waterline. Upon hatching, the embryos wriggle their way through the foam and suspended themselves, motionless, by a large cement gland below the nest. They remain in the vicinity of the spawning site until they reach a relatively advanced stage in their development. The adults exhibit territorial behaviour in the vicinity of the nests and add foam as necessary. Surface foam nests guarded by males are also known for *Ctenopoma damesi*.

Brood mixing

Broods may accidentally become mixed as a result of attacks by predators or by territorial fights between parents of adjacent broods, which result in local turbulence. Displaced fry may readily approach or be retrieved by unrelated adults. However, various field workers have mentioned large schools of fry of one species being guarded jointly by two or three pairs of adults. Such communal parental care has been documented for *Tilapia rendalli* in Lake Malawi (Ribbink *et al.*, 1981). Moreover, mixed-species schools of fry being guarded by adults of one or both parent species have also been observed in Nigerian rivers (Sjolander, 1972).

In Lake Tanganyika, an unusual behaviour among fish has been reported for the endemic catfish, *Synodontis multipunctatus*, that is a brood parasite of different species of mouthbrooding cichlids (Sato, 1986). It spawns its eggs into the mouth of the host together with the host's eggs. A number of different hosts have been identified, e.g. *Simochromis diagramma*, *Simochromis babaulti*, *Tropheus moorii*, *Gnathochromis pfefferi* and *Haplochromis horei*. The eggs of the catfish, which have the same mean size as those of the host species, are incubated together with the host's eggs but hatch earlier. The larval yolk sac

is absorbed within three days and the catfish fry start to feed upon the host fry while still in the parent's mouth, gradually sucking their yolk when small, or swallowing the whole host fry when larger. This unique development strategy allows the catfish eggs and fry to exploit the host parent as a refuge, and the host fry as food resource, and suggests a long-established association between the catfish and the cichlids in Lake Tanganyika (Sato, 1986).

Cuckoo behaviour has also been reported among cichlids from Lake Malawi (Ribbink, 1977). *Nimbochromis polystigma*, *Tyrannochromis macrostoma* and *Serranochromis robustus* were found guarding mixed broods including their own offspring, and fry of another more pelagic and open-water species that has been identified as the widespread *Copadichromis chrysonotus*. Although *C. chrysonotus* brood their own eggs in their mouths, they are also cuckoos in the sense that foster parents guard their young. Further investigations in Lake Malawi showed that at least 15 cichlid species will carry heterospecific fry in their mouths. Some mixed broods contained young of one, two or three foreign species, and a few had as many as 50% foreigners (Ribbink *et al.*, 1980).

Active transfer of fry from one brood to another has also been described for the biparental mouthbrooder *Perissodus microlepis* in Lake Tanganyika (Yanagisawa & Nshombo, 1983). The female broods the eggs but the male remains near by, and both parents guard the fry when they become free-swimming. If one parent deserts while the brood is at the fry stage, the remaining parent may take the fry into its mouth a few at a time and release them into a neighbouring conspecific brood. What has been called 'farming out' of the young apparently occurs commonly among *P. microlepis* (Yanagisawa, 1985). It may be the best option for a single parent protecting its brood.

A peculiar phenomenon observed in the Lake Tanganyika substrate brooder cichlids of the genus *Lamprologus* and *Julidochromis*, is called 'helpers at the nest' (Taborsky & Limberger, 1981). In *Neolamprologus brichardi*, helping has been observed in the field. Juveniles of different size classes may be found in the breeding territory of their parents, actively assisting in brood rearing by chasing predators, cleaning and occasionally fanning the eggs (Taborsky, 1984). This behaviour, instead of direct reproduction, may be the result of high local population densities leading to intense competition for shelter and breeding sites.

Social regulation of reproduction in cyprinodonts

In the tropical rain forest of Gabon, eight small sympatric species of cyprinodont (from the genera *Epiplatys*, *Aphyosemion*, *Raddaella*, *Hylopanchax*) live in the small creeks and pools, and feed on terrestrial insects falling from the overhanging foliage (Brosset, 1982). Their reproductive strategies differ according to the species but all cyprinodonts are 'K' strategists, spawning a few large-sized eggs and these particular cyprinodonts provide a well-documented example of the influence of environmental factors on social structure.

In the very small and shallow creeks, less than 3 cm deep, groups of various species may be observed. They are small adults belonging to all species of the fish assemblage, mixed together. The social groupings of these sympatric cyprinodonts are extremely variable, ranging from monogamous pairs to unstructured promiscuous groupings, as well as harems and multimale groups, the latter being most common. When moved to aquaria and well fed, these dwarf adults develop 'normally'. In the main stream of the creek (0.5–6.0 m wide and 1–50 cm deep), there is a patchy distribution of small groups of large adults (2–3 species) at low density. Finally, in residual pools and elephant prints, there are dense monospecific populations including both juveniles of all sizes and adults, structured in harems.

For each type of habitat, there is, therefore, a type of social and demographic structure that is developed by the species occupying this habitat. Moreover, laboratory experiments also demonstrated the existence of interspecific regulation. In aquaria, the presence of one species inhibits the reproductive activity of another species, an interaction probably mediated by pheromones that inhibit breeding. This inhibition does not exist between species belonging to different genera. However, in closed systems, it seems that one species may eliminate all the others, the inhibitor species always being the one established first. This may help to explain the existence of monospecific assemblages in the residual ponds (Brosset, 1982).

Parental care: a way for optimising reproduction success?

Many species of oviparous fish leave their eggs to the fate of currents and waves while others display more or less intensive parental care for their eggs. Fish of the second group have developed a number of specialisations for guarding their eggs. Some of them stick their eggs to a substrate, others build a nest that could be a simple pit or made of plants, or foam, while others have elaborate systems of parental care. Some form of parental care is known in only 22% of teleost families (Blumer, 1982) and is more prevalent in freshwater than in marine families. It has been recorded in 50, that is to say 60% of current freshwater fish families.

Parental care is a non-gametic contribution that enhances reproductive success and the survival of offspring. It can be defined as any aid, behavioural or material, given by either parent after zygote formation that results in increased survivorship of the zygotes of the current reproductive effort. Such care is a diverse phenomenon among fish and manifests itself in a wide variety of activities involving one or both sexes.

Types of parental care

Types of parental care range from pre-fertilisation activities, such as nest building, to guarding of eggs or fry. Parental care also includes ventilation of eggs to provide a flow of oxygen and to clear away any accumulation of silt. Parental care by males (11% of all teleost families) is more common than female parental care (7%), and biparental care occurs in 4% of families (Sargent & Gross, 1986).

Two main criteria have been used to classify patterns of parental care (Keenleyside, 1991 for review). The first is the physical relationship between parents and their offspring. Substrate brooders maintain their brood on or close to the substrate, from spawning until the young become independent, whereas mouthbrooders are species that carry their offspring in their mouths either from spawning or egg hatching, until independence. Mbuna offspring become independent as soon as released from the mother's mouth, while for many other cichlids (*Oreochromis* and non-Mbuna haplochromines), young continue to return to the female's mouth when danger threatens, for a further period. The second criterion for classifying parental care is the sex of the caring adult. The long period (up to 3 to 4 weeks) of parental care of the young after spawning, which is characteristic of the family Cichlidae, is an unusual feature among the teleosts.

The main functions of parental care are to protect the young from predators, and to promote favourable conditions for their growth and development (Keenleyside, 1991), and a variety of different behaviours have evolved in both substrate and mouth brooders. In the latter, mouthing cleans the eggs or fry, whereas fanning is the ventilation and aeration of the eggs to remove metabolic products and to increase the oxygen level near them. The entire brood may be frequently moved from one pit to another, possibly as a way of escaping from predators. In mouthbrooders, fanning and mouthing are replaced by churning that moves the brood around inside the parent's mouth, and may play a role in supplying their respiratory needs. The Syngnathidae, which are of marine origin, have a brood pouch that is only present in males and contains the developing eggs, deposited there by the female. This special behaviour enables Syngnathidae species to provide parental care without the need for a nest or attachment to a permanent site.

Parental care is widely encountered in many African fish families (Breder & Rosen, 1966; Blumer, 1982) (Table 8.5), and has been well documented for African cichlids (Keenleyside, 1991). Substrate brooding is considered to be the ancestral pattern, while mouthbrooding is the more specialised and more recently evolved. Many substrate-brooding species have a wide range across northern and western Africa, indicating a long history on that continent, but several species also occur in Lake Tanganyika. Conversely, the majority of mouthbrooding cichlid species (over 70% of all cichlids) are restricted to the East African Great Lakes. However, there are also many mouthbrooders elsewhere, such as *Serranochromis*, *Oreochromis* and *Sarotherodon* species. The paternal mouthbrooder, *Sarotherodon melanotheron*, is common in brackish waters all along the West African coast, and the biparental mouthbrooder, *Sarotherodon galilaeus*, is widely distributed in north tropical Africa. Oral incubation has evolved independently in phylogenetically distinct fish groups, such as the ariid catfish, various anabantoids, osteoglossids, cichlids.

Table 8.5. *Known occurrence of different forms of parental care in freshwater fish families with representative in Africa. This does not imply that all species in a family exhibit parental care or all the form of care listed*

Care given by	Form of care	
Lepidosireniformes		
Lepidosirenidae	male alone	GE, GF, N, F, GEC
Polypteriformes		
Polypteridae	sex not known	GF
Elopiformes	no parental care	
Anguilliformes	no parental care	
Clupeiformes	no parental care	
Osteoglossiformes		
Osteoglossidae	male or female alone, biparental	GE, GF, N, ORB
Pantodontidae	sex not known	GE
Notopteridae	sex not known	GE, N, F
Mormyriformes		
Mormyridae	male alone	GE, GF, N, ME
Gymnarchidae	sex not known	GE, N
Salmoniformes		
Galaxiidae	biparental	N
Gonorynchiformes	no parental care	
Cypriniformes		
Hepsetidae		N, GF
Characidae	male or female alone, biparental	GE, GF, N, F, RD, ME
Citharinidae	male alone	GE
Cyprinidae	male alone	GE, N, F, RD, EB
Siluriformes		
Bagridae	male alone and female alone	GF, N, EEC, EF
Clariidae	male alone	GE, N
Malapteruridae	sex not known	ORB
Ariidae	male alone and female alone	ORB
Plotosidae	male alone	GE, N, F
Atheriniformes		
Cyprinodontidae	male alone	GE, F, RD, EB
Channiformes		
Channidae	biparental	GE, GF, N
Gasterosteiformes		
Gasterosteidae	male alone	GE, GF, N, F, RD, RE, RF, CE, ME
Syngnathidae	male alone	EEC, BP
Synbranchiformes		
Synbranchidae	male alone	GE, N
Perciformes		
Serranidae	biparental	

Nandidae	male alone	GE, N, F, RD, RF
Cichlidae	male or female alone, biparental	GE, GF, N, F, RD, ORB, RF, CE, ME, MF, EF
Gobiidae	male or female alone, biparental	GE, N, F, IG, RD
Anabantidae	male alone and biparental	GE, N, RE, RF
Pleuronectiformes	no parental care	

From Blumer, 1982, with a few additions. Male alone means that when parental care occurs, it is given only by the male. GE, guarding eggs; GF, guarding fry; N, nest building; F, fanning and aeration; IG, internal gestation; RD, removal of dead or diseased eggs; ORB, oral brooding; RE, retrieval of eggs; RF, retrieval of fry; CE, cleaning eggs; EEC, external egg carrying; EB, egg burying; ME, moving eggs; MF, moving fry; GEC, coiling around eggs while guarding; EF, ectodermal feeding of fry; BP, brood pouch.

It is usually recognised that among biparental substrate brooders, the two parents do not share all parental duties equally. Females commonly concentrate on activities directly associated with the brood (fanning, mouthing), while males mainly patrol the brood-rearing territory and repulse predators. Quantitative studies support the generalisation that in biparental cichlids, the female is more involved than the male in the direct care of immobile young. Moreover, males are typically larger than their mates in biparental species, which could be useful for inhibiting potential predators. When the young are fully mobile, the parents share the duty of protecting the fry from predators more or less equally (Keenleyside, 1991). In *Tilapia mariae*, males spend less time than females in close care of the embryos. They can leave the brood more readily than females when disturbed and do not counterbalance their longer absence from the embryos by a more intensive or more specialised defence. In nature, females both guarded for a longer time and defended more than males (Schwanck, 1989).

The prevalence of male parental care may be a consequence of the relationship between body size and fecundity (Sargent & Gross, 1986). This parental behaviour is costly, both in the allocation of time and in energy expenditure. It reduces energy available for growth, and this reduction in the future production of offspring is

assumed to be greater for females than for males. Nevertheless, Baylis (1981) questioned the prevailing idea that females invest more than males in the reproductive process, as suggested by investment models of parental care that only rely on energetic gamete discrepancies. He suggested that a distinction should be made between gametic investment (the average material investment each sex must make in all the gametes necessary to produce one zygote) and zygotic investment (the investment each sex makes in caring for or transporting the zygote that results from fertilisation).

In a study of the proportion of eco-ethological sections of reproductive guilds of fish in some African inland waters, Bruton & Merron (1990) provided evidence of a higher percentage of bearers and guarders in the African Great Lakes that are characterised by relatively predictable abiotic regimes. This is partly a consequence of the large proportion of cichlids that have been highly successful in these lakes. For instance, 92% and 39% of mouthbrooders were observed in fish communities of Lake Malawi and Tanganyika respectively. The lower percentage of mouthbrooders in Tanganyika than Malawi is because Tanganyika also has substratum-brooding cichlids that are absent from Lake Malawi. But the results also suggest that fish from non-guarding reproductive guilds are better suited to the variable or less predictable environments of rivers and floodplains. This does not support the hypothesis that parental care is an adaptative response to unpredictability (see below). In fact, there is a greater variety of different individual reproductive guilds among fish communities inhabiting these unpredictable habitats. Thus, while the species richness may be higher in predictable environments that have evolved for a long time, these species are typically assembled into a relatively small number of reproductive guilds when compared to more variable and possibly more recent biota.

Evolution of parental care

For externally fertilising fish species the problem is that the zygotes must be left somewhere to develop after gamete release. They can be scattered in the water, or some kind of protection can be developed by the parents. The risk of predation on the adults may limit their activities, and predation on the fry can influence the selection of care behaviour. It has been suggested that parental care evolved in fish occupying spatially and temporally unpredictable environments for their eggs and larvae (Wootton, 1990), since it may reduce the dangers of that unpredictability, but this theory needs to be supported by data.

Many non-guarders, egg-scattering pelagic spawners, produce a large number of small eggs, poorly endowed with nutrients, and they have delayed embryonic differentiation and a long larval period. Pelagic eggs in freshwater fish tend to occur in species inhabiting large water bodies or which migrate upstream to spawn. In the latter case, the eggs drift downstream as they develop. After hatching, the larvae make their way to the river margins, into backwaters and side lakes where they find food and shelter. There is little danger that pelagic eggs will experience anoxic conditions, but they are more likely to suffer high mortalities as a result of predation. This reproductive style could reflect the ancestral conditions of reproductive strategies in fish. It might also have been best suited for dispersal, a feature of excellent survival value, as long as they are able to survive in unstable environments.

It can be assumed that the ability to disperse lost its initial value during the evolution of derived guilds, after new habitats were invaded. The release of eggs more or less at random by pelagic fish does not allow for the direct evolution of parental care, because the eggs are dispersed and hence are impossible to care for. In freshwater, it has been suggested that the diversity of microhabitats favours the production by fish of demersal and/or adhesive eggs. Selective spawning of demersal eggs, which are localised in space and time, provides a focus for parental care but it is not a sufficient condition for parental behaviour that could originate in the need to search for suitable sites for spawning (Baylis, 1981). This is particularly true in patchy microhabitats encountered in freshwaters that produce discrete optimal spawning sites. Once the suitable site was located, there could be an adaptive advantage over conspecifics in remaining at such a site. Evolution should then favour the appearance of territorial behaviour as well as defence behaviour to monopolise the optimal sites.

Young fish should not become independent of parental care until they are able to capture and process food for themselves, but the size of the available food items is likely to differ between habitats. For the rock-frequenting 'Mbuna' species from Lake Malawi, which are narrowly stenotopic (Ribbink et al., 1983b), large yolked eggs produce juveniles that are large enough to feed on approximately the same food items as the adults, thereby cutting out the need for a planktonic open-water stage. This reproductive strategy reinforces stenotopy.

Remaining with the eggs after spawning could be justified for a variety of reasons. One of them is to prevent disturbance by conspecifics and to deny them access to a resource. If modifications of the spawning site were required prior to spawning, then continued occupancy by the parents would also serve to maintain the optimal quality of the site until the eggs hatch. Lastly, if a male monopolises the optimal site, one can expect multiple spawnings with many females, and that would favour his remaining on the spawning site. All these advantages of the adult remaining with the nest and the eggs probably favour longer and more persistent site constancy and guarding behaviour. A special case is that of guarding species with floating, non-adhesive eggs. The parents place the eggs in a floating nest either made of bubbles or vegetation (e.g. Hepsetus odoe). This behaviour appears to be an adaptation to development in anoxic waters (eggs being at the air/water interface), but could also be viewed as an adaptation of reproductive behaviour in fluctuating environments.

The most primitive parental care pattern in the Cichlidae, monogamous biparental substrate brooding, probably evolved from an earlier pattern of male-only care (Keenleyside, 1991), which is recognised as the most primitive form of care in fish (Gross & Sargent, 1985). The substrate brooders attach eggs to the substratum and both parents care for the offspring. Substrate brooding is found among both the lacustrine and riverine species, which, apart from a few Tilapia species, occur mainly in West Africa and in the Zaïre basin. This behaviour also occurs in South American and Asian species, while mouthbrooding species are found mostly in Africa, and many are restricted to the East African Lakes. The highly adhesive eggs of substrate brooders, compared to mouthbrooders, also support the hypothesis of their greater phylogenetic age according to Keenleyside (1991), who implicitly assumed that the adhesiveness has been lost.

The last step in the mouthbrooding form of parental care probably derived from an ancestral substrate brooder, in which parental activities, such as mouthing of the eggs and moving the embryos to a new site by transporting them in the mouth, were incorporated into the mouthbrooding syndrome (Oppenheimer, 1970; Baylis, 1981). Such behaviour selection could have been derived from competition for nest sites or to avoid zygote predation, but in the latter case, the embryos are exposed to adult predators. In addition, carrying the eggs also has the adaptive advantage of liberating the African mouthbrooding cichlids from dependence on a substrate for spawning when benthic living space is at a premium (Balon, 1978). Competitive selection pressures associated with changes in the environment could conversely favour the loss of parental behaviour. The process is not unidirectional.

If we assume that mouthbearing strategies originated from guarding behaviour, only male mouthbrooding would be expected in families where only the males guard. But the primitive pattern of parental care is probably biparental guarding, and either sex should be equally likely to mouthbrood. In the great majority of teleost families in which parental care has evolved, it is the male that is the primary, and usually exclusive, care giver (Breder & Rosen, 1966; Blumer, 1982). Beyond courtship and spawning, females play little or no role in the care of their progeny. Unusually among care-providing teleosts, female cichlids adopt a major role in parental care of embryos and free-swimming larvae, often to the exclusion of the male. Maternal investment in the care of offspring seems to be an innovation of the family Cichlidae (Stiassny & Gerstner, 1992). One explanation (Baylis, 1981) is that while male residence at a particular site disposes the species toward male parental care, when reproduction is not tied to a site, the female will tend to be the parental sex because the relative gain from multiple mating will be least in that sex. This relies on the assumption that the male is able to produce sufficient fertile gametes to fertilise a clutch of eggs in less time than it takes a female to produce a new clutch of eggs. Other authors (Barlow, 1984; Gross & Sargent, 1985) make the assumption that the plesiomorphic state of biparental care arose, in cichlid ancestry, from a system of primary male care of eggs and embryos, with a gradual increase and ultimate usurping of care provision by the female. This is not the case for *Paratilapia polleni*, a phylogenetically primitive cichlid, where the male and female roles are already strongly differentiated with the female as the primary care giver (Stiassny & Gerstner, 1992). The evolution of parental care in this context is viewed as a strategy that maximises the production of offspring for members of each sex in the environmental situation encountered by the species. Nevertheless, the option that male mouth-bearing is derived from an antecedent male-guarding condition is still true. It could apply, for example, to the Syngnathidae.

The evolution from external to intrabuccal insemination has been tentatively explained by Ribbink (1990) for Lake Malawi haplochromines. He noted that the ancestral-type behaviour, as observed in *Pseudocrenilabrus philander* and *Haplochromis callipterus*, implies that gravid females lay batches of eggs upon the substratum that the male begins to inseminate. The female collects the eggs almost immediately and takes them into her mouth where intrabuccal fertilisation occurs. However, predation pressure is high in the rocky habitats of Lake Malawi, and most piscivorous fish are adapted to prey mainly on fry. Some paedophagic species evolved, eating mainly larvae and juvenile cichlid fish, and many other cichlids, regardless of their trophic specialisation, feed opportunistically upon eggs and fry. Faced with this severe predatory pressure, haplochromines developed different alternative evolutionary strategies in order to improve the chance of survival of offspring in their specific microhabitats. Offspring of lacustrine species are larger and more developed when released than their riverine relatives. Eggs are laid in very small batches and retrieved immediately by the parent. As a consequence, intrabuccal insemination becomes essential as the male has little opportunity to fertilise the eggs before they are collected. In such a case, the fish becomes almost independent of the substratum spawning site, and that could help to explain the appearance of open-water or semipelagic species of mouthbrooder fish. In Malawi, the Utaka, *Copadichromis chrysonotus*, spawns in mid-water, and is independent of the substratum (Eccles & Lewis, 1981).

Costs and trade-offs in reproduction

The evolution of life-history traits in fish is constrained by trade-offs between the biological traits, their compatibility, the amount of genetic variability in the population, and phylogenetic inertia (Wootton, 1990). The central concept is that an increase in the time or resources invested in one activity may be traded-off against a decrease or resources invested in others (Schaffer, 1979; Sibly & Calow, 1983). In that context, trade-offs are benefits from one process that are bought at the expense of another. It has been suggested that for fish, optimal reproductive effort is a trade-off between the present and the future. This principle, developed by Williams (1966), assumes that reproduction has a cost that can be modelled into a general framework for predicting and studying the variety of problems in the evolution and behaviour of parental care (Sargent & Gross, 1986). A parent that continues to invest in its offspring does so at the expense of its potential future reproduction. As animals attempt to optimise their remaining lifetime reproductive success, the cost of reproduction will limit the amount that they can invest in future reproduction. Animals that optimise this trade-off obtain equal rates of return on investments in present and future reproduction. A very high reproductive effort may endanger future spawning by decreasing the parent's chances of survival, through diverting too much of its energy resources away from maintenance and somatic growth. Thus William's principle is that natural selection favours animals that optimise their remaining lifetime reproductive success. As a consequence, there are two conflicting selection pressures: one to increase the size of the eggs to reduce larval starvation; and the other one to increase the number of eggs to overcome the effects of predation. The resulting compromise is not uniform among all fish populations. Fish inhabiting quickly changing environments may have to utilise a wider range of reproductive possibilities to ensure their survival.

Allocation of resources to gonads and eggs

The proportion of the available resource input that is allocated to reproduction over a given period of time is called 'reproductive allocation' or 'reproductive effort'.

Vitellogenesis is the process of provisioning the eggs with yolk that will be utilised by the zygote during its development. Growth of the ovaries during the reproductive cycle reflects the growth of the ovocytes as they accumulate yolk. The **gonadosomatic index** (GSI, the percentage of total body weight due to the weight of the gonad) is a common and simple method used to describe the relative size of the gonads. Some authors also use the percentage of gonad weight to somatic body weight (gonads not included).

There is a wide interspecific range for the GSI of ripe females. In some species, the ovaries just before spawning may represent 20 to 30% of the total body weight. In others, the ovaries form less than 5% (see Table 8.6). These differences partly reflect different strategies. Fish that spawn once a year often invest in large gonads with a high energy content. On the other hand, fish with small gonads in comparison to their total body size usually spawn many times during the reproductive period (partial and successive spawners). Altogether, the energetic investment in the reproductive effort could be the same, but the strategies of gamete release are different. For example a mature female of the killifish, *Nothobranchius guentheri*, weighting 1 g produces 20 eggs per day and would achieve a reproductive effort value of 27% of its weight in one month (Simpson, 1979). In such a case, the total egg production would approach maternal weight in a breeding season of 4–5 months.

Spermatogenetic activity varies greatly from one species to another (Billard, 1986), but the reasons for these differences are not clear and have been little discussed. Sperm motility is very short in freshwater species with external fertilisation. The gametes are released into a particularly hostile medium and have only a brief time to fertilise the eggs. In many tropical fish species the testes are much smaller than the ovaries, and in many African fish (Table 8.6), the testes of sexually mature males form less than 1% of their weight. It is likely, therefore, that the allocation of energy by males to sperm production is much lower than that allocated to eggs by females.

There is no relationship between the male and female GSI. Neither is there any between female fecundity or egg diameter and spermatogenetic production, i.e. the ratio of spermatozoa produced per ovum does not tend to be constant (Billard, 1986), but males often reach maturity earlier and remain mature for longer than females.

Fecundity

Fecundity could be defined as the number of ripening eggs found in the female just prior to spawning. This

Table 8.6. *Comparison of female and male gonadosomatic index (GSI) for different African species. Maximum observed values quoted for the species*

Species	River	Females max. GSI	Males max. GSI
Polypterus bichir	Senegal	23.4	0.2
Hyperopisus bebe	Senegal	6.3	0.2
Mormyrus rume	Senegal	11.8	0.3
Marcusenius senegalensis	Senegal	18.7	0.5
Petrocephalus bovei	Senegal	7.3	0.4
Hydrocynus forskalii	Senegal	9.4	2.1
Alestes baremoze	Senegal	13.4	1.3
Brycinus nurse	Senegal	26.1	2.0
Brycinus leuciscus	Senegal	17.0	1.2
Brycinus macrolepidotus	Senegal	19.8	6.7
Labeo senegalensis	Senegal	17.3	2.0
Labeo coubie	Senegal	12.0	0.8
Schilbe intermedius	Senegal	23.4	2.1
Chrysichthys auratus	Senegal	27.5	0.7
Synodontis schall	Senegal	16.7	2.1
Synodontis ocellifer	Senegal	26.3	1.5

Paugy, unpublished data.

contrasts with 'fertility', which is the number of eggs shed (Bagenal, 1978). Given the large range of reproductive habits in fish, a universal definition acceptable in all circumstances has not yet been devised.

The absolute fecundity (F = number of developing eggs per female) usually increases with the size of the fish, expressed as length, weight or age. A general relationship of the form $F = aX^b$ is often observed, where X could be the length (L) or the weight (W). When using weight as X the relationship is more often $F = a + bX$.

For partial or successive spawners, the number of eggs that a female spawns over a defined period depends on the number of eggs per spawning and the number of spawnings. It is therefore necessary to distinguish among:

1. **Batch fecundity,** or the number of eggs produced per spawning, that is a function of body size. The typical relationship with length is F batch $= aL^b$. There is intraspecific variability, and even the same female can have different batch fecundities within the breeding season. An important ecological question is whether it is possible to predict variations in batch fecundity, as a result, for instance, of the effect of environmental factors on oocyte development, or if variations are simply the consequences of random, unpredictable events during the maturation of ovaries (Wootton, 1990). For *Sarotherodon melanotheron*, in experimental tanks, whether the interval between two successive spawnings is short (6–8 days) or long, there is no significant variation in the absolute number of eggs in batches, which means that for shorter cycles there is an accelerated rate of vitellogenesis (Legendre & Trebaol, 1996).

2. **Breeding season fecundity** is the same as batch fecundity for total spawners, but for partial spawners it will depend on the number of times a female spawns during the season. An estimation of breeding season fecundity is usually difficult, because the number of yolked ovocytes at the beginning of the season may not be a reliable indication, since previously unyolked ovocytes may be recruited to vitellogenesis during the breeding season.

3. **Lifetime fecundity** depends on breeding season fecundity and life span. The annual fecundity of different species of cyprinodont successive spawners was studied in the Ivindo River (Gabon). Brosset (1982) estimated the number of eggs spawned annually by an individual female to be between 200 and 300, whatever the strategy adopted by the species during the breeding season: from 1 egg a day for *Diapteron* sp., to 8–15 eggs every 8–10 days for *Aphyosemion herzogi*, and 30–75 eggs every month for *Aphyosemion cameronense* and *Aphyosemion punctatum*.

In order to compare fecundities of fish of different sizes, or from different places, or different species, many authors calculate the relative fecundity, or the number of eggs per unit weight. The use of relative fecundity assumes that the regression coefficient b does not differ significantly from 1 in the relationship $\log F = \log a + b \log W$. But in many cases, the relative fecundity is clearly related to length, so it cannot be used alone to compare the fecundity of fishes.

Table 8.7. *Selected data on mean female size at first reproduction, maximum gonadosomatic index, egg diameter (mm) and relative fecundity expressed as number of eggs per kilogram of female body weight (or occasionally per g when species are small) for different African fish species*

Species	L50 size 1st maturity (mm)	Max GSI	Egg diameter (mm)	Relative F per kg	Author
Polypteridae					
Polypterus endlicheri	320 SL		2.45	15 000	Albaret, 1982
Polypterus senegalus			1.75	50 000	Albaret, 1982
Polypterus bichir		23.4			Paugy, unpub. data
Clupeidae					
Pellonula afzeluisi	28 SL	10.5	0.25–0.46		Otobo, 1978
Sierrathrissa leonensis	19 SL	12.5	0.12–0.34		Otobo, 1978
Notopteridae					
Papyrocranus afer		2	3.6	500	Albaret, 1982
Mormyridae					
Brienomyrus niger	110 SL		1.3–1.5	80 000	Lek, 1979
Gnathonemus longibarbis		15			Plisnier *et al.*, 1988
Hyperopisus bebe		8			Paugy, unpub. data
Mormyrus hasselquisti	190 SL		1.8–1.9	24 300	Albaret, 1982
Mormyrus rume		11.8			Paugy, unpub. data
Mormyrops longiceps	210 SL		2.65	11 300	Albaret, 1982
Marcusenius bruyerei	130 SL		1.75	51 800	Albaret, 1982
Marcusenius furcidens			1.80	39 250	Albaret, 1982
Marcusenius senegalensis		18.7			Paugy, unpub. data
Petrocephalus bane	110 SL		1.0–1.2	46 000	Lek, 1979
Petrocephalus bovei	67?	25.0	1.5–1.6	91 230	Albaret, 1982
Petrocephalus bovei	70 SL		1.0–1.1	133 000	Lek, 1979
Petrocephalus simus		16.3	1.6	76 440	
Pollymirus isidori	65 SL		1.2	141 000	Lek, 1979
Hepsetidae					
Hepsetus odoe	140 SL		2.3	18 250	Albaret, 1982
Characidae					
Alestes baremoze	175 SL	18	1.1	224 100	Paugy, 1978
Brycinus imberi	67 SL	17	1.0	191 000	Paugy, 1980*a*
Brycinus leuciscus	65 SL	18.9	0.9		Paugy, unpub. data
Brycinus nurse	80 SL	33	1.05	368 000	Paugy, 1980*b*
Brycinus longipinnis	45	17	0.95	166 000	Albaret, 1982
Brycinus macrolepidotus	185		1.20	182 400	Albaret, 1982
Hydrocynus forskalii	100 SL	8	1.05	127 300	Albaret, 1982
Micralestes acutidens	34	12	0.6	183 000	Lek & Lek, 1977
Distichodontidae					
Icthyborus besse	140 SL		0.6–0.9	111 500	Lek, 1979
Cyprinidae					
Raiamas senegalensis	100 SL		1.35	47 700	Albaret, 1982
Barbus sublineatus	59 SL	25		677/g	Albaret, 1982
Barbus trispilos	50 SL		0.90	449/g	Albaret, 1982
Labeo coubie	220		1.25	181 500	Albaret, 1982
Labeo senegalensis	175		1.0	122 500	Albaret, 1982
Labeo parvus	100	22	0.95	347 000	Albaret, 1982
Bagridae					
Chrysichthys maurus	140	25	2.6–3.2	19 360	Albaret, 1982
Chrysichthys auratus	70 SL	27.5	2.3	19 000	Paugy, unpub. data

Table 8.7. (*cont.*)

Species	L50 size 1st maturity (mm)	Max GSI	Egg diameter (mm)	Relative F per kg	Author
Clariidae					
Clarias gariepinus		17			Plisnier *et al.*, 1988
Schilbeidae					
Schilbe mandibularis	180	15.3	0.95	217 000	Albaret, 1982
Schilbe intermedius	100	12	0.8–0.9	253 700	Albaret, 1982
Schilbe mystus		16.2			Paugy, unpub. data
Schilbe uranoscopus	180	7	1	207 000	Mok, 1975
Mochokidae					
Synodontis afrofisheri		26			Plisnier *et al.*, 1988
Synodontis schall	155	17	1.2	156 600	Albaret, 1982
Synodontis ocellifer		26.3			Paugy, unpub. data
Amphiliidae					
Amphilius atesuensis	40		1.7	41/g	Albaret, 1982
Mastacembelidae					
M. nigromarginatus	150 SL	15	2.15–2.45	19 800	Albaret, 1982
Anabantidae					
Ctenopoma kingsleyae			1.05	103 000	Albaret, 1982
Cichlidae					
Hemichromis fasciatus	80 SL		1.65	30 000	Albaret, 1982
Chromidotilapia guentheri	60 SL	4.9	2.25	8100	Albaret, 1982
Oreochromis niloticus eduardianus		5.2	2		Plisnier *et al.*, 1988
Oreochromis macrochir		4	2		Plisnier *et al.*, 1988
Tilapia rendalli		6			Plisnier *et al.*, 1988
Tilapia zillii	70 SL		1.65	38 600	Albaret, 1982
Sarotherodon galilaeus	224 TL		2.1–2.9		
Oreochromis niloticus	296				Harbott & Ogari, 1982
Centropomidae					
Lates niloticus	520 SL	4.5	0.7	86 000	Loubens, 1974

SL, standard length; Max GSI, maximum gonadosomatic index; F, fecundity; TL, total length.

Trade-offs between egg size, egg number and juvenile survival

In fish, the range of egg sizes is restricted when compared to the range of adult body sizes. For comparison, the size of the eggs must be measured at the same developmental stage. This is often difficult, and authors either give the size of ovocytes in the ovary during the assumed pre-spawning stage, or the size just after fertilisation (see Albaret, 1982).

From an evolutionary perspective the question of egg size is a matter of debate. It should be stated that the optimal egg size is that which optimises the number of offspring surviving to become reproductively active. In other words, egg size could be considered as a trade-off between fecundity and juvenile survival (Sibly & Calow,

1986). Bigger eggs producing bigger larvae is likely to be an adaptative advantage, allowing increased juvenile survival if food supply is limited or variable. On the other hand, minimisation of egg size would tend to maximise fecundity.

The quality of parental care appears to correlate positively with egg size both among and within species. But this correlation has been difficult to test, and different models proposed fail to predict this relationship. Sargent *et al.* (1987) provided a model based on three assumptions about the dependence of offspring survival on egg size: (i) offspring from larger eggs develop more slowly and take longer to resorb their yolk sacs and become juveniles; (ii) egg size (amount of yolk reserves) determines initial juvenile size; and (iii) larger juveniles that hatch from larger eggs have lower mortality, experience faster growth, and take less time to become adults. Under these assumptions, as parental care reduces instantaneous egg mortality, the optimal egg size increases. This could be illustrated by comparing egg size of substrate spawners and mouthbrooding cichlids (Table 8.8). Mouthbrooders have larger eggs (Table 8.8), but conversely lower fecundity than substrate spawners. Eggs of the Tanganyika cichlids, *Bathybates* and *Hemibates*, measure up to 7 mm diameter in the ovary before expulsion, and are the largest eggs known for any African cichlid (Coulter, 1991*b*).

In most teleost fish, both fecundity and egg size usually increase with age or size (length, weight) of fish. However, when account has been taken of such trends, an inverse correlation between fecundity and egg size can often be demonstrated, and would be expected if selection pressures limit the overall reproductive effort.

Peters (1963) observed that the size of tilapia eggs is not constant in every species: it varies greatly within spawns but for some species it is also related to the size of the spawning females, in that larger fish generally produce larger eggs. For *Sarotherodon melanotheron* (ex *Tilapia macrocephala*) for instance, mean egg weight is less than 5 mg for females of 50 g body weight, whereas it rapidly reaches 20 mg for 200 g female body weight, and then stabilises. For the same species, Legendre & Ecoutin (1996) observed an increase in egg size with female size up to 100 g females, but above this fish size, the egg size remained stable. However, whatever the tactic, the relationship between spawn weight and body weight remained a simple function of body weight. A similar, but less spectacular tendency is also observed in *Oreochromis mossambicus*, and to a lesser extent for *Sarotherodon galilaeus*. A significant correlation between female size and egg size was also found for two haplochromine species from Lake Victoria (Goldschmidt & Witte, 1990). The food supplied to the female can also affect egg size, but little is known about the importance of this factor in African fish under natural conditions.

The above rule also applied when comparing different populations of the same species. For the same female weight and the same spawn weight, populations kept in enclosures had more but smaller ovocytes than natural populations from the Ebrié lagoon. The size at first reproduction was also smaller (Table 8.9). In so-called 'acadja-enclos', which could be considered as ecological intermediates between enclosures and the natural environment, intermediate biological characteristics were also observed (Table 8.9).

The ecological factors responsible for changes in reproductive tactics are for the moment unsolved. Probably many should be considered (food quantity and quality, space), including genetic ones.

For mouthbrooders, the number of young that can be mouthbrooded depends of the size of the buccal cavity. The number of brooded eggs and embryos is proportional to the size of the male, in the male mouthbrooder *Sarotherodon melanotheron* (Legendre & Trebaol, 1996), and the 'brooding efficiency' (number of young brooded/number of eggs produced) depends of the respective sizes of both sexes. It has been demonstrated that the volume of the buccal cavity increased faster with male size than brood volume with female size. Therefore, small males cannot receive all the eggs spawned by females of the same size. The percentage of the buccal cavity occupied by a brood can reach up to 90% for small males, whereas it is around 40–50% for large ones. Incubation is therefore much more efficient when a female mates with a male of larger size than herself.

The trade-off between egg number and egg size in a community of related species is well illustrated in the reproductive strategies adopted by 21 haplochromine species, belonging to different trophic groups and occurring in the different habitats of Lake Victoria, studied by Goldschmidt & Goudswaard (1989). They pointed out the existence of two distinct groups: one, the egg-number group, which invested heavily in higher fecundity while growing, but little or not at all in egg size; two, the egg-size group, which invested in egg size and much less in egg number with increasing body size. There are no major differences in the gonadosomatic indices of ripe females belonging to both groups, and a trade-off between egg number and egg size is therefore suspected. Species belonging to the different groups occurred in different habitats. During the day, adults of all species belonging to the 'egg-number' group frequented the mud–water interface; in contrast, most of the species of the 'egg-size' group are column-dwelling or surface-bound. The selective factors that have generated the two groups are not known, but tentative explanations may be suggested. One is that juveniles of the

Table 8.8. *Biological reproduction characteristics in mouthbrooder cichlids from East African Lakes*

Species	Origin	Care	Mean SL ripe females (mm)	Mean body weight (g)	Mean fecundity no. eggs	Length major egg axis (mm)	Egg weight (mg)	Relative fecundity no. eggs/g body weight	Max GSI	Parental care period (days)	Authors
Haplochromis 'kibensis'	Lake Victoria	m.b.	66		22	3.5	10.0				Goldschmidt & Goudswaard, 1989
Haplochromis pyrrhocephalus	Lake Victoria	m.b.	59	4.7	23	3.1	6.6	5.3	4.4		Goldschmidt & Witte, 1990
Haplochromis heusinkveldi	Lake Victoria	m.b.	63	7.4	20	3.6	11.4	3.5	5.1		Goldschmidt & Witte, 1990
Haplochromis laparogramma	Lake Victoria	m.b.	55	6.2	13	3.4	9.5	3.2	3.9		Goldschmidt & Witte, 1990
Haplochromis 'argens'	Lake Victoria	m.b.	61	6.1	21	3.4	9.4	3.9	4.4		Goldschmidt & Witte, 1990
Haplochromis piceatus	Lake Victoria	m.b.	62	4.3	33	3.2	7.6	7.2	5.7		Goldschmidt & Witte, 1990
Haplochromis 'profundus'	Lake Victoria	m.b.	86		94	3.2	7.5				Goldschmidt & Goudswaard, 1989
Haplochromis 'reginus'	Lake Victoria	m.b.	52	4.1	23	3.1	6.7	6.5	5.5		Goldschmidt & Witte, 1990
Haplochromis teegelaari	Lake Victoria	m.b.	86		87	3.2	7.6				Goldschmidt & Goudswaard, 1989
Astatoreochromis alluaudi	Lake Victoria	m.b.	98		170	2.9	5.8				Goldschmidt & Goudswaard, 1989
Melanochromis joanjohnsonae	Lake Malawi	m.b.	54	4.6	15	4.3	21.0	4.0			Marsh et al.,1986
Melanochromis auratus	Lake Malawi	m.b.	62	6.4	21	4.0	25.5	3.9			Marsh et al.,1986
Pseudotropheus zebra	Lake Malawi	m.b.	72	11.9	26	4.2	26.9	2.6			Marsh et al.,1986
Pseudotropheus tropheops	Lake Malawi	m.b.	78	16.0	34	4.5	22.8	2.5			Marsh et al.,1986
Cyrtocara taeniolata	Lake Malawi	m.b.	72	11.1	29	3.9	17.2	3.0			Marsh et al.,1986
Petrotilapia spp.	Lake Malawi	m.b.	109	44.3	78	4.7	25.9	2.5			Marsh et al.,1986
Labeotropheus fuelleborni	Lake Malawi	m.b.	74	14.4	23	5.0	42.0	1.9			Marsh et al.,1986
Tilapia grahami	Lake Magadi	m.b.	59		34	2.8					Coe, 1969
Haplochromis anastodon	Lake Kivu	m.b.	68		36	3.5		3.4			Losseau-Hoebeke, 1992
Haplochromis olivaceus	Lake Kivu	m.b.	67		28	3.7		3.1			Losseau-Hoebeke, 1992
Haplochromis crebidens	Lake Kivu	m.b.	69		31	3.6		3.0			Losseau-Hoebeke, 1992
Haplochromis paucidens	Lake Kivu	m.b.	70		25	3.6		2.5			Losseau-Hoebeke, 1992
Petrochromis polyodon	Lake Tanganyika	m.b.	134		12	7.1					Kuwamura, 1986
Petrochromis famula	Lake Tanganyika	m.b.	92		17	6.4					Kuwamura, 1986
Tropheus moorei	Lake Tanganyika	m.b.	70		11	5.7			4.6	35	Kuwamura, 1986
Simochromis diagramma	Lake Tanganyika	m.b.	75		20	5.2			5.5	28	Kuwamura, 1986
Gnathochromis pfefferi	Lake Tanganyika	m.b.	75		52	3.1			5.6	19	Kuwamura, 1986
Xenotilapia longispinnis	Lake Tanganyika	m.b.	69		19	3.2					Kuwamura, 1986
Xenotilapia ochrogenys	Lake Tanganyika	m.b.	68		35	3.1					Kuwamura, 1986

Table 8.9. *Comparison between size at first maturity (L50), mean egg weight and fecundity for different populations of* Sarotherodon melanotheron *in Côte d'Ivoire*

Type of environment	Sex	L50 (mm)	L95 (mm)	Mean egg weight (mg)	Fecundity for a 200 g body weight female
Lagoon	F	176	223	28	329
Enclosure	F	140	180	12	726
	M	138	200		
Acadja 1	F	189	220	19	495
	M	170	199		
Acadja 2	M	166	205	15	517
	F	170	203		

From Legendre & Ecoutin, 1996.

egg-size group have never been caught in open waters, but grow up in shallow waters where the availability for food may be smaller and predation pressure higher than in deeper and more exposed areas. Similar trade-offs have been observed for tilapiines (Peters, 1963) and for mouthbrooding cichlids from Lake Tanganyika (Kuwamura, 1986).

Life-history styles

Tropical freshwater fish exhibit great diversity in morphological, physiological and ecological attributes, and provide excellent systems for the evaluation of life-history patterns, that is their lifetime pattern of growth, differentiation, storage and reproduction. This raises the question: is the great variety of lifetime patterns a random assemblage or is it the result of adaptations to the environment? According to Harper & Ogden (1970) 'the expression of the life-cycle is the outcome both of the genotypic strategy and the particular tactics followed in response to the environment'. This raises another key question: what are the environmental factors that favour the evolution of particular life-history patterns? and how does a life-history pattern change when a species is faced with a new environmental situation?

Life-history theory deals therefore with constraints among the demographic variables and traits associated with reproduction and the manner in which these con-

straints shape strategies for dealing with different kinds of environments. This assumes that biological systems contain flexible mechanisms which allow them to respond to fluctuations of the environment. The challenge is to identify which are the factors and mechanisms responsible for the selection of a particular combination of biological or demographic traits (Barbault, 1981).

The response of individuals, in terms of their allocation of time and energy in the face of environmental changes over time and/or space, will determine, through the effects on survival and reproduction, the lifetime reproductive success of these individuals. This is usually called fitness, which can be considered as the proportionate contribution that an individual makes to future generations: 'the fittest individuals in a population are those that leave the greatest number of descendants relative to the number of descendants left by other, less fit individuals in the population' (Begon et al., 1986). Those individuals that leave the greatest proportion of descendants in a population have the greatest influence on the heritable characteristics of that population. This optimality approach has been very successful in aiding understanding of the observed patterns of life-history variation (Roff, 1992). Natural selection is assumed to favour those individuals with the greatest fitness.

Life-history theories

One of the fundamental challenges of modern ecology is to understand what are the similarities and differences in life histories of different organisms, despite the fact that every life history is to some extent unique (Begon et al., 1986), and to analyse the relationships between the characteristics of an environment and the life-history patterns of populations experiencing that environment. A major goal is to predict the direction of evolutionary change in life-history traits in response to an environmental change that has defined effects on age-specific schedules of mortality and/or fecundity. In order to ensure the continuity of their existence, species use different mechanisms, which can be divided into characteristics that determine the number of young born (e.g. growth, fecundity, etc.) or those which influence the survival of offspring. Both aspects are closely related, but the same result (e.g. the number of offspring) could be obtained by various combinations of characteristics.

An early life-history theory is that of r- and K-selection (MacArthur & Wilson, 1967). The terms r and K refer to the parameters of the logistic growth curve for populations, where r is the slope representing the growth rate of the population, and K the upper asymptote, which represents the carrying capacity of the environment. The theory suggests that high density-independent mortality (r-selection) would be expected in harsh or unpredictable environments, and selects for early reproduction, high fecundity and short life expectancy. Conversely, high density-dependent mortality (K-selection) would be expected in a clement environment in which populations are close to equilibrium densities, and selects for delayed reproduction, low fecundity and a long life expectancy. In other words, an r-selected species will have a life-history strategy that tends towards productivity, whereas the K-selected species are more geared to efficient exploitation of resources.

This r/K scheme is of course an over-simplification. Very few aquatic organisms are completely r- or K-selected, and most of them are somewhere between the two extremes. But this relatively simple concept can help to make sense of a large proportion of the multiplicity of life histories. The ecological attributes of r- and K- selected fish communities in Africa were tabulated by Ssentongo (1988). Communities in marshes, floodplains and swamps are characterised by high, catastrophic, density-independent natural mortality, and r-selected species predominate (Table 8.10). On the other hand, for populations from stable littoral and benthic areas of the East African Great Lakes, natural mortality is mostly due to predation. It is presumed that fish species in the East African Great Lakes were initially dominated by r-selected species but, during the course of evolution, each niche was filled through selective speciation, resulting in a progressive shift from r- to K-selection in the species assemblages (Ssentongo, 1988).

The terms generalist and specialist (or eurytope and stenotope) have also been used. A specialist may be considered as an organism adapted to its environment. A rich milieu where resources occur in abundance will tend to favour specialists, whereas an impoverished milieu favours generalists. In particular cases, specialisation may cause the explosive formation of evolutionary units that may be many trophic styles within one breeding mode. Such is the case for cichlid species flocks of the East African Great Lakes.

Table 8.10. *Distinctive features of 'r'- and 'K'-selected fish communities of tropical Africa*

Ecological attributes	'r'-selection	'K'-selection
Habitat	floodplains, marshes, swamps, estuaries, etc.	littoral and benthic areas of the East Africa Great Lakes
Species group	mostly pelagic fish: *Stolothrissa, Limnothrissa,* etc.	mostly demersal fish: *Cichlidae, catfish,* etc.
Diversity index	less diverse with dominant species	very diverse and species with more or less equatable distribution
Environmental stability	fluctuates seasonally	negligible fluctuations
Survivorship curve of fish species	sharply falling and concave	distinctly rectangular
Longevity	prey fish with short life-span	longer life-span
Natural mortality	usually high, sometimes catastrophic	relatively low and mostly due to predation
Protection cover	schooling behaviour	formation and defence of territories
Reproduction mode	numerous pelagic eggs, no parental care, some prolific substrate spawners (e.g. tilapias)	spawning displays, pairing and highly developed parental care
Feeding adaptations	specialists at low trophic levels few trophic levels	adaptative feeding types more trophic levels
Other favoured factors	rapid turnover early reproduction small body	slow turnover delayed reproduction large body size

From Ssentongo, 1988.

Instead of r- and K- strategies, Balon (1985, 1989) preferred the terms 'altricial' and 'precocial', which were already used by ornithologists to identify two life-history trajectories. Altricial animals produce small, incompletely developed young, with only a small volume of yolk that is not sufficient to produce the definitive

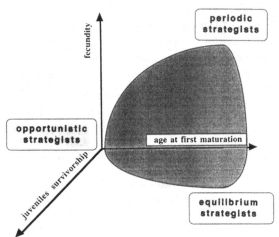

Fig. 8.2. Different life-history strategies for fish (from Winemiller, 1992)

phenotype. They are generalists and able to survive in unpredictable and uncrowded environments where they are mainly subjected to density-independent mortality. On the other hand, precocial forms produce large and well-developed young, are specialists, inhabit stable and crowded environments, and are subjected to density-dependent mortality. *Labeotropheus*, a mouthbrooder cichlid from Lake Malawi, provides a good example of a precocial fish that releases a relatively self-sufficient juvenile, 14% of the adult fish size, 31 days after fertilisation (Balon, 1977). Altricial species are more or less similar to 'r' strategists, while precocial species are equivalent to 'K' strategists.

As a result of comparative studies of fish in different environments, Winemiller (1992) identified three similar strategies (Fig. 8.2) as endpoints of a triangular continuum: (i) small, rapidly maturing, short-lived fish (opportunistic strategists); (ii) larger, highly fecund fish with longer life spans (periodic strategists); and (iii) fish of intermediate size that often exhibit parental care and produce fewer but larger offspring (equilibrium strategists).

The opportunistic strategy of repeated local recolonisations through continual and rapid population turnover is well exemplified in the killifish inhabiting temporary or highly variable habitats. It maximises the intrinsic rate of population growth (r) through a reduction in the mean generation time which is achieved via early maturation,

which in turn diminishes its capability to produce large clutches and large eggs. However, owing to their small size, the relative reproductive effort of opportunistic strategists is high, and the production of multiple batches sometimes results in annual fecundities that greatly exceed female body biomass in these small species.

Equilibrium strategists optimise juvenile survivorship by apportioning a greater amount of material into each individual egg, and/or the provision of parental care. That is the case for Balon's precocial forms with direct development. Trade-offs between body size, egg size and clutch size are probably inherent in the requirement for an adult body size sufficient to permit successful implementation of parental behaviour (nest, brooding, etc.). Equilibrium strategy is often associated with sedentary local populations, relatively stable adult food resources and prolonged breeding seasons. Cichlids, as well as many other fish species exhibiting parental care, are good examples of this strategy.

The periodic strategy maximises age-specific fecundity (clutch size) at the expense of optimising turnover time and juvenile survivorship. In environments which exhibit spatial or temporal variations that are predictable to some extent, selection will favour the strategy of synchronous reproduction in phase with the periodicity of optimal environmental conditions and the production of large numbers of small offspring that require little or no parental care. Large body size enhances adult survivorship during suboptimal conditions and permits storage of energy for future bouts of reproduction. This strategy is often associated with long-distance spawning migrations to productive, wet season floodplains. Species such as *Alestes baremoze* and *Schilbe mystus* are examples of periodic strategists, as are (to varying degrees) many of the species using floodplains as nurseries.

The ideal fish would have a long life, reach maturity at an early age, then reproduce often and produce numerous young at each breeding. But there are constraints that limit the capacity of fish to achieve all these desirable biological characteristics simultaneously (Wootton, 1990). For example, there are a number of broad-scale constraints within which the evolution of traits can take place: (i) genetic constraints, given that evolution requires some degree of genetic variation; (ii) phylogenetic constraints, resulting from the evolutionary history of the species, may limit possible directions, or make some directions more likely, given that all options are not simultaneously available (Southwood, 1988); (iii) physiological constraints, resulting from processes that act internally within the organism but are above the level of the genes; (iv) mechanical constraints, setting limits to the design of all types of animals; and (v) ecological constraints, also called external constraints, forming the backdrop of all life-history analyses, are those that are a function of the particular environment in which the animal lives (Roff, 1992).

Alternative reproductive strategies

Teleost fish have successfully colonised fresh and marine habitats, and proved their ability to adapt their life histories in response to environmental conditions. Major biological events may occur at varying periods of the life and the amount of energy allocated to reproduction or other physiological requirements can vary greatly between species, but also between populations experiencing different ecological circumstances. They have developed a wide range of life-history tactics to ensure that the maximum number of progeny reach sexual maturity, and to increase survival in order to maintain populations large enough to ensure their continuance. Slobodkin & Rapoport (1974) illustrated the central biological problem of organisms' response to a changing environment as follows: they can be seen as playing a game against Nature; and the measure of the success is how long they can stay in the game. In other words, an individual loses the game when it fails to reproduce, and a population loses the game when it becomes extinct, that is to say when all individuals are unable to maintain the gene pool. The optimal response to environmental change would therefore be, to minimise its cost and to answer through an adaptative response to the change.

The adaptation of individual populations to local ecological conditions is likely to result in alterations in genetic frequencies. However, the adaptive capacity of an organism is not entirely restricted to the ability of the genetic material to undergo spontaneous mutation, and living systems seem to have the potential to follow different trajectories in ontogeny or evolution in response to prevailing environmental conditions. Organisms may be regarded as scenario choosers that keep open two or more life-history options and allow the coevolving environment to determine which one will be most suc-

cessful (Bruton, 1989). In that case, unspecialized fish may have a wide range of phenotypic expression, and Mann et al. (1984) provided evidence that temperate fish species occupying a wide range of habitats, developed different life-history tactics. Stearns & Crandall (1984) also demonstrated that fish do not mature at a fixed age or size, but mature along a trajectory of age and size that depends on demographic conditions. Moreover, there are possibly different types of trajectories that correspond to particular demographic conditions. Finally, the model developed by Stearns & Crandall (1984) also suggested that age and size at maturity were not determined only by genetics or only by environment, but by both. In conclusion, different traits can be conceived as having a plastic trajectory in response to stress, and the way they interact to provide a co-ordinated response to stress reflects the demographic history of the population.

The existence of different reproductive patterns in subpopulations of the same species has been mentioned for African fish. One fairly well-documented example is given for Lake Chad by Bénech & Quensière (1985). In *Alestes baremoze*, one part of the adult population breeds in riverine habitats after an anadromous migration, whereas the other part breeds in the lake. The riverine habitat and its associated floodplains offer food and shelter for juveniles, but this reproductive pattern has a cost that is the energy expended on the migration of mature fish. The lacustrine reproductive pattern has a lower energetic cost, but the successful survival of demersal eggs is expected to be lower as a result of temporary deoxygenation of the lake water. It is difficult to decide at the moment if the existence of these two breeding patterns in *A. baremoze* is governed by genetics (two genetically distinct populations), or could be viewed as alternative tactics for one individual during its life history. If the latter, it may be that young females breed in the lake, whereas older ones undertake spawning migrations. There is, nevertheless, some evidence that this reproductive polymorphism may have physiological aspects, since migratory fish seem to accumulate more fat and to produce less gonad material than sedentary ones (Bénech & Quensière, 1985).

In Lake Kinnereth (Israel), *Tilapia zillii*, a monogamous biparental guarder, shows a remarkable range of nesting, brood care and courtship behaviour, according to the nature of the substrate and its exposure to wave action (Bruton & Gophen, 1992). On rocky substrates, exposed to wave action, the pair simply cleaned a horizontal rock surface and fertilised their eggs there. Adult fish did not regularly guard the young, which quickly became independent of the adults. In exposed sandy sites, the nests were simple, shallow saucer-shaped, and parental care was also abbreviated; but in partially sheltered lagoons with clay substrates, the nests consisted of a hollow with one to three brood chambers, which ranged in depth from 5 to 85 cm, and parental care was extensive. These observations support the idea that fish may express the full potential of their genotype to a greater or lesser extent depending on the nature of the environment. As a result of the plasticity of the phenotype, many major biological functions may differ markedly in different environments. This could be considered as an ecological advantage in a widely fluctuating tropical environment (Bruton & Gophen, 1992).

The occurrence of alternative reproductive behaviour has been reported in many species of fish (Taborsky *et al.*, 1987). In a number of cichlid populations, territorial males attract females into their territories and stimulate them to spawn within a nest or a prepared spawning site. However, the number of sexually mature males is usually greater than the number that can establish and defend territories in the space available. Supernumerary, sexually mature males that are excluded from arenas have few chances to successfully court females, and can develop an alternative mating pattern and intrude upon a spawning pair within a territory to steal insemination (McKaye, 1983). Such tactics are usually referred to as sneaking, and sneakers are usually inferior to territorial males. In the case of the riverine maternal mouthbrooding cichlid, *Pseudocrenilabrus philander*, any sexually mature male may alternate between the reproductive alternatives offered by territoriality and sneaking (Chan & Ribbink, 1990). Three types of sexually mature males were recognised: non-territorial males resembling females in colour; semiterritorial males that periodically attempted to defend a small area between existing territories and were dull in colour but with discernible male markings; and territorial males, brightly coloured and with defended territories. Studies in the laboratory showed that as the males grow, they pass progressively through these three stages and their reproductive behaviour therefore changes, from that of a sneaker to that of

a territorial male. During mating the female lays batches of eggs on the substratum and the male moves over these to inseminate them. Non-territorial and semiterritorial males also attempt to inseminate eggs by approaching and joining mating pairs. When eggs are present, they attempt to move over them, presumably to inseminate them (Chan & Ribbink, 1990). Mating has high costs (defence of the territory) for territorial males, but also high benefits as long as they have considerable success with the females. Territorial males contribute to 93% of fertilisations. Semiterritorial and non-territorial males have lower energy costs but also less success, since they contribute to 5 and 2%, respectively, of fertilisations.

Sneaking is an effective alternative strategy for at least some riverine haplochromines laying large batches of eggs on the substrate, but it is also found among the maternal mouthbrooders of the African Great Lakes: *Copadichromis eucinostomus* in Malawi (McKaye, 1983, 1984); and *Aulonocranus dewindti* in Tanganyika. In Lake Malawi, where there is intense predation pressure on eggs and juveniles (see above), haplochromine fish have developed other alternative evolutionary breeding tactics. One of them is the increased size of eggs (4–7 mm), which result in the release of larger and more developed offspring. Another is intrabuccal insemination, which avoids the eggs being exposed to predators when laid on the substrate for the males to fertilise. There are also other means of protecting precocial young. For example, in the case of the Mbuna, the parents do not practise parental care in nature but release their offspring into cracks and crevices among rocks (Trendall, 1988a; Ribbink, 1990). They find food and shelter in these habitats, emerging only when it is safe to do so. Other fish, such as semipelagic zooplanktivores, have a cuckoo-like behaviour in which other species care for their offspring (Ribbink, 1977).

Ribbink (1990) also reported two different alternative strategies for two related species from Lake Malawi. *Dimidiochromis kiwinge* protects its young for the few days necessary to bridge the gap between the protection afforded in the mouth and the protection conferred by schooling, the young then congregate in schools in inshore surface waters or macrophyte beds. In contrast, *Tyrannochromis macrostoma* guards its offspring for more than three weeks, and the juveniles become independent at 40 mm size. In this case, parental care gives the opportunity for the offspring to grow large enough to become independent, presumably with a better chance of survival.

It has been suggested that tilapias master both strategies of precocial and altricial life-history styles, under different environmental conditions (Noakes & Balon, 1982) . It is also possible to assume that tilapias have developed the same adaptation towards biotic pressures. The optimal foraging theory of predation suggests specific selective pressures on given sizes (Kolding, 1993). Mortality factors are usually size dependent (the predator fields or windows), and predation pressure is generally considered to affect the juveniles preferentially. In that case, the observed plasticity in size at maturation could very well be a response to variations in mortality rates. The situation observed in Ferguson Gulf in Lake Turkana offered the opportunity for a holistic approach, taking into account the impacts and interrelationships of both the changes in environmental parameters and the various predators (Kolding, 1993). One of the goals was to explain differences in population structure between those outside the Gulf, where samples did not contain *Oreochromis niloticus* below 17 cm TL, and those inside the Gulf where large numbers of juveniles occur. Another goal was also to explain the 13 cm decrease observed over 30 years in the median size of maturity of *O. niloticus* in Lake Turkana.

In Lake Turkana, there is no fish predator in the shallow areas inhabited by tilapias, but these areas are intensively predated by pelicans, crocodiles and man, and this predation is assumed to fall mostly on the larger specimens. Moreover, ecological conditions are more unstable in the marginal habitats, and must, in theory, be responded to by a rapid turnover rate. Lastly, early

maturation size might be a direct consequence of reduced oxygen levels (Pauly, 1984), and/or rapid diurnal fluctuations. This may help to explain why, during periods when the shallow marginal habitats provide a relatively long-term refuge, the overall selection pressure would favour small sizes. Conversely, during periods of decreasing water level, tilapias are forced into the deeper part of the lake where the predation upon juveniles would favour large sizes. Growing big and fast should be the way to escape a given predator.

Ecological conditions prevailing inside and outside the Ferguson Gulf of Lake Turkana represent 'forced alternating selective pressures towards which tilapias have become well-adapted. In contrast to the migratory type floodplain species, which counteract the changes by mobility, the tilapias simply follow the changes by their immense versatility' (Kolding, 1993). This example illustrates the application of the alternative life-history theory as a function of different size-selective mortality rates in different environments of Lake Turkana. In the deep central lake, the habitat is constant and predictable, mortalities are high among the small sizes and density dependent, longevity is long, and selection favours large body size, delayed reproduction, efficiency and survival. In contrast, in shallow marginal areas where the habitat is much more variable and unpredictable, mortality is high for large sizes, sometimes catastrophic, longevity is short, and selection favours small body size, early reproduction, high growth rates and productivity.

The more detailed studies of fish life-history strategies that are conducted, the more alternative reproductive tactics are discovered, and at an increasing rate. This field of research has great promise for future studies.

9 Diversity of responses to environmental constraints

Fish display a wide range of behavioural, physiological and biochemical adaptations to meet the challenges imposed by the changes that can occur within aquatic environments. These adaptations may range from avoidance reactions, in which the fish swim away from the source of the environmental stressor, to the initiation of complex suites of physiological changes requiring the integration and coordination of input signals from a variety of receptors.

Jobling, 1995

Unlike birds and mammals, fish do not regulate their body temperature, but can exhibit great flexibility in their responses to environmental changes and in traits such as growth, age at first reproduction, life span and other life-history traits. They may respond to this change through an adaptative response, its nature depending on the time-scale of the change compared to the generation time, or it may be in part a behavioural response to avoid a short-term change, or a biochemical and physiological response if change persists.

Speciation is fundamental to the evolution of taxa, one component of which is adaptation, the evolution of traits (Brooks & McLennan, 1991). Adaptation is related to notions of functional fit between an organism and its environment, and the processes that improve this fit, or which can maintain the functional efficiency of organisms, have not been well studied (Endler & McLellan, 1988). They have frequently been mistakenly equated with natural selection, whereas they are in fact the result of both mutation and selection in the past. Moreover, we know very little about the ecological, physiological or genetic mechanisms involved and the time-scales of the processes. The concept that species are in a continual race to maintain their existence by adaptation is exemplified by the Red Queen Hypothesis (Van Valen, 1973), named thus because the Red Queen said to Alice ' here, you see, it takes all the running you can do to keep in the same place'. Van Valen considered that in a constantly decaying environment, natural selection operates only to enable the species to maintain its current state of adaptation rather than improve it.

The nature of the responses of organisms to changing environmental conditions is a central problem in biology and the meaning of 'adaptation' has been widely discussed (Endler & McLellan, 1988; Løvtrup, 1988). One can distinguish between adaptative traits that were selected for one function (adaptations) and those which may originally have been developed for a completely different purpose, or may have arisen fortuitously, and could be used in a new way (exaptations).

This chapter is concerned with some of the major environmental constraints to which tropical freshwater fish have become adapted since diversity of conditions enhances biodiversity. In aquatic environments temperature changes more slowly than in terrestrial habitats but, nevertheless, has a direct influence on all aspects of life. It also has indirect effects, particularly through the influence of temperature on the amount of oxygen that can be dissolved in water. This is even more marked in tropical than temperate waters, since warm water, even when fully saturated, holds less oxygen than cold water. Many tropical freshwater fish must therefore be adapted to relatively low oxygen concentrations. Moreover, many inland waters are not fresh waters, but contain widely varying concentrations of salts. Salinity influences density as well as the oxygen-carrying capacity of the water. Finally, light is absorbed very rapidly by water and is frequently attenuated by turbidity. Thus many fish live and hunt in low light intensities, and sometimes darkness.

Temperature

Because of its effect on the velocity of chemical reactions, environmental temperature is perhaps the most pervasive of the abiotic factors. It has major influences on many biological processes, such as maturation and spawning, growth, development rate, metabolism, etc. Fish are faced with a simple choice in responding to temperature changes: either to move to another habitat when possible (a behavioural response); or to adapt their metabolic processes to the new situation (metabolic responses).

Temperature preferences and thermal tolerance ranges

Changes in temperature occur less abruptly in water than in air, but fish can, nevertheless, be exposed to large temperature variations. In shallow waters, for instance, temperature can fluctuate by several degrees during the diel cycle. There may also be major seasonal changes, especially in tropical waters, whereas in typically equatorial situations, the range of temperature variation over the year is small.

Adult fish can survive over a limited range of temperatures bounded by the upper and lower incipient lethal temperature (UILT and LILT). Philippart & Ruwet (1982) gave a valuable review of thermal tolerance ranges in tilapiines. The LILT for *Tilapia zillii* that occur in Lake Huleh (Israel) is 6.5 °C, and the UILT is 42.5 °C, but mortality begins when the temperature exceeds 39.5 °C. *Oreochromis niloticus* can also tolerate a temperature of 8 °C for several hours and an upper tolerance limit of 42 °C has been reported. These species, and probably also *Oreochromis aureus* and *Sarotherodon galilaeus* that tolerate a wide range of temperature, are typically eurythermal species. On the other hand, species, such as *Tilapia guineensis* (14–33 °C) and *Sarotherodon melanotheron* (18–33°C), that live in a narrower range of temperature are more stenothermal.

It is perhaps not the thermal tolerance limits that are so important for wild populations, but rather the tem-

perature preferenda at which maximum biological performances can be expected. In laboratory studies, fish typically exhibit a frequency distribution of temperature occupied over time that has a clear mode (the 'preferred temperature'), and upper and lower avoidance temperatures where the numbers of occurrences drop off markedly (Countant, 1987). It is generally accepted, for instance, that tilapias cease growing significantly at temperatures below 20 °C, and feeding stops completely around 16 °C. Although they may be able to tolerate short-term exposure to temperatures ranging from 7 to 10 °C, death can occur in some species at temperatures as high as 12 °C after long-term exposure. On the other hand, maximum swimming performance occurs at 28 °C in *T. zillii*, 28–32 °C in *O. niloticus* and 32 °C in *S. galilaeus* (Fukusho, 1968).

The effect of temperature has an important consequence for the egg stage. At low temperature, the development time from fertilisation to hatching is extended, and fish therefore spend longer in this most vulnerable period of their life history. For *Heterobranchus longifilis*, the thermal optimum for incubation and egg hatching, estimated experimentally, ranges from 25 to 29 °C (Legendre & Teugels, 1991), which closely corresponds to temperatures in natural habitats during the reproduction period of this species. No hatching was obtained experimentally at 21 or 33 °C. In contrast, the range of optimal temperatures for egg development is much wider for *Clarias gariepinus*: 19–31 °C according to Bruton (1979a). This better tolerance of low temperature may be explained by (or may explain) the wider distribution range of this species, which is recorded from South Africa to Israel.

Temperature and metabolism

For ectothermic living systems like fish, the temperature of the body will closely track that of the water. Whatever the geographical area, the standard metabolism of fish increases with temperature (Brett & Groves, 1979), and any form of thermal homeostasis exerts a significant cost through a higher rate of metabolism.

Caulton (1977, 1978, 1982) evaluated the routine metabolism, that is the energy required by unfed fish exhibiting spontaneous rather than directed movement, of two African cichlids. A relationship between temperature and metabolism in the range 16 °C to 38–

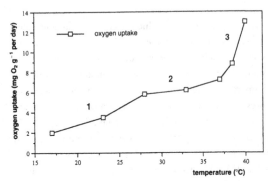

Fig. 9.1. The effect of temperature on oxygen uptake by young *Tilapia rendalli* (from Caulton, 1977). Numbers indicate the three phases described in the text.

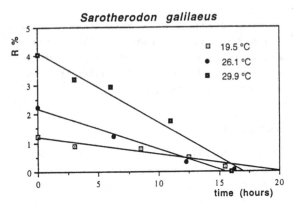

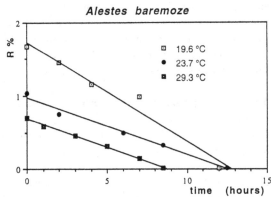

Fig. 9.2. Effect of temperature on gastric evacuation rates for two species from Lake Chad, *Alestes baremoze* and *Sarotherodon galilaeus* (from Lauzanne, 1969, 1977). R is the repletion coefficient expressed as % of the body weight.

40 °C was established for a single size class of *Oreochromis mossambicus* and *Tilapia rendalli*. For *O. mossambicus*, the routine metabolic energy demand can be described by the equation: $R_T = 0.0086\ t^{2.0783} M^{0.652}$, where R_T is the energy of metabolism expressed in J hr^{-1}, t the temperature in °C, and M the fresh mass of the fish in grams. For *T. rendalli*, three distinct phases of oxygen consumption were identified within the temperature range of 17–40 °C (Fig. 9.1). In the first phase (17–28 °C), the metabolic energy demand followed the normal logarithmic increase with increasing temperature. In the second phase, however, there is a distinctive and unexpected plateau between 28 and 37 °C. Over this range the metabolic demand of *T. rendalli* increased by only 22%, whereas that of *O. mossambicus* increased by 62%. Similar unusual metabolic curves have been reported for other species of fish, including *Oreochromis niloticus*, and are believed to illustrate a significant energy saving. This may help to explain the behaviour of juveniles that move into the warmer shallows during the day, but do not use much more energy than they would during the night in much cooler water. The final phase (phase 3) shows a return to the original logarithmic increase in oxygen consumption.

Temperature and feeding

Fish can also adjust their physiology to temperature changes by changes to their biochemical mechanisms that minimise the energetic cost and buffer, at least partially, the effect of the environmental change (see Hochachka &

Somero, 1984). Temperature affects both the rate of food consumption and the rate of gastric evacuation. At low temperatures the fish may cease to feed, but the rate of consumption increases with increasing temperature.

For *Sarotherodon galilaeus* in Lake Chad, the gut repletion coefficient and rate of gastric evacuation increased with temperature (Fig. 9.2), but a quite different situation was observed in the same lake with *Alestes baremoze* (Fig. 9.2). For this latter species gut repletion and gastric evacuation rates were negatively correlated with temperature (Lauzanne, 1969) . There is no clear explanation for these contrasting physiological behaviours. Lauzanne (1977) tentatively suggested that it could be related to differences in spawning periods, but this explanation is not convincing. What is clear is that the daily food consumption rate (expressed as the ratio of the weight of ingested food to fish weight × 100) decreases with temperature for *Alestes* (2.77% at 19.6 °C, 1.43% at 21.6 °C, and 1.18% at 30.6 °C), whereas it increases for *Sarotherodon* (1.5% at 19.5 °C, 3.3% at 26.1 °C and 6.1% at 30 °C). Caulton (1982) also reported a positive relationship between food intake and temperature for juvenile (50 g) *Tilapia rendalli* bred in tanks. This relationship holds over the range 18 to 30 °C, but between 30 and 35 °C consumption, like metabolism, is little affected by temperature, while at temperatures in excess of 35 °C, food intake declines and ceases at about 37–38 °C.

Behavioural responses to environmental temperature changes

Fish have thermosensory receptors that can detect a gradient in temperature and allow them to make decisions about the temperature range in which to live in natural environments. Behavioural thermoregulation is, therefore, a mechanism by which they can control the effect of temperature on metabolism. It is usually assumed, but apparently not really tested with quantitative data, that fish are able to select the temperature regime in which they would maximise their lifetime production of offspring (Wootton, 1990).

When juveniles of *Oreochromis niloticus* and *Tilapia rendalli* are submitted experimentally to a thermal gradient in tanks from 24 to 40 °C, they respond positively to the warm water and swim actively towards water that is only slightly cooler than their lethal temperature tolerance (Caulton, 1977, 1982). Such a response indicates that these fish are strongly thermophilic and that the critical temperatures (30–36 °C), demonstrated by the species tested, are fairly common in shallow tropical waters. According to Caulton (1982), the strong thermophily of tilapias could explain their daily inshore/offshore movements noted by many authors. Actually, in tilapias living in lakes, the juveniles, and above all the alevins, make daily cyclical movements between the shallower littoral zones, occupied during the day, and the deeper zones towards the open water, occupied by night. For instance, in Lake McIlwaine (Zimbabwe), the juveniles move in the morning from the deep homothermal

waters where they stay overnight, into shallow coastal waters where the temperature exceeds that of the homothermal waters (28 °C at midday). Large schools of juveniles are eventually found in warm waters at midday. In the late afternoon, a reversed movement is observed, and the juveniles move back into the deeper, homothermal water. The ecological significance and the survival value of these daily migrations have been much discussed. They may be a tactic for avoiding predators present in deep water during the day, or to reduce food competition by the successive exploitation of different habitats in the course of the day. An elegant explanation based on careful energy budgets was proposed by Caulton (1982), who claimed that thermal oscillations were beneficial to the growth of juvenile cichlids. The offshore–inshore migration tends to maximise growth: faster growth rates are achieved when fish move into warm inshore areas during the day (faster feeding and digestion rates), while at night, when they retreat to cool deeper areas, the energy demands are less than if the fish remained at constant temperature. Experimental data tend to support this hypothesis. In aquaria, 10 g *Tilapia rendalli* fry subjected to a thermal oscillation (18 °C night, 30 °C day) had a very similar growth rate to a control group of fry maintained at 30 °C. However, the major difference between the two groups was that fry maintained at a constant temperature consumed up to twice as much food as those fry subjected to daily temperature changes. For 50 g fish, Caulton (1982) also calculated that groups maintained at a constant temperature required 1.6 times more food than the group subjected to thermal oscillations. These data may indeed help to explain the observed fish behaviour.

Temperature creates distinct boundaries that may act as material barriers to fish movement. Details of thermal structure contribute to defining the amount of suitable habitat available at a given time. Knowledge of the thermal structure of a waterbody is therefore important for understanding fish distribution and abundance. The partitioning of aquatic environments among species and life stages along temperature gradients can be viewed as conveying survival value for individuals and populations. Spatial segregation results in utilisation of different food resources, and leads an individual to exploitation of a host of resources different from those used by either another species, or another life interval of the same species, and thus would seem to enhance the individual's survival.

Oxygen

With only a few exceptions, fish are dependant on aerobic aquatic respiration. But water has several disadvantages as a respiratory medium. Oxygen has a low solubility in water (1/30th of the oxygen content in an equivalent volume of air) and this solubility decreases with temperature, while the metabolic rate and thus the oxygen demand tends to increase. Moreover, due to the high viscosity of water compared to air, fish have to work harder to move it over their gills than do terrestrial vertebrates (Lindsey, 1978). For instance up to 10% of the metabolic rate may go to supporting gill ventilation in a fish at rest. As a result, there is the possibility that, under natural conditions, a fish will experience situations where oxygen may, at times, be a more important limiting factor than food (Kramer, 1987).

Oxygen availability can limit the activity of fish, and this element could, therefore, be considered as a 'limiting factor' (Fry, 1971). When oxygen availability is reduced, more energy must be allocated to breathing in order to maintain the same level of oxygen supply to the tissues. Ventilation frequency and amplitude increase with falling oxygen and feeding slows down, e.g. longer search, slower digestion and assimilation, which are all major components of the energy budget of many fishes. 'Changes in the activity of fish in hypoxic water suggest that they may experience reduced fitness. This results from increased energy allocation to ventilation and other activities and from decreased energy intake and ability to escape from predators' (Kramer, 1987).

Variations in the oxygen concentration in water depend on physical factors, such as temperature or turbulence, as well as biological ones. Primary production and photosynthesis during the day help to oxygenate the water, whereas the oxygen demand from decaying organic material could lead to anoxic situations. In flowing turbulent water, oxygen concentration is usually near its maximum, but in standing waters and particularly in deep tropical lakes, only the superficial water layer (epilimnion) is oxygenated, the hypolimnion often being permanently anoxic. When the oxygen content is low,

fish are faced with hypoxia, but fish actually exhibit a large variety of behavioural, physiological and morphological adaptations for coping with deoxygenated waters.

Morphological and physiological adaptations

Oxygen acquisition in all fish is achieved mainly by the active process of extracting dissolved oxygen from the water, but in waters where low oxygen concentrations occur for long periods of time, the fish fauna usually includes air breathing species. Such fish are categorised as bimodal breathers, although they do not necessarily use both modes simultaneously.

Most air breathers possess respiratory organs appropriate to both aerial and aquatic respiration, and a variety of adaptations for air breathing are found (Table 9.1), e.g. the swimbladder lungs of *Protopterus* and Polypteridae that originate on the ventral face of the oesophagous, the arborescent organs in the branchial cavity of Siluridae (*Clarias* and *Heterobranchus*), the modified swimbladder in *Gymnarchus* and *Heterotis*, and the pharyngeal diverticula in *Parachanna* (Bertin, 1958). In the catfish *Clarias*, part of the gill has become modified for air breathing. The primitive ostariophysan *Phractolaemus*, inhabiting swampy regions of the Lower Niger, also possesses an accessory respiratory organ.

Graham & Baird (1982) distinguished species that permanently use aerial respiration, from facultative air breathers using accessory organs only in cases of emergency. The partitioning of oxygen uptake between air and water in air-breathing species varies according to species, and depends on the activity and age of the individuals. In *Clarias gariepinus*, for example, more than 90% of juvenile oxygen consumption is of dissolved oxygen, whereas in adults 40 to 50% of the oxygen uptake is atmospheric (Babiker, 1979). Similar results (aerial respiration accounting for 40% of the total oxygen requirement) have been obtained for *Calamoichthys calabaricus* (Sacca & Burggren, 1982). For *Polypterus senegalus*, immature fish less than 22 g cannot utilise aerial oxygen, but older fish exhibit age-dependent reliance on aerial respiration in hypoxic and hypercarbic waters (Babiker, 1984). In the range of 3.75–2.50 mg dissolved oxygen l^{-1} mature *P. senegalus* partitioned its consumption of oxygen almost equally between air and water. Under adverse conditions, such as unavailability of dissolved oxygen or great activity, pulmonary oxygen con-

Table 9.1. *Different types of adaptation to aerial respiration among African freshwater fish*

Adaptations for air breathing

Adaptation of the branchial or pharyngeal cavity
Branchial modifications
    Synbranchid eels such as *Monopterus*, *Mastacembelus* spp.
Vascularised epithelium in a suprabranchial chamber
    Channidae: *Parachanna*
Labyrinth organ
    Anabantidae: *Ctenopoma*
Arborescent organs
    Clariidae: *Clarias*
Modified swimbladders
Osteoglossidae: *Heterotis niloticus*
Notopteridae: *Papyrocranus*
Gymnarchidae: *Gymnarchus niloticus*
Phractolaemidae: *Phractolaemus ansorgii*
True lungs
Polypteridae: *Polypterus senegalus*, *Calamoichthys calabaricus*
Dipnoi: *Protopterus*

sumption could account for nearly the total oxygen requirement (Babiker, 1984).

Aerial respiration appears to be a compensatory mechanism for *Clarias* (Magid & Babiker, 1975) and for *Polypterus senegalus* (Babiker, 1984) when branchial respiration is not sufficient; while for *Protopterus aethiopicus*, their aerial respiration accounts for 90–95% of total respiration, even in well-aerated waters (Lenfant & Johansen, 1968).

Some physiological adaptations were recorded for the endemic cichlid, *Konia dikume*, in Lake Barombi-Mbo (Cameroon). The relative blood volume of that species is noticeably higher than that of the other cichlids in the lake, as well as the haemoglobin concentration in its blood (Green *et al.*, 1973). This could function as an oxygen store, which would allow the fish to dive down into the deoxygenated hypolimnion and extend the time available for feeding on the *Chaoborus* larvae that spend the daytime in the hypolimnion. Various authors have attempted to link the affinity of haemoglobins for oxygen to the characteristics of the biota (Fish, 1956; Dusart, 1963). Green (1977) observed that *Synodontis* species living in poorly oxygenated waters tend to produce more heamoglobin than species living in well-aerated waters. The oxygen binding properties of *Oreochromis alcalicus*

grahami's haemoglobin are well adapted for extracting oxygen from water under very anoxic conditions (Lykkeboe *et al.*, 1975). The haemoglobin of *Oreochromis niloticus* has a high oxygen affinity, which allows this fish to survive hypoxic conditions (Verheyen *et al.*, 1985a).

Resistance to hypoxia

Oxygen requirements vary according to species, and tolerance of hypoxia has been studied experimentally, for 16 fish species from the Chad basin, in order to tentatively explain the selective mass mortalities that were observed at some periods during the drying up of Lake Chad (Fig. 9.3). Bénech & Lek (1981) pointed out that some strictly aquatic-breathers clearly exhibited a resistance to anoxia (*Tilapia zillii*, *Oreochromis niloticus*, *Schilbe intermedius*, *Brachysynodontis batensoda*), whereas other species need well-oxygenated waters (*Brienomyrus niger*, *Brycinus nurse*, *Micralestes acutidens*, *Labeo senegalensis*). Facultative air-breathing species, such as *Clarias* and *Polypterus*, do not survive in oxygen concentrations lower than 0.6 mg l⁻¹ when they are prevented from breathing air, which was the case in this experiment.

The oxygen consumption of *Oreochromis niloticus* is independent of oxygen concentration above a critical concentration of 2 mg l⁻¹ (Verheyen *et al.*, 1985a). This critical oxygen concentration is comparable to results obtained for *Tilapia rendalli* and *Tilapia macrochir* (Dusart, 1963). On the other hand, *Oreochromis niloticus*

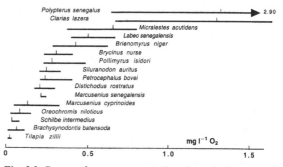

Fig. 9.3. Ranges of oxygen concentrations determined experimentally at 26 °C, at which fish species from Lake Chad suffer hypoxia. Upper and lower limits, oxygen concentration respectively at the first fish dead and at the last fish dead; |, oxygen concentration at 50% mortality (from Bénech & Lek, 1981).

is able to survive for up to five hours under anoxic conditions (0–0.1 mg O$_2$) according to Magid & Babiker (1975). However, its mean tolerance limit is generally much higher (1.4 mg l^{-1}) (Mahdi, 1973). According to Bénech & Lek (1981), *O. niloticus*, which has the ability to extract oxygen from very low concentrations, can survive a few hours of drastically hypoxic situations (less than 0.5 mg l^{-1}), but their behaviour is modified and the fish become lethargic, although they quickly recover normal behaviour when placed in well-oxygenated water.

Such physiological characteristics have an adaptive advantage for survival in natural environments where anoxia occurs regularly at night. This situation occurred frequently in Lake Chad during 1974, when it was drying up and shifting from a well-oxygenated lacustrine environment to swampy conditions. In the experimental fisheries (see Chapter 10), it appeared that some exclusively water-breathing species (*Alestes dentex* and *Labeo senegalensis*) rapidly disappeared from the lake, while species adapted to survive in waters with low oxygen concentration maintained their populations throughout the period of ecological changes, or only appeared in the community when ecological conditions became harsh. *Brienomyrus niger*, which required high dissolved oxygen levels to survive in experimental conditions (Fig. 9.3), is, nevertheless, well adapted to swamp environments and their populations increased in the marshy habitats of Lake Chad. In fact, this species might be a facultative air breather, but the adaptive mechanism is not known (Bénech & Lek, 1981).

Tilapias appear to be very resistant to low dissolved oxygen. This enables some species to live and reproduce in swamps and shallow lakes that become deoxygenated. These species also became proportionally more abundant in catches during the drought in the Sahel.

Behavioural responses to reduced oxygen availability

When fishes that normally live in well-oxygenated water are faced with oxygen-deficient water from which they cannot escape, suffocation could occur. But this does not usually happen in nature, because fish inhabiting such habitats frequently possess alternative modes of oxygen uptake, or may move to more oxygenated alternative habitats. Nevertheless, in most species, this avoidance behaviour is only observed when oxygen concentration

falls below 2–3 mg l^{-1} (Beitinger & Pettit, 1984). As a result, the survival of strictly aquatic breathers partly depends on their ability to find more suitable habitats during periods of hypoxia.

Some fish inhabiting shallow waters can survive low oxygen levels by using the surface film of water, which is relatively rich in oxygen because of diffusion from the atmosphere (Kramer *et al.*, 1978). This behavioural adaptation is part of a strategy, developed during periods of hypoxia, that allows fish to survive until the return of better oxygen conditions (Kramer & McClure, 1982). In Africa, *Hemisynodontis membranaceus* and *Brachysynodontis batensoda* possibly exhibit this type of behaviour (Green, 1977), as well as *Sarotherodon* (Bénech & Lek, 1981; Dusart, 1963) and cyprinodonts (Lewis, 1970). The exploitation of the oxygen-rich surface film is probably improved by anatomical features, such as a flattened head (cyprinodonts), small body size, upturned mouth (Kramer, 1983a). When the concentration of oxygen is near zero, fish spend most of their time at the surface. Such behaviour may nevertheless increase the risk of predation from fish-eating birds and other predators. This surface respiration has also been observed at oxygen levels where fish can survive without it and fish exhibit behavioural flexibility in the extent to which it is used (Kramer, 1987).

A few species not only use the superficial water layer, but also catch air bubbles in their mouths. As mentioned by Roberts (1975), 'almost all fish living in water of shallow to moderate depths will probably swim to the surface and "gulp air" if they have difficulty obtaining oxygen from the water. This behaviour is undoubtedly adaptive in itself and occurs in some fish in which one might not expect it'. American catfishes are also known to swallow bubbles of air. They pass through the alimentary canal, which has an area specialised for respiration.

Habitat shifts and the use of alternative breathing modes can change other ecological parameters, such as predation risk and food availability. The capacity for sustained swimming is reduced at low oxygen concentration and, consequently, predator avoidance capabilities might be reduced. However, there is little direct evidence to support this hypothesis. Laboratory studies have demonstrated that the risk of predation increases as oxygen decreases for fish attacked by herons (Kramer *et al.*, 1983) and for air-breathing fish (Wolf, 1985), but these effects are more likely to be related to changes in the use of cover than to reduced ability to escape.

Eggs and larvae are particularly at risk because they cannot escape hypoxic conditions, and parents have commonly developed behavioural adaptations to minimise this risk. Eggs are spawned in well-oxygenated waters, or parents ventilate their eggs. It has been said that under hypoxic conditions, the South American cichlid, *Heterotilapia multispinosa*, could move its newly hatched embryos up onto vegetation rather than into the substrate pits used under normal conditions (Courtenay & Keenleyside, 1983). Among the Channidae for instance, *Ctenopoma muriei* produce floating eggs, whereas *Ctenopoma damasi* spawn in a floating foam nest, guarded by males, and this may be to ensure adequate oxygenation of the eggs.

The African pike, *Hepsetus odoe*, has developed a specific behaviour to ensure successful breeding in oxygen-poor waters (Merron *et al.*, 1990). Eggs are deposited at the air/water interface in a floating foam nest among vegetation. One likely adaptative advantage is to provide the young with an hospitable oxygen environment, which is important for species living in swampy habitats were oxygen depletion may occur frequently and unpredictably. This spawning behaviour helps to explain the success of *H. odoe* in marshy, seasonally flooded habitats.

Mass fish mortalities resulting from anoxia

Mass fish mortalities have been reported from many waterbodies as a result of severe hypoxia occurring under particular environmental circumstances. In shallow lakes for instance, the disturbance of the superficial layer of the sediment, rich in decomposing organic matter, may result in a significant oxygen demand in the water column. Such disturbances, which generally occur after storms, were reported from Lake Chad in 1975 and 1976 (Bénech *et al.*, 1976). A marked decrease in the oxygen content of the water, as well as an unusually high CO_2 partial pressure, were responsible for the death of many fish species. It is also possible that disturbance was associated with the release of reduced substrates and/or toxins resulting from the anaerobic decomposition of the sedimented organic compounds. Indeed, the oxygen uptake resulting from the oxidation of reduced com-

pounds greatly contributed to the deoxygenation of the water column.

Mass fish mortalities, which were probably due to a mixing of the water column down to the mud, have also been reported from Lake George (Burgis *et al.*, 1973). Ganf & Viner (1973) estimated that the mean oxygen uptake for the five superficial centimetres of the muddy sediments of Lake George was 5 g O_2 m^{-2} during the first hour after being disturbed, while the oxygen content of the water column was about 13 g m^{-2}. If there was sufficient disturbance to resuspend the ten superficial centimetres of the mud, the oxygen uptake would be 17 g O_2 m^{-2} within one hour. No air-breathing species were observed among dead fish when this happened in Lake George.

Similarly, mass mortalities of *Oreochromis shiranus chilwae* and *Barbus paludinosus* but not of the air-breathing *Clarias gariepinus*, were observed in Lake Chilwa in 1966 and 1967, following heavy winds. These mortalities were ascribed to deoxygenation of the water column as a result of the resuspension of bottom sediment and excessively high turbidity levels (Furse *et al.*, 1979). This phenomenon was exacerbated in 1967 by blooms of blue-green algae, resulting in high levels of oxygen during the day but leading to hypoxia during the night. A similar situation following a bloom of blue-green algae was also observed in Lake Natron (Coe, 1966).

Mass fish mortalities occur occasionally in Lake Victoria under particular climatic conditions that lead to deoxygenation of the water column (Fish, 1956). In the shallow Nyanza Gulf, heavy storms can stir up the organic bottom sediment and distribute it throughout the water column. In such a situation, the nutrients from storm runoff can support algal blooms, which eventually lead to oxygen deficiencies. These adverse effects are now being enhanced by an increasing pollution load resulting from urban population growth, agricultural activities and the use of pesticides. In 1984 large numbers of *Lates niloticus* and *Oreochromis niloticus*, both introduced species, were found dead along the shores of the Nyanza Gulf (Ochumba, 1990). This massive kill of about 2400 tonnes occurred during a season of strong storms and algal blooms. Low pHs were observed, and could be considered as an indicator of the presence of carbon dioxide produced by the bacterial breakdown of the algal biomass. The dissected fish appeared to be in good physical condition, but flocculent mud and semi-gelatinous algal masses were found on the gills.

Cases of mass mortality have also been noted in Lake Albert. Greenwood (1966) quoted the experimental work of Fish (1956), which proved that *L. niloticus* has relatively high oxygen demands compared with other species of African freshwater fish. The lower ratio of gill area to body weight in large *Lates* may render them particularly susceptible to the effects of lowered oxygen concentrations that occur spasmodically under natural conditions.

Mass mortalities are not always associated with deoxygenation. For instance, Riedel (1962) has reported claims by Egyptian fisherman that periodic mass mortality of large *Lates niloticus* in the Nile is the result of the water temperature falling below 12 °C. Mortalities due to low temperatures have also been observed in South Africa.

Salinity

Most fish live either in freshwater or seawater, and only experience restricted changes in the osmotic characteristics of their environment. Adaptation to salinity involves physiological regulatory mechanisms and for species living in waters with fluctuating salinity, osmotic and ionic regulation has a energetic cost. For migratory fishes, moving along a salinity gradient may change the energy costs for osmoregulation, but could also result from a change in food availability, so that higher rates of food consumption may more than compensate for the extra costs of osmoregulation.

Fish species richness is low in saline inland waters and, apart from cichlids and some cyprinodonts, most fish species are restricted to fresh or slightly brackish waters. Some species are able to survive temporary increases in salinity, but, only a few species of primary freshwater fishes were able to evolve and to adapt to extreme saline conditions. For several fish species, salinity may appear to be an ecological barrier. Mormyridae, for example, live only in freshwater and avoid even slightly brackish waters. It seems that conductivity affects the electrosensory systems of those species. In Lake Chad, mormyrids are restricted to waters with a conductivity lower than 400 µS cm^{-1} (Bénech *et al.*, 1983). In Lake Turkana, where the mean conductivity is

3500 µS cm^{-1}, which is high compared with many other north tropical aquatic ecosystems, the fish fauna is apparently typically Nilo–Sudanian, with the exception of mormyrids. Hopson (1982) only recorded *Hyperopisus bebe* and *Mormyrus kannume* from the vicinity of the Omo Delta, and no mormyrids are known from the actual lake. The one species of mormyrid (*Hippopotamyrus discorhynchus*) presently occurring in Lake Tanganyika may be the result of increased salt concentrations in the lake water at some periods of the lake's history, which could have eliminated other mormyrid species (Lowe-McConnell, 1987).

Clarias gariepinus is able to adapt to brackish waters. It has been recorded living in Lake Chilwa at 10 000 µS cm^{-1} (Furse *et al.*, 1979). In Crater Lake, which probably became isolated from Lake Turkana at the beginning of the century, the conductivity is presently above 11 000 µS cm^{-1}. The presence of *Oreochromis niloticus*, *Synodontis schall*, *Clarias gariepinus* and *Haplochromis rudolfianus* suggests that these fish are to some extent able to adapt to increasing concentrations. The reverse is also true. A population of *Tilapia guineensis*, which normally prefers brackish waters, has been isolated in the man-made Lake Ayamé on the Bia River (Côte d'Ivoire), where it has succeeded very well in freshwater.

Some secondary freshwater fish such as the Cichlidae, can tolerate full seawater, at least for short periods. It is assumed that tilapias evolved from a marine ancestor and this origin may account for the fact that some species are markedly euryhaline: *Tilapia zillii* (0.16–44‰) and *Oreochromis mossambicus* (0–120‰) (Whitfield & Blaber, 1979). *Tilapia zillii* was found to reproduce in Lake Qarun (Egypt) in salinities between 10 and 26‰ (El Zarka *et al.*, 1970). This species was also found in the Red Sea at salinities of 42‰ and in hypersaline waters (Chervinski, 1972), but like *Oreochromis aureus*, *T. zillii* does not reproduce in seawater. The euryhaline species are usually widely distributed as long as they can move from one estuary to another during the floods. Several tilapias live and reproduce at salinities higher than 30‰, as is the case for coastal lagoons containing *Tilapia guineensis* and *Sarotherodon melanotheron* in West Africa, as well as *Oreochromis mossambicus*, *Oreochromis hornorum* and *Oreochromis placidus* in East African coastal rivers. *Oreochromis mossambicus* is euryhaline and grows and reproduces in fresh-, brackish and seawater. Hora & Pillay (1962) reported that reproduction occurred in seawater up to a salinity of 35‰ and Popper & Lichatowich (1975) showed that it was able to reproduce in ponds at salinities of 69‰. Fry of *O. mossambicus* were found to be in a good healthy condition at salinities of 69‰ (Pott *et al.*, 1967). *Oreochromis amphimelas* lives in the hypersaline Lake Manyara, Tanzania (salinity 58‰). *Oreochromis alcalicus grahami* is endemic to Lake Magadi (40‰), and *O. a. alcalicus* to Lake Natron (30–40‰). Among the Cyprinodontidae, *Cyprinodon fasciatus* inhabits waters up to 40‰ in North Africa (Beadle, 1943).

The reproductive performance of yearling *Oreochromis niloticus* evaluated at various salinities and under laboratory conditions shows that spawning occurs from freshwater to full seawater (32 ppt), but that extremely poor hatching success was obtained with eggs spawned in full seawater (Watanabe & Kuo, 1985). The total number of spawnings for yearlings was greater in brackish salinities of 5–15 ppt than in either full seawater or freshwater. This result has to be compared with other findings that tend to demonstrate a better physiological adaptation of *O. niloticus* to brackish waters. For instance, the lowest metabolic rates of *O. niloticus* at a given swimming speed are at a salinity of 11.6 ppt, and the highest at 30 ppt (Farmer & Beamish, 1969).

It has also been demonstrated that fish exposed to low concentrations of seawater at early stages of their life cycle are better adapted to subsequently tolerate higher salinities than those spawned in freshwater. Fry salinity tolerance of *O. niloticus* increased with increasing salinity of spawning, hatching or acclimatisation (Watanabe *et al.*, 1985).

So far, little is known about the osmoregulatory mechanisms of tilapias and their energy costs (Prunet & Bornancin, 1989). It is known that the branchial epithelium is the major salt transport site in fish, and in seawater-adapted fish, chloride cells are located on the filaments of the gills (Maetz & Bornancin, 1975), but they could also occur on the opercular membrane. For *O. mossambicus*, after adaptation from freshwater to one-third seawater or full seawater, the rate of effluxes increased 15 and 206 times, respectively, and a large increase in Na^+ turnover was also observed (Dharmamba *et al.*, 1975). For *O. niloticus*, the metabolic rate is higher

at salinities of 0.0, 7.5 and 22.5 ppt than at 11.6 (Farmer & Beamish, 1969). If a salinity of 11.6 is approximately iso-osmotic with the blood plasma, the increases in oxygen consumption at higher and lower salinities indicates the energy costs of regulation. Thus for *O. niloticus*, the proportion of total oxygen consumption required for osmoregulation is 29% at 30 ppt and 19% at 0.0, 7.5 and 22.5 ppt.

Drought

The ultimate catastrophe for aquatic animals must be for their habitat to dry up. Nevertheless, some species of fish are adapted to living in temporary waters. The major characteristic of temporary-water ecosystems is that they dry up seasonally. The extent of the desiccation period depends on the climate, the water budget, and the morphometry of the waterbody. The frequency of drying up varies greatly. In many cases, such as for tropical rain pools or floodplain ponds, there is seasonal cycle of filling and drying. But in more arid regions, such as the central Sahara, the dry season is longer and in extreme cases there are basins in which water appears only at rare intervals. Conversely, small puddles, as in the hoofmarks of cattle or elephants, may appear and disappear several times in the year, even in the humid tropics.

It is therefore not surprising that few fish species are adapted to survive complete desiccation. The ability of the lungfish, *Protopterus annectens*, to survive prolonged dry periods and seasonal desiccation of their habitats has been well known since the last century, but only really well investigated since the classical work of Johnels & Svensson (1954). With a fall in the water level, the fish excavates a vertical burrow into the soft mud, makes a bulb-shaped chamber (Fig. 9.4), and secretes large quantities of mucus into the chamber. As the water dries up, the mucus dries, and forms a cocoon surrounding the dormant lungfish in the so-called sleeping nest. The fish lies motionless and folded on itself, breathing air through a small aperture at the top of the cocoon. In nature, *P. annectens* can aestivate for about seven to eight months, depending on the length of the dry season. The cocoon formation does not appear to be the usual method of aestivation in other African *Protopterus* species (Greenwood, 1986). During the aestivation of *Protopterus aethiopicus*, the rate of oxygen consumption is progressively reduced

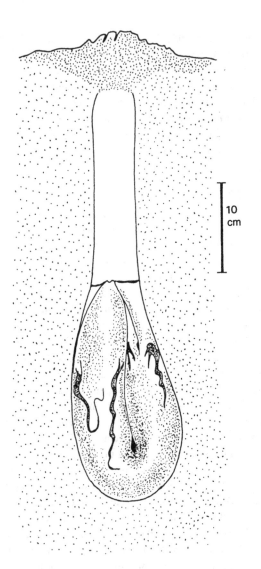

10
cm

Fig. 9.4. Schematic view of estivating and cocooned *Protopt-erus annectens*, as viewed from the ventral side of fish (Greenwood, 1986, modified from Johnels & Svensson, 1954).

to about 10% of that of the active fish, and the blood circulation is much retarded until the heartbeat falls to about three per minute (Beadle, 1981).

The behaviour of catfish when the habitat dries up is less well known. They are reputed to survive habitat desiccation by remaining dormant under dry mud or sand, in similar way to the lungfish. Survival of *Clarias* in

moist sand in an otherwise dry river bed has been reported by Donnelly (1978). A literature review and recent evidence indicate that clariids may survive whilst buried in moist sand, or in burrows with a water–air interface, but that survival is unlikely to occur in dried mud or sand (Bruton, 1979*d*). The recolonisation of newly flooded rivers and pools by *Clarias* spp. after a period of drought, probably results from their ability to move up to a few hundred metres over moist grassland areas. Various reports of Clarias 'walking' occur widely in the literature (Bruton, 1979*d*).

Oreochromis mossambicus is able to survive extreme reduction of temporary waterbodies and Donnelly (1978) collated several reports of the survival of this species in a layer of moist sand covered by up to 3 m of sand, the upper 2 m of which was quite dry.

Adaptation to temporary habitats is particularly well developed among the tiny fish of the family Cyprinodon-tidae. Species of the genera *Nothobranchius* and *Aphyose-mion* are small annual fish that inhabit temporary pools in tropical areas. These pools usually support a rich and abundant invertebrate fauna that provides excellent food for small fish. The adults can live only as long as water is available in the ponds, a time period that may be no more than a few months in some dry areas. Aestivation takes place in drought-resistant eggs, buried in the bottom of the pools. The embryonic period can last from a few weeks to several months. Pienaar (1968) reported that eggs of one population of *Nothobranchius* hatched after being dried for 18–24 months, and Scheel (1968) has collected viable eggs after a period of five and a half years. The eggs of the killifishes have developed thick-ened chorionic membranes, forming a hard incompress-ible 'shell' that can withstand severe mechanical stress and desiccation (Peters, 1963).

The reproductive behaviour of *Nothobranchius* in sea-sonally available habitats is adapted to getting the highest number of eggs into the substrate, with minimal energy expenditure, throughout the period of water availability (Haas, 1976*a*). Reproduction occurs daily throughout life and the number of eggs produced by females is pro-portional to their size and the amount of food consumed. Males are ready to spawn at any time, and most females can produce 45 eggs per day for extended periods (Turner, 1964). In many cyprinodonts, there is a daily spawning pattern with production of large eggs

(1–2 mm). The daily fecundity range from a few eggs to at least 50 depending on species, size and food availability.

While in most vertebrates there is a more or less fixed time from fertilisation to hatching or birth, in annual fish there is arrested development, or diapause, of varying duration that can occur at three specific stages of its normal embryonic development (Peters, 1963; Wourms, 1972; Simpson, 1979). It is a period when morphological growth and development is suspended, or at least retarded and, as such, differs from quiescence that is only a temporary inhibition directly influenced by environmental conditions (Simpson, 1979). In annual killifish development, Diapause I occurs when individual cells are dispersed throughout the periphery of the egg. Diapause II occurs just prior to heart contraction and is characterised by 38 somites. Diapause III occurs at full development when the heartbeat slows down and circulation decreases.

Diapause is either facultative and/or obligatory, depending on genetic and environmental conditions. When diapause occurs spontaneously, without changes in the surroundings, it is termed obligate. When external factors have a modifying effect, it is said to be facultative. Annual species of *Aphyosemion* or *Nothobranchius* can undergo facultative arrest at Diapause I and II and an obligate arrest at Diapause III. It should be noticed that Bailey (1972) found no evidence of recruitment of young fish into a population of *Nothobranchius guentheri* observed in seasonal pools of Tanzania, and it may be inferred that eggs laid in that specific year entered a period of prolonged diapause.

Eggs that have completed Diapause III have apparently an almost immediate response to the rains, and Peters (1965) suggested that the stimulus is given by a decrease in the oxygen supply to the embryo, caused by decomposition in the soil after the first rainfall. Growth after hatching is very rapid, and sexual maturity may be reached in less than 28 days for *Nothobranchius* (Haas, 1976a). For wild populations of *N. guentheri*, Bailey (1972) observed ripe eggs 7–8 weeks after hatching. The typical age at sexual maturity is earlier than in most other teleosts, and Simpson (1979) mentioned females of *Nothobranchius rachovii* that may spawn at four weeks at a size of 0.9 cm. Spawning as early as possible could be considered a selective advantage in case of drought occurring prematurely.

Acid waters

Most natural waterbodies have a pH close to neutrality, but some lakes and reservoirs have an acid pH sometimes as low as 3. The effect of extreme pH on fish physiology is poorly documented for African species. It has been observed in temperate fish that continuous exposure to acid stress can lead to an alteration of gill membranes and/or coagulation of gill mucus. In extremes cases, this may be the primary cause of death of exposed animals, while damage to the gill epithelium may affect gas exchange and gill ion exchange (Fromm, 1980). When subjected to debilitating acid stress, blood pH decreases and may precipitate an acidemia by preventing excretion of metabolically produced H^+ ions and CO_2, and/or interfere with ionic regulation (excessive loss of salt). In less extreme situations, low pH may result in reproductive failure resulting from acid stress related to an upset in calcium metabolism and to faulty protein deposition in developing oocytes.

Some cichlid species tolerate high pH waters. *Oreochromis alcalicus grahami* (pH 10.5 in Lake Magadi), can withstand a pH range of 5 to 11 for at least 24 h but dies after 2 to 6 hours at pH <3.5 or >12 (Reite *et al.*, 1974). *Oreochromis niloticus* seems to have only limited ability to adapt to very low pH, with a threshold at about pH 4 (Wangead *et al.*, 1988). All fish exhibited behavioural manifestations of physiological stress, rapid swimming and opercular movements, surfacing and gulping of air, and inability to control body position, almost immediately at pH 2 and 3. The major cause of death seemed to be respiratory failure as a result of acid water destroying gill tissues, increasing mucus secretion and causing redness and swelling of the gills.

Light attenuation and darkness

High suspensoid loads (solid or colloidal particles) are a common feature of many African inland waters, and result from decomposition of rocks and organic material. Soil erosion is an important source of suspended material in arid and semiarid areas, and contributes to increased silt loads of many rivers. In shallow lakes subject to wind-mixing, resuspended sediment also constitutes an important source of suspensoids (e.g. Lake Chad and Lake Chilwa). But turbidity is also caused by natural organic material such as phytoplankton and zooplankton. Whatever the causes, the turbidity of some freshwater systems may be such that water transparency becomes very low. During the drying up of Lake Chad, for example, the measured water transparency did not exceeded 10 to 20 cm (Carmouze *et al.*, 1983). Similar observations were made in Lake Chilwa (Kalk *et al.*, 1979). In many water systems, there is a seasonal change in water turbidity. In tropical rivers, waters are usually more turbid during the flood, as a result of the washing in of allochtonous material by rains and runoff in the river catchment.

The most likely effect of suspensoids on fish is reduced visibility, which means that it is more difficult for visual predators to identify their prey, but in turn it also reduces predation risk, both from other fish or from aerial predators. In addition, high turbidity may result in the clogging of gill rakers and gill filaments, reduced egg and larval survival, reduced habitat diversity and altered breeding behaviour (Bruton, 1985).

For surviving in low visibility habitats, fish have to use strategies other than visual orientation. However, this is probably not very different from strategies that are developed by fish active only at night. For example, in the case of mormyrids, which spend the day in cover and feed at night, electrolocation is a means of replacing visual location by a system more or less similar to the sonar of bats (Fig. 9.5). Electric organs are specialised for generating an electric field in the external environment, which is usually waters of low ion content and high resistivity. Mormyrids, contrary to most animals that exclusively depend on environmental signals, emit their own energy to test the environment and to locate objects, making them independent of the fortuitous and often unpredictable signals that the environment provides (Bullock, 1982; Kramer, 1990). The adaptive advantage of the electric system and active electrolocation, as found in weakly electric fishes, lies in the gain of specific information on position, size, conductivity properties and movement of surrounding objects. Fish use their electrolocation abilities to maintain their position and posture relative to the substrate. In riverine species that are mostly active at night, this mechanism allows them to monitor and direct their movements relative to the surrounding environment. Obviously, such a location mechanism is particularly useful for movement at night as well

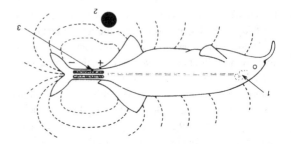

Fig. 9.5. The principle of active electrolocation in a mormyrid electric fish. The electric organ (3), which is monitored by electroreceptors (1) found in pores of the anterior body surface, produces electric discharges. The dotted lines give the current flow associated with electric organ discharge. Each object (2) with conductivity different from that of the surrounding water distorts the current pattern and thus modify the fish sensory information.

as in high turbidity waters, and this could explain why mormyrids are well represented in such habitats.

Extreme adaptations to light attenuation are encountered in species living in permanently dark environments. The major environmental characteristic of caves is the complete darkness, and to some extent a more or less constant temperature. Some animals use caves only occasionally to avoid unfavourable conditions outside. Others live permanently in caves and are called 'troglobionts', as is the case, for example, for *Glossogobius ankaranensis*, the most recently discovered blind fish in Madagascar (Banister, 1994). It is usually recognised that blind species evolved independently from fully-eyed epigean species, and they are of great intrinsic value to illustrate evolutionary processes. The striking morphological differences in comparison with their epigean equivalents are marked trends toward eyelessness, lack of pigment and low metabolic rate. The degree of eye reduction in different species that have been studied seems to be connected with the phylogenetic age of cave colonisation (Parzefall, 1993). Some species of blind fish have also been discovered in rapids. It is assumed that they live under stones, and that they lost their eyes and pigmentation as a result of lack of light, but this hypothesis has yet to be tested.

Very few species of blind fish (belonging to 6

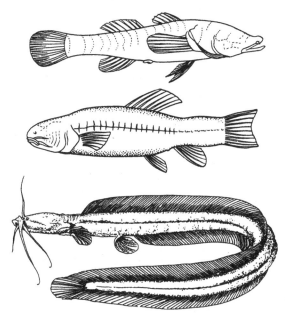

Fig. 9.6. Selected blind fish species from Africa: (top to bottom) *Typhleotris madagascariensis*, *Phreatichthys andruzzii* and *Platyallabes tihoni*.

Table 9.2. *Blind fish species recorded from Africa*

Family/species	Location
Cyprinidae	
Barbopsis devecchii	Somalia
Caecobarbus geertsi	Zaïre
Phreatichthys andruzzii	Somalia
Clariidae	
Channallabes apus	Zaïre
Clarias cavernicola	Namibia
Dolichallabes microphthalmus	Zaïre
Platyallabes tihoni	Zaïre
Uegitglanis zammaranoi	Somalia
Gobiidae	
Glossogobius ankaranensis	Madagascar
Eleotridae	
Typhleotris madagascariensis	Madagascar
Typhleotris pauliani	Madagascar
Synbranchidae	
Monopterus boueti	Liberia
Mastacembelidae	
Caecomastacembelus brichardi	Zaïre

Data from Thines, 1969.

families) (Fig. 9.6; Table 9.2) have been recorded in Africa. They are adapted to life in caves or honeycombed rocks (aquifers). A series of caves in the surrounding of Thysville (Zaïre), belonging to the Kwilu system, is the most extensive African system where the biology and ecology of blind fish have been investigated (Heuts, 1951; Heuts & Leleup, 1954). *Caecobarbus geertsi* is known from 10 caves out of 61 investigated. In some caves, there were also troglophile Clariidae, predators of *Caecobarbus*.

Caecomastacembelus brichardi and *Platyallabes tihoni*, which are not strictly cave fish, have been collected in the Stanley Pool. They live in riffles under flagstones or in crevices, and were captured with indigenous 'medicines'. *Barbopsis devecchii* and *Uegitglanis zammaranoi* were captured in wells. Not all species exhibit the same degree of non-development of eyes and pigment. In *Channallabes apus*, for instance, a few specimens only were eyeless among a large number of fish examined (Thines, 1969).

IV Dynamics of fish assemblages

10 Fish assemblages in tropical Africa

A calm evening on a lonely shore – the surface water near at hand is suddenly broken and a silvery form shoots six feet high into the air. It turns on its side and falls on the water with a flop and splash, disappearing into an ever-increasing circle of ripples. Another follows and yet another, and, as the eye becomes trained to distance, fish seem to leap continuously, perhaps to escape from the threatening jaws of the great Nile perch below, perhaps to attract the notice of some possible mate, perhaps for the sheer joy of feeling the cool air of evening on their scales . . . .The setting of these incidents is Lake Albert, almost in the middle of Africa, which, like all the African lakes, has its own particular assortment of living creatures.

S. & E. B. Worthington, 1933, Inland Waters of Africa

Any individual fish is usually part of a population of one species. It lives within a complex web of interactions with other individuals of its own species or other species, and can both affect or be affected by predator–prey relationships, host–parasite interactions, competition for food and reproductive activities, competition for space and habitats. The frequency and nature of the interactions experienced by an individual will, therefore, depend on the nature of its biotic environment, and particularly the structure of the assemblage of fish species. Moreover, the fish species diversity in a particular habitat is constrained by the nature of the abiotic environment, and is likely to change as conditions alter (see Chapters 1 and 6).

Many of the ideas and theories elaborated by community ecologists, particularly ornithologists, apply directly to fish. However, fish assemblages are often more complex than bird or mammal assemblages, because differences in size between juveniles and adults are more pronounced.

The community concept

There is no consensus definition of the concept of **community**, and many authors have expressed concern that communities rarely exist as naturally definable units. 'The differences in definitions relate to different emphases on organisational as against locational aspects, on spatial and taxonomic scale, and on the diversity of resource utilisation' (Giller & Gee, 1987). Southwood (1987) tentatively defined the concept of an ecological community as: 'a group of organisms (generally of wide taxonomic affinities) occurring together in a location; many of them will directly interact with each other within a framework of both horizontal and vertical linkages'. A major idea in this definition is the recognition of vertical or trophic relationships (predation, parasitism) or horizontal relationships (competition) between species of the same trophic level. Some other authors have suggested more functional definitions. A simple definition of community ecology is the study of the 'organisation' of communities including the number, identity and relative abundance of the component species, as well as their ecological attributes, their interrelationships, and how all of these vary over space and time (Brown, 1995).

Many studies concern taxonomic groupings that are part of a whole community (they form a subcommunity). The term assemblage describes a taxonomically or functionally restricted group of species occurring together in a given area, interacting or not (Giller & Gee, 1987). It is used, for instance, to describe all the fish species in a defined area. One can also recognise another level of organisation, the guild that is a group of co-occurring species populations utilising the same type of resources in similar ways and interacting with horizontal linkages.

One of the key questions in community ecology is to understand whether the collection of species populations observed at one site results from chance and hazard, or is the result of a selection based on criteria that allow these populations to coexist. We need to know whether community structure is controlled by deterministic interactions between species, such as competition or predation, or by stochastic factors, such as disturbance (Townsend, 1989). According to Wiens (1984), the goals of community ecology are to identify patterns of ecological communities, to determine the causal processes that underlie these patterns, and to generalise these explanations as far as possible. In other words, one of the major goals of ecological research is to synthesise information available in order to be able to forecast possible structural and functional responses to environmental changes.

During the 1960s and 1970s, community ecology relied mostly on recognition of patterns susceptible to analysis by rules governing species distribution and abundance, in a framework of equilibrium models involving biological processes, such as competition, predation and diseases. Much emphasis was therefore given to biotic interactions and local processes, with the belief that communities are structured in a predictable way that could be expressed in mathematical models (McIntosh, 1987; Brooks & McLennan, 1991).

A more pluralistic conception of ecology has been propounded during the last decade (see McIntosh, 1987), including what is called historical ecology (Brooks & McLennan, 1991), which is concerned with studying macroevolutionary components of ecological associations and the effects of evolutionary processes, such as speciation and adaptation in the production of both evolutionary groups of organisms and multispecies ecological associations. Historical ecology is primarily interested in incorporating the origins of diversity, and historical constraints on that diversity, into causal evolutionary explanations of ecology and behaviour.

The problem is further complicated by the effect of disturbance and the response time of the community. Fish ecologists are faced with a key problem: how to distinguish between shifts in species composition attributable to alteration in biological interactions (e.g. competition) mediated by temporal variability in the physical environment, and those attributable to individualistic responses to environmental changes (Schlosser, 1985). Dunson & Travis (1991) noticed that too little appreciation was given to abiotic influences and physiological tolerances on patterns of habitat segregation, especially among closely related species. This neglect of the role of abiotic influences may be partly due to the 'cultural gap' between community ecologists and physiologists.

Another major topic in community ecology, when studying organisms with multiphasic life cycles, is to understand the dynamics and interactions that all stages can play in the regulation of adult populations and species assemblages. The roles of adult phases have been emphasised partly because they are usually easier to sample and to manipulate, but much attention is now being given to the environmental requirements of larvae and juveniles, which may be different from those of the adults (see Chapter 12).

To summarise, 'the current uncertainties and controversies in community ecology fall into two broad areas' (Southwood, 1987): (i) what is our concept? Are communities mere assemblages of organisms in a location or are they tightly linked and structured groups of interacting species? (ii) to what extent is the nature of the community 'organism driven' or 'environment driven'? and (iii) what are the relative roles of speciation, colonisation (the biological history) and habitat preference compared with seasonality, physical conditions and disturbance frequency?

Sampling: the ichthyologists' inaccessible dream?

The study of fish assemblages requires an estimate, as accurate as possible, of the fish species composition and abundance in the habitat. This is a particularly difficult task, rarely achieved, that makes difficult the understanding of fish species relationships. Moreover, fish differ greatly in size, in behaviour and in their use of space during their life span. One is therefore unlikely to sample quantitatively both juveniles and spawning adults. As fish grow, sampling methods have to be modified, but the choice of sampling methods will also depend on the target fish and on the selectivity of the fishing gear used, which in turn affect the composition of samples obtained.

Sampling methods

The subject of fishing techniques is a huge field of knowledge that is outside the scope of this volume. Relatively few sampling techniques have been used for scientific purposes, either for the study of communities or for estimates of the abundance of fish populations. The most frequently used passive methods are netting and hook-and-line fishing. Gill nets are now widely used, and different mesh sizes are generally combined in order to sample a large range of sizes and species. One should notice that traditional local fisheries may be of particular interest for sampling fish populations, and have been used, for instance, in the study of the fish populations of the Yaéré floodplain in Northern Cameroon (Bénech & Quensière, 1982).

Active techniques include cast nets, but also a wide range of seines (beach seines, purse seines). Electrofishing has been used widely in shallow streams to collect a number of small-sized species that are difficult to sample with other methods and has the advantage of allowing collection of living specimens. Electric trawls have also been tested successfully in lakes and rivers (Bénech *et al.*, 1978).

Acoustic methods, which have been developed for sea fish populations, have been used for stock assessments in deep lakes (Turkana, Tanganyika, Malawi). They give only global biomass estimates, but they could also be useful for studying the distribution of fish in the water column and all over the lake.

Another set of active methods includes direct underwater observations using scuba, when water transparency allows it. This technique has been widely applied in large African lakes, such as Malawi and Tanganyika, where a lot of information about fish biology and ecology was gathered from underwater observations. Some quantitative data about fish composition and biomass have also been obtained from the drying up of a reach, for example, taking the opportunity of engineering works during the building of bridges or dams.

One of the most traditional scientific sampling techniques is still the use of ichthyotoxins, which are presumed to be less selective than most other fishing methods.

Whatever the equipment used, the fish catch does not exactly reflect the fish assemblage that has been sampled. There have, therefore, been many attempts to understand how far does the 'picture' given by the sample reflect the 'real' fish assemblage structure. A number of indices have been proposed to characterise the bias introduced by the fishing gear.

Catchability (q) is the probability of capturing one fish from the standing stock by one unit of effort ($q = C/N$ where C is the catch in number per unit of effort, and N the total standing stock). It may be divided into three elements (Laurec & Le Guen, 1981): the accessibility, vulnerability and efficiency (E).

Accessibility (p_A) is the probability of the presence of one fish in the fishing area (A). Local accessibility (p_a) is the probability of the presence of one fish in the area (a) of the fishing area (A) that has been swept by the gear in one fishing operation. It is associated with spatial distribution and migrations of the fish. *Efficiency* (E) is the ratio of the number of fish caught to the number of vulnerable fish that were present in the area swept in one set of the gear. *Vulnerability* is the probability of one fish being present in a given place to be caught by one unit of effort.

Selectivity, for a given gear, describes changes in catchability in relation to size of individuals of a given species. Selectivity S_{ij} of a mesh size i on fish of size j is the ratio of the number of fish of size j caught by the mesh i (C_{ij}), to the product of the fishing effort of the gear (f_i) by the absolute number of fish of size j (N_j) (Hamley, 1975): $S_{ij} = C_{ij}/f_i . N_j$

Despite their major importance in the study of population dynamics, all these parameters are generally difficult to estimate, and only very few studies have been conducted for that purpose.

Representativeness of data obtained with different fishing gear

For fisheries biologists, or for ecologists, it would be useful to estimate the total abundance of fish in a system, either for management purposes or for ecological studies. One way to achieve this would be to convert catch rates, which can be considered as relative abundance indices, into absolute abundance measurements. But it is a difficult process because the efficiency of fishing gear depends upon various factors, such as the behaviour of the fish, environmental conditions and characteristics of the gear (Charles-Dominique, 1989).

The main difficulty in obtaining reliable data lies in knowing the selectivity of fishing gear. Each type of gear has its own selectivity to different fish species, and results obtained with different fishing techniques can hardly be compared. To illustrate this difficulty, fish catches obtained with different fishing gear have been compared, as in the Nangoto experiment conducted in Chad (Bénech & Quensière, 1989). This residual temporary pool in the Chari floodplain is 400 m long by 55 m wide with a mean depth of 1.8 m. It is colonised by some 60 species of fish, mainly juveniles. This pool was fished with gill nets (from 20 to 60 mesh size) on 30 successive days, then with an electric trawl, followed by beach seining for 5 days and, finally, with ichthyotoxins in order to obtain an inventory of the remaining fish community (Bénech & Quensière, 1989). From the sum of these sampling data, an estimation of the population sizes at the beginning of the study was possible. Such an estimation is obviously an underestimate, if the natural mortality during that period is not considered.

For each fish population, the proportion caught after fishing with gill nets over 30 successive nights was very different depending on the species (Table 10.1). Predatory piscivore species were the most vulnerable, whereas the nets had a low efficiency for several other species, which is an illustration of the bias that may be introduced when studying fish assemblages using raw sampling data. This experiment also provided evidence that many species apparently learn how to avoid fishing gear, as seen in the sharp decrease in their vulnerability with time, after the beginning of the fishing period.

Another assessment of gear efficiency was conducted in the Côte d'Ivoire lagoons where Charles-Dominique (1989) studied catch efficiencies of purse and beach seines. The purse seine was 305 m long by 14 m deep and the beach seine 1100 m long and 8 m deep. For both these nets, the mesh size was 14 mm (bar measure). Net avoidance was estimated from the retention rates of marked fish released within the closed seines in shallow water. Reliable upper estimates of the catch efficiency for

Table 10.1. *Catch efficiency after 30 nights fishing in the Nangoto pool with a set of gill nets from 20 to 60 mm mesh size*

Fish species	C	N	E in %
*Lates niloticus	553	643	86.0
*Ichthyborus besse	677	948	71.4
*Gymnarchus niloticus	5	9	55.6
*Polypterus endlicheri	36	70	51.4
*Polypterus bichir	5	12	41.7
*Schilbe intermedius	205	496	41.3
*Hydrocynus forskalii	24	82	29.3
Petrocephalus spp.	107	367	29.0
*Malapterurus electricus	6	22	27.3
Distichodus rostratus	89	362	24.6
*Polypterus senegalus	86	380	22.6
*Bagrus bajad	55	247	22.3
Labeo coubie	19	90	21.1
Synodontis eupterus	55	298	18.5
Labeo senegalensis	587	3511	16.1
Brycinus nurse	311	2680	11.6
*Clarias spp.	53	486	10.9
Synodontis schall	213	2245	9.5
*Hydrocynus brevis	19	212	9.0
Marcusenius spp.	46	575	8.0
Heterobranchus spp.	2	27	7.4
Brachysynodontis batensoda	155	2161	7.2
Brycinus macrolepidotus	22	335	6.6
Synodontis clarias	13	228	5.7
Auchenoglanis spp.	18	334	5.4
Chrysichthys auratus	9	174	5.2
Hyperopisus bebe	5	96	5.2
Synodontis nigrita	153	3366	4.5
Alestes baremoze	759	16 721	4.5
Mormyrus rume	5	131	3.8
Hemisynodontis membranaceus	78	2041	3.8
Citharinus spp.	345	7136	4.8
Alestes dentex	43	3047	1.4
Tilapia spp.	12	2795	0.4

From Bénech & Quensière, 1989. C, number of fish caught; N, estimation of the initial population size; E, efficiency (C/ N × 100); *, piscivore predators.

the purse seine ranged from 47% for *Tylochromis jentinki*, through 54–58% for *Tilapia guineensis* and *Chrysichthys nigrodigitatus*, to 71–79% for *Chrysichthys maurus*, *Chrysichthys auratus* and *Hemichromis fasciatus*. For the beach seine, it was 35% for *Tilapia guineensis* and 53%

Table 10.2. *Vulnerability of four Mochokidae species to different fishing gear in the Nangoto pool*

Fish species	Original population size	Gill Net		Trawl		Seine	
		N	%	N	%	N	%
Brachysynodontis batensoda	2161	155	7.2	114	5.7	1110	58.7
Hemisynodontis batensoda	2041	78	3.8	59	3.0	1154	60.6
Synodontis schall	2245	213	9.5	0	0	109	5.4
Synodontis nigrita	3366	153	4.6	0	0	95	3.0

From Bénech & Quensière, 1989. N, number of fish caught; %, ratio between the number of fish caught (× 100) and the assumed abundance of the fish population at the beginning of the experiment.

Table 10.3. *Relationship between the mode (Y) of the selectivity curve for fish species considered and mesh size (X) of gill nets. Values in mm, mesh expressed as knot to knot measure*

Species	Regression
Alestes baremoze	$Y = 6.89\,X + 28.21$
Alestes dentex	$Y = 7.73\,X + 3.77$
Schilbe mystus	$Y = 8.28\,X - 20.84$
Hydrocynus forskalii	$Y = 7.60\,X + 24.47$
Hyperopisus bebe	$Y = 7.84\,X + 21.90$
Labeo senegalensis	$Y = 7.09\,X - 13.45$
Marcusenius cyprinoides	$Y = 7.04\,X - 9.64$
Brachysynodontis batensoda	$Y = 4.93\,X - 14.65$
Synodontis clarias	$Y = 4.50\,X - 6.73$
Synodontis schall	$Y = 5.50\,X - 18.27$

From Bénech & Quensière, 1989.

for *Chrysichthys* spp. The catchability of the purse seine was close to 1% for *Chrysichthys* spp. and 1.4% for *T. guineensis*. The efficiency was 12 and 18%, respectively, which is quite low. The purse seine is an efficient technique for pelagic fish schools, but apparently not for 'blind' fishing of demersal species. The main reason is likely to be avoidance during the surrounding phase of the operation.

The comparative vulnerability of species to different sampling gear may also vary according to the type of gear involved. In the Nangoto pool for instance, four morphologically similar species of the *Synodontis* group were equally abundant at the beginning of the fishing period. Their vulnerability to gill nets was also similar, but strong differences appeared in their vulnerability to other types of active gear (Table 10.2). In contrast to pelagic species (*Brachysynodontis batensoda* and *Hemisynodontis membranaceus*) the benthic species, such as *Synodontis schall* and *Synodontis nigrita* do not appear in the catches of the surface trawl and are not very abundant in the seine catch, possibly as a result of some avoidance behaviour.

The situation is still more complex when gear selectivity is also a function of mesh size. In the SE archipelago of Lake Chad for example, gill-net selectivity was estimated by comparing gill-net with beach-seine catches on the same fishing site (Bénech & Quensière, 1989). A linear relationship was established between the mode of the selectivity curve and mesh size. This relationship is different for each species (Table 10.3) and depends on their morphology.

Applying multilinear regression to the species under study, it was possible to express the mode of the selectivity curve as a function of mesh size (t) and of three morphological parameters (X, Y, Z):

$$\text{mode} = (0.077t + 25.42)\,X + (1.21t - 9.32)\,Y \\ + (0.24t + 11.42)\,Z + (0.47t - 76.87),$$

where X = the ratio of standard length to body depth; Y = the ratio of standard length to head length; Z = the ratio of head length to head width.

The selectivity of fishing gear explains why the use of fisheries catch statistics is of limited use for the study of fish assemblages. Moreover, fishermen also focus their activity on the catch of fish species that have a high commercial value, and they may ignore other species. The catch statistics may, nevertheless, be used with caution for the study of commercial species biology, and long-term trends in the catch composition may also be useful for understanding fish community dynamics in relation to environmental changes.

Fish communities in rivers

Rivers are the most important aquatic habitat in tropical Africa, with combined lengths amounting to some 13 million kilometres distributed mainly among small, lower-order streams (Welcomme & de Mérona, 1988). The research effort on lotic systems, which has sharply increased in temperate countries during the last two decades, has been comparatively low in tropical Africa where priority has been given, for example, to reservoirs and endangered Great Lakes. Despite the existence of a rich and diverse fish fauna, and an active commercial fishery, rivers remain poorly studied in comparison with lakes.

River zonation and fish assemblages

The basic idea of zonation is that there are large gradients to which the fauna must respond. The unidirectional flow of water, and the existence of a gradient in flow rate and in environmental conditions from source to estuary, is a basic pattern of river systems that may be modified by local geological or geomorphological features.

A stretch of river is commonly classified in terms of stream order (Strahler, 1957; Leopold et al., 1964). A first-order stream has no tributaries, and two first-order streams form a second-order stream when they connect, and so on. Usually, stream width is positively correlated with stream order, whereas flow rate is negatively correlated. This classification method, used in the United States and in Europe, has not been applied in Africa, where the main difficulty lies in the lack of maps sufficiently accurate to allow the identification of first-order streams. Species diversity increases with stream order (Horwitz, 1978), mainly through the addition of species. The reasons are the increase in the number of types of habitat, as well as the diversification of energy sources from upstream to downstream and the stability of the physical environment that tends to increase with stream order (Wootton, 1990). According to Zalewski & Naiman (1985), the importance of biotic interactions (such as competition) increases with stream order as the abiotic conditions become more stable.

Among the many attempts to provide a model of biological zonation in rivers (review in Wasson, 1989) Ilies & Botosaneanu (1963), using faunistic criteria, proposed to divide the river course into headwater (or creon), middle reach (or rhithron) and lowland (or potamon) zones. In Africa, this classification has only been applied successfully in some South Africa rivers, and in the Luanga, a high altitude tributary of the Zaïre

River system (Malaisse, 1976) where a succession of rhithron and potamon reaches was identified.

Vannote et al. (1980) developed the theory of the River Continuum Concept (RCC) that provided a conceptual functional framework, stressing changes in trophic resources with increasing river size. Briefly, this hypothesis holds that relative abundances of various types of food vary predictably with stream size, and that relative abundances of consumer guilds are correlated with those of their major food resources. The models therefore emphasise the importance of trophic function and food availability to the distribution and abundance of stream organisms. African riverine fish communities have not actually been studied in that context but, nevertheless, Lowe-McConnell (1975) suggested some general tendencies: the abundance in the upper course of surface eating insectivores and omnivores consuming riparian allochtonous material, and the presence of herbivores and benthic detritivores in the lower course. A distribution such as this was found in the Niandan River (Hugueny, 1990b) and the Mono River (Paugy & Bénech, 1989), but it has not been demonstrated that it is related to changes in food availability.

Fish species may exhibit physiological and morphological responses to changes in the riverine environment. The rhithron and the potamon, for instance, impose different environmental requirements on fish life, and one can expect that species will have developed specific adaptations (Bayley & Li, 1992). Rhithron fish are more constrained by habitat size and hydraulics than fish of the potamon. They are usually small-sized fish, in which small body mass favours agility in a turbulent environment (Webb & Buffrénil, 1990). Well-known morphological adaptations to swift currents also include enlarged and stiffened pectoral spines (genera *Amphilius*, Synodontis), the buccal suckers (genera *Garra*, *Chiloglanis*, *Synodontis*, *Labeo*), elongated body form and ventral flattening with a humped dorsal profile, etc. Small body size is encountered in fast-flowing habitats, as well as in seasonally inundated floodplains, in the form of small species and young of larger species that move upstream to spawn. But small size also places limits on trophic specialisation, life span and reproductive capacity (Bayley & Li, 1992). Rhithron fish are primarily adapted to consume small aquatic and terrestrial invertebrates, while more diversification is possible in the potamon where greater habitat and food diversity confers advantages to both large and small fish (Welcomme, 1985). The life span is also relatively short for rhithron fishes while large potamon fishes have a greater longevity.

In tropical Africa, the distribution of macroinvertebrates and/or fish along a longitudinal gradient has been studied in the Bandama basin (Mérona, 1981; Lévêque *et al.*, 1983; Gibon & Statzner, 1985), the Mono (Paugy & Bénech, 1989), the Ogun River (Sydenham, 1977), the Ebo stream (Lelek, 1968), the upper Niger (Hugueny, 1990*b*) and the Luongo River in the Zaïre system (Balon & Stewart, 1983).

In the Bandama basin, three main zones were distinguished (Lévêque *et al.*, 1983): (i) headwaters and small tributaries that are temporary streams running for part of the year, but dry out or eventually remain as pools during the dry season; (ii) a long relatively uniform middle reach, but with several successive alternations between slow-flowing and rapid reaches; and (iii) a relatively short estuarine zone where saline waters may penetrate a few tens of kilometres upstream. Based on a detailed study of the taxonomy and distribution of three insect groups, Hydropsychidae and Philopotamidae (Trichoptera) and the *Simulium damnosum* complex (Diptera), Gibon & Statzner (1985) provided evidence that the long mid-reaches are not a uniform faunistic stream zone. They found that, in general, from the furthest upstream to the lowest downstream riffles, there appeared to be no clear replacement and almost no loss of upstream species. Rather a steady increase of species numbers was observed, because of the occurrence of additional species. This odd zonation pattern, as compared to those found in other geographical areas, is assumed to be typical for streams of gentle slope and without a well-defined source in this part of Africa (Gibon & Statzner, 1985). Distribution of freshwater shrimps in the Bandama River did exhibit a clearer zonation pattern (Lévêque *et al.*, 1983). Among the Atyidae, *Atya* spp. occurs only in the estuarine zone, whereas *Caridina* colonises mainly the upper reaches. Two *Macrobrachium* species (*M. volenhoveni* and *M. felicinum*) are present in the lower and middle reaches, and one in the upper course (*M. raridens*).

The low-order streams in the Bandama basin are usually inhabited by a small number of fish species (de Mérona, 1981) with a small adult size and short life span

zonation	habitat	characteristic species
spring zone	brooks or pools	small *Barbus* small characids Cyprinodontidae *Neolebias sp.*
upper course	small pools	*Brycinus longipinnis* *Hepsetus odoe* *Schilbe intermedius* *Synodontis schall*
middle course	alternating pools and riffles	**pools** *Alestes baremoze* *Brycinus nurse* *Brycinus macrolepidotus* *Hydrocynus forskalii* *Schilbe mandibularis* **shallow running waters** *Labeo parvus* *Afromastacembelus nigromarginatus* *Nannocharax sp.* *Brycinus imberi* *'Tilapia' spp.*
lower course	large pools salinity influence	**continental species** *Brycinus longipinnis* *Hepsetus odoe* *Schilbe intermedius* *Synodontis schall* **estuarine species** *Elops lacerta* *Gobius guineensis*

Fig. 10.1. Schematic longitudinal zonation of fishes in the Bandama basin (from de Mérona, 1981).

(small *Barbus*, Cyprinodontidae, small Characidae) (Fig. 10.1). In the long middle reach, *Alestes baremoze*, *Brycinus nurse*, *Brycinus macrolepidotus*, *Hydrocynus forskalii* and *Schilbe mandibularis* are characteristic of the calm and deep reaches, whereas genera adapted to turbulent conditions prevail in the riffles (*Mastacembelus*, *Amphilius*, *Nannocharax*, etc.). The existence of a long midcourse with little physical change was also observed in the Ogun River (Sydenham, 1978). Such a situation is probably the fate of rivers with gentle slope, and differs from results obtained in rivers of the Zaïre (Malaisse, 1976) or Zambezi basins (Balon, 1974b), where there is a more distinct zonation pattern comparable to classifications developed in north temperate zones. A peculiar zonation pattern has been observed for a few species of fish (*Brycinus longipinnis* and *Hepsetus odoe*) that occur in the upper reaches, disappear in the mid-course zone where they are replaced by other species (*Brycinus nurse*, *Brycinus imberi*, *Hydrocynus forskalii*) and reappear in the estuarine zone.

In fact, the above scheme is more complicated when applied to the middle course of the Bandama and other surrounding rivers, where one can find a succession of riffles with fast-flowing waters, and pools with still and deeper water. Fish assemblages in the riffles have been sampled by electric fishing (Table 10.4) and are mostly constituted of small species adapted to life in a turbulent environment (*Amphilius*, *Phractura*, etc.) or juveniles of larger species (*Labeo parvus*, *Synodontis bastiani*), as well as species inhabiting rocky habitats where they find shelter in crevices (*Afromastacembelus*, small mormyrids). It is clear that when moving from one river system to another, the species composition may change, but one species is usually replaced by another one with more or less similar morphology and ecological function.

The general pattern emerging is of the existence of a clear longitudinal zonation in the upper course of rivers, in relation to its size, whereas in the middle course, other environmental factors predominate with apparently no clear change in community structure. At a small spatial scale, those factors could be the local characteristics of the stations such as current speed or substrate, which may be independent of river size. At a larger spatial scale, the river can cross different climatic or vegetation zones, such as savannah or rain forest that can influence community composition. This was observed on the Bandama River (Lévêque et al., 1983), as well as on the Niger River (Hugueny, 1989; Lévêque et al., 1991).

The zonation pattern may be modified as a result of river morphology. For example, Balon & Stewart (1983) described an 'unusual gradient' in the Luongo River, a tributary of the Luapula River that enters Lake Mweru. This river has upper and lower sections with a steep gradient, rapids and waterfalls, separated by a long

Table 10.4. *Species composition of fish assemblages sampled with electric fishing gear, in middle course riffles of different coastal river systems from Côte d'Ivoire*

	Rivers			
	Leraba	Bandama	Nzi	Sassandra
Mormyridae				
Petrocephalus bovei		*	*	
Characidae				
Brycinus nurse	*	*	*	*
Brycinus imberi	*	*	*	*
Micralestes occidentalis	*			
Distichodontidae				
Nannocharax occidentalis	*	*		*
Cyprinidae				
Labeo parvus	*	*	*	*
Barbus macinensis	*			
Barbus macrops	*			*
Barbus sublineatus	*		*	*
Barbus bynni waldroni		*	*	*
Barbus wurtzi		*	*	*
Raiamas senegalensis	*	*	*	*
Amphilidae				
Amphilius atesuensis	*	*	*	
Phractura intermedia	*			
Mochokidae				
Synodontis bastiani	*	*	*	
Synodontis comoensis	*			
Chiloglanis occidentalis				*
Cichlidae				
Tilapia spp.	*	*	*	
Hemichromis spp.	*	*	*	
Mastacembelidae				
Aethiomastacembelus nigromarginatus	*	*	*	*

Only characteristic and abundant species are mentioned.
ORSTOM, unpublished data.

floodplain section of low gradient. The fish faunas in the headwater zone and in the exit zone are characterised by the large number of species inhabiting flowing waters, but differ in species composition, a situation that could be compared to the upper and middle course of a more 'normal' hydrological gradient.

In the tropics, the upper reaches or tributaries of large rivers are seasonal and dry up or break up into chains of pools during the dry season. Most of our knowledge about them comes from studies carried out during the dry season, in these residual waterbodies.

There is very little information on the distribution of fish during the flooded phase. During the flood, the whole system becomes intractable and difficult to study, but it is known that many fish species migrate upstream at the beginning of the flood season and spawn in the smaller tributaries or the inundation zones when they are present. Thus the fish assemblages are fairly different during the flood season, but rather poorly documented. Such changes in the structure of fish communities of the seasonal Upper Ogun River were investigated by Adebisi (1988). Piscivorous fishes (*Hydrocynus forskalii, Hepsetus odoe, Bagrus docmak, Mormyrops deliciosus*) were particularly abundant and diverse in gill-net catches at the beginning and during the flood. Some of them (*Hydrocynus* spp., *Mormyrops* spp.) probably migrate up the river to spawn in the upper reaches, and migrate down as the water recedes. Towards the end of the flood, the omnivores (*Schilbe intermedius, Clarias gariepinus, Heterobranchus longifilis, Synodontis schall*) became increasingly more preponderant until the beginning of the lentic phase. In residual pools, the fish community consisted primarily of herbivores (mainly *Brycinus macrolepidotus, Brycinus nurse, Tilapia zillii*, but also *Labeo senegalensis, Chromidotilapia guentheri, Sarotherodon galilaeus*) and secondarily of insectivores (mormyrids, *Chrysichthys auratus*).

The floodplain 'ecotones'

The definition of Holland (1988) ' an ecotone is a zone of transition between adjacent ecological systems, having a set of characteristics uniquely defined by space and time scales, and by the strength of the interactions between adjacent ecological systems' applies well to the seasonally inundated areas bordering tropical rivers, which are a transition zone between a lotic system and the adjacent terrestrial system, where allogenic processes dominate.

The role of floodplains in the functioning of river systems has been extensively investigated in Africa where well-developed floodplains are present in nearly all savannah basins, e.g. Senegal, Niger, Chari, Logone, Nile, etc. These land–water ecotones play a major role in the water budget, the chemical and physical erosion-sedimentation cycle, the biological diversity and productivity of the lotic systems.

The hydrological buffering capacity of large catchments associated with floodplains results in rather smooth flood curves. However, in the arid tropics, where evaporation is high and exceeds precipitation, the net outflow may be very much lower than the inflow. From the 7.11×10^{10} m³ entering the Central Delta of the Niger, only 3.82×10^{10} m³ emerges, and the net loss is 46% (Welcomme, 1979). In the Logone floodplain (Yaéré) that covers 8000 km², the river input is 3.2×10^9 m³, and the outflow 1.1×10^9 m³. Rainfall contributes up to 8.5×10^9 m³, and evaporation 10.6×10^9 m³, to the water balance (Gac, 1980). A similar situation occurs in the Chari floodplain (Gac, 1980).

Floodplains are traps for suspended solids and thus alter the dissolved solids and water chemistry of the river. In the floodplains of the Chari basin, 250 000 tons of clay sediment out each year in the Massenya depression along the Chari, and 500 000 tons in the Ba-Illi floodplain (Gac, 1980; Burgis & Symoens, 1987). In the Yaéré floodplain, 900 000 tons of suspended material enter the system with the river water; 870 000 tons sediment each year on the floodplain while only 27 000 tons (3%) is exported through the outflow (Gac, 1980). For dissolved elements, 34 000 tons are lost in the Yaéré out of 185 000 tons entering with the floodwater.

Floodplains are a source of fish foods of both plant and animal origin, and the aquatic vegetation provides cover and refugia for the young fish from the many piscivores. In the river and its floodplain, there is a large variety of habitats, ranging from small temporary pools to large permanent lagoons and swamps, which vary between the flood and dry season phases (Welcomme, 1979; Welcomme & de Mérona, 1988). As a result, any attempt to describe fish communities inhabiting the main channel and associated floodplains can only be superficial.

From his own observations in the Oueme River (savannah type system) during the dry season, Welcomme (1979, 1985) suggested a general scheme for the distribution of fish in floodplains, linked to major trophic categories. The division of flood river communities into 'white fish' and 'black fish' components was an attempt to distinguish two major behavioural and ecological assemblages, but it was not wholly satisfactory, and a third assemblage, 'grey fish', has been proposed (Table 10.5) (Regier et al., 1989).

The first of these groups (white fish) is strongly dependent on the main channel for breeding, although many species move into the floodplain to breed and feed

Table 10.5. *Black–grey–white: three ecological assemblages of riverine fish and some African taxa and characteristic features*

	White fish	Grey fish	Black fish
Families	Characidae Cyprinidae Mormyridae (Mormyrops)	Cichlidae Citharinidae Cyprinidae (Labeo) Distichodontidae Mochokidae Mormyridae (Gnathonemus) Schilbeidae	Anabantidae Channidae Clariidae Gymnarchidae Mormyridae (Pollimyrus) Notopteridae
Respiratory organs	Gills	Gills with some physiological adaptations to low oxygenation	Gills and air-breathing organs with physiological adaptations to low oxygenation
Respiratory tolerance	Highly oxygenated waters	Medium to low oxygenation	Low oxygenation-anoxic
Muscle fibre type	Red	Red/white	White
Migratory behaviour	Long distance longitudinal	Short distance longitudinal often long lateral migrations	Local movements
Body form	Round, fusiform	Laterally compressed, spiny; often heavily scaled	Laterally compressed or soft; elongated and flabby; scales reduced or absent
Colour	Silvery or light	Dark, frequently ornamented or coloured	Very dark, often black
Reproductive guild	Non-guarders; open substratum spawners; lithophils; pelagophils	Guarders; nest spawners; open substratum spawners; phytophils	Guarders; external and internal bearers; complex nest builders
Dry season habitat	Main channel, lake	Backwaters or main channel fringes	Floodplain waterbodies
Wet season habitat	Main channel or flooded plain	Floodplain	Floodplain or marshy fringes

From Regier *et al.*, 1989.

during the flood. Extensive upstream migrations can take place, towards the headwaters, to breed just before or at the start of the floods.

The second group (black fish) inhabits the floodplain or the marshy fringes. Movements are limited to some lateral migrations, and species are adapted to resist adverse environmental conditions, principally deoxygenation. Such species tend to be partial spawners, with a breeding season starting in the pre-flood period and persisting over peak flood.

The third group (grey fish) inhabits backwaters, fringing vegetation, edges of floodplain lakes and the main channel during the dry season. There are usually lateral migrations from the main channel to the floodplain for breeding and feeding during the floods. Species in this group have a more flexible behaviour than species of the

other two groups, and they adapt more readily to changing hydrological conditions.

The Sudd wetlands in southern Sudan occupy about 30 000 km² and comprise a permanent complex of rivers, lakes, *Papyrus* and *Typha* swamps, as well as seasonal river-flooded grasslands. Four main aquatic fish habitats were recognised by Bailey (1988). The spatial and temporal distribution of species relates to their ecological tolerance and requirements for food, shelter, breeding, oxygenation, etc.

1. **Flowing river channels** that are relatively harsh habitats compared with lakes. The commonest species are characids (*Hydrocynus forskalii, Alestes dentex, Brycinus nurse, B. macrolepidotus*), bagrids (*Bagrus bajad, Auchenoglanis biscutatus*), mochokids

(*Synodontis schall*, *S. frontosus*), schilbeids (*Schilbe intermedius*) and the Nile perch, cichlids (*Oreochromis niloticus*). The tiny species, *Micralestes acutidens* and *Chelaethiops bibie*, are abundant in the water hyacinth fringe, the latter exhibiting a preference for running waters.

2. **Lakes and khors** (Arabic name for drainage channel) provide the greatest diversity of habitats and fish species. In many lakes investigated (Hickley & Bailey, 1986) the most numerous species caught were, in descending order: *Alestes dentex*, *Hydrocynus forskalii*, *Synodontis frontosus*, *S. schall*, *Schilbe mystus*, *Auchenoglanis biscutatus*, *Clarotes laticeps*, *Oreochromis niloticus*, *Labeo niloticus*, *Distichodus* spp., *Citharinus* spp., *Mormyrus cashive*, *Heterotis niloticus* and *Lates niloticus*.

3. **Shaded swamps** of *Cyperus papyrus* and *Typha domingensis* are rather inhospitable habitats with low levels of dissolved oxygen. Many air-breathing fish are potential colonisers of this biota in which only 23 species were recorded out of 62 identified in the Sudd: *Protopterus aethiopicus*, *Polypterus senegalus*, *Heterotis niloticus*, *Gymnarchus niloticus*, *Brienomyrus niger*, *Clarias gariepinus*, *Ctenopoma petherici*, *C. muriei*, *Hemichromis fasciatus* and *Parachanna obscura*.

4. **Floodplains** are predominantly *Oryza longistaminata* grasslands. Twenty-three species of fish were collected, including in descending order of abundance: cyprinodonts, anabantids (*Ctenopoma muriei*), *Clarias gariepinus*, *Oreochromis niloticus*, *Nannaethiops unitaeniatus*, *Barbus stigmatopygus*, *Polypterus senegalus*, *Parachanna obscura* and small mormyrids (*Brienomyrus niger*). It should be noticed that, in contrast to the general pattern observed in many floodplains, fish which enter the south-eastern floodplain grasslands become concentrated in isolated pools, with little return to the permanent system. Some stranded fish may survive in permanent waterbodies, but most of them will die or will be fished-out by piscivorous birds and fishermen (Hickley & Bailey, 1987a).

Response of fish communities to interannual changes in flow regimes

Quantitative studies on the response of fish assemblages to stream flow changes are rare, but results from studies on fish–habitat relationships in stream communities provide some information on community-level responses to flow regime. Moreover, some information is available on the effects of unusual natural streamflow events, such as extended droughts.

The reduction of flooding as a result of natural climatic changes, such as the severe Sahelian drought of 1970–7, has important consequences for fish biology and floodplain fisheries. The population of *Alestes baremoze* decreased dramatically in the Chari River and southern Lake Chad (Bénech *et al.*, 1983) as a result of the strongly reduced flooding of the North Cameroon Yaéré floodplains where the young of that species normally spend their juvenile phase. After a drought period from 1974 to 1978, the study of the catch composition of the traditional fishery on the El Beid River, an effluent of the Yaéré draining towards Lake Chad, showed a slow recovery of fish species richness, but a successive and apparently random appearance of abundant species (Bénech & Quensière, 1982, 1983a, b). Dansoko *et al.* (1976) also found that juvenile growth in two species of *Hydrocynus* was poor during two years of bad floods, and Reizer (1974) discerned great year-to-year differences in the growth of *Citharinus citharus* in the Senegal, corresponding to flood intensity. Long duration floods allowed a longer spawning period, which resulted in the production of a second cohort of juveniles by some species (Bénech & Quensière, 1983a).

The relationship between flood intensity and duration, and fisheries production is well known (Welcomme, 1979). In a detailed study of the Logone floodplain, Bénech & Quensière (1983b) found a positive correlation between fisheries production and flood volume. They also observed for different fish species, a significant correlation between growth of juveniles and the intensity of flood. The fish catch in the Cross River is also strongly correlated with the flood regime of the previous year (Moses, 1987). Such observations could be used as models to forecast the detrimental effect of suppressing flood after dam construction.

The close relationship between different aspects of fish life history and changes in the annual flood strength allows anticipation of year-to-year variations. In 1934 Wimpenny had already concluded that fisheries production in Egypt was proportional to the strength and duration of the floods. The same year-to-year effects usually result in a negative correlation between the amount of water remaining in the river during the dry season and the fishery catch (see Vidy, 1983; Welcomme et al., 1989). Assuming a constant fishing effort, the higher catches associated with low water levels arise from the greater vulnerability of the fish.

The delayed effects are more complex. Welcomme (1979) found a strong correlation between catch and flooding in the previous one or two years in the Shire, Kafue and Niger floodplains. Two components of the flood regime could influence the catch (Welcomme, 1986): (i) the magnitude of the high-water phase interferes with the size of the stock through reproductive success, survival and growth of fry, which are supposed to be better during the years of extensive flooding; and (ii) the degree of drawdown affects survival and ease of capture during the low-water phase. Using computer simulations, Welcomme & Harborg (1977) pointed out that magnitude and duration of flooding were important factors in explaining year-to-year differences in ichthyomass, but that the greater the amount of water remaining in the dry season the more such differences are transmitted to following years. The way in which a fish community reflects the 'hydrological memory' of a system also depends on its age structure. When most of the community is composed of long-lived species, many year classes are present and such an age structure tends to average out the hydrological regimes of several years and to reflect only long-term trends. Conversely, when the fish community consists of only a few age classes, it will be highly correlated with flood intensity in previous years (Welcomme, 1986).

For the Niger River, an analysis of a 22-year series (1963–84) of discharge data and the records of fish landings in the Inner Delta show catch in any year to be highly correlated with flood intensities (Welcomme, 1986). The overall regression was :

$$C_y = 151.73 \log (0.7HI_{y-1} + 0.3HI_{y-2}) - 428.26$$

where C_y is catch in a particular year and HI an index of flood intensity the previous year ($y - 1$) or two years ago ($y - 2$). Fishes six months old constituted 50% of the catch in 1974 and 90% in 1984. It is difficult to know exactly whether the correlation between falling floods and falling catches over the past 20 years is solely a result

Table 10.6. *Contribution of fish species to the biomass in the Bandama River, above and below Kossou before damming*

Species	Upper Kossou % biomass	Below Kossou % biomass
Brycinus macrolepidotus	19.2	7.9
Labeo coubie	16.2	49.2
Chrysichthys spp.	9.3	7.5
Polypterus endlicheri	7.4	1.8
Mormyrops longiceps	5.7	4.8
Heterobranchus longifilis	4.7	0.7
Auchenoglanis occidentalis	3.7	3.1
Citharinus eburneensis	3.5	
Labeo parvus	3.4	*
Papyrocranus afer	3.0	*
Lates niloticus	2.2	1.5
Tilapia zillii	2.2	3.5
Mormyrops elongatus	1.8	*
Synodontis schall	1.7	1.0
Hydrocynus forskalii	1.5	1.1
Chromidotilapia guentheri	1.4	*
Mormyrus rume	1.3	3.7
Labeo senegalensis	1.3	1.3
Brycinus nurse	1.2	*
Hepsetus odoe	1.1	
Petrocephalus bovei	*	
Pollimyrus isidori		*
Marcusenius ussheri	*	*
Marcusenius furcidens	*	*
Alestes baremoze	*	*
Brycinus imberi	*	*
Distichodus rostratus	*	*
Raiamas senegalensis	*	
Barbus bynni waldroni	*	*
Barbus wurtzi	*	*
Schilbe mandibularis	*	*
Synodontis bastiani	*	3.3
Malapterurus electricus	*	1.4
Clarias lazera	*	
Hemichromis fasciatus	*	*
Hemichromis bimaculatus	*	
Sarotherodon galilaeus	*	*
Ctenopoma kingsleyae	*	*
Parachanna obscura	*	
Aethiomastacembelus nigromarginatus	*	*
Area sampled	2.6 ha	1 ha

Data from Daget et al., 1973. *, contribution to biomass less than 1%.

Table 10.7. *The relative abundance of fish species caught in the 18 ha coffer-dammed lake in the Niger River at Kainji, 1966. Specimens of genera* Polypterus, Heterotis, Gymnarchus, Malapterurus, Lates, Oreochromis, Tilapia *and* Tetraodon, *were also collected.*

Genera	No. of species	No. of individuals	% of individuals	Total weight (kg)	% weight
Mormyridae	19	1198	20.7	219	19.5
Hyperopisus bebe	1	166		49	
Mormyrus	3	180		47	
Mormyrops	3	122		55	
Campylomormyrus	1	366		38	
Marcusenius	4	292		27	
Hippopotamyrus	3	56		3	
Others	4				
Characidae	8	2103	36.3	136	12.1
Hydrocynus	2	28		11	
Alestes	2	1447		100	
Brycinus	4	628		25	
Citharinidae	3	288	5.0	94	8.8
Distichodontidae	2	66	1.1	118	11.0
Cyprinidae	5+	192	3.3	48	4.3
Labeo	2	183		47	
Barbus	2+	8			
Raiamas	1	1			
Bagridae	7	422	7.2	204	18.2
Bagrus	2	187		157	
Chrysichthys + Clarotes	5	235		47	
Schilbeidae	3	463	8.0	40	3.6
Schilbe	1	350		34	
Eutropius	1	98		6	
Mochokidae	18	1064	18.0	209	18.7
Total catch	65	5796		1068	

From Motwani & Kanwai, 1970.

of poor climatic conditions (the drought) or also reflects some overfishing.

In a more recent study on the effects of hydrology on the evolution of the fisheries of the Inner Delta of the Niger from 1966 to 1989, Lae (1992) concluded that two main factors, not mutually exclusive, may explain the decrease in production from 87 000 tons in 1969–70 to 37 000 in 1984–5: the advent of drought in Africa; and/ or the increase of fishing effort. It seems that the considerable increase in fishing effort since 1966 has had little negative impact on the total catches, as long as fish less than one year old contributed up to 69% of the production. Lae (1992) also obtained a high correlation between the annual catch and the flood intensity for the same year, but he also tried to use another hydrological index for the particular case of the Inner Delta, which is the amount of water lost in the delta during an annual cycle. This amount of water lost through evaporation or infiltration may be estimated as the difference between the amount of water entering and leaving the delta. This water loss probably reflects more accurately the duration and extent of the inundation. Actually, there is a much better correlation between the annual fish catch and the loss in water volume (*w.l.*) from the Inner Delta for years y and $y - 1$ than with the catch and the index of flood intensity :

$$\log C_y = 0.26 \log w.l._y + \log w.l._{y-1} + 9.82$$
$$(\text{with } r^2 = 0.92).$$

It appears that competition for food resources is not

Table 10.8. *Relative abundance by weight of 18 species of commercial fishes at high water (June–July) and at low water (August–September) in the Kafue Flats in 1970*

Species	High water biomass % total	Low water biomass % total
Oreochromis andersonii	26.6	31.7
Oreochromis macrochir	18.4	19.1
Tilapia melanopleura	17.4	5.0
Tilapia sparrmanii	13.0	2.1
Serranochromis angusticeps	3.0	2.3
Serranochromis macrocephalus	0.3	0.2
Serranochromis robustus	0.2	0.3
Serranochromis thumbergi	0.6	0.1
Haplochromis carlottae	3.2	0.8
Haplochromis codringtoni	0.2	
Haplochromis giardi	0.1	1.5
Hepsetus odoe	4.0	6.0
Marcusenius macrolepidotus	0.7	2.2
Schilbe intermedius	1.8	15.6
Clarias gariepinus	5.6	8.0
Clarias ngamensis	3.5	4.4
Synodontis macrostigma	0.4	0.4
Labeo molybdinus	0.6	0.3
Total ichthyomass of commercial species	84%	84%
Total ichthyomass of non-commercial species	16%	16%
Mean ichthyomass	435 kg ha^{-1}	339 kg ha^{-1}

From Lagler *et al.*, 1971.

a serious factor in determining the composition and abundance of African riverine fish communities in the potamon (Welcomme, 1989; Lae, 1995). Rather, it would seem that limitations to the extent of breeding areas caused by flood restriction during droughts may play a greater role in determining relative abundance of species. In the Central Delta of the Niger, the fish catches for the period 1969–91 were characterised by a depletion of species, such as *Gymnarchus niloticus* and *Polypterus senegalus*, whose reproduction are linked to the floodplains, while families such as the Cichlidae and Clariidae increased to constitute 30% and 20%, respectively, of total annual catches during the drought (Lae, 1995). Similar changes were described in Lake Chad by Bénech *et al.* (1983).

Biomass estimates

Community structure and dynamics may be examined through estimates of the abundance (numbers) of individuals, or through the relative contribution of species to the biomass and, therefore, to the flow patterns of matter and energy in the ecosystem. Only a few such data, often from biased sampling, are available for assemblages of African riverine fish.

Such data as there are range from 50 kg ha^{-1} (Daget *et al.*, 1973) in the low-water season (May) of the Bandama River (Côte d'Ivoire) above the Kossou Dam, before it was closed, to 5260 kg ha^{-1} in backwaters of the Lower Chari River (Loubens, 1969). The latter, extremely high figure is probably rather exceptional, since the mean for the Chari backwaters was 1430 kg ha^{-1} and similar values (1210 kg ha^{-1}) were obtained in the Logone River. Biomasses in the region of 100–500 kg ha^{-1} are more common. The mean for the Maraoue was 100 kg ha^{-1} (Daget & Iltis, 1965) and that for the Nzi 102 kg ha^{-1} (Lévêque *et al.* 1983), both tributaries of the Bandama in the main stream of which values of 125 kg ha^{-1} and 177 kg ha^{-1} were obtained upstream in the high-water season and downstream in the low-water season, when 305 kg ha^{-1} was recorded in backwaters with no flow (Daget *et al.*, 1973). These figures are rather lower than those recorded on the Kafue Flats, which ranged from 339 at low water to 435 kg ha^{-1} during the flood (Lagler *et al.*, 1971), but even these are only about half those recorded on the Chari.

The use of methods such as poisoning and pumping out an isolated area, as on the Niger at Kainji, also allows an analysis of species contribution to the biomass. These are recorded in Tables 10.6, 10.7 and 10.8 from which it is evident that in each case relatively few species comprises the bulk of the biomass. This is particularly evident in the Kafue data where only four species of tilapias represented 75% of the biomass. The most abundant species in the Chari River were *Oreochromis niloticus*, *Sarotherodon galilaeus*, *Lates niloticus*, *Heterotis niloticus*, *Synodontis nigrita* and *Polypterus endlicheri*.

Fish communities in shallow lakes

Although most limnologists know what they mean by a 'shallow lake', there can be no universally recognised definition because the behaviour of the lake depends not only on the depth of the water, but the size and shape of the basin in which it lies. Nevertheless, one can choose to define them simply as lakes less than 10 m deep that are unlikely to have a permanent or long-term stratification of the water column (Lévêque & Quensière, 1988). This definition applies to many natural African waterbodies throughout Africa.

One of the principal characteristics of shallow waters is that fluctuations in depth may dramatically change the nature of the aquatic environment. Indeed, the composition and structure of fish communities in shallow lakes strongly depends on their water budget and hydrology, both of which play a major role in controlling water-level changes (Lévêque & Quensière, 1988). In general terms, two main categories of shallow lakes may be distinguished: (i) endorheic lakes with no surface outlets; (ii) and open lakes that are usually connected to larger riverine systems.

Water-level fluctuations in endorheic lakes are entirely dependent on the rainfall over the lake catchment. Depending on the morphology of the basin, the interannual changes in climate may result in drastic changes in area and depth, sometimes associated with changes in water quality. In extreme situations, small lakes may almost dry up, the aquatic biota being restricted to marshy swamps or even to the tributary river beds. During drought periods, the consequences may be dramatic for the fish fauna, which have little chance to survive unless they have developed special adaptations to harsh situations. The low number of fish species observed in endorheic lakes is therefore likely to be the result of past climatic events, during which the lake almost dried out, and tolerance to water chemistry. In such circumstances, only marshy species were able to survive and continue in the lake. In the absence of connections with other aquatic systems, the species richness of the fish community remained drastically reduced. This situation may continue even when the lake has been recovered for a long time and in the absence of any introduction of alien species by man.

Water-level fluctuations in shallow lakes connected to river systems are usually less extreme in the long term, and the fish fauna is largely of riverine origin. During drought, the fish remain in the river, and they recolonise the lake when the environmental situation is improved. The fish species richness in this case,

depends on the richness observed in the whole hydrographic basin.

Endorheic shallow lakes

In endorheic lakes without surface outlet, the water inflow (tributaries and rain) is almost completely counterbalanced by evaporation and, in a few cases, by underground seepage. Evaporation of incoming waters should result in the accumulation of salts, and a significant rise in salinity, as is obviously the case for some lakes such as Nakuru, Magadi, Turkana, and some lakes of the Ethiopian Rift. However, there are also a few endorheic lakes in Africa that remain fresh. Lakes Chad, Baringo and Naivasha are good examples. In the case of Lake Chad, the regulation of water salinity has been satisfactorily explained by chemical precipitation in the lake and underground seepage of water from the lake to the Kanem area (Roche, 1970; Carmouze, 1976; Carmouze et al., 1983). Lake Chilwa, apparently, also has a regulation mechanism, but its waters usually remain slightly saline.

Saline lakes

Only a few fish species are adapted to survive the extreme conditions of lakes with highly saline water and their fish fauna is extremely poor. The cichlid fish, *Oreochromis alcalicus grahami*, is endemic to Lake Magadi, where it lives in pools fed by alkaline hotsprings (Coe, 1966) and has been found in water of salinity up to about 40‰. *Oreochromis alcalicus* occurs in Lake Natron (Eastern Rift), also in water of 30–40‰ salinity, and the related species *Oreochromis amphimelas* in Lake Manyara, where it has been recorded from water of salinity as high as 58‰.

Lake Nakuru is an endorheic, highly alkaline lake in the Gregory Rift that was originally devoid of fish. The cichlid, *Oreochromis alcalicus grahami*, was introduced in the 1950s from Lake Magadi, and in its new environment has a very patchy distribution, with an increasing mean size from inshore to offshore regions; the smaller fishes concentrate near the shore whereas the larger fishes occur beyond 250 m (Vareschi, 1979). This type of pattern is known to occur in tilapia species of other lakes. At noon the fish concentrate near the shore and at night they move offshore, a migration pattern probably reflecting a preference for higher temperatures. Such a pattern has also been observed for *O. a. grahami* in Lake

Magadi, where the fishes move at sunset from the shallow waters to the deeper parts of the springs (Coe, 1966). The vertical distribution pattern is stable with almost 75% of the fish concentrated in the top 50 cm where oxygen concentration and light intensity are high.

An inshore to offshore gradient in fish biomass has been observed in Lake Nakuru (Vareschi, 1979). The average fish numbers per square metre decreased significantly from 20 at 50 m to 18 at 150 m 9 at 400 m and 8 at 1000 m. Conversely, the average fish biomass increased from 2.5 g dry weight m^{-2} at 50 m from shore to about 5.25 at 250–400 m and 4.2 at 600–1000 m. The mean biomass of $O.$ $a.$ $grahami$ was estimated in 1972 at 80 kg ha^{-1}, but reached 425 kg ha^{-1} in 1973. In 1976 the mean biomass was 300–400 kg ha^{-1} (Vareschi, 1979).

Lake Turkana is the fourth largest (by volume) of the African Great Lakes and its waters are slightly saline, with a mean conductivity $c.$ 3500 µS (Kolding, 1992), because, although its main tributary is the Omo River, it has no surface outlet. The lake level fluctuates with an annual amplitude of $c.$ 1–1.5 m but it also undergoes long-term variations. Low levels were recorded in the early 1940s, and in the early 1990s (Kolding, 1992).

Lake Turkana has retained a basically Nilotic riverine fish fauna. Of the 48 species recorded in the vicinity of Lake Turkana, 36 occur regularly within the lake, the other 12 being restricted to the area of the Omo Delta (Hopson & Hopson, 1982). A few species were found to occur over a wide range of habitats, e.g. *Lates niloticus*, *Synodontis schall*, *Barbus bynni bynni* and *Neobola stellae*, but most species tended to be restricted to a particular type of habitat and four major assemblages were recognised in the main lake.

1. The **littoral assemblage** restricted to an inshore belt between the lake margin and 4 m depth. *Oreochromis niloticus* and *Clarias gariepinus* occur everywhere. *Tilapia zillii* and *Raiamas senegalensis* prefer rocky or stony shores, whereas *Sarotherodon galilaeus*, *Brycinus nurse*, *Micralestes acutidens* and *Chelaethiops bibie* are more abundant on soft substrates.

2. The **inshore demersal assemblage** includes bottom-living fish restricted to inshore areas of the lake between 4 and 10–15 m depth. *Labeo horie*,

Citharinus citharus, *Distichodus niloticus* and *Bagrus docmak* are characteristic species.

3. The **offshore demersal assemblage** is distributed in a narrow layer rising 3–4 m above the bottom, in the deeper waters: *Bagrus bajad*, *Barbus turkanae* and *Haplochromis macconneli*.

4. The **pelagic assemblage** is spread over the water column from the upper limits of the demersal assemblages to the surface and encompassing both inshore and offshore areas of the lake. *Hydrocynus forskalii* and *Alestes baremoze* are the dominant species living in the superficial layers, from the fringe of the littoral zone into mid-lake. *Brycinus minutus* and *Brycinus ferox* preferred mid-waters, as do the predators – the endemic dwarf perch, *Lates longispinis*, and the catfish, *Schilbe uranoscopus*.

A large proportion of the species consists of highly seasonal anadromous spawners migrating into the Omo River to spawn. Among the species spawning in the lake, a few have demersal eggs (*Bagrus* spp.), some pelagic eggs (*Lates* spp. and the small *Brycinus* spp.), while the rest exhibit different degrees of parental care (*Oreochromis*, *Sarotherodon*, *Haplochromis*, *Tilapia* and *Heterotis*).

An experimental gill-net fishery was carried out in 1986–7 and compared with the results obtained during the earlier 1972–5 survey (Bayley, 1977; Hopson, 1982). A general reduction in catch per unit effort was observed since the mid-seventies, but unevenly distributed among communities and fish species. There was relatively little change among the inshore demersal community, whereas in the pelagic community some major species were drastically reduced in abundance (Kolding, 1992). For all species there was a decrease in mean size, and for some the maturation size apparently also decreased. It seems that the open-water pelagic zooplankton-based community have been the most affected by the lowering of the lake levels during the last decade. Populations of *Hydrocynus forskalii*, *Alestes baremoze* and *Schilbe uranoscopus* had all declined in abundance by at least 70% since the 1972–5 survey. As long as these species have not been commercially exploited, the decline must be due to natural fluctuations. According to Kolding (1992) this could be an indication that pelagic communities are very unstable, which contradicts Hopson's (1982) statement predicting more stable pelagic fisheries than inshore demersal fisheries.

Endorheic freshwater lakes

In freshwater endorheic lakes, the fish fauna is much more diversified. The most spectacular case is that of **Lake Chad** that occupies a vast basin of about two and half million square kilometres in the centre of Africa. The lake itself covered as much as 25 000 km² in the 1960s with a mean depth of 4 m. Since that time, the lake has been through a period of severe drought and was reduced in the 1970s to a remnant (5000 km²) in the south basin directly fed by the Chari flood (Carmouze *et al.*, 1983).

The ecology of fish in Lake Chad has been studied over a long period (1966–78), including a period of relative stability with a fairly well-developed lake (1966–71) and a period of drought (after 1972) associated with drastic ecological changes. The first period will be called 'Normal Chad' and the second 'Lesser Chad' according to Carmouze *et al.* (1983). Data were mostly obtained from experimental fisheries using standard gill nets or beach seines. But data from the local fishery were also available for the western Nigerian coast of the north basin.

During the so-called 'Normal Chad' period, the species distribution in the lake was strongly influenced by two main factors: (i) distance from the river system; and (ii) type of aquatic biota. Three main types of fish assemblages were observed in the south basin, in relation to the lacustrine landscapes (Bénech *et al.*, 1983; Bénech & Quensière, 1989). Three species were exclusively caught on the southern coast of the lake, in the vicinity of the Chari Delta: *Ichthyborus besse*, *Siluranodon auritus* and *Polypterus senegalus*. *Tetraodon lineatus* was also common in this area, along with juveniles of *Schilbe uranoscopus* and *Hyperopisus bebe* that were not found elsewhere in the lake. In the open waters, *Labeo coubie*, *Citharinops distichodoides* and *Synodontis clarias* were abundant, as well as large-sized *Hemisynodontis membranaceus*, but there did not appear to be any species exclusively associated with this biota. In contrast, many species common in the surrounding areas were rare or absent from open waters: cichlids (*Oreochromis niloticus*, *O. aureus*, *Sarotherodon galilaeus*, *Tilapia zillii*), characids (*Alestes baremoze*, *A. dentex*, *Brycinus macrolepidotus*), mormyrids

(*Marcusenius cyprinoides*, *Petrocephalus bane*) and *Heterotis niloticus*. No species appeared to be characteristic of the archipelago, but *Brachysynodontis batensoda* was particularly abundant. Many ubiquitous species were abundant everywhere in the south basin: *Hydrocynus forskalii*, *Schilbe mystus*, *Lates niloticus*, as well as *Hydrocynus brevis*, *Synodontis schall*, *Labeo senegalensis* and *Distichodus rostratus*.

Many ubiquitous species, which were abundant in the south, were also found in the north basin, with the exception of *Hydrocynus brevis*. In the western coast fisheries based on large-meshed gill nets, *Lates niloticus*, *Heterotis niloticus*, *Citharinus citharus* and *Citharinops distichodoides* represented more than 90% of the total catch. *Distichodus rostratus*, *Bagrus bajad* and *Labeo* spp. were also abundant but *H. membranaceus* was less common than in the south basin. Cichlids were abundant in all areas of reed islands and archipelagos where *Synodontis frontosus* was also widespread. Several species appeared to be less abundant or rare in the north: *Brachysynodontis batensoda*, *Hemisynodontis membranaceus*, *Schilbe uranoscopus*, and different mormyrids, such as *Pollimyrus isidori*, *Petrocephalus bane*, *Marcusenius cyprinoides* and *Hyperopisus bebe*. A progressively decreasing species diversity was observed towards the north-east (Carmouze *et al.*, 1972).

The Sahelian drought, which was one of the major climatic events of the last two decades, started in 1972 and resulted in drastic changes in the water level of the lake. It started with an overall drying up from 1972 to 1974, with a marked change from a stable lacustrine appearance to an unstable marshy appearance. This first phase was followed by the drying of the north basin in 1975. The lake was therefore restricted to the south basin ('Lesser Chad') and moved towards a new equilibrium, mixing at the same time the lacustrine and marshy characteristics.

This modification of the lacustrine hydrology caused serious disturbances to fish populations in the lake, but also to those in the associated lower reaches of the Chari and Logone Rivers and their floodplains. Reduction of lake area had various consequences on the environment for fish.

1. The decrease of the water volume increased fish con

centration. The consequences were an increase in inter- and intraspecific competition, and a greater vulnerability to fishing gear.

2. The lacustrine landscape was completely modified with an extreme reduction of open waters without any vegetation, and conversely the development of marshy biota resulting from the decreasing depth.

3. With decreasing depth, the wind disturbance allowed resuspension of sediment and consequently high water turbidity. At the same time, storms caused severe sediment disturbance resulting in mass mortalities of fish, as a result of deoxygenation of water (see Chapter 9).

4. The drying of the shallows caused isolation of parts of the lake, preventing the supply of flood water and fish movements. An abundant marshy vegetation then developed, and when decaying this partly contributed to the deoxygenation of the water.

During the drying-up phase, those fish species that were physiologically adapted to the marshy or swampy conditions developing in the lake selectively survived, as was illustrated by the changes in species composition of the experimental catches in the SE archipelago (Table 10.9). In 1973–4 this archipelago of dune-crest islands was isolated for part of the year from the open water of the south basin, but communication was then re-established. In 1973, which was the time of the lowest water level (Fig. 10.2), some species disappeared before the arrival of the flood (October–December), e.g. *Hydrocynus forskalii*, *Citharinus citharus*, *Hemisynodontis membranaceus*, *Lates niloticus*, *Alestes dentex* and *Labeo senegalensis*. A further sudden decrease in the number of species coincided with the period of storms (June, July 1973) (Fig. 10.2), and then the 1974 flood submerged a large amount of decomposing plants before reaching the archipelago, causing an oxygen deficit over at least three months. Most of the common species that had survived then disappeared: *Labeo senegalensis*, *Schilbe mystus*, *Lates niloticus*, *Polypterus bichir*, *Synodontis clarias*, *Marcusenius cyprinoides*, *Hyperopisus bebe*, *Petrocephalus bane* and *Pollimyrus isidori*. The number of species decreased sharply in the gill-net catches (Fig. 10.2) and by December 1974 almost the only species surviving were those with accessory aerial respiration.

The persistence of the communication with the south-eastern open waters in 1975, allowed some recolonisation by species that developed in the archipelago after the drastic environmental changes of 1973, e.g. *Sarotherodon galilaeus*., *Oreochromis* spp., *Schilbe intermedius*, *Siluranodon auritus*, *Brienomyrus niger*, *Heterotis niloticus*, *Polypterus senegalus*, and the true pelagic species from the 'Normal Chad', such as *Alestes baremoze* and *Alestes dentex*, also quickly recolonised the archipelago. The fish assemblage characteristic of the SE archipelago during the 'Lesser Chad' phase appeared more or less stable between 1975–77, in spite of the disappearance of three species that had survived from the 'Normal Chad' period: *Schilbe uranoscopus*, *Brachysynodontis batensoda* and *Synodontis frontosus*. After 1975, this assemblage had less than 20 species compared with the 34 species in the 1973 catches.

In the southern open waters, a constant connection with the river system was maintained, and after 1974 the biota evolved towards a new appearance with the expansion of open waters surrounded by a thick vegetation belt. The experimental sampling effort was less intensive in this area, but in 1973 *Brachysynodontis batensoda* and *Pollimyrus isidori* were abundant in the catches. After 1975, experimental catches indicated a rejuvenation and a progressive return to equilibrium of the age structure of some fish populations, such as *Hydrocynus forskalii*, *Synodontis clarias*, *Labeo senegalensis*, despite the fact they were deeply affected by the drought.

In mid-1973, the drying up of the Great Barrier isolated the north basin from the Chari floods, its only inflow coming from the Yobe River whose very low floods from 1973–4 did not reach the north of the lake. An important marshy vegetation developed on the exposed Great Barrier, between north and south basins, and hindered the flow of flood waters from the south basin towards the north in 1974–5. The evaporation loss from the north basin was not compensated for and it dried up completely in November 1975. At the time of the division of the lake, no migrations or mass movements towards the south basin were observed. The concentration of fish in the shrinking north basin attracted many professional fishermen, resulting in a high fishing mortality of most of the species that were present during the 'Normal Chad' period. *Heterotis niloticus*, *Hydrocynus brevis*, *Citharinus citharus*, *Mormyrus rume*, *Pollimyrus isidori* and *Tetraodon lineatus* progressively disappeared in the second half of 1974, and mass mortalities due to

Table 10.9. *Changes in fish species composition in the Lake Chad SE archipelago during a period of drought*

	Species	1971		1972		1973		1974		1975	1976	1977
		J–J	A–D	J–J	A–D	J–J	A–D	J–S	O–D			
M	Mormyrus rume	*	*	*	*	*	*					
	Mormyrops deliciosus	*	*	*	*	*						
	Hippopotamyrus sp.	*	*	*	*	*	*					
	Bagrus bajad	*	*	*	*	*	*					
	Chrysichthys auratus	*	*	*	*	*						
	Labeo coubie	*	*	*		*						
	Brycinus macrolepidotus	*	*	*		*						
M	Hydrocynus brevis	*	*	*	*	*						
	Hydrocynus forskalii	*	*	*	*	*	*	*				
M	Citharinus citharus	*	*	*	*	*						
M	Hemisynodontis membranaceus	*	*	*	*	*	*					
	Lates niloticus	*	*	*	*	*	*	*				
M	Hyperopisus bebe	*	*	*	*	*	*		*			
M	Marcusenius cyprinoides	*	*	*	*	*	*		*			
M	Petrocephalus bane	*	*	*	*	*	*		*			
M	Pollimyrus isidori	*	*	*	*	*	*		*			
M	Labeo senegalensis	*	*	*	*	*	*		*			
M	Schilbe mystus	*	*	*	*	*	*		*			
	Synodontis clarias	*	*	*	*	*	*		*			
MR	Polypterus bichir	*	*	*	*	*	*		*			
R	Polypterus endlicheri	*	*	*	*	*	*		*	*		
	Auchenoglanis spp.	*	*	*	*	*	*		*	*		
M	Schilbe uranoscopus	*	*	*	*	*	*		*	*	*	
M	Brachysynodontis batensoda	*	*	*	*	*	*	*	*	*	*	
	Synodontis frontosus	*	*	*	*	*	*		*	*	*	
M	Synodontis schall	*	*	*	*	*	*		*	*	*	*
M	Alestes dentex	*	*	*	*	*	*		*	*	*	*
M	Alestes baremoze	*	*	*	*	*	*		*	*	*	*
	Brycinus nurse	*	*	*	*	*	*	*	*	*	*	*
M	Distichodus rostratus	*	*	*	*	*	*	*	*	*	*	*
R	Gymnarchus niloticus	*	*	*	*	*	*	*	*	*	*	*
R	Clarias spp.	*		*		*	*	*	*	*	*	*
	Tilapia zillii	*	*	*	*	*	*	*	*	*	*	*
	Sarotherodon galilaeus	*	*	*	*	*	*		*	*	*	*
	Oreochromis niloticus	*	*	*	*	*	*	*	*	*	*	*
	Oreochromis aureus		*	*	*		*		*	*	*	*
R	Polypterus senegalus			*	*	*	*		*	*	*	*
	Schilbe intermedius				*	*	*		*	*	*	*
R	Heterotis niloticus	*					*	*	*	*	*	*
R	Brienomyrus niger	*					*	*	*	*	*	*
	Siluranodon auritus						*	*	*	*	*	*

From Bénech & Quensière, 1989. M, large-scale migratory fishes; R, aerial respiration, J–J, January–June; A–D, August–December; J–S, January–September; O–D, October–December.

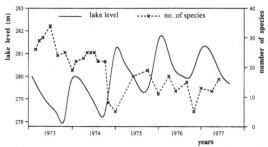

Fig. 10.2. Changes in water level and number of fish species caught with experimental gill nets in the SE archipelago of Lake Chad during the drought period (from Bénech *et al.*, 1983).

tornadoes were also observed (Bénech *et al.*, 1976). At the end of 1974, *Synodontis schall* and *Brachysynodontis batensoda* did not reproduce but survived, *Polypterus senegalus* increased in gill-nets catch, as well as *Sarotherodon galilaeus*, *Oreochromis aureus* and *Oreochromis niloticus* which exhibited a considerable increase in abundance. With the appearance of *Clarias* spp. in some localities, a marshy fish community was, therefore, established by the end of 1974 and developed in 1975 until the total drying up of the basin. After 1975, the north basin was partially flooded each year and a marshy fish assemblage, with abundant Clariidae, colonised this swampy area, which could be compared to a flooded zone.

During the drying phase, changes in lacustrine assemblages were the result of specific reactions to new environmental situations. The open-water species, for instance, moved to more suitable habitats, either from north to south basin, as for example *Alestes baremoze* probably did in 1971, or to the lower reaches of the Chari River, as did different Mochokidae in 1972–3 (Quensière, 1976). For some truly lacustrine species, their disappearance was accelerated by either natural or mass mortalities, fishing, or lack of recruitment due, for instance, to the isolation or the drying of spawning areas.

During the 'Normal Chad' period, the abundance of open-water species that are also migratory may result from the abundance of food resources in the lake compared to the river and from the existence of floodplains, which allow spawning in the lower reaches of the Chari and Logone Rivers and also provide food and shelter for juveniles. In the 'Lesser Chad' period after 1974, the floodplains again played their role as nursery, but the

trophic capacity of the lake was less than during the previous period and the risks of hypoxia increased dramatically as long as the lake was invaded by aquatic vegetation. It should be noticed that there are no endemic lacustrine species in Lake Chad, which is considered to be an extension of the river system. This means that all the species were able to survive in the river system and to recolonize the lake when ecological conditions improved. In fact, there was a richer ichthyofauna in the lower reaches of the Chari River than in the lake itself. A similar situation occurs in Lake Tumba (Zaïre) (Matthes, 1964).

Lake Chilwa, south-east of Lake Malawi, is another outstanding example of an endorheic lake that is subject to large fluctuations in level, area and salinity. There is basically a large open-water area bordered by wide swamps, which are more extensive at the north end of the lake, and dominated by *Typha domingensis*. A few small tributaries enter Lake Chilwa, and water level usually fluctuates by up to 1 m between dry and wet seasons, the lowest level occurring in November–December. As a result of long-term climatic fluctuations, the area and the depth of the lake are occasionally drastically reduced. For example, in 1968, following a series of dry years, the level reached the lowest recorded since 1920, and the lake almost dried (Kalk *et al.*, 1979). Such water-level fluctuations are associated with changes in salinity. During five years of investigations (1966–70), conductivity values ranging from 276 to 16 720 µS were recorded, with of course, higher values during the very low level in 1968 (Kalk *et al.*, 1979).

Altogether, some 30 fish species were recorded from the lake basin, many of them being only recorded from the affluent streams. Even at high lake levels the open water of Lake Chilwa, which is very turbid and slightly saline, is totally dominated by only three species, *Clarias gariepinus*, *Barbus paludinosus* and *Oreochromis shiranus chilwae* which provide the basis for the fishery. In the dry season, when the open water is particularly inhospitable, and during periods of major recession, the fish migrate into the surrounding swamps, which provide a refuge and where they find less saline and less turbid waters. When the water rises again, there is a restocking of the lake from this reservoir. But the swamps and river beds are inhabited by a more diverse fish fauna that only occasionally enters the open water during flooding, e.g.

Table 10.10. *Total fish landing from Lake Chilwa and its swamps, and the contribution of the three main species during normal, pre-drying, drying, filling and high level years*

Year	Minimum depth (m)	Total catch (tonnes)	*Oreochromis shiranus chilwae*	*Barbus paludinosus*	*Clarias gariepinus*
1962		3500			
1963		3262			
1964		5255			
1965	1.08	8820	4234	2293	2293
1966	0.43	7200	1368	2808	3024
1967	0.35	3139	220	157	2762
1968	0	97	4	7	86
1969	1.18	3326	13	67	3246
1970	1.05	4166	333	1458	2375
1971	1.36	3595	899	1258	1438
1972	0.90	5246	1312	2203	1731
1973	0.02	1903			
1974	2.21	3171			
1975	1.55	2809			
1976	2.00	19 746	2764	10 268	6714
1977		20 800			
1978		17 800			
1797		25 800			
1980		19 400			

From Furse *et al.*, 1979; Chaika, 1982.

Petrocephalus catostoma, Marcusenius macrolepidotus, Brycinus imberi, Labeo cylindricus, Barbus trimaculatus, Tilapia rendalli, Pseudocrenilabrus philander, etc. (Kirk, 1967; Furse *et al.*, 1979).

Changes in salinity and turbidity accompanying the fluctuations of lake level resulted in major changes in the composition of the flora and fauna in Lake Chilwa. The total fish landing decreased dramatically from 1965 to 1968, following the major recession, and then increased when the lake level rose (Table 10.10). By 1976 the fisheries had recovered. The open-water fish species each responded differently to the worsening situation. *Oreochromis shiranus chilwae* was most sensitive to change and their abundance declined three years before dryness. *Barbus paludinosus* only declined drastically in the fishery in 1967–8, whereas the *Clarias gariepinus* catch only failed in the year of complete dryness. The recovery occurred at an unexpectedly rapid rate. *Clarias* appeared in fish catches only one month after the onset of the rains

in December 1968, and *Barbus* was recorded early in 1969. The recovery of the *Oreochromis* fishery was much slower, which may have been a consequence of the heavy marginal fishing of cichlid stocks during the drought period (Furse *et al.*, 1979).

Open shallow lakes

The fish assemblages of open shallow lakes connected to large river systems would be expected to be more diverse than the fauna of endorheic lakes because of periodic reinvasion from rivers. There are several studies dealing with this habitat.

The hydrology of **Lake Liambezi** (Namibia) is complex. It receives water from different sources including the Zambezi River, which, on occasions, overflows southwards (Seaman *et al.*, 1978). Outflow from the lake via the Chobe River is intermittent and depends on lake level. The depth is less than 5 m and fairly constant in the open water (100 km²), which is surrounded by large swampy areas (200 km²). Up to 43 species, including 15 cichlids, were recorded in Lake Liambezi itself, all of them also occurring in the Upper Zambezi River where there are 73 fish species (van der Waal & Skelton, 1984). In the Chobe River and its floodplains, 56 species, which are also shared with the Zambezi River, have been identified. Most fishes encountered in the lake and swamps are palustrine species, some which are extremely tolerant of deoxygenation and can inhabit waterbodies completely covered by the floating exotic plant *Salvinia molesta*.

The shallow **Lake George** (Uganda) situated just on the Equator in the Western African Rift Valley has an average depth of 2.4 m. Most of the inflow water comes from the Ruwenzori mountains, and the single effluent is the Kazinga Channel that flows towards Lake Edward (Viner & Smith, 1973). Lake Edward itself flows to Lake Albert through the Semliki River. The isolation of Lake George occurred after the river drainage of western Uganda had been separated from the Lake Victoria drainage by tectonic movements. Lake George in its present form is rather young, having originated *c.* 4000 years ago (Beadle, 1981) after a period of volcanic activity about 8000–10 000 years ago, which is thought to have been fatal for many of the fish species inhabiting Lake Edward. The existence of falls and rapids, probably insurmountable to fish, apparently prevented any reinvasion from Lake Albert (Beadle, 1981). The origin of the invading fish fauna for the early Lake George was either from the riverine assemblage of the inflows, or from Lake Edward itself.

In Lake George, two main groups of fish were distinguished according to their distribution (Gwahaba, 1975). The first group of species includes fish caught at the inshore and offshore sites, such as *Protopterus aethiopicus*, *Haplochromis nigripinnis*, *Haplochromis pappenheimi*, *Haplochromis angustifrons* and *Aplocheilichthys eduardensis*. Nevertheless, a decreasing density pattern from inshore to offshore regions was observed for various species, such as *Bagrus docmak*, *Clarias gariepinus*, *Haplochromis squamipinnis*, *Oreochromis niloticus* and *Oreochromis leucostictus*. This pattern could be explained by the numerous juveniles of the latter species living in the shallow waters near the shore.

The second group includes 15 fish species caught only at sites near the shore, e.g. *Astatoreochromis alluaudi*, *Barbus kerstenii*, *Barbus neglectus*, *Ctenopoma muriei*, *Marcusenius nigricans* and many species of *Haplochromis*. This limitation to the distribution of some inshore species may be due to their food requirements. *Haplochromis aeneocolor* feeds mainly on macrophytic detritus and *Haplochromis limax*, which feeds on aufwuchs, is found near emergent vegetation such as *Vossia cuspidata*. *Haplochromis mylodon* eats the gastropods *Melanoides tuberculata*, *Haplochromis taurinus* feeds on eggs and embryos, and *Haplochromis petronius* is associated with rocky inshore habitats of which there is almost none in Lake George. Three species, *Oreochromis niloticus*, *Haplochromis nigripinnis* and *Haplochromis angustifrons* represents nearly 80% of the total fish biomass (Gwahaba, 1975). The former two (60% biomass) are herbivores and feed directly on the blue-green algae.

In Lake George, the pattern of fish abundance determined by purse seining (Gwahaba, 1975) is strongly influenced by the pattern of distribution of *Haplochromis*, and the pattern of biomass is determined mainly by the distribution patterns of two *Haplochromis* species (*H. nigripinnis* and *H. angustifrons*) as well as that of *O. niloticus*, since the biomass of these three species averages 80% of the total fish biomass. There is a clear trend of decreasing biomass from inshore (highest value 900 kg ha^{-1}) toward the centre of the lake (lowest value 60 kg ha^{-1}), the overall mean biomass being 290 kg ha^{-1}.

Fish assemblages in deep African Great Lakes

Much attention has been given to the fish fauna of the deep African Great Lakes during the last decade, as a result of the dramatic increase in the anthropogenic impacts on their aquatic environments (fisheries, introduction of alien species, eutrophication) and the consequences for their highly endemic fauna. The speciation processes, which resulted in the highly diversified cichlid fauna of these lakes, have already been discussed in Chapter 4, as well as aspects of the life-history strategies of many species (Chapter 8). This section is an attempt to give an overview of the distribution of multispecies communities according to broad habitat characteristics.

The communities recognised are a simplification of a much more complex reality. Each community has its own characteristics and is interlinked with neighbouring communities through transition areas. In each community there are narrowly stenotopic species strictly confined to their specific habitat, coexisting with eurytopic members that move from one habitat to another. Thus certain lithophilous species may make excursions onto sand or over mud (Ribbink, 1991). In addition to permanent residents, there are occasional visitors. Moreover cyclical activities specific to every species, daily and seasonal, both in space and time, also strongly influence the composition of fish assemblages at any one time.

In the littoral regions of Lakes Tanganyika and Malawi, the coastline is an alternation of rocky, sandy or muddy stretches, which is thought to have had profound evolutionary effects in isolating populations of fish restricted to a particular biota. The main sections, rocky and sandy, are discontinuous and alternate with each other. The rock-slope faunas have been compared with coral-reef fish faunas. The abundance of cover on rocky shores probably explains the persistence of a diverse cichlid fauna, despite the numerous predatory fishes.

Lake Tanganyika assemblages

With its 230+ fish species, Lake Tanganyika, the oldest of the East African Lakes (*c.* 20 Myr) has the most differentiated fish fauna. Seventy-two per cent of the species are endemic and 90% (or more) belong to the family Cichlidae, most of them being exclusively endemic to the lake. It is generally thought that the number of cichlids is probably much greater, many species still being undiscovered.

Compared to the other African Great Lakes, the fish fauna of Lake Tanganyika is unique in terms of :

the existence of species flocks other than cichlids, e.g. Mastacembelidae, Bagridae, Mochokidae, Centropomidae, etc.;

the existence of a true pelagic community with clupeids and large predators;

whereas in other African lakes most cichlids mouthbrood their young, Lake Tanganyika is unique in having, apart from tilapiine, substrate-spawning cichlids in addition to mouthbrooders; and

the absence of fish families such as the mormyrids.

The fish communities of Lake Tanganyika have been most intensively investigated during the last two decades. Brichard (1978) and, more recently, Konings (1988) provided the basic knowledge for the geographical distribution of cichlids. A Japanese team studied not only the biology and ethology of many cichlid species, but also conducted research on lithophilous communities (Kawabata & Mihigo, 1982; Hori et al., 1983; Yamaoka, 1983; Kuwamura, 1987). Much emphasis has also been given to the pelagic fish community (Coulter, 1970, 1981, and review in Coulter, 1991b).

The **rocky habitat communities** are analogous in many respects to those of Lake Malawi. Many species are strongly lithophilous, with microhabitat preferences for depth and substratum. The fish assemblage comprises numerous species, many with Mbuna-like characteristics but the contribution from other species may be large. The coastal fish communities include species flocks of different families. The catfishes, which are generally the largest fish in the area, include Mochokidae (10 species of *Synodontis*, most endemic), Malapteruridae (the widespread *Malapterurus electricus*), Clariidae (*Heterobranchus*) and Bagridae (such as *Chrysichthys* and *Bagrus*). The latter family also includes the small endemic genera, *Phyllonemus* and *Lophiobagrus*, that hide in crevices. Most of the *Afromastacembelus* live in rocks, as well as some species of cyprinids, such as *Varicorhinus* and juvenile *Labeo*.

The populations of rock-frequenting cichlids in Lake Tanganyika exhibit striking examples of convergent evolution with those of Lake Malawi. There are strong morphological parallels, and numerous species that appear to fill specialised ecological roles are also usually restricted to particular geographic localities isolated from one another (Ribbink et al., 1983b). In both lakes, cichlid communities are distributed according to depth and bottom characteristics, but the taxa are completely different in the two lakes. Well-known examples of parallel evolution may be found between the following genera (see Chapter 4): *Petrochromis* in Lake Tanganyika and *Petrotilapia* in Lake Malawi, *Tropheus* and *Pseudotropheus tropheops* species-complex, *Tanganicodus* and *Labidochromis*.

The aufwuchs-eating community is dominated by two territorial species, e.g. *Tropheus moorii* and *Petrochromis polyodon*, but many other species also share the same food resource, e.g. *Petrochromis*, *Simochromis*, *Telmatochromis*, *Ophthalmotilapia* and *Asprotilapia*.

The number of predatory species in the lithophilic communities seems higher in Lake Tanganyika than in Lake Malawi (Ribbink, 1991). The 'lamprologines', the most speciose cichlid taxon in the lake, may constitute more than 50% of the species in the littoral community. The genus *Lamprologus* (substratum spawners) includes many carnivorous species coexisting on the rocky shores (Hori, 1983). Some are benthic feeders, others piscivorous, their body shapes varying in accordance with their feeding habits.

While all the lithophilous cichlids of Lake Malawi are maternal mouthbrooders, a high proportion of those in Lake Tanganyika are substrate spawners.

Other microhabitats include extensive beds of gastropod shells that occur from 10 to 35 m depth, where the shore slopes gently. A community of shell dwellers, including more than ten cichlid species, principally lamprologines, utilise these shells as refuges (Ribbink, 1991). In the surge zone, a community of goby-like cichlids has evolved, including species of the genera *Eretmodus*, *Spathodus* and *Tanganicodus* adapted to life in the wave-washed area (Yamaoka et al., 1986).

The **sand and mud communities** are less well known, but the general trends demonstrated by the rock-dwelling fishes are paralleled by fish living in other habitats although these are not as clearly demarcated (Ribbink, 1991). Again, water depth and the nature of the substrate strongly influence species composition. Over sandy patches, typical sand-dwelling species, belonging to the genera *Callochromis*, *Xenotilapia* and *Cardiopharynx*, live in schools of several hundred.

The **deepwater communities**. Below 20 m rocky areas are rare and the bottom is muddy or sandy. Below 100 m the bottom is mainly mud and a remarkable fauna has evolved there. Species composition varies according to depth and nature of the bottom, and the lower limit of this deep-demersal fish community is determined by oxygen concentration. Below 60 m depth, the dissolved oxygen falls rapidly in the north basin, but there may be oxygen down to 250 m in the south basin. Many of these species occupy wide ranges of depth. Some remain all their lives in deep water while other make periodic migrations inshore (Coulter, 1991*b*).

A few species were caught in poorly oxygenated waters at 120 m: the cichlids *Hemibates stenosoma*, *Bathybates ferox*, *Bathybates fasciatus*, *Xenochromis hecqui*, *Gnathochromis permaxillaris*, two *Chrysichthys* that fill the deepwater scavenging role that the clariid *Dinotopterus* has in Lake Malawi, *Dinotopterus cunningtoni* and *Lates mariae* (Coulter, 1966). These species may be habituated to oxygen concentrations between 1 and 3 g m^{-3}. It is significant that three species, *Hemibates stenosoma*, *Chrysichthys stappersii* and *Lates mariae*, have been caught below 200 m. These species, and possibly also some others, can apparently survive temporarily without oxygen and are probably adapted to cope with this (Coulter, 1991*b*)

A group of at least ten species, all cichlids living above the bottom, has been also recognised as a bathypelagic community (Coulter, 1991*b*). They are small in size (less than 200 mm) and all are zooplanktivores. Most are feeble swimmers and they are not known to undertake migrations to the surface. They exhibit specialisations, such as big eyes and weak spines, and all are mouthbrooders. This group includes representatives of the genera *Trematocara*, *Greenwoodochromis*, *Haplotaxodon*, *Cyprichromis*, *Gnathochromis*, *Tangachromis*. *Bathybates fasciatus* and *Bathybates leo* that feed mainly on clupeids also occur in this bathypelagic community.

The **pelagic community** has a unique assemblage of fish species. Two endemic clupeids, *Stolothrissa tanganicae* and *Limnothrissa miodon*, occupy the pelagic zone where they feed on both phyto- and zooplankton. This clupeid stock supports piscivorous fishes, including four endemic species of the centropomid *Lates*. The young of *Lates mariae*, *Lates microlepis* and *Lates angustifrons* live among littoral macrophytes eating prawns and insects until maturity. Afterwards, they move partly offshore to feed on the clupeids. *Lates* (*Luciolates*) *stappersii* is completely pelagic all its life living in the upper 30 m. Lake Tanganyika, as well as Lake Kivu, lacks the *Chaoborus* larvae, predators of zooplankton, that are abundant in most of the other African Great Lakes.

The Lake Tanganyika fauna that evolved over millions of years is still intact. The extensive speciation of the Tanganyikan cichlids probably results from rapid speciation in both the *Lamprologus* and *Haplochromis* basic strains. The fact that the ancestral fluviatile *Lamprologus* stock comprises an important part of the endemic lacustrine cichlids is unique for the African Lakes, as this group does not appear to be present in any of the other lakes.

Investigations on patterns in resource utilisation in the cichlid fish communities of Lake Tanganyika showed that many species, with superficially similar ecological requirements, are actually well segregated in their utilisation of commonly required resources either for breeding or feeding (Nakai *et al.*, 1994). For example, several piscivorous species exploit similar food items but each employs a different hunting technique. This segregation is surely a mechanism by which biodiversity is formed and maintained, and may explain the coexistence of many closely-related species.

It should also be stressed that much attention has recently been paid to the indirectly beneficial relationships which have been found to be prevalent among species (Hori, 1983, 1987; Takamura, 1984; Nakai, 1993; Yuma, 1993). While the coexistence of many related fish species has been traditionally explained by ecologists in the context of resource segregation, other mechanisms, such as mutually or commensally beneficial, relationships, may be involved (see also Chapter 11)

Lake Malawi

Ecological research on Lake Malawi goes back to the pioneer works of Lowe (1952, 1953) on the tilapias, and Fryer (1959) on the Mbuna ecology of the rocky shore of Nkhata Bay. Then papers from Lewis (1981) and the detailed work of Ribbink *et al.* (1983*b*) added substantially to our knowledge of the fish populations.

Lake Malawi is also a Rift Lake, in many respects fairly similar to Lake Tanganyika but assumed to be a good deal younger (2–10 Myr). The same broad habitats and fish communities may therefore be recognised. The distribution and ecology of rock-dwelling cichlid species has been well investigated, but the communities of other habitats are not so clearly understood.

In the **pelagic community**, zooplankton is exploited by a range of closely-related haplochromine species belonging to the 'Utaka' group, which are found mainly inshore but usually occur in large shoals in places where upwellings lead to plankton abundance. These are supplemented by a cyprinid, *Engraulicypris sardella*, that is apparently a visual predator and is the only pelagic fish to venture far offshore, and by a number of catfish of the genus *Dinotopterus* (previously *Bathyclarias*), such as *Dinotopterus loweae* that eats both fish and plankton. *Dinotopterus* species have much reduced accessory respiratory organs and lateral eyes. The associated predators are cichlids of the genera *Rhamphochromis* that are essentially pursuit predators, and *Diplotaxodon*. The catfish, *Bagrus meridionalis*, is also a predator of zooplankton feeders.

Several fish species have been recorded from deep benthic regions, between 200–300 m around the lower limit of oxygen, and it is likely that they spend some time in poorly oxygenated waters. The **deepwater community** includes few species of pelagic cichlids (Utaka), but also *Bagrus meridionalis*, *Mormyrus longirostris*, *Synodontis njassae* and *Buccochromis heterotaenia* have been caught from these depths (Jackson *et al.*, 1963 cited *in* Beadle, 1981). At least four species of *Dinotopterus* have been recorded down to almost 300 m and it is interesting to note that their suprabranchial air-breathing organs have entirely disappeared.

The **rocky shores** provide rich and varied habitats colonised by small brightly coloured cichlids adapted to different niches. The surface of the submerged rocks is covered with a carpet of blue-green algae mixed with other algae that harbours a large number of invertebrates. This carpet, which is also called aufwuchs, serves as food for most of the rock-dwelling fish species.

Ribbink *et al.* (1983*b*) provided the results of detailed investigations of a group of ten closely related endemic genera known in Malawi as **Mbuna**. More than 200 species have been recorded, and most of them are stenotypic, that is to say that they are narrowly adapted to a specific habitat, and usually restricted to a particular sub-

stratum and a narrow depth range. Moreover, many Mbuna species are endemic to a single small part of the lake, sometimes restricted to a tiny rock outcrop, and the Mbuna community differs from one site to another in species composition and the relative abundance of the constituent species (Ribbink & Eccles, 1988).

Fryer (1959) had already pointed out that many fish species inhabiting rocky habitats are closely associated with the rocky zones, to such a point that some of them never occur more than one metre from the rocky substrate. The sedentary nature of the Mbuna has been demonstrated experimentally by tagging. Furthermore, there is evidence of species transferred within the lake that remained near the point of introduction and apparently reproduced there (Ribbink *et al.*, 1983*b*). Sexually mature fish live, feed and breed within their preferred habitat throughout the year. Moreover, the large-yolked eggs enable the young to hatch at a size large enough to use more or less the same food as the parents, which means that there is no need for a planktonic open-water stage as observed for most reef fishes (Lowe-McConnell, 1987), and the species may spend all their life in the same biota. Some species may have their entire distribution restricted to a few thousand square metres (Ribbink *et al.*, 1983*b*).

In Lake Malawi, a typical permanent rocky habitat community includes mainly Mbuna species, but also some other cichlids and non-cichlids. The non-Mbuna component also varies from one site to the next in both species composition and abundance. Nevertheless, a few species have a lake-wide distribution: *Nimbochromis polystigma*, *Labeo cylindricus*, *Barbus johnstonii*, *Opsaridium microcephalum* and *Dinotopterus worthingtoni*. But as a whole, there is high intralacustrine endemicity and this results in unique communities around the lake, a feature that has also been demonstrated for fish communities of other habitats (Ribbink *et al.*, 1983*b*). The rock communities also include non-permanent species, such as sand-dwellers that release and guard fry among the rocks, as well as fry and juveniles of many other species which shelter among rocks before becoming large enough to live in more open habitats.

Superficially, there is a considerable overlap between species in resource partitioning and space requirements. But Mbuna appear to be opportunistic feeders rather than rigid specialists. Regardless of their trophic specialisations, most of the species also feed on zooplankton when it is abundant. Furthermore, many species are adapted to live within a particular depth range and show a preference for rocks of a certain type. Syntopic species that ingest the same food collect it from different parts of the habitat (Marsh, 1981; Sharp, 1981). How and where they feed (zonation, depth and microhabitat) seems more important for resource partitioning than the items on which they feed (Lowe-McConnell, 1987).

The complexity of the rocky shoreline communities in Lake Malawi may be ascribed to the wide variety of microhabitats they offer: rock type, slope, depth, exposure to wave action, but also to the isolation of rocky areas. **Sedimented areas** (sandy and muddy substrates) appear to offer less diversity of situations, yet they too contain large numbers of species apparently adapted to them. But the presence of macrophytes, depth, organic content and grain size of the sediment are parameters that could affect the distribution of species (Ribbink & Eccles, 1988). Within the lake, there are also marked differences between the benthic communities in relation with their geographic situation. For instance, there are few species in common in comparable habitats situated in the northern and the southern extremities of the lake. *Lethrinops longipinnis*, a major component of catches in the south, is absent from the north where *Lethrinops argenteus* occurs. The genus *Gephyrochromis* from the north is replaced by *Pseudotropheus elegans* in the south.

As a result of demersal-trawl surveys (Lewis, 1981), it has been shown that cichlid species have restricted distributions, and exhibit strong depth preferences: even if the faunal list may be closely similar for adjacent stations, the proportion of species differs significantly with depth. The replacement of one species by another exploiting a similar ecological niche as depth increases has been exemplified by a series of molluscivorous species (Ribbink & Eccles, 1988). *Barbus eurystomus* and *Trematocranus placodon* exploit small gastropods down to 10–15 m depth. They are replaced at 20 m by *Maravichromis anaphyrmus*, which is replaced in turn by *Lethrinops mylodon* below 60 m. Similar examples may be found for other groups. *Clarias gariepinus*, common down to about 10 m, is replaced by various species of the genus *Dinotopterus*, the species *D. atribrancus* being confined to depths of over 90 m.

The nature of the sediment also influences these communities. The structure of the teeth of benthic species seems to be associated with the coarseness of the sediment, with a tendency to become smaller and more numerous as the substrate becomes finer with increasing depth (Ribbink & Eccles, 1988).

Fryer (1959) and later Ribbink & Eccles (1988) recognised an 'intermediate zone' between rocky and sandy areas, but most of the species apparently belong to the sandy communities, though a few species of cichlids seem to be more common in this intermediate zone than in sand–rock or sand–macrophyte habitats, e.g. *Cyathochromis obliquidens*, *Protomelas kirkii* and *Haplochromis callipterus*. These, usually very narrow, habitats could be referred to as ecotones, and the community of such an area is partly governed by the circumstances.

Lake Victoria

Until the 1950s, research on the fish of Lake Victoria mainly concerned species descriptions, but in the 1960s, investigations undertaken by scientists linked . with EAFFRO (East African Freshwater Fisheries Research Organisation) based in Jinja (Uganda) also dealt with conservation, ecology and fisheries. At the beginning of the 1970s, fisheries research was conducted towards a better exploitation of the cichlid stock. The pioneer work on Lake Victoria suggested a fairly uniform distribution of cichlids, which was in sharp contrast to the observations on Lakes Malawi and Tanganyika (Fryer & Iles, 1972). The lake-wide distribution of Lake Victoria species was also assumed in the publication of a study conducted in the Mwanza Gulf in southern Lake Victoria (van Oijen *et al.*, 1981). Nevertheless, this study demonstrated that most cichlid species had very restricted depth ranges, which were related either to the reduced oxygen concentration below 10 m or to the nature of the sediment. However, more recent data tend to support the hypothesis that many species actually have clear, geographically restricted distributions (Lewis, 1981; Witte, 1984*b*; Ribbink, 1991). Quite a lot of information has been obtained since 1975 by the Dutch *Haplochromis* Ecology Survey Team (HEST) that operated in southern Lake Victoria (Witte, 1981; Witte & van Oijen, 1990; Barel *et al.*, 1991)

The fish fauna of Lake Victoria, includes quite a lot of feeding specialisations, both non-cichlid (Corbet, 1961) and cichlid (Witte & van Oijen, 1990) (see also

Chapter 7). As in Lake Malawi, all endemic cichlids in Lake Victoria are maternal moutbrooders.

Although most of the bottom substrata of Lake Victoria are muddy or sandy, **rocky areas** that are inhabited by Mbuna-like fishes do occur (van Oijen *et al.*, 1981). In the Mwanza Gulf, a community of at least 16 species of rock-dwellers has been described along with their associated predators: a few large haplochromines, *Bagrus* and fish-eating birds. Parallel evolution in coloration, squammation, teeth and jaws has been observed between these Victorian rockfishes and Malawian ones, and this is illustrated by *Paralabidochromis victoriae*, which is similar to *Labidochromis vellicans* from Lake Malawi. Little migration apparently occurs between isolated rocky patches, and fish species appear to be geographically restricted.

The cichlid community of **sedimented areas** includes many trophic groups and resembles those of Lake Malawi, in that it contains unique assemblages of sediment-associated species. With 70% of the catch, insectivores clearly dominated over sandy substrates in shallow waters (2–6 m). This was also true to a lesser extent in shallow littoral waters over mud, but with increasing depth the detritivores/phytoplanktivore group became dominant (over 80% of the catch) (Witte & van Oijen, 1990).

According to the trawl catches, 90% of the cichlid community consists of detritus/phytoplanktivores, zooplanktivores or insectivores. Piscivores, while represented by the largest number of species (40%), contributed only 1% of the biomass (van Oijen *et al.*, 1981). This small contribution may be a result of selective fishing pressure on the larger fish. In a detailed study of the piscivores, van Oijen (1982) pointed out the importance of substrate type, depth range and exposure to wind in their habitat partitioning. Moreover, although there are numerous piscivores they are specialised on different prey taxa: different species, different prey sizes, different ontogenic stages and even different parts of the prey. Restricted depth distribution and substrate preferences for sand or mud occur among molluscivores that occur almost exclusively over sandy substrates (Witte, 1981).

Lake Victoria is mostly shallow, and stratified, and a thermocline develops at between 7 to 10 m in the rainy season (Beadle, 1981). Consequently, the **open water community** is not as distinct in Lake Victoria as in

Lakes Tanganyika and Malawi. Four cichlid zooplanktivore species appear to coexist within a feeding population. There is a spatial segregation as their distribution in the water column differs, and a temporal segregation occurs during the breeding season. The piscivore cichlids are habitat restricted and harvest particular food components. They also have species with specific depth distributions (van Oijen, 1982). The small cyprinid, *Rastrineobola argentea*, appears to be the only non-cichlid specialised as a pelagic feeder on zooplankton (Lowe-McConnell, 1987), but a few species also feed on surface insects, e.g. *Alestes* spp., *Barbus* spp., *Schilbe* and *Synodontis afrofischeri*. *Schilbe* is also a predator of *Rastrineobola*.

The effects of introduced species on the indigenous fish fauna of Lake Victoria are considered in Chapter 14.

Lake Kivu

Lake Kivu is a natural dam lake created on a northward flowing tributary of Lake Edward by an eruption of the Virunga volcanoes. As a result of this past but fairly recent volcanic event, Lake Kivu has a depauperate fish fauna compared with other Great Lakes (Chapter 6). Twenty-four species are currently recorded, including 13 species of *Haplochromis* and 3 tilapiines (2 introduced). The remaining families comprise 2 *Clarias* species, 5 cyprinids (*Raiamas moorei* possibly introduced from Tanganyika) and one clupeid, *Limnothrissa miodon*, introduced from Lake Tanganyika (Ulyel, 1991).

The fish assemblages are, therefore, much more simple than in any of the other Great African Lakes, but similar main types of habitats can be recognised (Ulyel, 1991).

1. A littoral zone, where rocks are covered with epilithic algae inhabited by numerous insect larvae. The aquatic vegetation is a source of detritus, but also offers shelter to fish and serves as support for aquatic insects. This zone is colonised by juveniles of most of the species, but also by the insectivore, *Haplochromis adolphifrederici*, and the omnivore, *Clarias liocephalus*. The tilapiine species, *Tilapia rendalli*, *Tilapia macrochir*, *Oreochromis niloticus*, also live on rocky littoral habitats, as well as the microphytophage, *Haplochromis olivaceus*, and a few *Barbus* species.

2. A benthic community of the sedimented substrates (mud and sand), where numerous juveniles are also present, as well as many other *Haplochromis* species and *Clarias gariepinus*.

3. A pelagic zone, about 70 m depth, with only a few species: *Limnothrissa miodon* (zooplanktophage), *Haplochromis kamiranzovu* (microphage), *Raiamas moorei* (insectivore and piscivore) and *Hydrocynus vittatus* (piscivore).

General conclusion

Fish assemblages observed at one time in a waterbody are no more than 'instantaneous samples' of a long-term dynamic process. They are the result of different major driving forces. For the East African Great Lakes, it is, for example, the result of a unique process of speciation, which occurred over millions of years in permanent but, nevertheless, always changing aquatic environments (see Chapter 6). Lake Tanganyika was colonised by direct invasion via the Lukuga outlet, Upper Zaïre basin; Lake Malawi was probably colonised from east coast and westward-flowing rivers (Skelton *et al.*, 1991) and now drains to the Zambezi, but has only nine fish families; Lake Victoria lies in the Nile basin. Despite the differences in their origin, the original faunas of these lakes responded with the evolution of a large number of endemic species, and a surprisingly high degree of convergence in morphology and ecological specialisation. The Great Lakes have evolved the most diverse fauna of endemic fish of any of the world's lakes (see Chapters 2 and 4). The biological resources have been repeatedly exploited in a similar way, but different groups of species may be involved. For instance, the pelagic zooplankton is consumed mainly by clupeid fishes in Lake Tanganyika, whereas in Lake Malawi several species, including the haplochromines of the Utaka group (local name for zooplanktivore cichlids) are involved. Turner (1982) observed that while the pelagic primary production in both lakes was broadly similar, the pelagic fish biomass of Lake Tanganyika is 1.6 to 10 times greater. The existence of small clupeids in Lake Tanganyika (*Limnothrissa*

and *Stolothrissa*) may be one of the major explanations.

In contrast to offshore fish communities with a rather simple structure dominated by several pelagic species of Clupeidae (see Chapters 7 and 13), littoral fish communities have far more components and a more complex structure. Most fishes of all three lakes for which data are available have a particular vertical depth range, sometimes extremely narrow, and habitat preferences for the nature of the substrate. For lithophilous fish, the habitat is similar for adults and fry, but for other species that need to find shelter for their fry, adults may occupy different habitats at different stages of their life history, and are therefore more eurytopic than the strictly lithophilous species.

In the short term, fish assemblages, such as those found in Lake Chad or Lake Chilwa, are strongly influenced by the nature of habitats provided in waterbodies, including their seasonal or year-to-year changes in ecological factors (see Chapter 1). It is important to stress that species assemblages should be examined and interpreted from a multiscale perspective, including both spatial and temporal scales (see Chapter 12). Moreover, taking into account the sampling constraints and the difficulties of obtaining a non-selective sample of fish communities, it is not surprising that fish assemblage studies are usually slightly boring and frustrating for ichthyologists. It takes much time in the field for limited scientific results. While for the moment studies of fish assemblages are much too descriptive and generally limited to the interpretation of gill-net (or any other gear) catches, which are no more than 'catch-assemblages', we really need, in the future, to develop a more dynamic approach. An improvement in sampling techniques is required, and at the moment that is a very difficult task.

The Great African Lakes are of world-wide interest for biologists, as natural laboratories to study evolution at work. Their colourful fishes, and the possibilities for making direct underwater observation of various aspects of their behaviour, such as feeding and breeding in natural conditions, excited aquariologists and scientists. But this valuable international heritage could be easily damaged irrevocably, either by the introduction of exotic species as happened in Lake Victoria, or by pollution.

11 Equilibrium processes of species richness and diversity

> There are two aspects to evolution: the creation of living beings, and their survival. Living organisms do not persist, but they survive in the form of offspring . . . Nature has created survival machines and natural interspecific selection has ensured that better and better ones survived.
>
> *Løvtrup, 1988*

Ecological theories based on equilibrium, such as the theory of island biogeography, argue that the actual number of species is the result of a balance between opposing forces occurring over a wide range of scales, so that care must be taken to distinguish between them. One of the major issues in ecology is to identify what causes changes in diversity. Does it tend towards an equilibrium, or increase without limits, or fluctuate randomly? Are there trends that could be identified and to what factors could they be related? If faunal lists are usually very large for tropical waterbodies, one of the questions is how many of these species can occur altogether within an assemblage or community in one biotope.

Actually, species richness and diversity in a particular region are determined both by historical and contemporary determinants operating at different time and spatial scales. Historical processes, including speciation as well as colonisation and extinction, are responsible for the number of species that may be encountered in a given biogeographical area (species pool). Contemporary processes will determine the species composition of communities encountered in particular localities within that area, each assemblage being a subset of the total pool. The relative role of contemporary abiotic (e.g. physical, chemical, climatic) and biotic (e.g. competition, predation, diseases) factors on community structure and diversity are the subject of an impassioned debate.

The understanding of species diversity is complicated because it is likely to be the outcome of many contribu-

ting factors. The question is, therefore, whether it is possible to formulate 'rules' about species diversity. According to Diamond (1988), determinants of species diversity can be grouped into four broad sets of factors, which are nothing more than a convenient empirical checklist.

1. **Resource quality**, consisting of the habitat and resource factors that determine the 'number and diversity of niches'. There is a general feeling that habitats with a more complex spatial structure contain more species than do simpler habitats. A related phenomenon is that the number of species increases with diversity of habitats encountered. Habitat structure and habitat diversity also have a time dimension, and temporal variability can provide opportunities for niche differentiation (e.g. 24-hour cycle or seasonal cycle). Moreover, consumers can develop different foraging strategies or adopt different life-history strategies with the spatial and/or temporal niche.

2. **Quantity of consumer individuals** as related to resources, including factors determining availability of resources and those determining the number of consuming individuals. The number of consumer individuals (N) or the population size, depends on the quantity of resources available (R). The maximum potential size of the population (Nmax) is affected by the ratio R/Rn where Rn is the quantity of resources required to sustain one individual.

Resource quantity (R) in turn, equals the product of area (A) times productivity per unit area (P). Resource requirement obviously increases with body size. Thus, the quantity of consumer individuals depends on three factors: (i) the increase in species diversity with the size of the area; (ii) the productivity of the area under consideration; and (iii) the body size of the consumers.

3. **Species interactions** that may boost or lower species richness through effects on number of individuals or their fitness. Predation and/or resource competition are assumed to be key factors controlling species diversity, in a complex array of cascading effects. Frequency and intensity of disturbances are considered to be the key elements in influencing species interactions. Species interaction through interference, competition and disease tends to decrease species richness by decreasing individual fitness, while species interaction through mutualism tends to increase species diversity by increasing individual fitness.

4. **Dynamics**: because even at equilibrium, dynamic processes, such as extinction, and emigration, tend to decrease species richness, whereas speciation and immigration tend to increase it. In non-equilibrium situations, pulse disturbances may transiently decrease species diversity below the equilibrium value until the resources or consumer populations are restored by immigration or speciation.

Some of these factors suggested by Diamond (1988) as being responsible for species richness may be controversial, and many of them will be discussed in this or the following chapters, where we shall attempt to illustrate both facts and controversies taking examples from African freshwater fishes.

The assessment of species diversity

For a long time, a measure of diversity has been seen by ecologists as an indication of the well-being of ecological systems. Consequently, they have devised a huge range of indices and models for measuring diversity, with considerable debate about their meaning. However, the term diversity has been used in different ways. It could relate to the number of species in a group, the relative number of individuals of different species in a given community or ecosystem, or the number of species in an assemblage. The latter is obviously a confusion with species richness.

Indeed, to describe the species composition of communities and assemblages, ecologists now widely use the following two definitions.

1. Species **richness** is simply the number of species occurring in a given biotic assemblage or area. It is not a very sophisticated index, but it has proved useful for fish where quantitative samples are usually difficult to obtain. Nevertheless, it should be remembered that there are difficulties in estimating richness, because estimates depend heavily on sample size, which may be compensated for by a plot of the number of species versus the sample size (area). Species density (number of species per unit area) is a commonly used measure of species richness.

2. Species **diversity** (also called evenness) considers the relative abundance of the species present. No community consists of species in equal abundance, and a variety of species abundance distributions have been proposed to describe the observed patterns.

Species diversity measures belong to two main categories (see Magurran, 1988, for review and details).

1. Models describing patterns of species abundance. A species abundance distribution that utilises all the information gathered in a community is the most complete mathematical description of the data. The main models used are the log normal distribution, the geometric series, the logarithmic series and MacArthur's broken stick model.

2. Indices based on the proportional abundances of species provide an alternative approach to the measurement of diversity. They take both evenness and species richness into account. The indices most commonly used are:

   the Simpson index: $D = \Sigma \{ni(ni - 1)/N(N - 1)\}$ with N the total number of individuals and ni the number of individuals of species i; and

   the Shannon diversity index: $H = -\Sigma\, pi \ln pi - (S - 1/N)$ where $pi = ni/N$ and S is the number of species.

The large number of indices available means that it

may be difficult to select the most appropriate method of measuring diversity. On the one hand, indices that reflect the species richness are better at discriminating between samples, but on the other hand, they are more affected by sample size than the other category of indices that express the degree of dominance (evenness) in the data. One of the weaknesses of diversity indices is that they do not contain much biological information. It should also be clear that, particularly for fish, their values are directly related to the sampling techniques, and one should bear in mind that most of these techniques are highly selective and provide a distorted picture of the fish community (see Chapter 10). One exception may be chemofishing, which has occasionally been used in different habitats.

The major applications of diversity measurement are in nature conservation and environmental monitoring. In both cases diversity is held to be synonymous with ecological quality. However, fish conservationists concentrate almost exclusively on measures of species richness and consider it as a more adequate measure of diversity, especially for comparisons between studies, because of the huge difficulties encountered in sampling correctly fish communities.

Abiotic determinants of local and regional species richness

For many decades, ecologists have sought to explain differences in local diversity by the influence of the physical environment and local interactions among species that are generally believed to limit the number of coexisting species. They now realise that local diversity also bears the imprint of more global processes, such as dispersal of species and historical events. A species cannot occupy an ecosystem that it has not been able to reach. Because most freshwater systems are geographically separated by terrestrial barriers uncrossable for primary division fishes, the occurrence of species in any one of these catchments may only be explained either by local speciation, or by historical accidents and past connections between catchments that allowed fishes to move from one system to another. Ecologists must therefore broaden their concept of community processes and incorporate data from systematics, biogeography and palaeontology into analyses of local patterns of diversity. In other

words, community ecologists are challenged to expand the geographical and historical scope of their concepts and investigations (Ricklefs, 1987). More recently, man has dramatically modified this situation by introducing alien species into many aquatic systems.

From continental to local scale: a conceptual framework

Tonn (1990) proposed a conceptual framework that views local fish assemblages, including the number, identity and relative abundance of species, as products of a series of filters operating at different spatial and temporal scales, and through which species of a given assemblage must pass. Figure 11.1 summarises this framework, the processes and features involved in each filter as well as their associated ranges of spatial and temporal scales. For Africa, the continental filter started with the break up of Gondwanaland. The continental fish fauna is the result of subsequent selective extinction and speciation, combined with distinct refuges and routes of dispersal, which shaped the continental patterns of species distribution, and selected the fish families currently encountered in African waters (Chapter 2).

A region (or province) may be defined as an area with relatively uniform climate and similar geomorphological

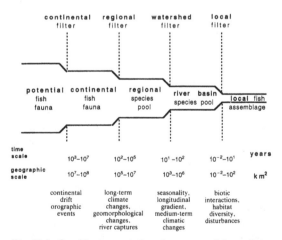

Fig. 11.1. Graphic representation of a conceptual framework of the structure and organisation of fish assemblages as a result of processes operating through different filters having spatial and temporal scales (based on Tonn, 1990).

history, so that its fauna experienced a common evolutionary history. Different ichthyoprovinces have already been identified in Africa on the basis of fish distribution (see Chapter 2). Within each province, there are different catchments, most of which are currently isolated and may be compared to islands. Connections have occurred, and will certainly occur in the future, through river capture, but also as a result of orogenic events and climatic changes causing, for example, high floods. The fish species pool in a catchment is the result of an equilibrium between past colonisations and extinctions, depending on the history of the catchment and its size, but also on speciation processes that are likely to be more or less related to the length of isolation of the catchment, and to the potential of fish families to speciate. It is also well known that fish species composition may change along a longitudinal river gradient, from small, first-order streams to large, lower courses.

Finally, at the local level, fish assemblages are shaped by the type of habitat available, including habitat diversity, and the frequency and amplitude of disturbances. The local level is also the spatial scale at which interactions between species are likely to be most important. The species richness of biological communities at local scales balances regional processes of species formation and geographical dispersal, which add species to communities against local processes of predation, competitive exclusion, adaptation and stochastic variation that may promote local extinction (Ricklefs, 1987).

This conceptual framework focuses attention on the influence that larger-scale phenomena have on local assemblages, but when changing scales the driving parameters also change. One major question, once patterns of fish assemblage organisation have been defined, is to identify the relative role of different filters in creating these patterns (Tonn, 1990). In other words, what are the causal processes behind the currently observed community level patterns ?

Why is species diversity higher in the tropics?
It is usually assumed that species diversity increases from high to low latitudes for most of the major groups of plants and animals, and that highest values occur at low latitudes in tropical and equatorial Africa, South America and Asia. Until now this assumption has received little analytical attention with regard to freshwater species,

compared to the large number of papers dealing with tropical rainforests. However, a comparison between species richness in African and European rivers (see Fig. 11.3) shows that the fish species richness is significantly lower for the same catchment area in European than in African rivers.

In fact, Wallace (1878) had already discussed this point more than a century ago. He wrote that the reasons for the peculiarities of tropical fauna were in the uniformity and constancy with which environmental and geographical factors exerted their influence without change in their essential features both throughout the year and during long time periods. 'While in the temperate zone the glaciations produced their destructive effects the equatorial zone was constantly full of life and, within its limits, that interaction of organisms did not cease, which is apparently the main factor in the appearance of the infinite variations and attempts to fill in all vacant places in nature'. In other words, Wallace assumed that in the temperate and cold zones, the struggle against the irregular and constantly recurring severity of the climate must have necessarily limited the success of speciation, whereas in equatorial zones the struggle for existence was less severe, and food was available in great abundance and constancy. Moreover, in equatorial zones, evolution and selection were probably adjusted to slow changes in physical conditions, favouring the emergence of countless number of organisms. In fact, Wallace expressed in a general form the various hypotheses that were used to explain the latitudinal gradients in diversity during the following century. As an example, MacArthur (1969) noticed that 'plausibly a greater historical accumulation of species in an area where production is greater and seasonality is less, should, coupled with effect of competition and predation, produce all of the characteristics of tropical communities'.

Since Wallace's time, different hypotheses have been proposed to explain the increasing diversity of species from the Pole to the Equator that has been observed in most of the higher taxonomic categories (Schoener, 1986). The predation hypothesis assumes that greater predation intensity in tropical zones results in an increase in the number of prey species by reducing the density of efficient competitors amongst the prey species. Similarly, the competition hypothesis assumes a greater intensity of competition in the tropics, predicting a greater number of species there. However, few, if any data have been available until now to test the above assumptions. Alternative explanations may be historical. Until the late Oligocene (30 Myr BP), the world had a tropical or sub-tropical climate, and the latitudinal temperature gradient developed further. The higher latitudes are impoverished in species because they have not yet recovered from Pleistocene glaciations. Possibly this period did not last long enough for different groups to evolve the adaptations necessary for surviving harsh climatic conditions (Futuyma, 1986).

May (1973) inferred that tropical communities contain more species because their environments are more stable than temperate ones, a hypothesis that is not supported by Lovelock (1992), who pointed out that biodiversity is greatest when rapid change, well within the limits of toleration, is taking place in the ecosystem. He argued that we should regard the greater diversity in the tropics as an indication that the ecosystem was healthy but has recently been perturbed.

Mina (1991) noticed that the tropical fish fauna is phenetically more diverse than that of the European rivers. What impresses an ichthyologist is the abundance of fishes in the tropics that differ sharply in their external characters, particularly coloration, in a way which is not observed in Holarctic fish. Also, there could be a considerable indirect effect of the temperature regime of northern waterbodies on the degree of morphological diversity of the ichthyofauna, especially as it is apparently more difficult for small fishes than for large fishes to adapt to life at low temperature in winter (Mina, 1991). As a result, the proportion of small fish (less than 5 cm length) would clearly decrease from south to north, and this may contribute to the fewer species and the lesser morphological diversity of fishes in the Holarctic. Actually, the number of small fish is very high in tropical Africa (see Fig. 11.10).

Relation between species richness and area or discharge in African river basins

The relationship of species number to area containing those species is a well-known empirical observation (Williamson, 1988) and a power function is widely used to describe this pattern mathematically: $S = cA^z$, where S is the number of species, A the area, Z the slope of the regression line and c a constant. It can also be expressed

as $\log S = c + Z \log A$. Some authors believe that the species–area relationship is no more than a statistical artefact, while others consider it a deterministic causal structure. The latter refer for instance to the theory of island biogeography put forward by MacArthur & Wilson (1967) who, according to Williamson (1988), 'have rather little to say about the shape of the species–area relationship'.

Whatever the reason for it, the relationship between catchment area and species richness (see Fig. 11.3) is a good illustration of a filter that may occur on a regional and catchment basis, as long as there is a reduction in the regional species pool in smaller areas. Since rivers are separated from each other by barriers impassable to strictly aquatic animals and can be considered as biogeographic islands, it is possible to analyse their species–area relationships. For 13 rivers, mostly from Côte d'Ivoire, the species richness of their fish communities increased in proportion to a power function of the catchment area (Daget & Iltis, 1965). Welcomme (1979) also described species–area relationships for 25 African rivers using a similar power function model, but Livingstone et al. (1982) demonstrated for 26 African rivers that a better prediction of species richness was obtained using discharge at the mouth rather than length or catchment area.

Using an updated set of data for 39 rivers, mostly from West Africa (Table 11.1), Hugueny (1989) was able to show that species richness was positively related to discharge (Fig. 11.2) and catchment area (Fig. 11.3) and in the latter case a power function with an exponent of 0.32 gave the best fit. Nevertheless, a multiple regression analysis using data collected on 26 rivers indicated that a model explaining species richness, as a function of both mean annual discharge at the mouth and catchment surface area, was an even better predictor: ln(species richness) = 0.231 ln(discharge) + 0.123 ln(area) + 1.695.

A subsample of 11 river tributaries in connection with the main stream shows a higher species richness than that predicted by the model from surface and discharge for isolated basins (Hugueny, 1989). Such a result could be explained by the existence in large river systems, where the tributaries have been investigated, of a richer fauna that can serve as a source of inoculum for recolonisation of the tributaries after disturbed periods. In river basins of similar area but isolated and without a connec-

Table 11.1. *Species richness, area and mean annual discharge for different African river basins*

Rivers	Species richness	Surface (km²)	Mean annual discharge (m³ s⁻¹)
Zaïre	690	3 457 000	40 487
Nile	127	3 349 000	2640
Niger	211	1 125 000	6100
Senegal	110	441 000	687
Volta	137	398 371	1260
Ogowe	184	205 000	4758
Sanaga	124	135 000	2060
Bandama	91	97 000	392
Comoe	88	78 000	206
Gambia	84	77 000	170
Sassandra	74	75 000	513
Niari	87	56 000	913
Oueme	91	50 000	220
Cross	111	48 000	
Ntem	94	31 000	348
Cavally	59	28 850	384
Nyong	77	27 800	443
Tomine	62	23 200	
Ogun	68	22 370	
Mono	50	22 000	104
Sewa	56	19 050	
Moa	44	18 760	
St Paul	61	18 180	
Konkoure	83	16 470	353
Loffa	37	13 190	
Nipoue	63	11 920	
Wouri	51	11 500	308
Bia	44	9730	81
Agnebi	56	8520	50
Mano	38	8260	
Jong	67	7750	
Kolente	63	7540	
Boubo	38	4690	32
Mungo	32	4570	
Me	52	3920	32
San Pedro	22	3310	31
Lobé	28	2305	102
Nero	21	985	16
Dodo	18	850	

From Hugueny, 1989.

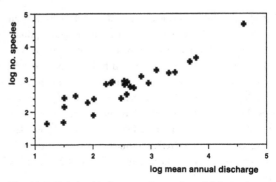

Fig. 11.2. Relationship between the number of fish species recorded in one river catchment and the corresponding mean annual discharge at the river mouth (expressed in m³ s⁻¹)

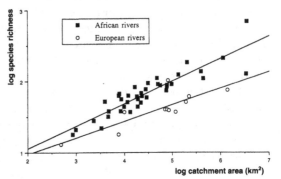

Fig. 11.3. Relationships between the number of fish species and the catchment area (in km²) (from Hugueny, 1990*b*). A comparison is made between African and European rivers.

tion to any other river system, there is no source of reinvasion when a species disappears.

Although the species–area relationship can be demonstrated by data, its explanation is not easy. The lower species richness in small isolated basins could result from a higher extinction rate. The hypothesis of area-dependent extinction rates is one element of the so-called dynamic equilibrium theory (MacArthur & Wilson, 1963, 1967). However, it is also assumed (Williams, 1964) that with larger surface, the probability increases that a certain type of habitat will be represented and that the associated fauna will be present too. Such an explanation means that surface area would only play a role via habitat diversity. This idea was supported by Williamson (1988) in a tentative explanation of species–area curves based on environmental heterogeneity. The effect of discharge is less clear. Livingstone *et al.* (1982) claimed that discharge and primary productivity (of the drainage area) are closely related, but such a relationship was not tested by Hugueny (1989) in the absence of reliable data on primary production. One of the reasons for the discrepancy between these results is that many figures for fish species richness used by Livingstone *et al.* (1982) for different river systems were much lower than those published later, after more detailed investigations of the fish fauna (Teugels *et al.*, 1988; Lévêque *et al.*, 1989, 1991, 1992). The interpretation of the results has, therefore, to be used with caution, because there was a large bias in the basic data. Possible explanations for the role of discharge are the volume of water available for the fishes and the extension of the inundation zone.

The River Nile, as already pointed out by Greenwood (1976a), has considerably fewer fish species than the Zaïre (127 against 690 for a comparable catchment area), even if such a difference is reduced when related to the differences in discharge (2640 m³ s⁻¹ for the Nile against 40 487 m³ s⁻¹ for Zaïre). In this particular case, and as suggested in the previous chapter, an additional historical explanation is rather more convincing: the Nile discharge was greatly affected by dramatic climatic changes during the Quaternary, and during dry periods a number of fish species, unable to survive extreme conditions, could have disappeared. Recovery of the presumed original richness needs either time for speciation, or opportunities for immigration from other river basins, which implies physical connections. Probably the time has been too short and colonisation has not occurred since the last severe drought. This explanation of the relatively poor fish species richness of the Nile could also apply to other major Sahelian rivers under similar climatic conditions, such as the Chari, the Niger or the Senegal. They have, in general, a lower number of fish species than would be expected from their size, when compared to coastal rivers from Upper or Lower Guinea.

Species richness along a longitudinal gradient in rivers

The identification of environmental factors influencing species distribution is of major importance in understanding the changes in species richness of fish assemblages along a longitudinal gradient. One of the major problems is to identify pertinent environmental variables. For stream ecosystems, the depth, current velocity, temperature and the nature of substrate have been

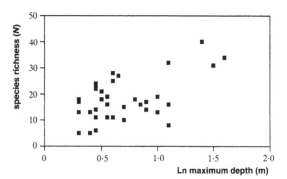

Fig. 11.4. Relationship between species richness and the maximum depth in the Niandan River (from Hugueny, 1990b).

the most widely used to characterise microhabitats (Gorman & Karr, 1978; Kinzie, 1988).

Angermeier & Schlosser (1989) pointed out that species richness in temperate and tropical rivers was better explained by the volume sampled than by the diversity of the habitat. In a study of the Niandan, an upper tributary of the River Niger in Guinea, Hugueny (1990b) also established a positive relationship between species richness and the habitat volume. One possible explanation is that the probability of occurrence of each species increases with increasing size of the sample, and this positive relationship between species richness and volume of the sampled habitat is known as the passive sampling hypothesis (Connor & McCoy, 1980). But an effect of habitat diversity was also observed. There was a positive relationship between richness and maximum depth, a characteristic related to river size (Fig. 11.4). River size, increasing along an upstream–downstream gradient, is equivalent to the stream-order approach, and similar results have been reported in temperate rivers (Beecher *et al.*, 1988). Similarly, in small tropical Panamanian streams, Angermeier & Karr (1983) reported a positive relationship between fish richness and width. An increase of fish richness along a longitudinal gradient has also been reported in north tropical Africa from the Mono (Paugy & Bénech, 1989) and the Ogun Rivers (Sydenham, 1977). In the Mono, fish species richness increases rapidly with increasing distance from the source, and then becomes asymptotic. It is therefore important to carefully investigate the headwaters in order to show an upstream–downstream gradient. This was not the case for the Bandama, where no clear gradient was observed (Mérona, 1981) because the headwaters were all temporary waters.

There were different attempts to explain the above observations. Gorman & Karr (1978) explained the increasing richness with river size by increasing habitat diversity. Schlosser (1982) pointed out that deep 'pools' are absent or rare in the upper streams, and this explains the absence of the fish species inhabiting these deep and quiet habitats. Mahon & Portt (1985) considered that each species, according to its size and behaviour, needs a minimal water depth to survive. As a result, the shallow waters could only be colonised by small species or juveniles of larger ones (Angermeier & Karr, 1983). This hypothesis was supported by a positive correlation between the length of the largest individual collected and maximal depth of the sampling station in the Niandan (Hugueny, 1990*b*). Another hypothesis is that the greater hydrological variability upstream results in the extinction of populations and thus in a lower species richness (Horwitz, 1978).

The current speed diversity, which is another way of estimating habitat diversity, also explains the local richness in the Niandan. Such results point out the major role of hydraulics in habitat selection by species (see Chapters 12 and 13). A significant correlation between both diversity and species richness and diversity of velocities has been demonstrated in the Sabie River, a tributary of the Limpopo River in the Kruger Park, Southern Africa (Gore *et al.*, 1992). Richness increases rapidly when velocity diversity is above a value of 2.75.

Biological determinants of local and regional species richness

As emphasised above, local or regional species richness is the result of an evolutionary process under changing environmental conditions. It could be viewed as a balance between speciation and colonisation processes on the one hand, and extinction or selection processes through selective filters on the other. All these processes are scale-dependent. Speciation contributes to species diversity over long time periods in comparison with immigration. Moreover, current knowledge on the modes of speciation suggests the importance of local processes,

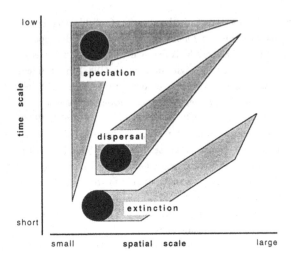

Fig. 11.5. Variation in the relative importance of extinction, speciation and dispersal on species diversity over a range of spatial and temporal scales.

resulting in the high endemicity of freshwater fish. The immigration of a species into small areas where it was absent, or into adjacent catchments, inevitably occurs much more frequently than the immigration of a species into a large area. The step-by-step colonisation of river catchments from a centre of speciation is likely to be the most common means of dispersal. Extinction, which is the factor reducing species diversity, can result from different processes operating at different scales (geological events, climate changes, biotic competition, human activities). However, due to the high endemicity of freshwater fish, and the island nature of watersheds, it is likely that local extinctions may occur more frequently than large-scale ones. In fact, the patchy distribution of fish species is a regulatory mechanism, which could explain why species disappearing from one watershed may hold on in another one. Figure 11.5 tentatively summarises the different biotic processes involved in the regulation of species diversity of freshwater fish.

Species richness and fish phylogeny

An important issue in relation to phylogeny is that speciation does not occur at equal speed and intensity in every fish family. There is, for example, empirical evidence that some families are more species-rich than others and that evolutionary explanations could be evoked for these differences. It is tempting to postulate that species-rich families have been more successful than groups with low numbers of species (Brooks & McLennan, 1991), but there are methodological difficulties in explaining the observed differences. For instance, what temporal scale should be selected?

Groups with unusually low diversity could generally be considered as evolutionary relicts (Simpson, 1944). Geographical relicts were once widespread and are now restricted in distribution. Phylogenetic relicts are something like living fossils, which have existed for a long time without speciating very much. By contrast, numerical relicts are surviving members of species-rich groups that have been depleted by extinction and are now represented by only a few species. Establishing relict status is not easy and needs fossil records, which are not always available. It can be suggested, for example, that several contemporaneous African fish species, belonging to so-called old families recorded in fossil remains and represented only by one or two species, are phylogenetic relicts, e.g. species of Malapteruridae, Protopteridae, Gymnarchidae, Denticipitidae. The species of Osteoglossidae on the other hand could be classified as numerical relicts, given the existence of several fossils that have been attributed to this family (see Chapter 2).

Conversely, unusually species-rich groups have experienced higher speciation rates (and possibly also less extinction) than sister-groups. This could result from abiotic factors, such as geological or geomorphological changes associated with vicariant speciation. Cold climates in temperate habitats could also be responsible for higher extinction rates than in tropical areas (Cracraft, 1985). But intrinsic attributes of particular groups of organisms, which allow them to invade and exploit new habitats, could also result in high speciation rates. The most obvious illustration of such speciation is given by the extraordinary haplochromine speciation in the African Great Lakes, where flocks of a hundred species have evolved in a fairly short time. Species flocks are not limited to cichlids, and in Lake Tanganyika flocks of mastacembelids, clariids or bagrids are also recorded (Coulter, 1991a; de Vos & Snoeks, 1994). For these families, it is not clear why speciation occurred to such an extent in lakes, bearing in mind that it did not happen to

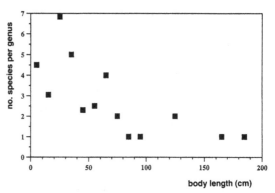

Fig. 11.6. Number of species in a genus in relation to body size for West African freshwater fishes (Hugueny, personal communication).

ancestor sister-groups remaining in rivers. Similar flocks of Cyprinodonts (*Orestias*) are known from Lake Titicaca in South America (Lauzanne, 1982), and flocks of cyprinids are known from Lanao in Philippines.

Speciation rate is expected to be higher for small, short-lived species than for large species with longer generation times. Assuming that generation time increases with body size, the number of species should be higher for small-bodied genera than for large-size ones. That is corroborated in West Africa where the mean number of species in a genus is actually much higher for small body sizes (Fig. 11.6).

The question as to why tilapias have not undergone the same explosive speciation as haplochromines has not received a clear answer. According to Fryer & Iles (1969) their 'generalised' state allowed *Tilapia* to adapt to new conditions without diversification. On the contrary, Trewavas (1983) pointed out that tilapias were probably too specialised trophically to do so.

The question has also been raised as to why the radiation of characoids generated an incomparably greater variety of forms in South America than in Africa and, on the contrary, why the radiation of cichlids led to their greater diversity in Africa than in South America, irrespective of the fact that both groups are present in both continents. According to Roberts (1975) the radiation of characoids in Africa was hindered by the fact that before their radiation took place the main food resources in this region had already been captured by representatives of other taxa, particularly the mormyrids and cyprinoids. Such an interpretation is based on the assumption that the result of radiation (diversity of forms) is mainly determined by the conditions in which the radiation proceeds and not by the properties of the radiating group.

It has been postulated, that adaptive radiation could be an explanation for differences in species richness among groups caused by unusually high speciation rates in the more speciose groups. The consensus view of adaptive radiation could be summarised by Futuyma's (1986) statement: 'a lineage may enter an adaptive zone and proliferate either because it was pre-adapted for niches that became available, or because it evolves "key innovations" enabling it to use resources from which it was previously banned'. Studies of adaptive radiations should, therefore, focus on assessing the degree of diversification in behavioural traits and reproductive strategies. But one can also consider that adaptive changes in the ancestry of a group could lead to unusually high speciation rates and survival in descendant species. Such 'key innovations', which are any novel features proposed to be correlated with the adaptive radiation of a clade, give advantage to the descendant lineage compared to its sister-group.

A high degree of faunal endemism usually suggests a long isolation period. It is also admitted that the more time there is for speciation to occur in an isolated system, the richer will be the indigenous fauna. But this statement does not apply equally among fish families, some families having obviously higher speciation rates than others. Lakes are certainly suitable sites for the evolution of species flocks, because they are often more stable than streams and rivers in terms of abiotic environmental factors. Moreover, deep lakes often occupy ancient geological sites and their catchments are often isolated, preventing faunal exchanges. Spectacular examples of cladogenesis are seen in the evolution of species flocks, where numerous closely-related species have evolved from a common ancestor in a restricted geographic area (see above).

There are also several examples of speciation and endemism in river basins. The African genus *Synodontis* for instance, which is distributed throughout tropical river basins and lakes in Africa, may offer a good biological model in which many species are strictly endemics (see Poll, 1971; Paugy, 1986). Out of some 110 species,

40 are known only from the Zaïre system, 7 are endemic to the Niger basin, 6 to Lake Tanganyika, 5 to the Nile, 3 to the Ogoue, and some 15–20 species are endemic to other river systems. Among African cyprinids, the polyphyletic group *Barbus* is represented by 300 species, many of them being geographically restricted to particular river basins (Lévêque & Daget, 1984; Lévêque et al., 1990)

Extinction

Extinction is part of the evolutionary process, an ultimate population catastrophe serving the primary role of making room for new species. It could be considered as an inevitable and irreversible end of species existence (Gould, 1983). For Darwinists, extinction is the fate of species that lose in the struggle for survival and a way of eliminating obsolete species (Marshall, 1988). There is much evidence that unpredictable environmental catastrophes occurred repeatedly in the past, leading to extinctions of living forms that had not had enough time or ability, or luck, to escape extinction. At present, there is a great deal of evidence that man is directly or indirectly responsible for most vertebrate extinctions in recent times.

Causes and processes of extinction

The process of extinction is central to both population and community ecology. When placed in unfavourable abiotic situations, organisms face the options of moving to other geographic locations or of undergoing biological changes, permitting them to adapt to new ecological conditions prevailing in the original habitat. If these alternatives fail, the organisms will become extinct.

Causes of extinction are generally assumed to be biotic or abiotic aspects of the environment of the organisms under consideration. Marshall (1988) tried to classify the proposed trends and scenarios, keeping in mind that in many cases extinctions do not result from a single process but from an array of interrelated processes, which should be considered at different time scales.

1. **Pseudo-extinction** is an evolutionary change, when a taxon evolves into one or more new species.
2. **Climate-related** extinctions are associated with environmental changes and have been considered as one of the major causes of extinction. Climatic

changes could dramatically affect habitats causing reduction or local extinctions. In north tropical Africa, past climatic changes have certainly caused extinction of fish species either from individual river basins, or from the African continent, when they were restricted to areas devoid of potential refuge zones (see below). Catastrophic extinction events, such as complete desiccation, were probably largely responsible for the present relatively low diversity and low endemism in Lake Turkana compared with other East African lakes (Coulter, 1994*b*). Similarly, in Lake Victoria, the recession about 14 000 years ago probably forced numerous cichlids species to extinction.

3. **'Prey naïvety'** means that any introduction or immigration of a predator into a fauna may result in extinction of prey species if they have not had enough time to adapt to this new situation. The disappearance of many haplochromines from Lake Victoria following the introduction of *Lates niloticus* is a dramatic example of such a process.

4. **Competition** between species to share space or resources may follow speciation, or immigration of a taxon with similar requirements. Specialisation presumably hastens a population's extinction by narrowing its range of acceptable food or habitats, but there is little evidence for this inference.

5. **Coextinctions, trophic cascades** and the **domino effect** could occur when the disappearance of a key taxon from the base of a specialised food chain causes extinction of 'dependant' taxa higher in the food chain.

6. **Species–area** effects resulting from decrease of living space and environmental heterogeneity and resources may also cause extinctions.

7. **Volcanic eruptions** and other **geological** events may eliminate local biotas and their associated endemic species. The paucity of cichlids in Lake Kivu is apparently the result of the destruction of the native fauna following volcanic events (Thys van den Audenaerde *et al.*, 1982).

In the long term, Marshall (1988) also identified other causes, the best documented being the **impact theory** which considers that one or more mass extinctions resulted from the impact of a large meteorite. Alvarez *et al.* (1984) provided evidence of such an event at the Cretaceous–Tertiary boundary that may have produced conditions similar to a 'nuclear winter'. This scenario is not widely accepted.

A number of ecological features that make species susceptible to extinction are as follows (Marshall, 1988): (i) large body size; (ii) position in the upper trophic levels within communities (i.e. carnivores); (iii) diet or habitat specialists; (iv) poor dispersal abilities; and (v) restricted geographic ranges (i.e. endemics). Pimm (1991) claimed that small populations are prone to extinction, but when environmental disturbances are the principal cause, large population size may offer no protection against extinction. Moreover, slowly growing populations should be more susceptible to extinction, because they recover less quickly after disturbance and could be badly affected if the disturbance frequency is high compared with the rate of recovery.

On the contrary, extinction-resistant taxa exhibit: (i) small body size; (ii) position in lower trophic levels; (iii) dietary and habitat generalists; (iv) good dispersal capabilities and ability to recolonize; (v) broad geographic ranges; (vi) short longevity of individuals; (vii) large population size and density; and (viii) low metabolic rate (Diamond, 1984).

Effect of environmental harshness: the 'last three'
All organisms have a defined physiological capacity and are only able to live or survive in a specific environmental regime, within limited ranges of abiotic gradients. These chemical and physical attributes of the environment are known to affect physiological functions, and therefore demographic parameters, such as fecundity and survivorship. When environmental characteristics shift away from the so-called optimal conditions for the species, populations decline. When abiotic factors are close to or beyond the tolerance limits for life, species disappear. Harshness means that abiotic factors such as temperature, oxygen concentration or salinity are close to the limits at which life is possible. After local extinction, fish can only recolonise aquatic systems when reintroduced or when physical connections exist between systems.

An example of the influence of environmental harshness on species richness can be demonstrated by changes observed in the estuarine fish community of the Casamance River (Senegal). Following a period of drought, the

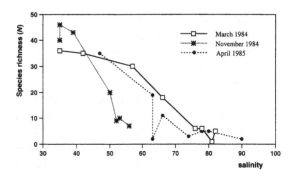

Fig. 11.7. Changes in fish species richness along an upstream salinity gradient in the Casamance River (Senegal), at three different seasons (from Albaret, 1987).

salinity of the Casamance River in 1984–5 was higher than in the open ocean for the first time since the beginning of the century. Salinity followed an increasing upstream gradient at the end of the rainy season, and values above 90‰ were observed in the upper reaches (Albaret, 1987). Along this gradient, the total fish species richness, estimated from commercial and experimental catches, decreased dramatically and only the two brackish water cichlids, *Sarotherodon melanotheron* and *Tilapia guineensis*, were recorded in the most extreme environmental conditions (Fig. 11.7).

Another drastic impact of this new ecological situation was the assumed disappearance from the Casamance River of many freshwater species that were previously recorded. This was the case for *Hepsetus odoe*, a few cyprinids including *Labeo* and *Barbus*, mormyrids, characids (*Alestes* and *Brycinus*), citharinids, *Malapterurus electricus.*, etc, which were all recorded by Pellegrin (1904) and still observed by fishermen in the river before the drought (Albaret, 1987). It is unlikely that those species were able to survive in refuge zones, given that most of the upper reaches dried out during the dry season. Only a few species, such as *Hemichromis bimaculatus*, some cyprinodonts and *Clarias*, were sampled in the upper reaches. The low fish species richness observed in the Nile River today (see Chapter 2) may also be attributed to environmental harshness, as a result of a drastic reduction in flow at some period during the Quaternary.

In many African shallow waterbodies, the triumvirate, '*Tilapia–Clarias–Barbus*', is the ultimate evolutionary stage in fish assemblages when freshwater systems are

invaded by vegetation, a characteristic of marshy conditions. They are the last species to survive until just before the drying up. During the drydown, the numbers of species and individuals progressively diminish, but if the system does not dry up completely, and when the other species have all been eliminated, the 'last three' are the only ones to survive these adverse conditions. This situation has been observed many times. In Lake Ngami (Jackson, 1989), *Barbus paludinosus*, *Clarias gariepinus* and *Oreochromis andersonii* were the last species recorded before drying. *Barbus paludinosus*, *C. gariepinus* and *Oreochromis shiranus chilwae* are also the only species recorded in the swampy waters of Lake Chilwa (Furse *et al.*, 1979). In Lake Mweru Wa'Ntipa, the fish community consists of over 12 species at high water, whereas during the dry phase only *Clarias gariepinus* and *Oreochromis macrochir* are recorded. Apparently, no *Barbus* have been observed in that lake (Bowmaker *et al.*, 1978; Jackson, 1989).

In the lakes mentioned above, the species temporarily eliminated were able to recolonise the system when hydrological conditions improved from populations that survived in an adjacent perennial stream or pools. The situation is rather different for isolated lakes with limited tributary systems where only the 'last three' occur today. In these cases, it is most certainly the result of historical dry climatic events during which the lake almost dried out. All fish species were eliminated with the exception of species best adapted to the harsh abiotic environment. Good examples are Lake Baringo, where *Clarias mossambicus*, *Oreochromis niloticus*, *Barbus intermedius australis* and *Labeo cylindricus* are present (Worthington & Ricardo, 1936), and the Ethiopian Rift Valley lakes (Ziway, Awassa, Langano, Abyata, Chala), which are also inhabited only by *C. mossambicus*, *O. niloticus* and some *Barbus* (Lévêque & Quensière, 1988). In contrast, Lakes Abaya and Shamo, situated southwards and at a lower altitude, are inhabited by a more diverse Nilotic fauna, including the predatory species *Lates niloticus*, *Bagrus docmac*, *Hydrocynus forskalii* and *Clarias gariepinus* (Riedel, 1962). *Clarias* possesses an accessory breathing organ and is thus well adapted to deoxygenated environments. The species of the other two genera must be flexible in their life-history styles and be able to succeed in an unstable, uncrowded environment, from which more specialised species have disappeared. Such phenotypic plasticity is well known in 'tilapia' (Noakes & Balon, 1982), but apparently not reported for cyprinids.

The effect of episodic environmental harshness on fish species richness, as presented here, may be compared to the effect of a disturbance of low frequency but high intensity. In extreme situations, and in the absence of recolonisation, species may disappear. It is usually difficult to provide evidence of such an extinction process, but the example of the Casamance River may be a contemporary demonstration that it has probably happened in historical or geological times.

The role of diseases

Until now, the effects of diseases on fish assemblage structure have probably been underestimated. However, at the moment, very few data support the hypothesis that diseases (including parasites, viruses, bacteria and fungi) may affect fish species richness in the tropics. There are documented examples from temperate countries of introduced parasites that have dramatically affected the native fish populations. In the absence of field research on the dynamics of parasites in wild African fish communities, their role is unknown, but should not be ignored. Long-term quantitative studies on the infestation cycles of fish populations in relation to environmental parameters are also lacking. Moreover, there is very little information about the life histories of most fish pathogens. From the experience in temperate regions, there should be an impact of diseases on fish populations that possibly lowers their abundance temporarily or, in extreme cases, may result in the disappearance of species with small populations. The impact of parasites can be subtle, as long as there is not an obvious adaptative advantage for the parasite to kill the host population. At least, sublethal consequences of pathogens for their fish hosts would be expected, such as effects on reproduction, behaviour, energy allocation, etc. One of the most dramatic outcomes is that parasitism may modify fish behaviour and increase risks of predation.

Guégan *et al.* (1992) pointed out that the maximal size of the host species accounted for most of the variation in the observed number of monogenean parasite species encountered in 19 West African cyprinid species (Fig. 11.8). The ecology of the host species was the second factor. Cyprinid species living in restrictive environments (turbulent waters, rapids) or near the sur-

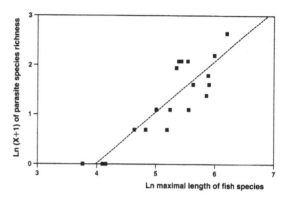

Fig. 11.8. Relationship between parasite species richness and maximal length (mm) of host species in 19 West African cyprinid species (from Guégan *et al.*, 1992).

face (inaccessible to swimming parasite larvae) carried fewer parasites than species living in the middle of the water column. This result may suggest that small species are comparatively less parasited than larger ones and, therefore, less exposed to capture pathogens.

Competition or mutualism?

There is a wide variety of intra- and interspecific relationships among organisms in biological communities, and communities exist as whole entities composed of interrelating elements. In most cases, the analysis of community structure and organisation has concentrated almost exclusively on the negative relations between species through predation and interspecific competition, which has been one of the greatest concerns in community ecology since Elton (1927) proposed the concepts of the food chain and the ecological niche.

Competition was defined as 'the demand of more than one organism for the same resource of the environment in excess of the immediate supply' (Darwin, 1859). This definition suggests that for competition to occur, two organisms must share the same resource which must be in short supply. The principle of competitive exclusion has been used to explain the disappearance of species. It assumes that members of two taxa cannot share the same niche and that competition brings about an ecological separation of closely related species or results in the extinction of one competitor: the dominant taxon exterminated the non-dominant one (Løvtrup, 1986). The latter is, nevertheless, able to survive in isolation, avoiding contact with possible dominant competitors, either in developing specialisations or protections, or in spatial segregation in a patchy environment.

One of the common approaches used to investigate competition has been to describe patterns of resource use among species coexisting in a community. Certain patterns of niche segregation or resource partitioning are consistent with the competition hypothesis, but species may differ in resource use for several reasons other than competition, including historical and phylogenetic constraints. Thus, resource partitioning provides only circumstantial evidence that competition is a major factor organising interaction in an assemblage (Werner, 1986). Actually, it is almost always possible to detect differences in habitat use, diet, or time of activity among coexisting species, if one looks closely enough. Many recent studies on African Great Lake cichlids support this idea (see Chapters 4, 8 and 9). In the case of fish populations, ontogenic, size-related changes in diet and habitat make the interpretation of field data even more complex.

Several ecologists who were not entirely satisfied with the competitive exclusion theory, posed the question: are competitive interrelationships sufficient to explain the high diversity of ecological communities ? In fact, direct mutualism and commensalistic relationships have been more or less neglected in the discussion of community organisation (Kawanabe *et al.*, 1993), because such positive relationships were regarded as 'mere fascinating topics' with less importance in population dynamics. However, during the past two decades, many examples of direct mutualism and commensalism between cohabiting species in Lake Tanganyika have been reported, even among species sharing similar resources in fish communities: increasing food quality by using feeding sites used by other species of algae eaters (Takamura, 1983); increasing feeding efficiency by following and coforaging among small-fish eaters, scale eaters and benthos eaters (Hori, 1983, 1987; Yuma, 1993); restriction of brood predators foraging (Nakai, 1993); and codefending the territorial border in algal feeders (Kohda, 1991). These positive relationships should promote the existence of many species with diverse ecological niches (Kawanabe, 1986, 1987), and enable a greater density of species and/or individuals (Yuma, 1993).

However, the distinction between mutualism, commensalism and competition is not always unambiguous. Yuma (1993), for instance, reviewed research on competitive and cooperative interspecific interactions in fish communities in Lake Tanganyika and discussed the idea that the substratum type of the lake is, to a large extent, responsible for the variation in the interactions: a community on a sandy bottom is characterised by fewer species, fewer threats by predators and fewer cooperative relationships; while a community on a rocky substrate has more species, more threats by predators and more co-operative relationships, depending on the topographic diversity.

Ability of fish to disperse

In studying the geographic range of species, one can assume that there have been barriers to their dispersal, and that all species cannot therefore occupy the whole of the geographic area under study because they are simply not able to cross these barriers (see previous section of this chapter). Another approach is to consider that all species, at some time, had potentially the same opportunity to spread all over the area. Therefore, the present-day geographical distribution may only be a question of chance. In fact, the current geographic range could also result from differences in intrinsic biological characteristics of species and their ability to disperse, to compete or to avoid extinction. The first two approaches are not antagonistic but complementary, and one or the other is more or less pertinent according to the level of perception used.

In north tropical Africa, the fish fauna of the Nilo–Sudan province displays considerable homogeneity. It is assumed that there were frequent contacts between river catchments in the past, and that the whole of this region was potentially colonisable by most of the species presently recorded in the province. Therefore, the question was to investigate the size of the geographic range (expressed as the number of river catchments in which the species is present) of the different species in order to test if the species all have the same colonisation ability, and what are the biological characteristics that may explain differences among species (Hugueny, 1990*a*).

There is a significantly positive correlation between fish body size and the geographic range of that species' distribution estimated as the number of catchments

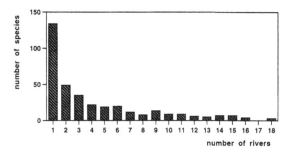

Fig. 11.9. Distribution of Nilo–Sudanian fish species according to their range size (number of rivers inhabited) (from Hugueny, 1990c).

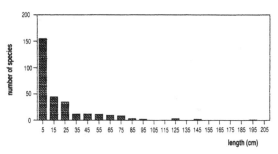

Fig. 11.10. Distribution in body length class of 295 fish species in the Nilo–Sudanian region (from Hugueny, 1990c).

where the species is present (Hugueny, 1990a). This agrees with the results obtained by McAllister *et al.* (1986) on North American fishes, in which large-sized fish tend to have a larger geographical distribution range. Given that the swimming speed of fish is related to size (Bainbridge, 1958), large fish are indeed better prepared than small ones to travel long distances, and have the ability to more rapidly cross ephemeral or hostile environments (Hugueny, 1990a). Only 2 out of 23 long-distance migrators in the Chad basin are less than 20 cm long (Bénech & Quensière, 1989). In fish of a similar length, micropredators have the largest geographic distributions, followed by top predators, omnivores and finally herbivore–detritivores.

The distribution of range sizes observed in Nilo–Sudan fishes is well represented by a negative power function. There is a clear predominance of species with a small geographic range (Fig. 11.9). This 'hollow curve' type of frequency distribution is similar to distributions observed for other vertebrates. It is possible that the hollow curve type distribution may result from another macroscopic law which states that small species are more numerous than large ones (Van Valen, 1972) and in fact, the length distribution of 295 Nilo–Sudan fish species (Fig. 11.10) shows a very similar trend to that of the curve representing range sizes (Hugueny, 1990b).

The possession of an accessory respiratory device does not affect range size. Therefore, this result does not support the hypothesis that fishes possessing accessory respiratory organs may be more successful dispersers, because they should be able to migrate through poorly oxygenated flooded and marshy areas that form exchange

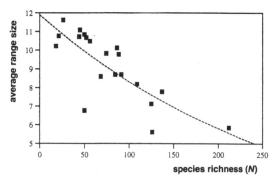

Fig. 11.11. Average range-size of resident species in relation to species richness in 20 Nilo–Sudanian rivers (from Hugueny, 1990a).

paths between catchments (Lowe-McConnell, 1988). The idea that accessory respiration accounts for the large geographic range of *Clarias*, including dispersion between Africa and South-East Asia, remains to be validated quantitatively. Similarly, data gathered in West Africa do not confirm that secondary division fish species have an advantage for greater dispersal through saline waters. Nevertheless, *Brycinus longipinnis*, which belongs to the primary division family Characidae, is tolerant of seawater and probably colonised rivers on the West African coast from the sea (Paugy, 1986).

One of the main results of the above study (Hugueny, 1990a) was that there is an inverse relationship between the species richness of a river and the average range size of the species inhabiting the river. The fewer species in a river, the more it contains species distributed among a large number of rivers (Fig. 11.11). This too has been

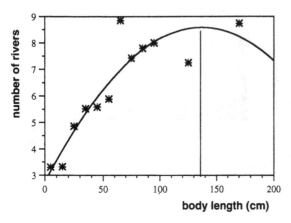

Fig. 11.12. Mean number of rivers inhabited as a function of body size, in West African river basins (Hugueny, personal communication).

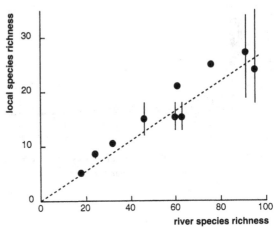

Fig. 11.13. Relationship between average local species richness by river and number of fish species present in the catchment area (from Hugueny & Paugy, 1995).

observed for different vertebrate groups. The data as a whole lead to the conclusion that the effect of extinction-related factors is negligible in comparison to that of immigration-related factors in explaining the distribution of ranges sizes in Nilo–Sudanian fish.

The consequences of the above observations are that, within a homogeneous biogeographic area, a model combining the different relationships should be a dome-shaped function of body size (Hugueny, personal communication). The number of rivers inhabited by a small fish species should be low as a result of low dispersal and high speciation rate. For large species, the high probability of extinction also reduces the number of inhabited rivers, mainly the small ones where populations are small. Consequently, species of medium body size should occur in a higher number of rivers than small or large species. This hypothesis was tested (Fig. 11.12), and the distribution of freshwater fishes in West Africa is in good agreement with the suggested model, at least for the small sizes.

Saturated or unsaturated assemblages?

In saturated assemblages in which strong biotic interactions take place among species within a local habitat, it is not possible to add a species unless it replaces, and eliminates, another (Ricklefs, 1987). Conversely, in non-interactive assemblages, local biotic interactions are feeble or absent, and species richness must increase pro-

portionately with the species richness of the regional pool of potential colonists. This pattern is called 'proportional sampling' (Cornell & Lawton, 1992).

The question has been repeatedly raised: are local communities ever saturated with species? In other words, do they ever reach a point where species from the regional pool are unable to invade the local habitat because of exclusion by resident species (Cornell & Lawton, 1992). That local richness of communities is not only dependent on local conditions but is also affected by regional richness has been stressed recently, despite the few examples available. In a review of the theoretical evidence for saturation in various community models, Cornell & Lawton (1992) concluded that the principal direction of control for species richness is from regional to local. The interplay between local and regional richness is not simply that local communities are samples of the regional pool of species; by definition, the regional pool of species is the sum of the species in local assemblages. In this chicken and egg problem, it is unclear whether local richness derives mainly from regional richness or vice versa. Dependent and independent variables are therefore hard to define, but Cornell & Lawton (1992) concluded that many ecological communities should not be saturated and that the size of local assemblages should be strongly dependent on the size of regional pools. If this assumption is correct, 'then the key to community structure may lie in extrinsic biogeography rather than in intrinsic local processes, making ecology a more historical science'.

The above view has been verified in rivers from Côte d'Ivoire (Hugueny & Paugy, 1995). Data on ten river fish assemblages sampled in the river pools were analysed. As expected, the average local species richness, the species richness of the river and area of the river catchment are positively related. Path analysis of the data led to the conclusion that the area of the river catchment governs the number of species present in the catchment, which in turn governs the average local species richness of the pool assemblages. The relationship may well be represented by a linear model (Fig. 11.13).

The positive and linear relationship between local and regional species richness suggests that the assemblages are not locally saturated with species in the rivers studied, but are shaped by processes operating on a broader spatial scale, such as the river as a whole. One of these processes is probably recolonisation from neighbouring sites, which is an important phenomenon in river systems where seasonal or interannual changes in hydrology may cause the extinction of local subpopulations.

These results emphasise that factors which affect the overall richness of a river are dominant in accounting for local assemblage structure. The catchment area plays an important role here, and it is, therefore, necessary to find out what mechanisms are responsible for the species–areas relationships observed in aquatic systems.

12 Diversity of habitats, temporal change and assemblage dynamics

Streams evidently vary in their disturbance regimes and in the provision of refugia for their inhabitants. Ultimately it is environmental heterogeneity (well matched by the 'perception' of habitat by the organisms involved), at all spatio-temporal scales up to the drainage network and above, that seems to explain the resilience of lotic communities.

Hildrew & Giller, 1994

A central issue in community ecology is to determine the various factors underlying spatial and temporal variations in community structure, and to place this knowledge into a predictive framework. The search for relationships between scales of biological processes and habitat is fundamental to the understanding of fish community dynamics in freshwater habitats.

For many authors, there is a close relationship between the physical and chemical characteristics of the habitat and the organisation of fish assemblages. From a deterministic perspective, the structure of fish assemblages may therefore be predicted from habitat characteristics. It has been demonstrated that the structural diversity of habitats is expected to increase the number of potential niches and, therefore, the coexistence of a greater number of species. A major share of this habitat heterogeneity arises from the consequences of disturbances operating at various temporal and spatial scales.

In identifying the factors responsible for the community patterns observed and their temporal changes, many papers since the end of the 1970s have also focused on habitat variability as a driving factor. Most freshwater fish have to live in a frequently changing world, both temporally and spatially. We now recognise a close cause-and-effect relationship between the spatial and temporal variability of the habitat and the structure and dynamics of communities. This environmental patterning, which refers to the non-uniform spatial and temporal distribution of resources and abiotic conditions that influence species or species interactions (Addicott et al., 1987), has a profound influence on biological processes, such as resource partitioning (food and habitat), life histories, demography, regulation of predator–prey systems, or the coexistence of potentially competitive species.

From the organism's point of view a habitat can be constant, that is to say the conditions remain favourable or not indefinitely; it can be predictably seasonal with possibly an alternation of favourable and unfavourable periods, or unpredictable, or ephemeral. A habitat can be continuous in space, or patchy, or isolated (Southwood, 1977). Fish have developed adaptive strategies that allow them to cope with periodic changes in the environment, but in some cases also with unexpected events, provided the disturbance does not drastically affect their basic biological needs. One major adaptive strategy may be viewed as the way they maximise the use of space, and therefore of resources available at a particular time of the year, bearing in mind that fish exhibit ontogenetic changes in food and habitats during their life cycles. How organisms respond to environmental patterning is becoming a central focus of current ecological research (Addicott et al., 1987). The question may therefore be formulated as, how to be at the right place at the right time in order to ensure optimal use of the resources available for a short period of time and for a specific ecophase, in a particular and sometimes temporary habitat? (Lévêque, 1995*b*).

Typology of habitat scales and environmental variability

The ecosystem concept, which has been used for a long time, was non-dimensional. However, during the last decade, the emphasis of landscape ecology on the heterogeneity in nature has increased interest among ecologists in the nature of patchiness. Authors, such as Kolasa (1989) or Kotliar & Wiens (1990), have explored the possibility of direct applications of the concept of hierarchy to problems traditionally assigned to community ecology. For Kolasa (1989), any habitat is considered to be hierarchically heterogeneous, composed of subunits, which are themselves composed of even smaller subunits, and this structure may be responsible for the observed patterns of species abundances. Various biotic and abiotic factors can be considered as specific mechanisms, sorting biological components into respective levels and compartments of the hierarchical structure of the environment. Kotliar & Wiens (1990) apply a similar conceptual hierarchical model of heterogeneity to patches.

If freshwater habitats, and particularly fluvial systems or shallow lakes, are hierarchically structured, they also exhibit a high spatial and temporal variability that mainly results from their overall geomorphology, as well as from hydrological patterns and budgets (see Chapters 1 and 6). In these freshwater systems, community dynamics clearly depends on the habitat dynamics that determine species distribution and the nature of ecological processes (Lévêque & Quensière, 1988). In other words, physical characteristics, that is to say the spatial pattern of heterogeneity and its change in time, will strongly influence community structure and dynamics. This is usually known as the bottom-up concept.

Spatial and temporal scales

Obviously, community composition at one scale is often influenced by processes that operate at a different scale (see Chapter 11). Patterns in ecological communities, such as abundance, distribution or diversity of species, depend on a complex interplay between processes acting

Table 12.1. *Theoretical relationships between time scales, physical events and biological phenomena*

Years	Scale	Physical events	Biological phenomena
100 000	100 millenniums		evolution of species
10 000	10 millenniums	glaciations, climate changes, drought–pluvial	species extinction, refuge zones
1000	millennium		exchanges between watersheds
100	century	long-term hydrological changes	community changes
10	decade	anthropogenic impacts	fish lifetimes
1	year	flood cycle	reproduction cycle
0.1	'month'	lunar cycle	gonad maturation
0.01	'day'	diel cycle	feeding cycle
0.001	'hour'		physiological processes

Modified from Magnuson, 1990.

at different spatial and temporal scales. Thus, the spatial and temporal scales of community investigations determine the range of patterns and processes that may be detected and, therefore, the level of understanding and explanation that can be achieved. They also influence our perception of the nature of the community. Events may seem to be random at one scale, but to be predictable at other scales where new patterns and new phenomena may appear. In other words, while changing scales, the nature of the parameters involved as well as the nature of the patterns observed also change.

The definition and use of time scales are particularly complex. The composition and formation of species assemblages depends on biogeographical limits derived during evolutionary time scales (see also Chapter 6). Distribution limits in ecological time scales are also constrained by morphological and physiological preadaptations developed during these evolutionary time scales. Table 12.1 shows a tentative simplification of the relationships between times scales, physical events and biological phenomena. The decade is the period of time covering the lifetime of most fish species, corresponding to a period from one to less than ten years in the tropics. For changes occurring over decades, which in many instances cover several generations, a major difficulty is to get the basic information and to interpret the cause-and-effect relationships in the data. At this time scale, humans are inclined to think of the world as static, and have difficulty in sensing slow changes. This is the world of the 'invisible present' (*sensu* Magnuson *et al.*, 1983).

Another difficulty lies in time lags that may occur between cause and effect, i.e. the time required before ecological responses to a disturbance permeate natural systems to the level at which they can be recorded in terms of a significant change. Time lags may occur because biological or physical process simply take time. The time lag for many fish may be a few years, whereas for long-lived trees it may be 200 years or more. There are also examples of biological relicts persisting long after the environmental conditions changed for fish, such as *Tilapia* in the Banc d'Arguin, off the Mauritanian coast, which is a biological relict of the past existence of mangroves in that area. Time lags in ecosystems are the rule (Magnuson, 1990) rendering it more difficult to separate cause and effect and thus, they may confuse our interpretation of the natural world and the search for ecological rules.

Fish habitats in freshwater systems

River systems have been described as hierarchically organised, and incorporating a number of levels nested at successively smaller spatio-temporal scales (Frissell *et al.*, 1986). This hierarchy can be easily visualised at the extremes as ranging from the complete drainage basins to a single substratum particle.

Hierarchy theory suggests that higher levels of organisation incorporate and constrain the behaviour of lower levels (O'Neill *et al.*, 1986), and there have been various attempts to impose a hierarchical approach on the habitat units of freshwater systems. The modern view is to consider ecological systems as collections of entities and processes occupying a continuum of temporal and spatial scales. Thus, physical characteristics from microhabitat, to reach, to zoogeographical areas persist on time scales from 10^{-1} to 10^9 years (Frissell *et al.*, 1986). This hierarchi-

cal scaling implies that the larger, more stable environments impose limits on the smaller, more variable environmental units, but many ecological phenomena operate over large spatial and temporal scales and may not be completely understood at small scales of investigation. Larger spatial scales are usually associated with increases in the relative influence of variations in 'climatic' condition whereas on local scales, biotic factors, such as predation and competition, may exert a major influence.

Within the framework of the spatial and temporal scales recognised by Frissell *et al.* (1986) in small rivers, Bayley & Li (1992) identified four main ranges of activity behaviour. The diel and microhabitat scales correspond to foraging activity, social behaviour (territoriality, gregarism), or the selection of better abiotic conditions implying small-scale displacements. At the month or seasonal scales, the behavioural activity concerns a larger spatial scale, the home range, and eventually the stream or drainage scale. This eventually includes large-scale migrations to the spawning areas, but also the use of different biotopes in relation to ontogenetic changes in ecology and shifts in resource use. Finally, the larger scales deal with the evolutionary processes that led to the present-day species composition: speciation, extinction, colonisation, climatic and geomorphological events.

There are no general procedures or criteria for determining how organisms respond to, or are affected by, environmental patterning. To arrive at a general clear-cut typology of spatial and temporal behaviours is not an easy task, and greatly depends on the species involved. Without reasonable means of scaling, it is difficult to compare the behaviour of the same species in different environments or with different species in the same environment. A classification of fish assemblages needs, therefore, to identify the hierarchical structure of the environmental units defined according to the spatial and temporal scales.

Partly based on the Bayley & Li (1992) approach modified by Lévêque (1995*b*), it is possible to propose four main patterns (Fig. 12.1).

1. The **microhabitat** is the lower scale at which the fish responds to a more or less complex array of biotic and abiotic stimuli, by showing a preference for living in an area that can provide the best trade-off between the constraints. The microhabitat may

be restricted to the resting area, that is to say the area where fish find shelter against predators and abiotic environmental factors. However, until now, the microhabitat has been defined more broadly on the basis of morphometric characteristics of the river, and not in reference to the resting area.

2. In a restricted area, biological and behavioural events are rhythmically driven by the diel and possibly the lunar cycles. For some territorial fish the territory is the spatial scale of reference for this category, whereas for fish that limit their activity to a restricted area, which is not defended but shared with other fish, the scale is the so-called 'home-range', that is to say the habitat distribution often limited by physico–chemical factors. The home-range concept could be extended to species that undergo small-scale migrations between refuges and feeding habitats. In some cases the home-range may be diffuse, when fish move from site to site. However, the weak point is that the home-range may vary following changes in the environment, and it should be recognised that it is not a static definition, but that it applied for a short period of time. This 'small-scale' category may nevertheless be called the **home-range activity scale** despite the uncertainty arising from the definition of the concept of home-range.

3. Some fish complete their life cycle within the limits of the home-range defined above, while others undergo large-scale seasonal migrations within the river basin. As long as these migrations are usually spawning migrations, and that juveniles often occupy different habitats from the adults, it is possible to identify an ontogenetic niche. Schematically, the spatial limits should include the spatial range of migration, as well as the habitats used by the different developmental stages. Usually, the migration and spawning patterns occur seasonally so that the temporal scale should be the year. This is the **ontogenetic scale**.

4. Finally, the last category to consider is that of the biogeographical distribution of species. Populations may be geographically isolated within a basin or between basins, with a low probability of meeting individuals from different patches. This is typically the **metapopulation scale**.

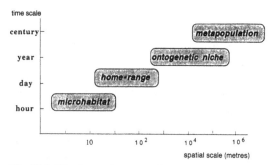

Fig. 12.1. The four main suggested patterns of spatial and temporal models of habitat use.

Effect of disturbances on species diversity and patch dynamics

To understand the dynamics of fish assemblages it is also important to consider the influence of heterogeneity in time and space. The heterogeneity in space results in the patchy distribution of habitats and species in the freshwater systems. The heterogeneity in time may be observed through the temporal patterns, as well as through disturbances that are non-cyclic events.

The recognition of a few patterns of habitat is complicated by the fact that most natural ecosystems are mosaics of subsystems exposed to different ecological conditions. The term 'patch' implies: (i) a relatively discrete spatial pattern but does not establish any constraint on patch size and discreteness; and (ii) a relationship of one patch to another in space (White & Pickett, 1985). The smallest scale at which an organism responds to patch structure by differentiating among patches is its 'grain', a term that may be compared to microhabitat. The upper limit, the 'extent', is the largest scale of heterogeneity to which an organism responds, and may be defined by the ontogenetic niche of the individual. The particular scales of grain and extent will vary with the type of organisms studied, and between the limits of grain and extent there may be several levels in the hierarchy of heterogeneity. Habitat selection by individuals involves patch choice and thus responses to patch structure at a series of hierarchical levels. The problem also remains here of how to develop methodologies in order to quantify the 'grain' and 'extent' scales.

Perturbation or disturbance may be defined as 'any

relatively discrete event in time that disrupts ecosystem, community, or population structure and changes resources, substrate availability, or the physical environment' (White & Pickett, 1985), or 'any relatively discrete event in time that removes organisms and opens up space which can be colonised by individuals of the same or different species' (Townsend, 1989). For Sousa (1984) disturbance is 'a discrete, punctuated killing, displacement, or damaging of one or more individuals (or colonies) that directly or indirectly creates an opportunity for new individuals (or colonies) to become established'. It is possible to distinguish the more or less cyclical environmental changes from exceptional and destructive events. Disturbances may be defined by several descriptors: the type (physical, biological), the regime (spatial distribution, frequency, intensity) and the regional context.

The majority of both theoretical and empirical works in ecology has been dominated for a long time by an equilibrium perspective: communities evolve naturally towards an equilibrium state (climax). An alternative hypothesis is that populations exposed to the effect of natural disturbances, e.g. periods of unfavourable abiotic conditions or overpredation, rarely reach densities characteristic of the so-called equilibrium stage, at which interspecific competition should appear as one of the major driving forces. Disturbances have demonstrable effects on community richness: they open gaps which are colonised by pioneer species. In the absence of further disturbance, communities are expected to evolve towards an equilibrium or climax state, with many of the pioneer species driven to extinction. But if another disturbance intervenes, its effect will be to knock the community back to an earlier stage of succession, and the process of recolonisation starts again. This situation may be observed, for example, in intermittent rivers that are recolonised seasonally.

Depending on the scale of the disturbance it may either increase or decrease environmental heterogeneity. Some disturbances (called in this case catastrophes, e.g. extreme floods or drying out) may be so severe as to decrease resource availability or to obliterate the system completely. Highly unpredictable environments usually have low species diversity. But several studies have concluded that disturbance may enhance species diversity by lowering the dominance of one or a few species, and thereby relax the competition sufficiently to allow other

species to develop (Denslow, 1985). The interaction of disturbance and competition seems to suggest that species richness will be greatest in communities experiencing some intermediate level of disturbance (Connell, 1978; Caswell & Cohen, 1991). The intermediate disturbance hypothesis states that if disturbance is too rare, the competition will result in some equilibrium and fugitive species will be eliminated, reducing diversity to minimal levels. If disturbance is more frequent or too intense, equilibrium species will be eliminated; only few populations of pioneer species could establish themselves after each disturbance event, and this would also lead to minimal diversity. If disturbances are of intermediate frequency and/or intensity, there will be repeated opportunities for the re-establishment of pioneer populations that would otherwise be outcompeted, and the populations of the successful competitors could withstand the disturbance without completely taking over the community. Thus a peak of diversity should be found at intermediate frequencies and intensities of disturbance.

Within a community, recolonisation patterns in every patch following a disturbance depend on interactions between disturbance regime and species biology (biological cycle, behaviour, physiology). For a single individual, perceived heterogeneity depends on the temporal and spatial scale at which the individual operates, and the physiological time scales of an organism become shorter with decreasing size. Diversity in community species composition is the result, therefore, of both the frequency and intensity of disturbances compared to the life-cycle length of species. In a hierarchical approach, it is possible to construct a continuum of habitat sensitivity to disturbances (natural or man-induced) and recovery time (Fig. 12.2) (Naiman et al., 1992). Large-scale disturbances usually directly influence smaller-scale features, while events that affect smaller-scale habitat characteristics may not affect larger-scale system characteristics.

The importance of disturbance to the structure of cyprinodont fish assemblages living in the forest creeks and pools of the Ivindo basin (Gabon) has been stressed by Brosset (1982). The eight sympatric, closely-related species living in the shallow waters are strongly dependent on food of the forest invertebrates falling from trees. In creeks and in the mainstream, assemblages include all the species. But residual pools in the inundation area are

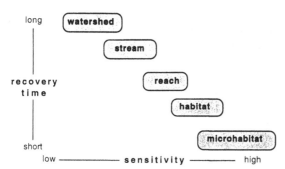

Fig. 12.2. Relation between recovery time and sensitivity to disturbance for different spatial scales associated with stream systems (from Naiman *et al.*, 1992).

inhabited by monospecific populations of cyprinodonts including adults and juveniles. In such a patchy habitat, catastrophic events such as sudden and violent swellings of the streams, or disturbance of the habitat by forest elephants, result in habitat changes at random. Community structure can change drastically in the same habitat at different times, and vary in similar habitats at a given period. In one year, the cyprinodont fish assemblages may be completely modified. But in the absence of catastrophic events, the community remains stable, without marked seasonal changes. Obviously the principle of competitive exclusion does not operate in that case which is rather close to 'the interspecific lottery for living space' suggested for coral-reef fishes by Sale (1978).

Another important concept linking patch distribution of species and habitats, and temporal changes, is the concept of 'patch dynamics', which emphasises changes in the patches such as colonisation, or dispersal, as well as flux and exchanges of individuals, matter and energy between patches. To understand interactions between patches, it is necessary to examine changes at different spatial and temporal scales. An important role of disturbance is to create and increase spatial and temporal heterogeneity (Denslow, 1985), because their effect and intensity vary in a site. As long as species are able to exploit that heterogeneity, disturbance can favour the existence of species with different environmental requirements in different patches. On the temporal scale, the most obvious role of disturbance in ecosystems is in the deflection of a community from otherwise predictable successional paths. These successions may involve sev-

eral changes in dominant species. The history of disturbance at a site and its implicit role as a reset mechanism have important implications for the rate and pattern of succession at this site. Townsend (1989) argued that this concept of patch dynamics of community ecology provides a unifying theme in stream ecology, where one can generally observe a marked spatial heterogeneity in conditions such as flow regime, substratum, and resources.

Examples of patchy environments may be found in the middle course of savannah rivers, where a series of riffles separated by large pools are usually observed during the low-water period. This is the case for the Bandama River in Côte d'Ivoire (Lévêque *et al.*, 1983) and for many other coastal rivers in West Africa. Each habitat unit has its own specific fish assemblage, but there are also differences in community structure between riffles (e.g. size of the biota, species composition) as well as between pools. Exchanges of individuals may occur, mainly during flood. Because of the multiscale organisation of metapopulations, coexistence of species at the regional level may result from mechanisms that do not apply to local populations (Caswell & Cohen, 1991); the interaction of competition and disturbance can maintain fugitive species persisting regionally, whereas they could be excluded locally whenever they are faced with successful competitors.

How to investigate fish activities?

Direct observation, using free-diving systems, is the simplest way of obtaining information on fish activities and movements. Since the main restriction on this method is the transparency of the water, it is only used in large, clear lakes where it is possible to observe the fish without disturbing them.

The different tags or marks used for tagging fish, which have been developed for temperate fishes, are still of limited use in the tropics, where one problem is the low probability of recapturing the tagged fishes. In any case, this method gives no information on the behaviour of fish between tagging and recapture.

During recent decades, biotelemetry has proved to be a very useful technique for the study of behavioural patterns of animals, particularly for birds and mammals, but its application to aquatic organisms is increasing only slowly (see Mitson, 1978; Winter, 1983). This technique relies on tagging the fish with a radiotransmitter that can be attached either externally to the dorsal musculature, or inserted into the stomach, or surgically implanted into the peritoneal cavity. Electronic tags transmit data through the water to receivers, and triangulation techniques using two or more receivers allow the monitoring of individual movements by plotting the successive positions of the fish, both horizontally and vertically in the water column. Sensors have also been developed to measure and transmit physical variables at the position of the fish (e.g. temperature, light, salinity) as well as physiological data on the fish itself (e.g. heart rate, tail beat). The present size of electronic tags still restricts their use to large fishes, but we can expect technical improvements in the future. Thus, biotelemetry is still of limited use, and very few data are available for African fish, apart from the paper from Hocutt (1989). Development of the technique is expected in the future and it will probably improve our knowledge of the use of time and space by African freshwater fishes.

Finally, echosounding devices are used to locate shoals of fish and to record patterns of movement, including vertical migrations in lakes. This technique is still mainly directed at stock assessment rather than to the study of fish behaviour, but has proved to be efficient for the latter purpose. The disadvantage is that the species of fish under observation is not known for certain.

The microhabitat

A fish exhibits a preference for living in an area that can provide a more or less complex set of environmental stimuli which may, of course, change during its lifetime. A microhabitat is the place, and associated environmental conditions, where a fish spends most of its time. It could be defined as patches within pool/riffle systems that are relatively homogeneous in substrate type, water depth, and velocity (Frissell et al., 1986).

Microhabitat conditions are, of course, closely related to the overall macrohabitat characteristics, such as temperature and water quality that define the limits of suitability for different species. These macrohabitat conditions determine the streams, or the length of a stream, that could potentially be inhabited by a species. Other macrohabitat characteristics, such as slope, elevation or water supply, create longitudinal changes in the shape, pattern and dimension of the river channel. These in turn are major determinants of the types of microhabitat

that occur at any location in the stream. Fish do not respond directly to physical macrohabitat characteristics but to the microhabitat conditions associated with the macrohabitat.

The most obvious characteristics associated with microhabitats are velocity, depth, substrate and cover. The microhabitat preference of many species based on water velocity is a well-known phenomenon, and many authors have for a long time distinguished between rheophilous species inhabiting rapids and riffles, and lentic species inhabiting pools, but until now few studies have been conducted on African fishes in order to accurately assess their microhabitat preferences. Instream Flow Incremental Methodology (IFIM), which is widely used to investigate the flow/habitat relationships of fish (Bovee, 1982), is a set of techniques that assesses the flow patterns of typical reaches and attempts to correlate these patterns with the hydraulic preferences (for velocity, depth, substrate and cover) of target species of concern. Microhabitat data are collected by selecting a species/life stage of interest and gathering sufficient information to describe patterns of habitat use accurately. Frequency distributions of the range of conditions selected by sampled fish are then calculated to determine the suitability of various microhabitat conditions.

Habitat suitability curves have been established for a few species of fish from the Sabie River (a tributary of the Limpopo) in the Kruger Park (Southern Africa) (Gore et al., 1992). Juveniles of Serranochromis meridianus and Barbus viviparus are pool dwellers with B. viviparus being slightly more tolerant of moderate velocities (up to 60 cm s^{-1}). Both species preferred shallow pools (up to 2.5 m), and their substrate preferences were also quite similar with a preference for sand and gravel having significant overhead cover. Both species were classified as pool- and edge-dwelling species. Conversely, Chiloglanis swierstrai is a typical riffle dweller found primarily in shallow water (<50 cm) with velocities between 35 to 150 cm s^{-1} (Fig. 12.3) passing over substrates ranging from small to large cobbles and boulders.

Combining preference data with hydraulics, it is possible to predict the available microhabitats at critical flow levels in a reach. The Weighted Usable Area or WUA (an index used to evaluate the microhabitat available) for a 300 m length of river, increases sharply from 2 to 8 m^3 s^{-1} discharge for C. swierstrai and S. meridianus, while for B. viviparus it remains constant above

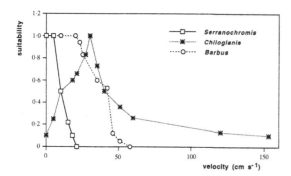

Fig. 12.3. Habitat suitability in terms of mean water current velocity for three species from the Sabie River, Kruger National Park (from Gore *et al.*, 1992).

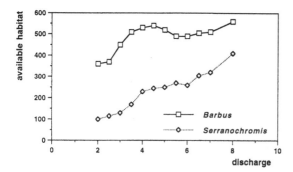

Fig. 12.4. Habitat available for *Serranochromis meridianus* and *Barbus viviparus* as a function of discharge at the Skukuza site on the Sabie River (from Gore *et al.*, 1992). Available habitat is expressed as usable area in m² for 300 m river length and discharge in m³ s⁻¹.

3.5 m³ s⁻¹ (Fig. 12.4). These results illustrate changes in the spatial pattern of microhabitat availability, as a function of temporal changes in discharge at the macrohabitat level. Such changes occur regularly in most rivers with a tropical discharge regime.

Territoriality and home-range: temporal and spatial patterns of fish activity

Some fish species restrict their activities to a limited space that can provide protection from enemies and predators, facilities for spawning and parental care, as well as food. The size of the defended area depends on the costs and benefits of holding a territory of a given size. That is a question of trade-off between the time and energy spent either on foraging and/or defence. The size of the territory may also vary according to the presence of conspecifics and the abundance and spatial distribution of food. For example, living space may be more limiting than food resources in littoral and benthic communities. In the littoral communities of the African Great Lakes, many fish are territorial and stay in one area, which generates very stable communities in which both species composition and number of individuals stay relatively constant throughout the year and from year to year in the absence of major disturbance. Territorial behaviour is probably important in limiting littoral fish populations in African lakes, and in Lake Malawi this keeps numbers below the level of competition for food.

Many fish, including territorial fish that have not yet found a territory, do not exhibit territorial behaviour. Several occupy restricted home-ranges, generally including a refuge habitat, where they spend the inactive period of the daytime, and a feeding habitat. This pattern is well known for many riverine fishes, such as mormyrids that spend most of the day in cover (rock crevices, roots, etc.) and feed at night in the open water on plankton or on bottom invertebrates. Most of these species with restricted areas of activity may also reproduce locally. In such cases, the extant home-range concept includes the habitats occupied by the species during their life cycle (including feeding and spawning grounds) to achieve their main biological functions, but in a restricted spatial range. This broad concept would apply to a large number of African freshwater fish, even if the spatial boundaries of this concept are somewhat vague.

Among the main driving factors at the home-range scale are the temporal rhythms of the environment. Fish respond to rhythmic signals received from the environment and to their changes. The dominant temporal pattern is the diel pattern of light and dark. This simple diel cycle imposes on the behaviour and activity of fish a dramatic, overriding set of predictable constraints. However, in tropical regions, freshwater organisms can also be influenced by the lunar cycle or the succession of wet and dry seasons. It is important to understand such rhythms in the context of animal behaviour, to identify how they are integrated into the life habits, and to relate the patterns to the different strategies developed by the species to cope with environmental changes. Most of these aspects have so far been largely neglected for tropical fish.

Diel activity patterns

Daily rhythms are primarily endogenous and are inherent characteristics of living systems. Changes in fish activity during a 24 hour period is a well-known phenomenon among fishermen and ecologists. These patterns can be recognised in such fundamental activities as the ability of fish to forage, to rest, to aggregate or to attract mates, all of which may vary within the daily cycle of light and dark. There are species that are mainly active during the day, or at night, or during periods of rapidly changing light intensity (dawn and dusk). It has been estimated that in temperate as well as in tropical shallow-water fish assemblages roughly one-half to two-thirds of the species are diurnal, one-quarter to one-third are nocturnal, and about one-tenth are crepuscular (Helfman, 1986).

Diel pattern of drift

The importance of diurnal light–dark changes upon locomotor activity patterns has been demonstrated for many aquatic invertebrates. Decreasing light intensity results in increasing activity and/or changes of the microhabitat, and the natural drift of invertebrates in rivers exhibits a diel periodicity (Statzner et al., 1984). This has been observed in running waters all around the world and is a reflection of the rhythmic nature of the animals' locomotor activity. The drift of invertebrates is undoubtedly important as a food source for many fish species that consume all types of drifting organisms, i.e. insect larvae, emerging or emerged insects, ovipositing insects and, to a large extent, terrestrial insects fallen into the river.

The existence of a similar diel periodicity in fish drift has been documented in Côte d'Ivoire (Elouard & Lévêque, 1977). At the end of the flooding period, juveniles of Schilbe mandibularis were absent from drift nets during daytime, but were caught in great numbers at twilight, between 18.00 and 20.00 h. The catch decreased over night and the species disappeared from samples at dawn, after 06.00 h. It was assumed that the drift activity of S. mandibularis just after dusk might result from the abundance of drifting and emerging insects serving as potential food at that time of the day.

Diel feeding patterns

In general, most fish species forage primarily either during the day or night, with a few species particularly active during crepuscular periods of dawn and dusk. This is also the case for African species (Table 12.2).

Major foraging activities include movements between resting sites and foraging areas as well as movements to follow, if necessary, daily activity patterns of prey. For example, vertical migrations of Lake Victoria haplochromines are assumed to follow the migration of the prey Chaoborus larvae which stay in the muddy bottom during the day and migrate to the water surface during the night to feed on the zooplankton (Witte, 1984b). Similarly, in Lake Kariba, adults and copepodids of Mesocyclops leuckartii exhibit a clear vertical migration pattern, moving upward to the surface of the lake at night where they remain until dawn, when they descend rapidly (Begg, 1976). Mesocyclops leuckartii feeds on the smaller forms of zooplankton, such as the rotifers near the surface. In the same lake, Bosmina longirostris also shows detectable diurnal vertical migrations to feed on phytoplankton at night. Field observations and echosounding surveys demonstrated that the clupeid, Limnothrissa miodon, that feeds principally on B. longirostris shows a pattern of diurnal vertical migration similar to that of its prey (Begg, 1976).

For species feeding continuously, there should be a striking difference in diet composition between day and night. For four zooplanktivore Haplochromis species in Lake Victoria, microcrustaceans and rotifers are the main prey during the daytime. At night, in all species except H. 'reginus', the fraction formed by planktonic Chaoborus increased considerably while microcrustaceans and rotifers decrease in the diet. In contrast H. 'reginus' switches from zooplankton during the day to mainly phytoplankton at night (Goldschmidt et al., 1990).

In Lake Sibaya, the predator, Clarias gariepinus, feeds in inshore areas at night, although its main night prey, Sarotherodon mossambicus, inhabits these areas mainly during the day. During the day, C. gariepinus ignores fish prey and feeds mainly on invertebrates, which are easier to catch (Bruton, 1979b). At night, or if hunting in packs in shallow water by day, fish prey are more easily caught. Experimental data have shown that low light intensities, shallow water and high predator densities may increase the predation efficiency of this fish species (Bruton, 1979c).

Table 12.2. *Feeding activity patterns of selected African fish species*

Species	Localities	Authors	Food	Diel feeding activity
Labeo parvus	Bandama	de Mérona, unpub. data	epilithic algae	diurnal feeder – stop feeding at night
Oreochromis niloticus	L. Turkana	Harbott, 1982	phytoplankton	diurnal feeder – marked diel feeding pattern
	L. George	Moriarty, 1973	phytoplankton	*idem*
Sarotherodon galilaeus	L. Chad	Lauzanne, 1977	phytoplankton	diurnal feeder
Alestes baremoze	L. Chad	Lauzanne, 1977	zooplankton	diurnal feeder – peak stomach fullness in the afternoon
Alestes baremoze	L. Turkana	Hopson, 1982	zooplankton	chiefly diurnal feeder
Syndontis schall	L. Turkana	Lock, 1982	zooplankton	peak stomach fullness in the dawn and morning – little feeding in the afternoon
Brachysynodontis batensoda	L. Chad	Lauzanne, 1977	zooplankton	nocturnal feeder
Bagrus bajad	L. Turkana	Lock, 1982	prawns and fish	peak stomach fullness at night but feeding occurs also during the day
Schilbe uranoscopus	L. Turkana	Lock, 1982	prawns and fish	peak stomach fullness at night – most active at sunset and during the early part of night
Lates longispinis	L. Turkana	McLeod, 1982	prawns and fish	non-feeding period between sunset and midnight – peak fullness in the afternoon
Malapterurus electricus	R. Niger	Belbenoit *et al.*, 1979		nocturnal feeder – peak hunting and feeding period of 4–5 h immediately following sunset
Aphyosemion geryi	Guinea	Pandare & Roman, 1989	invertebrates	feeds both day and night, but a peak of feeding between 10:00–19:00 h
Trachinotus teraia	Ebrié lagoon	Trebaol, 1991	molluscs	diurnal feeder
Hydrocynus forskalii	L. Turkana	Hopson *et al.*, 1982	piscivore	diurnal feeder – peak fullness in the afternoon
Hydrocynus brevis	Lake Kivu	Losseau-Hoebeke, 1992	piscivore	feeds on haplochromines at night and usually hide behind rocks during the day

Diel breeding patterns

Fish with a marked diurnal activity pattern of feeding may have somewhat different behaviour during the spawning period. For example, four species of normally day-active cichlids fan their eggs more actively at night than during the day (Reebs & Colgan, 1991). This may be an adaptive behaviour since egg respiration does not stop after sunset: parental care cannot be limited to just one period of the day, particularly if eggs or young are in relatively exposed environments.

In the killifish, *Nothobranchius guentheri*, diel periodicity occurs in both agonistic and reproductive behaviours and is linked to male hierarchies (Haas, 1976a). Reproductive activity is low early in the morning, begins to increase at about two and a half hours after dawn, peaks at about six hours after dawn, and then decreases to morning levels by day's end. Male agonistic behaviour rises and drops in similar fashion, and the reproductive activity of dominant males is much greater than that of the subordinate males. Time of ovulation appears to be the proximal cause of reproductive behaviour and periodicity for females.

Differences in rhythms of spawning activity may avoid, or limit, interspecific competition. The species of the closely-related genera of Tanganyikan cichlids *Tropheus* and *Simochromis* live in the same habitat (between rocks at depths of a few metres), and feed upon the aufwuchs covering the rocks. However, they do not spawn at the same time of the day: the former spawns in the early morning while the latter spawns throughout the light period (Nelissen 1977a). Moreover, the different species occurring together in the same habitats exhibit different rhythms of activity with respect to peaks of activity which do not overlap.

Circadian activity patterns of fish in different freshwater systems

In their study of fish migrations in the El Beid River, Bénech & Quensière (1983*a, b*) pointed out the existence of a diel activity pattern for most of the species investigated. Four typical activity patterns were observed (Fig. 12.5).

1. Essentially diurnal species, rare in night catches but abundant during daytime, e.g. *Alestes baremoze, Alestes dentex, Hydrocynus brevis, Distichodus brevipinnis, Distichodus rostratus, Citharinus citharus, Labeo senegalensis, Tilapia zillii, Sarotherodon galilaeus* and *Oreochromis niloticus.*

2. Usually diurnal species but with a peak at dusk, such as *Polypterus senegalus, Polypterus bichir, Brycinus nurse, Heterotis niloticus, Siluranodon auritus, Schilbe uranoscopus, Synodontis nigrita* and *Lates niloticus.*

3. Species more abundant at dusk and at night, rare at the end of night and during the day: *Schilbe intermedius, Brachysynodontis batensoda, Synodontis clarias, Auchenoglanis* spp. and *Ctenopoma* spp.

4. Essentially nocturnal species: *Brienomyrus niger, Hyperopisus bebe, Marcusenius cyprinoides, Pollimyrus isidori, Petrocephalus* spp., *Synodontis schall, Clarias anguillaris, Clarias albopunctatus, Clarias lazera* and *Auchenoglanis occidentalis.*

The diel activity patterns observed in the El Beid River for *A. baremoze, S. galilaeus* and *B. batensoda* are similar to the feeding activity patterns observed for these species in the SE archipelago of Lake Chad (Fig. 12.6) (Lauzanne, 1969, 1978*b*; Gras *et al.*, 1981). In both cases the sympatric zooplanktivore species *A. baremoze* and *B. batensoda* have completely different feeding periods.

There can be seasonal changes in these circadian rhythms, as, for example, in the El Beid River where the maximum catches were obtained over night during the flood and high-water period, but over the day at low water. *Hyperopisus bebe*, as well as most mormyrids, exhibits typical nocturnal activity during floods, but there is a behavioural change during the post-flood period (Fig. 12.7), with the apparent disappearance of the circadian rhythm. This drastic change also occurs at the beginning of the post-flood for other mormyrids, *Brachysynodontis batensoda, Clarias lazera, Schilbe intermedius* and *Siluranodon auritus*. These changes in behaviour could be interpreted as the fish escaping from the floodplain to the lake at the first signs of falling water levels. In the case of *H. bebe* this hypothesis is supported by the disappearance of the species after that period.

Another illustration of a diel cycle is the circadian activity of the haplochromines of Lake Kivu, which is characterised by two periods of synchronised activity: one in the early morning and another in the early evening (de Vos *et al.*, 1987; Losseau-Hoebeke, 1992). Most of these fish are active during the day and rest on the lake bottom over night, and the activity peaks observed could be related to the haplochromines dispersing from the lake bottom in the morning, while concentrating there

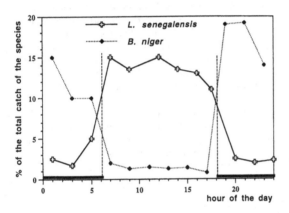

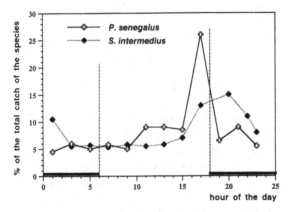

Fig. 12.5. Typical periods of activity for different species in the El Beid River, Chad basin (expressed as % of the total catch of the species) (from Bénech & Quensière, 1983*a*).

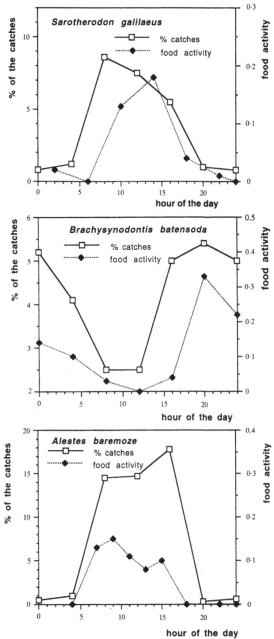

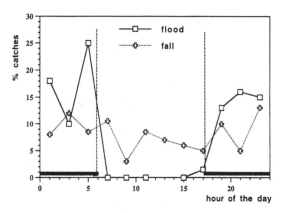

Figure 12.7. Variation of the diel cycle in the catches of *Hyperopisus bebe* in relation to the hydrology (flood or fall) of the El Beid River (from Bénech & Quensière, 1983*a*).

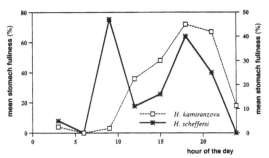

Fig. 12.8. Daily changes in the mean stomach fullness of two species of haplochromines from Lake Kivu (data from Ulyel, 1991).

Fig. 12.6. Comparison between the diel feeding activity patterns of *Sarotherodon galilaeus*, *Alestes baremoze* and *Brachysynodontis batensoda* in Lake Chad, and the rhythms of activity observed in the El Beid River for the same species (Lake Chad data from Lauzanne, 1969, 1978*b*; El Beid data from Bénech & Quensière, 1983*a*).

over night (Losseau-Hoebeke, 1992). The circadian activity observed (Fig. 12.8) is apparently a reflection of the major resting and swimming periods of most haplochromines, which may be related to their daily feeding cycle in Lake Kivu (Ulyel, 1991). Some species exhibit a unimodal cycle with a peak at the beginning of the night (18.00–21.00 h), e.g. *H. nigroides* (omnivore) and *H. kamiranzovu* (phytoplanktivore); but most species have a bimodal feeding cycle, with two longer periods of intensive feeding after dawn (from 6.00 to 12.00 h) and before dusk (from 15.00 to 19.00 h), e.g. *H. scheffersi*, *H. olivaceus*, *H. gracilior* and *H. graueri*. In Lake Kivu, haplochromines are therefore mainly active during daytime with the exception of *H. kamiranzovu*, which apparently migrates to offshore open waters and feeds on zooplankton from the end of the afternoon to around 21.00 h. These data support the observations made on vertical migrations of haplochromines in Lake Kivu (Thys van den Audenaerde, 1986; de Vos *et al.*, 1987), where it was pointed out that most species exhibit a maximum of activity between 6.00–8.00 h in the morning, and 16.00–18.00 h in the afternoon.

The circadian activity patterns of fish species investigated by gill netting in Lake Kariba (Mitchell, 1978) showed that all the species examined exhibited particular activity periods and used particular depth zones. *Hippopotamyrus discorhynchus*, *Mormyrus longirostris* and *Marcusenius macrolepidotus* were essentially nocturnal species. The predator *Hydrocynus vittatus*, the most commonly caught fish in the study area, was one of the few fish species to be caught around noon, but catches at that time were significantly lower than catches in the evening and at night. The catfish *Schilbe intermedius* and *Synodontis zambesensis*, as well as the cichlid, *Oreochromis mortimeri*, were also caught in greatest abundance during the night. The pattern observed in Lake Kariba has been observed in many other African lakes. It is well known that gill netting is much more efficient during the night than during the day, but the question remains open as to whether this pattern is the result of fish avoidance of nets during daytime or of a biological rhythmic activity shared by most fish species in Africa.

The development of other techniques may help to improve our knowledge of rhythmic activities in fish. For *Malapterurus electricus*, a field study in a tributary of Lake Kainji (Belbenoit *et al.*, 1979) demonstrated the existence of a daily activity rhythm of electric organ discharge. This activity (measured as the number of discharges per unit of time) remains at a constant level throughout the day from 07.00 to 18.00 h. Immediately after sunset, it increased five fold to its peak value at 19.00 h. The occurrence of discharges declined throughout the night with a noticeable drop to the daytime low after sunrise between 06.00 and 07.00 h. In the same location, mormyrid locomotor activity and electric organ discharge patterns also exhibit a daily activity rhythm, where changes in light intensity affect the onset and cessation of the fish's nocturnal activity (Moller *et al.*, 1979). But other environmental variables, such as temperature,

probably interact with light. In the Ivindo basin, Hopkins (1981) also provided evidence of nocturnal activity in mormyrids. During the day, most species are found in tree rootlets, inside hollows, sunken logs or under rocks, along banks and within vegetation, while at night they move to open waters.

Bruton (1978, 1979c) made extensive behavioural observations of the daily activity patterns of *Clarias gariepinus* in Lake Sibaya, a clearwater coastal lake in South Africa. He used different methods, such as tagging, gill netting, gut analysis, and SCUBA observations, and concluded that *C. gariepinus* exhibited diel migrations. This species moved inshore at night, the period of peak activity related to feeding, and offshore into deeper water in daytime. The higher catches of *C. gariepinus* during the night compared to the day in Lake Sibaya, and also in Lake Kariba (Mitchell, 1978), support the idea of a peak of activity occurring at night. However, the monitoring of diel behaviour of six radio-tagged *Clarias gariepinus* in Lake Ngezi, Zimbabwe (Hocutt, 1989), led to very different results. These fish exhibited individualistic behaviour and diel patterns were not confirmed. A primary feature was that major movements were not restricted to night-time, because all radio-tagged fish exhibited daytime activity with little indication of inshore/offshore movements being a function of time of the day. Such differences in observed behavioural patterns could result from different environmental situations. It is also possible that *Clarias* avoids nets more easily during the day than at night.

Lunar cycles

An influence of the lunar cycle on fish activity has been demonstrated in many parts of the world, but it is not a universal and well-identified phenomenon. In West Africa, a relationship between high catch levels and the lunar cycle has been observed for migrating fish (Daget, 1957). In the Central Delta of the Niger, *Brycinus leuciscus* forms extensive shoals during moonlit nights at the time of their upstream dispersal migrations. The shoals dissociate when there is no moon (Daget, 1952).

In the course of their study of juvenile migrations in the El Beid River towards Lake Chad, Bénech & Quensière (1983a) tried to demonstrate the influence of the lunar cycle on the migration patterns. The results were very different depending on the species and only some of them seem to be sensitive to this factor. Catches of *Labeo senegalensis*, *Brycinus nurse*, *Brachysynodontis batensoda* and *Oreochromis aureus* are maximum at full moon. Conversely, catches of *Schilbe intermedius*, *Brienomyrus niger*, *Marcusenius cyprinoi*des, *Clarias* spp., and to some extent *Pollimyrus isidori* and *Petrocephalus* spp., exhibit a maximum in abundance during the new moon, and a minimum during the full-moon period. Apparently, diurnal or crepuscular species are more sensitive to full-moon phases than nocturnal species that are sensitive to the new-moon phase. For instance, night catches of *Labeo senegalensis*, a typical diurnal species, are usually low. They are, however, more important during full-moon period when the nocturnal luminosity is higher. For *Hyperopisus bebe*, a nocturnal species whose activity period is maximal at the beginning and end of the night, maximum catch occurs during the first half of the night at new moon, and during the second half at full moon. In the latter case, the luminosity due to the full moon would inhibit the activity of *H. bebe* early in the night.

Lunar periodicity in the breeding activity of *Tilapia mariae* has also been demonstrated in the Ethiop River, a clearwater Nigerian stream (Schwanck, 1987). Most egg clutches were laid during the last quarter of the moon's cycle. The author suggested that the lunar cycle might be used as a cue for synchronised breeding. Spawning before full moon could also enhance the effectiveness of their parental care by allowing a maximal amount of moonlight during the most critical phases in the development of the young. Okorie (1973), working in the northern part of Lake Victoria, also found that the proportion of female *Oreochromis niloticus* with ovaries in an advanced stage of development was greater during the full moon than during new moon.

Ontogenetic niche and migrations: to be at the right place at the right time?

The need for increased empirical and theoretical attention to the role of size-mediated biological interactions in regulating community structure is widely recognised by aquatic biologists (Werner & Gilliam, 1984; Schlosser, 1987). Since body size has major effects on competitive and predator–prey interactions, ontogenetic changes in these processes greatly increase the complexity of species interactions and have important consequences for community dynamics.

Fishes utilise a variety of microhabitats at different times and at different phases of their life history. Size, shape, swimming performance, feeding strategy, predation and competition all combine to define the suitability of a microhabitat, as well as the limits of tolerance. Changes in food or habitat use with respect to body size are the niche dimensions in which fish have usually been investigated.

Niche relationships are an excellent example of the critical role played by body size in regulating species interactions. Juveniles of different species frequently overlap extensively in habitat and food utilisation patterns, and diverge in niche characteristics as they increase in size and become adults. The result is a reduced niche overlap with larger-size classes, but increased niche overlap and reduced foraging profitability among smaller-size classes.

Spatial and temporal scales of the ontogenetic niche

In fish, the body weight of individuals within a species commonly spans four or more orders of magnitude during their ontogeny and many species will undergo extensive ontogenetic shifts in food and habitat use (Werner & Gilliam, 1984). Such shifts create complex ecological interactions in natural communities, while individuals face different competitors and/or predators as they grow. The stage-specific nature of these interactions is important in shaping species' life histories, the dynamics of species interactions and finally the structure of communities.

Ontogenetic niche refers to the patterns in an organism's resource use that develop as it increases in size from birth or hatching to its maximum (Werner & Gilliam, 1984). It includes changes in habitats used, and timing in the occupation of these habitats, keeping in mind that important factors such as predation risks or susceptibility to abiotic factors also scale with body size. The whole range of habitats required by fish during their development may be considered a key functional unit, because it is necessary to achieve the full life history of the species. Spatially, the ontogenetic niche may be either limited, when species complete their life cycle in a restricted area, or may include most of the river basin when adults undertake large-scale upstream spawning migrations.

Ontogenetic changes in resource use are nearly universal, and these shifts are often associated with, or caused by, shifts in habitat. Frequently, piscivorous fish undergo a few rather abrupt shifts in diet (see above). Positive correlations between food size and body size have been documented for a number of species (see Chapter 7). Juveniles of large species often negotiate juvenile periods in which individuals prey on resources that are similar to those consumed by smaller species which eventually serve as prey to the larger adults. This can result in dramatic competitive effects. Another consequence of food and body size interactions is that food availability for juveniles may be a limiting factor to reproductive success in fish. The developmental stage has to meet a specific prey size whatever the habitat. This is the match–mismatch theory that focuses on the importance of juveniles appearing at a time when their food resource is available.

Shifts in habitat have been recorded for many species. Small fishes are often found in shallow water whereas large fish inhabit more open water. In Lake George for instance, the juveniles of most species occur in the shallow water near the shore (Gwahaba, 1975). This is due to the presence of nurseries for cichlids and of breeding sites for some other species. This phenomenon is also well known in inundated floodplains, where the juvenile fish find food and shelter while parents stay on the edges or in the mainstream. Many riverine fish also move upriver to spawn and the young often remain upstream for some time, though many of them perish as headwater streams dry up in regions with extended dry seasons. Spatial segregation as a function of size has also been observed in lakes and the presence of predators may strongly affect the distribution of juveniles (Schlosser, 1987). High density and extensive overlap in habitat use of small fishes in shallow habitats may be related to the increased risks of predation in deeper areas.

Types and benefits of migrations

Migration can be defined as a 'movement resulting in an alternation between two or more habitats (i.e. a movement away from one habitat followed by a return again), occurring with regular periodicity (seasonal or annual), but certainly within the life span of an individual and

involving a large fraction of the breeding population' (Northcote, 1979). Myers (1949), Gross (1987) and McDowall (1987) have provided definitions to describe distinctive migratory patterns in fish.

1. *Diadromous*: truly migratory fish that use two ecosystem types and migrate between the sea and freshwater. Within this general term one can distinguish three types.
   (i) Anadromous fish that spend most of their lives in the sea. Mature to ripe adults migrate to freshwater to breed. Spawning could occur just above the tidal zone or far inland.
   (ii) Catadromous fish that spend most of their lives in freshwater and migrate to the sea to breed.
   (iii) Amphidromous fish that divide their life cycles between freshwater and marine habitats. Migrations are not for the purpose of breeding but occur at some regular phases of the life cycle. They involve larval or juvenile fish not close to reproduction.

   It is also convenient to refer to *euryhaline* fish that can move freely and frequently between fresh and salt waters, and to *amphihaline* fish that can do so only at carefully regulated stages of their lives (McDowall, 1987).

2. *Potamodromous*: truly migratory fishes whose migrations occur wholly within freshwaters. This term is especially used for numerous species of fluviatile fishes (particularly in the tropics) that migrate long distances upstream to spawn (Myers, 1949). But it could also be used for any truly migratory, permanently freshwater fish.

There is a general feeling that anadromous fishes are missing from the tropics, where eels are the only catadromous species (see Day *et al.*, 1981), but McDowall (1987) pointed out that whereas anadromy predominates in cool to cold temperate regions, it does also occur in the tropics. Catadromy predominates in the tropics and southern subtropics, but rarely occurs in both northern and southern cool temperate regions.

According to Gross (1987), diadromy occurs when the gain in fitness from using a second habitat, minus the migration costs of moving between habitats, exceeds the fitness from staying in only one habitat. The relative availability of food in the sea and freshwater habitats could therefore be the key factor driving the evolution of diadromy. In northern latitudes, marine productivity exceeds that of freshwater, while in tropical latitudes the reverse is true. As a result, selection for diadromy has probably been responsible for the colonisation of tropical freshwaters by certain marine fishes, in which case many amphidromous fishes may possibly be of marine origin.

As discussed in detail by Welcomme (1985), potamo-dromous migrations would appear to have an adaptive advantage, and in this way some species may avoid unfavourable conditions in the lower reaches. But most of these migrations seem to be directed at reaching localities suitable for reproduction or feeding. Jackson (1961) suggested that the potamodromous habit is a way of protecting the young from predation, whereas Fryer (1965) considered it as a mechanism to secure dispersal over the whole course of the river. Both are probably true. Spawning near the headwaters allows the fry to drift with the flood wave over hundreds of kilometres downstream, and to reach suitable feeding grounds in floodplains where they will find appropriate prey and also shelter from predation. Upstream migrants who time their arrival and spawning to coincide with the first flow of water on to the plain in systems with predictable floods have a similar strategy: they give the fry the opportunity to reach the floodplain as early as possible. Such a drift is possibly an adaptative advantage in ensur-ing considerable dispersal of the population and allowing mixing of stocks within large systems.

The best breeding and resting sites may differ from the best feeding sites, and many species therefore have to travel, sometimes over long distances, between them. In flood rivers, Daget (1960) and later Welcomme (1985) discriminated between the longitudinal migrations (usually for spawning) that take place within the river channel, and the lateral migrations (mostly for feeding, but also spawning migrations) when fish leave the main channel to spread over the floodplain.

Timing of migrations

The factors responsible for upstream migrations of trop-ical African fishes are not well known. Some species move during low water in order to arrive at the upstream spawning sites at the beginning of the flood, as is the case of *Alestes baremoze* in the Chad basin, which moves up the Chari and Logone Rivers at low water to reach the Yaéré floodplains, a few hundred kilometres upstream, at

the beginning of the flood. But other species or families, such as the mormyrids, initiate their upstream migration as the flood appears. In view of the great differences in timing among species, it is likely that there is a whole range of fine tuning stimuli which influence the various species in different ways (Welcomme, 1985).

There are many indications that adult fish leave the floodplain before the young-of-the-year. That is the case for older age groups in the swamps of the Niger and Benue (FAO/UN, 1970), and a similar phenomenon was observed on the Kafue Flats (University of Idaho, 1971; Williams, 1971). In the North Cameroon Yaéré flood-plain of the Chad basin, lacustrine breeding fish do not return through the El Beid River draining the floodplain to Lake Chad, but use the main channels of the Chari and Logone Rivers.

According to Welcomme (1985), it is possible that simple mechanical and chemical factors could be suf-ficient to initiate movement out of the floodplain habi-tats, in so far as the end of the flood tends to be less pre-dictable than its beginning. It is important for their survival that species be able to receive signals in order to anticipate the onset of conditions that would be lethal for the majority of the population. Such stimuli are not always completely effective, and large quantities of fish are trapped every year in shallow ponds that will dry up later in the season. The tendency for both the larger individuals and the larger species to leave the floodplain earlier than smaller fish indicates that depth is possibly one of the driving factors. Dissolved oxygen concen-tration also seems to be of major importance.

Spawning migrations in a large riverine system: the Chad basin

In the Chad basin (Bénech et al., 1983; Bénech & Quen-sière, 1989) many species undertake longitudinal and/or lateral migrations. The North Cameroon Yaéré (over about 8000 km²), a floodplain of the Logone River (see Fig. 12.9), serves as a nursery for many of them. As floodwaters subside at the end of the flood season, part of the flow does not recede to the Logone River channel but instead moves towards Lake Chad through the tempor-ary El Beid River. Significant numbers of juvenile fish leave the Yaéré and reach Lake Chad by this route. The catches of the El Beid fisheries were used to study the influence of hydrology on the fish production of the Yaéré.

Three groups of adult migrants have been recognised in the Lake Chad–Chari system, and similar groups of migratory species have been observed in the Senegal River (Reizer, 1974).

1. **Large-scale migrants.** Only the adults undertake large-scale longitudinal movements at the time of reproduction. *Alestes baremoze* has been particularly well studied (Durand, 1978 ; Bénech & Quensière, 1989). A large part of the lake population migrates, before the flood, into the Chari River and spawns on the nearby floodplain (the North Cameroon Yaéré) from August to September. The eggs and larvae disperse in the flooded areas where they spend a few months before returning to the lake via the El Beid or the main river channel. After spawning, the adults disperse in the margins of the flooded areas before returning to the lake through the Chari River at the subsidence. *Brachysynodontis batensoda*, another abundant lacustrine species, as well as *Synodontis schall*, *Schilbe intermedius* and *Hyperopisus bebe* have fairly similar patterns of migration and reproduction. *Schilbe uranoscopus* and *Schilbe mystus* also migrate upstream but reproduce in the main channel. *Alestes dentex* is a long-distance migrant that apparently spawns in the higher reaches of the rivers, further upstream than *A. baremoze*. Immature individuals were observed either in the lower reaches of the Chari River or in Lake Chad, and Blache & Miton (1962) recorded movements of up to 650 km upstream from Lake Chad. *Labeo senegalensis* and *Hemisynodontis membranaceus* probably have a similar behaviour pattern. *Polypterus bichir*, *Distichodus rostratus*, *Pollimyrus isidori* and *Marcusenius cyprinoides* also exhibit the characteristics of true migrants, but data are too scarce for safe conclusions to be drawn. In the River Niger, before the construction of the Markala Dam, *Brycinus leuciscus* migrated up to 400 km upstream in the dry season (Daget, 1952).

2. **Medium-scale migrants.** These species are also longitudinal migrants but cover much shorter distances in the Chad basin. Fluvial populations of *Hydrocynus forskalii* spawn in the main channel

during the flood, but juveniles were never observed in the Yaéré. The Lake Chad population apparently migrates to the delta and the lower reaches of the Chari to reproduce between November and March, at the time of the subsidence. The abundance of *Bagrus bajad* in the delta region in May–June could also be attributed to a medium-scale migration from the lake.

3. **Lateral migrants.** Many more or less sedentary species move from the main channel to inundated areas. The movements are not clearly related to reproduction but rather to feeding. They are usually multiple spawners exhibiting parental care of eggs and juveniles. Most cichlids, e.g. *Brienomyrus niger*, *Petrocephalus bovei*, *Gymnarchus niloticus*, *Heterotis niloticus*, *Ichthyborus besse*, *Clarias lazera* and *Siluranodon auritus* exhibit such behaviour.

The downstream movement of juveniles is an important phase of the migratory pattern of long-distance migrants. It generally occurs through the main channel, as a more or less passive drift that ensures recolonisation of the various reaches of the river. A detailed study of juvenile downstream movements has been conducted in the El Beid River, a temporary tributary of Lake Chad, draining a large part of the flood waters of the Yaéré floodplain to the lake (Durand, 1971; Bénech & Quensière, 1982, 1983a,b; Bénech et al., 1982). The existence of an extensive traditional fishery allowed the monitoring of downstream migrations of juveniles that constituted over 95% of the catch. This traditional fishery on the El Beid River consists of dams made of tree trunks and branches built across the river bed (Bénech & Quensière, 1982). In 1969 the number of dams was estimated at 270. Dams divert the fish to shallow waters of the minor bed, where they are caught with a triangular push net, the 'boulou'. This sampling technique is very efficient and the total catch, estimated to be 1200 tons during the 1968–9 fishing season, was made up almost exclusively of 3- to 5-month-old juveniles.

From data obtained before the 1972 drought for the 22 most common of the 74 species, Durand (1971) recorded that there was a marked change in catch composition in relation to hydrology and time (Fig. 12.9). A first group of species, including *Hyperopisus bebe*, *Marcusenius cyprinoides*, *Alestes dentex* and *Labeo senegalensis*,

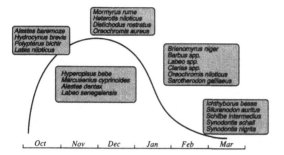

Fig. 12.9. Groups of migrating species observed at the Daga station on the El Beid River, for different phases of the flood (from Bénech *et al.*, 1983).

exhibit a very clear peak of abundance from mid-November to the end of December, during the high water period. Accompanying species related to this group were *Alestes baremoze*, *Polypterus bichir*, *Hydrocynus brevis* and *Lates niloticus* that appeared early in November, as well as *Heterotis niloticus*, *Distichodus rostratus* and *Oreochromis aureus* that appeared later, until January. *Mormyrus rume*, *Pollymirus isidori* and *Distichodus brevipinnis* were also abundant during the first two months of the flood. This first group corresponded to fish migrating at the end of the Logone floods.

A second group of species was characteristically very abundant by the end of January: *Sarotherodon galilaeus*, *Brienomyrus niger*, *Clarias* spp., *Barbus* spp., and, accompanying these species, *Oreochromis niloticus* and *Labeo coubie*.

The third group was the most homogeneous one with *Ichthyborus besse*, *Siluranodon auritus*, *Schilbe uranoscopus*, *Synodontis schall* and *Synodontis nigrita*. Many of these species were caught at the beginning of the flooding of the El Beid River and disappeared afterwards. They reappeared only in February when their maximum abundance occurred. The second and third groups moved from the floodplain towards the El Beid and Lake Chad with the water draining off the flooded plain.

Some species did not appear to have a clear period of abundance and were caught throughout the hydrological cycle: *Brycinus nurse*, *Citharinus citharus*, *Brachysynodontis batensoda*, *Hemisynodontis membranaceus*, *Synodontis clarias*. No juveniles of *Schilbe mystus* and *Hydrocynus forskalii*, two common species in the permanent aquatic environments, were recorded in the El Beid

catches. This confirms the observations of Daget (1954) in the Niger.

As a result of the Sahelian drought (1972/1973), the Yaéré was not flooded for two consecutive years. Residual, usually permanent, ponds on the floodplain dried up and the autochthonous species disappeared. At the same time, drastic ecological changes occurred in Lake Chad and its fish assemblages (see Chapter 10). During the relatively low flood of 1974, following two years of drought, the El Beid catches reflected the disappearance of autochthonous species and lacustrine migratory species (Bénech & Quensière, 1983b).

The sedentary species *Brienomyrus niger*, *Siluranodon auritus* and *Brycinus nurse* that were characteristic in the phase before the drought were absent in 1974. Adults of these species that usually spent the dry season in the permanent ponds of the Yaéré probably disappeared as a result of the drying up of permanent floodplain ponds during the two years of drought. On the other hand, tilapiines were particularly abundant in 1974–5. *Oreochromis niloticus* and *Tilapia zillii*, for instance, spend the dry season in the fluvial system where populations were able to survive during the drought. Their greater capacity for rapid recolonisation may be a consequence of their multiple spawning habits and the care given to the eggs and young, all allowing them to respond very quickly to favourable conditions, and to occupy the places left vacant by the fish populations that would normally have colonised the system.

The reduction of lacustrine stocks as a result of the drought, the disappearance of some fish populations from the southern basin of Lake Chad following the drop in the water level, and the appearance of marshy conditions also had important consequences for the species composition of migratory fish assemblages in the El Beid. The contribution of large-scale migrant juveniles, such as *Alestes baremoze*, *Schilbe intermedius*, *Synodontis schall*, *Citharinus citharus*, *Distichodus rostratus*, decreased sharply in the 1974–5 catch on the El Beid. Conversely, benthic mormyrid species resistant to hypoxia, such as *Mormyrus rume*, *Marcusenius cyprinoides*, *Pollimyrus isidori* and *Hyperopisus bebe*, maintained their populations. *Brachysynodontis batensoda*, a species also resistant to hypoxia and found in the Lake Chad basin but also able to colonise the fluvial system during the low-water period, contributed significantly to the El Beid juvenile stock in the years following the drought. There were also some indications from the catch of juveniles that the lacustrine population of *Citharinus citharus* recovered over the period 1974 to 1978.

The type of migration described in the Chad basin also occurs in other Nilo–Sudan basins. Monteil (1932) described in great detail the migration of *Brycinus leuciscus*, at the very beginning of the century (period 1900–3), in the inner delta of the Niger. These observations are of particular interest in relation to the fisheries, because Monteil was already claiming that many species were overexploited, which proves that problems encountered today already existed a century ago.

Migrations between the East African lakes and their tributaries

In the East African lakes, many fish species have retained the habit of spawning migrations to the inflowing rivers. In Lake Turkana, *Alestes baremoze*, *Citharinus citharus*, *Distichodus niloticus* and *Barbus bynni* migrate into the Omo River, and *Brycinus nurse*, *Labeo horie*, *Clarias gariepinus* and *Synodontis schall* into rivers and ephemeral affluents when these are in spate (Hopson, 1982). In the equatorial Lake Victoria, the flood of northern rivers usually occurs twice a year (April–May and September–December), while rivers in the south flood once a year in accordance with a well-marked annual rainy season. Whitehead (1959) recognised three distinct migration patterns for anadromous fishes under these conditions.

1. Long duration, with fishes, such as the large *Barbus altianalis*, that enter the river for an extended period and spawn 80 km or more upstream.
2. Medium duration, a behaviour common to most of the potamodromous fishes, such as *Labeo victorianus* and *Schilbe intermedius*, that migrate up to 25 km upstream and move laterally into floodwater pools to spawn.
3. Short duration species, such as the *Brycinus*, that ascend flooding temporary streams in huge numbers.

The migrations of *Labeo victorianus* and other small cyprinids have been studied by Cadwalladr (1965a) and by Balirwa & Bugenyi (1980). The upstream migration seems to be preceded by a preparatory movement to the river mouth within the lake itself. The arrival of the first

freshets of the flood seems to trigger longitudinal migratory behaviour. The small mormyrids (*Marcusenius victoriae*, *Gnathonemus longibarbis*, *Hippopotamyrus grahami*, *Pollimyrus nigricans*, *Petrocephalus catostoma*) move up to the northern affluents on both floods. The ripe fish remain near the river mouth until the flood arrives, then migrate upriver at night, with peak runs at dawn and dusk. They spawn in pools 8–24 km upriver and the young remain in river pools for 3–7 months (Okedi, 1969, 1970). Pre-migratory aggregations of *Clarias gariepinus* have been observed in different places. Bowmaker (1969) reported such aggregations at the mouth of a tributary of Lake Kariba about six weeks before the first floods; they migrated upstream as soon as the first flood waters reached the mouth. However, according to Cambray (1985), potamodromy is not obligatory in *C. gariepinus*, since they are able to perform lateral migrations along the shores in the absence of flowing water.

In Lake Malawi, some potamodromous species run up the North Rukuru River to spawn early in the flood (Tweddle, 1982). Many species, such as *Barbus eurystomus*, *Barbus johnstonii* and *Labeo cf molybdinus*, run up early in the rainy season, whereas *Opsaridium microlepis* tends to migrate towards the end of the rainy season, having a more prolonged breeding season than the other species.

The regional scale and the metapopulation range

The populations of many species are not distributed continuously, but are concentrated in patches. In the case of fish, one should consider two levels: (i) the patchy distribution within a watershed, e.g. fish populations inhabiting headwaters of different tributaries, or fish assemblages of riffles separated by pools; and (ii) the distribution of populations across different watersheds, having no present-day connections.

Originally, the term metapopulation was used to describe 'a population of populations', that is to say an abstraction of the population concept to a higher level (Hanski & Gilpin, 1991). Metapopulations are, therefore, ensembles of interacting populations with a finite lifetime, and the concept is closely linked with the processes of population turnover, extinction and establishment of new populations. The concept can also be defined as 'systems of local populations connected by dispersing individuals' (Hanski & Gilpin, 1991). The metapopulation scale is inbetween the local scale (scale at which individuals interact routinely with each other in breeding and feeding activities) and the geographical scale (individuals have no possibility of moving to most parts of the range). It is the scale at which individuals might move infrequently from one place to another, crossing unsuitable habitats. Like the 'islands theory', metapopulation dynamics emphasises the role of emigration and immigration, extinction and colonisation in population dynamics. At a still higher hierarchical level, one can define metacommunity as a community of populations occupying a set of habitat patches sharing the same species assemblage.

For African fishes the main ichthyoprovinces recognised in Chapter 2 are inhabited by species whose distribution is roughly restricted to that particular area. Many of these species occur in different, presently isolated watersheds, and therefore constitute regional metapopulations. These isolated populations have occasionally mixed together when climatic and/or geological events allowed connections between watersheds, and this might again happen in the future. During periods of drought, some of the populations probably disappeared, whereas others survived in refuge zones and were the source of recolonisation for other rivers when climatic and hydrological conditions improved (see Chapter 6).

The early interest in metapopulation dynamics stemmed from the idea that metapopulations should be more stable and less likely to become extinct than single local populations, even if the size of the latter is equivalent to the size of the metapopulation. Today, there is a slightly different approach (Hanski, 1991). Many environments are becoming increasingly fragmented, and 'newly created' metapopulations are not a guarantee of long-term persistence.

Flood patterns and disturbances as major determinants in fish assemblage dynamics?

Statzner (1987) has suggested that stream communities are systems more physically than biologically controlled. If this is true, then the analysis of river systems should focus on a better understanding of the role of the physi-

cal variables and on a proper description of variations in the physical environment. The role of hydraulics is particularly important in affecting the behaviour and the distribution patterns of organisms (Statzner & Highler, 1986; Statzner et al., 1988). Variations in discharge can change the mosaic of flow in a stream section, and changes in species assemblages may be correlated with changes in stream hydraulics. The extent of their variability is often viewed as regulating the relative contributions to community structure of abiotic and biotic processes, and can serve as an *a priori* criterion for predicting community pattern and process (Wiens, 1984). Stream ecologists are interested in this theory because of the great temporal variability within and between lotic environments, particularly with respect to stream flow, which plays a central role in stream ecology.

Similarly, stream hydraulics may also determine the sequence of species assemblages from source to mouth in a river system. Nevertheless, until now few studies have used a hydraulic approach, combining parameters, such as current velocity, depth and substrate roughness; and in Africa the only quantitative information available is on invertebrate distribution in relation to current velocity (Petr, 1970, 1986; Dejoux et al., 1981; Elouard, 1983, 1987; Elouard & Gibon, 1985).

The flood pulse concept

Welcomme (1979) and Junk et al. (1989) recognised that the principal driving force responsible for the existence and productivity of river floodplain systems is the flood pulse. The system responds to the amplitude, duration, frequency and regularity of the pulses. Short duration and often unpredictable pulses occur in low-order streams. Conversely, in most large rivers with large floodplains, regular pulses of long duration result in extensive but temporary lentic habitats. Organisms inhabiting these types of systems have had to adapt to spatial and temporal fluctuations.

Seasonal changes in flow volume (and to a lesser extent temperature), triggered by variable rainfall, appear, therefore, to be the main physical disturbance stimulating different aspects of fish life history (Karr & Freemark, 1985). In most African savannah rivers, the large seasonal variations in rainfall result in great fluctuations in water level causing a seasonal cycle of flood and drought over the lateral plains and low-lying areas,

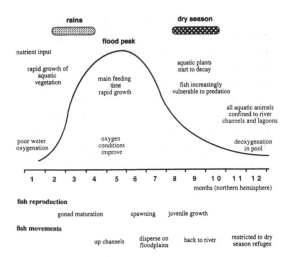

Fig. 12.10. The seasonal cycle of events in a floodplain river (from Lowe-McConnell, 1985).

although some permanent water does persist within the main river channels and the depressions of the floodplain itself. For rivers with a tropical hydrological regime, water-level changes associated with seasonal flooding appear to be the key factors in the biology of species and system functioning, rather than changes in water temperature or day length (Welcomme, 1979; Lowe-McConnell, 1985, 1988). Periodicity (or seasonality) of flood as well as intensity must be considered. The seasonal cycle of events in a floodplain has been summarised by Lowe-McConnell (1985) (Fig. 12.10). The inundation of savannah is associated with an enrichment of the water by nutrient salts from the breakdown of organic matter, decaying vegetation and the dung of animals (cattle or wildlife) grazing on the floodplain. This leads to the rapid development of bacteria, algae and zooplankton, later supporting a rich fauna of invertebrates. At the same time, aquatic vegetation grows rapidly. There is an extremely rapid increase in the production and biomass of different kinds of fish food. After the peak, the level declines and water flows back through numerous channels. Animals migrate to the main river bed or are trapped in isolated small lakes, ponds and swamps, where the water becomes deoxygenated. Some of these waterbodies persist throughout the year, but tend to become overgrown with vegetation. They play the role of refugia. The relationship between hydrology and fish biology is

quite well documented in West Africa. Many Sudanese fish species spawn at the beginning of the flood, often at the end of a long-distance migration of ripening adults to riverine breeding areas (Welcomme, 1979; Albaret, 1982; Bénech & Quensière, 1985).

Flood patterns and structure of fish assemblages
Most published works have failed to put fish studies into an adequate hydrological framework. LeRoy Poff & Wards (1989) developed a regional classification of stream communities based upon variation in streamflow patterns of streams in the USA. They demonstrated for temperate rivers that long-term, daily streamflow records are rich sources of information with which to evaluate temporal and spatial patterns of lotic environmental variability and disturbance across many physiographic and ecographic regions. Patterns of diversity of all major lotic assemblages have been related to patterns of temporal variation in flow, and extreme conditions, such as high flow (flood) and low flow disturbances, are primary sources of environmental variability that play a central role in structuring stream communities. The degree of physical control depends on combinations of streamflow variation (e.g. range and predictability), patterns of flooding (e.g. frequency and predictability) and how intermittent are the disturbances. These hydrological components establish a significant portion of the gross physical flow template (*sensu* Southwood, 1988) in most streams, and a general conceptual model may be proposed using variables such as degree of intermittency, flood frequency, flood predictability and overall flow predictability/variability (LeRoy Poff & Ward, 1989). It is assumed that in habitats with highly variable and/or unpredictable flow regimes, abiotic processes are of predominant importance in controlling ecological patterns, whereas in more benign or predictable flow environments stronger biotic interactions, such as competition or predation, predominate. However, in most lotic systems, flow regimes are intermediate between these extremes.

In low-order streams (rhithron), strongly influenced by precipitation, floods are erratic, unpredictable and of short duration. They correspond to catastrophic events and bottom materials, including detritus, are swept downstream. In the dry season, the system may become completely dry in the most extreme conditions. According to the conceptual (and speculative) model proposed,

streams that experience prolonged periods of intermittency are inhabited by a few species of fish that have
specific life-history characteristics, such as diapause or
physiological tolerance to low dissolved oxygen. Quite a
number of river systems are intermittent along the Saharan border for instance. During the dry season, such
rivers usually appear as a succession of isolated pools in
which dissolved oxygen content is poor, and only a few
species of animals are able to survive. Their colonisation
occurs during the flood as a result of upstream migration,
or from residual pools in which fish spend the dry
season. As intermittency becomes less prolonged, other
hydrologic forces, such as floods, become increasingly
important in structuring communities.

For perennial streams, flood frequency and flood predictability are likely to be of primary significance. As
flood frequency increases, selection pressure is presumably exerted in favour of certain behavioural and life-
history characteristics: small, vagile or colonising assemblages of fish and invertebrates. The more persistent fish
species are likely to be those morphologically adapted to
survive flood flows. Fish avoidance behaviour may reflect
an evolutionary history of frequent floods. A combination of high flood frequency and low flood predictability (perennial flashy streams) is likely to minimise
the contribution of biotic interactions to community
structure.

For fish that spend their early life in floodplains,
where they find food and shelter, an unexpectedly poor
flood may result in poor year-class survival. Wide variations in recruitment are typical of many such populations. A rare but long-lived species may be able to
maintain its presence if it enjoys occasional high recruitment. In contrast, short-lived species may be endangered
if the frequency of disturbances does not allow successful
spawning. Cichlids have developed parental care, such as
mouthbrooding, which in some way allows the species to
ensure spawning success even in drastic environmental
conditions.

Conclusion

The questions of spatial and temporal scales are of prime
importance in any fish community investigation. Four
main ranges of activity behaviour have been identified,
which may be considered along an increasing gradient of
complexity: (i) the microhabitat (or resting habitat); (ii)
the home activity-range (including the refuge and feeding habitats); (iii) the ontogenetic scale (the spatial limits
of the habitat required for the different developmental
stages); and (iv) the metapopulation scale. Migrations
may be regarded as a way to be at the right place at the
right time for the implementation of the different developmental stages.

The flood pattern is the principal driving force
responsible for the diversity of available habitats for fish.
Seasonal changes in flow volume stimulate most of the
aspects of fish life history, and may explain changes in
the patterns of species diversity in fish communities.

13   Fish diversity and ecosystem functioning

Many believe that species diversity is essential for the proper function of communities and for the emergence of new properties at the community level. ... If all cells had the same characteristics there could be no specialised organs. Consequently species diversity is clearly necessary for community structure. But is any amount of diversity sufficient, or are there specific mixes of species that are necessary for the proper function of communities and ecosystems? This is a very old question in ecology . . . .

Solbrig, 1991a

The functional relationships between species in ecosystems are still poorly understood, and one of the central questions dealing with biodiversity is: would it really matter if our living systems were greatly simplified? Given the assumed importance of biological diversity in ecosystem functioning, some major ecological questions have been identified.

1. What is the linkage between species diversity and ecosystem function: is species composition responsible for particular ecological characteristics and processes of the system under study, when compared with other systems? Another related question could be: is the redundancy in function among species useful and necessary for ecosystems or is there an overriding influence of certain species in controlling both the structure and function of ecosystems (dominant species and keystone species) (Paine, 1969; di Castri & Younès, 1990a, b; Lawton & Brown, 1993).

2. How is system stability and resistance affected by species diversity, and to what extent could the integrity and sustainability of ecosystems be maintained in spite of species deletions resulting from degradation of environmental conditions (Solbrig, 1991a). Could the ability to resist external episodic extreme perturbations depend on system heterogeneity and species diversity?

3. The relationship of species richness to such processes as biological productivity remains dubious and is an open question.

At present, there are few data with which to examine such questions, and even fewer for African freshwater systems. Traditionally, fishery biologists, particularly those working in tropical countries, have tended to consider fish in isolation, as a natural renewable resource, rather than as integral components of the aquatic ecosystem, interacting with other biotic and abiotic components of the whole system. However, the re-examination of the pool of data already available is a unique opportunity to discuss some results that may be relevant for this purpose (Lévêque, 1995a). Indeed, information on the role of species diversity in the functioning of ecosystems could have implications for environmental management policy. Some scientists contend that all species are useful, while others have argued that not all species can, or will, be saved and that choices have to be made. If it is demonstrated that diversity *per se* plays a major role in ecological functioning, then ecological research should contribute to decisions on what to preserve. Better information on the importance of biodiversity to ecosystem functioning could also potentially alter the perceptions of decision-makers, if the argument that biodiversity should be maintained in order to preserve the integrity and functioning of ecosystems can be made more convincing to a wider range of people than some of the reasons currently advanced for the conservation of biodiversity.

The top-down effect

In the classical limnological approach, it was usual to envisage freshwater ecosystems as operating in a physical–chemical milieu, which conditions the food chain from primary producers to top predators (Le Cren & Lowe-McConnell, 1980). In this 'bottom-up' control, competition between primary producers for limiting nutrients determines the state of higher trophic levels. More recently, the role of fish in regulating the structure and functioning of freshwater ecosystems has become a fashionable topic (Werner, 1986). 'The top-down view includes those various ways in which fish affect the function and structure of an ecosystem in contrast with the bottom-up view which considers the ways in which food limitation and related physical–chemical factors affect fish' (Northcote, 1988). Foraging activities of fish, for example, can directly affect water transparency and thereby primary and secondary production, either by stirring up bottom sediment or through intense phytophagous feeding. Fish also have direct effects on the abundance of phytoplankton, periphyton and macrophytes, as well as on plankton and benthic communities. The 'top-down control' approach argues that the effects of fish predation cascade down the trophic chain and are responsible for controlling the state of the entire ecosystem. Size-selective predation by fish may not only play a major role in the population dynamics of prey species (see Chapter 7) but also result in shifts in the relative abundance of species. The influence of fish on nutrient cycling and transport, through nutrient release in faeces and migration from one habitat to another, has certainly been underestimated in many studies. Diel vertical migration, for instance, may be an important means of transporting nutrients from deeper waters, perhaps below a thermocline, up into surface or pelagic waters. All the above effects are also time-related, and large seasonal changes may occur.

There is, of course, an important degree of com-

plexity in the top-down processes involving fishes, and in most situations both top-down and bottom-up processes are involved, although in some systems or at some times one or the other may be paramount.

The influence of fish predation on prey assemblages

The effect of fish on the abundance and composition of their food organisms has been widely documented (Lazzaro, 1987) and includes different aspects.

Various sympatric species may, of course, behave differently and have specific impact on the prey communities. In Lake Chad, for instance, the smallest prey captured by *Brachysynodontis batensoda*, a microzooplanktivore, are about 80 µm in length (Gras *et al.*, 1981). Nauplii and rotifers are progressively selected as a function of their size up to 260 µm. The larger microcrustaceans are mainly selected on the basis of their ability to avoid predation: the relatively big cladoceran, *Moina micrura*, with low motility is easily captured, whereas the highly vagile diaptomids are not. Large adults (230–250 mm SL) of *Alestes baremoze*, another zooplanktivore in Lake Chad, do not retain small-sized nauplii and rotifers, and the branchial filter retains particles in excess of 400 µm in length. All filtered planktonic items above a size of 880 µm are collected (Lauzanne, 1970). The two species feeding on zooplankton therefore have quite different behaviour. *Alestes baremoze* feeds on bigger prey and consumes more copepods than *B. batensoda*. The latter selects smaller prey, such as rotifers and nauplii, more efficiently.

Passive size-selective predation by some filter-feeders can dramatically lower the mean prey size and induce changes in the species composition of zooplankton communities. Where planktivorous fish are absent, invertebrate planktivores and large crustacean zooplankton predominate. In Lake Naivasha, for example, *Daphnia laevis* grows to between 2.0 and 2.5 mm, *Diaphanosoma excisum* occasionally attains 1.0 to 1.6 mm in length, whilst adults of *Tropodiaptomus neumanni* attain 2.6 mm (Mavuti, 1983) and the presence and abundance of these large cladocerans and copepods in the limnetic zone suggest a low predation pressure from fish (Mavuti, 1990). Actually, the present fish fauna of Lake Naivasha is composed of introduced species, and adults or juveniles are found almost exclusively in the littoral and rarely occur in the limnetic zone. Predation pressure is therefore high on the littoral zooplankton, while it is low on the limnetic populations and this is considered to be a major trophic gap in the pelagic food chain (Mavuti, 1990).

Evidence that the distribution and abundance of prey may be controlled by fish behaviour has also been gathered in other African lakes. Cladocerans are absent from the pelagic waters of Lake Tanganyika. According to Green (1967), the reason may well be the intense predation by the endemic clupeid fish, *Stolothrissa tanganicae* and *Limnothrissa miodon*. Conversely, in Lake Malawi, large numbers of *Bosmina longirostris* were recorded in the southern end of the lake, as well as fairly large numbers of the cladoceran *Diaphanosoma excisum*, in the open waters where *Daphnia lumholtzi* was also recorded (Turner, 1982). Therefore, planktivore fish species of Lake Malawi do not appear to be very effective predators in the open waters and Fryer & Iles (1972) stated that they were inshore species that rarely venture far toward the middle of the lake. The truly pelagic zooplankton feeder is *Engraulicypris sardella*, but it does not support an important fishery as do the clupeids in Lake Tanganyika. In Lake Albert, where another pelagic species (*Engraulicypris bredoe*) is present, Green (1967) also found that it concentrated inshore, and that it was not an important predator on the large cladoceran, *Daphnia lumholtzi*. There are also indications that the other haplochromine zooplanktivores of Lake Malawi, belonging to the Utaka group, are much more numerous inshore than offshore.

In the absence of any experimental evidence on which to assess the impact of African fish on plankton assemblages, one can alternatively monitor the effect of the introduction of alien species on the planktonic communities of natural systems. For example, after the introduction of *Limnothrissa miodon* into Lake Kariba in 1967–8, observations were made on the pelagic community (Begg, 1976). There was an obvious decline in the large zooplankton from 1968 to 1974, in particular *Ceriodaphnia*, *Diaphanosoma* and *Diaptomus*. This decline could be attributed to the introduced species. In 1973–4 the diet of *Limnothrissa miodon* included 80% *Bosmina longirostris*, whereas *Mesocyclops leuckarti* numerically dominated the large zooplankton in the open lake. By 1976, *M. leuckarti* was the dominant food item of *L. miodon*, and *Bosmina* contributed only to 5% (Cochrane, 1978).

Thus over a period of ten years, the largest forms of cladocerans had markedly declined, and the smallest ones such as *Bosmina* were disappearing. A small copepod later became the major food item, a situation approaching that in Lake Tanganyika (Turner, 1982). In Lake Kivu where the clupeid fish *Limnothrissa* was introduced, the original open-water zooplankton community, composed of large pigmented pond species of Cladocera and Copepoda, has also been modified with a drop in standing crop, a decrease in size of individuals and an increase in diversity (Dumont, 1986).

The impact of predation has also been advanced as an explanation for the absence of particular species from waterbodies, without being always well supported by data. For instance, the absence of *Chaoborus* from Lake Tanganyika and Lake Kivu, compared with other East African lakes, has been a subject for debate. Turner (1982) suggested that in Lake Malawi, where *Chaoborus edulis* is very abundant, the pelagic cyprinid *Engraulicypris*, which feeds only during the day, is relatively inefficient in utilising this food item compared with the Tanganyika clupeids, while *Chaoborus* must play a significant role as a predator of zooplankton. Night feeding by *Limnothrissa* is a possible mechanism by which the *Chaoborus* could have been eliminated from Lake Tanganyika, assuming that they were formerly present. This species is known to exhibit strong vertical migrations, being benthic during daylight in lakes with planktivorous fish, but in Lake Malawi there is no evidence for diel vertical migrations of *Chaoborus* (Degnbol & Mapila, 1982). Turner's hypothesis was discussed by Ribbink & Eccles (1988) who suggested that the dominance of *Chaoborus* in the central part of Lake Malawi may be the result of a lower production and standing stock of zooplankton in this area. The passive predatory strategy of *Chaoborus* would be energetically more efficient than active hunting as practised by *Engraulicypris*. They also suggested, as an alternative explanation, that the absence of *Chaoborus* from Lake Tanganyika may lie in abiotic factors, such as the chemical characteristics of the waters. *Chaoborus* are also absent from Lake Kivu, and the Ruzizi River that drains Lake Kivu is the main inflow to Lake Tanganyika.

The influence of fish predation upon benthic communities has been observed in Lake Chad where molluscivore fish (*Synodontis clarias*, *S. schall*, *Hyperopisus bebe*) feed selectively upon young benthic molluscs, as demonstrated by the prey size frequencies in their gut contents (Lauzanne, 1975*a*). This strong predation pressure results in a truncated size frequency in the benthic populations: despite sustained reproduction throughout the year, the size frequency distribution exhibits a domed shape, rather than that of a negative exponential (Lévêque, 1972).

Cascading trophic interactions

The concept of cascading trophic interactions, reflects an elaboration of long-standing principles of fishery management based on logistic models. In simple words, a rise in piscivore biomass results in decreased fish zooplanktivore biomass, which in turn allows an increase in herbivore zooplankton biomass, and decreased phytoplankton biomass (Fig. 13.1). Productivity at a given trophic level is maximised at intermediate intensities of predation at all trophic levels. The concept of cascading trophic interactions links the principles of limnology to those of the fishery biologists: potentially, variation of primary productivity is mechanistically linked to variation in piscivore populations. If piscivore populations can be controlled, it should therefore be possible to control the cascade of trophic interactions that regulate algal dynamics, through programmes of stocking and harvesting (Carpenter *et al.*, 1985). For these authors, 'altering food webs by altering consumer populations may be a promising management tool'.

In natural environments, sequences of cascading trophic interactions will be propagated from stochastic fluctuations in piscivore year-class strength and mortality. Fish stocks, reproduction rates and mortality rates exhibit enormous variance, and lags in ecosystem response occur because generation times differ among trophic levels. Examples of lags may be found in the development of the predator, when it acts first as a zooplanktivore and then as piscivore. As a planktivore, it drives the ecosystem toward small zooplankton and higher phytoplankton abundance, which is the reverse of when the fish grows and becomes a predator of planktivorous fish. If potentially a system may be managed by increasing or decreasing the intensity of piscivore predation, the system responses are non-equilibrium, transient phenomena that are difficult to detect using long-term averages. Nevertheless, in a few cases, following for

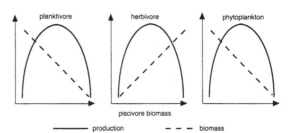

Fig. 13.1. Theoretical changes in biomass and production of vertebrate zooplanktivores, large herbivores and phytoplankton in relation to piscivore biomass (from Carpenter *et al.*, 1985).

instance the introduction of a new predator in an aquatic ecosystem, examples of trophic cascades have been documented.

A striking example of a trophic cascade is known from Lake Victoria, where the depletion of haplochromines by Nile perch shows how a predator can affect other trophic levels. The food webs in Lake Victoria have changed considerably during the past 20 years. The pre-*Lates* food web was dominated by the haplochromines, encompassing many trophic specialisations (Ligvoet & Witte, 1991). The detritivore/ phytoplanktivore group of haplochromines, as well as the zooplanktivores group, constituted most of the total demersal fish biomass. The major food chains (Fig. 13.2) starting from phytoplankton and bottom deposits were: (i) directly via detrivorous/phytoplanktivorous haplochromines to piscivorous fish; (ii) via zooplankton, and then zooplanktivorous haplochromines to the piscivorous catfishes and haplochromines; (iii) via molluscs to various fish taxa and piscivorous fish; and (iv) via insect larvae and *Chaoborus* to various fish taxa (haplochromines, Mormyridae, *Barbus*, *Alestes*, *Synodontis*) to piscivores.

Haplochromines disappeared in the early 1980s simultaneously with the explosive increase of the introduced Nile perch. Some 13+ detrivorous/phytoplanktivorous haplochromines seem to have been replaced by the native detritivorous atyid prawn, *Caridina nilotica*, and 20+ zooplanktivorous haplochromines by the native zooplanktivore cyprinid, *Rastrineobola argentea* (Witte *et al.*, 1992*a*; Goldschmidt *et al.*, 1993). The two latter species became major prey for the Nile perch after the decline of the haplochromines. *Lates* also replaced the 109+ species of

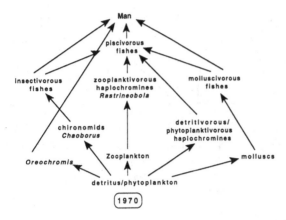

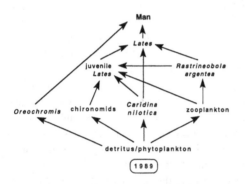

Fig. 13.2. Simplified diagrams of the food web in the sub-littoral area (6–20 m deep) of Lake Victoria in the 1970s and in 1989 (modified from Witte *et al.*, 1992*a*).

fish community has shifted from primary consumers (detrivorous haplochromines) to the top predator (*Lates*) operating mostly as a secondary and tertiary consumer (Ligtvoet & Witte, 1991).

The algal grazing was reduced by the disappearance of haplochromine phytoplanktivores and detritivores, and this situation may have contributed to recent algal blooms of cyanobacteria (Goldschmidt *et al.*, 1993). Another consequence of the introduction of *Lates* into Lake Victoria, and the subsequent crash of the native cichlid populations feeding on invertebrates, was a population explosion of emergent insects, which at times resembled a cloud over the lake. This insect population in turn is now supporting a huge population of the sand martin, *Riparia riparia*, which winters in Africa, and has increased massively from small numbers to hundreds of thousands (Sutherland, 1992). It should be noted, however, that there were always fly swarms over Lake Victoria (Beadle, 1981). Other changes were also observed at the top of the food pyramid. The pied kingfisher (*Ceryle rudis*) shifted from a diet of mainly haplochromines to one that consists exclusively of *Rastrineobola* (Wanink & Goudswaard, 1994), but more fish need to be caught to meet daily energy demands (from 17 haplochromines in former years to 55 nowadays) and hunting time is increased (Goudswaard & Wanink, 1993). A similar shift in diet was also found for the Great Cormorant, *Phalacrocorax carbo*, and the Long-tailed Cormorant, *Phalacrocorax africanus* (Goudswaard & Wanink, 1993).

It may be of interest to compare the food web suggested by Witte *et al.* (1992*a*) with that proposed some 60 years ago by S. & E. B. Worthington (Fig. 13.3). One of the main differences is the disappearance of the crocodile from Lake Victoria, which was a crocodile-infested lake according to Worthington & Worthington (1933): 'Crocodiles act as an important link between the larger water creatures and man, unless they take it into their heads to reverse the topmost link'. The disappearance of crocodiles some decades ago led to profound modifications in the lake fish assemblages, but unfortunately these are not documented by data. An interesting question to ask: would the introduced *Lates* have been better controlled in Lake Victoria if the crocodiles were not extinct?

Another example of a trophic cascade is provided by the introduction of *Oreochromis alcalinus grahami* to Lake Nakuru. In this saline lake, previously devoid of fish, the

original haplochromine piscivores as well as the piscivorous catfishes (*Bagrus docmak*, *Clarias gariepinus*). The introduction of the Nile perch therefore severely disrupted the ecosystem, with a simplification of the food web through virtual eradication of haplochromines from many interactions, and the development of short food chains to the Nile perch (Fig. 13.2). Whereas the haplochromines converted numerous protein sources (algae, zooplankton, insect larvae, molluscs, etc.) into fish protein for consumption by higher trophic levels, the Nile perch now eats quite a lot of its own juveniles, which in some way feed on the same food items as the haplochromines did previously. In comparison with the pre-*Lates* situation, the bulk of the biomass within the

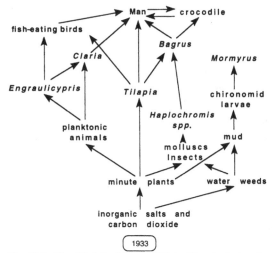

Fig. 13.3. Simplified food chains of Lake Victoria during the 1930s (from Worthington & Worthington, 1933).

cichlid introduced in the 1950s developed quickly, feeding on the high standing crop of the cyanobacteria, *Spirulina platensis*. All primary consumers in the lake, Lesser Flamingos, fish, copepods, rotifers, chironomid larvae and corixids, together consume about 1% of the algal biomass per day (Vareschi, 1978). It is, therefore, very unlikely that the grazing *Tilapia* compete with other grazers and/or significantly reduce the algal biomass whose turnover rate may even be accelerated by additional grazing. But the most striking effect of the introduction on the Lake Nakuru ecosystem has been a substantial increase in diversity by the extension of the food chain to fish-eating birds, of which the Great White Pelican is dominant (85% of the biomass of fish-eating birds) (Vareschi, 1979). The bird life of the lake was poor in the 1950s since there were few other water birds beside the huge number of flamingos: this later increased to more than 50 water bird species. Fish-eating birds started to invade Lake Nakuru around 1963, about two years after the fish were introduced. It was estimated that breeding pelicans harvested some 16 000–20 000 kg fresh weight of fish each day, removing some 72 kg of phosphorus and 486 kg nitrogen every day (Vareschi, 1979; Vareschi & Jacobs, 1984).

If we agree with the idea that it is possible to manage lake ecosystems through trophic manipulations and the introduction of species, one could understand that fish diversity may find a wide field of application. Selection of species according to their trophic behaviour needs better knowledge of their life cycles and their ability to adapt to new ecological situations. A good knowledge of the ecological requirements of wild fish populations may be useful in the process of selecting candidates for biomanipulation. The above example of the introduction of *Oreochromis alcalinus grahami* to Lake Nakuru is an application of that knowledge, but less obvious and probably more sophisticated choices might be expected.

Intraguild predation

Another level of complexity arises when species may be in a position to compete for prey at one stage of their ontogeny. Intraguild predation (IGP) 'is the killing and eating of species that use similar resources and are thus potential competitors' (Polis & Holt, 1992). It often occurs among species that eat the same food resource but differ in body size, such that the smaller species is a potential prey for the larger. Piscivorous fish species, for instance, usually begin their life by feeding on zooplankton or small invertebrates, in competition with species that will serve as important prey when they are large enough to become piscivorous. Whereas many studies have analysed systems in which predators and competitors are considered as different species groups that interact, intraguild predation, combining predation and competition, has been less well integrated into the conceptual framework of ecology.

Intraguild predation interferes with trophic cascade scenarios that involve webs having between three and five links (e.g. plant–herbivore–intermediate-level predator–top predator). A change in one trophic level could affect other levels. For instance, an increase in a top predator could result in a decrease of the intermediate-level predator numbers, allowing herbivore abundance to increase, and ultimately increasing plant predation (Fig. 13.4). This scenario could function in the absence of strong intraguild predation links (i.e. top predators that also eat herbivores). If this link is strong, changes in the top predator will not cascade as predicted, or may even act in an opposite direction to that predicted above (Polis & Holt, 1992). The dynamic complexity of intraguild predation could possibly explain the difficulties (sometimes disasters) frequently encountered when exotic species are introduced into aquatic systems.

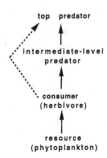

Fig. 13.4. The geometry of a cascading trophic interaction with and without intraguild predation (IGP). Trophic cascades are only possible if the IGP link is not strong compared to the link between the top and intermediate-level predators (from Polis & Holt, 1992).

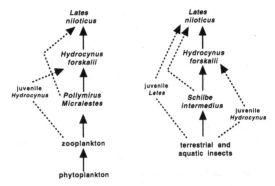

Fig. 13.5. Food webs with intraguild predation, illustrated with data from Lake Chad open-water food chains (from Lauzanne, 1976). In the case of *Lates* and *Hydrocynus*, feeding on small pelagic zooplanktivores or on *Schilbe intermedius* (consumers), the top predator (*Lates*) also eats the other consumer. Both predator species when juveniles, also potentially compete with the consumer.

An example of intraguild predation may be suggested from food chain data obtained in the open waters of Lake Chad (Lauzanne, 1976) (Fig. 13.5). The producers (phytoplankton) are consumed by zooplankton, itself serving as the main food source for small zooplanktivore species (*Micralestes*, *Pollymirus*). The latter constitute the bulk of the food of *Hydrocynus forskalii*, a piscivore when adult but whose juveniles are at some stage zooplanktivores. *Hydrocynus forskalii* is itself one of the few major

prey species of the top consumer, *Lates niloticus*, which also consume smaller planktivore fish. In this case, intraguild predation occurs at two different levels in the food chain.

A similar pattern occurs in the open waters of Lake Chad, with another food source. Indeed, Lauzanne (1976) stressed the importance of falling terrestrial insects in the food chain: they serve as the main food source of the pelagic *Schilbe intermedius*, itself one of the major prey species for *H. forskalii* and *Lates niloticus*. Juvenile *Lates* and *Hydrocynus* also consume terrestrial insects and, therefore, theoretically compete for food with *S. intermedius* (Fig. 13.5).

Keystone species

The question has been raised as to whether certain species are more important to global ecosystem functioning than others, and may have disproportionate influences on the characteristics of an ecosystem. The term 'keystone species' was originally applied by Paine (1969) to a predator in the rocky intertidal zone. There are two hallmarks of keystone species: (i) their presence is crucial in maintaining the organisation and diversity of their ecological communities; and (ii) it is implicit that these species are exceptional, relative to the rest of the community, in their importance. The loss of such keystone species may transform or undermine the ecological process, or fundamentally change the species composition of the community, even though they may appear numerically unimportant either in space and time.

According to Solbrig (1991b), three general classes of keystone species are recognised: (i) keystone predators, herbivores or pathogens that allow the maintenance of diversity among competing organisms by controlling the abundance of dominant species, thus preventing competitive exclusion; (ii) keystone mutualists that link the fate of many partner species; and (iii) species that provide keystone resources which are critical for the survival of dependant populations during bottlenecks of low resource availability. Keystone resource species are expected to be found in conditions where resource availability is low, whereas keystone predators should play a significant role in situations where competitive exclusion is not prevented by disturbances. This is a phenomenon similar to the intermediate disturbance hypothesis, but in one case the species richness is maintained or increased by intermediate levels of disturbance involving abiotic factors, whereas in the other case, predation and competition, through control of dominant competitors, opens up room for maintenance of other species.

Long-term observations and/or experimentation are necessary in order to determine the existence of and to identity keystone species (Solbrig, 1991*b*). At the moment, this concept of keystone species has not yet been investigated for African fish, but as discussed earlier in this chapter (see top-down effect), keystone predators have been convincingly demonstrated in aquatic communities. There is some empirical evidence that predators exert a control on prey populations, and examples are available for plankton, benthos and fish populations. In Gebel Aulia Reservoir (Sudan), Hanna & Schiemer (1993*a*, *b*) pointed out that *Alestes baremoze* and *Brycinus nurse*, exhibiting opportunistic foraging behaviour, fill the niche of zooplankton feeders in the absence of a specific zooplanktivorous fish, but can shift to insects and plant material in other circumstances. As a result of their high population densities, it is likely that they can function as keystone species influencing zooplankton composition and in turn phytoplankton productivity and nutrient dynamics in Gebel Aulia Reservoir.

It should be stressed that piscivores are more vulnerable to fishing gear than benthivores or planktivores (see Chapter 10). This explains why the proportion of fish predators, which is usually high in the catch at the start of fishing a virgin stock, decreases rapidly when the fishing effort is maintained. The depletion of the predator stock following exploitation should have feedback consequences on the abundance of prey, which could initiate a host of indirect effects, including intense competition between species previously coexisting at lower densities. Unfortunately, documented data are not available for African inland waters.

There are also experimental demonstrations of the existence of keystone predators using either fish removal experiments or species introductions. The most obvious demonstration is given by the introduced Nile perch in Lake Victoria. The cascading effect, associated with other man-induced changes, has resulted in the disappearance of many haplochromines and possibly in the increase of phytoplanktonic production (Goldschmidt *et al.*, 1993; Hecky, 1993; Mugidde, 1993). One can conclude therefore *a contrario* that the disappearance of

Lates from any freshwater ecosystem, where it presently occurs, would also result in changes in the overall functioning of that system.

The introduced *Oreochromis alcalicus grahami* in Lake Nakuru might now be considered as a keystone resource species. Indeed, the disappearance of that species, supporting a huge population of piscivore birds, would dramatically affect the ecosystem as a whole. Such a disappearance could result from long-term climatic cycles, or for example temporary drying out of the lake, resulting in the elimination of the fish. This situation has certainly occurred in the past and explains the poor fish fauna currently present in various shallow lakes. In the case of Lake Nakuru, high water salinity also explained the absence of fish before the introduction of *Oreochromis alcalinus grahami* .

The concept of keystone species has been questioned by Mills *et al.* (1993) and, while the term keystone species is very popular among both scientists and managers, it nevertheless lacks clear definition, and means different things to different people. Moreover, it is very difficult to define objectively which species are keystones, and it is likely that subjectively chosen groups of species will be so labelled, whereas other species of similar importance will be ignored (Bender *et al.*, 1984). Finally, the keystone role of a given species is particular to a specific set of environmental conditions and species assemblages, and that does not take into account the spatial and temporal variability of those associations (Gauthier-Hion & Michaloud, 1989). While recognising that the concept has been useful in demonstrating that under certain conditions some species have particularly strong interactions, Mills *et al.* (1993) claimed that emphasising strengths of interactions within communities, instead of a keystone/non-keystone dualism, is a better way to recognise the complexity as well as the temporal and spatial variability of interactions. In the long run, and despite the danger of making communication between biologists and policy makers more difficult by the abandonment of a popular and evocative concept (keystone species), this attitude will favour the development of management guidelines that will explain more explicitly the complexity of interactions in natural systems.

Trophic interactions have been used by most authors to identify keystone species, but other kinds of interaction exist. Actually, we know nothing about the potential control of African freshwater fish populations by keystone pathogens although their role has been demonstrated in temperate waters.

Fish biodiversity and responses of ecosystems to perturbation

The terminology for properties of populations that determine their response to perturbations is rather confused. Underwood (1989) provided a review and tried to suggest a more standard terminology. The property of a population that determines whether or not it will respond to a given type of perturbation of known frequency is variously known as 'persistence' (Margalef, 1969), 'resistance' (Connell & Sousa, 1983 ; Begon et al., 1986) and 'inertia' (Orians, 1974). The lack of response to perturbation should be called inertia (Underwood, 1989). Persistence should be simply the observed property of a population able to continue to exist without changes. Theoretically, the inertia of a population should be measured as the greatest magnitude of a particular type of perturbation that causes no response.

When a population reacts to a perturbation, the capacity of the system to recover includes two components. The first important one, called stability, is the rate of recovery after the stress has disappeared. There has been some confusion between stability and inertia, and the term 'elasticity' (Orians, 1974) has also been proposed to mean rate of return to equilibrium. Stability must be measured in terms of a defined magnitude of stress. A population may recover at different rates with respect to stresses of different sizes. It should be noticed that the word 'resilience' has been used by some authors to mean the rate of recovery (see Begon et al., 1986). The second component is the ability of a population to recover from different magnitudes of stress, which has been called 'amplitude' (Orians, 1974) or 'resilience' (Holling, 1973). Amplitude should be reserved for the magnitude of resilience of a particular population (Underwood, 1989). The magnitude of resilience could be measured by recording the amplitude of the largest response to stress from which a population can regain its equilibrium. Pimm (1991) defined population resilience as 'the rate at which population density returns to equilibrium after a disturbance away from equilibrium'.

There is no single answer to what determines resilience, but there are probably many factors operating. In the short term, the rate of recovery depends on the reproductive rate: the more young produced, the faster the population can recover its former level. It is likely that more resilient populations are to be found among species with high reproductive rates, which are usually small-bodied species, with a short life span. But resilience in the long term also depends on that species interactions with other species in the community, and this could be explored through the length of food chain. Long food chains may be expected to reduce the resilience of the constituent species according to Pimm (1991). From this point of view, resilience may also depend on the availability of food or the nutrients necessary for the species' growth, which is partially related to abiotic processes. At the ecosystem level, the energy flow through the system and its availability for species could have consequences for the overall resilience of the system. There are, therefore, sets of factors at different hierarchical levels (population, community and ecosystem) that could be involved in resilience. Any one of these factors may predominate depending on the temporal or spatial scale under consideration.

'Perturbations' or disturbances are changes that occur in environments. The word 'stress' has also been used in a more or less similar way (Underwood, 1989). Some perturbations could cause no response. Others could be described as 'catastrophes', for example, when the habitat of some population is destroyed and the population becomes locally extinct. In between these extremes, perturbations usually cause some response by the population under stress. A general definition of 'disturbance' has been given by Pickett & White (1985) and modified by Resh et al. (1988): 'any relatively discrete event in time that is characterised by a frequency, intensity, and severity outside a predictable range, and that disrupts ecosystems, community, or population structure, and changes resources or the physical environment'.

Detection of perturbations, and responses of communities, refers more or less implicitly to the concept of equilibrium. The definition of equilibrium in natural populations is not an easy task. Organisms fluctuate widely in time and space, as do their food or their pathogens (Connell & Sousa, 1983).

Biodiversity and stability
The relationships between community complexity and stability have received some attention. For a long time,

the so-called 'conventional wisdom' in ecology (Begon *et al.*, 1986) was that increased complexity within a community leads to increased stability. Complexity is used here to mean more species, more interactions between them, and more pathways. Some authors like Elton (1958) supported the view that more complex communities are more stable, and MacArthur (1955) suggested that the greater the complexity, the greater is the ability of the community to respond to a perturbation. The basic assumption is that if the number of pathways increases, any blockage at one point of the network would be compensated for by the opening of another pathway. However, until now this conventional wisdom has not received any support from field or experimental work, and mathematical simulation models have led to different conclusions.

May (1972) constructed a model of randomly assembled food webs, including a number of interacting species. He defined as 'connectance' of the web, the fraction of all possible pairs of species which interacted directly. The overall result was that increase in number of species, increase in connectance and increase in interaction strength tended to increase instability. Other models (May, 1981) also suggested that complexity leads to instability, which is in opposition to the ideas of Elton and MacArthur. However, some criticisms arose concerning the characteristics of the model communities used, asking if the connection between complexity and instability was not an artefact. For instance, models often refer to randomly constructed communities, whereas in natural communities, interactions are probably not random but are in part the result of coevolution, which will have selected the more stable associations.

Another important field in which complexity and instability could be associated is the range and predictability of environmental conditions that can vary from place to place. In a stable and predictable environment, one expects to find an assemblage of species that is stable only within a narrow range of environmental constraints (a so-called dynamically fragile community). This should be true if the relatively stable conditions have lasted for a long time as, for instance, in the Great African Lakes were fish species have become highly specialised. On the other hand, in a variable and/or unpredictable environment, a dynamically robust community (which is stable over a wide range of conditions and characteristics) will be able to persist. Such is the case for fish communities of Sahelian rivers, which are characterised by the more or less predictable alternation of flood and severe low-waters, with superimposed and unpredictable long-term climate changes. It seems likely that communities from stable and predictable environments are much more susceptible to man-made disturbances than the more robust communities from variable environments, which are already used to adapting to changing situations (Begon *et al.*, 1986).

It should also be pointed out, that properties of communities, and their ability to respond to disturbances, also depend on the biological characteristics of their component populations. Life-history strategies differ according to environmental conditions. In stable environments, the selection of K strategies would be expected, and populations will have high inertia, but once perturbed will have low stability or resilience. By contrast, the reverse (r selection, low inertia, high stability or resilience) is expected in variable environments.

To conclude, while most ecologists feel that an ecosystem is stable if the number of its biological constituents return after perturbation to their original equilibrium values, the present evidence suggests that diversity does not contribute significantly to this kind of stability

Diversity and niche structure

According to ecological theory, every species in an ecosystem occupies a specific niche that is 'a region in a factor space whose axes represent the critical resources or environmental variables to which species in the community respond differentially' (Colwell, 1979). The niche theory should, in principle, provide the theoretical framework to explain the number and type of species that exist in a community (Solbrig, 1991*a*). According to Diamond (1988), the diversity of niches in a community could be determined by four sets of factors; quantity of resources, quality of resources, species interactions, and community dynamics. But application of niche theory to the prediction of the number of species in a community or to explaining why communities differ in number of species has proved very difficult in practice (Solbrig, 1991*a*).

One of the key questions is whether two species can have the same niche, which could also be: is there species redundancy in a community? The corollary question is: how similar can two coexisting species be in the same community? Present theory predicts that two species with similar niches cannot coexist, but this seems to be verified by few empirical studies (Roughgarden, 1989) and for tropical freshwater fish, very few data are available. If many species are present in tropical aquatic systems, are there a larger number of niches, or are there different ways of exploiting the system other than those found in temperate waterbodies? From various detailed studies on the very specialised cichlid fish communities of the East African Lakes, it has become clear that species exhibiting broadly similar behaviour and requirements, actually have developed specific strategies so that they avoid niche overlap.

Various authors have raised the issue of 'functional redundancy' (di Castri & Younès, 1990a; Solbrig, 1991b). According to this hypothesis, within some functional types there are multiple species that perform the same function. Is it necessary to maintain all these species to ensure the ecosystem functions, or could it be possible to remove some species without affecting the response potential of the ecosystem? This question is central for ecosystem management and conservation purposes. It also has to be investigated in relation to the number of functional types in the ecosystem. An ecosystem in which species diversity is associated with many functional types, so that only a few functionally analogous species belong to each type, is likely to respond differently to disturbances or environmental changes than an ecosystem in which high species richness results from relatively few functional types, with many functional analogs within each type (Solbrig, 1991b).

Rare species

Species differ greatly in abundance, and some are extremely rare and highly localised. Using the word 'rare' is a statement about the distribution and abundance of a particular species, but there is a large number of causes that may explain this rarity. Some of them relate to the geological and evolutionary history of the taxon in question, while others are the result of human activities. Different causes of rarity also result from phenomena that are distinguished by, at minimum, different temporal and spatial scales (Fiedler & Ahouse, 1992).

Many fish species are rare species in one or more of the following sense: their distribution is clumped and

individual abundance is low; their distribution is broad but they are found only in very low numbers compared with other fish species; and the distribution is narrow yet populations are represented by many individuals where they are found. Rarity is undoubtedly a natural and common phenomenon, and causes include specialised habitat requirements, trophic position, poor dispersal abilities, and pattern of range expansion after speciation and of range contraction before extinction (Gaston & Lawton, 1990). For a long time, ecologists have considered that large-bodied animals occur at lower population densities than small-bodied animals, but recent studies show that the correlation between population density and body size is feeble and often non-existent (Lawton, 1989; Gaston & Lawton, 1990), and fish are no exception.

Do these rare species play a role in ecosystem function? For some authors they may be viewed as a record of the past. Rare species are those that have, as yet, avoided extinction. Actually, populations of a number of species have apparently persisted for thousands of years, even though many are small and very isolated, as is the case for the remaining isolated fish populations in the northern Sahara (see Chapter 6). For other authors, they are alternative components of the ecosystem, or they are insurance policies, so that roles will be fulfilled even though changes occur (Main, 1982). In fact, common organisms perform the bulk of essential ecosystem functions. However, biotic and abiotic environments change through time, leading to correlated changes in the relative abundance of species. One can assume that stable ecosystems are those in which essential functions are maintained in the face of disturbance. Thus rare species, which make up the bulk of the taxa in most ecosystems, provide an insurance of ecosystem stability, even though they might be considered as non-essential in terms of ecosystem function at the present moment (Pate & Hopper, 1993). This hypothesis may lead to the conclusion that the most stable ecosystems in terms of key functions are those richest in species. However, well-documented studies of rare species substituting for declining common species in the maintenance of key ecosystems functions following disturbance are scant. It is possible, nevertheless, to consider the changes in the Lake Chad fish assemblages following a drastic fall in water level (see Chapter 10) as an illustration of this hypothesis. Some species, which were obviously rare during the period of high water (*Polypterus senegalus, Brienomyrus niger, Schilbe intermedius, Siluranodon auritus*, etc.), became dominant after the drought because they were able to develop in more marshy environments, probably as a result of better physiological adaptation, but also in a situation of relaxed predation constraints. In many riverine fish assemblages in Nilo–Sudanian river basins, one can observe the coexistence, alongside the pool of truly rheophilic species, of a pool of pelagic species able to colonise newly created environments, such as reservoirs, as well as of a pool of palustrine species able to colonise marshy biota. In current situations, most pelagic and palustrine species may be considered as rare species, but they demonstrate high colonisation potential as soon as they are faced with suitable environments. This ability of Nilo–Sudan fish assemblages to adapt to changing environmental situations is certainly both a result and a necessity when considering the long-term climatic changes that have prevailed in this region for million of years.

'At present a general theory of rarity does not exist although one is sorely needed, both as part of a broader understanding of the dynamics of populations, and as a basis for determining conservation priorities' (Gaston & Lawton, 1990). Much can be learnt from the classical 'compare and contrast' approach using assemblages of species from a variety of taxa (including fish) that have contributed little until now to the improvement of this theory, but which may prove to be a very valuable biological model.

Species diversity and productivity

The question has been raised as to what extent are patterns of biodiversity important in determining the behaviour of ecological systems (Lubchenko *et al.*, 1991; Solbrig, 1991*b*). Surprisingly, there are very few theories or empirical studies that might help to answer this fundamental question (Lawton & Brown, 1993). One can state that all species contribute to the integrity of the ecosystem and some ecologists consider that the greatest species richness appears to occur in highly productive habitats. Different investigations have suggested that mechanisms were involved by which increased energy availability tended to result in the proliferation of different species rather than increased populations of existing species (Wright, 1983). It has been suggested that the more productive environments support more but smaller populations of specialised species that are more susceptible to extinction and unable to persist in unproductive environments. However, empirical observations also provide evidence of a decrease in species richness in aquatic environments when overall productivity of the system increases. An alternative hypothesis is that species richness is irrelevant. What is important is the biomass, and the ecological processes in general may function perfectly well with very few species. Available information suggests that major patterns of energy flow may be broadly insensitive to the number of species involved and that is consistent with the redundant species hypothesis.

There is no simple relationship between biodiversity and ecological processes such as productivity. Simple systems, both natural and artificial, appear to be more productive than diverse ones. Moreover, a similar level of production can be observed in species-rich ecosystems and in ecosystems with a low level of diversity.

One way to investigate the role of fish species diversity in ecosystem production is to compare the productivity of different assemblages (expressed for instance by an index such as the mean annual catch per hectare of commercial fisheries) in aquatic ecosystems with similar ecological characteristics. Such an attempt was made for three shallow tropical African lakes: Lakes Chad, George and Chilwa (Lévêque & Quensière, 1988). Fisheries data available (Table 13.1) must, of course, be used with caution, but it appeared that the annual catch per hectare was more or less in the same order of magnitude, i.e. 100–200 kg, in the three lakes under comparison, whereas the fish diversity and trophic chains were rather different. In Lake Chad, there are about 100 species, belonging to different trophic levels from detritivores to top consumers, that are actively fished. In Lake George, 21 of the 30 species recorded are cichlids, and four of them that are phytoplanktivores and exploit the cyanobacteria make up 60% of the fish biomass. In Lake Chilwa, the fish community is still more reduced, the bulk of the catch being made up of three species (*Barbus, Clarias, Oreochromis*), which are more or less opportunists. It should be noted that zooplanktivores are poorly represented in Lake George, in spite of a large biomass

Table 13.1. *Fish diversity, feeding groups and fish production estimated by fishery catches (Lakes Chad, Chilwa, George), or by estimation of fish consumption by birds (Lake Nakuru)*

	Nakuru	Chilwa	George	Chad
No. species	1	3	30 including 21 cichlids	100
Feeding	*Oreochromis* phytoplanktivore	*Clarias–Barbus, Oreochromis* detritivores, zooplanktivores	biomass: 64% phytoplanktivores, 20% ichthyophage	every types
Fishery production kg/ha/year	625–2436	80–160	100–200	100–150

of zooplankton similar to the biomass observed in Lake Chad.

The fishery production of the above lakes may be compared to the fish production of Lake Nakuru, a shallow hypereutrophic equatorial soda lake in East Africa whose area covers up to 50 km². The only fish species is *Oreochromis alcalicus grahami*, introduced from Lake Magadi around 1960 (Vareschi, 1979; Vareschi & Jacobs, 1984). This fish feeds almost exclusively on the blue-green alga, *Spirulina platensis*. There is no commercial fishery, but these fish support a huge population of fish-eating birds (90% *Pelecanus onocrotalus*). The total fish yield taken by the birds was estimated to be between 2700–9500 metric tons per year or 625–2436 kg ha⁻¹ yr⁻¹, which is one of the highest fish yield evaluations for natural lakes, and this high production results from a single species. The short food chain may be one of the reasons, as well as the very high phytoplankton productivity and the short life cycle of the cichlid. On present evidence, species richness is not a major determinant of basic production trends.

More or less similar conclusions were made in the Great Lakes of North America. They have been subjected to over a century of pollution and intensive fishing, and they have lost several native species but some other fish species have been introduced. Despite the dramatic changes in fish species contribution to total catch, and the observation that the lakes no longer function in the same way as they did in the past, the secondary productivity of these ecosystems has changed remarkably little during that period (Pimm, 1993).

625–2436 kg ha^{-1} yr^{-1}

v Conservation of biodiversity

14 The threats to fish biodiversity

I see the world as a living organism of which we are part; not the owner, nor the tenant, not even a passenger. To exploit the world on the scale we do is as foolish as it would be to consider our brains supreme and the cells of other organs expendable. Would we mine our livers for nutrients for some short-term benefit?

Lovelock, 1988

The roughly 3000 species of African freshwater fishes which have been identified are undoubtedly the modern-day survivors of the several thousands of species that have ever existed. Extinctions have occurred in the past, and probably only a very small proportion of the numerous species that have existed over millions of years are known from the generally poor fossil record of Africa. Natural processes have been responsible for these extinctions, but today, human activities are increasingly modifying freshwater habitats and threatening fish species. Over the past 20 years there has been a rapid growth in awareness among freshwater ecologists of the anthropogenic threats facing the endemic biotas of the East African ancient lakes.

Inland waters are particularly sensitive to degradation as a result of the complex properties of water itself, as well as from interactions between the aquatic and terrestrial environments, and the demographic pressure resulting from the acceleration of human population growth since the middle of this century. This acceleration is associated with a reduction in mortality rate, fertility having changed little in sub-Saharan Africa.

One of the major causes of change in biodiversity has been, and probably will be in the long term, climatic changes. In addition to natural changes, human-induced climate changes, resulting for instance from the greenhouse effect, are expected to have severe and rapid consequences for the biosphere and its living compo-

nents. However, at the local or regional scale most impacts come from the use of water and aquatic ecosystems for human purposes. Freshwater systems tend to be the first habitats to experience degradation because people congregate beside water. As a consequence of anthropogenic pressure, they are subjected to a considerable range of stresses including damming and canalisation, reclamation of floodplains, water abstraction, pollution, loading with organic matter and detritus. In addition, activities pursued within the drainage basin (e.g. land use or deforestation) may have consequences for the aquatic ecosystem. It should be stressed that aquatic environments are recipients of virtually every form of human waste and this results in their rapid and continuous degradation. The increasing demand for freshwater resources generated by population growth will probably result in further declines of freshwater biotas in the absence of strict regulation.

The four direct major causes of loss of diversity, called the 'evil quartet' (Diamond, 1984) or the 'four horsemen of the environmental apocalypse' (Pimm & Gilpin, 1989) are:

fragmentation or destruction of some habitats, degradation or pollution of others;
overkilling of plants and animals by humans;
introduction of alien species; and
secondary effects of extinction – the extinction of one species caused by the extinction of another.

According to Reid & Miller (1989), habitat loss and the introduction of exotic species are by far the major threats or causes of extinction for fish species, followed by overexploitation. Whatever the causes, aquatic species are especially vulnerable to changes that result from development, since many are confined to individual watersheds and cannot readily disperse to undisturbed areas. Distinct species assemblages, highly adapted and specialised, have evolved in a wide array of habitats now threatened, and it has been estimated that one-fifth of the world's freshwater fish are either endangered or recently extinct.

The effects of a single perturbation may interact with and multiply the consequences of other disturbances (Fiedler & Jaine, 1992), and the incremental effects of different types of disturbance have a cumulative impact on an ecosystem. Such cumulative, long-term effects are evident in many African aquatic ecosystems, but in most cases, neither the effects nor their causes have been clearly identified or quantified. These include natural long-term changes in water discharge or water level, immediate but also delayed effects of system management, effects of fishing activities and of changes in technology, introduction of alien species, pollution and/or increase in nutrient loading resulting in lake eutrophication.

Theoretically, there should be a unique response to each stress, but it is difficult to identify the particular response to each one. There is considerable evidence that some stresses mimic each other in their effects (Rapport et al., 1985), and it may be valid to propose generalised heuristic models of the behaviour of ecosystems under a family of human-induced stresses (Welcomme et al., 1989). While the various threats to biodiversity tend to be cumulative in their effects, it is nevertheless informative to examine the manifestations of these individual threats on species and habitats.

Global environmental change

As we reach the end of the twentieth century, we have come to realise that our everyday activities may affect people elsewhere on Earth. Global change relates to changes in the Earth's biosphere, including atmospheric, geological, hydrological and biological systems (Price, 1992). The scientific community recognises that such changes have always occurred throughout the history of the Earth, but many scientists are also concerned about the increasing effects of the activities of billions of people on these systems, and some of them consider that these effects have now become equal to or greater than the effects of natural processes.

Two main types of global environmental change are usually recognised. In the first, changes that take place at discrete locations around the world can have a global importance when combined. The cumulative effects of deforestation and changes in land use on regional climates, as well as the release of toxic pollutants, can affect natural cycles at the biosphere scale. The anthropogenic causes of this type of global change lie in processes occurring at regional scales. The second type, systemic changes that occur throughout global systems, includes modifications of the stratospheric ozone layer and of the climate system as a result of increasing concentrations of certain atmospheric gases. These factors are likely to lead to a further systemic change, namely rising sea-levels. Of course, global systemic changes have, in turn, regional and local consequences for the environment.

Global change is, therefore, not always restricted to greenhouse warming, but must also include large-scale alterations in patterns of land and water use and anthropogenic changes in environmental chemistry.

Climate changes and greenhouse warming

The environmental problems of today occur on various scales, and climate change is a global problem with global dimensions (Wyman, 1991; Adams & Woodward, 1992). There has never been any such thing as a stable climate, and climates have varied throughout the Earth's history in response to natural events, such as volcanic activity and continent drift, changes in the Earth's orbit, etc. Periods of glaciation in northern latitudes, corresponded to hyperarid periods in the tropics, whereas the interglacial phases corresponded to more humid tropical climates. There is no evidence at present of major climate change having occurred in the last 2500 years, despite minor fluctuations that have taken place. But within the next few decades it is forecast that the increasing concentrations of greenhouse gases, resulting from human activities (carbon dioxide, chlorofluorocarbons, methane and other anthropogenic polyatomic gases), will lead to a warming of the planet (also called 'global warming'). The recent historical record of atmospheric carbon dioxide concentrations shows an unmistakable upward trend from 280 ppb (parts per billion by volume) during the pre-industrial period, to 350 ppb for the present era. Analyses of air samples during the past two to three decades have corroborated the observed trend. The anticipated result will be a warming of the Earth's troposphere, because these 'greenhouse gases' allow the passage of short-wave radiation from the sun but trap infrared radiation emitted by the Earth. This warming is predicted to direct regional effects on patterns of temperature, and indirect effects on precipitation and evaporation rates.

Rising sea levels would involve saltwater intrusion further into estuaries, higher water tables, and would undoubtedly modify the structure and landscape of coastal wetlands. A rise in sea level may create new aquatic habitat while flooding terrestrial habitat. It could be forecast that in the absence of human interference, coastal marshes will reconstitute along the new shorelines as sea levels advances, but the response will be site-specific.

Much of the research dealing with climate change is co-ordinated through the International Geosphere-Biosphere Program (IGBP) established in 1986 with the objective 'to describe and understand the interactive physical, chemical, and biological processes that regulate the total Earth system, the unique environment that it provides for life, the changes that are occurring in this system, and the manner in which they are influenced by human actions' (ICSU, 1986).

Consequences of climate warming for the water cycle

In order to study the climatic consequences of increased carbon dioxide, a number of general circulation models have been developed (Wetherad, 1991). There are many areas of disagreement between the various models, but it is usually recognised that a doubling of carbon dioxide would result in an increase of 3.5 to 5.2 °C in global surface temperature. This increase would be greater at higher latitudes than in the tropics. Because of a time lag caused by thermal inertia of the oceans, there could be a delay of 30–40 years beyond the time that a doubling equivalent of carbon dioxide is reached, but substantial warming could occur sooner (Peters, 1991).

With 3 °C of global warming we should be faced with a warmer world than has ever been experienced in the past 100 000 years. With a 4 °C rise, it would be the warmest climate since the Eocene, 40 million years ago (Peters, 1991). Moreover, the expected warming would be very fast compared to natural fluctuations in the known past. For the moment we have only the vaguest notion of how climate will change in any given region. If the climate becomes warmer, the Earth's hydrology will certainly change, but one of the major uncertainties is the response of precipitation over the continents during summer. Warmer temperatures will increase evapotranspiration, leading to more clouds and more precipitation (Rind, 1988), and many other components of the hydrological cycle will be affected, such as surface runoff, soil moisture content, river flows, lake levels and groundwater recharge (Gucinski et al., 1990). Current velocity and turbidity of streams will change and if that leads to altered flow during spawning periods, spawning and nursery areas of migratory fishes may be unavailable. Warming may result also in faster drying out of temporary waters, increasing the challenge they present to aquatic life. Water temperature will also increase and thus decrease dissolved oxygen content, and this will have an immediate and direct effect on fish, particularly those living in shallow warm waters. Moreover, changes in the terrestrial ecosystems surrounding the rivers are likely to increase sediment runoff. Climatic effects and multiple human uses of waterbodies may exacerbate these problems.

Forecasting climate change effects on fish biodiversity

One can infer the possible response to climate change by observing the present and past distributions of ecosystems and aquatic organisms. Present distribution ranges are closely linked to precipitation regimes and temperature patterns. Roughly speaking, when temperature and rainfall patterns are changing, species' ranges will shift. But extreme events like droughts or floods may, in some cases, have more effect on ecosystems and species distribution than average climate. Most studies have focused on species response to climate change as independent units, not as assemblages, but from a conservation point of view, the way in which assemblages will respond to climate change is important.

The information that is most relevant to predicting future changes comes from the detailed study of climatic and biotic changes in the relatively recent past of the Quaternary, in which the geographical and biological setting is most closely related to the present. The fairly well-known palaeoclimatic sequence of events in the Sahara (Chapter 6) could, for instance, suggest future scenarios. The average temperature was lower some 20 000 years ago when drought prevailed in that region, but it increased during the late pluvial, when heavy rains allowed extension of freshwater ecosystems far to the north. The fish fauna, restricted to refuge zones during drought, recolonised the fluvial valleys. Since about the end of the pluvial period, the drier climate has led to the shrinking of the Chad and Niger basins, which today occupy about half of their catchment of 8000 years ago. A depauperate relict fauna now occurs in isolated water bodies in the Sahara (Lévêque, 1990b), and is probably endangered due to pollution or destruction of habitats by humans.

The shift of aquatic habitats and species ranges, either contracting or expanding according to the prevailing climate, is not surprising. Nevertheless, the ability of species to adapt to changing climatic environments will depend upon their ability to track shifting specific habitats by dispersing colonists. Any barrier to dispersal will increase the probability of extinction, and species are more likely to become extinct when they occupy limited geographic ranges, such as localised endemics. On the other hand, an overall increase of global precipitation would be beneficial for arid zones and so-called fossil rivers could recover, for instance in the Sahara.

For some authors who argue that the climate may be quite different, the past is no longer a reliable guide to the future. However, we can probably draw the best inferences from the effects of historical changes in climate on freshwater systems, fish assemblages and fisheries. The effect of drought and recovery after drought are, for instance, well documented for Lake Chad (Carmouze et al., 1983) and Lake Chilwa (Kalk et al., 1979) (see also Chapter 10).

In order to evaluate the anticipated effects of climate change on freshwater fish and their habitats, Regier & Meisner (1990) sketch out an iterative assessment process that uses water temperature, water quantity and water quality variables. Temperature affects all vital processes, including activity, feeding, growth and reproduction. According to recent studies (Hill & Magnuson, 1990),

growth of fish is expected to increase with climate warming if other factors now limiting growth also change with climate. But climate warming is also expected to increase productivity at all aquatic trophic levels by about 10–20% per 1 °C increase in temperature (Regier et al., 1990). In his examination of the tendency for species richness to increase with decreasing latitude, Stevens (1989) postulated that low-latitude species have narrow climatic tolerance. Thus, even a small increase in temperature may be deleterious to low latitude organisms. But such speculations are hardly supported by the information on climatic changes in Africa over the last twenty thousand years (cf Chapter 6).

Another likely long-term consequence of global climate change is the modification of the genetic composition of populations and species. Selection for tolerance to heat can quickly lead to the evolution of tolerant genotypes that may have altered life-history traits (Parsons, 1988).

Perspectives

At the moment, there are too many unknowns regarding the degree and extent of potential changes to allow accurate predictions of likely effects of global climate change. It is unclear how the African tropical freshwater biota would be affected. Most simulations predict a relatively small temperature increase for the lower latitudes, and we know that the history of that part of the world has been a complex succession of droughts and humid periods. Nevertheless, there are reasons to pay particular attention to possible future changes (Peters, 1991).

1. The past natural changes were quite slow compared to changes predicted for the near future (the change to warmer conditions at the end of the last ice age spanned several thousand years), and the speed of the projected change may exceed the ability of many species to develop the necessary physiological adaptations, or to disperse.

2. The conjunction between climate changes and anthropogenic impacts could threaten more species than either factor alone. Habitat destruction and artificial barriers will obviously prevent many fish from colonising new habitats when their previous one is threatened by climate changes.

Opinions regarding the responses necessary range

from those economists who think mankind should simply use his technology to adapt to changes as they occur, to those who advocate the expensive option of reducing carbon dioxide emissions and thus prevention of further global change. Research on the ecological aspects of global change should contribute to basic ecological understanding of processes regulating the Earth's biota. Basic questions, such as what regulates the large-scale dynamics of plant and animal populations or the fluxes of energy and pollutants within and between ecosystems, need collaboration between different disciplines including atmospheric science and environmental toxicology. But the ultimate goal for ecologists is to be in a position to assist decision-makers in devising policies to anticipate, ameliorate or respond to global environmental change.

Flow regulation in Africa

Habitat alteration is one of the major causes of loss of diversity of aquatic life, and degradation or destruction of habitats are particularly threatening in rivers. Flow regulation occurs on almost every large river in tropical Africa. Man-made lakes associated with hydroelectric dams or built for irrigation purposes prevent fish migrations, and alter the flow pattern downstream, one of the major consequences being the disappearance of floodplains and spawning grounds for different species. Canalisation of the stream bed reduces diversity of habitats. Catchment changes caused by large-scale land-use practices associated with deforestation give rise to erosion and increase the sediment load in river waters, as do mining-related activities. In this context, fish communities are affected not only by events occurring in the channel, but also by several external influences occurring in the catchment.

The multidimensional nature of streams

Many human activities interfere with the natural dynamics of river ecosystems in ways that can be examined from four different perspectives: (i) the longitudinal dimension from upstream to estuary; (ii) a lateral dimension extending beyond the channel and including associated wetlands; (iii) a vertical dimension encompassing the groundwater system; and (iv) each of these has a temporal dimension (Ward & Stanford, 1989). It should be

stressed that the functioning of freshwater systems must be considered in this multidimensional space.

The upstream–downstream linkages in a river system may be modified by hydraulic engineering within the basin and different impacts can be identified.

1. Changes in flow patterns (quantity and timing of discharge). Flow regimes are radically altered upstream and downstream after the building of large dams, as illustrated by the River Nile (Fig. 14.1). Loss of floodplain fertility induces a decline in species diversity and productivity. The numerous small dams built during the last two decades on the tributaries of many large rivers in the Sudano–Sahelian zone, for irrigation development or water supply, also have an overall impact on the hydrology by reducing the discharge in the main rivers.

   The distribution of flow in time may also be altered with different biological consequences: spawning success and growth rate of river species are known to be related to flood strength (Chapter 8). Therefore, changes in discharge and velocity usually result in a shift and simplification of fish community structure.

2. Blocking of the channel: interruption of migratory pathways after the building of the dam walls could have dramatic consequences for migratory fish species that can no longer find suitable habitats for spawning.

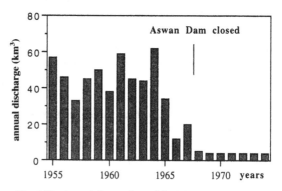

Fig. 14.1. Annual flow regime of the Nile River (from Balarin, 1986).

3. Changes in habitats: establishment of a man-made lake behind a dam creates a novel environment that is colonised only by species able to adapt to lentic conditions. Conversely, impoundments could eliminate specific habitats such as rapids, which are usually inhabited by fishes highly adapted to torrential environments. Below dams, the reduction in flow results in a narrower river bed and a concomitant loss of habitats. The most drastic change is undoubtedly the disappearance of downstream floodplains.

The interactions between the river and its catchment are also strongly modified by hydraulic works.

1. Floodplain interactions: alteration of the flood regime results in major disruptions in exchanges between the channel and the floodplains including side arms, lakes, swamps and other wetlands. Change in species composition and loss of obligate floodplain spawners are observed, as well as a general diminution in productivity. Moreover, because of the fertility of their soils, floodplains are regarded has having high development potential and there are many efforts to 'reclaim' them by flood control.
2. Deforestation: disappearance of forested areas in river catchments leads to erosion and siltation of rivers and lakes, as well as changes in the patterns of water discharge.

Human activities or natural climatic events may change the height of the groundwater table, with sometimes serious ecological consequences. Groundwaters play a major role in sustaining the river flow during low-water periods, and a lowering of the water table could result in only temporary flow or even the drying up of rivers. Consequences for off-channel habitats are also severe.

Changes in land-use and their consequences for aquatic systems

As indicated above, the quantity and quality of water in aquatic systems may be affected by changes in the terrestrial components of the watershed. One well-known phenomenon is the degradation of wetlands that are highly complex productive systems and that serve important functions for fish communities. World-wide, wetlands are rapidly being lost to drainage, mainly to increase agricultural production. In many areas, the traditional practices of flood recession agriculture on the floodplains has extended in response to increase in human population. Such practices, which were not so destructive to the environment, are being replaced by irrigation schemes as, for instance, in the Senegal valley where floodplains are being used for irrigation agriculture and the growth of swamp rice after damming of the Senegal River.

There is widespread concern about rapid deforestation in the tropics partially because tropical rain forests, occupying 7% of the land surface, contain more than half the species in the entire world biota (Wilson & Peters, 1988). Barnes (1990) provided an analysis of published statistics on deforestation in Africa. The East African countries have only small areas of forest, while much larger tracts still exist in West Africa. The human population is dense in both regions, and deforestation rates are high, especially in Côte d'Ivoire and Nigeria (see Table 14.1). In the unbroken block of the central African forests, human populations are sparse and deforestation rates are low. Highest values of per capita deforestation rates (area deforested per million people, see Table 14.1) are found in West and Central Africa. Data from three selected countries illustrate the dramatic reduction of forested areas. If present trends continue, 70% of West African forests, 95% of forest cover in the eastern countries, and 30% of the Zaïre forest will disappear by 2040. Deforestation is linked to growing human populations, and correlated with increasing timber exports and fuelwood use.

In Madagascar, the 3.8 million hectares of rain forest remaining in 1985 represented only 50% of what existed in 1950, and 34% of the estimated original extent. In those 35 years, the rate of deforestation averaged 111 000 ha per year. Deforestation was more rapid in areas with low relief and high population density (Green & Sussman, 1990). Soil no longer protected is subject to rapid erosion from hurricanes and rainstorms. Erosion rates up to 250 tonnes per hectare have been reported (Helfert & Wood, 1986), much of which ends up in freshwater ecosystems.

The destruction of aquatic habitats is a major consequence of deforestation and the possible disappearance of

Table 14.1. *Annual deforestation rates in 1980 for different African countries. All figures for 1980*

Countries	Original forest area (km²)	Mean % deforested 1976–80	Human population density (per km²) 1980	Mean area deforested per million people (km²)
West Africa				
Guinea	20 500	1.76	22.1	66
Sierra Leone	7400	0.78	48.8	17
Liberia	20	2.05	16.8	219
Côte d'Ivoire	44 580	6.95	25.6	375
Ghana	17 180	1.57	48.3	23
Togo	3040	0.66	45.4	8
Nigeria	59 500	4.79	91.7	34
Central Africa				
Cameroon	179 200	0.45	17.8	95
Gabon	205 000	0.07	2.5	227
Congo	213 400	0.10	4.7	137
Zaïre	1 056 500	0.16	12.3	57
East Africa				
Uganda	7500	1.33	53.4	8
Tanzania	14 400	0.69	19.7	5
Kenya	6900	1.59	28.6	7

From Barnes, 1990.

endemic species adapted to life in clear forest waters. Another major consequence is a change in the water quality of the associated rivers. In undisturbed tropical forest the streams that leave the forest contain almost pure rainwater with low mineral content. As soon as the forest is cut, nutrients are washed out by rains and the mineral content of the water increases rapidly.

Deforestation severely affects aquatic communities, because massive quantities of sediment eroded from clear-cut watersheds are eventually discharged into freshwater systems where they can be extremely destructive to aquatic organisms. Soil erosion can be particularly dramatic following deforestation on steep slopes and can result in siltation that has adverse effects on river biology, by covering spawning sites with sediment, destroying benthic food sources and reducing water clarity for visual-feeding fish (Bruton, 1985). Excess sediment loading in standing waterbodies reduces light penetration and thereby photosynthetic rates both planktonic and benthic. Herbivorous fish may be directly affected through reduced foraging efficiency, while zooplanktivores may also be indirectly affected by a decrease in the herbivorous zooplanktonic organisms on which they prey. Suspended sediments also affect filter-feeding

organisms by interfering with their feeding apparatus. In South Africa, sedimentation of estuaries has contributed to the decline of many native fishes (Skelton, 1987).

Cohen *et al.* (1993) investigated the impact of excess sediment on the biodiversity of Lake Tanganyika. Landsat image analysis has illustrated the severity of deforestation in many parts of the Lake Tanganyika drainage basin. Forest clearing by massive, uncontrolled fires is followed by conversion of original forest/woodland areas to grazing land or for use in subsistence agriculture. Between 40–60% of the original woodland or forest land has been cleared in the central portions of the lake's drainage basins, and almost 100% around the northernmost portion of the lake. Steep slopes, heavy rainfall, generalised cultivation without protecting the land and slopes, and the abundance of torrential mountain streams are all factors generating rapid headwater erosion and stream incision, which generate a massive increase of suspended sediment and sedimentation rates in the nearshore region of the lake. For example, the rate of outbuilding of the Ruzizi River Delta, the major drainage in the northern end of the lake, has probably increased by an order of magnitude in the past 20 years over its rate prior to deforestation (Cohen, 1991). There are also

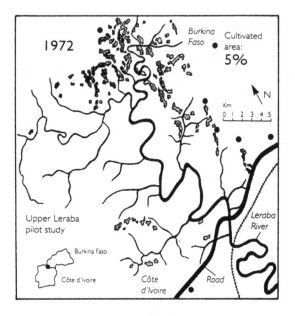

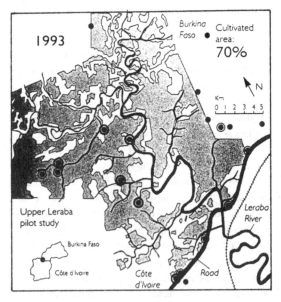

Fig. 14.2. Enhanced aerial photographs of the pilot study
area taken in 1972 and 1993 show the extension of
land use along the Upper Leraba River (from
Liese, 1994).

many fewer species of diatoms, ostracods and fish, in
highly disturbed portions of Lake Tanganyika than in
moderately disturbed or undisturbed areas (Cohen *et al.*,
1993).

Changes in land-use as a result of the elimination of
diseases also has an indirect effect on the aquatic
environment. In the Onchocerciasis Control Programme
in West Africa (see below), the use of aerial photographs
allowed us to document the change in settlement, land-
use and the environment of the Leraba River (north of
Côte d'Ivoire). In 1972, only 5% of the land of the pilot
area was under cultivation, while in 1983 it was 30%, and
by 1993, 20 years after the start of the control programme, 70% of the land was being farmed (Fig. 14.2)
(Liese, 1994). The gallery woods bordering the small
streams had been cut down for building materials and
firewood, and the remaining savannah woodland is less
than 25%. It appears that the fish community in the
areas around the settlements has also been disturbed and
is less diverse and abundant than in habitats located a few
kilometres downstream. The future spread of this situation in the pilot zone is a matter for concern.

Man-made lakes

Dams are costly, prestigious and controversial structures.
They can be regarded as central to national development,
or as disasters as a result of unintended ecological, economic or social side-effects. Most major rivers have at
least one impoundment on them, and almost every river
on the continent of Africa has been interfered with,
mainly by the construction of man-made lakes. This is a
fairly recent situation. Most of the largest dams were
built after the mid 1950s, on large rivers and for elec-
tricity supply (see Table 14.2). A useful bibliography was
prepared by Ita & Petr (1983). More recently, probably
thousands of smaller reservoirs have been established,
principally on smaller rivers and tributaries, to meet
other water demands including domestic use, irrigation,
stock-farming, fish production, etc. The impact of
African tropical dams is quite different from that of
north temperate dams. Often what is considered an
undesirable environmental effect in the temperate situation (e.g. in Canada) is either unimportant or even beneficial in the tropical situation. The transferability of
scientific and engineering expertise is therefore low
(Bernacsek, 1984).

Table 14.2. *Approximate data relating to major tropical African dams (above 500 km²)*

Lake	River	Country	Closure	Area (km²)	Max depth (m)
Volta	Volta	Ghana	1964	8270	80
Kariba	Zambesi	Zambia–Zimbabwe	1958	4300	125
Nasser	Nile	Egypt	1964	6000	80–90
Cahora Bassa	Zambezi	Mozambique	1974	2660	36
Kossou	Bandama	Côte d'Ivoire	1971	1630	60
Kainji	Niger	Nigeria	1968	1270	60
Buyo	Sassandra	Côte d'Ivoire	1980	900	
Kafue Gorge	Kafue	Zambia	1972	809	58
Lagdo	Benue	Cameroon	1982	700	20
Mtera	Great Ruaha	Tanzania	1980	610	7
Gebel Aulia	White Nile	Sudan	1937	600	
Mbakaou	Djerem	Cameroon	1968	500	

For a few major impoundments, the sequence of events after damming has been studied. Soon after the closure of the dam, there is a substantial increase in fish populations favoured by the new lacustrine conditions. This expansion does not, however, last very long and the fish biomass decreases sharply as predators reduce the inflated population that first dominated the reservoir's ecosystem. It is therefore difficult to predict the ultimate nature of the fish communities that become established. They have distinctive features depending on the physical and chemical characteristics of the lake and the original fish fauna. Nevertheless, the different lakes have certain trends in common (Jackson *et al.*, 1988). The most obvious change is that reservoirs behind dams do not provide habitats for most riverine fish, and the transformation of the ecosystem from fluviatile to lacustrine results in the disappearance of fish populations adapted to running waters and the development of others. A well-documented example is reported by Jackson *et al.* (1988). *Labeo congoro* and *Labeo altivelis* were important commercial fishes in the Zambezi and abundant at the Kariba site before closure of the dam. They used to have well-marked annual spawning migrations up the tributary rivers. The decline in the *Labeo* stocks after closure of the dam was attributed to the flooding of its favourite habitat. This hypothesis was supported by observation of the progressive reduction in standing stocks of *Labeo* as distance from the more riverine end of the reservoir

increased. In West Africa, a sharp decline of Mormyridae was noticed both in Lake Kainji (Turner, 1970) and Lake Volta (Petr, 1968), which has been attributed to the inundation of their preferred benthic habitats by deoxygenated hypolimnetic waters.

However, with the creation of a pelagic zone, a new environment is offered to fish capable of adapting to life in open waters. In many instances, small clupeid fishes became abundant after impoundment: *Pellonula afzeluisi* and *Sierrathrissa leonensis* in Lake Kainji (Otobo, 1978), *P. afzeluisi* and *Cynothrissa mento* in Lake Volta (Reynolds, 1971), and *Limnothrissa* was introduced into Lake Kariba (Cochrane, 1978). Predators moving out to feed on the schooling pelagic species, also became adapted to reservoir open waters, for example the tiger fishes of the genus *Hydrocynus* in Kariba. Although the number of species in a reservoir may be equivalent to the number inhabiting the original river at the reservoir site, native forms often disappear (Moyle & Leidy, 1992).

Case study: Lake McIlwaine

Lake McIlwaine is a reservoir created in 1953, close to the city of Harare (Zimbabwe). It is a recreational centre, with angling as a major attraction, and a commercial fishery was established in 1956. Marshall (1982) gave an overview of the changes in the fish communities as a consequence of different anthropogenic events.

Twenty-one indigenous fish species had been

recorded and five species were introduced. *Oreochromis macrochir* was introduced with *Tilapia rendalli* in 1956 to improve the commercial fishery and to reduce weed growth. *Serranochromis codringtoni* was introduced from Lake Kariba in 1978 as a snail predator and to improve angling. *Cyprinus carpio* reached the lake from fish ponds, but has probably never bred there so that it is now likely to be very rare or extinct. *Micropterus salmoides* also reached the lake in small numbers and appears to have increased since 1978.

Between 1953 and 1959 the quality of water in Lake McIlwaine was good. From 1960, periodic cyanobacteria blooms appeared (*Anabaena* and *Microcystis*), as a result of eutrophication caused by the drainage of sewage effluents rich in nitrogen and phosphorus from the urban area. By 1970, there was a massive build-up of water hyacinth and extensive deoxygenation of the water, especially at the time of overturn, which caused fish kills. Increasing concentration of nutrients led to a hypereutrophic state in 1971. By 1972, water hyacinth was virtually eradicated by chemical control, and waste water was beginning to be diverted to irrigation schemes. This continued until 1977 when almost 100% of the municipal waste water was being treated. The last fish kill was observed in 1976 and by 1980 the lake was close to mesotrophy (Thornton, 1982).

Commercial fishing data could be used to provide an assessment of population changes for the five most important species over the 25 years, 1956–79. Several typically riverine cyprinid species, such as *Barbus marequensis*, were unable to adapt to lacustrine conditions, and have now virtually disappeared from the lake, whereas this fish was taken in early commercial catches. *Clarias gariepinus* was the major component of the commercial catch until 1966 (more than 50% by weight) and then declined to only 15% in 1979. *Hydrocynus vittatus*, the major fish predator, made up 40% of the catch during the first year, but later declined to about 5%. The most striking changes occurred amongst the cichlids. *Tilapia rendalli* increased until 1962 (20% of the catch) and destroyed the marginal vegetation, which could explain its subsequent decline, although this coincided with increasing intensity of algal blooms that may have also contributed to its decline (to about 2% in 1979). *Oreochromis macrochir*, introduced in 1956, did not appear in catches before 1963, but it rapidly increased and by 1972

it comprised 60% of the commercial catches. It feeds on blue-green algae, which it possibly digests in a similar way to *Oreochromis niloticus* (Minshull, 1978). The increasing water quality by the end of the 1970s was probably detrimental to *O. macrochir*, while on the contrary, *Labeo altivelis* has shown a steady increase since 1970 and made up 60% of the catches in 1979. This may be the result of improved water quality and an increase in benthic diatoms, a major food item for *L. altivelis*.

Case study: Lake Kainji

This large reservoir (1800 km^2) on the River Niger was closed in 1968, and has been much studied. Preimpoundment studies were made by Banks *et al.* (1966), and a census of fish in a stretch of the River Niger was possible during the construction of the dam (see Table 14.3). About 90 species representing 24 families occur in the lake, most of them being widespread Nilo–Sudan species. The fish community started to change as soon as the dam was closed. Changes in composition of the fish communities were investigated by studying the commercial fisheries, and the more striking features observed during the first years were as follows.

1. A decrease of mormyrids, which were previously very abundant in the river where they constituted 40% of gill-net catches. There could be several different causes for this decline. Most of them are insectivores and utilise the same prey organisms, so the paucity of benthos in the new lake, reflected in the limited range of prey in their stomachs, certainly led to feeding limitations. However, it is unlikely that this was the only cause of the decline (Blake, 1977c). The need for riverine conditions for their reproduction may have been another important factor. Many of the mormyrids appeared to be restricted to sheltered shallow water with extensive reed beds and *Hippopotamyrus pictus* was the dominant species in open water gill nets.

2. The dramatic increase in *Citharinus citharus* in gillnet catches, probably because of successful spawning as the lake filled (Lowe-McConnell, 1987).

3. A boom of small clupeids and the schilbeid catfish, *Schilbe mystus*, during the first year. Among the zooplankton-feeding schilbeids, *Physailia pellucida* inhabited the open waters, whereas the clupeid *Sier-*

Table 14.3. *The relative abundance of fish families caught before the closure of the dam in the 18 ha coffer-dammed lake in the Niger River at Kainji, 1966[1], and after the closure in 21 rotenone samples in Kainji Lake between 1975 and 1976[2]. Specimens of* Polypterus, Heterotis, Gymnarchus, Malapterurus, Lates, Oreochromis, Tilapia *and* Tetraodon *were also collected*

| Fish families | R. Niger before damming | | | Kainji Lake | | |
	No. of species	% of individuals	% weight	No. of species	% of individuals	% weight
Mormyridae	19	20.7	19.5	12	1.4	1.4
Characidae	8	36.3	12.1	8	18.6	10.3
Citharinidae + Distichodontidae	5	6.1	19.8	4	4.7	15
Cyprinidae	5+	3.3	4.3	7	8.6	5.5
Bagridae	7	7.2	18.2	8	18.5	15
Schilbeidae	3	8.0	3.6	3	0.1	*
Mochokidae	18	18.0	18.7	11	2.3	2.5
Clariidae	2	*	*	2	*	2.2
Malapteruridae	1	*	*	1	0.1	2.5
Centropomidae	1	*	*	1	1.7	1.2
Cichlidae		*	*	7	36.1	43.6

[1]From Motwani & Kanwai, 1970; [2] from Ita, 1984. *, very low %.

rathrissa leonensis was more numerous in the shallows (Otobo, 1974). A specific fishery developed for these small clupeids, which were caught in 'atalla' lift nets and by a mid-water trawl dragged between two boats.

4. An increase in the catch of large predators such as *Lates niloticus* and *Hydrocynus forskalii* feeding on the clupeid stock. Before impoundment, the predators represented 15–17% by weight of the catches, increasing to 18–31% after impoundment (Lelek & El Zarka, 1973).

Immediately after impoundment, annual commercial catches rose to a maximum of 28 000 tons in 1970–1, far above the expected 10 000 tons yr^{-1}. But the boom was short and followed by a rapid decline, stabilising at 4500 tons by 1975. This was associated with further changes in the fish communities.

1. *Citharinus citharus* declined rapidly.
2. Cichlid populations, which had been fairly numerous along the edges of the river, increased slowly, in contrast to what had been observed in Lakes Volta and Kariba. *Sarotherodon galilaeus* populations built up in the flooded bush and by 1972 were an important element of the cast-net catches, together

with *Oreochromis niloticus* and *Tilapia zillii*. By 1976, tilapias had become the main component of the inshore community, in waters less than 7 m deep, and were also fished by beach seines.

3. Mochokid catfish fall into two main trophic groups: (i) pelagic zooplankton feeders that also eat surface insects and; (ii) benthic feeders eating mainly insect larvae or molluscs. Eighteen species were recorded, but the species composition changed as riverine gave way to lacustrine conditions. Such changes apparently tied in very well with changes in food availability (Willoughby, 1974). Changes in habitats and in the oxygenation of the bottom water could also explain the observed shifts, but we have very little information on the breeding sites and habits of these fish. Under lacustrine conditions, numbers of the plankton feeder, *Brachysynodontis batensoda*, increased and it became abundant in gill-net catches. This was also the case for the bottom feeder, *Synodontis schall*, while other benthic species, such as *Synodontis violaceus* and *Synodontis clarias*, became rare.

Downstream from the lake, there was an overall decline in the fish catch upstream of the Benue River confluence. The Mormyridae almost disappeared from commercial catches in the two downstream stations nearest the dam. In contrast, *Lates niloticus* increased in the catches at all stations. Adeniyi (1973) reported problems for fishermen in the clearer water below Kainji where the fish were apparently able to see and avoid nets. In addition, productive pools and swamps, previously flooded, were now dry.

Case study: Lake Kariba

Lake Kariba was formed by the damming of the Zambezi River at Kariba Gorge in 1958. Prior to the creation of the dam, Jackson (1961) and Harding (1964) listed up to 31 species from the Middle Zambezi. A few years after closure, Balon established a list of 39 species, the most abundant being *Brycinus lateralis*, which started to explore the new open-water environment where almost no fish were present, except for the eel, *Anguilla bengalensis labiata*. Most of the fish population of riverine origin was concentrated in shallow inshore waters.

One interesting modification after the dam closure was the balance between *Brycinus imberi* and *Brycinus lateralis*. The latter was a fish of the Upper Zambezi before the construction of the dam, whereas *B. imberi* was the most abundant fish of the Middle Zambezi, unknown in the Upper Zambezi. Lake Kariba was inhabited exclusively by *B. imberi* until 1962. The first specimens of *B. lateralis* were caught in the lake in 1963, but in 1964–5, Matthes (1968) noted that *B. lateralis* was largely dominant, and had eliminated *B. imberi*. *Brycinus lateralis* reached the Middle Zambezi area via Victoria Falls. It did not colonise the Middle Zambezi earlier, because *B. imberi* was apparently better adapted to spawning in the floodplains while *B. lateralis* preferred lentic conditions (Balon, 1971). However, in 1975, *B. imberi* became more abundant again, and later remained an important component of the catch. The reasons for such changes are not clear. Marshall (1984a) suggested that *B. imberi* was unable to compete with the tiger fish *Hydrocynus vittatus*. The abundance of that species increased until 1974 when commercial purse seining led to a rapid decline. Concurrently, the *B. imberi* population increased (Table 14.4). According to Jackson *et al.* (1988), *B. lateralis* was supplanted by the introduced clupeid *Limnothrissa*, which efficiently utilised pelagic zooplankton.

Several other fish species present at Victoria Falls also

invaded the newly created Lake Kariba, including *Marcusenius macrolepidotus*, *Labeo forskalii lunatus*, *Schilbe intermedius*, *Serranochromis macrocephalus*, *Serranochromis robustus jallae*, *Sargochromis giardi* (Balon, 1974a). Nevertheless, other species recorded from Victoria Falls were never found in Lake Kariba. The cyprinids, *Labeo altivelis* and *Labeo congoro*, which were initially abundant before the dam was closed, declined rapidly and none were taken in 1967 (Table 14.4). A similar trend was observed for the two *Distichodus* species (*D. mossambicus* and *D. schenga*). The river cichlids, *Tilapia rendalli* and *Oreochromis mortimeri*, expanded rapidly after the lake formed (Harding, 1966).

The lack of fish in the open water, despite the abundance of zooplankton, justified the decision to stock the lake in 1967–8 with *Limnothrissa miodon*, a zooplanktivorous clupeid endemic to Lake Tanganyika. In 1969, a survey revealed that the introduction was a success, and in 1971 there were indications that the tiger fish, *Hydrocynus vittatus*, was moving offshore to exploit the Tanganyika sardines, which were abundant in its stomach contents.

Other impacts

Large water-diversion projects in Africa have also caused reductions in estuarine and marine fish populations. Reduction of freshwater inflows reduces the amount of nutrients flowing into downstream areas and increases salinities. The Aswan Dam, for instance, impounds 50 to 80% of the flow of the River Nile. Its closure in 1965 was associated with a drastic decline in pelagic fish populations in the eastern Mediterranean (Ryder, 1978), where the annual landing of sardines (*Sardinella aurita* and *Sardinella maderensis*) and other fish has declined roughly 96% and 36% respectively.

Channelisation involves the realignment, clearing and lining of the stream channel, usually for flood control. Its major physical consequences are to create uniform habitat conditions (reducing habitat heterogeneity), to drain adjacent wetlands and to destroy riparian vegetation. Until now there were few large projects involving the channelisation of river systems in Africa. However, the Jonglei Canal, a huge project of channelisation of the Nile where it crosses the marshy Sudd system in Sudan was planned (Howell *et al.*, 1988) for many years. The sudden announcement in 1974 of its impending

Table 14.4. *Lake Kariba: changes in the abundance of major fish species, lakeside station. Expressed as numbers per 50 standard gill-net fleet settings*

Taxa	1960	1961	1962	1963	1964	1967	1969	1970	1971	1972	1973	1974	1975	1976	1977	1978	1979	1980	1982
Mormyridae																			
Hippopotamyrus discorhynchus					13	20	225	318	556	1090	1028	532	159	378	389	500	286	565	270
Mormyrus longirostris						90	602	294	800	371	324	266	68	96	165	275	112	184	33
Mormyrops deliciosus			4					24	16	68	141	277	413	460	443	159	91	122	284
Marcusenius macrolepidotus					32			6				20	6	11	7	9	8	18	26
Characidae																			
Hydrocynus vittatus	2900	2000	2773	3258	2058	1370	2876	2718	5936	6042	4466	7020	4219	2906	1979	1808	1332	1646	2095
Brycinus imberi	3150	2200	968	1233	180	110	6	6	36	50	30	206	994	1093	1379	827	1649	1364	878
Distichodontidae	400	650	1884	109	114	50	5	6		6	6	12	2	4	1	4	1		1
Cyprinidae																			
Labeo spp.	11 650	4600	2096	1384	656		4			6	6	4	22	9	45	29	90	87	18
Siluriformes																			
Schilbe intermedius			46	146	18	80	664	900	1260	694	295	604	212	160	601	299	324	735	162
Clarias gariepinus	600	1450	828	667	45	90	209	42	48	172	104	458	345	231	168	294	73	78	204
Heterobranchus longifilis								6		3	4	6	12	12	16	2	1	2	4
Synodontis zambezensis					26		63	18	36	50	99	167	88	103	125	80	57	103	84
Cichlidae																			
Oreochromis spp.	1900	50	1050	408	139	480	915	192	356	1039	757	2684	2046	1632	1493	1178	150	171	332
Serranochromis condringtoni		1	50	21	6	140	401	162	312	1635	1145	1967	2439	1860	1118	723	571	534	1567
Tilapia rendalli		150	23	46	9	40	83	18	80	141	64	193	304	314	144	132	248	80	117

From Marshall, 1984a.

implementation caused unfavourable reactions among
southern Sudanese, partly due to unease over lack of
consultation on a matter that was clearly of deep concern
to them. This project started in 1978, but has come to a
halt, incomplete, because of political trouble since 1983.
Among the expected effects of the canal, a decrease in
permanent swamp area would certainly have happened,
as well as a loss of habitat diversity and the selective disappearance, or at least the drastic reduction, of many
floodplain-dependant species.

Pollution

Most detailed work on aquatic pollution has been carried
out in temperate countries with concern for temperate
zone organisms and ecosystems. However, many of the
principles demonstrated in temperate regions are certainly applicable in the tropics, modified primarily by the
effects of higher water temperatures. Very little detailed
work has been carried out in Africa, despite the fact that
sources of all types of pollution are rapidly increasing on
that continent and their impact on freshwater communities is clearly becoming apparent (Dejoux, 1988).
Because there is so little information on the specific
effects of oil, heavy metals and eutrophication in Africa,
these topics will be briefly considered first. Little more
has been done on pesticides in general, but one marked
exception is the Onchocerciasis Control Programme in
West Africa, which will be discussed as a case study.

Eutrophication
Eutrophication results when nutrients from agricultural,
industrial and human sewage sources are released into
fresh waters in excess of quantities normally required by
the ecosystem. With increasing human populations and
development in Africa the eutrophication of natural
waters is becoming serious. In particular, excess phosphorus and/or nitrogen leads to enhanced growth of
plants and increase in other forms of organic productivity. The blue-green algal blooms that frequently
result from eutrophication often have consequences
which can include the deoxygenation of deeper water and
sometimes of the whole water column, in extreme cases
leading to massive mortalities of fish. Some species of
cyanobacteria also produce poisons, which have been
shown to be responsible for fish deaths.

The eutrophication of Lake Victoria during the last
25 years is quite well documented (Hecky & Bugenyi,
1992). Enhanced quantities of nutrients appear to have
been entering this lake for many years, both through
rivers and from aerosols (ash particulates), as a result of
human activities in its watersheds. The observed increase
in algal biomass in this case might be attributed to an
increase of nitrogen and sulphur, as the phosphorus concentrations changed little from 1961 to 1988. The
eutrophication could lead to increased oxygen demand in
the lake's deep water and thus decrease the hypolimnetic
volume habitable by fish during seasonal stratification.
This phenomenon is partly responsible for the threatening, or disappearance of cichlid species belonging to the
haplochromine flock. The eutrophication of Lake Victoria may be the result of the cumulative effects of the
increasing urbanisation around the lake, as well as the
use of fertilisers and pesticides for agriculture or tsetse
fly control. It is also possible that the drastic reduction in
the herbivorous haplochromine stocks, resulting from
intense fishing activities and predation by the introduced
Nile perch, may contribute to the phenomenon by lowering the consumption of plant material and therefore
increasing oxygen demand through decomposition of this
material.

A slow eutrophication of Lake Kivu has also been
reported, with lower average oxygen levels and higher
nitrate levels. Population growth, deforestation and erosion are likely to be the main causes of these changes.
Symptoms of eutrophication, such as cyanobacteria
blooms, have been detected in Lake Naivasha as a consequence of human population increase and agricultural
intensification in its hinterland (Harper et al., 1993).
Eutrophication appears to be promoted by the destruction of papyrus and ploughing of lake-edge land.

Domestic and industrial organic loads
During recent decades, a considerable population growth
has taken place in many African countries, accompanied
by a steep increase in urban, industrial and agricultural
land use. Inland waters often became the recipients of
organic matter in amounts exceeding their natural
capacity of self-purification. Sewage and other effluents
rich in decomposable organic material cause primary
organic pollution, which is highly oxygen-demanding
and causes, in turn, anoxia. Many African towns have

open drain systems for municipal waste waters and they are flooded during the rainy season, leading to high organic discharge to the receiving waters over short periods of time (Dejoux, 1988; Saad *et al.*, 1994). Organic wastes from food processing industries, such as sugar refineries, breweries, palm-oil industries, etc., are also major sources of organic pollution in African freshwater.

Industrial pollutants and pesticides are used in the Bujumbura area, and their discharge into Lake Tanganyika is largely unregulated. However, several recent studies of the lacustrine fauna and flora have failed to reveal significant levels of toxic wastes in fish. DDT concentrations in fish tissue have showed a decline in recent years (Sindayigaya, 1991). The impact of increased dissolved pollutants on the lake's phytoplankton around Bujumbura appears to be limited to a zone of a few kilometres around the urban area (Caljon, 1992).

The effects of organic pollution, resulting in mass mortalities of the aquatic fauna (including fish) and of whole aquatic communities, have been reported from different areas (see Calamari, 1985; Dejoux, 1988; Saad *et al.*, 1994). Considering that little attention has been given to such events in an age where development takes precedence, it must be obvious that, as a result of urban and industrial organic pollution, fish biodiversity is under threat, and the case is much more serious than current published works would indicate. The future must be to facilitate the introduction and use of technologies to control this type of pollution.

Heavy metals

Heavy metals occur naturally in low concentrations, but recently the occurrence of metal contaminants in natural aquatic ecosystems has become of increasing concern. This is the result of increasing urbanisation, expansion of industrial activities, mining, atmospheric sources, e.g. burning of fossil fuels and incineration of wastes, and modernisation of agricultural practices. Unlike other pollutants, trace metals may accumulate, unnoticed, to toxic levels. This explains why problems with trace metal contamination were first highlighted in the industrially advanced countries with their larger industrial discharges. However, with the increase in industrial and urban activities, there is a growing awareness of the need for the control of wastes discharged into African environments.

Some heavy metals, such as zinc, copper, manganese and iron, are essential for the growth and well-being of living organisms, but they are likely to cause toxic effects when the organisms are exposed to levels higher than normally required. Other elements, such as lead, mercury and cadmium, are not essential for metabolic activities and exhibit toxic properties. Heavy metals are taken up by both flora and fauna. If the excretion phase is slow, this can lead to bioaccumulation. A few metals, such as mercury, have been shown to undergo biomagnification through the food chain. The existing information on various environmental problems in Africa has been reviewed by Dejoux (1988) and Biney *et al.* (1992).

While available data on heavy metals in the African aquatic environment are still scattered, the occurrence of trace metals is quite low when compared to some other areas of the world. On the whole, the levels of such metals in inland water fish muscle were below WHO limits (Table 14.5). Despite a few hot spots, such as Lake Mariut (Egypt), Lagos lagoon (Nigeria), Ebrié lagoon (Côte d'Ivoire), concentrations in inland and coastal environments (sediments, water organisms) for the moment pose no environmental concern and exhibit no significant differences at a continental level (Biney *et al.*, 1992).

Oil pollution

Accidental oil pollution has been possible for a long time in African inland waters, mainly from ships running on fuel oil, but also from storage tank leaks, accidents to road or rail tankers, pipeline failures, etc. Chronic discharges (including those of refinery or industrial effluents, sewage and municipal wastes) are considered to be the major threat (Baker, 1992) because they are usually deliberate and long term.

Concern about oil pollution has increased since 1982 because exploration for oil has spread over a number of freshwater lakes in East Africa (Hartman & Walker, 1988). Oil exploitation in or near lakes may conflict with other activities, such as fisheries and tourism, and their confined waters provide less scope for dilution and dispersion than the open sea. Oil leaks or spills might therefore have catastrophic consequences (Coulter *et al.*, 1986).

Baker (1992) gave an overview of the impact of oil and breakdown products in tropical lakes. Crude oils are

Table 14.5. *Mean metal concentrations in inland water fish* ($\mu g\ g^{-1}$ *fresh weight*)

	Hg mercury	Cd cadmium	Pb lead	As arsenic	Cu copper	Zn zinc	Mn manganese	Fe iron
Lake Mariut, Egypt		0.15			3.7	7.6	0.9	11.2
Lake Idku, Egypt	0.01	0.004	0.67	0.031	1.77	7.4		
River Wiwi, Ghana	0.37	0.19	0.47		0.18	3.0		
Niger Delta, Nigeria	0.034	0.03	0.48		0.70	4.8	1.1	5.4
Lake Nakuru, Kenya	0.044	0.05	0.17	0.36	2.0	22	1.8	
Lake Victoria, Kenya		0.04–0.12	0.4–1.1		0.15–0.53	2.21–7.02	0.22–0.74	0.53–4.65
Lake McIlwaine, Zimbabwe		0.02	0.17	0.28	1.08	9.6	5.4	
Hartbeesport Dam, SA		0.02	<0.02	0.40	0.30	6.6	0.24	
WHO limits	0.05	2.0	2.0		30	1000		

From Biney *et al.*, 1992.

mixtures of complex compounds, mainly thousands of different hydrocarbons. The lighter, relatively toxic, hydrocarbons are both the most soluble in water and the quickest to evaporate. Remaining compounds eventually either degrade or become associated with sediments. Degradation occurs through oxidation/reduction, but many bacteria and some fungi are hydrocarbon degraders. Degradation rates may vary according to physical and chemical characteristics of the water, including the oxygen concentration.

There are probably oil pockets in the sediment of Lake Malawi and drilling is a possibility. If oil extraction is planned, it should be remembered that in this seismically active area earthquakes could rupture oil pipelines or drilling platforms, the result being oil pollution on a devastating scale, which could eliminate the fauna (Tweddle, 1992). Similarly, the oil potential of the sediments of Lake Tanganyika are being investigated, and their exploration and exploitation carry pollution risks (Coulter, 1992). Lake Tanganyika is also very vulnerable, and the social as well as ecological effects of pollution would be extremely serious. Substantial investments in environmental assessments and safeguards are needed to be able to quickly respond to any adverse event.

Pesticides

Since the late 1950s, the scientific community, through the chemical and biological monitoring of various environmental parameters, has become aware of the dangers that the massive use of chemical substances may

create in ecosystems. The perception of environmental degradation led, in the 1960s, to a number of environmental regulations and legislative interventions, the most extreme example of which was the banning of certain individual substances, e.g. DDT, or categories of chemicals.

One of the most important sources of contamination in African lakes and rivers is certainly the increasing use of pesticides in the public and animal health sectors to curb endemic diseases, or for agriculture purposes (Calamari, 1985; Dejoux, 1988). What is known of the fate and effects of these chemicals on the environment is based essentially on knowledge acquired in temperate zones but, in general, there appears to be little basic difference between the effects seen in terrestrial temperate and tropical ecosystems (Bourdeau *et al.*, 1989).

The effects of chemical pollution on fish communities could be schematically summarised as:

lethal toxicity, killing fishes at some stage of their life history;

sublethal effects that alter the fish's behaviour (e.g. foraging, predation), and could affect their physiology (e.g. reproduction) thereby making it difficult to complete the normal life cycle (but these kinds of insidious effects are usually difficult to detect and to prove); and

accumulation of pollutants in tissues that can render fish unsafe for human consumption.

It is likely that a species' sensitivity to pollution might

result in overall deterioration of the ecosystem if that species is critical in maintaining the ecological balance. However, except for the Onchocerciasis Control Programme in West Africa (see below), there are very few long-term monitoring programmes carried out to monitor the impact of pesticides on fish communities.

Whatever the effects, the community response is a reduction in diversity and change in species composition, leading to dominance by smaller, short-lived forms. Graded responses to the damaging agent are recognised: (i) rare or sensitive species are eliminated; (ii) demographic changes occur in some long-lived species; (iii) there is reduced species diversity; and (iv) opportunist species become dominant (Howells et al., 1990)

Information available on the toxicity of substances to fish often relates to data obtained with one chemical, in laboratory conditions. In field situations, several harmful substances are usually present, and the effect of such mixtures on fish needs appropriate tests. It is possible to account for the harmful effect of a mixture of chemically similar poisons in water by summation of their individual toxic fraction, but some results, dealing with the acute lethal toxicity to fish of mixtures of pesticides and other substances, have indicated that more-than-additive effects (or synergism) could occur (Alabaster et al., 1988).

The evidence for and the significance of long-term changes associated with exposure to low-level contamination is frequently questioned, both by scientists and policy-makers. In many cases, the process of change is insidious and the possible agents present only in low concentration or intermittently (Howells et al., 1990). In order to detect such processes that could lead to irreversible damage to ecosystems, the need for an 'early warning' system that could invoke protective measures at an early stage has often been stressed. But the task is difficult because of the complexity in space and time of the ecosystems, and the methodological difficulties of identifying the nature and source of contaminants, as well as the type of exposure. A major challenge in natural environments is to detect and interpret the long-term effects of low-level contamination as distinct from those of other stressing agents, such as climatic and seasonal changes. The challenge is, of course, more difficult in situations where the ecosystem and species biology are poorly understood.

The pollution resulting from biocides applied to agricultural crops and leaching into streams can be an important source of toxin, which gradually builds up in the bodies of fish or other lacustrine organisms. This problem is causing concern in several countries bordering the East African lakes (Alabaster, 1981). Pesticides residues have been detected in fish caught in northern Lake Tanganyika where the total quantity of DDT and its metabolites in whole dried fish varied between 0.45 and 2.4 mg kg^{-1} in *Stolothrissa*, with a maximum after the cotton harvest (Deelstra et al., 1976; Deelstra, 1977). A marked accumulation of DDE in the flesh of certain fish species from the Chari River has also been observed by Everaerts et al. (1971).

The impact of pesticides on the functioning of natural aquatic ecosystems and on fish biology is very poorly documented in Africa. However, there is a strong feeling that problems are likely to occur, but it is very difficult to demonstrate real effects in the field.

Case study: the Onchocerciasis Control Programme (OCP)

Human Onchocerciasis is a dermal filariasis that causes an irreversible blindness in people long exposed to infection, and is widespread throughout tropical Africa. The parasite, *Onchocerca volvulus*, is strictly specific to man and is transmitted by the female blackfly of the *Simulium damnosum* complex. In the absence of effective treatment (prophylaxia, chemotherapy) suitable for mass application, vector control is the only way to prevent spread of this disease. Adult blackflies are widely dispersed and their control is difficult, but it is possible to attack the aquatic larvae, which develop in river rapids, with larvicides. The duration of the larval stage is short (8–10 days) so, in order to efficiently control the *Simulium* population, weekly application was necessary (Yaméogo et al., 1988).

In 1974, a 20-year Onchocerciasis control campaign was initiated. The overriding objective was to reduce the impact of Onchocerciasis to a sufficiently low level so that it no longer represents either a public health problem or an obstacle to socio–economic development. The initial control area of 764 000 km^2 covered parts of seven West African countries and about 18 000 km of rivers. Most of the rivers in the OCP area are of the savannah type, with a tropical water regime characterised by a flood period from June–July to November. River discharge is very low during the dry season and the upper course is sometimes reduced to a series of pools. During the flood, the rivers spill over from the main channel, invading the fringing plains.

The insecticide selected for a large-scale campaign due to last for about 20 years had to have often contradictory properties, such as an effective impact on the larvae of *Simulium damnosum*, ease of application, low cost, little residue but far-reaching effect, harmless for man and mammals, and the lowest possible toxicity for the rest of the aquatic environment. Temephos (Abate), an organophosphate larvicide, more or less meet these criteria. It was the only larvicide used from 1975 to 1980, until temephos resistance developed in larvae of some cytospecies of the *Simulium damnosum* complex. To control resistant strains, other insecticides had to be used. This resulted in large-scale application of *Bacillus thuringiensis* H14 (Teknar) during the dry season in the areas of resistance, together with chlorphoxim, another organophosphate, during the wet season. But very soon, a resistance to chlorphoxim was discovered and intensive screening of alternative insecticides was necessary. Carbosulfan and permethrine appeared to be promising, provided they were used only during the rainy season, and for a limited number of applications.

Since such prolonged and intensive use of insecticides could present risks, it was necessary to evaluate the possible short-term effects of their application on the aquatic flora and fauna. An aquatic environmental monitoring programme was therefore devised before the beginning of OCP to be sure that the insecticides used did not excessively disturb the biotic systems, and to provide warning to those carrying out treatments should toxic effects be noted (Lévêque *et al.*, 1979).

An Ecological Group, of five independent experts, controls and directs the Programme's activities regarding environmental protection. The Ecological Group meets each year to evaluate aquatic environmental data collected by OCP and national teams of hydrobiologists involved in monitoring activities.

The monitoring programme was primarily concerned with two major categories of organisms (Lévêque *et al.*, 1988*b*).

1. The benthic invertebrates that abound in the rivers and that are directly threatened by the insecticide in the same way as *S. damnosum* larvae, or through disruption of food chains.

2. Fish and shrimps that represent a major natural resource for the people living along the river. Insecticides could affect the food chain by killing fish prey, and repeated long-term treatments could also affect the reproductive cycle of fish, either by a direct effect on eggs and juveniles, or indirectly through the reproductive physiology of adults.

Immediate toxicity in laboratory tests
The larvicides used for this large scale campaign were tested in the laboratory to determine the toxicity levels and lethal doses (LC50), as well as to compare the relative toxicity of different compounds in controlled situations. The toxicity of the insecticides tested (permethrine, cyphenotrin, pyraclofos and carbosulfan) is not very different for African fish species and fish from temperate areas. African fish react more or less like warm-water fish (Yaméogo *et al.*, 1991). Small pelagic species appeared to be less tolerant than benthic ones. For many species, the 24 h LC50 is not far from the operational dose, but it should be kept in mind that the exposure time in the river is much less (10 minutes).

Long-term monitoring

Ecotoxicological studies The effect of organophosphates in laboratory experiments showed that fish were able to accumulate temephos (Matthiessen & Johnson, 1978). As an example, *Oreochromis mossambicus* exposed weekly to 0.05 mg l^{-1} for 10 minutes, accumulated 3–4 mg kg^{-1} by direct absorption. An affinity of temephos for fatty tissues was observed. In the OCP and in field conditions (Quelennec *et al.*, 1977), fish captured during the dry season just below the application point exhibited traces of temephos: 14.3 mg kg^{-1} one day after and 7.1 mg kg^{-1} six days after treatment for *Tilapia zillii*; 1.30 and 1 mg kg^{-1} respectively for *Labeo parvus*; and 0.8 and 0.3 mg kg^{-1} for *Brycinus nurse*. At 1 km below the point of application, contamination was much less: 0.14 mg kg^{-1} one day after treatment for *Tilapia zillii*. During the rainy season, accumulation of temephos is much lower than in the dry season. DDT residues were also found (0.01–0.35 mg l^{-1}) in the fish studied.

Biochemical monitoring Exposure to organophosphates reduces acetylcholinesterase activity. In laboratory tests, inhibition by temephos was about 25% for *Tilapia guineensis* at the operational dose (0.05 mg l⁻¹ for 10 minutes). But no effects on fish were noted after repeated weekly exposures (Gras *et al.*, 1982). When the length of exposure is 24 hours, the inhibitory effect is much higher: 38% after one exposure, 69% after three weekly exposures. In the latter case, the fish did not survive (Pélissier *et al.*, 1982). Inhibition of acetylcholinesterase appears to be more important with chlorphoxim.

In field conditions, there seemed to be no significant difference in acetylcholinesterase activity of fish in rivers treated with temephos or untreated. Below spraying points, when chlorphoxim was used, a 20% reduction in enzymatic activity was observed but is reversible (Scheringa *et al.*, 1981; Antwi, 1984).

Changes in natural fish communities Assuming that there was no direct negative impact on adult fishes, the long-term monitoring programme considered two fundamental hypotheses (Lévêque *et al.*, 1988*b*): (i) insecticides could affect the food chain of aquatic invertebrate feeders if prey are seriously reduced; and (ii) repeated long-term treatments could change the reproductive cycle of fish, either by affecting their physiology or by direct effects on eggs and juveniles.

Several monitoring sites were selected on different rivers using criteria, such as accessibility at all seasons, suitability for sampling, availability of hydrological data and abundance of fish stocks (Lévêque *et al.*, 1979, 1988*b*). Experimental fishing was carried out every two months, using sets of gill nets with different mesh sizes. The following biological parameters were recorded: total catch for the set of gill nets, number of species caught, structure of the catch, coefficient of condition and gonad maturation.

Overcoming many scientific and practical problems, the monitoring programme on invertebrates and fish has operated since 1975. Results of the monitoring and also of short-term research are available (see Yaméogo *et al.*, 1988; Lévêque, 1989*c*; Yaméogo, 1994). In most of the monitoring sites the number of fish species caught shows seasonal fluctuations with a maximum at low-water periods. There is no clear evidence of reduced species richness after 20 years of larviciding in the original area of the Programme (Fig. 14.3).

For long-term changes in fish abundance (Fig. 14.4), the situation is more complex. In some cases, there is no evidence of a significant decrease in experimental catches, and fluctuations may be related mainly to interannual changes in hydrological regimes. In other cases, a decrease is clearly noticeable, which could be attributed both to the increasing use of unregulated fishing gear (including the forbidden use of pesticides) and to the resettlements in the vicinity of the monitoring stations, creating more disturbance to the fish fauna (WHO-Oncho, unpublished data).

Therefore, after 20 years monitoring, it has not been possible to identify the effects on fish populations due to the use of larvicides in West African rivers treated for *Simulium* control. In operational conditions, the treated rivers seem to have fairly strong resilience and a great capacity for recovery. However, it should be stressed that this is the result of the implementation of a strict protocol starting with a drastic selection of larvicides in relation to their environmental impact (Lévêque *et al.*, 1988*b*; Hougard *et al.*, 1993).

Impact of fisheries

In the late 1970s the yearly catch of inland fisheries in Africa amounted to 1.5 million tonnes. It dropped back to about 1.3 million tonnes, but from 1981 onwards the trend was upwards until it reached 1.8 million tonnes in 1988, of which about 10% came from reservoirs. The total catch is again likely to increase slowly during the 1990s as long as there is no return to severe drought conditions. The increase will come from new reservoirs, opening up previously isolated areas, and from better management of large lake systems. On the other hand, these improvements will be offset by the general degradation of water quality and quantity in rivers, as well as habitat deterioration (particularly on floodplains).

As long as they continue to deplete stocks of primary consumers or remove predators, fishing activities will produce changes in fish assemblages. Destructive fishing methods include such activities as use of nets with excessively small mesh sizes, intensive beach seining, fishing with explosives, poisons and electrical devices. The use of multiple types of gear and changes in the fishing effort may also put different pressures on different species. Alteration in the species composition itself may, in turn, result in changes in catch and fishing activities. Overfishing is

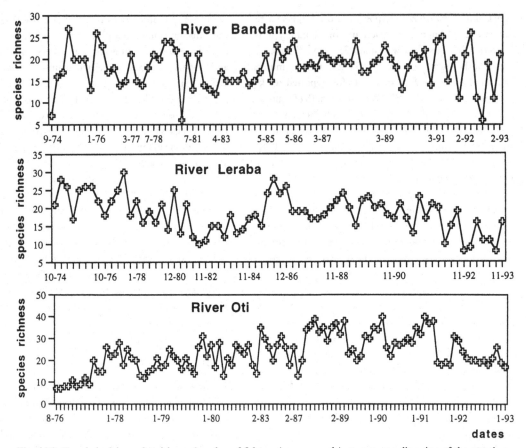

Fig. 14.3. Trends in fish species richness (number of fish species per sample) at some sampling sites of the aquatic monitoring programme of the Onchocerciasis Control Programme in West Africa (data from WHO-Oncho).

another recognised form of perturbation, but there are actually very few documented examples of species that have definitely disappeared entirely following overexploitation of a freshwater habitat. Taking samples of ornamental fishes is another type of impact, similar to conventional fisheries, but more selective.

Impact on individual species

The most striking effect of fisheries is observed on large species of low reproductive capacity and low resilience. Daget *et al.* (1988) cited the case of the endemic catfish, *Arius gigas*, in the Niger basin. At the beginning of the century, specimens 2 metres long were reported by travellers, and could be inferred from the special fishing gear especially devised for their capture (Monteil, 1932). The

biology of the species is unknown, except that males are mouthbrooders of few but large eggs. At the end of the 1940s, this fish was already very rare, and nowadays most fishermen ignore the existence of this large species.

The impact of fisheries on the cyprinid *Labeo* is quite well documented. *Labeo victorianus* was abundant in the 1950s in fisheries of the affluent rivers of Lake Victoria, but declined drastically, before *Lates* was established, as a result of unregulated fishing practices. While the indigenous fishing methods did not have a significant deleterious effect on the stocks, the introduction of more efficient gill nets and fishing methods, associated with the search for gravid fish (regarded as a delicacy), contributed to the decline (Ogutu-Ohwayo, 1990*b*). The decline of *Labeo altivelis* in the Luapula River was also attributed to the

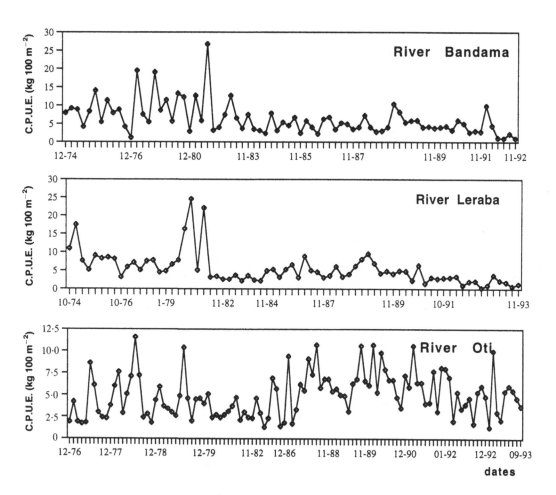

Fig. 14.4. Trends in mean catches per 100 m² gill nets per night (C.P.U.E. [catch per unit effort] in kilograms) at different monitoring sites of the aquatic monitoring programme of the Onchocerciasis Control Programme in West Africa (data from WHO-Oncho).

development of a large-scale fishery in the late 1940s (Jackson, 1961). For *Labeo mesops* in Lake Malawi, the decline is assumed to result from deterioration in stream conditions, which became unfavourable for spawning and egg development (Skelton *et al.*, 1991). However, according to Turner (1994), it seems likely that *L. mesops* suffered from overfishing by fences or gill nets set across the mouths of spawning streams. In Lake Chad, the catch of *Labeo coubie*, which was 7 kg per 100 m² of gill net per night in 1963, fell to less than 1 kg in 1966 and the fish almost disappeared from the fishery in 1969 (Durand,

1980). In this case it is also likely that overfishing was responsible for the disappearance of the species.

This sensitivity to fishing is not restricted to cyprinids. *Lates niloticus*, *Heterobranchus* and *Bagrus docmac* also disappeared from the heavily fished lower reaches of the Oueme River (Welcomme, 1979), and before the drying up of Lake Chad, there was a continuous decrease in the catch of *Citharinus* and *Heterotis* (Durand, 1980).

As pointed out in Chapter 10 predatory piscivore species are particularly vulnerable to gill nets in comparison to other species. As a result, the higher contribution of predators to the total catch, which is generally observed when a fishery starts on an unexploited stock, is partly the result of their greater vulnerability. Many predators are also large animals, with a longer turnover time than many other food species, and this too may partially explain the diminution of their relative contribution to the catch of such fisheries when they are well established.

Impact of fishing on fish communities

African Great Lakes

Ribbink (1987) stressed the fragility of the cichlid communities endemic to the African Great Lakes, which showed overt signs of overfishing even in response to low levels of fishing pressure. The structure of fish communities can change dramatically within a few years when trawlers and other such fishing gear are used. For example, the commercial bottom trawl fishery started in Lake Malawi in 1968, and experimental surveys conducted from 1971 to 1974 showed a rapid decline in abundance of the larger species (individuals over 190 mm) from 79% of the catch in 1971 to 18% in 1974, at one sampling site in the south-east arm of the lake. Cichlid species, such as *Lethrinops stridei* and *Lethrinops macracanthus*, declined drastically (Turner, 1977). Over 20% of the species disappeared from the catch in the four years 1971–4. In the early 1990s, the large benthic feeder, *Lethrinops macracanthus*, had disappeared, while *Lethrinops microdon*, formerly the mainstay of this fishery, had declined from 16% by weight of the 1971–2 catch to less than 2% in 1991–2 (Turner, 1994). Demersal trawl surveys have also shown that fish communities in the most heavily fished areas in the south of Lake Malawi have been severely disrupted by trawlers; three large haplochromine species have been eliminated and a further eight show statistically significant declines (Turner, 1994). Ribbink (1987) suggested that it is almost impossible to regulate a multispecific fishery in which as many as 120 closely related species are caught by trawling, and that it is therefore difficult to develop a sustainable fishery based on the African Great Lakes cichlids. It would also be expected that the lacustrine cichlids would be slow to recover from exploitation because they invest heavily in parental care and produce only a few large eggs. Moreover, the fishing gear destroys their nesting arenas, and many species that are habitat

restricted cannot benefit from recolonisation from elsewhere since no conspecific populations exist. In general, cichlid recruitment is poor, and yield estimates were probably far too optimistic and should be re-evaluated.

Overfishing by small-meshed seines, which has increased tenfold in Lake Malombe since 1980, has affected the tilapias and the larger haplochromine cichlids. Nine of these large species that were previously reported from Lake Malombe were not found in 1991–2 (Turner, 1994).

A number of authors have recorded the effects of overfishing in Lake Victoria, from the decline of some species to the virtual disappearance of others (Garrod, 1960, 1961; Cadwalladr, 1965*b*; Jackson, 1971), and the history of the fishery has been briefly reviewed by Barel *et al.* (1991), Craig (1992) and Witte *et al.* (1992*a*). An endemic tilapia fishery was already established for *Oreochromis esculentus* and *Oreochromis variabilis* at the beginning of the century (Acere, 1988), and in the following decades catches decreased and fish became smaller (Graham, 1929). Nylon gill nets and outboard motors were introduced in 1952. The improved technology and increased fishing effort resulted in a severe drop in the catches of endemic *Tilapia* (Fryer & Iles, 1972). In the 1960s, the use of small mesh nets (38–46 mm) and an increase in the use of beach seines resulted in a further reduction of the small fish, including haplochromines and juveniles from other genera, such as *Synodontis* spp. and *Schilbe intermedius* (Marten, 1979*b*). Meanwhile, light fishing for the cyprinid, *Rastrineobola argentea*, developed in both Kenyan and Tanzanian waters.

By the beginning of the 1970s, well before the Nile perch was established, a significant decline of *O. esculentus* was noted, as well as an increase in the catches of the large predators *Bagrus* and *Clarias*. Fish that migrated into rivers to spawn had been heavily exploited for some time with traps and by blocking the river, and since the 1950s fishing pressure had been increased by the use of gill nets across the river mouth. The valuable *Labeo victorianus* for instance, which was caught during its spawning migration, was reduced to very low population levels in the late 1950s (Fryer, 1973), and a marked decline was noted in the migratory species of the lake community (Cadwalladr, 1965*b*). As the catch of the non-haplochromine species declined, fishermen started to concentrate on haplochromines, which constituted at least 80% of the demersal biomass at the end of the 1960s. In the early 1970s they became an important source of protein around Lake Victoria. Meanwhile, a trawl fishery developed in the Tanzanian part of the lake, which led to a reduction of the haplochromine stocks. The highest yields were recorded in 1977, but by then the haplochromines already showed signs of overfishing (Witte & Goudswaard, 1985; Witte *et al.*, 1992*b*).

A great deal of attention has been paid to the pelagic community of Lake Tanganyika, where three different categories of fishery exploit fish resources: (i) the traditional fishery (scoop net, gill net, beach seine); (ii) the artisanal fishery (lift net); and (iii) the commercial fishery (purse seine) (Roest, 1992). Most of these fisheries use lights to attract clupeids and to capture them at night, together with *Lates* spp. The initial effect of the commercial fishery was a steady reduction in all *Lates* species, whose catches fell to less than half of those recorded initially, but an increase in clupeids (Coulter, 1991*b*). In the second phase, the small pelagic *Lates stappersii* increased considerably, but fluctuations of this predator and its prey species (*Limnothrissa*) appear to alternate in abundance with periods of 6–8 years between successive peaks. Thus the system became reduced to a single predator–prey system interacting under intensive fishing pressure (Roest, 1988). The years that favour clupeid production are also suitable for juvenile *L. stappersii*. It takes 2–3 years for these juveniles to become adult, which accounts for the time lag in abundance cycles reflected in the catches of the two species (Coulter, 1991*b*). The decline of the clupeid fishery observed in recent years may result from overfishing.

The benthic fish fauna of Lake Tanganyika has been exploited by gill nets and the top benthic predator, *Lates mariae*, has diminished considerably. There has been a marked trend over the past 25 years towards a relatively greater abundance of small fish and a decline in the total biomass, while no alternative benthic predator has emerged (Coulter *et al.*, 1986). However, the fish population did not decline in the unexploited parts of the south-west arm.

In Lake Barombi-Mbo (Cameroon) it is not clear whether fisheries are responsible for the decline of rare endemic cichlid species, such as *Stomatepia mongo* and *Konia dikume*, but water management has probably had some influence on these species (Reid, 1991).

According to Coulter *et al.* (1986), the collective experience in recent years on the African Great Lakes seems to show that large-scale mechanised fishing is incompatible with the continued existence of the highly diverse cichlid communities. Cichlids appear especially vulnerable to unselective fishing because of their particular reproductive characteristics. According to Turner (1977), it is not possible to maximise the yield of protein without causing a change in species composition and a decline in the number of endemic species. A number of suggestions have been made that could afford at least partial protection to sections of the cichlid communities. These include increasing the mesh size of nets and the replacement of mechanised fishing gear with labour intensive methods (Fryer, 1984). An alternative to large mechanised fisheries would be to increase the number of artisanal units, which should be preferable for the conservation of species (Coulter *et al.*, 1986). But smaller units are more difficult to manage and involve different social and economic factors. It has been suggested that the haplochromines of Lake Victoria may be partly saved only if the Nile perch densities are severely reduced.

Another suggestion has been that parks should be developed (Coulter *et al.*, 1986), and that fishing should be rendered impossible in certain areas by placing obstructions on the bottom that would snarl trawl nets (Ribbink, 1987). Lake Malawi National Park will very probably afford protection to a wide range of species, but no data are at present available to confirm this hypothesis. It is not known yet whether these reserves can adequately preserve the integrity of populations, and it may be that only stenotopic populations whose distribution coincides with the park area will be protected. The size of the reserves, the intensity of fishing in nearby areas, and the possible influence of pollution or introduced alien species should also be taken into account (Coulter *et al.*, 1986).

Maximisation of yield has highest priority in developing countries with expanding populations and increasing food requirements. Tweddle (1992) posed the question about the cichlids of Lake Malawi: 'does it matter if, in the process of maximising yields, a few cichlid species are forced into extinction? The rate of speciation in cichlids is such that a 'species', at least in the haplochromine group, can be regarded as a temporary taxonomic category ... Lake Malawi is a continuously evolving system, and if man's fishing activities cause a few more changes, including species extinctions, is this necessarily a disaster?' He pointed out that when overfishing of both tilapiine and haplochromine stocks has occurred, there is evidence of rapid recovery (within two years) when fishing pressure was reduced (Tweddle & Magasa, 1989). In the case of random trawl fisheries, Tweddle (1992) also questioned the supposed impact of fishing gear on breeding areas. Destroyed nests are rapidly rebuilt if the males avoid the gear. In the south of Lake Malawi, shallow-water nesting arenas were totally flattened twice in 48 hours by earth tremors, but resumption of breeding activity started within a day, and the nests were restored to their previous species-specific diameters in two days. Experience in Lake Malawi suggests that cichlids do not require unique management because of their breeding behaviour. Providing good catch data are available, the fisheries could be effectively managed, ensuring that sufficient adults survive to maintain steady recruitment. Tweddle assumed, however, that commercial fisheries are currently closely controlled in Lake Malawi, and information on the impact of traditional fisheries on stock is now available.

During the 1970s, Marten (1979b) had already noticed that the inshore fishery of Lake Victoria showed symptoms of severe overfishing. 'Despite the numerous species and general abundance of *Haplochromis*, inshore *Haplochromis* in the heavier fished areas of lake Victoria already seem to be cropped at the limits of their potential'. The problem was that heavy fishing was associated with the use of small mesh gill nets and seines to catch small fish, a practice that does not seem to justify the damage to catches of the larger species by the same gear. Actually, since the three countries bordering the lake achieved their independence in 1961–3, no restriction on fishing gear has been in effect in Tanzania and Uganda, whereas nets below 13 cm mesh size were prohibited throughout the lake before independence. A great deal of 'noise' has been made about the consequences of the introduction of the Nile perch, a so-called 'ecological disaster', but it is obvious that fishing pressure in the absence of regulation has also played a large part in the disappearance of the haplochromines in Lake Victoria.

Nilo–Sudan ecosystems

In north tropical Africa, a number of highly productive aquatic systems are heavily exploited by fisheries. Monod

(1928) and Blache & Miton (1962) described the numerous traditional fishing techniques that are well adapted to the many situations encountered in the complex and varied network of the Lower Chari and Logone Rivers. Many of these techniques disappeared during the 1960s with the introduction of nylon gill nets. At the same time, the fishery, which was restricted to the lower reaches of the Chari and Logone, expanded to the lake itself where the fishing effort dramatically increased from 1963 to 1970. In the northern basin of Lake Chad, the total fishing effort increased about 40 fold between 1963 and 1970 (Durand, 1973), while the catch per unit effort (kg per $100 \, m^2$ gill net per night of fishing) fell from 18 kg in 1963 to less than 1 in 1970. During subsequent years, the Sahelian drought and resultant shrinking of Lake Chad resulted in temporarily very high catches and the disappearance of many species from the lake itself, whereas they were able to survive in the lower reaches of the main tributary rivers.

In Lake Turkana, the commercial fishery started around 1950. In the absence of standardised landing statistics and based on available information, Kolding (1989, 1992) has provided a reconstructed series of yield and effort data for the western side of Lake Turkana since 1962. In the 1950s, the main target fish was *Oreochromis niloticus* caught with a 4 inch beach seine. By 1956, a strong decline in tilapia was observed, compared with other fish. With a reinitiation of the commercial fishery in the early 1960s, tilapia was again predominant, but *Lates* also made a significant contribution to the catch, with smaller quantities of *Clarias* and *Bagrus*. The exploitation of the large stocks of *Citharinus* and *Distichodus* available was at first discouraged because of the lack of a market, but around 1963–4, an 8 inch gillnet fishery was established and the Zaïre market opened up. *Citharinus* became the dominant species in the catch between 1966 and 1972, but thereafter it declined sharply. The fishery then shifted from a nearly single species fishery to a multispecies fishery based on previously unexploited species, such as *Labeo horie*, *Barbus bynni* and *Bagrus bajad*. In 1975–6, there was a tilapia boom that markedly increased the total fish yield, but this situation was ephemeral, although another small tilapia boom also took place in 1982. An increasing use of hooks might explain the relative rise in the contribution of *Lates* to the total catch, from around 1985.

Yields from Lake Turkana have fluctuated greatly

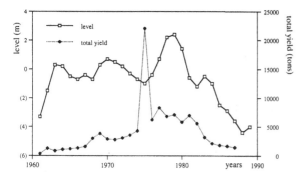

Fig. 14.5. Estimated total yields and lake level on the western side of Lake Turkana, 1962–88 (from Kolding, 1989).

(Fig. 14.5) and, since 1984, there has been a general decline, partly attributed to the lowering of the lake level and the drying up of areas such as Ferguson's Gulf, which provided a large tilapia population. As stressed by Kolding (1989), the many attempts at calculating the so-called 'sustainable yield' as a management strategy for Lake Turkana have not been very fruitful, but this has only been acknowledged by a few past researchers. 'It has even had adverse effects in inducing an over-capitalisation that could not be sustained ... Still the concept is deeply rooted in planners, administrators and research personnel and several attempts have been made to adapt relationships to confirm or predict the potential yields of Lake Turkana' (Kolding, 1989). Actually, there is a fairly strong correlation between the mean annual lake level and the commercial catch rates expressed as annual catch per boat. The rate of change in fish production with variations in water inflow is apparently very fast, and it appears that tilapias make a major contribution to yields when lake levels are rising.

Captures for aquarists
For almost a century, African freshwater fish have been exported for the aquarium trade. However, the capture of live fish for aquarists does not conform to the same rules as capture for food fisheries. The constraints are mainly economic, as the fishery is heavily dependent on demand. Prices can only be maintained sufficiently high to cover freight costs, when demand exceeds supply.

Cyprinodonts, and to some extent cichlids, received far more attention than other groups, probably as a result of their bright colours and the relative ease of breeding them. One of the positive consequences of the interest in cyprinodonts is a much better knowledge of their taxonomy and biology, and owing to their short life cycles they are also very good material for genetic studies. They are, therefore, one of the fish groups for which detailed inventories and distribution maps are available, and these should be used for conservation purposes.

For many species, such as easily bred cyprinodonts, the impact of sampling appears to be low, but for other rare taxa with known limited stocks, there is a real danger of eradication. Daget et al. (1988) drew attention to Caecobarbus geertsi, a blind cyprinid inhabiting caves near Thysville in Zaïre and of special interest to aquarists. The stock was estimated at 4000 individuals and, in the absence of regulation, collectors would be tempted to exhaust this stock for short-term profits.

The African Great Lakes cichlids are very attractive and have became popular abroad. One fish exporter is running a profitable concern from Malawi (Tweddle, 1992). Demand for specific wild fish is, nevertheless, a potential threat. Many of the fish in greatest demand are very rare, with populations probably restricted to a few hundred individuals inhabiting specific localities. Conservation of such a resource should be wisely managed to avoid the danger of achieving only short-term profits.

It should be clearly recognised that the fishing and trading of fish for aquarists is not subject to the same regulation that applies to other vertebrates. It is difficult, if not impossible, in Europe to obtain any reliable information about species, origin and quantities of fish imported. Most traders are not willing to provide such information, and this opens the way to suspicion about the origin and nature of fish imported.

Ichthyotoxins and misuse of chemicals
The use of ichthyotoxins to catch fish is fairly traditional in tropical areas. In Africa, the use of the Papilionaceae, Tephrosia vogeli, has long been reported from Cameroon (Monod, 1928; Stauch, 1966), Zaïre (Claus, 1930; Malaisse, 1969) and Nigeria (Reed, 1967). One of the earliest reports is probably that made by Goffin (1909), who wrote about the Congo (Zaïre): 'Les Upotos emploient le Tephrosia de Vogel. Les feuilles de cet arbuste sont écrasées puis mises dans un récipient; après plusieurs jours de macération dans l'eau, la décoction est jetée dans les petits ruisseaux ou le mare. Aussitt que le pois-

son a absorbé cette substance, il devient malade et flotte à la surface'. Tephrosine, the active substance, is chemically close to rotenone and has similar physiological effects. Traditional *Tephrosia* fisheries were usually collective and involved the participation of whole villages to prepare the poison and collect the fish. The fishery usually took place once a year, at low-water level.

The long-term environmental effect of such so-called 'traditional fishery' is unknown, but the natural ichthyocides are being widely superseded by chemicals that can kill every fish over kilometres of running waters. The appearance of industrial insecticides in the 1950s, and later their widespread use in agriculture, opened the door to uncontrolled fishing activities using industrial products. Stauch (1966) already mentioned the use of aldrin and lindane in the Benue River on the Nigeria boundary in Cameroon in the early 1960s. The use of chemical pesticides should be strictly forbidden, yet it is difficult to control.

Impact on genetic resources

The effects of fishing and related activities on genetic resources have been stressed. Fishing pressure reduces population numbers and this may have an effect on genetic resources, through drift and other random events, but so far there is little clear evidence that fishing irreversibly changes the genetic structure of populations. The classical effects of overfishing on population structure and life-history characters, e.g. early maturity and small size for age, have been seen to reverse themselves when the fishing pressure was removed (Bartley, 1993). Stocking and the introduction of exotic species can, nevertheless, have a significant influence, both positive and negative, on aquatic genetic diversity. The elimination of races or local stocks, either through overfishing or through environmental degradation, is another cause of reduced heterozygosity.

Species introductions

The introduction of alien fish into inland waters has occurred all around the world (Welcomme, 1988). The main goals of deliberate introductions by fishery officers were initially to improve sport fisheries and aquaculture, or to develop biological control of aquatic diseases, insects and plants, or else to fill supposed 'vacant niches'

and improve wild stocks in old or newly created waterbodies. More recently, transfers of ornamental fish for the aquarium trade have also increased.

Introductions of exotic species may displace indigenous fish, some of which are known to be threatened by alien competitive fish introduced either directly into freshwater bodies by fisheries officers, or indirectly when they escape from fish ponds. They could also hybridise with local species. One of the major problems in species introductions is their irreversibility. Once established, it is impossible, given current technology, to deliberately eradicate a fish species from a large natural waterbody.

Purposes of the introductions

The fish assemblages in African inland waters are usually sufficiently diverse to fill most of the available niches (trophic and spatial). But in a few systems there is, for historical reasons, a poor or depauperate fauna, and during recent decades there have been several attempts to introduce alien species into some natural waterbodies or new impoundments in order to improve local fisheries (Moreau *et al.*, 1988). Most of these introductions occurred in southern and east tropical Africa (Moreau *et al.*, 1988; Welcomme, 1988) and cichlids were particularly involved (Table 14.6). Tilapias have been introduced into many African waters for various purposes (see Philippart & Ruwet, 1982):

stocking natural lakes where no tilapias occurred (e.g. *Oreochromis alcalicus grahami* into Lake Nakuru, or *Oreochromis spilurus niger* and *Tilapia zillii* into Lake Naivasha);

filling a supposed vacant ecological niche (e.g. *Oreochromis niloticus* and *T. zillii* into Lake Victoria);

development of new fisheries in artificial waterbodies based on introduced cichlids (e.g. *O. niloticus* in Lakes Ayame and Kossou in Côte d'Ivoire); and

biological control of aquatic vegetation (e.g. *Tilapia rendalli* into Sudanese irrigation channels or some artificial lakes of Shaba).

Fish introduction has periodically been suggested for disease control. For example *Astatoreochromis alluaudi*, a 'mollusc crushing' cichlid under threat of extinction in East Africa, has been proposed for introduction to Cameroon for snail control (Slootweg, 1989). This fish has already been introduced into different places in

Table 14.6. *Some selected records of fish species introductions in intertropical African freshwater systems*

Species	To	From	Year	Purpose	Source
Aristichthys nobilis	Egypt	USA	1975	weed control	Moreau *et al.*, 1988
Astatoreochromis alluaudi	Côte d'Ivoire	Uganda	1969	snail control	Lazard, 1990
Clarias gariepinus	Côte d'Ivoire	CAR	1973	aquac.	Welcomme, 1988
Ctenopharyngodon idella	Egypt	Hong Kong	1969	aquac.; weed control	Welcomme, 1988
	Ethiopia	Japan	1975	aquac.; weed control	Welcomme, 1988
	Kenya	Japan	1969	aquac.; weed control	Welcomme, 1988
	Côte d'Ivoire	France	1979	weed control; small dams	
	Sudan	India	1975	irrigation channels	Moreau *et al.*, 1988
Cyprinus carpio	Egypt	Indonesia	1934	aquac.; man-made lakes	Welcomme, 1988
	Ethiopia	Italy	1936	aquac.; man-made lakes Koka and Akoki	Moreau *et al.*, 1988
	Ghana	?	1962	aquac.	Welcomme, 1988
	Côte d'Ivoire	Italy	1976	aquac.	Welcomme, 1988
	Nigeria	Austria/Israel	1954/1976	aquac.	Welcomme, 1988
	Sudan	India	1975	?	
	Togo	?	?	aquac.	Welcomme, 1988
Gambusia affinis	CAR	?	1958	mosquito control	Welcomme, 1988
	Egypt	?	1929	mosquito control	Welcomme, 1988
	Sudan	Italy	1929	mosquito control	Welcomme, 1988
	Ghana/Côte d'Ivoire	?	?		
Heterotis niloticus	CAR	Cameroon	1956	introduced in the Oubangui River	Moreau *et al.*, 1988
	Côte d'Ivoire	Cameroon	1958	introduced in Ayame (1962) and Kossou (1972) dams	Moreau *et al.*, 1988
Hypophthalmichthys molitrix 'silver carp'	Egypt	Japan	1962	experiment	Welcomme, 1988
	Ethiopia	Japan	1975	stocking impoundments	Welcomme, 1988
Ictalurus punctatus	Nigeria	USA	1970/1976	aquac.	Welcomme, 1988
Lates niloticus	L. Victoria	L. Mobutu	1960	fill vacant niche	Welcomme, 1988
	L. Kyoga	L. Mobutu	1955	fill vacant niche	
Limnothrissa miodon	L. Kivu	L. Tanganyika	1958–1960	fill vacant niche	Spliethoff *et al.*, 1983
	L. Kariba	L. Tanganyika		fill vacant niche	
Micropterus salmoides	Egypt	Europe	1949	control tilapia in ponds	Moreau *et al.*, 1988
	L. Naivasha	USA	1929	aquac.; fill vacant niche	Welcomme, 1988
	Nigeria	USA	1976	?	Welcomme, 1988
Oreochromis aureus	Côte d'Ivoire	Israel	1981	aquac. lagoons	Lazard, 1990
	Côte d'Ivoire	Egypt	1988	aquac. lagoons	Lazard, 1990
	Benin	Israel	1983	aquac. brackish waters	Lazard, 1990
Oreochromis alcalicus grahami	L. Nakuru	L. Natron?	1953–1960	fill vacant niche	
Oreochromis leucostictus	L. Victoria	L. Mobutu	1954		Welcomme, 1988
	L. Naivasha	L. Victoria	1956		Siddiqui, 1979
Oreochromis macrochir	Benin	?	?	aquac.	Welcomme, 1988
	Ghana	Kenya	1962	aquac.	Welcomme, 1988
	Côte d'Ivoire	Cameroon	1957	aquac.; hybridisation	Lazard, 1990
Oreochromis mossambicus	Benin		1980s	aquac.	Welcomme, 1988

Table 14.6. *(cont.)*

Species	To	From	Year	Purpose	Source
	Egypt	Thailand	1954	aquac.	Welcomme, 1988
	Côte d'Ivoire	Mozambique	1966	hybridation	
	Côte d'Ivoire	Mozambique	1982	aquac. lagoons; hybrids	Lazard, 1990
	Benin	Côte d'Ivoire	1984	aquac. brackish waters	Lazard, 1990
	L. Victoria		1961		Welcomme, 1967
Oreochromis niloticus	CAR	Congo	1957	aquac.	Welcomme, 1988
	Côte d'Ivoire	Burkina	1957	aquac.; introduced in many reservoirs	Welcomme, 1988
	Côte d'Ivoire	Uganda	1968		
	Lake Ayame		1962		Welcomme, 1988
	Lake Kossou		1971		Welcomme, 1988
	Côte d'Ivoire	Burkina	1987	aquac.	
	Côte d'Ivoire	Egypt	1988	aquac. lagoons	
	Lake Naivasha		1965	fill vacant niche	
	Sierra Leone	Volta	1978	aquac.	
	L. Kyoga	L. Albert	1957	improve fisheries	Moreau *et al.*, 1988
	Benin	Côte d'Ivoire	1979	aquac.; hybrids	Lazard, 1990
	L. Victoria		1954, 1956	Welcomme, 1967	
Oreochromis spilurus	Benin	Kenya	1986	aquac.; brackish waters	Lazard, 1990
O. spilurus niger	L. Naivasha	River Athi	1925	fill vacant niche	Siddiqui, 1979
Oreochromis urolepis hornorum	Côte d'Ivoire	Malaysia	1967	aquac.	Welcomme, 1988
	Côte d'Ivoire	USA	1982	aquac. lagoons	Lazard, 1990
	Benin	Côte d'Ivoire	1983	aquac. brackish waters	Lazard, 1990
Osphronemus gouramy	Côte d'Ivoire	Singapore	1957	?	Welcomme, 1988
Poecilia reticulata	Kenya	Uganda	1950	mosquito control	Welcomme, 1988
'guppy'	Nigeria	UK	1972		Welcomme, 1988
Oncorhynchus mykiss	Ethiopia	Kenya	1967	sport; rivers Danka and Welo	Welcomme, 1988
	Kenya	South Africa	1910	sport	Welcomme, 1988
	Sudan	Kenya	1947	?	
Salmo trutta	Kenya	UK	1921/1949	sport; rivers Danka and Welo	Welcomme, 1988
Tilapia rendalli	CAR		1953	aquac.	
	Côte d'Ivoire	Zaïre	1957	aquac.	Lazard, 1990
	Sudan		1960	weed control	
	L. Victoria	Kenya	1952		
	River Tana		1955	aquac.; fisheries	
	Liberia	Zaïre	1960		
Tilapia zillii	Ethiopia	Uganda	1974	aquac.	Welcomme, 1988
	Côte d'Ivoire	Congo	1957	aquac.	Welcomme, 1988
	L. Victoria	L. Mobutu	1954	fill vacant niche	Welcomme, 1988
	L. Kyoga		1956	fisheries	
	L. Naivasha		1956		Siddiqui, 1979

Africa (not always well recorded), but adverse ecological impacts have not so far been recorded. *Gambusia affinis* and *Poecilia reticulata* were introduced to control mosquito vectors of malaria, but the impact of these introductions is not well documented.

Many introductions of alien fish into natural environments appear to have been more or less accidental, and occurred because of confusions in identification, e.g. with batches of stock-fish contaminated with unnoticed juveniles of another species. But involuntary introductions certainly occur on a wider scale as a result of the escape of specimens from culture ponds. A striking example is that of *Ophicephalus* in Madagascar. More unusual is the transport of fish in ship's ballast water. This was the hypothesis put forward by Miller *et al.* (1989) to explain the spread to the mangroves of Lower Guinea of *Prionobutis koilomatodon*, an Indo-Pacific eleotride goby.

Translocations of geographically endemic species have been reported from Lake Malawi, where aquarium fish exporters have returned fish species to the lake in areas where they did not previously exist (Ribbink *et al.*, 1983*b*). Where such translocated species have become established, they have often usurped habitat from local endemics (Trendall, 1988*b*). It is currently unknown whether such translocations have occurred in Lake Tanganyika.

Consequences of introductions for the aquatic environment

Impact on the aquatic environment

Introduced fish species have a variety of impacts on their environment: predation on autochthonous (sometimes endemic) species, for life when ecological requirements overlap, hybridisation, introduction of parasites and diseases, etc. One indirect effect of introductions could be changes induced in aquatic habitats (Welcomme, 1988). The introduction of herbivorous species, such as *Tilapia rendalli* or *Tilapia zillii*, led to a significant reduction of aquatic vegetation in Lakes Victoria and Naivasha.

Oreochromis alcalicus grahami, an endemic species from the saline Lake Magadi (Kenya), was introduced into Lake Nakuru, a salt lake previously devoid of fish, in 1959 and 1962 to combat mosquito breeding. It is now one of the main primary consumers of the lake, feeding on cyanophytes and supporting a wide variety of fish-eating birds, such as *Pelecanus onocrotalus* (Vareschi, 1979).

Interactions with native species

Competition with the indigenous fish fauna may play a major role in determining the success or not of an introduced species. Nevertheless, this is difficult to evaluate even in simple fish faunas, and almost impossible in diverse and complex fish assemblages.

There are four ways in which introduced fish can interact with native ones.

1. The introduced species could be rejected because there is no 'vacant niche', or because the population is eradicated by predators at an early stage. Exotic fish must be introduced in sufficient number and must be large enough to avoid predation in order to become established. The introduced fish may have to learn to avoid newly encountered predators, or how to feed on new prey and eventually to compete with indigenous predators.
2. They can hybridise with very closely related stocks.
3. Introduction fish can eradicate or suppress stocks that are either an easily available prey or an 'ecological homologue'.
4. They can find a 'vacant niche' and adapt to resources that are not fully exploited by native fish species.

Another way that alien species affect native species may be through the introduction of diseases, but this hypothesis is not well documented for African fishes.

The mode of reproduction plays a major role in determining the establishment or otherwise of an introduced species. The tilapiine, *Oreochromis niloticus*, introduced into Lakes Kyoga and Victoria is more fecund and grows to a larger size than the endemic *Oreochromis* species. But eggs, juveniles and other developmental stages, also need to find suitable environmental conditions, including food and shelter, to ensure reproductive success.

An example of the negative impact of a cichlid introduction upon the native fish fauna is illustrated in Lake Luhondo (Rwanda) (de Vos *et al.*, 1990). Until 1934, three species of cyprinids were known from this lake: a small barbel (*Barbus neumayeri*) and two large cyprinids

(*Barbus microbarbis* and *Varicorhinus ruandae*). Young *O. niloticus* were introduced in 1935 from the nearby Lake Bunyoni. The introduced *Oreochromis* adapted well and was the dominant species at the beginning of the 1950s. After that period, the more slowly-reproducing *Haplochromis* sp., probably introduced accidentally as a contamination with *Oreochromis*, increased in numbers and is now the most abundant fish in the lake. The small barbel survives only in some small tributaries of the lake, and *V. ruandae* in the nearby effluent rivers; the large *Barbus* has disappeared.

Impacts of introduced predators

The introduction of the Nile perch, *Lates niloticus*, and several tilapiine species (*Oreochromis niloticus, O. leucostictus, Tilapia zillii*) to Lakes Victoria and Kyoga during the 1950s and early 1960s is one of the most significant and well-documented examples of the impact of introduced species. The aim of these introductions was to improve the fisheries.

Lake Kyoga In Lake Kyoga, where the Nile perch was introduced between 1954 and 1957 (Gee, 1969), the effects resemble those in Lake Victoria. The formerly abundant haplochromine species are now rare, with relict populations restricted to inshore areas or to areas with aquatic macrophytes (Ogutu-Ohwayo, 1990b), and presumably almost nothing is known of those species that may have disappeared completely. The changes in the relative composition of the commercial fish catches are of particular interest for an understanding of the possible future trend of the Lake Victoria fisheries. At the time of the introduction of *Lates*, the total fish catch was low (4500 t) and native species constituted the bulk of the catch. In 1963, the commercial catch had already increased to 20 000 t but *Lates* was still poorly represented, as was *O. niloticus* (Table 14.7). By the end of the 1960s *Lates* was well established and in 1977 the commercial catch rose to a peak of 167 000 t, *Lates* and *O. niloticus* contributing over 80% of the total catch. At the same time, native tilapiines had disappeared from the fisheries (Table 14.7). In 1978, *Lates* and *O. niloticus* contributed almost equally to the catch, but after that the proportion of Nile perch decreased while that of the Nile tilapia increased to the point where it constituted 80 to

90% of the catch in the 1980s (Table 14.7), concomitant with a decrease in the total catch.

It can be assumed that establishment of the Nile perch in Lake Kyoga was initially supported by the large quantities of native haplochromines, which were their main prey. It is possible that *Lates* exploited the small species of the shallow, extensively swampy areas that are difficult of access to fishermen (Ribbink & Eccles, 1988). With the increasing *Lates* stock, the haplochromines declined rapidly and were drastically reduced to the point where they are now virtually absent from Lake Kyoga. However, increased fishing pressure also contributed to the decline of the haplochromines and, later, to the overfishing of juvenile Nile perch. Moreover, with the disappearance of haplochromines, the large Nile perch started to feed on juvenile *Lates* (cannibalism), also contributing to the depletion of the *Lates* breeding stock. Altogether, these different parameters, contributed to reducing the number of adults that could support the fishery, and this has been worsened by beach seining of juveniles (Ogutu-Ohwayo, 1990b). The observation of the changes in the fish catch from Lake Kyoga tends to suggest that although *Lates* can support a commercial fishery of some magnitude, it might not maintain a high yield over a long period. A survey in 1985 (Ogutu-Ohwayo, 1990b), showed that *O. niloticus* contributed 78% of landings and *L. niloticus* only 17%. Comparatively, after the establishment in the early 1960s of a commercial fishery in the north basin of Lake Chad, there was a dramatic decline of *Lates* in the gill-net catch per unit effort, from 6–7 kg per 100 m^2 per night in 1963 to less than 1 in 1970 (Carmouze *et al.*, 1983). In Lake Kyoga, the Nile perch is a commercially valuable species and has been fished for a long time, but the total landing remains at less than 3000 t (Cadwalladr & Stoneman, 1966).

One explanation for the decrease of *Lates* in the Lake Kyoga fishery is that an explosive development of the predator population was possible due to the availability of large quantities of vulnerable haplochromine prey. Later on, the depleted prey stock was insufficient to sustain large stocks of *Lates*, which consequently declined. At the same time, the Nile perch exhibited changes in feeding habits in Lake Kyoga, switching from haplochromine and small mormyrid prey soon after its introduction, to a predominance of tilapiine and then of *Rastri-*

Table 14.7. *Commercial fish landings and catch composition (%) for selected years in Lake Kyoga, from 1963 to 1988*

Years	Total catch (tonnes)	L. niloticus	O. niloticus	Native tilapiine	Others
1963	20 000	2	3	37	58
1965	21 000	24	3	20	53
1967	26 000	48	19	11	22
1970	63 000	52	26	4	18
1973	97 000	57	32	*	11
1975	121 000	43	42	*	15
1978	162 000	42	41	*	17
1985	103 000	17	78	*	5
1986	57 000	4	91	*	5
1988		13	86	*	1

Data from the Uganda Fisheries Department in Ogutu-Ohwayo, 1990*b*. *, very low.

neobola argentea. An increase in cannibalism was also noticed.

The Lake Victoria story One of the major examples of well-documented impact following the introduction of alien species is Lake Victoria, where *Lates niloticus* and several tilapiine species (*Oreochromis niloticus*, *O. leucostictus*, *Tilapia zillii*) were introduced during the 1950s and early 1960s to improve the fisheries. The question of whether or not to introduce this large predatory fish (*L. niloticus*) was controversial from the beginning (Fryer, 1960; Anderson, 1961). Opponents pointed out that the harvest to be expected from a top predator should be less than the yield of their prey. Others argued that *Lates* could improve the local fishery by feeding on the large but relatively unexploited stock of small haplochromines. The circumstances of the introduction are not clear, given that the fish was first noticed in the Ugandan part of the lake in 1960, even as the debate on its translocation continued. Nevertheless, and after subsequent deliberate introductions in 1962 and 1963 (Gee, 1965), the Nile perch (*L. niloticus*) established itself within one decade (Fig. 14.6) as a major component of the fish population, as did *O. niloticus*, to the extent that they now dominate the commercial fisheries in parts of Lake Victoria as they do in Lake Kyoga (Ogutu-Ohwayo, 1990*b*, *c*). The other tilapiine species, which were originally important in the commercial fisheries, have almost disappeared, as well as many native endemic haplochromine cichlids.

The controversy about such an introduction led to impassioned debate, popularised in the press, between 'those who regard the Nile perch introduction as a great calamity, and those who take a milder or more positive view, noting certain benefits that the new population of fish may bring over time' (Reynolds & Greboval, 1988).

In the early 1980s the impact of the introduced *Lates* upon the indigenous fish fauna was considered an ecological and conservation disaster (Balon & Bruton, 1986; Coulter *et al.*, 1986; Ligtvoet, 1989). 'Fisheries have been not merely damaged but destroyed ... The establishment of *Lates* is not only an economic and ecological tragedy but also an enormous loss to evolutionary biology' (Barel *et al.*, 1985). Discussing the concept of 'enemy-free space' for endemic species, Fryer (1986) stated that 'the arrival of enemies to whose form of attack prey species have had no time to adapt also sometimes indicates that enemy-free space exists in complex communities. A dramatic example is provided by the depredations of the criminally introduced Nile perch *Lates niloticus*, on the endemic haplochromine cichlid fishes of Lake Victoria, many species of which have virtually disappeared from areas in which this predator has become established'. It was also forecast that future prospects were poor. After eating and exhausting all available prey, including its own progeny through cannibalism, *Lates* will move the system from 'a richly polyspecific, trophically diverse system to a virtually monospecific, trophically restricted system, dominated by *Lates* and its predatory and competitive interactions' (Hughes, 1986). Nile perch stock would ultimately reach a breakdown point resulting in

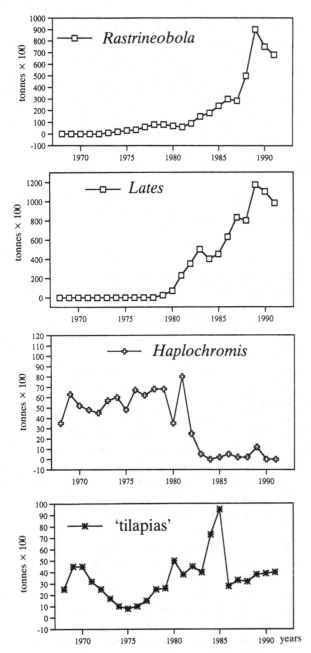

Fig. 14.6. Fisheries in the Kenyan part of Lake Victoria: species annual catches (metric tons) vs. year regressions (data from Gophen *et al.*, 1995).

collapse of the ecological system (Ribbink, 1987). As a result, it was recommended by many experts that selective fishing of *Lates* should be encouraged in order to reduce pressure upon the small cichlid stocks, and to establish conservation sanctuaries (Coulter *et al.*, 1986).

It was later recognised, that predation by *Lates* may not be solely responsible for the depletion of haplochromine stocks, and that marked changes in the endemic cichlid assemblages were partly the result of overfishing. *Lates*, which is an opportunistic predator feeding upon a wide size range of most fish species occurring in Lakes Kyoga and Victoria, is obviously directly responsible for the decline in haplochromine populations that were abundant previously. But it has also been shown that the haplochromine stock was already affected by fisheries before the establishment of *Lates*, and particularly by unregulated fishing or by trawling techniques introduced in the Tanzanian part of the lake (Ogutu-Ohwayo, 1990*b*). The collapse of the *Labeo victorianus* fishery, which started before *Lates* was established, is also attributed to overfishing of juveniles and gravid individuals.

Experience gained from other lakes has also shown that large-scale mechanised fishing threatens a Great Lakes fish community: in Lake Malawi for instance, 20% of the species disappeared from the catch between 1971–4 (Coulter *et al.*, 1986). Harrison *et al.* (1989) pointed out the surprising results of a field survey in Lake Victoria, which showed that haplochromines were still common in some areas, particularly in reserves in which no commercial or subsistence fishing was allowed. In reserves of the Kenyan waters of the lake, haplochromines occurred with large numbers of juvenile tilapiines, and representatives of several other families reported as virtually non-existent. Similarly, in Tanzania where there are no designated reserves, large catches of haplochromines were only made in less populated regions where the fishing pressure was low. Some scientists suggested a long time ago that haplochromines were being exterminated by overfishing, and believe this to be the major cause of their decline (Acere, 1988).

Witte *et al.* (1992*b*) provided quantitative data on experimental catches from 1979 to 1990 in the Mwanza Gulf, which showed that the decline of haplochromines had already started prior to the Nile perch upsurge. Of the at least 123 species recorded at a series of sampling stations, more than 80 (*c.* 70%) disappeared from the catches after 1986. In the deepwater and in the sublittoral regions, haplochromine catches decreased to virtually zero after the Nile perch boom. Haplochromines were still caught in the littoral regions where Nile perch densities were lower, and this finding corroborated to a certain extent the observations made by Harrison *et al.* (1989). However, in spite of the local abundances of haplochromines, the data provided by Witte *et al.* (1992*b*) clearly demonstrated that a considerable decrease of species had occurred in this region too. Their data also suggested that some species were more prone to extinction than others, and this appeared to depend on their abundance and adult size, and on their degree of habitat overlap with Nile perch. Thus, many piscivores species, which were relatively large compared to other species, were the first to disappear from the catches, probably as a result of fishing gear selectivity. Simultaneously, a decrease in mean size of many small zooplanktivorous and detritivorous species was observed, but the reasons for this are obscure, since surviving zooplanktivorous species increased again in size after 1986 in the presence of the Nile perch (Wanink, 1991). Moreover, pelagic zooplanktivorous and phytoplanktivorous species were less seriously affected by the demersal Nile perch than benthic haplochromines. A reduced overlap in horizontal distribution may also explain the presence of larger numbers of haplochromines in the littoral areas, where large Nile perch were either relatively rare or absent. Molluscivorous and insectivorous species, which are full-time residents in the littoral habitats, were also less seriously affected than other trophic groups.

The abundance of native cichlids in Lake Victoria may also have been reduced through competition with the introduced *O. niloticus* and other alien cichlids that have similar ecological requirements, overcrowding, and possibly the introduction of parasites and diseases (Bruton, 1990*b*). *Oreochromis niloticus*, which feeds on phytoplankton and minute water plants, grows faster and to a larger size than most tilapiine species, has a longer life span, a wider food spectrum, and is less habitat restricted (Fryer & Iles, 1972). Achieng (1990) also noticed that about 40% of the fish species before introduction of *Lates* were at least partly piscivorous (*Clarias, Bagrus, Schilbe,* etc.) and could have contributed, along with the fishing pressure, to the disappearance of haplochromine species.

There is evidence that changes also occurred in the stocks of fish species other than haplochromines: the most important catfish in the lake, the piscivorous, *Bagrus docmac,* and the omnivorous, *Clarias gariepinus,* decreased strongly after the Nile perch explosion (Witte *et al.,* 1992*a, b*). Both competition with and predation by Nile perch may have played a role in their decline. *Synodontis victoriae* and *Synodontis afrofischeri,* catfish feeding on molluscs and insects, also declined in the catches. *Schilbe intermedius* seems least affected, and has escaped the predatory impact of Nile perch because of its pelagic habits.

The impact of human activities in the vicinity of Lake Victoria could also explain the decline of the haplochromine stock, as well as various natural phenomena. There is no evidence of drastic pollution. Algal blooms, and deoxygenation following the mixing of stratified layers during storms have been reported (mimeo papers quoted in Reynolds & Greboval, 1988), but new results also provide evidence that the limnology of Lake Victoria has changed over the last few decades, and a possible effect of lake eutrophication is suspected.

The lake has experienced profound changes over the last 30 years since it was studied by J. F. Talling (1965, 1966): increases in algal biomass (3–5×) and in primary production (2×), and dramatic decline in hypolimnetic oxygen, with seasonal anoxia below 40 m (Lehman & Branstrator, 1993; Muggide, 1993). Paleolimnological studies show that eutrophication of the lake began early this century, but accelerated rapidly after 1960 as the phytoplankton community changed significantly from a diverse community with large diatoms to one dominated by filamentous cyanobacteria. There has been a shift in the dominant diatom genera from *Melosira* to *Nitzschia,* probably as a result of a decrease in silicon concentrations (Mugidde, 1993). After 1986, intense algal blooms were observed in the Mwanza Gulf, and this had never been observed previously (Witte *et al.,* 1992*a*).

It is assumed that watershed disturbance, which may have resulted in nutrient enrichment of the drainage basin, is the primary cause of this eutrophication. According to Witte *et al.* (1992*a*), the cyanobacteria were one food source of the detritivorous/phytoplanktivorous haplochromines, and the disappearance of these haplochromines may have contributed to the algal blooms.

When large Nile perch started to appear in great

numbers in the local catches, different stories emphasised its voracity, and it was even said that parts of human bodies had been found in the fish's stomach. The absence of traditional cooking methods for this species, which is more oily than tilapia, and the damage caused by large specimens to gill nets, contributed to the bad reputation of the Nile perch all around the lake, and its selling price remained low compared to other species (Achieng, 1990). It has also been condemned because, after its establishment, populations of other species, some of which were considered a delicacy, declined drastically or even disappeared. Fish traders (mostly women), therefore, faced a serious crisis and the fishing industry appeared for some time to be in jeopardy.

But the socio-economic structure quickly evolved (Reynolds & Greboval, 1988) and, with the introduction of new methods of cooking, the Nile perch was better accepted by consumers, its price rose and the fishermen reacted quickly, adopting new gear such as large-mesh nets to catch it. Overall, local fish yields have, in fact, increased almost threefold from the levels prevailing in the 1970s, and the higher catch levels have resulted in greater economic returns. Different processing methods were also introduced, such as smoking, frying and filleting. Large cold-storage trucks owned by industrial traders collect Nile perch and transport the fish to centres from which fillets are exported to large hotels or supermarkets for local consumption, and also to some overseas markets. Use of Nile perch by-products for cooking fat, medicines and soap making has become common (Bwathondi, 1985). On the other hand, these processing methods create an environmental impact due to increased demand for scarce fuelwood supplies.

'In the final analysis, Nile perch has become the most important commercial fish species, supporting a major and thriving industry on a scale not anticipated either by those who introduced it into Lake Victoria, or by those who opposed its introduction into the lake' (Achieng, 1990). '*Lates* fishery has been an exceedingly positive development from an economic benefit and food resource point of view'(Reynolds & Greboval, 1988).

Controversies over the Nile perch introduction to Lake Victoria obviously involve a complex of issues, perspectives and interests (Reynolds & Greboval, 1988). In the absence of baseline data on the original haplochromine community of Lake Victoria, it is difficult to assess precisely the impact of the introduction of *Lates* on that

community. *Lates niloticus* is an indigenous widespread African species that probably occurred in Lake Victoria during the Miocene, but in quite a different context. It should be noted that cichlid flocks can coexist with large centropomid predators, as is the case in Lake Tanganyika where four endemic species of *Lates* have coevolved with a large cichlid flock estimated at over 300 species. The vulnerability of Lake Victoria cichlids to *L. niloticus* could therefore be explained by the fact that they did not coevolve in the same lake (Bruton, 1990*b*). It has also been observed that the impact of this introduction differed strikingly according to the haplochromine trophic groups. The large species, e.g. piscivores and molluscivores, were the first to decline, being affected by the trawl fishery. The impact of Nile perch on the algal grazers and trophic groups inhabiting shallow waters was less than on the deepwater bottom dwellers, such as the prawn eaters (Witte & van Oijen, 1990). The control of the predator by selective fishing may allow the reappearance of small endemic cichlids, as long as their populations have not been completely exhausted. However, the shift in diet from haplochromines to prawns and *Rastrineobola* has allowed the predator to maintain high biomass levels in the absence of cichlid prey elsewhere.

'The introduction of alien fish into Lake Victoria constitutes a large-scale experiment that is still in progress, and the extraordinary fluctuations in the fortunes of the lakes's fishes may contain further surprises' (Achieng, 1990). The Lake Victoria ecosystem is currently unstable, and changes in the flora and fauna can be expected as a result of a process of eutrophication. Whether or not these changes will affect (positively or negatively) the survival of the haplochromines is an open question. Remnants of the cichlid fauna will probably survive in the littoral areas, but may be threatened by fishing pressure with small-meshed nets (Witte *et al.*, 1992*b*). However, there is little hope of restoring the original fish species diversity. Meanwhile, it appears that the introduction of *Lates* to Lake Victoria has provided a basis for a useful fishery and is not the only cause of decline among the haplochromines. The future of both remains uncertain.

Cumulative impacts of introductions
Madagascar lakes The freshwater ichthyofauna of Madagascar evolved independently from that of main-

Table 14.8. *Introduced species in Madagascar*

Years	Fish name	From	Adapted	References
1857	*Osphronemus gouramy*	Far East	Pangalanes	Kiener, 1963
1861	*Carassius auratus*	France	widespread	Kiener, 1963
1914	*Cyprinus carpio*	France	widespread	Kiener, 1963
1922	*Salmo irideus*	USA	Ankaratra	Kiener, 1963
1926	*Salmo fario*	France	Ankaratra	Kiener, 1963
1929	*Gambusia holbrooki*	USA	widespread	Kiener, 1963
1940	*Oncorhynchus mykiss*	Réunion		Moreau *et al.*, 1988
1951	*Tilapia melanopleura*	Congo	widespread	Kiener, 1963
1951	*Micropterus salmoides*	North America	middle altitudes	Kiener, 1963
1951	*Tilapia rendalli*	Congo		Moreau, 1982
1955	*Tilapia zillii*	Kenya	in a few places	Kiener, 1963
1955	*Oreochromis macrochir*	Congo	widespread	Kiener, 1963
1956	*Oreochromis mossambicus*	Mozambique	widespread	Kiener, 1963
1956	*Oreochromis niloticus*	Egypt	widespread	Kiener, 1963
1963	*Heterotis niloticus*	CAR	Pangalanes	Moreau *et al.*, 1988
1975	*Ophicephalus striatus*	Asia	many lakes	Rabelahatra, 1988

From Kiener, 1963; Moreau, 1979.

land Africa, and is much less diverse but highly endemic. At present, 28 freshwater species and 10 euryhaline ones are known (Kiener, 1963; Reinthal & Stiassny, 1991). Several endemic African freshwater families are not naturally present in Madagascar: mormyrids, characins, cyprinids, polypterids, catfishes, etc. The importance of Madagascar as a centre of endemism is obvious, and representatives from the families Cichlidae and Bedotiidae occupy a basal phylogenetic position of utmost importance for evolutionary studies, within their respective clades (Reinthal & Stiassny, 1991). The main threats to Madagascar's endemic fish species are undoubtedly exotic species introductions and the effects of deforestation, but the conservation of native fishes has not received much attention while efforts have focused on the mammalian fauna.

The introduction of alien species started in the last century (Table 14.8) (Kiener, 1963; Moreau, 1979). Many species were introduced for aquaculture and others to improve fish production in the natural waterbodies, but breeding fishes often escaped from fishponds and succeeded well in wild habitats. Managers largely ignored the possible impact of exotics on endemic species, and few, if any, studies were conducted in order to assess this impact or to ensure the conservation of endemics.

The introduction of exotics and the decline of native fishes are coincidental. In Lake Itasy, for instance, the decline of *Ptychochromoides betsileanus* is attributed to the progressive introduction of different species, among which tilapiines are powerful competitors. The recent introduction of *Ophicephalus striatus* has also had a detrimental effect on *Paratilapia polleni* (Raminosoa, 1987). It should, nevertheless, be stressed that many introduced species have managed very well in their new environments. Lake Alaotra and to some extent Lake Itasy are probably unique examples of natural waterbodies where species originating from Eurasia, North America, Africa and Asia coexist and succeed. An artisanal fishery has developed for those species and is doing well.

The case of Lake Alaotra has been studied in detail (Moreau, 1980). Before any introduction of exotic species, the restricted fish community was dominated by *Paratilapia polleni*, *Rheocles alaotrensis* and *Ratsirakia legendrei*, and two eels (*Anguilla mossambica* and *Anguilla marmorata*) were also recorded. A limited subsistence fishery had developed based on the endemic species. *Carassius auratus*, introduced into Madagascar in 1865, was recorded from Lake Alaotra between 1900 and 1910 and led to a slight depletion of *Paratilapia polleni*. In 1925, there were approximately 75% of *P. polleni*, 20% of *C. auratus* and 5% of others in catches. *Cyprinus carpio* was imported to Madagascar in 1916 and was introduced into Lake Alaotra in 1925, but only became significant in

Table 14.9. *Fishing activities, introductions and changes in commercial catches in Lake Alaotra (Madagascar)*

	1925	1935	1954	1957	1960	1963	1966	1969	1972	1975
Total catches (tons)			1800	2200	2700	3000	3200	2500	2100	2100
No. fishermen	50?	50?	100	230	720	1000	1400	1250	1250	1200
No. senies			6	8	30	110	130	135	120	110
No. cast nets			0	20?	80	150	240	310	270	240
No. gill nets			0	0	5	30	350	800	700	600
% in catches										
Anguilla spp.	5	5	1	1	1	1	1	1	1	1
Paratilapia polleni	75	40	5	3	2	1				
Carassius auratus	20	25	14	10	4	2				
Cyprinus carpio (1925)		30	80	40	15	14	5	11	21	21
Tilapia rendalli (1954)				46	23	8	10	20	16	16
Oreochromis macrochir (1958)					55	74	85	66	56	50
Oreochromis niloticus (1961)								1	3	7
Micropterus salmoides (1961)							1	1	3	4

From Moreau, 1980.

catches in 1935. It increased up to 1954, and in 1952 made up 80% of catches estimated at 1800 tonnes. The other species, *P. polleni* and *C. auratus*, were present but not abundant. This was the situation when *Tilapia rendalli* was introduced into the lake in 1954. It proliferated quickly and made up 46% of the fishery in 1957. This could be explained by its high fecundity, and the existence of an empty ecological niche that it occupied (being predominantly a phytophagous species). But the introduction in 1958 of *Oreochromis macrochir* led to a drastic change and this species quickly became dominant, whereas *T. rendalli* was depleted (Table 14.9). *Oreochromis macrochir* also occupied an empty niche (planktophagous species) and reproduced at a fast rate. The introduction in 1961 of *Oreochromis niloticus* and *Oreochromis mossambicus* did not modify the situation much, but *O. macrochir*, still abundant in 1975 (50% in catches), was less dominant than ten years previously. Overexploitation of this species following the introduction of gill nets in open waters (1967–9) could explain the observed trend. The black bass, *Micropterus salmoides*, introduced in 1962, represented only 4% of the catches in 1975. From 1966 to 1972, juveniles of *C. carpio* have been reintroduced into the lake, which could explain a slight increase in catches.

To summarise, introductions of exotic species to Lake Alaotra have induced a drastic reduction in the native fauna, but apparently no native species have actually gone extinct. Successive introductions and increasing fishing effort are responsible for a very unstable fish community in a constant state of change.

A similar story happened to Lake Itasy (Moreau, 1979) (Table 14.10). At the beginning of the century, the lake was inhabited by native species: *Anguilla mossambica*, *Anguilla bengalensis labiata*, *Anguilla marmorata*, *Ptychochromoides betsileanus*, *Ratsirakia legendrei* and *Chonophorus macrorhynchus*. *Carassius auratus* was introduced in 1899 and *Paratilapia polleni* (an endemic Malagasy species) in 1924, but they did not proliferate. In 1930, *Ptychochromoides betsileanus* made up 40% of the catch. Later, from 1950–5, *Cyprinus carpio*, which was introduced in 1925–30, contributed up to 85% of the total catch, and *P. betsileanus* only 10%. *Tilapia rendalli* introduced in 1955, quickly replaced *Cyprinus carpio*, and by 1958 made 70% of the catch. But *T. rendalli* itself was replaced by *Oreochromis macrochir*, which was introduced in 1958 and which constituted the bulk of the fishery in 1963. In 1965, the importance of *O. macrochir* decreased following the appearance in catches of a hybrid population called *Tilapia 3/4*, and by 1972 *O. macrochir* had disappeared. *Tilapia 3/4* is the result of a natural hybridisation between *O. macrochir* and *O. niloticus*, which was introduced in 1961. Such spontaneous hybridisation gave rise to fertile hybrids that were much more fecund at a

Table 14.10. *Fishing activities, introductions and changes in commercial catches in Lake Itasy (Madagascar)*

	1930	1950–1955	1958	1963	1964	1965	1966	1967	1968	1969	1970	1971	1972	1973	1974	1975	1985
Total catch (tons)				1200	850	500	300	280	84	830	450	300	290	280	280	275	625
No. fishermen				700	650	550	450	320		320	420	400	350	330	350	300	993
No. senies				63	60	45	22	10		3	0	0	0	0	0	0	
No. cast nets				70				90		93	80	80	80	80	80	80	
No. gill nets				45	100	200	300	400		570	550	550	500	450	400	400	
% in catches																	
Ptychochromoides betsileanus	40	10	0	0													
Anguilla spp.	50?			3	2	2	2	1	1	1	1	1	1	1	1	1	
Cyprinus carpio (1925)		80–85		6	6	5	4	4	4	3	3	3	3	3	3	3	1.5
Tilapia rendalli (1955)			70	5	4	4	3	3	2	1	1	1	1	1	1	*	0.4
Oreochromis macrochir (1958)				85	86	78	56	30	10	1	1	0					
Oreochromis niloticus (1961)				1	2	5	8	10	15	18	24	32	39	45	50	55	57
Tilapia 3/4				0	0	5	25	50	66	74	68	61	54	48	43	39	38
Micropterus salmoides (1961)				0	0	1	1	2	2	2	2	3	3	3	3	3	1.5

From Moreau, 1979.

similar size than *O. macrochir*. But the *O. niloticus* population developed slowly and in 1975 represented 55% of the catch, followed by *Tilapia* 3/4, which exhibited signs of degeneration. The contribution of species to the catch was roughly similar in 1985 to the situation observed in 1975 (Matthes, 1985). The history of Lake Itasy is therefore similar to that of Lake Alaotra until 1964, when *Tilapia* 3/4 appeared. In addition to the successive fish species introductions, overfishing and changes in fishing gear probably played a role in the changing population dynamics in this lake (see Table 14.10).

Some other species have also been introduced to Lake Itasy, e.g. *Micropterus salmoides* in 1962, and *Gambusia holbrooki* around 1950. *Oreochromis mossambicus* was also introduced in 1961, but this species, as well as *O. macrochir*, seems to have disappeared (Matthes, 1985). *Ophicephalus striatus* was recorded in the catches in 1985, and is expected to have an impact on the fish fauna in the future (Matthes, 1985). The fish fauna of Lake Itasy does not include any catfish, whereas this group is well represented in most inland waters in Africa.

The aquatic ecosystems of Madagascar are today a curious melting pot of fish species from Africa, Asia, Europe and even North America, which have been introduced into and apparently coexist in many lakes. Several of these fish are widespread species that have already been introduced into other parts of the world, but Madagascar is unique in having such a wide variety of exotics coexisting in its natural waters. The native species are undoubtedly threatened and probably some of them have already disappeared. There is an urgent need for a better understanding of the status of these endemic species and an international programme is planned.

Lake Naivasha The present-day fish fauna of Lake Naivasha (Kenya), a shallow equatorial lake, consists of only three introduced species: *Oreochromis leucostictus* introduced from ponds in 1956 and dominant in the mid-1970s; *Tilapia zillii* also introduced in 1956; and *Micropterus salmoides*, which was introduced as sport fish in 1929 (Siddiqui, 1977, 1979). The only indigenous and endemic fish species (*Aplocheilichthys antinorii*), which was found in large numbers in the surface waters in 1962, is no longer recorded.

The history of species introductions into Lake Naivasha began in 1925 when the mouthbrooding cichlid, *Oreochromis spilurus niger*, was introduced from the River Athi. This species became well established during the 1950s and the 1960s, but has not been reported since 1971. This disappearance may be related to the introduction in 1956 of *Oreochromis leucostictus*. Hybrids of *O. spilurus niger* and *O. leucostictus*, two mouthbrooding fishes, were abundant during the 1960s, but their numbers have declined considerably. They lost their intermediate characters described by Elder et al. (1971), and acquired O. leucostictus characters by the mid-1970s.

The large black bass, Micropterus salmoides, was introduced successfully from Europe in 1929 to improve sport fishing in East Africa. This population presumably disappeared around 1950 as a result of a long period of falling water level. It was reintroduced in 1951. A second cichlid, Tilapia zillii, was introduced in 1956 with O. leucostictus. In 1965, Oreochromis niloticus, a third species of cichlid, was introduced but disappeared after 1969. Other cyprinodonts (Poecilia sp. and Gambusia sp.), which were introduced at unknown dates for mosquito control, have also been eliminated, possibly through predation by M. salmoides.

The tilapiine commercial fishery started in 1959 and the catch increased sharply, reaching a peak of 1100 tons per annum in 1970, followed in the next two years by greatly reduced catches. From 1973 to 1977 the fishery remained depressed, but catches have steadily improved since 1977. The data available for the commercial landings of M. salmoides show a similar pattern to that of the tilapia fishery. The decline of the Naivasha fishery since 1971 is probably partly attributable to overfishing, but also appears to be strongly dependent on lake level fluctuations. Rising water levels correlate well with increased fish production, while catches increase approximately two years later when the increased stock has grown to a commercially exploitable size. The 1.2 metre drop in lake level between 1971 and 1975 undoubtedly contributed to the collapse of the fishery during the same period, along with overfishing (Litterick et al., 1979). The fish population of Lake Naivasha is essentially unstable and unpredictable, as a result of the relationship between fish population dynamics and lake level that drastically modifies the morphology of the littoral zone. It is also sensitive to a commercial fishery that could overexploit target fishes.

Perspectives

Are introductions a game of chance? Most scientists more or less agree that the introduction of exotic species is risky, and many examples provide evidence of the potential threats. But there are also examples of apparent success, which demonstrates the need for very careful impact assessments prior to any introduction.

The success of the Lake Tanganyika fishery, based largely on catches of clupeids, gave rise to proposals for the introduction of those clupeids, which could feed on the abundant planktonic organisms, into other lakes lacking pelagic fish species. In 1967 and 1968 Limnothrissa miodon was introduced into the man-made Lake Kariba where it undoubtedly produced a sustainable fishery (Marshall, 1984b). By 1970, L. miodon was found all over the lake and became the most important prey of the native characid tiger fish, Hydrocynus vittatus. It also invaded Lake Cahora Bassa, downstream of Kariba and no disastrous consequences were reported. However, it was thought that L. miodon could have prevented the native Brycinus lateralis from colonising the open waters of Lake Kariba (Balon & Bruton, 1986), but this species did not succeed in doing so before the introduction.

The Tanganyika clupeids, L. miodon and Stolothrissa tanganicae, were also transplanted to Lake Kivu in 1958–60, and an artisanal fishery started to exploit the clupeid stock at the beginning of 1980. This plankton feeder occupied a vacant niche and this introduction has been presented as a biological and economical success (Spliethoff et al., 1983).

From time to time, it has been suggested that Tanganyika sardines should be introduced to Lake Malawi, but this has been hotly debated. Iles (1960) concluded that the endemic Engraulicypris sardella occupies a similar niche in Lake Malawi and that another clupeid introduction might result in a reduction in the populations of this endemic planktivore that supports a large fishery. Turner (1982) suggested that an alien introduction might boost the declining local fishery, arguing for instance that Tanganyika sardines are more efficient zooplanktivores than the Lake Malawi clupeids. Eccles (1985) provided evidence that Turner's biological assessment in support of this proposed introduction was either erroneous or based on poor information (see also Coulter et al., 1986). For the moment the three riparian countries of Lake Malawi are opposed to this introduction.

The need for better communication between the scientific community and managers or administrators has been raised many times (Balon & Bruton, 1986), as well as the need to establish formal protocols for the evaluation of risks, both for indigenous species and for ecological balance, prior to any introduction. Such a protocol has already been drawn up in the USA by Kohler & Stanley (1984) to avoid the harmful effects of fish transfers. One important aspect is to gather all the information available on the proposed species to be introduced and to call for the experts' professional advice before making regulatory decisions.

Balon & Bruton (1986) clearly pointed out the difficulties of such procedures. Are decision-makers willing to accept advice that does not conform to their schemes? Do scientists have sufficient knowledge to give clear and unequivocal advice? A number of international organisations formulated a code of practice to be considered for any new candidate species for introduction to inland waters, but it has not been used yet in Africa.

Conclusion

It is clear that although the biodiversity of inland waters in tropical Africa is threatened in a number of ways, the most serious and immediate threats are river regulation and the introduction of alien species. However, as the countries of Africa develop and industrialise, pollution from the catchments will become even more serious than the present effects of soil erosion and agricultural runoff. These latter causes of habitat degradation need immediate remedial action, and it is urgent that African governments are assisted in learning the lessons from other tropical areas, so that their efforts to enhance socio-economic development will be compatible with the sustainability of biodiversity. This need is particularly pressing in cases such as Lakes Tanganyika and Malawi that, so far as is known, still retain most of their pristine, unique biodiversity.

15   The economic role of fish biodiversity

> Our planet's essential goods and services depend on the variety and variability of genes, species, populations and ecosystems. Biological resources feed and clothe us and provide housing, medicines and spiritual nourishment.
>
> *Agenda 21, 15.2*

Fishing and fish trading have provided food for centuries and contribute substantially to the supply of animal proteins in many countries. In Africa, the aquatic harvest still consists of wild rather than farmed species, and the diversity found within a small number of fish, such as the tilapiines, that form the basis of an important world tropical aquaculture production remains a small but vital part of the rich biodiversity of African fish. Cultured fish will certainly provide a major source of food in the future, even if aquaculture has not yet been adopted widely in rural Africa, primarily because it has no traditional base in smallholder agriculture. Moreover, the new field of research opened up by gene technology in relation to the development of aquaculture will probably stimulate greater interest in the fish genetic resources of Africa, as it already has for domesticated plants. In this case intraspecific genetic variability will be very useful for the improvement of selected strains.

More recently, there have also been attempts to use various fish species for the biological control of pests, and meeting the demands of the ornamental fish industry is another way to increase the economic value of the diversity of fish species. Species richness and diversity is also a criterion for assessing the health of freshwater ecosystems.

Assessing the value of biological diversity

Despite the past neglect of environmental problems and externalisation of environmental costs, since the 1970s economists have taken a broader interest in environmental issues than previously. This change was largely due to scientists who argued that there were limits to human population and development. Fortunately, not all the pessimistic scenarios foreseen, for instance by the 'Club of Rome' and others, have yet to come to pass, but the decisive question raised by the debates about the 'limits to growth' is still unanswered: how large can the human population become and how long can it be sustained with the given available set of resources?

Economic and environmental issues

Because human welfare ultimately depends on natural resources, there are strong interactions between ecological, economic and social systems. Many of the critical questions at the interface of ecology and society involve economics (Ehrlich, 1989), and the word bioeconomics has been coined for this complex of interactions (Clark, 1989). In general biologists have tended to oversimplify the economic side of the relationship and have displayed greater concern for the health and persistence of ecosystems as a foundation for human well-being. On the other hand, economists have largely ignored natural systems and resources and are not sufficiently well equipped to deal with them. Fishery management illustrates the complexity of biological and economic systems and the difficulties that specialists encounter in trying to understand each other.

Growing environmental problems have not led to abandonment of the idea that the maximisation of human welfare is an inherent goal of economics, but have deprived it of its exclusivity. To achieve global sustainability we need to take account of the fact that economic systems are dependent on their ecological life-support systems. Until recently, time scales were not given enough attention in the use of natural resources, but this attitude has to be modified because we know that many resources are finite. Economists, therefore, have to extend their theoretical horizons and to consider the environment as an extra factor in production costs (Hohl & Tisdell, 1993).

Assessing the costs of environmental deterioration (or benefits of an unspoilt environment) is fraught with difficulty. Conceptual models include the impossibility of measuring values that are intrinsic (i.e. values of the natural world independent of its value to mankind), as well as those that are anchored in a culture. Empirical problems include the difficulty of estimating costs of environmental trends (such as the accumulation of greenhouse gases) in the presence of great uncertainty regarding their potential impact. Nevertheless, an attempt to assess costs and benefits is an essential ingredient of the process of setting priorities and designing policies and programmes.

Confusion over the exact meaning of biodiversity and the assertion that biological resources are the physical manifestation of biodiversity (McNeely et al., 1990) may have contributed to the idea that biological resources and biodiversity are synonymous and that their values must also be synonymous. Even if biological resources depend on biodiversity, it does not mean that the economic value of the former is the value of the latter. For example, the presence of many species in a lake means valuable biodiversity, but not necessarily high resource value if all the fish are too small for fishing. Conversely, low diversity could still be economically valuable if the few species were highly edible. The perverse result would be that biodiversity may become a redundant factor and that development projects may fail to distinguish between conserving valuable resources and conserving valuable biodiversity. The urgent need to act on conservation measures, combined with a lack of consensus on definitions, could, nevertheless, explain why scientists and economists choose the easier route of considering the value of biological resources rather than that of biodiversity per se.

In order to counter this approach, Barbier (1989) and

Table 15.1. *Ecological and economic concepts compared*

System concepts	Ecosystem concepts	Ecological variable	Economic concepts
stocks	structural components	matter, space	goods
flows	environmental functions	time, energy	services
organisation	biological and cultural diversity	diversity	attributes

From Aylward & Barbier, 1992.

Aylward (1991) proposed a conceptual framework for understanding the relationship between biodiversity and biological resources, in order to formulate a methodology for the valuation of biodiversity. They distinguished between structural components, such as genes, water, species, etc., and environmental functions, such as groundwater recharge by wetland areas. Moreover, they (Aylward & Barbier, 1992) proposed a conceptual linkage between ecology and economics, using three basic system concepts – stocks, flows and organisation of these stocks and flows – that can be applied to both ecology and economics (see Table 15.1).

Structural components of the ecosystem are species populations and non-living matter. When they are appropriated for use, e.g. fish for consumption, economists call them goods. Environmental functions, such as hydrological and nutrient cycles or energy flows, correspond to change in stocks over time. When they produce benefit flows (e.g. watershed protection by vegetation cover), they provide economic services. Goods and services are the outputs that affect human welfare through the economic process of production and consumption. Attributes are not outputs per se but indicate how goods and services (e.g. the components and functions of an ecosystem) are organised. The attribute of diversity in ecological components, environmental functions and ecosystems may affect the value arising from these outputs into the economic system. Diversity of an ecosystem is, therefore, a qualitative attribute, which in a broader sense comes in many guises including cultural, geological and biological diversity.

Nevertheless, discussions about biological diversity and biological resources should not mask the need to assess the total economic value of ecosystems, given that the diversity of ecosystems is also part of the biodiversity concept.

Economic value of biodiversity

Valuation is a fundamental step in informing planners and resource managers about the economic importance of biodiversity in national development objectives, and in demonstrating the importance of different areas for the biological resources they contain. A major problem in resource management is the conflicting demands on resources, as well as on the environment. Current methods of evaluation used in decision-making, such as cost–benefit analysis, inadequately reflect the true environmental and socio–economic value of natural resources and ecosystems. For a long time, natural ecosystems were considered as unproductive areas whose benefits could only be realised by conversion to some other use. Consequently, many systems have been altered to serve other purposes, simply because their value to society was not adequately demonstrated, and because traditional evaluations favoured short-term benefits. Many rivers, for instance, were dammed to provide electric power, but the long-term consequences of damming were not properly evaluated before any impoundment project.

A variety of methods have therefore been devised for assigning values to natural biological resources that in some way reflect the different values which can be recognised for natural resources. The mainstream economic approach today is to compile a utilitarian calculation expressed in money value, and to include the commercial values that are expressed in markets. Fish production may be one of the many economic uses of waters; others include, for instance, domestic use, drinking water supply, energy (hydroelectric dams) and agriculture (irrigation). Decisions on land and water use have a great impact on biodiversity but political considerations are often paramount in these decisions, whereas the value of biodiversity should be introduced as a major component in the evaluation of alternative land/water uses.

The next step is to determine how to protect an area and its species and how much it will cost. Conserving biodiversity in its present condition would require far more funds than are assumed to be available in the near future. Difficult decisions have to be made in order to

Table 15.2. *Types of economic values usually recognised in order to evaluate biological resources*

Direct use values	
consumptive	harvesting of food, firewood, fish, game meat
productive	entering into market: fishery, forestry
non-consumptive	recreation, tourism
Indirect use values	ecological functions
Option value	expected, still unknown use
Existence value	intrinsic value; knowledge that something exists

identify which species, genetic resources and habitats should be conserved, and how to manage conservation programmes. Economic tools have an important role to play in such an approach, but coherent conceptual and methodological frameworks for measuring the economic value of the benefits of biodiversity have to be developed. The most important obstacle is that it is difficult to discuss the value of nature and to formulate the necessary measures in specific concrete terms.

The total economic value of an ecosystem is made up of its direct and indirect uses (Table 15.2). The direct and indirect use values of biodiversity are clearly part of the total economic value of the components and functions of an ecosystem as traditionally calculated in the cost-benefit analysis of alternative land/water uses. Some of the biological resources can be transformed into revenue, while others provide flows of service that do not carry an obvious price-tag. The economic value derived from the structural components and environmental functions of ecosystems is often classified by economists into use and non-use components (McNeely, 1989; Aylward, 1991, Angel et al., 1992).

Direct and indirect use values

Direct use values describe the benefits of the goods and services that enter directly into the human economy. Assessing the value of direct use of ecosystem outputs (nature's products or structural components) is the impact, which is commonly measured in valuation exercises and is a long-established area in resource economics. The 'consumptive use value' deals with products, such as fish, firewood and game meat, that are consumed directly without passing through a market, whereas the 'productive use value' concerns products that are commercially harvested (fish and game meat sold in a market, timber, etc.). A 'non-consumptive use value' is sometimes identified, e.g. tourism, recreation, bird-watching, etc.

Attributes such as biodiversity may also have different use values (Aylward, 1991; Angel et al., 1992). It has a direct use value when diversity of any type enters directly into consumer or producer preference. For example, people could exhibit preferences for different species of fish, and these preferences could vary according to cultures and food habits.

The productive use value, which is the most powerful of the measurable economic values, shows, in fact, that loss of biodiversity can be economically advantageous for mankind. A few fish species with high growth rates may produce, in intensive culture, the majority of fish food requirements. This could be a substitute for harvesting natural populations, but the establishment of large fish-culture systems may have negative impacts on water quality. On the other hand, greater specialisation in fish strains may also increase risks of catastrophic failures, and this could provide an argument for the conservation of wild strains of domesticated species, which could eventually be used to improve the genetic diversity of cultivated strains.

The 'indirect use values' mainly concern the ecological services provided by biological resources. Although it is generally accepted that the economic benefits derived from the ecological functions of ecosystems and the biosphere are dependent on biodiversity, ecologists have enormous difficulty in explaining the true importance and role of biodiversity in the functioning of nature. Moreover, calculation of the economic benefits of such ecosystem services are difficult to assess and to quantify.

In Africa, inland-water fish make a considerable contribution to human welfare. Part of the catch is directly consumed by villagers, while a commercial fishery is active in most countries. Fish capture for aquariology is another direct use of the resource. The indirect use value of fish biodiversity is revealed when changes in diversity influence the biological production of a good or service that itself enters directly into people's preference struc-

ture. Hogson & Dixon (1988) illustrated, for instance, how changes in both the coral species diversity and the gross amount of coral cover could affect the fish biomass in a coral reef ecosystem. An underestimated role of fish biodiversity in the ecosystem is that they are a source of food for many piscivorous bird populations. The introduction of a tilapia into Lake Nakuru (Kenya) allowed the establishment of a huge colony of pelicans and supports a wide diversity of other fish-eating bird species that are attractive to tourists (Vareschi & Jacobs, 1984). Indirect non-consumptive use through bird-watching and tourism is far from negligible in this case and is beginning to be more widely appreciated in Africa, as elsewhere.

It should also be noted that direct values often derive from indirect values, for example, when harvested fish species are supported by the goods and services provided by their environments. Species without consumptive or productive values may support valuable species, and therefore play an important role in the ecosystem.

Option and existence values

An assessment of the indirect use values of environmental functions of an ecosystem that supports economic activity needs to identify the 'option values' which are associated with the future of a resource when options are kept open, and the 'existence values' which simply recognise that certain species or resources exist but are not used.

The option value amounts to 'a willingness to pay' for preserving access to a diverse range of ecosystems, species and genes that might later reveal economic benefits. Some authors see option value as part of 'insurance value' or protection against risks.

The empirical measurement of indirect values, such as option and existence values, is still a largely unexplored subject (Aylward, 1991; Aylward & Barbier, 1992). An emerging consensus that option values for biodiversity are positive needs to be extended by a debate over their order of magnitude. Estimation of option values, beyond an affirmation that they are positive, is in fact very difficult and no methods are really available. Methods that have been developed to measure the existence value are also based on the willingness to pay principle, and usually relate to charismatic species rather than to diversity itself. The value of future use of genetic material, or pharmaceutical compounds for instance, could be compared to the high costs of product identification and research. The increasing development of biotechnology on the other hand may contribute to decreasing interest in wild strains. In general, option values may not be as large as was first suggested, but another approach is to consider that natural habitats preserve a reservoir of continually evolving genetic material that enables the various species to adapt to changing conditions. In this case, option value is a means of assigning a value to risk aversion in the face of uncertainty, and society may be willing to pay for it.

Conversely, the existence values of species and ecosystems are likely to increase along with growing public interest and increased media coverage (Aylward, 1991). Many people in the industrialised nations attach particular value to the existence of species or habitats, or expect that future generations may derive some benefits from them. The ethical dimension is therefore important in determining the 'existence value' (McNeely, 1989), which reflects the concern that some people may feel towards species and ecosystems. Nature has considerable symbolic attraction for people stressed by urban life and a variety of economic activities have developed from this, including books, films, photographs, TV, etc. A large existence value for biodiversity, when diversity is threatened and well publicised, could be potentially an important component of conservation projects. Therefore, they should not be included again in the environmental economic assessment, but could be used separately for evaluation of impacts in the case of conservation projects on biodiversity.

The different economic values that are today associated with biodiversity are, in most cases, a valuation of biological resources (individual, species, ecosystems) rather than a valuation of biodiversity itself. The latter can only be measured by assessing the impact on economic values if that diversity is reduced – the social cost of biodiversity loss. From this utilitarian perspective, biodiversity loss seems in many cases to have no measurable economic impact. Thus, if the economic argument for conservation of biological resources cannot provide powerful conservation incentives, the argument in favour of the diversity of these resources is much less straightforward and convincing. Therefore, the task would be to optimise a complex relationship between the level of

diversity and the socio-economic benefits to be derived from it, bearing in mind that cost/benefits might change over time, might be different according to the chosen geographical perspective and might be influenced by other social objectives.

One great danger of the economic valuation arguments is that we could be increasingly forced into piecemeal argument over every individual species or individual policy or action, instead of looking at biodiversity as a whole. Ehrenfeld (1988) cautions that arguments for conservation should not be based simply upon economic considerations, and that by assigning value to biodiversity we merely legitimise the process which says that the tangible magnitude of the dollar costs and benefits is the first thing that matters in any important decision. 'It is certain that if we persist in this crusade to determine value where value ought to be evident, we will be left with nothing but our greed when the dust finally settles. I should make it clear that I am referring not just to the effort to put an actual price on biological diversity but also to the attempt to rephrase the price in terms of a nebulous survival value. . . . As shown by the example of the faltering search for new drugs in the tropics, economic criteria of value are shifting, fluid, and utterly opportunistic in their practical application. This is the opposite of the value system needed to conserve biological diversity over the course of decades and centuries' (Ehrenfeld, 1988).

Case studies: economic value of an African tropical wetland

There are many components of tropical wetlands that are directly exploited to support people, e.g. through fishing, hunting, fuelwood extraction, etc. But wetlands also perform ecological functions that may benefit economic activity, within, or beyond, the wetland area, e.g. downstream flood control, groundwater recharge and sediment or nutrient retention. These functions of wetlands appear to have indirect use value. Barbier (1989) discussed approaches for incorporating environmental functions into the valuation process for South American wetlands. The value of wetland environmental functions arises through their support or protection of economic activity. Where economic production is being supported, values can be measured in terms of the value of change in productivity attributed to these functions operating nor-

Table 15.3. *Economic values of wetlands characteristics:
the Hadejia-Jama'are floodplain in northern Nigeria*

	Economic values		
	Direct	Indirect	Non-use
Components			
forest resources	*		
wildlife resources	*		
fisheries	*		
forage resources	*		
agricultural resources	*		
water supply	*		
Functions			
groundwater recharge		*	
flood and flow control		*	
sediment retention		*	
nutrient retention		*	
water quality maintenance		*	
shoreline stabilisation		*	
nursery (fish, shrimps)		*	
recreation/tourism	*		
water transport	*		
Attributes			
Biological diversity	*	*	*
uniqueness to culture/heritage			*

Modified from Barbier (1989); Aylward & Barbier (1992).

mally. In other words, one should estimate the damage costs to the economic activity that are currently being avoided. Estimating non-use/preservation values is extremely difficult. Valuation of the direct uses, indirect uses and non-use/preservation values of wetlands relates to the consumer's willingness to pay (WTP) for these various uses.

Barbier *et al.* (1991) calculated the benefits of the direct uses of the Hadejia-Jama'are floodplain in northern Nigeria (Table 15.3). Looking simply at the uses for agricultural, fishing and fuelwood production, they found that the wetland netted a higher economic return per cubic metre of water, than water developments being considered upstream. The estimated annual net economic benefits from fishing in the floodplain was 2 million dollars for the 100 000 ha of flooded land (24 US$ per ha). The net economic benefits of agriculture were estimated at around 32 US$ per ha per annum. However, adding in the indirect benefits of other environmental functions whose value does not overlap

with those of the direct uses (household water use, other wildlife benefits, non-timber forest products) increases the value of the total benefits to a level equalling or perhaps exceeding that of agriculture, strengthening the case for conservation.

Direct estimation of WTP for water use could be used to value environmental functions. For example, an important function of the Hadejia-Jama'are floodplain is to recharge the Chad formation aquifer, and this groundwater is in turn drawn off by numerous small village wells for domestic use and agricultural activities. By combining information on the groundwater recharge supplied by the wetlands, the impact on the water table of the water extraction and the WTP of villagers, it might be possible to give a value to the groundwater recharge function provided by the floodplain, or to the cost of digging deeper wells.

The economic importance of the floodplain suggests that the benefits it provides cannot be excluded as an opportunity cost of any scheme that diverts water away from the floodplain system. Policy-makers should be aware of this problem when designing water-development projects in the Hadejia-Jama'are river system. In the valuation of this Nigerian floodplain, Barbier *et al.* (1991) ignored the health factor, that is to say the cost of endemic diseases, such as malaria or bilharziasis, associated with marshy systems and the cost of control and/or treatment would need to be offset against the benefits.

Fisheries

In most African countries, fish are important natural renewable resources (Table 15.4). Whereas in the past certain tribes, particularly those with a cattle-owning tradition, did not eat fish for reasons of custom or taboo, catching fish is now a major source of protein all over tropical Africa and the economy of many rural communities is heavily dependent on inland fisheries. With increasing human populations, this source of food is of even greater value, and fisheries science, including fishing technology, regulation and management of fisheries, is therefore of great significance.

Much has been written about African fisheries (for example Balon & Coche, 1974; Welcomme, 1979, 1985, 1989; Hopson, 1982; Carmouze *et al.*, 1983; Kapetsky &

Table 15.4. *Fish catches and fish culture production in inland waters in 1987 in tropical Africa*

Country	Catches inland water fisheries (tons)	Production inland water aquaculture (tons)	% inland fish to total fish production
Angola	8000	0	10
Benin	31 973	14	76
Botswana	1900	0	100
Burkina Faso	6964	36	100
Burundi	4984	25	100
Cameroon	19 863	137	24
Central Africa	8800	88	100
Chad	110 000	0	100
Congo	13 385	115	43
Côte d'Ivoire	27 353	847	27
Egypt	141 700	60 000	81
Equatorial Guinea	400		10
Ethiopia	3500	0	87
Gabon	1897	3	9
Gambia	2700	0	19
Ghana	53 614	386	14
Guinea	1999	1	7
Kenya	124 096	210	95
Liberia	3997	3	21
Madagascar	45 806	194	72
Malawi	88 485	103	100
Mali	55 690	12	100
Mozambique	246	21	0.7
Namibia	150	0	0.03
Niger	2386	14	100
Nigeria	103 209	5528	44
Rwanda	1565	65	100
Senegal	14 966	34	5
SierraLeone	15 982	18	30
Sudan	22 757	43	95
Tanzania	265 735	35	85
Togo	705	9	5
Uganda	200 000	387	100
Zaïre	163 300	700	99
Zambia	66 980	1020	100
Zimbabwe	17 344	156	100
Totals	1 632 431	70 214	

Data from Vanden Bossche & Bernacsek, 1990, 1991.

Petr, 1984; Reynolds & Greboval, 1988; Vanden Bossche & Bernacsek, 1990, 1991; Coulter, 1991a; Kolding, 1994; Quensière, 1994). It is not within the scope of this book to discuss this topic in detail and only a general overview will be provided.

It is difficult to get a reasonable long-term assessment of fishery catches in any particular water system due to the large number of landing points and the difficulties encountered in sampling a very scattered fishery. Furthermore, standing stocks and production are seasonal and variable, and in river systems are related to flood intensity. As a result, most fishery statistics must be used with great caution, assuming that in many cases they are no more than a very rough estimate. Official statistics from FAO indicate that African inland waters alone yield almost 2 million tonnes of fish annually (Fig. 15.1)

In African inland waters, a broad spectrum of species are caught. In the Niger River, for instance, the catch may contain over 50 fish species. However, in some cases, there is a temptation for managers to introduce alien species into water systems in order to increase fish yield. This has been successful in a few cases, such as in the newly created man-made lakes (the introduction of the clupeid *Limnothrissa* into Lake Kariba is probably a good example). However, the recent controversy about the Nile perch introduction into Lake Victoria (see Chapter 14) has resulted in a general feeling that all introductions have negative effects. This is not always true, and there are many examples of past introductions that have been highly successful in meeting ecological and economic objectives. A central problem is the lack of good predictive models and codes of practice for introductions into African waters (Bernacsek, 1987). Research into modelling introductions and assessment of their impacts on native fauna and flora is badly needed.

Catches from fisheries based on wild stocks are supplemented by traditional, indigenous technologies designed to increase fish production: the acadja brush parks of Benin (Welcomme, 1972; Kapetsky, 1981); the garse system in Cameroon (Stauch, 1966); and the hoshas of Egypt are examples.

Until the middle of this century, African inland fisheries were traditionally pursued with a variety of locally fabricated gear such as baskets, spears, seine nets, etc. Traditional fishing methods, ranging from simple harpoons to basket-work fishtraps, are typically selective for both size and species and are adapted to the diversity of fish capture possibilities under particular environmental conditions. Fishing strategies adapted to the hydrological cycle have been documented (see Blache & Miton, 1962; Welcomme, 1979). After the Second World War, the

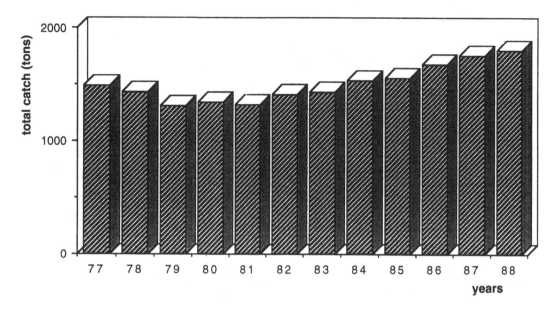

Fig. 15.1. Annual total fish catch in inland waters of tropical Africa (data from FAO).

introduction of nylon ensured the nylon gill net's pre-eminence in most African fisheries. Later, in the 1960s, mechanised fishing and trawling were developed in the East African Lakes. Commercial gear often has a by-catch of unwanted species and this has contributed to the overexploitation of resources.

African fishermen still use a wide variety of gear. In the Inner Delta of the Niger for example, fishermen still use harpoons, cast nets, various types of fish trap, barrage traps, dip nets, lift nets, fixed or drifting gill nets, hooks and lines, etc. (Lae *et al.*, 1994). A similar diversity of fishing gear existed in the Benue River (Stauch, 1966), the Chad system (Blache & Miton, 1962) and Madagascar (Kiener, 1963).

The huge diversity of fishing gear, which has been developed over centuries, should be considered as a cultural heritage that is rapidly disappearing. With the introduction and spread of new fishing technologies many traditional practices have become extinct.

Fish culture

Aquaculture is the farming of aquatic organisms under controlled or semicontrolled conditions. This implies some form of intervention in the rearing process to enhance production, such as regular stocking, feeding, etc. There is a growing feeling that inland aquaculture is more akin to agriculture than to fisheries and will become integrated into mixed farming systems.

Although aquaculture is of considerable antiquity, especially in the Far East, there is apparently very little tradition of fish culture in tropical Africa, despite considerable effort by many bilateral or international agencies to promote fish farming, particularly since World War II. There are various reasons for this apparent failure (Jackson, 1988). One is that fish culture is still often uneconomic, artificial food being too costly in comparison with the price of the fish on the local markets, so that fish farming does not pay. But a major obstacle seems to be social, and this has resulted, up to now, in the absence of any real interest in domesticated fish. As a result, fish culture production in tropical Africa only amounts to 10 000 tonnes per annum, whereas it is 6 to 7 million tonnes in fresh and brackish waters world-wide, mostly produced in Asia.

Species used for fish culture
Many cichlids have excellent attributes for aquaculture and they are easily bred in captivity without complex hatchery equipment. Undoubtedly, tilapias are more

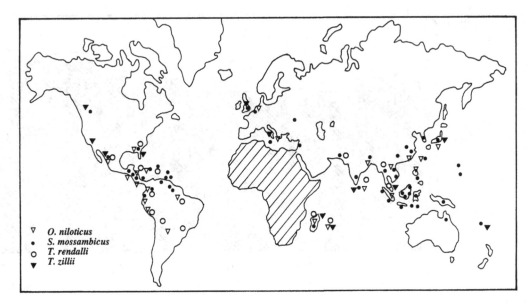

Fig. 15.2. Introductions of tilapias (4 principal species) outside Africa. From Philippart & Ruwet, 1982; Welcomme, 1988).

popular in African aquaculture than any other fish and various species are of value in fish culture (Balarin & Haller, 1982; Pullin & Lowe-McConnell, 1982). *Oreochromis niloticus* was the original tilapia to be cultured in Africa, and remains the most widely used. However, various other species, such as *Oreochromis aureus*, *Oreochromis macrochir*, *Oreochromis mossambicus*, *Tilapia rendalli*, *Tilapia guineensis*, and *Sarotherodon melanotheron*, have also been used in African aquaculture.

Although endemic to Africa, many tilapiine species are now firmly established in warm-water aquaculture throughout the world (Fig. 15.2), and as a result of their excellent attributes for aquaculture, tilapias have been dubbed 'aquatic chickens'. Nevertheless, while their culture has made great advances during the last two decades in some Asian countries, it remains poorly developed in Africa. FAO Statistics for 1990, show world production of tilapias over the period 1985–8 was between 250 000 and 270 000 tonnes a year. Over 95% of the current production is *Oreochromis* spp. and their hybrids (*O. niloticus* being the dominant species), the major part of which comes from Asia, only 6000 to 7000 tonnes (2.5 to 3% of

world production) being produced in Africa, the native habitat of tilapias.

Interest in the culture of African catfish is relatively recent, and was probably stimulated by the success of the Asian *Clarias batrachus*. Three species have been particularly studied for this purpose: the clariids, *Clarias gariepinus* and *Heterobranchus isopterus*, and the bagriid, *Chrysichthys nigrodigitatus*.

Many other species, including *Schilbe mystus* (Kruger & Polling, 1984), *Gymnarchus niloticus* and *Distichodus rostratus*, have been used or studied for use in African aquaculture, but did not give sufficiently good results to be considered as good candidates.

While cyprinids are naturally found in very diverse habitats and have many advantages for fish culture, native species have not been used in African aquaculture to date and only a few attempts have been made to use exotic species. However, various Chinese carps have been introduced, such as the common carp, *Cyprinus carpio*, and also *Hypophthalmichthys molitrix* (silver carp), *Aristichthys nobilis* (bighead carp) and *Ctenopharyngodon idella* (grass carp). The common carp was first introduced into

Madagascar where the native fish fauna is depauperate, and later into Kenya, Cameroon, Rwanda, Malawi, Côte d'Ivoire, Central African Republic, etc.

Fish genetics applied to aquaculture
Genetic diversity is the basic material of evolution. Conservation and sustained utilisation of genetic resources is especially important in the field of aquaculture and fisheries.

Sex control
The reason for controlling the proportion of sexes in fish breeding is that in fish sexes often differ in characteristics of economic relevance, such as growth rates. In many cichlids, for example, males have a more rapid growth than females, and the use of monosex male populations is one way of improving the productivity of tilapia cultures. Another reason for using only males is to avoid unwanted natural reproduction, which would lead to overpopulation to the detriment of the fishery. To obtain such monosex populations different techniques have been proposed, but most of them are not entirely satisfactory.

1. Individuals can be sorted manually (with 2.7–10% error) when 2–3 months old. This technique has the disadvantage of eliminating females; almost half the population is wasted after 2–3 months of rearing (see Baroiller & Jalabert, 1989).
2. In tilapiines, the crossing of *Oreochromis mossambicus* (female) with *Oreochromis hornorum* (male) produced all-male progeny, these hybrids being fertile (Hickling, 1960). Other *Oreochromis* crosses were later reported to produce all-male F1 hybrids, for instance: *O. niloticus* and *O. hornorum* (Pruginin, 1967), *O. niloticus* and *O. macrochir* (Jalabert *et al.*, 1971), and *O. niloticus* and *O. aureus* (Fishelson, 1962; Pruginin *et al.*, 1975). However, hybridisation between two parental species, such as *O. niloticus* (homogametic female) and *O. aureus* (homogametic male), does not always provide entirely 100% males (Wohlfärth & Wedekind, 1991).
3. Hormonal inversion, a widely used technique, consists of masculinising a population of tilapia fry by incorporation of a synthetic steroid into their food

(Pandian & Varadaraj, 1987; Baroiller & Jalabert, 1989). It is a very efficient technique, but there are questions about the fate of degradation products of the synthetic steroids and their potential ecological consequences (Wohlfärth & Wedekind, 1991). Moreover, the use of hormones to produce sex-reversed fish for food production remains forbidden in many north temperate countries.

4. Another approach is to produce homogametic males or females whose progeny are entirely all male. For species such as *O. aureus* where the male is homogametic (ZZ) and female heterogametic (WZ), sex-reversed female fish (ZZ) may be produced using feminising hormones. Such fish have a male genotype but a female phenotype. When bred with non-treated males, the sex-reversed females theoretically produce 100% animals of male phenotype. On the contrary, *O. niloticus* males are heterogametic (XY) and females homogametic (XX). A sex-reversed female (XY) may be obtained by hormonal inversion with oestrogen. When this sex-reversed female is crossed with a classic male (XY), the progeny partly contains a new genotype that is a homogametic male (YY). Theoretically, when manipulated homogametic males are crossed with naturally homogametic females, all progeny should be entirely males. But technical problems remain.
5. For fish, the gonadal differentiation of sex is plastic and environmental factors may also influence the appearance of phenotypic sex, independent of genetic sex, and the thermosensitivity of sex differentiation has been demonstrated in tilapias (Baroiller *et al.*, 1996). The progeny of sex-reversed males of *O. niloticus* (XX) crossed with homogametic females (XX) should produce all-female progeny. But when fry are exposed to raised temperatures above 34 °C, a very high proportion of males is produced (84–87%), indicating that warm temperatures appear to affect sex determination directly. The thermosensitivity of the fry is critical for the 10–13 day period after fertilisation. When a thermal treatment is applied after that time, sexual differentiation appears to be already genetically determined.

The introduction of techniques allowing for the selection of all-male populations is probably responsible for the considerable increase of tilapia production since 1980, estimated today at 500 000 tonnes per year worldwide.

Polyploid induction

In order to control reproduction in cultured fish, it is also theoretically possible to produce sterile males through induction of polyploidy. The economic potential of polyploidy led many investigators to study the conditions by which it could be artificially induced.

The induction of triploidy provides functional sterility and the development of gonads is reduced or completely inhibited, allowing for increased somatic growth. But methods used to produce triploids may have detrimental effects on embryonic development, and the mating of diploid and tetraploid organisms is an alternative method for inducing triploidy. Triploidy caused by inhibition of extrusion of the second polar body in fish has been achieved in a number of species (Don & Avatlion, 1986, 1988a). At the moment survival rates are low after triploidisation and only females are entirely sterile (Penman et al., 1987b).

Success in producing viable male and female tetraploid rainbow trout by using pressure shock was reported by Chourrout et al. (1986). For cichlids, tetraploid induction, using a combination of pressure and cold treatments, has been attempted on two species of tilapia, Oreochromis niloticus and Oreochromis mossambicus, and their hybrids (Myers, 1986), but survival was very poor among the tetraploids produced. Development abnormalities were also pronounced, and no tetraploid survived beyond seven days post-hatching. In another experiment, a cold-shock treatment of 1 h at 11 °C applied to embryos incubated at 25 °C at the zygotic age of 92 min resulted in the production of viable tetraploid Oreochromis aureus (Don & Avatlion, 1988b). The efficiency of tetraploidisation so far is very limited, but only a few pairs of mature tetraploid fish would be sufficient to create a whole tetraploid population if the 4N fish obtained were fertile.

Artificial hybridisation

Fish hybridise relatively easily and the interbreeding of different taxonomic groups has been widely practised by fish culturists. The production of new forms of fish that are more suited to specific needs, such as an improved productivity, is certainly the most purposeful approach. Hybridisation allows the production of monosex populations (see above) and makes possible uniform products, or new breeds or strains. A particular use is for the production of infertile hybrids to be stocked in natural water bodies where they will not be able to maintain a population.

Artificial hybridisation between two Clariidae, Clarias gariepinus and Heterobranchus longifilis, was reported by Hecht & Lublinkhof (1985) and Teugels et al. (1992a). Both species have a similar morphology and mainly differ in the adipose fin complex. Striking similarities were found in the karyotypes of these species and most pairs appear homologous, but the modal chromosome number is 2n = 56 for C. gariepinus and 2n = 52 for H. longifilis (Teugels et al., 1992a). The hybrid karyotype (2n = 54) is intermediate between that of the parental species and in addition, two unpaired chromosomes that originate from the C. gariepinus karyotype can be observed.

Genetic improvement programmes

Genetic improvement is the process of replacing a given population of genotypes with another exhibiting better phenotypic performance. This implies an understanding of the relative influence of a fish's genotype on its phenotype, and evaluation of which production traits can be improved by selection programmes.

On the whole, warm-water aquaculture has been very slow in recognising the scope for improvement of farmed breeds through applied genetics. As reviewed by Pullin & Capili (1988), most farmed tilapias derive from very small founder stocks and little has been done to improve farmed breeds.

In order to assist aquaculturists, a number of breeding centres have been developed where the principles of quantitative genetics are applied to the improvement of farmed species. The ICLARM Genetic Improvement of Farmed Tilapia (GIFT) programme collected different strains of tilapia and evaluated them for growth characteristics in several different environments. Significant differences in growth characters existed among the strains, but that their relative performance was nearly the same for all tested environments. This project is ongoing and should promote increased production of tilapia.

Genetically modified organisms

Molecular biology has revolutionised the study of genetics in recent years and has created the possibility of genetic engineering by gene manipulation. Gene transfer from one organism to another, or transgenics, is a very simple idea: DNA is inserted into a nucleus so that it takes part in chromosomal replication and becomes part of the hereditary material of that cell. However, the technology required to achieve it is very complex.

Gene transfer technology improves the economics of animal production for food and makes possible the development of lines of fish-bearing introduced genes (Kapuscinski & Hallerman, 1990; Chourrout, 1991a, b). However, in contrast to mammals, studies on transgenic fish are still in their infancy even though it is likely that such fish will be of considerable economic benefit (Hew & Fletcher, 1992; Okada & Nagahama, 1993). Fish species with short reproductive cycles may be excellent systems for transgenic studies (Powers, 1989) as they produce many eggs and fertilisation is external, which is an advantage compared with mammals.

Transgenic fish

Transgenic fish are defined as fish-bearing (within their chromosomal DNA) copies of novel genetic constructs introduced through molecular genetic techniques. Such fish are produced by the insertion of copies of the novel DNA, produced by recombinant DNA methods, into newly-fertilised eggs, or by reproduction of the individuals so produced (Chourrout, 1993).

The ability to transfer 'novel' genes to improve the genetic and phenotypic characteristics of fish opens up a new area in fish culture. Transgenic fish may contain genetic elements derived from conspecifics, derived from unrelated species, or from a combination of intra- and interspecific sources. The development, for instance, of transgenic fish that are faster growing or more disease resistant would be a significant advancement (Hew & Gong, 1992).

The direct method of DNA microinjection into fertilised eggs has been the most widely used for transfer of exogenous genes to fish. Rahman & Maclean (1992) successfully produced *Oreochromis niloticus* that have chromosomally integrated copies of transgenes in both blood and fin. But there is still a limited number of genes with well-defined functions that are available and most current investigations focus on the genetic control of growth hormone, to produce the 'super fish' that grow faster and larger. Different laboratories have succeeded in producing transgenic fish with a growth hormone (GH) gene inserted (Kapuscinski & Hallerman, 1990). For example, Zhu *et al.* (1985, 1986) reported the production of transgenic carp and loach, three to four fold larger than the control, using a mouse metallothionein promotor–human growth hormone gene construct. For African fish, transgenic Nile tilapia, *Oreochromis niloticus*, have been produced by incorporating into eggs a mouse metallothionein–human growth hormone fusion gene (Brem *et al.*, 1988).

Consequences of genetic manipulations of fish

Phenotypic changes created by the inserted DNA may lead to broader changes in life-history patterns. At least three conceptual classes of phenotypic changes might be anticipated for transgenic fish (Kapuscinski & Hallerman, 1990). The first includes changes in physiological rates affecting one or more components of the balanced energy budget for individual fish. An example is the increased growth rates of fish through the transfer of growth hormone genes, which will probably affect a number of life-history traits, such as age and size at maturity, maximum length, reproductive effort, etc. The second class of phenotypic changes among transgenic fish would entail alteration of species' tolerance of physical factor, such temperature or salinity. Changes could affect upper or lower lethal limits, or optimum values. In the third class, behavioural changes may occur, for example in seasonal migrations, habitat selection, territoriality, etc. Gene transfer could also be used to alter other traits and develop transgenic lines with increased disease resistance or drug resistance.

Any human intervention carries an element of risk and there are several concerns about the use of transgenic fish for aquaculture, particularly regarding the possibility of accidental release. Development, field testing and commercial use of transgenic fish introduces ecological and public policy uncertainties that may constrain progress. The possible ecological impact has to be fully understood as long as transgenic fish may invade and dominate wild types, destroying the ecological balance (Hallerman & Kapuscinski, 1991). Given the complexity of interactions in aquatic communities, including indirect effects and time lags, it is anticipated that there will be difficulty in predicting community-level impacts of transgenic fish.

The risks of introducing transgenic fish must be weighed against the perceived benefits and the risks of not making the introduction. This balancing is part of risk management. For some authors, there are no convincing scientific grounds for distinguishing engineered organisms from natural ones. However, some safety testing is desirable before considering the deliberate introduction of genetically modified organisms. A detailed evaluation of the traits of the organism and of the host environment are required before release. It is necessary to consider both the short- and long-term consequences of the introduction because once an introduction is successful there will be no easy recall.

Experience in the introduction and stocking of non-transgenic fish into new environments may be relevant. Current evidence suggests that the stocking of a piscivorous fish, such as *Lates niloticus* (see Chapter 14), destabilises aquatic ecosystems in the short term, with resultant trophic level changes occurring through top-down forces that could result in the extirpation of native aquatic species. But indirect effects of introductions on fish assemblages have also been mentioned for non-piscivore fish such as tilapiines, as well as direct effects on the aquatic zooplanktonic communities through selective predation of larger forms. The release of sterilised transgenic fish would reduce long-term environmental risks, in allowing the option of halting further releases and limiting ecological perturbations if ecological impacts are detected.

An ecological perspective is essential to the assessment of risks associated with any introduction. The introduction of any organism into the environment, modified or not, should be undertaken within a framework that maintains appropriate safeguards for environmental protection without discouraging innovation.

At the moment, the public are ambivalent about the introduction of transgenic food. Public confidence will prove critical in determining whether transgenic fish reach the stage of practical utilisation. Commercialisation of transgenic fish, as well as any other transgenic food, requires that the public is honestly informed about both the benefits and any potentially hazardous side-effects of the products.

However, commercialisation of transgenic fish will also be influenced by the patentability of novel forms of life, and it is likely that any transgenic lines developed will be protected by patents. An obligation to pay royalties to patent holders will affect the utilisation of transgenic lines by the aquaculture sector. There are differences among countries regarding what constitutes patentable material, and so far there is no international agreement, raising the question of the origin of the genetic material and to whom royalties will be paid. This question is particularly complex. For example, tilapias are cultured throughout the tropics and subtropics, but while the largest tilapia culture industries are in Asia, nearly all tilapia genetic resources are in Africa. Numerous transfers have been made within and between African countries, including strains from western universities and Israel. Many recorded and unrecorded introductions and transfers have also been made within Asia (Pullin & Capili, 1988). In such a situation it will not be easy to determine to which country royalties have to be paid for the use of original genetic material.

It is difficult to forecast what will be the future of research on transgenic fish. Their production has become a reality, and undoubtedly the subject is fascinating to many scientists. However, there are as yet no applications seriously foreseen that are likely to improve fish genetically with gene transfer, as is the case for cultivated crops (Chourrout, 1993), and huge methodological problems will probably hamper the production of animals suitable for large-scale production. Considerable amounts of research are still required in order to develop strains of transgenic fish for commercial aquaculture purposes.

Ornamental fish

The annual wholesale trade value of ornamental fish has been estimated at 900 million US$, but at present Africa only supplies approximately 3% of the total quantity of live fish (Bassleer, 1994).

Laws and rules have been developed for the aquatic trades, and we now have, in principle, a selection of bodies and legislation responsible, such as CITES (Convention on International Trade in Endangered Species of Wild Flora and Fauna), and IATA (International Animal Transportation Association). However, in most cases the laws do not apply to trade in fish. Even in Europe it is impossible to know which species are imported, and from where. There can be almost no control of species while customs do not have fish experts, and the aquatic industry is not willing to share its information with government officials or scientists, apprehensive that this information would be used against it or would help competitors. It should also be clear that only a few fish species are listed in the Red Data Book, because little is known about the status of most African fish (Chapter 16).

Biological control

One of the oldest methods of biological control for mosquito larvae is the use of predatory fish. Species of cyprinodont fish belonging to the genus *Nothobranchius* inhabiting temporary waterbodies, subject to varying periods of drying, are potential candidates for mosquito control.

Annual fish offer the advantages that their eggs may be relatively easily transported and dispersed into waterbodies by hand, thus avoiding the difficulties and expense of transporting live fish. Completely developed eggs remain viable after several months and ripe eggs hatch within a few hours after reintroduction to water.

The use of mollusc-eating fish has from time to time been suggested for controlling the vectors of schistosomiasis. Slooteweg (1989), following others, has discussed the possible translocation of the mollusc-crushing cichlid, *Astatoreochromis alluaudi*, from Lake Victoria into artificial and seminatural waterbodies in the north of Cameroon. This fish has already been introduced as a means of snail control in Kenya and Rwanda, with success in some cases. Compared to the use of molluscicides, fish have some disadvantages, such as the limited number of habitats that can be controlled and uncertainty about possible ecological impacts, but they are less expensive and, generally, ecologically safer than chemical products.

Fish diversity as an indicator of ecosystem health

The use of indicator species to assess or to monitor environmental conditions is a firmly established tradition in ecology, but it has also been widely criticised. In toxicity testing, for instance, it is difficult to claim that responses, universally, can be predicted by single-species toxicity tests. A number of other difficulties have also been pointed out. In general, indicator species have failed to predict population trends of other species, or trends in other environmental characteristics. The interest in biodiversity presents an opportunity to address environmental problems holistically rather than in a species-by-species or stress-by-stress fashion.

Indicators are measurable surrogates for environmental endpoints, such as biodiversity, that are assumed to be of value to the public (Noss, 1990). Ideally, an indicator should be sufficiently sensitive to provide: early warning of change; distributed over a broad geographical area, or otherwise widely applicable; relatively independent of sample size; easy and cost-effective to measure, collect, assay and/or calculate; able to provide information over a wide range of stress, and to differentiate naturally cyclical trends from those induced by anthropogenic stress; and relevant to ecologically significant phenomena. While no indicator will possess all these desirable properties, a set of complementary indicators is required (Noss, 1990). One way is to identify measurable attributes or indicators of biodiversity for use in environmental inventory, monitoring and assessment programmes.

Due to their ecological and economic importance, fish assemblages are among the best studied aquatic organisms all around the world. They are impacted by the cumulative effects of a wide array of anthropogenic stressors, and they integrate adverse effects of complex and varied stresses on other components of the aquatic ecosystem, such as habitat and macroinvertebrates, on which they depend for survival and growth. Many species utilise several different habitats during their life cycle (see Chapter 12) and integrate effects of stressors over large spatial scales. Therefore, fish are useful for basin-wide monitoring and should constitute an essential part of a comprehensive monitoring programme. Fish communities reflect watershed conditions (Fausch *et al.*, 1990), as was the case in monitoring the Onchocerciasis Control Programme (see Chapter 14) covering a vast area of West Africa .

Numerous measures of community structure have been used as indicators of the response of natural ecosystems to anthropogenic stress. They usually deal with a limited number of structural attributes: number of species (including species richness and species evenness), relative abundance/dominance, biomass, guild structure, size spectra and food web structure. However, in practice several problems may limit the usefulness of routine diversity measures as indicators of ecosystem deterioration and recovery. While changes in a diversity measure may provide an indication of changes in environmental quality, the measure provides no indication as to whether conditions are improving or deteriorating. In other words, changes in diversity do not reliably indicate changes in the degree of ecosystem impact, and empirical evidence does not support a consistent relationship between diversity and environmental stress (Cairns et al., 1993).

Given the inevitable limitations to the use of any single indicator for monitoring ecosystem conditions, various attempts have been made to combine a suite of biological indicators into a robust index of ecosystem health or integrity that is sensitive to several different types of stressors. Integrated indices that reduce information from several measures into a single value are advantageous for decision-makers. Indices of Biotic Integrity (IBIs) using fish are increasingly being used to assess and monitor ecosystem health (Karr, 1991; Karr et al., 1986).

The Index of Biological Integrity (IBI) is an attempt to provide a measure of stress or stream degradation that is more robust than richness, diversity or other methods using an indicator species (Karr, 1981; Karr et al., 1987; Fausch et al., 1990). IBI was defined as 'the capability of supporting and maintaining a balanced, integrated, adaptive community of organisms having a species composition, diversity, and functional organisation comparable to that of the natural habitat of the region' (Karr & Dudley, 1981). Three basic measures have generally been used to assess ecosystem health using fish communities: (i) species richness and composition including indicator taxa; (ii) trophic composition (proportion of species in different trophic groups); and (iii) overall abundance and condition (extent of hybridisation, proportion diseased or with tumors). Originally the IBI, which was primarily

applied to streams and rivers, was a sum of 12 parameters, each scored as odd numbers up to 5 that are estimated to represent the degree to which a particular stream locality is degraded from its natural state. The IBI could be modified to detect disruptions of longitudinal zonation patterns (Fausch et al., 1984).

The IBI approach is not without limitations (Steedman & Regier, 1990), and some authors suggest that it should be regarded as a useful management tool for preliminary diagnosis rather than as an index of ecological or heuristic value. IBI has not yet been applied to African fish, but there should be some attempts soon.

Thus, while fish have a very direct economic value as food they also have a number of indirect values as well as their intrinsic value as living organisms. Although there is almost no evidence that the direct value of the fish harvest must depend on the diversity of species present, indeed there is a good deal of evidence to the contrary, there is no doubt that natural fish harvests are underpinned by more species than those which dominate the catch. Moreover, taking a continent-wide view, loss of ecosystem diversity and/or loss of species diversity would probably reduce the present level of fish harvest, which is almost entirely dependent on natural stocks. The genetic diversity of African fish has already provided the foundations of a world-wide tropical aquaculture, and still has the potential to not only extend aquaculture within Africa but also to provide resources for the enhancement of aquaculture yields in general.

The aquarist trade values species diversity directly, and fish also have value as hidden components of food chains that form the basis of tourism, recreation and possibly disease control, in the same way that wetland ecosystems have long ignored values in such things as flood control and water supply. Although the evaluation of biodiversity per se is problematic, if not impossible, some strong economic arguments can be advanced for its preservation and it is increasingly important that economists and politicians are made to understand its less obvious values.

16 Conservation options

Administrators, policy makers, and managers have a right to ask for the bottom line. . . . And biologists have the right and sometimes the obligation not to give an oversimplified misleading answer to such a question.

Soulé, 1986

As we have seen, the range of anthropogenic impacts on tropical and subtropical aquatic systems includes deforestation, disturbance of the riparian zone, river diversion, impoundment and flow regulation, establishment of settlements, chemical and biological pollution, mining, overexploitation of endemic species and the introduction of exotic species (see Chapter 14). However, unlike the tropical rain forests, the threats facing inland water ecosystems and biodiversity have received very little attention in any part of the world. Although there are many programmes for conservation of endangered species, few deal with endangered fish species. Despite being the oldest and the largest group of vertebrates, fish seem to be less attractive than the mammals and birds. Nevertheless, the conservation of endangered African fish is at least as urgent and complex as the conservation of other forms of wildlife (Andrews, 1992). A major exception to this general indifference has been some international concern for the endangered endemic cichlid fauna of Lake Victoria and other East African Great Lakes.

Conservation biology

Human attitudes to living species and nature have changed considerably within a fairly short period. The idea of nature conservation has moved rapidly from preservation in zoos and 'nature reserves' to a more complex and comprehensive approach, the management of the biosphere. This change has resulted in the fundamental assumption that the human habitat is the biosphere as a whole, and yet our capacity to alter the biosphere is far greater than our current understanding of its functioning. Therefore, we have to ensure that the biosphere will maintain both its capacity to renew itself in order to provide a suitable environment for human life, and its capacity to produce the resources needed to ensure human survival. Humanity's future needs are unpredictable and could be very different from those of today, so one of the goals of conservation is to keep open as many options as possible in order to meet the potential needs of future generations.

Conservation of biodiversity: a crisis discipline

Conservation Science has evolved rapidly as an emergency attempt to provide tools for conservationists and managers who are faced with the enormous and urgent problem of an imminent, catastrophic loss of biodiversity. This has proved difficult, given that it has to synthesise results from several different biological sciences and to balance them with economic and social concerns. The danger for ecologists and biologists, in this emergency situation, is of providing managers with technical tools, or conceptual rules, that have not been properly tested. Thus, they have to be very circumspect in their recommendations, and they have to propose guidelines and plans of action to conservationists who cannot wait two further decades before protecting endangered species and areas. Soulé (1991) described conservation biology as a 'crisis discipline' in which 'one must act before knowing all the facts' and where 'tolerating uncertainty is often necessary'. It is therefore important to define the criteria to be used in making conservation decisions. These criteria include designation of key species, overall indices of species diversity, interactions between species, designation of important ecosystems and phylogenetic analyses.

There is still much debate in the field of fish conservation. Substantial work has still to be done even to establish the status of fish in each geographic area, to identify the specific conservation needs of the most endangered species and to implement appropriate measures as soon as possible. In many cases precise information is lacking on environmental threats and on the basic biology (or even taxonomy) of the fish at risk.

Goals and principles for biodiversity conservation

The Global Biodiversity Strategy (1992) identified three basic elements for successful action to conserve biodiversity.

1. '**Saving biodiversity** means taking steps to protect genes, species, habitats and ecosystems. The best way to maintain species is to maintain their habitats. Saving biodiversity therefore often involves efforts to prevent the degradation of key natural ecosystems and to manage and protect them effectively.'

2. '**Studying biodiversity** means documenting its composition, distribution, structure, and function; understanding the roles and functions of genes, species and ecosystems; grasping the complex links between modified and natural systems; and using this understanding to support sustainable development.'

3. '**Using biodiversity sustainably and equitably** means husbanding biological resources so that they last indefinitely, making sure that biodiversity is used to improve human condition, and seeing that these resources are shared equitably.'

The strategy document also lists ten principles (Table 16.1) for conserving biodiversity which outline an ideal relationship between the natural world and human society.

Table 16.1. *Ten principles for conserving biodiversity*

1. Every form of life is unique and warrants respect from humanity.
2. Biodiversity conservation is an investment that yields substantial local, national and global benefits.
3. The costs and benefits of biodiversity conservation should be shared more equitably among nations and among people in nations.
4. As part of the larger effort to achieve sustainable development, conserving biodiversity requires fundamental changes in patterns and practices of economic development world-wide.
5. Increased funding for biodiversity conservation will not, by itself, slow biodiversity loss. Policy and institutional reforms are needed to create the conditions under which increased funding can be effective.
6. Priorities for biodiversity conservation differ when viewed from local, national and global perspectives; all are legitimate and should be taken into account. All countries and communities also have a vested interest in conserving their biodiversity; the focus should not be exclusively on a few species-rich ecosystems and countries.
7. Biodiversity conservation can be sustained only if public awareness and concern are substantially heightened, and if policy-makers have access to reliable information upon which to base policy choices.
8. Action to conserve biodiversity must be planned and implemented on a scale determined by ecological and social criteria. The focus of activity must be where people live and work, as well as in protected wildland areas.
9. Cultural diversity is closely linked to biodiversity. Humanity's collective knowledge of biodiversity and its use and management rests in cultural diversity; conversely, conserving biodiversity often helps strengthen cultural integrity and values.
10. Increased public participation, respect for basic human rights, improved popular access to education and information, and greater institutional accountability are essential elements of biodiversity conservation.

From Global Biodiversity Strategy, 1992.

Ecological paradigms in conservation biology

What now complicates conservation is the demise of the paradigm of 'the balance of nature', which itself is an inheritance of the 'climax' concept: all ecosystems steer inexorably towards a climax state that, once reached, effectively persists for ever. This classical paradigm, with its emphasis on the stable state, its suggestion of natural ecosystems as closed and self-regulating and its resonance with the non-scientific idea of the balance of nature, is no longer an adequate concept to serve as a keystone for conservation (Pickett et al., 1992). Actually, the course of succession depends very much on circumstances, and each ecosystem is no longer assumed to be self-regulating but influenced by the surrounding context (see Chapter 14). We are, for instance, about to impose temperature rises on the entire planet, and this in itself will make a nonsense of many of the present-day reserves. The impact of global warming both on water budgets and temperatures will certainly be significant for aquatic ecosystems, but the overall use of surrounding terrestrial systems may also result in long-term changes in their functioning. The contemporary paradigm (Pickett et al., 1992) can be labelled the 'non-equilibrium paradigm' and emphasises processes rather than end points. The main concern now is with how systems actually behave, and how their structures and trajectories are actually determined. The main inputs to the new paradigm in conservation biology are its recognition of the role of episodic events, the multiplicity of regulation processes, the openness of ecological systems and people as part of the ecosystem.

Monitoring of biodiversity

Most of the questions asked of scientists by managers concern the ability to detect changes in the physical, chemical or biological states of the environment, and to distinguish cause from effect. The recognition of ecological changes due to adverse impacts, or the assessment of ecological improvement after a disturbance has been reduced or suppressed, needs quantitative, regional, long-term data. It is, therefore, necessary to set up long-term monitoring programmes based on a network of selected sites and the use of quantitative protocols. Furthermore, particularly in poorly known tropical ecosystems, research combined with monitoring is likely to be most useful. However, all long-term data collection programmes face problems of continuity of variables measured, continuity of funding and comparability of data as analytical methods change.

In particular need of monitoring are the long-term and large-scale environmental changes occurring slowly in a piecemeal fashion, such as climatic changes. Ecological processes operate at a broader range of temporal and spatial scales than is typically addressed in ecological

studies, and long-term research reveals processes and events that have often been invisible in the short term. Such is the case for slow changes occurring over years or decades, which are hidden in the so-called 'invisible present' (Magnuson, 1990), i.e. we frequently observe the response of an ecological system to a cause that occurred before the beginning of the monitoring, and in most impact studies we often see transition of the system rather than the new state it is likely to reach. Such time lags between cause and response to a disturbance contribute to the difficulties of interpreting short-term field records. Among other reasons for the lag is that the time for response is often related to the duration of the life cycle of species, which varies from less than a day for microorganisms, to a century or more for trees. Long-term ecological monitoring, which is primarily designed to capture the effects of the environment and biotic interactions in ongoing ecological processes, could also help to put the present situation into the context of temporal changes. There is no doubt that the need to provide decision-makers with convincing data on environmental changes, requires long-term studies.

A major problem in monitoring programmes, as well as for restoration purposes, is to have a baseline reference situation with which to compare the data collected. Any impact study should refer to a standard so-called natural or non-perturbed ecosystem. What constitutes a 'healthy' system, which could serve as reference, is a key question. In many places, the natural community has in fact already been disturbed by fishing, pollution, water management, or species introduction, and trying to ascertain its original characteristics is problematic.

In order to detect potential ecosystem damage, it would also be useful to use the point of no return (PNR) concept (Loehle, 1991). For a fish population under stress (fisheries, pollution, etc.), the PNR has been reached when irreversible damage is expected. This could be a drastic decline in population size leading to extinction even when protection is ensured. It could also result from habitat degradation and alteration of productive capacity. The eutrophication process in aquatic ecosystems or marsh reclamation could, for example, indirectly affect the existence of endemic or indigenous species to such a point that they disappear. For shallow water systems, small changes in the water input may be dramatic for the indigenous fauna. In the absence of refuge zones for fish, the drying up of an aquatic system as a result of anthropogenic activities is an extreme example.

Ethics

Faiths and ethical traditions give people their basic orientation toward the natural world. It is therefore important to understand how ethical norms, culture and religion may condition the commitment of societies when faced with the problem of conserving biodiversity. Motivation for nature conservation may differ greatly among people, but whatever the attitudes, there must be an overall goal to ensure that the biosphere can continue to renew itself and provide the means for all life, including human well-being. Many conservationist movements actually began as essentially moral crusades.

Ehrenfeld (1988) highlighted the notion of wrongness. If conservation is to succeed, the public must come to understand that it is wrong to destroy biodiversity. It is a powerful argument with great breadth of appeal to all manner of personal philosophies. The long-established existence of biodiversity confers a powerful right to continued existence. Moreover, for those who believe in God, diversity is God's property and we have no right to destroy it. It was God who caused this diversity to appear here in the first place.

The 'World Charter for Nature', adopted by the General Assembly of the United Nations on 28 October 1982, expresses absolute support by governments for the principles of conserving biodiversity. It recognises that mankind is part of nature, that life depends on the uninterrupted functioning of natural systems and that every form of life is unique and warrants respect regardless of its worth to mankind. Nature as a whole has to be respected and its essential processes must not be impaired. The Charter calls for strategies for conserving nature, scientific research and international cooperation in conservation action, but it has been all but forgotten by both governments and conservationists, and needs to be given far greater exposure in the future than it has received so far (McNeely, 1989). The ethical declaration contained in the World Charter for Nature is only a 'soft law', which does not bind governments to any commitments and has therefore seldom been invoked or quoted.

Building on the World Charter for Nature and the World Conservation Strategy (IUCN, 1980), the

Table 16.2. *An ethical basis for conserving biological diversity*

1. The world is an interdependent whole made up of natural and human communities. The well-being and health of any one part depends upon the well-being and health of the other parts.
2. Humanity is part of nature and humans are subject to the same immutable ecological laws as all other species on the planet. All life depends on the uninterrupted functioning of natural systems, which ensure the supply of energy and nutrients, so ecological responsibility among all people is necessary for the survival, security, equity and dignity of the world's communities. Human culture must be built upon a profound respect for nature, a sense of being at one with nature and a recognition that human affairs must proceed in harmony and balance with nature.
3. The ecological limits within which we must work are no limits to human endeavour; instead, they give direction and guidance as to how human affairs can sustain environmental stability and diversity.
4. All species have an inherent right to exist. The ecological processes that support the integrity of the biosphere and its diverse species, landscapes and habitats are to be maintained. Similarly, the full range of human cultural adaptations to local environments is to be enabled to prosper.
5. Sustainability is the basic principle of all social and economic development. Personal and social values should be chosen to accentuate the richness of flora, fauna and human experience. This moral foundation will enable the many utilitarian values of nature (for food, health, science, technology, industry and recreation) to be equitably distributed and sustained for future generations.
6. The well-being of future generations is a social responsibility of the present generation. Therefore, the present generation should limit its consumption of non-renewable resources to that level which is necessary to meet the basic needs of society, and ensure that renewable resources are nurtured for their sustainable productivity.
7. All persons must be empowered to exercise responsibility for their own lives and for the life of the Earth. They must therefore have full access to educational opportunities, political enfranchisement and sustaining livelihoods.
8. Diversity in ethical and cultural outlooks towards nature and human life is to be encouraged, by promoting relationships that respect and enhance the diversity of life, irrespective of the political, economic or religious ideology dominant in a society.

From McNeely, 1989.

IUCN's Working Group on Ethics and Conservation has produced an ethical foundation for conservation (Table 16.2), and concluded that the ethical basis for conserving biological diversity needs to be consistent with ecological principles and to promote activities that are sustainable in the long term (McNeely, 1989).

Threatened fish species in Africa

In order to select priorities for conservation action, it is necessary to recognise emergency situations and to identify several degrees of threat. The use of simple but standard scales is recommended for monitoring the status of species or ecosystems on a regional or global scale.

The major classification system used for assessing the status of threat to fish species is the Red Data Book Categories, which have been developed by IUCN.

1. **Extinct**: the species is no longer known to exist in the wild, but could survive in captivity.
2. **Endangered**: species that are in danger of extinction and whose survival is unlikely if the threats continue operating. This includes species with critically small populations as a result of habitat destruction or over exploitation. Many endemic species inhabiting only one system could be destroyed by one incident.
3. **Vulnerable**: populations that have been seriously depleted and are likely to move to the endangered category if the causal factors continue operating. This is also the case for populations that are under threat from serious adverse factors throughout their range.
4. **Rare**: species with small populations inhabiting restricted geographical habitats, not yet endangered, but which are at risk. These may be species that are seldom recorded or not easy to sample.
5. **Indeterminate**: taxa for which the information available is too restricted to say if they belong to the above categories.

Critical fish species in tropical Africa: a world of ignorance

Fish have received much less attention than other vertebrates, and the inventory of threatened species is still at an early stage of development in tropical Africa. Never-

theless, a few species have been recognised as in a critical situation (Stuart *et al.*, 1990).

1. In Madagascar, 14 of 29 endemic freshwater species are considered threatened. Habitat loss due to deforestation has resulted in the extinction of three of the five described species of *Rheocles*, and several nominal species of *Bedotia* (Stiassny, 1990). The cyprinodont, *Pachypanchax sakaramyi*, has not been collected since its description in 1928. The cichlid, *Ptychochromoides betsilianus*, has been extirpated by black bass, and other native fish have been virtually eliminated from the central plateau by exotic competitors and predators (Reinthal & Stiassny, 1991).
2. In Cameroon, crater lakes in the western part of the country contain a number of vulnerable endemic species. These fish are at particular risk by virtue of their small populations, low fecundity, and the limited size and isolated nature of the lakes. In Lake Barombi-Mbo for instance, the most important cichlid 'species flock' in West Africa is presently endangered by deforestation and cultivation of the inner face of the crater (Reid, 1991). Any increase in nutrient levels would be likely to result in significant eutrophication, with a concomitant loss of biodiversity. The species *Stomatepia mongo* seems to be currently absent from fishermen's catches (Reid, 1991).
3. In the East African Great Lakes, the endemic fauna, especially the cichlids, is potentially threatened by overfishing, pollution and the introduction of alien species. In Lake Victoria for example, some 200 taxa of over 300 haplochromine cichlids have now disappeared, a loss attributed to predation by *Lates* and overfishing (see Chapter 14) (Witte *et al.*, 1992*a*; Lowe-McConnell, 1993). Many species, endemic or not, from Lakes Nabugabo and Kyoga have either been depleted or have disappeared since *Lates niloticus* and *Oreochromis niloticus* were introduced in the 1960s (Ogutu-Ohwayo, 1993).
4. In southern Africa, a few species are considered to be endangered: *Afromasculatus vanderwaali* (Zambezi and Okavango Rivers in Namibia, Angola and Zambia); *Clariallabes platyprosopos* (Zambezi and Okavango Rivers in Namibia and Angola); *Clarias cavernicola* (Namibia, only known from Aigamas

Cave); *Austroglanis sclateri* (Orange River, Namibia); and *Tilapia guinasana* (from Lakes Guinas and Ojikota in Namibia).

This list is far from complete, and throughout tropical Africa there are localised communities of endemic species that are at least vulnerable without being recorded in a Red List. The major reasons are that we know little of their geographic distribution and, in the absence of any large-scale monitoring of the biodiversity of freshwaters, the fate of most species. In South Africa, which is not within the scope of this book, 24 of the 220 freshwater fish are considered to be globally threatened (Skelton, 1987).

Conservation priorities: which taxa to save?

Biologists are frequently faced with assessing conservation priorities, and fish conservation provides good examples of the difficulties involved. Resources and technical ability are too limited for the use of captive propagation to save all threatened taxa and it is therefore unrealistic to hope that everything can be conserved. Decisions regarding priorities are essential (Ribbink, 1986). It is impossible to breed all species of a complex fish community and to imagine that such communities could be restructured through reintroductions (Reid, 1990). Only fish that are truly threatened should be selected for captive breeding, and these should belong to as many different trophic levels as possible and to different phylogenetic lineages (Ribbink, 1986). In the absence of absolute biological criteria with which to judge the relative value of a given species or community, the problem of establishing priorities may become acute. Frequently taxa are chosen irrationally, depending on the personal whims of those making the decision (Ribbink, 1987).

In fact, many different criteria are available, ranging from the preservation of rare species to the protection of representative samples of ecosystems or biotas. An important factor for choosing a threatened species for a large-scale captive breeding programme is that it may be possible to reintroduce it into the wild. Other factors influencing choice are the taxonomic uniqueness of the species and their suitability for captivity. In some cases, certain species may be especially important candidates for breeding on account of their particular conservation

message that can be conveyed to the general public via zoo or aquarium displays (Andrews, 1992).

Frankel & Soulé (1981) proposed criteria by which to evaluate candidates for conservation within a community in which all taxa are equally threatened, and Ribbink (1987) examined them with reference to their applicability to the haplochromines of Lake Victoria. Their discussions could be summarised as follows.

1. Scientific interest is a criterion that cannot be applied realistically to a species flock, in so far as each species is unique and has played a role in the evolution of the community to which it belongs. Moreover, given the poor information available as a whole for tropical fish faunas, it is unlikely that consensus will be reached among scientists to provide agreed scientific criteria for selection on this criterion.

2. Economic importance: fish contributing substantially to the catch, and those favoured by consumers would be valuable candidates for fish biologists. Fish of ornamental value should also be selected under this criterion.

3. Aesthetics: many colourful fish and those with unusual shapes or behaviour are already kept by aquarists who constitute a potentially cost-effective system for maintaining fish gene pools provided they abide strictly by the protocols of captive propagation.

4. Ecological value: the ecological role of a species is theoretically a powerful criterion for judging whether or not it should be selected for captive propagation. Ignorance of the precise role of many species is a handicap, but in this case, species known to be abundant are likely to have greater ecological impact than rare species, and should be chosen. There are, nevertheless, exceptions to this generalisation. For instance, in Lake Victoria, piscivorous haplochromines are the most speciose group (40% of all cichlid species) and include members adapted to different microhabitats and specialised on particular prey. These predatory species are also among the largest but the least numerous members of the cichlid community, and are severely affected by commercial fishing pressure. They are, therefore, urgently in need of rescue and should be

included in the captive propagation list. By carefully selecting different ecological components of the fish community for captive propagation, there is hope of a partial restoration of the ecosystem through reintroduction.

Systematic biology should play a key role in conservation biology, in that the taxonomic composition and biogeographical distribution of the fish assemblages are basic information needed to establish inventories of threatened areas and endangered species lists, and to formulate conservation plans. Stiassny (1992) also emphasised the potential role of phylogenetic analysis, and how it could provide information on which to assess conservation priorities. Phylogenetic trees, which help us to understand the genealogical relationships and sequence of historical events uniting taxa, could be of particular interest when conservationists are faced with the 'quality versus quantity' dilemma. The African cichlids illustrate the kind of information that may be gained using phylogenetic analysis (Stiassny, 1992). Cichlid species richness in Madagascar is low (9 species) compared to the hundreds of endemic species in Lake Malawi for instance, but the Madagascan assemblage represents the phylogenetically oldest cichlid lineage. Extant representatives of the most basal of cichlid lineages should be of particular interest to evolutionists, for they provide a unique resource for baseline comparative studies within the family. The loss of the nine Madagascan species may, in that context, be more dramatic than the loss of several Malawian species. Similarly, the African cichlid species *Heterochromis multidens*, restricted to the central basin of the Zaïre River, is unique in being the sister-taxon of the remaining Afro-Neotropical radiation (Stiassny, 1991). Thus, the eventual loss of this taxon (as well as the Madagascan cichlids) should not be equated with the loss of just another African cichlid, since it would result in the extinction of an entire evolutionary lineage.

For Lake Malawi, Reinthal (1993) concluded that to conserve all Mbuna species the entire rocky shore of the lake should be protected, an unrealistic goal. Mbuna genera, rather than species, may be a more realistic taxonomic group upon which to base conservation decisions. Given the high levels of diversity and endemicity, all of the Mbuna genera should be protected within the Lake Malawi basin, rather than protecting species from allo-

patric localities in genera that are already adequately represented in conserved areas.

Ecosystem and resource management options

The well-being of human populations depends on the availability of a variety of renewable resources including freshwater fish, which may either be utilised sustainably, at rates that permit harvests over a long time, or exhaustively, at rates that in the short or long term lead to a decline in the total stocks and possible harvests therefrom. The success of fish biodiversity conservation will also depend on how well the aquatic landscape, and to some extent the terrestrial landscape, is managed to minimise the loss of diversity. It is unrealistic to imagine that protected areas can be large enough to meet all species' habitat needs, and the quality of aquatic habitats, even those in protected areas, depends on the overall management of the surrounding landscape.

The sustainable development concept

It is likely that one of the major environmental concerns of the 1990s will be the preservation of biodiversity in the context of sustainable development. Recognising that our natural resources are not being managed on a sustainable basis may lead to the conclusion that further loss of biological diversity will reduce our future options for sustainable biological resource management (Cairns & Lackey, 1992). This is, of course, closely related to our options in the management of lands and waters for the future, all of which have ecological, socio-economic and political constraints. Conflicts, or at least competition, between protection of biodiversity and production of resources are likely to occur. What might be in the interest of society is not always in the best interest of individuals. Poverty is a major cause of habitat and biodiversity loss because the urgent need for food, energy and money leads to overexploitation and degradation of terrestrial and aquatic habitats. Thus, actions to alleviate the loss of biodiversity must address the socio-economic cause of poverty (Schweitzer, 1992). The problem is still more complex because aquatic resources are usually owned and treated as public goods and services without costs. The result is a tendency toward their overexploitation.

Sustainability concepts are increasingly important to

policy-makers. The broad concept of 'sustainable development' was widely publicised by the World Conservation Strategy (IUCN, 1980) and became central to the relationships between development and environment during the following decade. We shall not discuss different definitions of sustainability and their implications (see Pezzey, 1989), but give some information about the main environmental issues. In the Bruntland Report (WCED, 1987), sustainable development is defined as a 'development that meets the needs of the present without compromising the ability of future generations to meet their own needs'. This report also pointed out that the loss (i.e. extinction) of plant and animal species can greatly limit the options of future generations. Goodland & Ledec (1987), gave a rather similar definition: 'it is a pattern of social and structural economic transformations, which optimises the economic and societal benefits available in the present, without jeopardising the likely potential for similar benefits in the future'. This implies using renewable natural resources in a manner that does not eliminate or degrade them, or otherwise diminish their usefulness for future generations (see also Pearce, 1988). At the same time, Norgaard (1988) claimed that 'if sustainable development is to be achieved, we will have to devise institutions, at all levels of government, to reallocate the use of stock resources towards the future, curb the pace and disruption of global climatic changes, reverse the accumulation of toxins in the environment and slow the loss of biological diversity. These are the key resource and environmental issues that must be addressed'.

The idea of sustainable development, however defined, has contributed greatly to increasing awareness of the ecological and economic importance of the environment. During the last two decades, the interaction between biodiversity conservation and sustainable socio-economic development has been increasingly recognised. Managers have discovered that natural capital may be depleted to the point where irreversible damage may occur. A condition for sustainability is, therefore, to conserve the natural stock. This can be done in various ways: taxes, laws, integrated national policies, international technical cooperation, etc. Today, many sustainable development programmes are designed to minimise losses of biological diversity, and efforts to conserve biological diversity are also expected to generate economic and social benefits, particularly for those who bear most of the opportunity costs associated with the preservation of areas for conservation.

Towards a holistic approach of aquatic systems management

Spatial interaction is very high in aquatic systems because of the continuity of water, so unwise land-use and lake- or river-use practices in one area may produce deleterious effects throughout the entire system. From the perspective of sustainable development, multiple-use management of aquatic ecosystems can provide the framework for achieving long-term sustainability of resources and for maintaining biodiversity, but the primary goal must be to maintain ecosystem integrity. Important ecosystem processes, such as energy and water flow, must be recognised, as well as natural disturbances, such as floods and droughts that help maintain the natural communities.

Aquatic conservation biologists should therefore give more emphasis to the wise management of natural systems than to establishing more reserves and protected areas. Nevertheless, habitat protection and restoration are the only major long-term means by which successful fish conservation will be achieved. Stream management requires a holistic, ecosystem approach. The ultimate determinant of the quality and quantity of water in a stream is the catchment from which it originates. It is difficult to manage an entire drainage basin, but particular attention must be given to the riparian zones that are the interface between running waters and terrestrial landscapes (Petersen et al., 1987). Riparian zones control the interactions between the stream and its surroundings in acting as nutrient filters and buffer zones. Riparian zones, such as floodplains, are also important refuges for a wide diversity of life, which is the main source of food for many riverine fishes. The loss of riparian zones as a result of agricultural practices, or fuelwood harvesting, may also cause soil erosion and siltation, modifying the water turbidity with subsequent impacts on fish biology.

Maintaining healthy aquatic systems contributes to providing harvestable fish populations, maintaining biological diversity and helping to control the flux of nutrients from the land to lakes and seas. However, stream management, apart from impoundment, is a neglected discipline in Africa. Aquatic resources management tends to concentrate on fisheries development and the engineering of water storage. Enormous damage has been done to many fish habitats and the situation is often not easy to reverse. It is certainly unrealistic to envisage the destruction of many of the large and small dams built during recent decades. It is probably also difficult to stop within a short time the on-going processes that have been observed in East Africa, where cultivation and deforestation on steep slopes will definitely increase in all the Great Lakes watersheds if current population growth trends continue. Conservation efforts in the African Great Lakes could therefore be jeopardised by eutrophication (Bootsma & Hecky, 1993) or sediment pollution (Cohen et al., 1993). It would be technically easier, but expensive, to control the industrial and urban pollution. It is certainly possible to drastically reduce the use of pesticides for fishing purposes as these are destroying the fish fauna along many kilometres of rivers.

The watershed management concept is not new. There is evidence that various societies developed such an ecosystem view long ago (Gadgil & Berkes, 1991). For example, The Sultan Mehmed II instituted watershed conservation measures in the river basin supporting Constantinople when he captured that city in 1453. He prohibited overgrazing and tree-cutting, and encouraged river bank stabilisation and the revegetation of the catchment area. The 'dina' system of Mali also provide for integrated resource management through specialisation of different groups of people, and by codifying the then existing resource-management systems into grazing, fishing and farming territories. In the Inner Niger Delta, for example, the Bozo people specialised in shallow-water fishing systems, whereas the Somono specialised in net fishing.

When a lake or other aquatic system is bordered by many countries, effective management requires international cooperation. It would therefore be useful to establish cooperation between riparian countries as a first step for an integrated regional approach to the conservation and management of the shared resource. Such organisations as the Lake Chad Basin Commission already exist, but are not always very efficient due to difficulties encountered in the implementation of recommendations. There is also a need at the regional level to harmonise national policies and legislations governing environmental management.

Common property

Most cultures emphasise responsibility to the community, and traditional community-based resource-management systems have recently become of major interest to international organisations (UNESCO, FAO, IUCN). As an example, according to some authors, small-scale fisheries are the key to long-term management success. Where local communities of fishermen can control access to fishing space and enforce regulations, exploitation levels can be managed, and this is an essential condition for sustainable exploitation (Berkes & Kislalioglu, 1991).

The term 'common property' is controversial. For some scientists it means resources that are not amenable to private appropriation, i.e. it is a free good, which is not owned by anyone, such as marine resources, including fisheries. They are open-access and freely available to any user. According to an alternative view, the term 'common property' should be restricted to communally owned resources that are managed through communal arrangements for allocation among co-owners.

The first interpretation resulted in the 'tragedy of the commons' model of Hardin (1968) which made the critical assumption that common-property resources are really open-access, and that such resources held in common are doomed to overexploitation since each resource-user places immediate self-interest above community interest. This implies that users are individualistic and unable to cooperate. The conclusion is that resources should be either privatised or controlled by a central government authority to ensure sustainable use. In equating communal property with open-access, the Hardin theory mislead a whole generation of fishery ecologists and managers by suggesting that absolute governmental controls need to be established over both the resource and the user.

This model still persists in the conventional wisdom of many resource managers, and is also dominant in models of development exported to Third World countries. However, this situation does not occur frequently, because valuable natural resources are almost never open-access but are (or were) managed under traditional rules governing their use. Many case studies indicate that cooperation for communal interest more frequently occurs (Berkes & Farvar, 1989). Recent literature on common property rejects a deterministic 'tragedy of the commons'. The common-property approach reverses the

Table 16.3. *Main types of property-rights regimes relevant to common property resources*

Open access (res nullius)	Access is free and open to all; rights are owned in common; no management regime at all.
State property (res publica)	Public resources to which access rights have not been specified; management control is held by the central governments; management decision by technical experts.
Communal property (res communes)	User-rights and management are controlled by an identifiable group; there exist rules on the ways to use the resource.
Private property	Allocation of exclusive rights to individuals or firms.

Modified from Berkes & Farvar, 1989.

traditional emphasis of fishery management, which has been on the resource rather than the people, and starts with an analysis of the local property rights regime.

Many different resources show characteristics of common property, e.g. water, fish, wildlife, forests and pastures, but all common-property resources share two important characteristics. The first is that exclusion (or control of access) of users to these resources is problematic, and the second is that each user is capable of subtracting from the welfare of other users (Berkes & Farvar, 1989).

Common-property resources may theoretically be held within four property-rights regimes (Table 16.3). In the real world, many resources are held under regimes that combine characteristics of two or more of these types. In many traditional systems, communal property resources are nationalised and turned into state property, a situation that complicates the ownership status of the resource and sometimes results in resource depletion.

Common-property systems have certain critical roles in local communities that are relevant to sustainable development.

1. Livelihood security: to guarantee that no one in the group starves, many societies have elaborated rules for sharing food through access rights to a vital resource.
2. Access equity and conflict resolution: rules mutually agreed serve to reduce conflict over resource use.

3. Mode of production: social roles and obligations are often defined in terms of one's participation in work teams. Common-property systems are an integral part of the local culture, including knowledge of the resources and knowledge of resource-use rules.
4. Resource conservation: common-property systems are basically conservative in the way resources are utilised. The emphasis is on taking what is needed, and there are social sanctions against excessive individual gain from a communal resource. The ecological wisdom of many common-property systems emphasises respect, responsibility and stewardship.
5. Ecological sustainability: the traditional use of resources often incorporates rituals to help synchronise harvesting with natural cycles. This, in reinforcing social controls, may help to maintain a productive resource from generation to generation.

In the Hadejia-Nguru wetlands (Nigeria), until recently, it was possible for individuals to own the rights to fish particular areas, stretches of river, or small ponds. These rights were handed down from father to son, and especially concerned prime sites for fish traps (Thomas *et al.*, 1993). In some cases, access to ponds and lakes could be 'bought' from the village head, acting on behalf of the local government. There was a great variety of mechanisms controlling access to the fishery and there was rarely 'open-access'. However, it would appear that the existing patterns of allocation of rights have broken down over the last 20 years and have not been replaced by any effective alternative at state or local government level. Moreover, sociological changes in the floodplain are now having an impact on the fishery. New roads open up the area for increasing numbers of migrant fishermen who are likely to show much less concern for the long-term welfare of the fisheries and enjoy short-term benefits. This may have an eroding effect on the behaviour of resident fishermen just as Hardin suggested. Instead, a combination of traditional ownership rights and local government licence fees and taxation should control access to the fishery and determine to a large extent who fishes.

Living resources management

A fishery is a complex array of human activities based on a renewable natural resource system whose main elements are the environment, the commercially exploited fish stocks, the communities of fishermen, the fish processing activities, the marketing, and finally the consumers.

African inland fisheries traditionally utilised a great variety of gear, but with the introduction of nylon, gill nets and cast nets became the most commonly used in inland waters. This change of gear has altered the way in which the resource needs to be managed to achieve its sustainable use. Should new management systems be based on the resource-management techniques of the industrialised countries or should they be developed by rehabilitating and adapting 'indigenous' resource-management systems and upgrading traditional local-level institutions? Is there any way to integrate scientific and traditional management systems? These are the central questions to be answered in order to propose development models to take care of the environment and serve the needs of the people who use the fish resource. This is a really new challenge because, until recently, scientists and policy-makers knew little about traditional management systems and accorded them little credibility.

Limitations of traditional fisheries management

It has been claimed that rational fisheries management is the only universal solution for a sustainable use of fish stocks. There is an extensive literature on this subject, and most policies for sustainable management of fish stocks derive from the concepts of equilibrium population dynamics and stock assessment, and aim to achieve a level of fishing effort at which the stock or population is conserved at its level of maximum yield. 'Although the principle of Maximum Sustained Yield has frequently been challenged, it is still *de facto* accepted as one of the main bases for management. Despite the apparent capacity to determine the levels of effort or access needed to conserve fish stocks, there has been an almost universal failure to do so. This failure may lie in part in the shortcomings of scientific advice, but for the major part lies in the difficulties of applying coherent management strategies for political and sociological reasons' (Welcomme, 1992).

In other words, fisheries management principles that have prevailed during recent decades are not very successful. The reasons for this failure in inland waters, lie in technical and social determinants.

1. As pointed out in other chapters, fish biology and

dynamics in many systems are closely dependent on the flood. There is also good evidence from various river systems that catches are more or less related to the flood of the year (see Chapter 12). While it is not possible to predict rains, and therefore flood intensity from year to year, there are doubts about the predictive value of models based only on biological and fish capture data.

2. Fisheries licensing exists in many countries, but is not in reality used as a means of regulating the number of fishermen. In Nigeria for example, anyone can obtain a licence, and licences are more or less a means of generating income for local government agencies (Thomas *et al.*, 1993).

3. The decision-making processes, by which technical advice is transformed into laws, are not efficient and information is largely ignored for economic or social reasons. Even when a fishery service exists, it is certainly unrealistic to expect good catch statistics from large areas where fishing activity is scattered. The disruptive effect of central fisheries authorities on rational community-based management has also been noted by Scudder & Conelly (1985).

4. Fisheries regulations frequently exist at the level of the state, local government and community, but can rarely be put into operation. For instance in Kano State (Nigeria), the Fisheries Edict of 1985 specifies the allowable mesh and hook sizes and other aspects of fishing gear, forbids the introduction of non-native species and the catching of immature fish. However, there is little implementation, mainly due to lack of funds, staff and transport. Many fishermen seem to be unaware of the existence of such regulations (Thomas *et al.*, 1993), and any kind of control is in fact very difficult in the large wetlands, inaccessible with vehicles.

Lessons from traditional community-based management

Efforts at central fisheries management in Africa over the last few decades have not been particularly cost-effective, and serious consideration needs to be given to reinstating community-based, traditional-type management structures. This idea that communities of fishermen ought to be able to control their resources is not new, although this renewed interest in traditional management systems stems partly from the past failures of development projects. In fact, rural populations have had the capacity throughout their history to manage their resources for a sustainable yield. It is only in recent times, beginning in the colonial period but greatly accelerated after 1950, that traditional systems for managing resources have been replaced by government agencies, which have proved to be less successful than they should be.

Pliya (1980) reviewed changes in the management of continental fisheries in Benin over the last century. Before the colonial period, access to water and fishing was controlled by religious authorities and the lakes were the common properties of the villagers. There were many social prohibitions, religious beliefs and local customs, the result of which was to forbid the use of some gear, to forbid access to some areas that were usually sacred places and to protect areas locally important for reproduction. Thus, the traditional society developed a set of social mechanisms that resulted in a wise and sustainable use of the common resource. During the colonial period, from the beginning of the century up to 1939, changes in social structures weakened the traditional authority in favour of central administration. But the 'Eaux et Forêts' service did not provide any legislation, and progressively the traditional rules were infringed. This resulted in an anarchic development of the fishery, rapidly leading to conflicts and overexploitation. After 1939, during the end of the colonial period and the beginning of independence, there were various attempts to improve the role of the fishery service. In favouring this service, the government continued to weaken the traditional authorities that controlled the common property of waterbodies and its management. The government also recognised the private ownership of some water systems. In general, this was a complete failure, resulting in the late 1970s in the absence of any control. This gave rise to social conflicts and to the proposal by the fishermen that traditional management should be rehabilitated. Presumably, the anarchy continues as the administration apparently has not heeded these proposals.

More or less similar trends were observed in Côte d'Ivoire (Verdeaux, 1986, 1989; Perrot, 1989) where the crash in the fish production of the Ebrié lagoon in 1982 is another example of the mismanagement (or absence of management?) of the fishery. Until the early twentieth century, the fishing organisation in such coastal lagoons,

and the control exerted on the environment, lay in the hands of decentralised and self-governing lineal powers who regulated fishing activities. The adequacy of the regulation was secured by religious interdicts, each collective fishing ground being under the authority of deities. The period 1935–82 was characterised in the Ebrié lagoon by a shift from collective ownership to individually appropriated means of production, and the use in the 1970s of gear (e.g. large seine nets) responsible for the overexploitation of the fish stock. Reinstatement of the traditional management systems would probably ensure social order and the preservation of the abundance of the resource.

In Mali, in the Middle Niger valley, the political organisation of space includes traditional inherited authorities and gives rise to a series of similar 'territories' occupied by groups that enforce homogeneous regulations (Fay, 1989a, b). Fishing is regulated by 'masters of the water' who supervise the use of allowable techniques, set opening dates for different fisheries, can extend fishing rights to outsiders and conduct ceremonies for water deities. This political and symbolic organisation has evolved during recent decades as a response to the introduction of new fishing gear that makes sampling homogeneous, and of new powers that have brought a new organisation of space and a new definition to identities and relationships. Thus, the current situation does not amount to a conflict between tradition and modernity.

In the Logone floodplains (Chad), the 'Chefs de Pêche' who have jurisdiction over a section of river belonging to a village enforce a ban on fisheries in certain stretches of the river from December until the end of March. Afterwards, they invite villagers and inhabitants of surrounding areas to join in the collective fishing and villagers pay 10% of their revenues to the chiefs. By maintaining the ban in certain parts of the river until water levels start to rise in April, the presence of spawning fish at the beginning of the flood is ensured. Under this system, overfishing is prevented (Dadnadji & Wetten, 1993).

It would appear that there have been human groups whose interests were strongly linked to the prudent use of their resource base and who have evolved appropriate conservation practices. These practices were based on some simple and approximate rules of thumb that tended to ensure the long-term sustainability of the resource base. There was probably a process of trial and error, with the continual acceptance of practices that appeared to keep the resource base secure, coupled with the rejection of those practices that appeared to destroy the resource base.

However, too much emphasis has probably been given to traditional community-based management systems. They were efficient in an historical context but are no longer adapted to the present-day situations where increasing populations, urbanisation, and changes in local and national economies have led to different social structures and relationships between people and nature.

The need for new management strategies

Most African inland fisheries are artisanal or subsistence in nature and their management is incredibly complex. Models based on fish population dynamics and fishing effort, which were developed for marine fisheries, do not work for inland fisheries (see also Chapter 15). 'Traditional management, sanctioned by local custom or religious belief, is still strong in many areas, and its replacement by modern methods necessitates widespread modifications of the whole socio–economic framework of the societies concerned. Management of fisheries in African river systems by governmental organisations is particularly difficult under these conditions and is still at a relatively early stage of development' (Welcomme, 1989). In fact, inland-fisheries management also has to be considered within the framework of water-systems management. During recent decades, priority has been given to water supply for irrigation or to hydroelectric plans. This, associated with a period of severe drought over 20 years in tropical Africa, has resulted in a decrease of fishery production in many waters.

If community controls are eroded, and state controls are ineffectual as a result of inadequate means to enforce them, the fishery resource is left without a framework to ensure that it is properly managed. But with the social and environmental changes that are now underway, some kind of regulatory mechanism is more important than ever. It is becoming increasingly clear that precise prescriptions for the prudent use of aquatic living resources are difficult to define, particularly if the entire prey population is continually subject to

harvesting. However, simple general prescriptions for averting collapse of the resources are feasible and yet have a significant effect in enabling sustainable use (Gadgil & Berkes, 1991). According to Berkes & Kislalioglu (1991), it is doubtful if multidisciplinary considerations play a role in fishery management decision-making in the real world. 'Decisions are made by whatever understaffed, underpaid and overworked agency happens to be responsible for the fishery, and are often not based on sound scientific knowledge of the stocks or sound economic and social information on the fishery. It is also likely that decisions are made on an *ad hoc* basis, modified by whatever political pressure that has been brought to bear on the agency'. These events, which everyone knows to be largely true, may have discouraged most attempts at better resource management based only on a technical approach. As already seen, the solution to environmental problems lies in the structure of human societies.

Modern science seems unable to halt and reverse the depletion of resources and the degradation of the environment. Resource management has not been designed for the sustainable use of resources, but for their efficient utilisation as if they were boundless. There is a need to develop a new resource-management science that is better adapted to serve the needs of ecological sustainability. The concept of 'people's participation' in national resources management is being voiced and increasingly recognised in international forum, and we can assume that the public will be more and more associated with the decision-making process on environment and development issues. We have to conserve the diversity of traditional resource-management practices and systems if we want to achieve the task of reconstructing this new resource-management science so that it is better adapted to the real world. However, rejection of the monolithic scientific vision of resource management does not mean an overall rejection of science. The task is to develop a flexible approach by conserving what is useful in science, and 'ecology is in a unique position to be the cornerstone of a new science of resource management, that synthesises the best of the old and new wisdom towards a more sustainable future. But ecology would first have to extricate itself from the older utilitarian, "control over nature" tradition of resource management' (Gadgil & Berkes, 1991).

In situ conservation options

Conservation science has developed gradually but has recently shifted focus (Simberloff, 1988). The older literature in this field, rests principally on autoecological studies and species requirements. Conservation efforts were usually reactive rather than preventive. The consequence was too much emphasis given to those rare or endangered species that are the most likely to become extinct. One major recent consensus is the need to promote a strategy of ecosystem protection over a species-by-species approach, which has been summarised as 'Saving species by saving ecosystems' (Noss, 1987). Protecting large and representative habitats is seen as less expensive and more effective than the previous species-centred strategy. Therefore, the conservation of freshwater ecosystems is vital and is often the most important step to be taken for the conservation of threatened fish species. It can be achieved through judicious resource management, but also by considering the ecosystem as a whole, including the abiotic and biotic components, as well as the needs of relevant human societies. Some of the basic concepts that are particularly relevant to the development of *in situ* conservation strategies are reviewed here.

Most of the concepts and strategies for conservation were designed for terrestrial ecosystems and little has been done for the conservation of aquatic ecosystems, particularly rivers where the specific problems due to their longitudinal structure must be addressed. A hindrance often experienced in aquatic conservation is the 'out of sight, out of mind' phenomenon. It is difficult to make legislators and public realise the importance of protecting the environment of organisms that they never see in their natural habitat. There is, therefore, an urgent need to improve our expertise in this field of conservation.

Some basic concepts for *in situ* conservation of biodiversity

A broad-scale perspective is necessary for decisions about the creation or protection of sustainable landscapes, which is a prerequisite for the long-term maintenance of biological diversity. This explains why there is increasing interest in the application of the concepts developed within the discipline of landscape ecology, which seeks to understand the relationships between spatial patterns and ecosystem processes, to environmental management. A landscape can be considered as a spatially heterogeneous area. Landscape ecology emphasises the influence of spatial heterogeneity on biotic and abiotic processes, and the interactions and exchanges across heterogeneous landscapes. Landscape ecological studies focus on the effects that spatial patterning and changes in landscape structure have on the distribution, movement and persistence of species, that is to say the spatial patterns of biological diversity. The connectivities within components of the landscape may be important for species persistence, while in a mosaic of habitat patches, interconnections allow movement of species among patches and recolonisation when a population becomes locally extinct. Modifications of habitat connectivity, or in sizes and shapes of patches, can have strong influences on a species abundance.

A widespread approach, developed in conservation science in the mid 1970s, is based on ideas deriving from the dynamic equilibrium theory of island biogeography (MacArthur & Wilson, 1963, 1967; Terborgh, 1974; Diamond, 1975). This theory originated general principles of refuge design, such as optimal shapes, sizes and connectivity (Goeden, 1979; Temple, 1981), and became very popular among the scientific community concerned with conservation. One of the major debates generated by the refuge design discussion was the size of the reserve. Is a single reserve better than several small ones of equal total size? (Simberloff, 1988). This debate proved to be somewhat academic because, even if there was empirical and theoretical evidence that in the short term several small sites would contain more species than one large one, habitat destruction, leading to fragmentation into small isolated patches, results ultimately in decreased species diversity.

In the case of fish communities, there is a relationship between the size of the watershed, and the number of species (Chapter 11). This is partly explained by the increase of habitat diversity when river size increases, as well as by specific environmental requirements along a longitudinal upstream–downstream gradient.

Out of the Man and the Biosphere Programme (MAB), launched by UNESCO in 1971, grew the Biosphere Reserve concept, which was designed to improve the management of protected areas so that they also promote the sustainable development of surrounding regions. The basic idea was that the establishment of many national parks usually imposed legal restrictions on the use of their resources by local people. In fact, the international definition of national parks stressed the elimination of settlements and exploitation of land and resources within parks. In Africa, the management of national parks, for instance, focused more on the needs of tourists than those of local people (Hough, 1988). The alternative Biosphere Reserve concept takes a more conciliatory approach and is characterised by three functions: (i) conservation of species and ecosystems; (ii) logistics (including research, monitoring of environmental parameters, training); and (iii) the development of regions immediately adjacent to the reserve (Ishwaran, 1992). The Biosphere Reserve, therefore, consists of a core zone comprising a minimally disturbed system, surrounded by a buffer zone, with legal protection and an outer transition zone, which is an area of active cooperation between local people and reserve management for promoting sustainable socio–economic development.

Aquatic parks and reserves

The protection of habitats in parks has been the main conservationist approach throughout the world, and there are several examples of successful underwater parks, mostly marine. Although the establishment of freshwater conservation areas is a relatively new concept in Africa, its necessity was realised as a result of the recent awareness of the potential threats to the African' Great Lakes cichlid communities. An example is the Lake Malawi National Park, located near the southern end of Lake Malawi. This was gazetted in 1980 with the objective of preserving 'a sample of the Lake Malawi biome with particular reference to the rocky lakeshore and its specialist cichlid communities' . . . and developing 'the tourist industry in a manner consistent with the preservation of the biological and aesthetic features of the area' (Croft, 1981). It is the world's first national park established specifically for the protection of freshwater fish. It consists of the waters within 100 m of the shoreline of several islands and disjunct portions of the mainland. Of special interest is the 'Mbuna', a group of ten cichlid genera that are more or less confined to rocky habitats, brightly coloured and exhibit peculiar behaviour (Bootsma, 1992). The Mbuna are not

exploited as major food fish, but taken for the aquarium trade. The park also serves as a breeding and feeding area for commercially important fish.

The extent to which parks will be effective in helping to conserve the fish fauna in lakes depends on whether selected areas are representative in species and habitat of the entire lake, and whether the natural integrity of the parks can be preserved if pervasive changes take place in the lake environment through human activity (Coulter & Mubamba, 1993). Criteria for the selection of underwater parks in Lake Tanganyika were developed by Cohen (1992). 'Reserves should attempt to preserve viable-sized areas of lake bottom in all habitat classes, preferably within a continuous region of lake margin, but if necessary in separated parcels. Reserve boundaries should incorporate both pelagic zones offshore from the shallow water regions of highest diversity and onshore watersheds to avoid lake margin disturbance by agricultural-induced siltation and changes in local nutrient input.'

A prudent policy would be to set aside without delay areas covering as many habitats as feasible in the light of present knowledge. Others should be added as our knowledge increases. Coulter & Mubamba (1993) suggested extending four existing terrestrial wildlife parks to incorporate underwater parks: the Rusizi Park in Burundi, the Gombe Park in Tanzania, the Mahali Mountains Park in Tanzania, and the Nsumbu Park in Zambia. These four underwater areas contain a variety of habitats and faunal complexes and the plan could be implemented quickly with few administrative or social difficulties.

In addition to the underlying purpose of conservation, a park must accommodate existing local activities, and should offer employment opportunities to lakeshore people involved until now in small-scale fishing and farming. Lakeshore residents are well placed to assist in park protection and development. However, to 'pay their way', underwater parks have also to provide facilities to attract visitors. Underwater parks should be great attractions, often quite different from those of land parks. Lakes Malawi and Tanganyika have been compared to coral reefs because of their colourful fish of the rocky littoral, and their cichlids are known world-wide.

Until recently, the villagers living in and around the Lake Malawi National Park could see no benefit in the establishment of the park (Bootsma, 1992). No fishing of any kind is permitted, and this restriction has soured relations between the fishermen and the park management. To solve this problem, it is necessary to promote public education and local involvement. It is only through education that the local inhabitants will understand that the protection of woodlands and fish is for their long-term benefit. This understanding is essential for the success of the park. An education centre is now being built with the assistance of the WWF. It will include display tanks with live fish, poster displays and audio–visual presentations. An underwater trail has been constructed where scuba divers can view fish in their natural environment.

Stock transfer

When a habitat has been completely destroyed, it may be extremely difficult or impossible to return a species to its natural range. One way to save the native species is to remove fry or individuals of endangered species from their habitat and to introduce them into other ecologically suitable habitats, to initiate reserve populations.

Translocation is likely to be less expensive and have a greater chance of success than habitat restoration plans. However, it is not easy to identify habitats where the threatened species may be introduced and translocation projects must meet some important criteria (Maitland & Lyle, 1992). Translocation activities and the removal of eggs or adults must pose no threat to the parent stock, and the introduction must pose no threat to the ecology of the introduction site. Special consideration should be given to the genetic diversity of the stock to be translocated, by selecting material widely in space and time, and transfer of undesirable diseases or parasites must be avoided.

In Africa, there are some examples of species introduced into new habitats that did not create major ecological problems, but probably none of these introductions were devised for conservation purposes. It should be important to focus efforts on the location of valuable sites, and to take a census of candidate sites for stock transfer. This implies that scientists should no longer consider species introduction as a Pandora's box (see Chapter 14).

Restoration ecology

With a good knowledge of the functioning of aquatic systems, it should be possible to rehabilitate degraded eco-systems. Restoration ecology is a relatively new applied science that requires an in-depth knowledge of the original ecosystem, as well as time, money and human resources. The primary goal is to protect an area from further degradation and, when feasible, restore the habitat to a state where it is able to sustain its original biodiversity. Simple protection may be sufficient but, more often, restoration will require more aggressive interventions such as planting trees, or re-establishing river systems or wetlands that have been disrupted due to an extensive water project.

The decision to begin a restoration effort depends upon several factors, the most important being scale, cost and knowledge (Langton, 1993). For aquatic systems, re-establishing historic water flow is often a major problem. For lakes, it is pollution, which disrupts the whole system. In the case of Lake Victoria (Chapter 14), where introduction of exotic species, eutrophication, overfishing and human population growth are responsible for the extinction of many haplochromines, the problems are enormous and a truly comprehensive effort to restore the lake is probably beyond reach (Langton, 1993). A more manageable situation is found in Lake Tanganyika (Coulter & Mubamba, 1993), where many fish species are affected by sediment pollution due to deforestation. Conservation measures must be implemented now, before the situation reaches a point where it will face insurmountable problems and costs.

There are very few examples to illustrate the successful recovery of African freshwater systems following deterioration. However, Lake McIlwaine (actually a reservoir) provides good examples of lake restoration. It was used for many years for dumping sewage from the city of Harare, and consequent eutrophication led to massive algal blooms in the 1960s (Thornton, 1982). A nutrient diversion programme in the early 1970s resulted in improvement of the lake's water quality a few years later and a return to a mesotrophic state.

Ex situ conservation options

Some endemic fish species are under intense and increasing pressure from a number of threats, and there are certain groups of species that are particularly endangered, especially forms with restricted distribution, those of large body size, those of high economic value and those at the top of food chains. Species in these categories are likely to be lost first, but a wide range of other forms are also at risk. For such species, habitat protection alone is not always sufficient and other supportive interventions, including captive breeding, will be needed to avoid the loss of many species, especially those at high risk in greatly reduced, highly fragmented and disturbed habitats.

Captive breeding and captive propagation

The establishment of self-sustaining captive populations has proved to be successful for several species of birds and mammals, and has also been used for fish but breeding programmes are not yet well established (Ribbink, 1986; Meffe, 1987; Ribbink & Twentyman-Jones, 1989). Cichlids are probably ideal candidates for captive propagation because they adapt well to aquarium conditions, they have stereotyped behaviour and mate readily in captivity, they grow rapidly to sexual maturity, and they can be propagated rapidly under artificial conditions (Ribbink, 1986). Phenotypic changes in response to laboratory conditions have been observed, but it appears that this is reversible, particularly in young fish.

Captive breeding should be regarded as a short-term emergency measure only, because a variety of genetic and other difficulties are likely to arise. Captive breeding programmes may best be considered as a way to save taxa threatened with extinction in the wild and to provide propagules for the repopulation of natural habitats. The ultimate goal of captive breeding should be the successful re-establishment of rescued taxa in their native environment. Captive propagation thus averts extinction by providing an 'ark' (Soulé, 1986) to carry endangered taxa through a crisis period.

The objectives of captive propagation are not always readily attainable. Restocking of captive-bred fish into their native habitats is not always realistic and cannot necessarily be considered as the ultimate goal of breeding programmes. Indeed, reintroduction should take place only where the original causes of extinction have been removed, and where habitat requirements of the species are satisfied. This excludes reintroduction into localities where significant habitat deterioration has occurred. We can only hope that habitat rehabilitation will become an important area of research in the next few decades. In the case of Lake Victoria for instance, reintroduction would be futile as long as the *Lates* threat persists, and the

eutrophication of the lake is not controlled. In such cases, captive breeding offers the only alternative to extinction.

Breeding programmes for endangered species

An aquarium working group, within the framework of the IUCN Captive Breeding Specialist Group and Species Survival Commission, has been created to study the need for co-ordinated breeding programmes for fish. Aquaria and aquatic zoos are far behind terrestrial zoos in the percentage of specimens displayed that were captive-born, and in the number of programmes undertaken for rearing endangered species to maturity (Van den Sande, 1991). This situation must change, and Faunal Survival Programmes are being developed, including conservation research, captive propagation and habitat preservation, and require sound understanding of the biology of relevant species.

A few zoos and aquaria have undertaken their own captive-breeding programmes, but the main role of zoos and aquaria will probably remain education. An exhibition of cichlids from Africa's Great Lakes can be used to illustrate fundamental biological phenomena, such as evolution and speciation, fish reproduction (mouthbreeding) and feeding habits. The educational value of displaying fish that are threatened or extinct in nature should not be underestimated, and such displays will help inform people of the value of biodiversity and may help to stimulate the conservation ethic (Langton, 1993). Modern aquaria can also offer good facilities for research as they permit observations and experimentations that are very difficult to achieve in the field, but unfortunately most are under-used in this respect (Andrews, 1992).

Most people involved in breeding fish for conservation purposes are aware of the problem of inbreeding, which has negative consequences in the form of defective offspring. One side effect of captive breeding is a possible decrease in genetic variability of the species as a result of a reduction in population size (bottleneck situation). A high level of genetic diversity will be essential when a species is to be reintroduced into a natural environment and genetically impoverished stocks may have a reduced ability to survive reintroduction. Further, Bruton (1990*b*) stressed the possibility that genetically impoverished, captive-bred species could reduce the fitness of wild, resistant individuals with which they may interbreed when reintroduced into the natural habitat.

Only time and experience will show if these obstacles can be overcome. Until now the general prescription for avoiding inbreeding is to maintain a large breeding stock and to exchange animals with other breeders.

The role of the aquarium industry

The aquarium industry has a key role to play in conservation by trading in an ethical and responsible way that promotes economic gain, but which does not create conservation problems. A greater emphasis on pond culture of ornamental fish would take the pressure off exploiting sensitive wild stocks, should prove economically beneficial to local communities (Reid, 1993), and should be possible throughout Africa (Bassleer, 1994). It should also be possible for the aquarium industry to support selected conservation initiatives. Moreover, individual aquarists could assist in particular conservation endeavours, and the Aquatic Conservation Network is currently preparing a guideline document for aquarists, outlining the potential contributions they can make to captive breeding programmes. As a whole, cooperation between scientists and the aquatic trade, on a basis of exchange of information, would develop a sound strategy for the conservation of fish fauna. Fish exporters usually have knowledge of fish biology, valuable information that ichthyologists seek and require.

Conservation strategy for Lake Victoria haplochromines

A great deal of attention has been given to the dramatic situation of the endangered native fish of Lake Victoria. Various institutions have developed a conservation strategy for the cichlids and one of the Faunal Survival Programmes (FSP) of IUCN deals with Lake Victoria cichlids (Andrews, 1992). It is suggested that these programmes will have three essential elements: (i) captive maintenance and propagation activities; (ii) active management of the captive population (aided by studies on the genetic structure of wild and captive populations); and (iii) restoration and recovery efforts in the wild.

The European Union of Aquarium Curators stressed that they would like to participate in a co-ordinated captive-breeding programme for these cichlids, with three main objectives (Van den Sande, 1991): (i) establish exhibits focusing on the importance of conservation of native fish related to habitat conservation and preservation; (ii)

to stabilise, conserve and propagate haplochromine stocks presently in North America and Europe; and (iii) to assess the status of Victoria cichlids in the wild, and to determine conservation priorities. For a maximum chance of success in such a conservation programme, key elements include:

creation of a network of the major organisations involved, including international institutions as well as specialised laboratories, private breeders and hobbyists;

acquisition of most of the stock of known provenance now in captivity;

begin to explore cryopreservation of gametes as a management tool; and

starting of long-term field studies with the countries involved, to assess the changing status of endemic species in the lake and possibilities for long-term on-site conservation measures.

A Lake Victoria Species Survival Plan was established in 1993. It is a cooperative population management and conservation programme for endangered species at zoos and aquaria in North America. Thirty-four species are being maintained by various institutions, and 30 species have reproduced in captivity (Norton, 1993). Similarly the European captive-breeding programme is managing approximately 24 species. The intention is to consolidate and focus support by the professional aquarium and academic communities on the conservation of representative remnants of the unique endemic fish fauna of Lake Victoria. For many species, the number of founders is small, usually less than 30, but populations have expanded quickly due to the high fecundity of these fish. Most species have reproduced at more than one institution.

Conserving fish genomes

Genetics is a key area in biological conservation literature. Apart from the immediate possibility of extinction for geographically restricted and endemic forms, there is a more subtle threat: erosion of the genetic diversity upon which long-term persistence and adaptability depend (Vrijenhoek et al., 1985).

Interest in the genetics of fisheries resources has increased recently. One of the reasons is concern about the potentially harmful genetic effects of selective harvesting on fish populations, since fisheries can alter the size and genetic structure of the target population. It is also recognised that genetically discrete races or stocks of many fish species should be harvested as distinct units, but there is little information on this topic for African inland waters. The capture of ornamental fish can also threaten the conservation of genetic resources, particularly since a fishery may concentrate on species that are already rare.

Habitat destruction can subdivide a species into a series of small, partially isolated local populations, and captive breeding also usually deals with limited fish populations. In both cases, fish conservation is faced with the problems of the genetics of small populations (Meffe, 1986; Soulé, 1986), which fall into two broad categories: random (bottlenecks and drift), and directional (inbreeding depression) effects (Meffe, 1987). A bottleneck occurs when a population goes through a sudden and usually drastic decline in effective size. Survivors possess only a part of the genetic variance contained in the pre-bottleneck population, and the genetic pool of subsequent generations will therefore be much poorer unless there is gene flow via migration or introduction from other sources. Genetic drift is a prolonged bottleneck that erodes genetic variance through change in gene frequency in small populations. This loss of variation is random, but selection may subsequently play a role in altering gene frequency in succeeding generations.

The deleterious effect of inbreeding, which is the mating of individuals sharing a common ancestor, results in increases in homozygous genotypes. Some authors believe inbreeding depression is likely to threaten refuge populations already impoverished by genetic drift. This situation may negatively affect fitness in characters such as fecundity and growth rate, but there is at present no empirical evidence that it automatically causes the collapse of a population.

Before developing rigorous programmes for the conservation of gene pools, deficiencies in our knowledge must be assessed (Meffe, 1987). The first is, of course, to collect genetic information on those fish judged to be threatened. Knowledge of the distribution of genetic variation within and between remnant populations of an endangered species is necessary if we are to design sound conservation programmes and make informed decisions. We also need a better understanding of the role that gen-

etic heterozygosity plays in the individual fitness of fish, and what really is the heterozygote advantage. The heterozygosity detected in electophoretic analysis may have nothing to do with fitness, and we need to learn more about it to better serve our interest in maintaining endangered populations of fish. Conservation programmes must also consider features of fish life histories that could affect the genetic structure of populations. Dispersal of individuals within and among populations can affect genetic diversity by increasing or decreasing gene flows. This is particularly true for freshwater fish living in isolated waterbodies that are more constrained than species in many terrestrial or marine situations where migratory pathways between suitable habitats are often available. Isolation has tended to increase genetic divergence among populations, and this differentiation must be considered an important aspect of genetic variation to be preserved in any recovery efforts.

To conclude, we can only agree with Meffe (1987) that the science of conservation genetics is in its infancy and we will make mistakes. They can be minimised by retaining a maximum amount of global genetic diversity for maximum flexibility in future research and management efforts.

Cryopreservation

Modern techniques for rapid freezing of gametes to very low temperatures have proved successful for a variety of animals, including fish. After freezing for many years and then thawing the material is still viable. Cryopreservation of gametes therefore opens up the possibility of large-scale gene banks, free from genetic contamination and at relatively low recurrent costs. Cryopreservation of embryos would also be useful for ecotoxicology bioassays.

Despite the economic importance of some African fish species, such as the tilapias, only a few papers deal with the technical aspects of cryopreservation of their spermatozoa (Chao *et al.*, 1987; Rana & McAndrew, 1989; Rana *et al.*, 1990). Cryopreservation of fish spermatozoa in sperm banks may yet provide long-term preservation and access to a wide range of genetic material for the animal breeding industry, and this could be a way to avoid genetic drift in small breeding populations.

Although rapid freezing to very low temperatures has been successful for fish sperm, the cryopreservation of fish eggs and embryos has proved more difficult. The eggs of most species are large, have a high yolk content and a low permeability. The size of the embryos and their more complicated structure may explain the difficulty of preservation, and methods of avoiding ice formation within the egg or embryo during cooling need to be developed in order to achieve higher viability. Nevertheless, in the past ten years there has been some limited success in the cryopreservation of eggs and embryos, e.g. of zebrafish (Zhang *et al.*, 1993). Even if this technique is still of limited value for the conservation of fish species, there is a realistic longer-term prospect of applying cryobiological techniques to cichlid zygotes and ova.

While efforts to contain and control the accelerating erosion of biodiversity should be strengthened, cryopreservation could become the ultimate conservation method. It seems likely that captive-breeding programmes and *in situ* protected areas can preserve only a tiny fraction of threatened species. So, a broad programme of freezing endangered species has been suggested in order to preserve biodiversity for future generations (Benford, 1992). This programme (Saving the 'library of life') proposes a combined strategy of preserving alive some fraction of each ecosystem type, its population represented intact at the genus level, and freezing as many species related to the preserved system as possible. This will allow future biologists to extract DNA from frozen samples, assuming that genes of interest could be expressed in living examples of the same genus, by systematic replacement of elements of the genetic code with information from the frozen DNA. This venture relies on future biotechnology greatly surpassing ours, but might resurrect then–extinct species.

Legal and international aspects of conservation

The position of biological diversity in national legal systems is one of the important judicial problems of today. Plants, animals, water, etc. are considered as matters or objects whose degree of protection depends on the value they represent for human beings. Even under seemingly sympathetic intent, this specifically anthropocentric view leads directly to the subordination of biological diversity,

and to its unilateral sacrifice in spite of modern people understanding its advantages.

International conventions for protection of biodiversity

Nevertheless, many treaties and conventions, as well as bilateral or multilateral agreements, have been adopted for the conservation of particular species or ecosystems. Only a few major ones are relevant for inland ecosystems and freshwater fish species.

1. The **Ramsar** Convention on Wetlands of International Importance (1971) is the only global nature conservation convention designed to cover a particular broad habitat type. Contracting parties undertake to use wisely all wetland resources under their jurisdiction and to designate for conservation at least one wetland of international importance under criteria provided by the convention. In 1990 the 61 contracting parties had designated 421 sites. A conservation fund for wetlands collects the voluntary contributions of contracting parties and may help those that are faced with economic difficulties in achieving their commitments.

2. The **Paris Convention** concerning the protection of the World Cultural and Natural Heritage (1972) recognises the obligation of all states to protect unique natural and cultural areas, as well as the obligation of the international community to help pay for them. An International Committee establishes and publishes the list of exceptional natural and cultural sites, which had reached 337 sites in 1991. Each country contributes to a World Heritage Fund for the conservation of exceptional sites and to promote research. The national contributions are currently calculated at 1% of the contributions to the annual budget of UNESCO.

3. The **Washington** Convention on International Trade in Endangered Species of Wild Fauna and Flora (CITES) (1973) has established lists of endangered species for which international trade is controlled via permit systems, as a means for combating overexploitation.

Many other treaties that deal entirely or in part with biological diversity have evolved in an uncoordinated manner. Some are global conventions, but most are regional conventions of little interest for inland waters and freshwater fish. The total obligations explicit in existing treaties fall far short of a comprehensive system. The Convention on Biological Diversity attempts to meet many of these demands, and is the first treaty to concentrate specifically on the conservation and use of global biodiversity.

The Convention on Biodiversity

The Rio Declaration on Environment and Development is intended to form the basis for an 'Earth Charter', comparable to the Universal Declaration of Human Rights. For the first time it has established in global terms the principle of prevention, and the internalisation of environmental costs. For the first time, world leaders publicly recorded their concern at the accelerating destruction of habitats, and the mass extinction of species. One of the main achievements of UNCED was to emphasise the interrelationships of world problems as well as existing chains of causality. A joint search for solutions to the problems of planetary management may be the starting point for a new ethical basis for relations between states.

The Rio Convention is the product of a long-term process that has resulted in different attempts to establish the legal status of biodiversity. The Convention was originally prepared by scientists wishing to promote *in situ* conservation. The World Conservation Union (IUCN) played a major role in considering biodiversity as a World Heritage. The main goal was the gathering of information through national surveys in order to identify threatened species and ecosystems. However, the original concept suggested open-access to biodiversity, both for scientists and industry. Moreover, while each country was responsible for conservation of biodiversity, an international scientific committee was proposed to control the application of the Convention. In the context of the North–South divide, this was considered by several developing countries as a threat to their national sovereignty as well as an impediment to their economic development. In fact, developing countries were concerned to achieve a greater share of the economic benefits arising from the use of their natural resources, which until now

have mainly accrued to the industries of developed countries. It was also recognised that the redistribution of benefits arising from the use of their natural resources would provide the developing countries with the necessary economic incentive to reinforce the conservation of their biodiversity. Conversely, the biotechnology industry was concerned about unrestricted and free access to biodiversity. As a consequence of the very different interests and expectations of developed and developing countries, the negotiations became increasingly polarised, and the result was a loose compromise.

The Convention on Biodiversity has now been signed by most states and it contains commitments relating to the conservation and sustainable use of the elements that make up biodiversity, as well as to a fair and equitable share-out of the advantages stemming from their use, particularly through the transfer of technologies.

The so-called Agenda 21, as well as the Convention on Biological Diversity, recognises the right of developing countries to sustainable socio-economic development based on the principle of national sovereignty over their own biodiversity. The contracting parties also recognised that the responsibility for conserving biodiversity rests with sovereign national governments. The principles recommend states to carry out regular surveys and inventories of biological resources and to identify those requiring urgent conservation measures. Strategies for long-term monitoring should be developed and adopted. Conserving biological resources *in situ* is highly desirable and should be promoted whenever possible. States were asked to avoid practices and policies, such as the introduction of alien species and unregulated removal of indigenous species, that undermine conservation efforts.

The Convention also pointed out that without active cooperation from local people, the conservation of biological diversity would be impossible. This cooperation should therefore be sought and made an integral part of policy and practice. The participating states also agreed that cooperation between countries was essential, particularly in promoting awareness among the public, in the carrying out of environmental impact assessments, and in building national capacity and capability.

So, the international legal framework exists to conserve biodiversity. The question now is whether these ideals will be put into practice soon enough to be successful.

The safe minimum standard and precautionary principle

A delicate balance has to be found between the obligation to advise decision-makers and the obligation to disseminate only proven findings. Scientists cannot wait indefinitely for better measurements or better methods while the erosion of biodiversity and the degradation of the environment is taking place today. It has been suggested that the establishment of safe minimum standards for conservation of wildlife could be one way to preserve biodiversity and genetic variability. Such standards are, for example, minimum habitat requirements or minimum viable populations, but their validity is very uncertain (Hohl & Tisdell, 1993). In fact, no standard ensures the continued existence of any species, and minimum standards need to be cast in terms of those required to achieve a particular probability of survival of the targeted species for a specified period. Consequently, even if it may sound unsatisfactory, decisions on environmental issues will have to be based mainly on ethical considerations for quite a long time to come, while the scientific knowledge necessary still remains inadequate. According to Bishop & Ready (1991): 'Society may choose to adopt the safe minimum standards not because it results from a rigorous model of social choice, but simply because individuals in society feel that the safe minimum standard is "the right thing to do" '.

The precautionary principle emerged during the last decade as a recognition of the uncertainty involved in impact assessments and management, including future consequences of present decisions. It is related to intergenerational equity (our responsibility towards future generations), one of the central issues of the concept of sustainable development. The precautionary principle emphasises growing awareness that fisheries management cannot be seen in isolation and must fit an integrated context that satisfies the requirement both for long-term resource sustainability and environmental conservation (Garcia, 1994). It puts the focus more clearly on uncertainty and related hidden costs of present decisions for future generations. It intends to promote fishery practices compatible with the requirements of ecosystems, resources and consumers, and therefore offers a good opportunity to progress towards sustainable fisheries development. The concept tends to promote fishery practices compatible with the requirements of ecosystems, and the guidelines needed for its implementation will have to give due consideration to the need for precautionary approaches (Garcia, 1994). Meanwhile, FAO is developing guidelines for 'Responsible Fishing'.

If the 'precautionary principle' is accepted, developers would then have to prove that the preservation of species is the wrong choice because of the unacceptably high cost of foregone benefits, instead of the present situation where conservationists have to prove that environmental costs outweigh economic benefits (Hohl & Tisdell, 1993).

Bibliography

Abban E. K. & Skibinski D. O. F., 1988. Protein variation in *Schilbe mystus* (L.) and *Eutropius niloticus* (Rüppel) (Pisces, Siluriformes) in the Volta Basin of Ghana, West Africa. *Aquaculture and fisheries management*, **19**: 25–37.

Acere T. O., 1988. The controversy over Nile Perch, *Lates niloticus* in Lake Victoria, East Africa. *Naga*, **11**(4): 3–5.

Achieng A. P., 1990. The impact of the introduction of Nile perch, *Lates niloticus* (L.) on the fisheries of Lake Victoria. *Journal of Fish Biology*, **37**, Supplement A: 17–23.

Adams J. M. & Woodward F. I., 1992. The past as a key to the future: the use of palaeoenvironmental understanding to predict the effects of man on the biosphere. *Advances in Ecological Research*, **22**: 257–314.

Adams W. M., 1985. River control in West Africa. In Grove A. T. (ed.), *The Niger and Its Neighbours*, pp. 177–228. Balkema, Rotterdam.

Adamson D. A. & Williams F. M., 1980. Structural geology, tectonic and the control of drainage in the Nile basin. In Williams F. M. & Faure H. (eds.), *The Sahara and the Nile*, pp. 225–52. Balkema, Rotterdam.

Adamson D. A., Gasse F., Street F. A. & Williams M. A. J., 1980. Late Quaternary history of the Nile. *Nature*, **287**: 50–5.

Addicott J. F., Aho J. M., Antolin M. F., Padilla D. K., Ridcharson J. S. & Soluk D. A., 1987. Ecological neighborhoods: scaling environmental patterns. *Oikos*, **49**: 340–6.

Adebisi A. A., 1981. Analyses of the stomach contents of the piscivorous fishes of the upper Ogun River, Nigeria. *Hydrobiologia*, **79**: 166–77.

Adebisi A. A., 1988. Changes in the structural and functional components of the fish community of a seasonal river. *Archiv für Hydrobiologie*, **113**: 457–63.

Adeniyi E. O., 1973. Downstream impact of the Kainji Dam. In Mabogunje A. L. (ed.), *Kainji Lake Studies*, 2. NISER, Ibadan.

Agnèse J. F., 1989. Differentiation génétique de plusieurs espèces de Siluriformes ouest-africaines ayant un intérêt pour la pêche et l'aquaculture. Thèse de l'Université des Sciences et Techniques du Languedoc, Montpellier.

Agnèse J. F., 1991. Taxonomic status and genetic differentiation among West African populations of the *Chrysichthys auratus* complex (Pisces, Siluriforme) based on protein electrophoresis. *Aquaculture and fisheries management*, **22**: 229–37.

Agnèse J. F. & Bigorne R., 1992. Premiéres donnèes sur les relations génétiques entre onze espèces ouest-africaines de Mormyridae (Teleostei, Osteichthyes). *Revue d'Hydrobiologie tropicale*, **25**: 253–61.

Agnèse J. F., Romand R. & Pasteur N., 1987. Biochemical differentiation between some genera of African Cyprinodontidae. *Zeitschrift für Zoologische Systematik und Evolutionforschung*, **25**: 140–6.

Agnèse J. F., Pasteur N. & Lévêque C., 1989. Différenciation génétique de quelques populations de *Chrysichthys nigrodigitatus* et de *C. johnelsi* (Pisces, Bagridae) de Côte d'Ivoire et du Mali. *Revue d'Hydrobiologie tropicale*, **22**: 101–6.

Agnèse J. F., Berrebi P., Lévêque C. & Guégan J. F., 1990*a*. Two lineages, diploid and tetraploid, demonstrated in African species of *Barbus* (Osteichthyes, Cyprinidae). *Aquatic Living Resources*, **3**: 305–11.

Agnèse J. F., Oberdoff T. & Ozouf-Costaz C., 1990*b*. Karyotypic study of some species of family Mochokidae (Pisces, Siluriformes): evidence of female heterogamety. *Journal of Fish Biology*, **37**: 375–81.

Alabaster J. S., 1981. *Review of the State of Aquatic Pollution of East African Inland Waters*. CIFA Occasional Paper, 9. FAO, Rome.

Alabaster J. S., Calamari D., Dethlefsen V., Konemann H., Lloyd A. & Solbé J. F., 1988. Water quality criteria for European freshwater fishes. *Chemistry and Ecology*, **3**: 165–253.

Albaret J. J., 1982. Reproduction et fécondité des poissons d'eau douce de Cte d'Ivoire. *Revue d'Hydrobiologie tropicale*, **15**: 347–71.

Albaret J. J., 1987. Les peuplements de poissons de la Casamance (Sénégal) en période de sécheresse. *Revue d'Hydrobiologie tropicale*, **20**: 291–308.

Albaret J. J. & Charles-Dominique E., 1982. Observation d'un phénomène de maturation sexuelle précoce chez l'Ethmalose *Ethmalosa fimbriata* Bowdich, dans une baie polluée de la lagune Ebrié (Côte d'Ivoire). *Documents Scientifiques de Centre de Recherches Océanographiques, Abidjan*, **13**: 23–31.

Almaça C., 1976. La spéciation chez les Cyprinidés de la Péninsule Ibérique. *Revuedes travaux de l'Institut des Péches maritimes*, **40**: 399–411.

Alvarez W., Kauffman E. G., Surlyk F. *et al.*, 1984. Impact theory of mass extinctions in the invertebrate fossil record. *Science*, **223**: 1135–41.

Amiet J. L., Poliak D. & Chauche M., 1987. Le genre *Aphyosemion* Myers (Pisces, Teleostei, Cyprinodontiformes). *Faune du Cameroun*, Vol. 2. Venette, Compiègne.

Anderson A. M., 1961. Further observations concerning the proposed introduction of Nile perch into Lake Victoria. *East African Agricultural and Forestry Journal*, **26**: 195–201.

Andrews C., 1992. The role of zoos and aquaria in the conservation of the fishes from Africa's Great lakes. *Mitteilungen-Internationale Vereinigung für Theoretische und Angewandte Limnologie*, **23**: 117–20.

Angel M., Glachant M. & Lévêque F., 1992. La préservation des espèces: que peuvent dire les économistes? *Economie et Statistiques*, **258-9**: 113–19.

Angermeier P. L. & Karr J. R., 1983. Fish communities along environmental gradients in a system of tropical streams. *Environmental Biology of Fishes*, **9**: 117–35.

Angermeier P. L. & Schlosser I. J., 1989. Species-area relationships for stream fishes. *Ecology*, **70**: 1450–62.

Antwi L. A. K., 1984. The effect of abate on the brain acetylcholinesterase activity of fish from two treated rivers in the upper Volta: rivers White Volta and Black Volta. Report OCP/VCU/Hybio/84.13.

Arai R. & Koike A., 1980. A karyotype study on two species of freshwater fishes transplanted into Japan. *Bulletin of the National science Museum (Tokyo)*, **6**: 275–8.

Arambourg C., 1963. Continental vertebrate fauna of the tertiary of north Africa. *Publications in Anthropology, Viking Fund*, **36**: 55–64.

Arambourg C. & Magnier P., 1961. Gisements de vertébrés dans le bassin tertiaire de Syrte (Libye). *Comptes rendus hebdomadaires des Séances de l'Académie des Sciences, Paris*, **252**: 1181–3.

Arawomo G. A. O., 1982. Food and feeding of three *Distichodus* species (Pisces, Characiformes) in Lake Kainji, Nigeria. *Hydrobiologia*, **94**: 177–81.

Arnold S. J., 1985. Quantitative genetic models of sexual selection. *Experientia*, **41**: 1296–310.

Arnoult J., 1959. *Poissons des eaux douces*. In *Faune de Madagascar*, Tananarive, **10**.

Arnoult J., 1963. Un Oriziiné (Pisces, Cyprinotontidae) nouveau de l'est de Madagascar. *Bulletin du Muséum national d'Histoire naturelle, Paris*, (2) **35**: 235–7.

Ashlock P. D., 1979. An evolutionary systematists view of classification. *Systematic Zoology*, **25**: 441–50.

Avise J. C., 1990. Flocks of African fishes. *Nature*, **347**: 512–13.

Axelrod H. R., Burgess W. E., Pronde N. & Walls J. G., 1986. *Dr Axelrod's Atlas of Freshwater Aquarium Fishes*, 2nd edn. T.H.F. Publications, Neptune City, NJ.

Aylward B., 1991. The economic value of ecosystems: 3. *Biological Diversity*. Gatekeeper Series No. GK 91–03. IIED/UCL London Environmental Centre.

Aylward B. & Barbier E. B., 1992. Valuing environmental functions in developing countries. *Biodiversity and Conservation*, **1**: 34–50.

Babiker M. M., 1979. Respiratory behaviour, oxygen consumption and relative dependance on aerial respiration in the african lungfish (*Protopterus annectens* Owen) and an air-breathing teleost (*Clarias lazera*). *Hydrobiologia*, **65**: 177–87.

Babiker M. M., 1984. Development of dependance on aerial respiration in *Polypterus senegalus* (Cuvier). *Hydrobiologia*, **110**: 351–63.

Baerends G. P. & Baerends-van Roon J. M., 1950. *An Introduction to the Study of the Ethology of Cichlid Fishes*. E. J. Brill, Leiden.

Bagenal T. B., 1978. Aspects of fish fecundity. In Gerking S. D. (ed.) *Methods of Assessment of Ecology of Freshwater Fish Production*, pp. 75–101. Blackwell, Oxford.

Bailey R. G., 1969. The non-cichlid fishes of the eastward flowing rivers of Tanzania, East Africa. *Revue de Zoologie et de Botanique africaine*, **80** (1–2): 170–99.

Bailey R. G., 1972. Observations on the biology of *Nothobranchius guentheri* (Pfeffer) (Cyprinodontidae), an annual fish from the coastal region of east Africa. *African Journal of tropical Hydrobiology and Fisheries*, **1**: 33–43.

Bailey R. G., 1988. Fish and fisheries. In Howell P., Lock M. & Cobb S. (eds.), *The Jonglei Canal: Impact and Opportunity*, pp. 328–49. Cambridge Studies in Applied Ecology and Resource Management. Cambridge University Press, Cambridge.

Bainbridge R., 1958. The speed of swimming of fish as related to size and the frequency and the amplitude of the tail beat. *Journal of Experimental Biology*, **35**: 109–33.

Baker J. M., 1992. Oil and African Great Lakes. *Mitteilungen-Internationale Vereinigung für Theoretische und Angewandte Limnologie*, **23**: 71–7.

Balarin J. D., 1986. *National Review for Aquaculture Development in Africa. 8. Egypt*. FAO Fisheries Circular, 770.17.

Balarin J. D. & Haller R. D., 1982. The intensive culture of tilapia in tanks, raceways and cages. In Muir J. F. &

Roberts R. J. (eds.), *Recent Advances in Aquaculture*, pp. 265–356, Croom Helm, London.

Balek J., 1977. *Hydrology and Water Resources in Tropical Africa*. Developments in Water Science, 8. Elsevier, Amsterdam.

Balirwa J. S. & Bugenyi F. W. B., 1980. Notes on the fisheries of the river Nzoia, Kenya. *Zoological Conservation*, 18: 53–8.

Balon E. K., 1971. Replacement of *Alestes imberi* Peters 1852 by *Alestes lateralis* Boulenger 1900, in Lake Kariba, with ecological notes. *Fisheries Research Bulletin, Zambia*, 5: 119–62.

Balon E. K., 1974a. Fish of the edge of Victoria Falls, Africa: demise of physical barrier for downstream invasions. *Copeia*, 3: 643–60.

Balon E. K., 1974b. Fish population of drainage area and the influence of ecosystem changes on fish distribution. In Balon E. K. & Coche A. G. (eds.), *Lake Kariba: A Man Made Tropical Ecosystem in Central Africa*, Monographiae Biologicae, pp. 459–96. Junk Publishers, The Hague.

Balon E. K., 1975. Reproductive guilds of fishes: a proposal and definition. *Journal of the Fisheries Research Board of Canada*, 32: 821–64.

Balon E. K., 1977. Early ontogeny of *Labeotropheus* Ahl, 1927 (Mbuna, Cichlidae, Malawi) with a discussion on advanced prospective styles in fish reproduction and development. *Environmental Biology of Fishes*, 2: 147–76.

Balon E. K., 1978. Reproductive guilds and the ultimate structure of fish taxocenes: amended contributions to the discussion presented at the mini-symposium. *Environmental Biology of Fishes*, 3: 159–62.

Balon E. K., 1981a. About processes which cause the evolution of guilds and species. *Environmental Biology of Fishes*, 6: 129–38.

Balon E. K., 1981b. Additions and amendments to the classification of reproductive styles in fishes. *Environmental Biology of Fishes*, 6: 377–89.

Balon E. K., 1984. Patterns in the evolution of the reproductive styles in fishes. In Potts G. W. & Wootton R. J. (eds.), *Fish Reproduction: Strategies and Tactics*, pp. 35–53. Academic Press, London.

Balon E. K., 1985. *Early Life Histories of Fishes: New Development, Ecological and Evolutionary Perspectives*. Developments in Environmental Biology of Fishes, 5. W. Junk Publishers, Dordrecht.

Balon E. K., 1986. Types of feeding in the ontogeny of fishes and the life history model. *Environmental Biology of Fishes*, 16: 11–24.

Balon E. K., 1989. The epigenetic mechanisms of bifurcation and alternative life-history styles. In Bruton M. N. (ed.), *Alternative Life-History Styles of Animals. Perspectives in Vertebrate Science*, 6, pp. 467–501. Kluwer Academic Publishers, Dordrecht.

Balon E. K., 1990. Epigenesis of an epigeneticist: the development of some alternative concepts on the early ontogeny and evolution of fishes. *Guelph Ichthyology Reviews*, 1: 1–42.

Balon E. K. & Bruton M. N., 1986. Introduction of alien species or why scientific advice is not heeded. *Environmental Biology of Fishes*, 16: 225–30.

Balon E. K. & Coche A. G., 1974. *Lake Kariba. A Man-made Tropical Ecosystem in Central Africa*. W. Junk Publishers, The Hague.

Balon E. K. & Stewart D. J., 1983. Fish assemblages in a river with unusual gradient (Luongo, Africa Zaïre system), reflections on river zonation, and description of another new species. *Environmental Biology of Fishes*, 9: 225–52.

Banarescu P., 1973. Origin and affinities of the freshwater fish fauna of Europe. *Ichthyologia*, 5: 1–8.

Banarescu P., 1990. *General Distribution and Dispersal of Freshwater Animals*. Vol. 1, Zoogeography of Fresh Waters. Aula-Verlag, Wiesbaden.

Banister K. E., 1994. *Glossogobius ankaranensis*, a new species of blind cave goby from Madagascar (Pisces: Gobioidei: Gobiidae). *Aqua-Journal of Ichthyology and Aquatic Biology*, 1(3): 25–8.

Banister K. E. & Bailey R. G., 1979. Fishes collected by the Zaïre River Expedition 1974–75. *Zoological Journal of the Linnean Society*, 66: 205–49.

Banister K. E. & Clarke M. A., 1980. A revision of the large *Barbus* (Pisces, Cyprinidae) of Lake Malawi with a reconstruction of the history of the southern African Rift Valley lakes. *Journal of Natural History*, 14: 483–542.

Banks J. W., Holden M. J. & McConnell R. H., 1966. Fishery report. In White E. (ed.) *The First Scientific Report of the Kainji Biological Research Team*. Liverpool University.

Barbault R., 1981. *Ecologie des Populations et des Peuplements*. Masson, Paris.

Barbier E. B., 1989. The economic value of ecosystems: 1. *Tropical Wetlands*. Gatekeeper Series No. LEEC 89–02. London Environmental Economics Centre.

Barbier E. B., Adams W. M. & Kimmage K., 1991. *Economical Valuation of Wetland Benefits: The Hejia-Jama'are Floodplain, Nigeria*. LEEC Paper DP 91–02. London International Institute for Environment and Development.

Barel C. D. N., 1983. Towards a constructional morphology

of cichlid fishes (Teleostei, Perciformes). *Netherlands Journal of Zoology*, 33: 357–424.

Barel C. D. N., Dorit R., Greenwood P. H., Fryer G., Hughes N., Jackson P. B. N., Kawanabe H., Lowe-McConnell R. H., Nagoshi M., Ribbink A. J., Trewavas E., Witte F. & Yamaoka K., 1985. Destruction of fisheries in Africa's lakes. *Nature*, 315: 19–20.

Barel C. D. N., Anker G. C., Witte F., Hoogerhoud R. J. C. & Goldschmidt T., 1989. Constructional constraint and its ecomorphological implications. *Acta Morphologica Neerlando-Scandinavia*, 27: 83–109.

Barel C. D. N., Ligvoet W., Goldschmidt T., Witte F. & Goudswaard P. C., 1991. The haplochromine cichlids in Lake Victoria: an assessment of biological and fisheries interests. In Keenleyside M. H. A. (ed.), *Cichlid Fishes: Behaviour, Ecology and Evolution*, pp. 258–79. Fish and Fisheries Series 2. Chapman & Hall, London.

Barlow G. W., 1961. Causes and significance of morphological variation in fishes. *Systematic Zoology*, 10: 105–17.

Barlow G. W., 1964. Ethology of the Asian teleost *Badis badis*. V. Dynamics of fanning and other parental activities, with comments on behaviour of the larvae and post larvae. *Zietschrift für Tierpsychologie*, 21: 99–123.

Barlow G. W., 1984. Patterns of monogamy among teleost fishes. *Archiv für Fischereiwissenschaft, Hamburg*, 35: 75–123.

Barlow G. W., 1991. Mating systems among cichlid fishes. In Keenleyside H. A. (ed.), *Cichlid Fishes: Behaviour, Ecology and Evolution*, pp. 173–90. Fish and Fisheries Series 2. Chapman & Hall, London.

Barnard K. H., 1943. Revision of the indigenous freshwater fishes of the S. W. Cape Region. *Annals of the South African Museum*, 36(2): 101–262.

Barnes R. F. W., 1990. Deforestation trends in tropical Africa. *African Journal of Ecology*, 28: 161–73.

Baroiller J. F. & Jalabert B., 1989. Contribution of research in reproductive physiology to the culture of tilapias. *Aquatic Living Resources*, 2: 105–16.

Baroillier J. F., Fostier A., Cauty C., Rognon X. & Jalabert B., 1996. Significant effects of high temperature on sex-ratios of progenies from an *Oreochromis niloticus* sex-reversed male broodstock. In Pullin R. V. S., Lazard J., Legendre M., Amon Kotias J. B. & Pauly D. (eds.), *Third International Symposium on Tilapia in Aquaculture*. ICLARM Conference Proceedings 41.

Bartley D., 1993. FAO expert consultation on utilisation and conservation of aquatic genetic resources. *Aquatic Survival*, 2: 15–17.

Barton N. H., 1988. Speciation. In Myers A. A. & Giller

P. S. (eds.), *Analytical Biogeography: An Integrated Approach to the Study of Animal and Plant Distributions*, pp. 185–218. Chapman & Hall, London.

Barton N. H., Jones J. S. & Mallet J., 1988. No barriers to speciation. *Nature*, 336: 13–14.

Baasleer G., 1994. The aquatic industry and conservation of African fish species. *OFI Journal*, February 1994: 12–13.

Bayley P. B., 1977. Changes in species composition of the yields and catch per unit effort during the development of the fishery at Lake Turkana, Kenya. *Archiv. für Hydrobiologie*, 79, 111–32.

Bayley P. B. & Li H. W., 1992. Riverine fishes. In Calow P. & Petts G. E. (eds.), *The Rivers Handbook*, Vol. 1, pp. 251–81. Blackwell Scientific Publications, Oxford.

Baylis J. R., 1981. The evolution of parental care in fishes, with special reference to Darwin's rule of male sexual selection. *Environmental Biology of Fishes*, 6: 223–51.

Beadle L. C., 1943. An ecological survey of some inland saline waters of Algeria. *Journal of the Linnean Society (Zoology)*, 41: 218–42.

Beadle L. C., 1981. *The Inland Water of Tropical Africa*, 2nd edn. Longman, Harlow, Essex.

Beaudet G. R., Coque R., Michel P. & Rognon P., 1977. Y a-t-il eu capture du Niger? *Bulletin de l'Association de Géographie*, 445–6: 215–32.

Beecher H. A., Dott E. R. & Fernau R. F., 1988. Fish species richness and stream order in Washington State streams. *Environmental Biology of Fishes*, 22: 193–209.

Begg G. W., 1974. The distribution of fish of riverine origin in relation to the limnological characteristics of the five basins of Lake Kariba. *Hydrobiologia*, 44: 277–86.

Begg G. W., 1976. The relationship between the diurnal movements of some of the zooplankton and the sardine *Limnothrissa miodon* in Lake Kariba, Rhodesia. *Limnology and Oceanography*, 21: 529–39.

Begon M., Harper J. L. & Townsend C. R., 1986. *Ecology: Individuals, Populations and Communities*. Blackwell Scientific Publications, Oxford.

Beitinger T. L. & Pettit M. J., 1984. Comparison of low oxygen avoidance in a biomodal breather, *Erpetoichthys calabricus*, and an obligate water breather, *Percina caprodes*. *Environmental Biology of Fishes*, 11: 235–40.

Belbenoit P., Moller P., Serrier J. & Push S., 1979. Ethological observations on the electric organ discharge behaviour of the electric catfish, *Malapterurus electricus* (Pisces). *Behavioral Ecology and Sociobiology*, 4: 321–30.

Bell-Cross G., 1966a. The distribution of fishes in Central Africa. *Fisheries Research Bulletin of Zambia*, 4: 3–20.

Bell-Cross G., 1966b. Physical barriers separating the fishes of the Kafue and Middle Zambezi River Systems. *Fisheries Research Bulletin of Zambia*, 4: 97–8.

Bell-Cross G., 1972. The fish fauna of the Zambezi River system. *Arnoldia, Rhodesia*, 5: 1–19.

Bell-Cross G., 1976. *The Fishes of Rhodesia*. National Museums & Monuments of Rhodesia, Salisbury.

Bender E. A., Case T. J. & Gilpin M. E., 1984. Perturbation experiments in community ecology: theory and practice. *Ecology*, 65: 1–13.

Bénech V., 1975. Croissance, mortalité et production de *Brachysynodontis batensoda* (Pisces, Mochocidae) dans l'Archipel sud-est du lac Tchad. *Cahiers ORSTOM, série Hydrobiologie*, 8: 23–33.

Bénech V. & Lek S., 1981. Résistance à l'hypoxie et observations écologiques pour seize espéces de poissons du Tchad. *Revue d'Hydrobiologie tropicale*, 14: 153–68.

Bénech V. & Ouattara S., 1990. Rôle des variations de conductivité de l'eau et d'autres facteurs externes dans la croissance ovarienne d'un poisson tropical, *Brycinus leuciscus* (Characidae). *Aquatic Living Resources*, 3: 153–62.

Bénech V. & Quensière J., 1982. Migration de poissons vers le lac Tchad à la décrue de la plaine inondée du Nord-Cameroun. 1- Méthodologie d'échantillonnage et résultats généraux. *Revue d'Hydrobiologie tropicale*, 15: 253–70.

Bénech V. & Quensière J., 1983*a*. Migration de poissons vers le lac Tchad à la décrue de la plaine inondée du Nord-Cameroun. 2- Comportement et rythme d'activité des principales espéces. *Revue d'Hydrobiologie tropicale*, 16: 79–101.

Bénech V. & Quensière J., 1983*b*. Migration de poissons vers le lac Tchad à la décrue de la plaine inondée du Nord-Cameroun. 3- Variations annuelles en fonction de l'hydrologie. *Revue d'Hydrobiologie tropicale*, 16: 287–316.

Bénech V. & Quensière J., 1985. Stratégies de reproduction des poissons du Tchad en période 'Tchad normal' (1966–1971). *Revue d'Hydrobiologie tropicale*, 18: 227–44.

Bénech V. & Quensière J., 1989. *Dynamique des peuplements ichtyologiques de la région du lac Tchad (1966–1978). Influence de la sécheresse sahélienne.* Travaux et Documents Microédités, No. 51. ORSTOM, Paris.

Bénech V., Lemoalle J. & Quensière J., 1976. Mortalité de poissons et conditions de milieu dans le lac Tchad au cours d'une période de sécheresse. *Cahiers ORSTOM, série Hydrobiologie*, 10: 119–30.

Bénech V., Franc J. & Matelet P., 1978. Utilisation du chalut électrifié pour l'échantillonnage des poissons en milieu tropical (Tchad). *Cahiers ORSTOM, série Hydrobiologie*, 12: 197–224.

Bénech V., Quensière J. & Vidy G., 1982. Hydrologie et physicochimie des eaux de la plaine d'inondation du Nord-Cameroun. *Cahiers ORSTOM, série Hydrologie*, 19: 15–35.

Bénech V., Durand J. R. & Quensière J., 1983. Fish communities of Lake Chad and associated rivers and floodplains. In Carmouze J. P., Durand J. R. & Lévêque C. (eds.), *Lake Chad: Ecology and Productivity of a Shallow Tropical Ecosystem*, pp. 293–356, Monographiae Biologicae 53. W. Junk Publishers, The Hague.

Benford G., 1992. Saving the 'library of life'. *Proceedings of the National Academy of Sciences of the United States of America*, 89: 11098–101.

Bennett M. V. L., 1971*a*. Electric organs. In Hoar W. S. & Randall D. J. (eds.), *Fish Physiology*, pp. 347–491. Vol. 5, Sensory Systems and Electric Organs. Academic Press, New York, London.

Bennett M. V. L., 1971*b*. Electroreceptors. In Hoar W.S. & Randall D. J. (eds.), *Fish Physiology*, pp. 493–574. Vol. 5, Sensory Systems and Electric Organs. Academic Press, New York, London.

Berkes F. & Farvar M. T., 1989. Introduction and overview. In Berkes F. (ed.), *Common Property Resources. Ecology and Community-based Sustainable Development*, pp. 1–17. Belhaven, London and Columbia University Press, New York.

Berkes F. & Kislalioglu M., 1991. Community-based management and sustainable development. In Durand J. R., Lemoalle J. & Weber J. (eds.), *La recherche face à la pêche artisanale*, pp. 567–574. ORSTROM, Paris.

Bernacsek G. M., 1984. *Dam Design and Operation to Optimise Fish Production in Impounded River Basins.* CIFA Technical Paper, 11. FAO, Rome.

Bernacsek G. M., 1987. *Policy Options for Development of the Fisheries Sector in Africa.* FAO, Fisheries Department, Fishery Policy and Planning Division, Rome.

Bernacsek G. M., 1990. Fisheries development and wetland conservation in Africa. *Journal of West African Fisheries*, 3: 121–35.

Berra T. M., 1981. *An Atlas Distribution of the Freshwater Fish Families of the World.* University of Nebraska Press, Lincoln.

Berrebi P., Lévêque C., Cattaneo-Berrebi G., Agnèse J. F., Guégan J. F. & Machordom A., 1990. Diploid and tetraploid African *Barbus* (Osteichthyes, Cyprinidae): on the coding of differential gene expression. *Aquatic Living Resources*, 3: 313–23.

Berruti A., 1983. The biomass, energy consumption and breeding of waterbirds relative to hydrological conditions at Lake St Lucia. *Ostricht*, 54: 65–82.

Bertin L., 1958. Systématique des poissons. In Grassé P. P. (ed), *Traité de Zoologie*, XIII, pp. 2204–500. Masson, Paris.

Beverton R. J. H., 1992. Fish resources; threats and protection. *Netherlands Journal of Zoology*, **42**: 139–75.

Bigorne R., 1987. Le genre *Mormyrops* (Pisces, Mormyridae) en Afrique de l'Ouest. *Revue d'Hydrobiologie tropicale*, **20**: 145–64.

Bigorne R., 1989. Les genres *Brienomyrus* et *Isichthys* (Pisces, Mormyridae) en Afrique de l'Ouest. *Revue d'Hydrobiologie tropicale*, **22**: 317–38.

Bigorne R., 1990a. Révision systématique du genre *Pollimyrus* (Teleostei, Mormyridae) en Afrique de l'Ouest. *Revue d'Hydrobiologie tropicale*, **23**: 313–27.

Bigorne R., 1990b. Mise en synonymie de *Gnathonemus brevicaudatus* Pellegrin, 1919 avec *Gnathonemus petersii* Günther, 1862 (Teleostei, Mormyridae). *Cybium*, **14**: 125–9.

Bigorne R. & Paugy D., 1991. Note sur la systématique des *Petrocephalus* (Teleostei: Mormyridae) d'Afrique de l'Ouest. *Ichthyological Exploration of Freshwaters*, **2**: 1–30.

Billard R., 1986. Spermatogenesis and spermatology of some teleost fish species. *Reproduction, Nutrition, Development*, **26**: 877–920.

Billard R. & Breton B., 1978. In Thorpe (ed.), *Rhythmic Activity of Fishes*, pp. 31–53. Academic Press, London.

Biney C., Amuzu A. T., Calamari D., Kaba N., Mbome I. L., Naeve H., Ochumba P. B. O., Osibanjo O., Redegonde V. & Saad M. A. H., 1992. Review of heavy metals in the African aquatic environment. *CIFA, FAO Fisheries Report*, No. 471, pp. 7–43. FAO, Rome.

Bishai H. M. & Abu Gideiri Y. B., 1965. Studies on the biology of the genus *Synodontis* at Khartoum. I–Age and growth. *Hydrobiologia*, **26**: 87–97.

Bishop R. C. & Ready R. C., 1991. Endangered species and the safe minimum standard. *American Journal of Agricultural Economics*, **73**: 309–12.

Biswas A. K. & Biswas M. R., 1976. Hydropower and the environment. *Water Power & Dam Construction*. May 1976, pp. 40–3.

Blache J. & Miton F., 1962. *Première contribution à la connaissance de la pêche dans le bassin hydrographique Logone-Chari-lac Tchad*. Mémoire ORSTOM, 4(1).

Blache J., Miton F., Stauch A., Iltis A. & Loubens G., 1964. *Les poissons du bassin du Tchad et du bassin adjacent du Mayo-Kebi. Etude systématique et biologique*. Mémoire ORSTOM, 4(2).

Blackman R. I. & Day M. C., 1981. Introduction. In Forey P. L. (ed.), *The Evolving Biosphere*, pp. 3–7. British Museum (Natural History) and Cambridge University Press, Cambridge.

Blake B. F., 1977a. Food and feeding of the mormyrid fishes of Lake Kainji, Nigeria, with special reference to seasonal variation and interspecific differences. *Journal of Fish Biology*, **11**: 315–28.

Blake B. F., 1977b. Aspects of the reproductive biology of *Hippopotamyrus pictus* from L. Kainji, with notes on four other mormyrids species. *Journal of Fish Biology*, **11**: 437–45.

Blake B. F., 1977c. The effect of the impoundment of Lake Kainji, Nigeria, on the indigenous species of mormyrid fishes. *Freshwater Biology*, **7**: 37–42.

Blanc M., 1954. La répartition des poissons d'eau douce africains. *Bulletin de l'Institut français d'Afrique noire*, **16**: 599–628.

Blanck J. P., 1968. Schéma d'évolution géomorphologique de la vallée du Niger entre Tombouctou et Labbezanga (Rép. du Mali). *Bulletin de l'Association française pour l'étude du Quaternaire, Paris*, **19–20**: 17–26.

Blondel J., 1986. *Biogéographie Évolutive*. Collection d'Écologie No. 20, Masson, Paris.

Blondel J., 1987. From biogeography to life history theory: a multithematic approach illustrated by the biogeography of vertebrates. *Journal of Biogeography*, **14**: 405–22.

Blumer L. S., 1979. Male parental care in the bony fishes. *Quarterly Review of Biology*, **54**: 149–61.

Blumer L. S., 1982. A bibliography and categorization of bony fishes exhibiting parental care. *Zoological Journal of the Linnean Society, London*, **76**: 1–22.

Bonnefille R., 1993. Afrique, paléoclimats et déforestation. *Sécheresse*, **4**: 221–31.

Bonnefille R., Roeland J. C. & Guiot J., 1990. Temperature and rainfall estimates for the past 40 000 years in equatorial Africa. *Nature*, **346**: 347–9.

Bonou C. A. & Teugels G. G., 1985. Révision systématique du genre *Parachanna* Teugels & Daget, 1984 (Pisces, Channidae). *Revue d'Hydrobiologie tropicale*, **18**: 267–80.

Bootsma H. A., 1992. Lake Malawi National Park: an overview. *Mitteilungen-Internationale Vereinigung für Theoretische und Angewandte Limnologie*, **23**: 125–8.

Bootsma H. A. & Hecky R. E., 1993. Conservation of the African Great Lakes: a limnological perspective. *Conservation Biology*, **7**: 644–56.

Boulenger G., 1905. The distribution of African fresh-water fishes. *Nature*, **72**: 413–21.

Boulenger G., 1907. *Zoology of Egypt. The Fishes of the Nile*, 2 vols. Published for the Egyptian Government. London.

Boulenger G., 1909–16. *Catalogue of the Freshwater Fishes of Africa in the British Museum (Natural History)*. Vol.

I-IV. Trustees of the British Museum (Natural History), London.

Bourdeau P., Haines J. A., Klein W. & Krishna Murti C.R. (eds.), 1989. *Ecotoxicology and Climate. With Special Reference to Hot and Cold Climates.* SCOPE38, IPCS Joint Symposia 9. John Wiley & Sons, Chichester.

Boureau E., Cheboldaeff-Salard M., Koeniguer J. C. & Louvet P., 1983. Evolution des flores et de la végétation Tertiaires en Afrique, au nord de l'Equateur. *Bothalia*, 14: 355–67.

Bovee K. D., 1982. A Guide to Stream Habitat Analysis Using the Instream Flow Incremental Methodology. Instream Flow Information Paper No. 12. USDI Fish and Wildlife Service, Office of Biological Services. FWS/OBS-77/63.

Bowen S. H., 1976. Mechanism for digestion of detrital bacteria by the cichlid fish *Sarotherodon mossambicus* (Peters). *Nature*, 260: 137–8.

Bowen S. H., 1979. A nutritional constraint in detrivory by fishes: the stunted population of *Sarotherodon mossambicus* in Lake Sibaya, South Africa. *Ecological Monographs*, 49: 17–31.

Bowen S. H., 1982. Feeding, digestion and growth-qualitative considerations. In Pullin R. V. S. & Lowe-McConnell R. H. (eds.), *The Biology and Culture of Tilapias*, pp. 141–56. ICLARM Conference Proceedings 7, Manula.

Bowen S. H., 1988. Detrivory and herbivory. In Lévêque C., Bruton M. N. & Ssentongo G. W. (eds.), *Biologie et écologie des poissons d'eau douce africains*, pp. 243–7. ORSTOM, Paris.

Bowers N., Stauffer J. R. & Kocher T. D., 1994. Intra and interspecific mitochondrial DNA sequence variation within two species of rock-dwelling cichlids (teleostei: Cichlidae) from Lake Malawi, Africa. *Molecular Phylogenetics and Evolution*, 3: 75–82.

Bowmaker A. P., 1963. Cormorant predation in two Central African lakes. *Ostricht*, 34: 2–26.

Bowmaker A. P., 1969. Fish population fluctuations associated with anadromesis at the Mwenda river mouth, Lake Kariba. *Newsletter of the Limnological Society of Southern Africa*, 1: 45–9

Bowmaker A. P., 1973. Potamodromesis in the Mwenda River, Lake Kariba. *Geophysical Monograph*, 17: 159–64.

Bowmaker A. P., Jackson P. B. N. & Jubb R. A., 1978. Freshwater fishes. In Werger M. J. A. & Van Bruggen A. C. (eds.), *Biogeography and Ecology of Southern Africa*, pp. 1181–230. W. Junk Publishers, The Hague.

Breder C. M. & Rosen D. E., 1966. *Modes of Reproduction in Fishes.* Natural History Press, Garden City, New York.

Brem G., Brenig G., Horstgen-Schark G. & Winnacker E.-L., 1988. Gene transfer in tilapia (*Oreochromis niloticus*). *Aquaculture*, 68: 209–19.

Brett J. R., 1979. Environmental factors and growth. In Hoar W. S., Randall D. J. & Brett J. R. (eds.), *Fish Physiology*, Vol. VIII, pp. 599–675. Academic Press, London.

Brett J. R. & Groves T. D. D., 1979. Physiological energetics. In Hoar W. S., Randall D. J. & Brett J. R. (eds.), *Fish Physiology*, Vol. VIII, pp. 279–352. Academic Press, London.

Brewster B., 1986. A review of the genus *Hydrocynus* Cuvier, 1819 (Teleostei, Characiformes). *Bulletin British Museum (Natural History)* Zoology, 50: 163–206.

Brichard P., 1978. *Fishes of the Lake Tanganyika*. T.H.F. Publications, Neptune City.

Brichard P., 1989. *Cichlids of Lake Tanganyika*. T.H.F. Publications, Neptune City, NJ.

Briggs J.C., 1979. Ostariophysian zoogeography: an alternative hypothesis. *Copeia*, 1979: 111–18.

Brooks D. R. & McLennan D. A., 1991. *Phylogeny, Ecology and Behaviour.* The University of Chicago Press, Chicago.

Brosset A., 1982. Le peuplement de cyprinodontes du bassin de l'Ivindo, Gabon. *Revue d'Ecologie (Terre et Vie)*, 36: 233–92.

Brown J. H., 1988. Species diversity. In Myers A. A. & Giller P. S. (eds.), *Analytical Biogeography: An Integrated Approach to the Study of Animal and Plant Distributions*, pp. 57–89. Chapman & Hall, London.

Brown J. H., 1995. *Macroecology.* The University of Chicago Press, Chicago.

Brown L. H. & Urban E. K., 1969. The breeding biology of the Great White Pelican *Pelecanus onocrolatus roseus* at Lake Shala, Ethiopia. *Ibis*, 111: 199–237.

Brundin L. Z., 1988. Phylogenetic biogeography. In Myers A. A. & Giller P. S. (eds.), *Analytical Biogeography: An Integrated Approach to the Study of Animal and Plant Distributions*, pp. 343–69. Chapman & Hall, London.

Bruton M. N., 1978. The habitats and habitat preferences of *Clarias gariepinus* (Pisces: Clariidae) in a clear coastal lake (Lake Sibaya, South Africa). *Journal of the Limnological Society of Southern Africa*, 4: 81–8.

Bruton M. N., 1979a. The breeding biology and early development of *Clarias gariepinus* (Pisces: Clariidae) in Lake Sibaya, South Africa, with a review of breeding in species of the subgenus *Clarias (Clarias)*. *Transactions of the Zoological Society of London*, 35: 1–45.

Bruton M. N., 1979b. The food and feeding behaviour of *Clarias gariepinus* (Pisces, Clariidae) in Lake Sibaya, South Africa, with emphasis on its role as a predator of cichlids. *Transactions of the Zoological Society of London*, 35: 47–114.

Bruton M. N., 1979*c*. The role of diel inshore movements by *Clarias gariepinus* (Pisces: Clariidae) for the capture of fish prey. *Transactions of the Zoological Society of London*, 35: 115–38.

Bruton M. N., 1979*d*. The survival of habitat desiccation by air breathing clariid catfishes. *Environmental Biology of Fishes*, 4: 273–80.

Bruton M. N., 1985. The effects of suspendoids on fish. *Hydrobiologia*, 125: 221–41.

Bruton M. N., 1989. The ecological significance of alternative life-history styles. In Bruton M. N. (ed.), *Alternative Life-history Styles of Animals*, pp. 503–33. Perspectives in Vertebrates Sciences 6. Kluwer Academic Publishers, Dortrecht.

Bruton M. N., 1990*a*. The conservation of alternative life-history styles: a conclusion to the second ALHS volume. In Bruton M. N. (ed.), *Alternative Life-history Styles of Animals*, pp. 309–13. Perspectives in Vertebrates Sciences 6. Kluwer Academic Publishers, Dortrecht.

Bruton M. N., 1990*b*. The conservation of the fishes of Lake Victoria, Africa: an ecological perspective. *Environmental Biology of Fishes*, 27: 161–75.

Bruton M. N. & Allanson B. R., 1980. Growth of *Clarias gariepinus* in Lake Sibaya, South Africa. *Suid-Afrikaanse tydskrif vir dierkunde*, 15: 7–15.

Bruton M. N. & Boltt R. E., 1975. Aspects of the biology of *Tilapia mossambica* Peters (Pisces, Cichlidae) in a natural freshwater lake (Lake Sibaya, South Africa). *Journal of Fish Biology*, 7: 423–46.

Bruton M. N. & Gophen M., 1992. The effect of environmental factors on the nesting and courtship behaviour of *Tilapia zillii* in Lake Kinnereth (Israel). *Hydrobiologia*, 239: 171–8.

Bruton M. N. & Kok H. M., 1980. The freshwater fishes of Maputaland. In Bruton M. N. & Cooper K. H. (eds.), *Studies on the Ecology of Maputaland*, pp. 210–44. Rhodes University, Grahamstown & Natal Branch of Wildlife Society of southern Africa, Durban.

Bruton M. N. & Merron G. S., 1990. The proportion of different eco-ethological sections of reproductive guilds of fishes in some African inland waters. *Environmental Biology of Fishes*, 28: 179–87.

Budgett J. S., 1900. The breeding habits of *Polypterus*, *Gymnarchus* and some other West African fishes. *Science*, 11: 1015.

Bullock T. H., 1982. Electroreception. *Annual Review of Neuroscience*, 51: 121–70.

Burgis M. J. & Symoens J. J., 1987. *African Wetlands and Shallow Water Bodies. Directory*. Travaux et Documents, No. 211. ORSTOM, Paris.

Burgis M. J., Darlington J. P. E. C., Dunn I. G., Ganf G. G., Gwahaba J. J. & McGovan L. M., 1973. Biomass and distribution of organisms in Lake George, Uganda. *Proceedings of the Royal Society of London*, B, 184: 227–346.

Buth D. G., Dowling T. E. & Gold J. R., 1991. Molecular and cytological investigations. In Winfield I. J. & Nelson J. S. (eds.), *Cyprinid fishes: Systematics, Biology and Exploitation*, pp. 83–126. Fish and Fisheries Series 3. Chapman & Hall, London.

Butler G. C. (ed.), 1978. *Principles of Ecotoxicology*. SCOPE 12. John Wiley & Sons, Chichester.

Butzer K. W., 1980. Pleistocene History of the Nile valley in Egypt and Lower Nubia. In Williams F. M. & Faure H. (eds.), *The Sahara and the Nile*, pp. 248–76. Balkema, Rotterdam.

Bwathondi P. O. J., 1985. The future of fisheries in the Tanzanian part of Lake Victoria in view of the wide spread of Nile perch (*Lates niloticus*). *FAO Fisheries Report*, 335: 143–5.

Bye V. J., 1984. The role of environmental factors in the timing of reproductive cycle. In Potts G. W. & Wootton R. J. (eds.), *Fish Reproduction: Strategies and Tactics*, pp. 187–205. Academic Press, London.

Cadwalladr D. A., 1965*a*. Notes on the breeding biology and ecology of *Labeo victorianus* Blgr (Cyprinidae) of Lake Victoria. *Revue de Zooloogie et de Botanique africaine*, 72: 109–34.

Cadwalladr D. A., 1965*b*. The decline of *Labeo victorianus* Blg fishery of Lake Victoria and an associated deterioration in some indigenous fishing methods in the Nzoia River, Kenya. *East African Agricultural and Forestry Journal*, 30: 249–56.

Cadwalladr D. A. & Stoneman J., 1966. A review of the fisheries of Uganda waters of Lake Albert, East Africa, 1928–1955/66, with catch data mainly from 1953. East African Freshwater Fisheries Research Organisation, Supplement to publication No. 3.

Cahen L., 1954. *Géologie du Congo belge*. Vaillant-Carmanne, Liège.

Cairns J., McCormick P. V. & Niederlehner B. R., 1993. A proposed framework for developing indicators of ecosystem health. *Hydrobiologia*, 263: 1–44.

Cairns M. A. & Lackey R. T., 1992. Biodiversity and management of natural resources: the issues. *Fisheries*, 17: 6–10.

Calamari D., 1985. *Situation de la Pollution des Eaux Intérieures de l'Afrique de l'Ouest et du Centre*. FAO,

Document Occasionnel du CPCA, No. 12, CPCA/OP 12.

Caljon A., 1992. Water quality in the Bay of Bujumbura (Lake Tanganyika) and its influence on phytoplankton composition. *Mitteilungen-Internationale Vereinigung für Theoretische und Angewandte Limnologie*, 23: 71–8.

Callicott J. B., 1991. Conservation ethics and fishery management. *Fisheries*, 16: 22–8.

Cambray J. A., 1983. The feeding habits of minnows of the genus *Barbus* in Africa, with special reference to *Barbus anoplus* Weber. *Journal of the Limnological Society of Southern Africa*, 9: 12–22.

Cambray J. A., 1984. Fish populations in the middle and lower Orange River, with special reference to the effects of stream regulation. *Journal of the Limnological Society of Southern Africa*, 10: 37–49.

Cambray J. A., 1985. Observations on spawning of *Labeo capensis* and *Clarias gariepinus* in the regulated lower Orange River, South Africa. *South African Journal of Science*, 81: 318–21.

Cambray J. A. & Teugels G. G., 1988. Selected annotated bibliography of early developmental studies of African freshwater fishes. *Annals of the Cape Provincial Museums, Natural History*, 18: 31–56.

Cappetta H., Jaeger J. J., Sabatier M., Sigé B., Sudre J. & Vianey-Liaud M., 1978. *Geobios*, Lyon, 11: 257–63.

Carmouze J. P., 1976. *La Régulation Hydrogéochimique du Lac Tchad*. Travaux et Documents No. 58. ORSTOM, Paris.

Carmouze J. P., Dejoux C., Durand J. R., Gras R., Lauzanne L., Lemoalle J., Lévêque C., Loubens G. & Saint-Jean L., 1972. Grandes zones écologiques du lac Tchad. *Cahiers ORSTOM, série Hydrobiologie*, 6: 103–69

Carmouze J. P., Durand J. R. & Lévêque C. (eds.), 1983. *Lake Chad: Ecology and Productivity of a Shallow Tropical Ecosystem*. Monographiae Biologicae 53, W. Junk Publishers, The Hague.

Carpenter S. R., Kitchell J. F. & Hodgson J. R., 1985. Cascading trophic interactions and lake productivity. Fish predation and herbivory can regulate lake ecosystems. *BioScience*, 35: 634–9.

Caswell H. & Cohen J. E., 1991. Communities in patchy environments: a model of disturbance, competition and heterogeneity. In Kolasa J. & Pickett S. T. A., *Ecological Heterogeneity*, pp. 97–122. Springer-Verlag, New York.

Cataudella S., Sola L. & Capanna E., 1978. Remarks on the karyotype of the Polypteriformes: the chromosomes of *Polypterus delhezi*, *P. endlicheri congicus* and *P. palmas*. *Experientia*, 34: 999–1000.

Caulton M. S., 1975. Diurnal movement and temperature selection by juvenile and sub adult *Tilapia rendalli*

Boulenger. *Transactions of the Rhodesian Science Association*, 56: 51–6.

Caulton M. S., 1976. A quantitative assessment of the daily digestion of *Panicum repens* L. by *Tilapia rendalli* Boulenger (Cichlidae) in Lake Kariba. *Transactions of the Rhodesian Science Association*, 58: 38–42.

Caulton M. S., 1977. The effect of temperature on routine metabolism in *Tilapia rendalli* Boulenger. *Journal of Fish Biology*, 11: 549–53.

Caulton M. S., 1978. The effect of temperature and mass on routine metabolism in *Sarotherodon (Tilapia) mossambicus* (Peters). *Journal of Fish Biology*, 13: 195–201.

Caulton M. S., 1982. Feeding, metabolism and growth of tilapias. Some quantitative considerations. In Pullin R. V. S. & Lowe-McConnell R. (eds.), *The Biology and Culture of Tilapias*, pp. 157–80. ICLARM Conference Proceeding 7, Manila.

Cavender T. M., 1991. The fossil record of the Cyprinidae. In Winfield I. J. & Nelson J. S. (eds.), *Cyprinid fishes. Systematics, Biology and Exploitation*, pp. 34–54. Fish and Fisheries Series 3. Chapman & Hall, London.

Cézilly F., Brun B. & Hafner H., 1991. Foraging and fitness. *Acta Oecologica*, 12: 683–96.

Chaika J. J., 1982. 'Planning development of the fisheries and implications of management: Malawi, a case study.' CIFA Technical Paper, FAO, Rome. No. 8, pp. 52–67.

Chan T. Y. & Ribbink A. J., 1990. Alternative reproductive behaviour in fishes, with particular reference to *Lepomis macrochira* and *Pseudocrenilabrus philander*. *Environmental Biology of Fishes*, 28: 249–56.

Chao H. N., Chao W. C., Liu K. C. & Liao C. I., 1987. The properties of tilapia sperm and its cryopreservation. *Journal of Fish Biology*, 30: 107–18.

Chapman D. W. & van Well P., 1978. Growth and mortality of *Stolothrissa tanganicae*. *Transactions of the American Fisheries Society*, 107: 26–35.

Charles-Dominique E., 1989. Catch efficiencies of purse and beach seines in Ivory Coast lagoons. *Fishery Bulletin*, 87: 911–21.

Chen F.Y., 1969. Preliminary studies on the sex determining mechanisms of *Tilapia mossambica* Peters and *T. hornorum* Trewavas. *Verhandlungen-Internationale Vereinigung für Theoretische und Angewandte Limnologie*, 17: 719–24.

Chervinski J., 1972. Occurrence of *Tilapia zillii* (Gervais) (Pisces, Cichlidae) in the Bardawil lagoon in northern Sinai. *Bamidgeh*, 24: 49–50.

Chourrout D., 1991a. Revue sur le déterminisme génétique du sexe des poissons téléostéens. *Bulletin de la Société zoologique de France*, 116: 123–44.

Chourrout D., 1991*b*. La transgénèse chez les poissons. *Bulletin de la Société zoologique de France*, **116**: 151–7.

Chourrout D., 1993. Transgenic technology in fish. *Biology International* (special issue), **28**: 99–106.

Chourrout D., Chevassus B., Krieg F., Happe A., Burger G & Renard P., 1986. Production of second generation triploid and tetraploid rainbow trout by mating tetraploid males and diploid females. *Theoretical and Applied Genetics*, **72**: 193–206.

Cichocki F. P., 1976. Cladistic history of cichlid fishes and reproductive strategies of the American genera *Acarichthys, Biuotodoma* and *Geophagus*. Ph.D. thesis, University of Michigan, Ann Arbor, MI.

Clark C. W., 1989. Bioeconomics. In Roughgarden J., May R. M. & Levin S. A. (eds.), *Perspectives in Ecological Theory*, pp. 275–86. Princeton University Press, Princeton, NJ.

Claus F., 1930. La capture des poissons au moyen de plantes toxiques. *Bulletin agronomique du Congo belge*, **21**: 1095–114.

Cochrane K. L., 1978. Seasonal fluctuations in the catches of *Limnothrissa miodon* (Boulenger) in Lake Kariba. *Lake Kariba Fishery Research Institute, Project Report*, **29**: 158–62.

Coe M. J., 1966. The biology of *Tilapia grahami* Boulenger in Lake Magadi, Kenya. *Acta Tropica*, **23**: 158–62.

Coe M. J., 1969. Observations on *Tilapia alcalica* Hilgendorf an endemic cichlid fish from Lake Natron, Tanzania. *Revue de Zoologie et de Botanique africaine*, **80**: 1–14.

Cogels F. X., 1984. Etude limnologique d'un lac sahélien. Le lac de Guiers (Sénégal). Dissertation pour obtention du grade de Docteur en Sciences de l'Environnement. Fondation Universitaire Luxembourgeoise, Arlon, Belgique.

Cohen A. (ed.), 1991. *Report on the First International Conference on Conservation and Biodiversity of Lake Tanganyika*. Biodiversity Support Program, Washington, DC.

Cohen A. S., 1992. Criteria for developing viable underwater natural reserves in Lake Tanganyika. *Mitteilungen-Internationale Vereinigung für Theoretische und Angewandte Limnologie*, **23**: 109–16.

Cohen A. S., Bills R., Cocquyt C. Z. & Caljon A. G., 1993. The impact of sediment pollution on biodiversity in Lake Tanganyika. *Conservation Biology*, **7**: 667–77.

Cohen J. E., 1989. Food webs and community structure. In Roughgarden J., May R. M. & Levin S. A. (eds.), *Perspectives in Ecological Theory*, pp. 181–202. Princeton University Press, Princeton, NJ.

Colwell R. K., 1979. Towards a unified approach to the study of species diversity. In Grassle J. F., Patil G. P. & Taillie C. (eds.), *Ecological Diversity in Theory and Practice*. International Cooperative Publishing House, Fairland, MD.

Connell J. H., 1978. Diversity in tropical rainforest and coral reefs. *Science*, **199**: 1302–10.

Connell J. H. & Sousa W. P., 1983. On the evidence needed to judge ecological stability or persistence. *American Naturalist*, **121**: 789–824.

Connor E. F. & McCoy E. D., 1980. The statistics and biology of the species area relationship. *American Naturalist*, **113**: 791–833.

Copley H., 1958. *Common Freshwater Fishes of East Africa*. H.F & G. Witherby Ltd, London.

Corbet P. S., 1961. The food of non-cichlid fishes in lake Victoria Basin, with remarks on their evolution and adaptation to lacustrine conditions. *Proceedings of the Zoological Society of London*, **136**: 1–101.

Cornell H. V. & Lawton J. H., 1992. Species interactions, local and regional processes, and limits to the richness of ecological communities: a theoretical perspective. *Journal of Animal Ecology*, **61**: 1–12.

Coulter G. W., 1966. The deep benthic fishes of the South of Lake Tanganyika with special reference to distribution and feeding in *Bathybates* species, *Hemibathes stenosoma* and *Chrysichthys* species. *Fishery Research Bulletin, Zambia*, **4**: 33–8.

Coulter G. W., 1970. Population changes within a group of fish species in Lake Tanganyika, following their exploitation. *Journal of Fish Biology*, **2**: 329–53.

Coulter G. W., 1981. Biomass, production and potential yield of the L. Tanganyika pelagic fish community. *Transactions of the American Fisheries Society*, **110**: 325–35.

Coulter G. W. (ed.), 1991*a*. *Lake Tanganyika and its Life*. Natural History Museum Publications and Oxford University Press, Oxford.

Coulter G. W., 1991*b*. Pelagic fish. In Coulter G.W (ed.), *Lake Tanganyika and its Life*, pp. 111–38. Natural History Museum Publications and Oxford University Press, London.

Coulter G. W., 1992. Vulnerability of Lake Tanganyika to pollution, with comments on social aspects. *Mitteilungen-Internationale Vereinigung für Theoretische und Angewandte Limnologie*, **23**: 67–70.

Coulter G. W., 1994*a*. Lake Tanganyika. In Martens K. F., Goddeeris B. & Coulter G. (eds.), *Speciation in Ancient Lakes*, pp. 13–18. *Archiv für Hydrobiologie*, **44**.

Coulter G. W., 1994*b*. Speciation and fluctuating environments, with reference to ancient East African lakes. In Martens K. F., Goddeeris B. & Coulter G. (eds.), *Speciation in Ancient Lakes*, pp. 127–37. *Archiv für Hydrobiologie*, **44**.

Coulter G. W. & Mubamba R., 1993. Conservation in Lake Tanganyika with special reference to underwater parks. *Conservation Biology*, **7**: 678–85.

Coulter G. W. & Spigel R. H., 1991. Hydrodynamics. In Coulter G.W (ed.), *Lake Tanganyika and its Life*, pp. 49–75. Natural History Museum Publications, Oxford University Press, Oxford.

Coulter G. W., Allanson B. R., Bruton M. N., Greenwood P. H., Hart R. C., Jackson P. B. N. & Ribbink A. J., 1986. Unique qualities and special problems of the African Great Lakes. *Environmental Biology of Fishes*, **17**: 161–84.

Countant C. C., 1987. Thermal preference: when does an asset become a liability. *Environmental Biology of Fishes*, **18**: 161–72.

Courtenay S. C. & Keenleyside M. H. A., 1983. Wriggler-hanging: a response to hypoxia by brood-rearing *Heterotilapia multispinosa* (Teleostei, Cichlidae). *Behaviour*, **85**: 183–97.

Cracraft J., 1985. Biological diversification and its causes. *Annals of the Missouri Botanical Garden*, **72**: 794–822.

Craig J. F., 1992. Human-induced changes in the composition of fish communities in the African Great lakes. *Reviews in Fish Biology and Fisheries*, **2**: 93–124.

Crapon de Crapona M-D., 1980. Olifactory communication in a cichlid fish *Haplochromis burtoni*. *Zeitschrift für Tierpsychologie*, **52**: 113–34.

Crapon de Caprona M-D. & Fritzsch B., 1984. Interspecific fertile hybrids of haplochromine cichlids (Teleostei) and their possible importance for speciation. *Netherlands Journal of Zoology*, **34**: 503–38.

Craw R., 1988. Panbiogeography: method and synthesis in biogeography. In Myers A. A. & Giller P. S. (eds.), *Analytical Biogeography: An Integrated Approach to the Study of Animal and Plant Distributions*, pp. 405–35. Chapman & Hall, London.

Croft T. A., 1981. *Lake Malawi National Park Master Plan*. Malawi Department of National Parks and Wildlife.

Crow J., 1976. *Genetics Notes*. Burgess Publishing Co., Minneapolis, MN.

Dadnadji K. K. & Wetten C. J. van, 1993. Traditional management systems and integration of small scale interventions in the Logone floodplains of Chad. In Davis T. J.(ed.), *Towards the Wise Use of Wetland*s. Report of the Ramsar Convention Wise Use Projects. Ramsar Convention Bureau, Gland, Switzerland.

Daget J., 1952. Mémoires sur la biologie des poissons du Niger. I. Biologie et croissance des espèces du genre

Alestes. *Bulletin de l'Institut français d'Afrique noire*, **14**: 191–225.

Daget J., 1954. Les poissons du Niger supérieur. *Mémoire de l'Institut français d'Afrique noire*, **36**.

Daget J., 1957. Données récentes sur la biologie des poissons dans le delta centra du Niger. *Hydrobiologia*, **9**: 321–47.

Daget J., 1959a. Notes sur les poissons du Borkou-Ennedi-Tibesti. *Travaux de l'Institut de Recherches Sahariennes, Univresité d'Alger*, **18**: 173–81.

Daget J., 1959b. Restes de *Lates niloticus* (Poissons, Centropomidae) du Quaternaire saharien. *Bulletin de l'Institut français d'Afrique noire (A)*, **21**: 1103–11.

Daget J., 1960. Les migrations de poissons dans les eaux douces tropicales africaines. *Proceedings IPFC*, **8**: 79–82.

Daget J., 1961a. Note sur les *Nannocharax* (Poissons Characiformes) de l'Ouest africain. *Bulletin de l'Institut français d'Afrique noire (A)*, **23**: 165–81.

Daget J., 1961b. Restes de poissons du Quaternaire saharien. *Bulletin de l'Institut français d'Afrique noire (A)*, **23**: 182–91.

Daget J., 1962a. Le genre *Citharidium* (Poissons Characiformes). *Bulletin de l'Institut français d'Afrique noire (A)*, **24**: 505–22.

Daget J., 1962b. Le genre *Citharinus* (Poissons Characiformes). *Revue de Zoologie et de Botanique africaine*, **66**: 81–106.

Daget J., 1962c. Les poissons du Fouta Djalon et de la basse Guinée. *Mémoires de l'Institut français d'Afrique noire*, **65**: 1–210.

Daget J., 1963. Sur plusieurs cas probables d'hybridation naturelle entre *Citharidium ansorgii* et *Citharinus distichodoides*. *Mémoires de l'Institut français d'Afrique noire*, **68**: 81–3.

Daget J., 1966. Taxonomie numérique des Citharininae (Poissons Characiformes). *Bulletin du Muséum national d'Histoire naturelle, Paris*, (2)38(4): 376–86.

Daget J., 1967. Le genre *Ichthyoborus* (Poissons Characiformes). *Cahiers ORSTOM, série Hydrobiologie*, **1**: 141–5.

Daget J., 1968a. Contribution à l'étude des eaux douces de l'Ennedi. IV- Poissons. *Bulletin de l'Institut français d'Afrique noire (A)*, **30**: 1582–9.

Daget J., 1968b. Le genre *Hemidistichodus* (Poissons Characiformes). *Cahiers ORSTOM, série Hydrobiologie*, **2**: 297–309.

Daget J., 1978. Contribution à la faune de la République Unie du Cameroun. Poissons du Dja, du Boumba et du Ngoko. *Cybium*, 3° série (3): 35–52.

Daget J., 1979. Contribution à la faune de la République

Unie du Cameroun. Poissons de l'Ayina, du Dja et du Bas Sanaga. *Cybium*, 3° série (6): 55–64.

Daget J., 1988. Systématique. In Lévêque C., Bruton M. N. & Ssentongo G. W. (eds.), *Biology and Ecology of African Freshwater Fishes*, pp. 15–34. ORSTOM, Paris.

Daget J., 1994. Aperçu historique sur l'ichtyologie africaine. In Teugels G. G., Guégan J. F. & Albaret J. J. (eds.), *Biological Diversity of African Fresh- and Brackish Water Fishes*, pp. 17–19. *Annales Sciences zoologiques*, Vol. 275, Musée royal d'Afrique centrale, Tervuren, Belgique.

Daget J. & Depierre D., 1980. Contribution à la faune de la République Unie du Cameroun. Poissons du Sanaga moyen et supérieur. *Cybium*, 3° série (8): 53–65.

Daget J. & Desoutter M., 1983. Essai de classification cladistique des Polyptéridés (Pisces, Brachiopterygii). *Bulletin du Muséum national d'Histoire naturelle, Paris*, (4)5A(2): 661–74.

Daget J. & Ecoutin J. M., 1976. Modèles mathématiques de production applicables aux poissons tropicaux subissant un arrêt annuel prolongé de croissance. *Cahiers ORSTOM, série Hydrobiologie*, 10: 59–69.

Daget J. & Iltis A., 1965. Poissons de Côte d'Ivoire (eaux douces et saumtres). *Mémoire de l'Institut français d'Afrique noire*, **74**.

Daget J. & Moreau J., 1981. Hybridation introgressive entre deux espèces de *Sarotherodon* (Pisces, Cichlidae) dans un lac de Madagascar. *Bulletin du Muséum national d'Histoire naturelle, Paris*, (4)3A(2): 689–703.

Daget J., Planquette N. & Planquette P., 1973. Premières données sur la dynamique des peuplements de poissons du Bandama (Côte d'Ivoire). *Bulletin du Muséum national d'Histoire naturelle, Paris*, (3) 151: 129–43.

Daget J., Gosse J. P. & Thys van den Audenaerde D. F. E. (eds.), 1984. CLOFFA 1. *Check-list of the Freshwater Fishes of Africa*. ISNB,MRAC, ORSTOM, Vol. 1.

Daget J., Gosse J. P. & Thys van den Audenaerde D. F. E. (eds.), 1986a. CLOFFA 2. *Check-list of the Freshwater Fishes of Africa*. ISNB, MRAC, ORSTOM, Vol. 2.

Daget J., Gosse J. P. & Thys van den Audenaerde D. F. E. (eds.), 1986b. CLOFFA 3. *Check-list of the Freshwater Fishes of Africa*. ISNB, MRAC, ORSTOM, Vol. 3.

Daget J., Gaigher I. C. & Ssentongo G. W., 1988. Conservation. In Lévêque C., Bruton M. N. & Ssentongo G. W. (eds.), *Biology and Ecology of African Freshwater Fishes*, pp. 481–91. ORSTOM, Paris.

Daget J., Gosse J. P., Teugels G. G. & Thys van den Audenaerde D. F. E. (eds.), 1991. CLOFFA 4. *Check-list of the Freshwater Fishes of Africa*. ISNB, MRAC, ORSTOM, Vol. 4.

Dansoko F. D., Breman H. & Daget J., 1976. Influence de la sécheresse sur les populations d'*Hydrocynus* dans le delta du Niger. *Cahiers ORSTOM, série Hydrobiologie*, **10**: 71–6.

Darlington P. J., 1957. *Zoogeography: The Geographical Distribution of Animals*. John Wiley & Sons, New York.

Darwin C., 1859. *On the Origin of Species by Means of Natural Selection, or the Preservation of Favoured Races in the Struggle of Life*. 1st edn. John Murray, London.

Darwin C., 1871. *The Descent of Man*. John Murray, London.

Davies B., 1986. The Zambezi river system. In Davies B. R. and Walker K. F. (eds.), *The Ecology of River Systems*, pp. 225–57. W. Junk Publishers, Dortrecht.

Davies B. & Gasse F. (eds.), 1988. *African Wetlands and Shallow Water Bodies. Bibliography*. Travaux et Documents, No. 211. ORSTOM, Paris.

Day J. H., Blaber S. J. M. & Wallace J. H., 1981. Estuarine fishes. In Day J. H. (ed.), *Estuarine Ecology – With Particular Reference to Southern Africa*, pp. 197–221. Balkema, Rotterdam.

de Kimpe P., 1964. Contribution à l'étude hydrobiologique du Luapula-Moero. *Annales du Musée royal d'Afrique centrale, in-8°*, **12**: 1–238.

de Vos L., 1984. Preliminary data of a systematic revision of the African species of the family Schilbeidae (Pisces, Siluriformes). *Revue de Zoologie africaine*, **98**: 424–33.

de Vos L. & Skelton P., 1990. Name changes for two common African catfishes. Rehabilitation of *Schilbe intermedius* Rüppell, 1832 (Siluriformes, Schilbeidae). *Cybium*, **14**(4): 323–6.

de Vos L. & Snoeks J., 1994. The non-cichlid fishes of the Lake Tanganyika basin. *Archiv für Hydrobiologie. Ergebnisse der Limnologie*, **44**: 391–405.

de Vos L., Janssens L. & Thys van den Audenaerde D., 1987. Etude préliminaire des migrations verticales et cycle d'activité de quelques espèces d'*Haplochromis* (Pisces, Cichlidae) du lac Kivu. *Revue de Zoologie africaine*, **101**: 265–70.

de Vos L., Snoeks J. & Thys van den Audenaerde D., 1990. The effect of *Tilapia* introductions in Lake Luhondo, Rwanda. *Environmental Biology of Fishes*, **27**: 303–8.

Deelstra H., 1977. Danger de pollution dans le lac Tanganyika (Bassins de Bujumbura). *Bulletin de la Société d'Etudes Géographiques*, **46**: 23–53.

Deelstra H., Power J. L. & Kenner C. T., 1976. Chlorinated hydrocarbon residues in various fish species in Lake Tanganyika. *Bulletin of Environmental Contamination and Toxicology*, **15**: 689–98.

Degens E. T., von Herzen R. P., Wong H. K., Deuser W. G. & Jannasch H. W., 1973. Lake Kivu: structure, chemistry and biology of an East African Rift lake. *Geologische Rundschau*, **62**: 245–77.

Degnbol P. & Mapila S., 1982. Limnological observations on the pelagic zone of Lake Malawi from 1970 to 1981. FAO Fishery Expansion Project, Malawi FI: DP/MLW/75/O19. *Technical Report*, I: 5–19.

Dejoux C., 1988. *La pollution des eaux continentales africaines. Expérience acquise, situation actuelle et perspective*. Travaux et Documents, No. 213. ORSTOM, Paris.

Dejoux C., Elouard J. M., Forge P. & Jestin J. M., 1981. Mise en évidence de la microdistribution des invertébrés dans les cours d'eau tropicaux. Incidence méthodologique pour la recherche d'une pollution à long terme par insecticide. *Revue d'Hydrobiologie tropicale*, **14**: 253–62.

Denny P., 1985 (ed.). *The Ecology and Management of African Wetland Vegetation*. Geobotany, 6. W. Junk Publishers, The Hague.

Denny P., 1993. Wetlands of Africa: Introduction. In Whigam D., Dykyjova D. & Hejny S., *Wetlands of the World: Inventory, Ecology and Management*, Vol. 1 pp. 1–31. Kluwer Academic Publishers, Dordrecht.

Denslow J. L., 1985. Disturbance-mediated coexistence of species. In Pickett S. T. A. & White P. S. (eds.), *The Ecology of Natural Disturbance and Patch Dynamics*, pp. 307–23. Academic Press, New York.

Denton T. E. & Howell W. M., 1973. Chromosomes of the African polypterid fishes, *Polypterus palmas* and *Calamoichthys calabaricus* (Pisces, Brachiopterygii). *Experientia*, **29**: 122–4.

Dharmamba M., Bornancin M. & Maetz J., 1975. Environmental salinity and sodium and chloride exchanges across the gill of *Tilapia mossambica*. *Journal of Physiology*, **70**: 627–36.

di Castri F. & Younès T., 1990*a*. Fonction de la biodiversité biologique au sein de l'écosystéme. *Acta Oecologica*, **11**: 429–44.

di Castri F. & Younès T. (eds.), 1990*b*. *Ecosystem Function of Biological Diversity*. Special issue Biology International, 22. IUBS, Paris.

Diamond J. M., 1975. The island dilemma: lessons of modern biogeographic studies for the design of natural reserves. *Biological Conservation*, **7**: 129–46.

Diamond J. M., 1984. "Normal" extinctions of isolated populations. In Nitecki M. N. (ed.), *Extinctions*, pp. 191–46. University of Chicago Press, Chicago.

Diamond J. M., 1988. Factors controlling species diversity: overview and synthesis. *Annals of Missouri Botanical Garden*, **75**: 117–29.

Diamond J. & Case T. J. (eds.), 1986. *Community Ecology*. Harper & Row, New York.

Diamond J. M. & Hamilton A.C., 1980. The distribution of forest passerine birds and quaternary climatic change in tropical Africa. *Journal of Zoology, London*, **191**: 379–402.

Dill I. M., 1983. Adaptative flexibility in the foraging behaviour of fishes. *Canadian Journal of Fisheries and Aquatic Sciences*, **40**: 398–408.

Din N. A. & Eltringham S. K., 1974. Ecological separation between White and Pink-backed pelican in the Rwenzorie National Park, Uganda. *Ibis*, **116**: 28–43.

Doadrio I., 1990. Phylogenetic relationships and classification of western palaearctic species of the genus *Barbus* (Osteichthyes, Cyprinidae). *Aquatic Living Resources*, **3**: 265–82.

Doadrio I., 1994. Freshwater fish fauna of North Africa and its biogeography. In Teugels G. G., Guégan J. F. & Albaret J. J. (eds.), *Biological Diversity of African Fresh- and Brackish Water Fishes*, pp. 21–34. *Annales Sciences zoologiques*, Vol. 275, Musée royal d'Afrique centrale, Tervuren, Belgique.

Dobzhansky T., Ayala F. J., Stebbins G. L. & Valentine J. W., 1977. *Evolution*. W. H. Freeman, San Francisco.

Dominey W. J., 1981. Anti-predator function of bluegill sunfish nesting colonies. *Nature*, **290**: 586–8.

Dominey W. J., 1984. Effects of sexual selection and life history on speciation: species flocks in African cichlids and Hawaiian Drosophila. In Echelle A. A. & Kornfield I. (eds.), *Evolution of Fish Species Flocks*, pp. 231–49. University of Maine at Orono Press, Orono, ME.

Don J. & Avatlion R. R., 1986. The induction of triploidy in *Oreochromis aureus* by heat shock. *Theoretical and Applied Genetics*, **72**: 186–92.

Don J. & Avatlion R. R., 1988*a*. Comparative study on the induction of triploidy in tilapias, using cold- and heat-shock techniques. *Journal of Fish Biology*, **32**: 665–72.

Don J. & Avatlion R. R., 1988*b*. Production of viable tetraploid tilapias using the cold shock technique. *Israeli Journal of Aquaculture*, **40**: 17–21.

Donnelly B. G., 1969. A preliminary survey of *Tilapia* nurseries on Lake Kariba during 1967/68. *Hydrobiologia*, **34**: 195–206.

Donnelly B. G., 1978. Evidence of fish survival during habitat desiccation in Rhodesia. *Journal of the Limnological Society of Southern Africa*, **4**: 75–6.

Douchement J., Romand R. & Pasteur N., 1984. Biochemical differentiation in West African Cyprinodontoid fish of the genus *Aphyosemion*. *Biochemical Systematics and Ecology*, **12**: 325–33.

Douglas M. E. & Matthews W. J., 1992. Does morphology predict ecology? Hypothesis testing within a freshwater fish assemblage. *Oikos*, **65**: 213–24.

Dubois A., 1988. Le genre en zoologie: essai de systématique théorique. *Mémoire du Muséum national d' Histoire naturelle, Zoologie*, **139**.

Dudley R. G., 1972. Biology of Tilapia of the Kafue floodplain, Zambia: predicted effects of the Kafue Gorge Dam. Ph.D. Dissertation, University of Idaho, Moscow, USA.

Dudley R. G., 1974. Growth of Tilapia of the Kafue floodplain, Zambia: predicted effects of the Kafue Gorge Dam. *Transactions of the American Fisheries Society*, **103**: 281–91.

Dumont H. J., 1979. *Limnologie van Sahara en Sahel*. Unpublished thesis, University of Ghent.

Dumont H. J., 1982. Relict distribution patterns of aquatic animals: another tool in evaluating late Pleistocene climate changes in the Sahara and Sahel. *Palaeoecology of Africa*, **14**: 1–24.

Dumont H. J., 1986. The Tanganyika sardine in lake Kivu: another ecodisaster for Africa ? *Environmental Conservation*, **13**: 143–48.

Dumont H. J., 1987. Region 2. Sahara. In Burgis M. & Symoens J. J. (eds.), *African Wetlands and Shallow Waterbodies. Directory*. pp. 79–154. ORSTOM, Paris.

Dunson W. A. & Travis J., 1991. The role of abiotic factors in community organization. *American Naturalist*, **138**: 1067–91.

Durand J. R., 1971. Les peuplements ichtyologiques de l'El Beid. 2ème note: Variations inter et intraspécifiques. *Cahiers ORSTOM, série Hydrobiologie*, **5**: 147–59.

Durand J. R., 1973. Note sur l'évolution des prises par unité d'effort dans le lac Tchad. *Cahiers ORSTOM, série Hydrobiologie*, **7**: 195–207.

Durand J. R., 1978. *Biologie et dynamique des populations d'Alestes baremoze (Pisces, Characidae) du bassin tchadien*. Travaux et Documents, No. 98. ORSTROM, Paris.

Durand J. R., 1980. Evolution des captures totales (1962–1977) et devenir des pêcheries de la région du lac Tchad. *Cahiers ORSTOM, série Hydrobiologie*, **13**: 93–111.

Durand J. R. & Loubens G., 1970. Observations sur la sexualité et la reproduction des *Alestes baremoze* du Bas-Chari et du lac Tchad. *Cahiers ORSTOM, série Hydrobiologie*, **4**: 61–81.

Dusart J., 1963. Contribution à l'étude de l'adaptation du *Tilapia* (Pisces, Cichlidae) à la vie en milieu mal oxygéné. *Hydrobiologia*, **21**: 328–41.

Eccles D. H., 1974. An outline of the physical limnology of Lake Malawi (Lake Nyasa). *Limnology and Oceanography*, **19**: 730–42.

Eccles D. H., 1985. Lake flies and sardines-cautionary note. *Biological Conservation*, **33**: 309–33.

Eccles D. H., 1992. *Field Guide to the Freshwater Fishes of Tanzania*. FAO species identification sheets for fishery purposes. FAO, Rome.

Eccles D. H. & Lewis D. S. C., 1981. Midwater spawning in *Haplochromis chrysonotus* (Blgr) (Teleosteri, Cichlidae) in Lake Malawi. *Environmental Biology of Fishes*, **6**: 201–2.

Eccles D. H. & Trewavas E., 1989. *Malawian Cichlid Fishes. The Classification of Some Haplochromine Genera*. Lake Fish Movies, Zevenhuizen, Holland.

Ehrenfeld D., 1988. Why put a value on biodiversity? In Wilson E. O. (ed.), *Biodiversity*, pp. 212–16. National Academy Press, Washington, DC.

Ehrlich P. R., 1989. Discussion: ecology and resource management – is ecological theory any good in practice. In Roughgarden J., May R. M. & Levin S. A. (eds.), *Perspectives in Ecological Theory*, pp. 306–18. Princeton University Press, Princeton, NJ.

El-Zarka S., Shaheen A. H. & Aleem A. A., 1970. Reproduction of *Tilapia nilotica* L. *Bulletin of the Institute of Oceanography and Fisheries, Cairo*, **1**: 193–204.

Elder H. Y., Garrod D. J. & Whitehead P. J. P., 1971. Natural hybrids of the African cichlid fishes *Tilapia spirulus nigra* and *T. leucosticta*: a case of hybrid introgression. *Biological Journal of the Linnean Society*, **3**: 103–46.

Eldredge N., 1971. The allopatric model and phylogeny in Palaeozoic invertebrates. *Evolution*, **25**: 156–67.

Eldredge N. & Gould S. J., 1972. Punctuated equilibria: an alternative to phyletic gradualism. In Schopf T. J. M. (ed.), *Models in Palaeobiology*, pp. 82–115. Freeman and Cooper, San Francisco.

Ellenbroek G. A., 1987. *Ecology and productivity of an African wetland system. The Kafue Flats, Zambia*. Geobotany, 9. W. Junk Publishers, Dortrecht.

Elouard J. M., 1983. Impact d'un insecticide organophosphoré (le témephos) sur les entomocénoses associées aux stades préimaginaux du complexe *Simulim damnosum* (Diptera: Simulidae). Thèse de Doctorat d'Etat, Université de Paris XI.

Elouard J. M., 1987. Rheopreferendums des stades préimaginaux de quatre espèces de simulies ouest-africaines. Incidences des épandages répétés de témephos. *Cahiers ORSTOM, série Entomologie Médicale et Parasitologie*, **25**: 3–11.

Elouard J. M. & Gibon F. M., 1985. Compétition interspécifique entre les stades préimaginaux de quelques espèces de simulies ouest-africaines. *Bulletin d'Ecologie*, 16: 223–9.

Elouard J. M. & Lévêque C., 1977. Rythme nycthéméral de dérive des insectes et des poissons dans les rivières de Côte d'Ivoire. *Cahiers ORSTOM, série Hydrobiologie*, 11: 179–83.

Elton C., 1927. *Animal Ecology*. Sidgwick and Jackson, London.

Elton C. S., 1958. *The Ecology of Invasions by Animals and Plants*. Methuen, London.

Endler J. A., 1977. *Geographic Variation, Speciation and Clines*. Monographs in Population Biology No. 10. Princeton University Press, Princeton, NJ.

Endler J.A., 1982. Pleistocene forest refuges: fact or fancy? In Prance G. T. (ed.), *Biological Diversification in the Tropics*, pp. 641–57. Columbia University Press, New York.

Endler J.A., 1986. *Natural Selection in the Wild*. Princeton University Press, Princeton, NJ.

Endler J. A. & McLellan T., 1988. The processes of evolution: toward a newer synthesis. *Annual Review of Ecology and Systematics*, 19: 395–421.

Estève R., 1949. Poissons du Sahara central. *Bulletin de la Société Zoologique de France*, 74: 19–20.

Estève R., 1952. Poissons de Mauritanie et du Sahara oriental. *Bulletin du Muséum national d'Histoire naturelle, Paris*, 24: 176–9.

Euzet L. & Combes C., 1980. Les problèmes de l'espèce chez les animaux parasites. In les problèmes de l'espèce dans le règne animal. *Mémoire de la Société Zoologique de France*, 40: 239–83.

Euzet L., Agnèse J. F. & Lambert A., 1989. Valeur des parasites comme critére d'identification de leur hôte. Démonstration convergente par l'étude parasitologique des monogénes branchiaux et l'analyse génétique des poissons hôtes. *Compte rendus hebdomadaires des Séances de l'Académie des Sciences, Paris*, 308: 385–88.

Everaerts J. M., Koeman J. H. & Brader L., 1971. Contribution à l'étude des effets sur quelques éléments de la faune sauvage des insecticides organophosphorés utilisés au Tchad en culture cotonniére. *Coton et fibres tropicales*, 26: 4.

Ewulonu U. K., Haas R. & Turner B. J., 1985. A multiple sex chromosome system in the annual killifish, *Nothobranchius guntheri. Copeia*, 1985(2): 503–8.

Falter U. & Charlier M., 1989. Mate choice in pure-bred and hybrid females of *Oreochromis niloticus* and *O. mossambicus* based upon visual stimuli (Pisces, Cichlidae). *Biology of Behaviour*, 14: 265–76.

Falter U. & Dolisy D., 1983. The effect of female sexual

pheromones on the behaviour of *Oreochromis niloticus*, *O. mossambicus* and hybrid males (Pisces, Cichlidae). Proceedings of the IV Workshop on biology, ecology and conservation of Cichlids. *Annales du Musée royal d'Afrique centrale (Sciences Zoologiques)*, Tervuren, 257: 35–8.

FAO/UN, 1970. *Report to the Government of Nigeria on Fishery Investigations on the Niger and Benue Rivers in the Northern Region and Development of a Programme of Riverine Fishery Management and Training*. Based on the work of M. P. Motwani. Report FAO/UNDP (TA), (2771).

Farmer G. J. & Beamish F. W. H., 1969. Oxygen consumption of *Tilapia nilotica* in relation to swimming speed and salinity. *Journal of the Fisheries Research Board of Canada*, 26: 2807–21.

Fausch K. D., Karr J. R. & Yant P. R., 1984. Regional application of an index of biotic integrity based on stream fish communities. *Transactions of the American Fisheries Society*, 113: 39–55.

Fausch K. D., Lyons J., Karr J. R. & Angermeier P. L., 1990. Fish communities as indicators of environmental degradation. *American Fisheries Society Symposium*, 8: 123–44.

Fay C., 1989a. Sacrifices, prix du sang, 'eau du maitre': fondation des territoires de péche dans le delta central du Niger (Mali). *Cahiers de Sciences humaines*, 25: 159–76.

Fay C., 1989b. Systèmes halieutiques et espaces de pouvoirs: transformation des droits et des pratiques de pêche dans le delta central du Niger (Mali) 1920–1980. *Cahiers de Sciences humaines*, 25: 213–36.

Fiedler P. L. & Ahouse J. J., 1992. Hierarchies of cause; toward an understanding of rarity in vascular plant species. In Fiedler P. L. & Jain S. K. (eds.), *Conservation Biology. The Theory and Practice of Nature Conservation, Preservation and Management*, pp. 23–47. Chapman & Hall, London and New York.

Fiedler P. L. & Jaine S. K., (eds.), 1992. *Conservation Biology. The Theory and Practice of Nature Conservation, Preservation and Management*. Chapman & Hall, London and New York.

Fink S. V. & Fink W.L., 1981. Interrelationships of the ostariophysian fishes (Teleostei). *Zoological Journal of the Linnean Society*, 72: 297–353.

Fish G. R., 1956. Some aspects of the respiration of six species of fish from Uganda. *Journal of Experimental Biology*, 33: 186–95.

Fishelson L., 1962. Hybrids of two species of the genus

Tilapia (Cichlidae, Teleostei). *Fishermen's Bulletin, Haifa*, 4: 14–19.

Fleger-Balon C., 1989. Direct and indirect development in fishes-examples of alternative life-history. In Bruton M. N. (ed.), *Alternative Life-history Styles of Animals*, pp. 71–100. Perspectives in Vertebrates Sciences 6. Kluwer Academic Publishers, Dortrecht.

Foerster W. & Schartl M., 1987. Karyotype and isozyme patterns of five species of *Aulonocara* Regan, 1922. *Courier-Forschungsinstitut Senckenberg*, 94: 55–61.

Fowler H. W., 1949. Results of the two Carpenter African Expeditions, 1946–1948. Part II- The fishes. *Proceedings of the Academy of Natural Sciences of Philadelphia*, 101: 233–75.

Fox L., 1975. Cannibalism in natural populations. *Annual Review of Ecology and Systematics*, 6: 87–106.

Franck J. P. C., Wright J. M. & McAndrew B. J., 1992. Genetic variability in a family of satellite DNAs from tilapia (Pisces, Cichlidae). *Genome*, 35: 719–25.

Frankel O. H. & Soulé M. E., 1981. *Conservation and Evolution*. Cambridge University Press, Cambridge.

Frissell C. A., Liss W. J., Warren C. E. & Hurley M. D., 1986. A hierarchical framework for stream habitat classification: viewing streams in a watershed context. *Environmental Management*, 10: 199–214.

Fromm P. O., 1980. Review of some physiological and toxicological responses of freshwater fish to acid stress. *Environmental Biology of Fishes*, 5: 79–93.

Fry F. E. J., 1971. The effect of environmental factors on the physiology of fish. In Hoar W. S. & Randall D. J. (eds.), *Fish Physiology*, 6, pp. 1–98. Academic Press, New York.

Fryer G., 1959. The trophic interrelationship and ecology of some littoral communities of Lake Nyasa with special reference to the fishes, and a discussion of the evolution of a group of rock-frequenting Cichlidae. *Proceedings of the Zoological Society of London*, 132: 153–281.

Fryer G., 1960. Concerning the proposed introduction of Nile perch into Lake Victoria. *East African Agricultural and Forestry Journal*, 25: 267–70.

Fryer G., 1961. Observations on the biology of the cichlid fish *Tilapia variabilis* Boulenger in the northern waters of lake Victoria (East Africa). *Revue de Zoologie et de Botanique africaine*, 64: 1–33.

Fryer G., 1965. Predation and its effect on migration and speciation in African fishes: a comment. *Proceedings of the Zoological Society of London*, 144: 301–22.

Fryer G., 1973. The Lake Victoria fisheries: some facts and fallacies. *Biological Conservation*, 5: 304–8.

Fryer G., 1984. The conservation and rational exploitation of the biota of Africa's great lakes. In Hall A. V. (ed.),

Conservation of Threatened Natural Habitats, pp. 135–54. South African National Scientific Programmes Report (92), CSIR, Pretoria.

Fryer G., 1986. Enemy-free space: a new name for an ancient ecological concept. *Biological Journal of the Linnean Society*, 27: 287–92.

Fryer G., 1991. The evolutionary biology of African cichlids fishes. *Annales du Musée royal d'Afrique centrale (Sciences Zoologiques)*, 262: 13–22.

Fryer G. & Iles T. D., 1969. Alternative routes to evolutionary success as exhibited by African cichlid fishes of the genus *Tilapia* and the species flocks of the great lakes. *Evolution*, 23: 359–69.

Fryer G. & Iles T. D., 1972. *The Cichlid Fishes of the Great Lakes of Africa: Their Biology and Evolution*. Tropical Fish Hobbyist Publications, Edinburgh and Oliver & Boyd, London.

Fryer G., Greenwood P. H. & Peake J. F., 1983. Punctuated equilibria, morphological stasis and the paleontological documentation of speciation: a biological appraisal of a case history in an African lake. *Biological Journal of the Linnean Society*, 20: 195–205.

Fryer G., Greenwood P. H. & Peake J. F., 1985. The demonstration of speciation in fossil molluscs and living fishes. *Biological Journal of the Linnean Society*, 26: 325–36.

Fukuoka H. & Muramoto J., 1975. Somatic and meiotic chromosomes of *Tilapia mossambica* Peters. *Chromosome Information Service*, 18: 4–6.

Fukusho K., 1968. The specific difference of temperature response among cichlid fishes genus *Tilapia*. *Bulletin of the Japanese Society of Scientific Fisheries*, 34: 103–11.

Furse M. T., Kirk R. C., Morgan P. R. & Tweddle D., 1979. Fishes: distribution and biology in relation to changes. In Kalk M., McLachlan A. J. & Howard-Williams C. (eds.), *Lake Chilwa: Studies of Change in a Tropical Ecosystem*, pp. 209–29. Monographiae Biologicae 35. W. Junk Publishers, The Hague.

Futuyma D. J., 1986. *Evolutionary Biology*, 2nd edn. Sinauer Associates, Sunderland, MA.

Gac J. Y., 1980. *Géochimie du bassin du lac Tchad: bilan de l'altération, de l'érosion et de la sédimentation*. Travaux et Documents, No. 123. ORSTOM, Paris.

Gadgil M. & Berkes F., 1991. Traditional resource management systems. In *Resource Management and Optimisation*, pp. 127–41. Hardwood Academic Publishers, Reading, Berks.

Gaemers A. M., 1989. The first cichlids (Perciformes, Pisces) from Europe: the new fossil genus *Eurotilapia*, evidence from otoliths and teeth. *Annales du Musée royal d'Afrique centrale (Sciences Zoologiques)*, **257**: 109–16.

Gaigher I. G. & Pott R. Mc. C., 1972. A checklist of indigenous fish in the east flowing rivers of the Transvaal. *Limnological Society of Southern Africa Newsletter*, **18**: 26–32.

Gaigher I. G., Hamman K. C. D. & Thorne S. C., 1980. The distribution, conservation status and factors affecting the survival of indigenous freshwater fishes in the Cape province. *Koedoe*, **23**: 57–88.

Ganf G. G. & Viner A. B., 1973. Ecological stability in a shallow equatorial lake (Lake George, Uganda). *Freshwater Biology*, **5**: 13–39.

Garcia S. M., 1994. The Precautionary Principle: its implications in capture fisheries management. *Ocean & Coastal Management*, **22**: 99–125.

Garrod D. J., 1960. The fisheries of Lake Victoria, 1954–1959. *East African Agricultural and Forestry Journal*, **26**: 42–8.

Garrod D. J., 1961. The history of the fishing industry of Lake Victoria, East Africa, in relation to the expansion of marketing facilities. *East African Agricultural and Forestry Journal*, **27**: 95–9.

Gashagaza M. M., 1991. Diversity of breeding habits in lamprologine cichlids in Lake Tanganyika. *Physiology and Ecology, Japan*, **28**: 29–65.

Gasse F., 1977. Evolution of Lake Abhe (Ethiopia and TFAI), from 70 000 BP. *Nature*, **265**: 42–5.

Gasse F., Rognon P. & Street F. A., 1980. Quaternary history of the Afar and Ethiopian Rift lakes. In Williams M. A. J. & Faure H. (eds.), *The Sahara and the Nile*. pp. 361–400. A. Balkema, Rotterdam.

Gaston K. J. & Lawton J. H., 1990. The population ecology of rare species. *Journal of Fish Biology*, **37**: 97–104.

Gauthier-Hion A. & Michaloud G., 1989. Are figs always keystone resources for tropical frugivorous vertebrates? A test in Gabon. *Ecology*, **70**: 1826–33.

Gauthier A. & Van Neer W., 1989. The animal remains from the Late Palaeolithic sequence in Wadi Kubbaniya. In Wendorf F., Schild R. & Close A. E. (eds.),*The Prehistory of Wadi Kubbaniya*, Vol. 2, pp. 119–61.

Gayet M., 1980. Hypothéses sur l'origine des Ostariophysaires. *Comptes rendus hebdomadaires des Séances de l'Académie des Sciences, Paris*, **290**: 1197–9.

Gayet M., 1981. Considérations relatives à la paléoécologie du gisement Cenomanien de Laveiras (Portugal). *Bulletin du Muséum national d'Histoire naturelle (sciences de la terre), Paris*, sér. 4 secteur C No. 4: 311–15.

Gayet M., 1982a. Découverte dans le Crétacé supérieur de Bolivie des plus anciens Characiformes connus. *Comptes rendus hebdomadaires des Séances de l'Académie des Sciences, Paris*, **294**: 1037–40.

Gayet M., 1982b. Cypriniformes Crétacés en Amérique du sud. *Comptes rendus hebdomadaires des Séances de l'Académie des Sciences, Paris*, **295**: 661–4.

Gayet M., 1982c. Considération sur la phylogénie et la paléobiogéographie des Ostariophysaires. *Geobos*, Mémoire spécial, **6**: 39–52.

Gayet M., 1983. Poissons. In Petit Maire N. & Riser J. (eds.), *Sahara ou Sahel? Quaternaire récent du bassin de Taoudénni (Mali)*, pp. 183–209. Laboratoire de géologie du quaternaire du CNRS, Marseille.

Gayet M. & Meunier F. J., 1992. Polyptériformes (Pisces, Cladistia) du Maastrichtien et du Paléocéne de Bolivie. *Geobios*, **14**: 159–68.

Gayet M., Meunier F. & Levrat-Calviac V., 1988. Mise en évidence des plus anciens Polypteridae dans le gisement sénonien d'In Becetem (Niger). *Comptes rendus hebdomadaires des Séances de l'Académie des Sciences, Paris*, **307**: 205–10.

Gee J. H. R. & Giller P. S. (eds.), 1987. *Organizations of Communities: Past and Present*. Blackwell Scientific Publications, Oxford.

Gee J.M., 1965. Nile perch investigation. *Annual Report of the East African Freshwater Fisheries and Research Organisation*, (1964): 13–17.

Gee J. M., 1969. A comparison of certain aspects of the biology of *Lates niloticus* (Linne) in some East African lakes. *Revue de Zoologie et de Botanique africaine*, **80**: 244–62.

Géry J., 1969. The freshwater fishes of South America. In Fittkau J. *et al.* (eds.), *Biogeography and Ecology in South America*, pp. 828–48. W. Junk Publishers, The Hague.

Géry J., 1977. *Characoids of the World*. T.H.F. Publications, Neptune City, NJ.

Ghaza M. A., Bénech V. & Paugy D., 1991. L'alimentation de *Brycinus leuciscus* (Teleostei, Characidae) au Mali: aspects qualitatifs, quantitatifs et comportementaux. *Ichthyological Exploration of Freshwaters*, **2**: 47–54.

Gibon F. M. & Statzner B., 1985. Longitudinal zonation of lotic insects in the Bandama River system (Ivory Coast). *Hydrobiologia*, **122**: 61–4.

Giller P. S., 1984. *Community Structure and the Niche*. Chapman & Hall, London.

Giller P. S. & Gee J. H. R., 1987. The analysis of community organization: the influence of equilibrium, scale and terminology. In Gee J. H. R. & Giller P. S. (eds.), *Organizations of Communities: Past and Present*, pp. 519–42. Blackwell Scientific Publications, Oxford.

Gjerstad D., 1982. Recent findings in the Lake Turkana

Fisheries. Paper presented at the Seminar on the Future of the Lake Turkana Fisheries. Kalokol, Kenya, September 20–25, 1982.

Glantz M. H., 1990. Does history have a future? Forecasting climate change effects on fisheries by analogy. *Fisheries*, 15(6): 39–44.

Global Biodiversity Strategy, 1992. *Guidelines for Action to Save, Study and use Earth's Biotic Wealth Sustainability and Equitability*. WRI, IUCN, UNEP.

Goeden G. B., 1979. Biogeographic theory as a management tool. *Environmental Conservation*, 6: 27–32.

Goffin A., 1909. *Les pêcheries et les poissons du Congo*. Verteneuil & Desmet (ed.). Bruxelles.

Goldschmidt R., 1940. *The Material Basis of Evolution*. Yale University Press, New Haven, CT.

Goldschmidt T., 1991. Egg mimics in haplochromine cichlids (Pisces, Perciformes) from Lake Victoria. *Ethology*, 88: 177–90.

Goldschmidt T. & Goudswaard K., 1989. Reproductive strategies in haplochromines (Pisces, Cichlidae) from Lake Victoria. In Goldschmidt T., An ecological and morphological field study on the haplochromine cichlid fishes (Pisces, Cichlidae) of Lake Victoria, pp. 93–119. Ph.D. Rijksuniversiteit Leiden, Netherlands.

Goldschmidt T. & Visser J., 1990. On the possible role of egg mimics in speciation. *Acta Biotheoritica*, 38: 125–34.

Goldschmidt T. & Witte F., 1990. Reproductive strategies of zooplanktivorous haplochromine cichlids (Pisces) from Lake Victoria before the Nile perch boom. *Oikos*, 58: 365–8.

Goldschmidt T., Witte F. & Visser J. de, 1990. Ecological segregation in zooplanktivorous haplochromine species (Pisces, Cichlidae) from Lake Victoria. *Oikos*, 58: 343–55.

Goldschmidt T., Witte F. & Wanink J., 1993. Cascading effects of the introduced Nile perch on the detrivorous/phytoplanktivorous species in the sublittoral areas of Lake Victoria. *Conservation Biology*, 7: 686–700.

Golubtsov A. S. & Krysanov E. Y., 1993. Karyological study of some Cyprinid species from Ethiopia. The ploidy differences between large and small *Barbus* of Africa. *Journal of Fish Biology*, 42: 445–55.

Goodland R. & Ledec G., 1987. Neoclassical economics and principles of sustainable development. *Ecological Modelling*, 38: 19–46.

Goody P.C., 1969. The relationships of certain Upper Cretaceous teleosts with special reference to the myctophoids. *Bulletin British Museum (Natural History)* Geology, Supplement 7.

Gophen M., Ochumba P. B. O. & Kaufman L. S., 1995. Some aspects of perturbations in the structure and biodiversity of the ecosystem of Lake Victoria (East Africa). *Aquatic Living Resources*, 8: 27–41.

Gore J. A., Layzer J. B. & Russell I. A., 1992. Non-traditional applications of instream flow techniques for conserving habitat of biota in the Sabie River of Southern Africa. In Boon P. J., Calow P. & Petts G. E., *River Conservation and Management*, pp. 161–77. John Wiley & Sons, Chichester.

Gorman O. T. & Karr J. R., 1978. Habitat structure and stream fish ecology. *Ecology*, 59: 507–15.

Gosline W. A., 1971. *Functional Morphology and Classification of Teleostean Fishes*. University Press of Hawaii, Honolulu.

Gosse J. P., 1963. Le milieu aquatique et l'écologie des poissons dans la région de Yangambi. *Annales du Musée royal d'Afrique centrale (Sciences Zoologiques)*, 116: 113–271.

Gosse J. P., 1966. Remarques systématiques sur quelques espéces de la faune ichtyologique congolaise. *Revue de Zoologie et de Botanique africaine*, 73: 186–200.

Gosse J. P., 1968. Les poissons du bassin de l'Ubangui. *Musée royal d'Afrique centrale, Documents Zoologiques* (13): 1–56.

Goudswaard K. & Wanink J. H., 1993. Anthropogenic perturbation in Lake Victoria: effects of fish introductions and fisheries on fish eating birds. *Annales du Musée royal d'Afrique Centrale (Sciences Zoologiques)*, 268: 312–18.

Gould S. J., 1983. *Hen's Teeth and Horse's Toes*. W. W. Norton, New York.

Gourène G. & Teugels G. G., 1989. Révision systématique du genre *Microthrissa* Boulenger, 1902 des eaux douces africaines (Pisces, Clupeidae). *Revue d'Hydrobiologie tropicale*, 22: 129–56.

Gourène G. & Teugels G. G., 1991a. Révision systématique des genres *Odaxothrissa* Boulenger, 1899 et *Cynothrissa* Regan, 1917 (Pisces, Clupeidae) des eaux douces et saumâtres de l'Afrique. *Revue de Zoologie africaine*, 105: 439–59.

Gourène G. & Teugels G. G., 1991b. Révision du genre *Pellonula* des eaux douces africaines (Pisces, Clupeidae). *Ichthyological Exploration of Freshwaters*, 2: 213–25.

Gourène G. & Teugels G. G., 1993. Position taxonomique de *Limnothrissa stappersii*, un clupeidé lacustre d'Afrique centrale. *Ichthyological Exploration of Freshwaters*, 4: 367–74.

Gourène G. & Teugels G. G., 1994. Synopsis de la classification et phylogénie des Pellonulinae de l'Afrique occidentale et centrale (Teleostei, Clupeidae). *Journal of African Zoology*, 108: 77–91.

Graham J. B. & Baird T. A., 1982. The transition to air breathing in fishes. I. Environmental effects on the facultative air breathing of *Ancistreus chagresi* and *Hypostomus plecostomus* (Loricaridae). *Journal of Experimental Biology*, 96: 53–67.

Graham M., 1929. *The Victoria-Nyanza and its Fisheries. A Reports on the Fishing Surveys of Lake Victoria*. Crown Agents for the Colonies, London.

Grant P. R., 1986. *Ecology and Evolution of Darwin's Finches*. Princeton University Press, Princeton, NJ.

Gras R., Lauzanne L. & Saint-Jean L., 1981. Régime alimentaire et sélection des proies chez les *Brachysynodontis batensoda* (Pisces, Mochocidae) du lac Tchad en période de basses eaux. *Revue d'Hydrobiologie tropicale*, 14: 223–31.

Gras G., Pelissier C. & Leung Tack D., 1982. Etude expérimentale de l'action du temephos sur l'activité acetylcholinestérasique du cerveau de *Tilapia guineensis*. 1° partie: étude expérimentale aux doses opérationnelles. *Toxicological European Research*, 4: 301–8.

Gras R. & Saint Jean L., 1982. Comments about Ivlev's electivity index. *Revue d'Hydrobiologie tropicale*, 15: 33–7.

Grassle J. F., Lasserre P., McIntyre A. D. & Ray G.C., 1991. Marine biodiversity and ecosystem function. A proposal for an international programme of research. *Biology International, Special Issue* No. 23.

Graves J. E. & Somero G. N., 1982. Electrophoretic and functional enzymatic evolution in four species of Eastern Pacific barracudas from different thermal environments. *Evolution*, 36: 97–106.

Green G. M. & Sussman R. W., 1990. Deforestation history of the Eastern rain forest of Madagascar from satellite images. *Science*, 248: 212–15.

Green J., 1967. The distribution and variation of *Daphnia lumholtzi* (Crustacea: Cladocera) in relation to fish predation in Lake Albert, East Africa. *Journal of Zoology, London*, 151: 181–97.

Green J., 1977. Haematology and habitats in catfish of the genus *Synodontis*. *Journal of Zoology, London*, 182: 39–50.

Green J., Corbet S. A. & Betney E., 1973. Ecological studies on crater lakes in West Cameroon. The blood of endemic cichlids in Barombi Mbo in relation to stratification and their feeding habits. *Journal of Zoology, London*, 170: 299–308.

Greenwood P. H., 1951. Fish remains from the Miocene deposits of Rusinga Island and Kavirondo Province, Kenya. *Annals and Magazine of Natural History*, (12)4: 1192–201.

Greenwood P. H., 1959. Quaternary fish-fossils. In Institut des Parcs nationaux du Congo belge; *Exploration du parc national Albert*. Mission J. de Heinzelin de Braucourt (1950), 4(1).

Greenwood P. H., 1960. Fossil denticipitid fishes from East Africa. *Bulletin British Museum (Natural History)* Geology, 5: 1–11.

Greenwood P. H., 1965a. Environmental effects on the pharyngeal mill of a cichlid fish *Astatoreochomis alluaudi* and their taxonomic implications. *Proceedings of the Linnean Society of London*, 176: 1–10.

Greenwood P. H., 1965b. The cichlid fishes of Lake Nabugabo, Uganda. *Bulletin British Museum (Natural History)* Zoology, 12: 313–57.

Greenwood P. H., 1966. *The Fish of Uganda*, 2nd edn. Uganda Society, Kampala.

Greenwood P. H., 1972. New fish fossils from the Pliocene of Wadi Natrum, Egypt. *Journal of Zoology, London*, 168: 503–19.

Greenwood P. H., 1973a. Fish fossils from the late Miocene of Tunisia. *Notes du Service Géologique de Tunis*, 37: 41–72.

Greenwood P. H., 1973b. Morphology, endemism and speciation in African cichlids. *Verhandlungen des Deutschen Zoologischen Gesellschaft, Mainz*, 1973: 115–24.

Greenwood P. H., 1973c. The interrelationships of the Osteoglossomorpha. *Zoological Journal of the Linnean Society*, 53 (Supplement 1): 307–32.

Greenwood P. H., 1974a. Review of Cenozoic freshwater fish faunas in Africa. *Annals of the Geological Survey of Egypt*, IV: 211–32.

Greenwood P. H., 1974b. The cichlid fishes of Lake Victoria, East Africa: the biology and evolution of a species flock. *Bulletin British Museum (Natural History)* Zoology, Supplement 6.

Greenwood P. H., 1974c. The *Haplochromis* species (Pisces, Cichlidae) of Lake Rudolf, East Africa. *Bulletin British Museum (Natural History)* Zoology, 27: 141–65.

Greenwood P. H., 1976a. Fish fauna of the Nile. In Rzoska J. (ed.), *The Nile, Biology of an Ancient River*, pp. 127–41. Monographiae Biologicae. W. Junk Publishers, The Hague.

Greenwood P. H., 1976b. A review of the family Centropomidae (Pisces, Perciformes). *Bulletin British Museum (Natural History)* Zoology, 29: 1–81.

Greenwood P. H., 1979. Towards a phyletic classification of the 'genus' *Haplochromis* (Pisces, Cichlidae) and related taxa. Part I. *Bulletin British Museum (Natural History)* Zoology, 35: 265–322.

Greenwood P. H., 1980. Towards a phyletic classification of

the 'genus' *Haplochromis* (Pisces, Cichlidae) and related taxa. Part II. The species from Lakes Victoria, Nabugabo, Edward, George and Kivu. *Bulletin British Museum (Natural History)*. (Zoology) 39: 1–101.

Greenwood P. H., 1981. Species flock and explosive evolution. In Greenwood P. H. & Forey P. L. (eds.), *Chance, Change and Challenge – the Evolving Biosphere*, pp. 61–74. Cambridge University Press, Cambridge and British Museum (Natural History), London.

Greenwood P. H., 1983a. The zoogeography of African freshwater fishes: bioaccountancy or biogeography. In Sims R. W., Price J. H. & Whalley P. E. S. (eds.), *Evolution, Time and Space: The Emergence of the Biosphere*, pp. 179–99. Systematic Association Special Volume No. 23. Academic Press, London and New York.

Greenwood P. H., 1983b. On *Macropleurodus, Chilotilapia* (Teleostei, Cichlidae) and the interrelationships of African cichlid species flocks. *Bulletin British Museum (Natural History)* Zoology, 45: 209–31.

Greenwood P. H., 1984. African cichlids and evolutionary theories. In Echelle A. A. & Kornfield I (eds.), *Evolution of Fish Species Flocks*, pp. 141–54. University of Maine at Orono Press, Orono, ME.

Greenwood P. H., 1986. The Natural History of African lungfishes. *Journal of Morphology*, Supplement 1: 163–79.

Greenwood P. H., 1987. The genera of pelmatochromine fishes (Teleostei, Cichlidae). A phylogenetic review. *Bulletin British Museum (Natural History)* Zoology, 53: 139–203.

Greenwood P. H., 1991. Speciation. In Keenleyside H. A. (ed.), *Cichlid Fishes: Behaviour, Ecology and Evolution*, pp. 86–102. Fish and Fisheries Series 2, Chapman & Hall, London.

Greenwood P. H., 1992. Are the major fish faunas well-known? *Netherlands Journal of Zoology*, 42: 131–8.

Greenwood P. H., 1994. Lake Victoria. In Martens K. F., Goddeeris B. & Coulter G. (eds.), *Speciation in Ancient Lakes*, pp. 19–26. *Archiv für Hydrobiologie*, 44.

Greenwood P. H. & Howes G. J., 1975. Neogene fossil fishes from the Lake Albert–Lake Edward Rift (Zaïre). *Bulletin British Museum (Natural History)* Geology, 26: 71–127.

Greenwood P. H., Rosen D. E., Weitzman S. H. & Myers G.S, 1966. Phyletic studies of teleostean fishes, with a provisional classification of living forms. *Bulletin of the American Museum of Natural History*, 131: 339–456.

Greenwood P. H., Miles R. S. & Patterson C., 1973. *Interrelationships of Fishes*. London, Academic Press.

Groenewald A. A. van, 1958. A revision of the genera *Barbus*

and *Varicorhinus* (Pisces, Cyprinidae) in Transvaal. *Annals of the Transvaal Museum*, 23: 263–330.

Gross M. R., 1987. Evolution of diadromy in fishes. *American Fisheries Society Symposium*, 1: 14–25.

Gross M. R. & Sargent R. C., 1985. The evolution of male and female parental care in fishes. *American Zoologist*, 25: 807–22.

Grove A. T., 1983. Evolution of the physical geography of the East African Rift valley region. In Sims R. W., Price J. H. & Whalley P. E. S. (eds.), *Evolution, Time and Space: The Emergence of the Biosphere*, pp. 115–55. Systematic Association Special Volume No. 23. Academic Press, London and New York.

Grove A. T. (ed.), 1985. *The Niger and its Neighbours. Environmental History and Hydrobiology, Human Use and Health Hazards of the Major West African Rivers.* Balkema, Rotterdam.

Grubb P., 1982. Refuges and dispersal in the speciation of African forest mammals. In Prance G. T. (ed.), *Biological Diversification in the Tropics*, pp. 537–53. Columbia University Press, New York.

Gucinski H., Lackey R. T. & Spence B. C., 1990. Global climate change: policy implications for fisheries. *Fisheries*, 15: 33–8.

Guégan J. F. & Agnèse J.F., 1991. Parasite evolutionary events inferred from host phylogeny: the case of *Labeo* species (Teleostei, Cyprinidae) and their dactylogirid parasites (Monogenea, Dactylogiridae). *Canadian Journal of Zoology*, 69: 595–603.

Guégan J. F. & Lambert A., 1990. Twelve new species of dactylogyrids (Platyhelminthes, Monogenea) from West African barbels (Teleostei, Cyprinidae) with some biogeographical implications. *Systematic Parasitology*, 17: 153–81.

Guégan J. F. & Lambert A., 1991. Dactylogyrids (Platyhelminthes, Monogenea) of *Labeo* in west African short coastal river systems. *Journal of the Helminthological Society of Washington*, 58: 85–99.

Guégan J. F., Lambert A. & Birgi E., 1988a. Observations sur le parasitisme branchial des Characidae du genre *Hydrocynus* en Afrique de l'Ouest. Description d'*Annulotrema pikoides* n.sp. (Monogenea, Ancyrocephalidae) chez *Hydrocynus vittatus* (Castelnau, 1861). *Annales de Parasitologie Humaine et Comparée*, 63: 91–8.

Guégan J. F., Lambert A. & Euzet L., 1988b. Etude des Monogénes des Cyprinidae du genre *Labeo* en Afrique de l'Ouest. I – Genre *Dactylogyrus* Diesing, 1850. *Revue d'Hydrobiologie tropicale*, 21: 135–51.

Guégan J. F., Lambert A. & Euzet L., 1989. Etude des

Monogénes des Cyprinidae du genre *Labeo* en Afrique de l'Ouest. II – Genre *Dogielus* Bykhowski, 1936. *Revue d'Hydrobiologie tropicale*, **22**: 35–48.

Guégan J. F., Lambert A., Lévêque C., Combes C. & Euzet L., 1992. Can host body size explain the parasite species richness in tropical freshwater fishes? *Oecologia*, **90**: 197–204.

Guégan J. F., Rab P., Machordom A. & Doadrio I., 1995. New evidence of hexaploidy in 'large' African *Barbus* (Cyprinidae, Teleostei) with some considerations on the origin of hexaploidy. *Journal of Fish Biology*, **47**: 192–8.

Guenero R. D., 1975. Use of androgens for the production of all-male *Tilapia aurea* (Steindachner). *Transactions of the American Fisheries Society*, **94**: 386–9.

Guillet A. & Crowe T. M., 1985. Patterns of distribution, species richness, endemism and guild composition of water birds in Africa. *African Journal of Ecology*, **23**: 89–120.

Guillet A. & Furness R. W., 1985. Energy requirements of a Great White pelican (*Pelecanus onocrotalus*) population and its impact on fish stocks. *Journal of Zoology, London* (A), **205**: 573–83.

Gwahaba J. J., 1973. Effects of fishing on the *Tilapia nilotica* populations of Lake George, Uganda, over the past 20 years. *East African Wildlife Journal*, **11**: 317–28.

Gwahaba J. J., 1975. The distribution, population density and biomass of fish in an equatorial lake, Lake George, Uganda. *Proceedings of the Royal Society of London*, B, **190**: 393–414.

Haas R., 1976*a*. Behavioral biology of the annual killifish, *Nothobranchius guentheri*. *Copeia*, **1**: 80–91.

Haas R., 1976*b*. Sexual selection in *Nothobranchius guentheri* (Pisces, Cyprinodontidae). *Evolution*, **30**: 614–22.

Haberyan K. A. & Hecky R. E., 1987. The late Pleistocene and Holocene stratigraphy and paleolimnology of lakes Kivu and Tanganyika. *Palaeogeography, Palaeoclimatology, Palaeoecology*, **61**: 169–97.

Haffer J., 1982. General aspects of the refuge theory. In Prance G. T. (ed.), *Biological Diversification in the Tropics*, pp. 6–24. Columbia University Press, New York.

Hallerman E. M. & Kapuscinski A. R., 1990. Transgenic fish and public policy: regulatory concerns. *Fisheries*, **15**: 12–20.

Hallerman E. M. & Kapuscinski A. R., 1991. Ecological and regulatory uncertainties associated with transgenic fish. In Hew C. L. & Flechter G. L. (eds), *Transgenic Fish*. World Scientific Publishing Company, Singapore.

Hamblyn E. L., 1966. The food and feeding habits of *Lates niloticus* (Linné) (Pisces, Centropomidae). *Revue de Zoologie et de Botanique africaine*, **74**: 1–28.

Hamilton A. C., 1982. *Environmental History of East Africa. A Study of the Quaternary*. Academic Press, London.

Hamilton A. C., 1988. Guenon evolution and forest history. In Gauthier-Hion A., Bourlière F. & Gauthier J. P. (eds), *A Primate Radiation. Evolutionary Biology of the African Guenons*, pp. 13–34. Cambridge University Press, Cambridge.

Hamley J. M., 1975. Review of gillnet selectivity. *Journal of the Fisheries Research Board of Canada*, **32**: 1943–69.

Hanna N. S. & Schiemer F., 1993*a*. The seasonality of zooplanktivorous fish in an African reservoir (Gebel Aulia Reservoir, White Nile, Sudan). Limnological cycle and the fish community dynamics. *Hydrobiologia*, **250**: 173–85.

Hanna N. S. & Schiemer F., 1993*b*. The seasonality of zooplanktivorous fish in an African reservoir (Gebel Aulia Reservoir, White Nile, Sudan). Part 2: Spatial distribution and resource partitioning in zooplanktivorous fish assemblages. *Hydrobiologia*, **250**: 187–99.

Hanski I., 1991. Single-species metapopulation dynamics: concepts, models and observations. *Biological Journal of the Linnean Society*, **42**: 17–38.

Hanski I. & Gilpin M., 1991. Metapopulation dynamics: brief history and conceptual domain. *Biological Journal of the Linnean Society*, **42**: 3–16.

Hara T. J. (ed.), 1992. *Fish chemoreception*, Chapman & Hall, London.

Harbott B. J., 1982. Studies on the feeding activities of *Sarotherodon niloticus* (L.) in Lake Turkana. In Hopson A. J. (ed.), *Lake Turkana. A Report of the Findings of the Lake Turkana Project 1972–1975*, pp. 1357–68. Overseas Development Administration, London. Published by the University of Sterling, Sterling.

Harbott B. J. & Ogari J. T. N., 1982. The biology of the larger cichlid fish of Lake Turkana. In Hopson A. J. (ed.), *Lake Turkana. A Report of the Findings of the Lake Turkana Project 1972–1975*, pp. 1331–55. Overseas Development Administration, London. Published by the University of Sterling, Sterling.

Hardin G., 1968. "The tragedy of the commons". *Science*, **162**: 1243–8.

Harding D., 1964. Hydrology and fisheries in Lake Kariba. *Verhandlungen-Internationale Vereinigung für Theoretische und Angewandte Limnologie*, **15**: 139–49.

Harding D., 1966. Lake Kariba. The hydrology and development of fisheries. In Lowe McConnell (ed.), *Man Made Lakes*, pp. 7–20. Symposium Institute of Biology, No. 15, Academic Press, London.

Harper D. M., Phillips G., Chilvers A., Kitaka N. & Mavuti K., 1993. Eutrophication prognosis for Lake Naivasha,

Kenya. *Verhandlungen-Internationale Vereinigung für Theoretische und Angewandte Limnologie*, **25**: 861–5.

Harper J. L. & Ogden J., 1970. The reproductive strategies of higher plants. I. The concept of strategy with special reference to *Senecio vulgaris*. *Journal of Ecology*, **58**: 681–98.

Harrison K., Crimmen O., Travers R., Maikweki J. & Mutoro D., 1989. Balancing the scales in Lake Victoria. *Biologist*, **36**: 189–91.

Harrison R. G. & Rand D. M., 1989. Mosaic hybrid zone and the nature of species boundaries. In Otte D. & Endler J. A. (eds.), *Speciation and its consequences*, pp. 111–33. Sinauer Associates, Sunderland, MA.

Hart P. B. J., 1986. Foraging in teleost fishes. In Pitcher T. J. (ed.), *The Behaviour of Teleost Fishes*, pp. 211–235. Croom Helm, London.

Hartman J. B. & Walker T. L., 1988. Oil and gas developments in Central and Southern Africa in 1987. *American Association of Petroleum Geologists Bulletin*, **72**: 196–227.

Havens K., 1992. Scale and structure in natural food webs. *Science*, **257**: 1107–9.

Heal O.W., 1991. The role of study sites in long-term ecological research: the U.K. experience. In Risser P. G. (ed.), *Long-term Ecological Research*, pp. 23–44. SCOPE 47, John Wiley & Sons, Chichester.

Hecht T. & Appelbaum S., 1988. Observations on intraspecific aggression and coeval sibling cannibalism by larval and juvenile *Clarias gariepinus* (Clariidae: Pisces) under controlled conditions. *Journal of Zoology*, **214**: 21–44.

Hecht T., & Lublinkhof W., 1985. *Clarias gariepinus* × *Heterobranchus longifilis* (Clariidae, Pisces): a new hybrid for aquaculture? *South African Journal of Science*, **81**: 620–1.

Hecky R. E., 1991. The pelagic ecosystem. In Coulter G. W. (ed.), *Lake Tanganyika and Its Life*, pp. 90–110. Natural History Museum Publications and Oxford University Press, Oxford.

Hecky R. E., 1993. The eutrophication of Lake Victoria. *Verhandlungen-Internationale Vereinigung für Theoretische und Angewandte Limnologie*, **25**, 39–48.

Hecky R. E. & Bugenyi F. W. B., 1992. Hydrology and chemistry of the African Great Lakes and water quality issues: problems and solutions. *Mitteilungen-Internationale Vereinigung für Theoretische und Angewandte Limnologie*, **23**: 45–54.

Hecky R. E. & Degens E. T., 1973. Late Pleistocene-Holocene chemical stratigraphy and Paleolimnology of the Rift Valley lakes of Central Africa.

Woods hole Oceanographic Institution, Technical Report, 73–28 (unpublished).

Hecky R.E, Fee E. J., Kling H. J. & Rudd J. W., 1981. Relationship between primary production and fish production in Lake Tanganyika. *Transactions of the American Fisheries Society,* **110**: 336–45.

Heiligenberg W., 1977. Principles of electrolocation and jamming avoidance in electric fish: a neuroethological approach. In Braitenberg, V. (ed.), *Studies of Brain Functions*, Vol. 1, pp. 1–85. Springer, Berlin.

Helfert M. R. & Wood C. A., 1986. *Geotimes*, **31**: 4.

Helfman G. S., 1986. Fish behaviour by day, night and twilight. In Pitcher T. J. (ed.), *The Behaviour of Teleost Fishes*, pp. 366–87. Croom Helm, London.

Hennig W., 1950. *Grundzüge einer Theorie der Phylogenetischen Systematik*. Deutscher Zentralverlag, Berlin.

Hermitte M. A., 1992. La gestion d'un patrimoine commun: l'exemple de la diversité biologique. In Barrère M. (ed.), *Terre, Patrimoine Commun*, pp. 120–8. La Découverte, Paris.

Hert E., 1992. Homing and home-site fidelity in rock-dwelling cichlids (Pisces, Teleostei) of Lake Malawi, Africa. *Environmental Biology of Fishes*, **33**: 229–37.

Heuts M. J., 1951. Ecology, variations and adaptation of the blind African cave fish *Caecobarbus geertsi* Blgr. *Annales de la Société royale zoologique de Belgique*, **82**: 155–230.

Heuts M. J. & Leleup N., 1954. La géographie et l'écologie des grottes du Bas-Congo. Les habitats de *Caecobarbus geertsi* Blgr. *Annales du Musée royal du Congo belge*, Tervuren, 35.

Hew C. L., 1989. Transgenic fish. *Fish Physiology and Biochemistry*, **7**: 409–13.

Hew C. L. & Fletcher G. L., 1992. *Transgenic Fish*. World Scientific Publishing, Singapore.

Hew C. L. & Gong Z., 1992. Transgenic fish. A new technology for fish biology and aquaculture. *Biology International*, **24**: 2–10.

Hewitt G. M., 1989. The subdivision of species by hybrid zones. In Otte D. & Endler J. A., *Speciation and Its Consequences*, pp. 85–110. Sinauer Associates, Sunderland, MA.

Hickley P. & Bailey R. G., 1986. Fish communities in the perennial wetland of the Sudd, southern Sudan. *Freshwater Biology*, **16**: 695–709.

Hickley P. & Bailey R. G., 1987*a*. Fish communities in the eastern, seasonal floodplain of the Sudd, southern Sudan. *Hydrobiologia*, **144**: 243–50.

Hickley P. & Bailey R. G., 1987*b*. Food and feeding

relationships of fish in the Sudd swamps. *Journal of Fish Biology*, **30**: 147–59.

Hickling C. F., 1960. The Malacca *Tilapia* hybrids. *Journal of Genetics*, **57**: 1–10.

Hildrew A. G. & Giller P. S., 1994. Patchiness, species interactions and disturbance in the stream benthos. In Giller P. S., Hildrew A. G. & Raffaelli D. G. (eds), *Aquatic Ecology. Scale, Pattern and Process*, pp. 21–62. Blackwell Scientific Publications, Oxford.

Hill D. K. & Magnuson J. J., 1990. Potential effect of climate warming on the growth and prey consumption of Great Lakes fishes. *Transactions of the American Fisheries Society*, **119**: 265–75.

Hinch S. G., 1991. Small-scale and large-scale studies in fisheries ecology. The need for cooperation among researchers. *Fisheries*, **16**: 22–7.

Hoar W. S. & Randall D. J. (eds.), 1978. *Fish Physiology*, Vol. VII, Academic Press, London.

Hochahchka P. W. & Somero G. N., 1984. *Biochemical Adaptation*. Princeton University Press, Princeton, NJ.

Hocutt C. H., 1989. Seasonal and diel behaviour of radio-tagged *Clarias gariepinus* in Lake Ngezi, Zimbabwe (Pisces: Clariidae). *Journal of Zoology, London*, **219**: 181–99.

Hogson G. & Dixon J. A., 1988. Logging versus fisheries and tourism in Palawan. East-West Environment and Policy Institute occasional Paper No. 7. East-West Center, Honolulu.

Hohl A. & Tisdell C. A., 1993. How useful are environmental safety standards in economics? – The example of safe minimum standards for protection of species. *Biodiversity and Conservation*, **2**: 168–81.

Holland M. M. (compiler), 1988. SCOPE/MAB technical consultations on landscape boundaries: report of a SCOPE/MAB workshop on ecotones. *Biology International*, Special Issue, **17**: 47–106.

Holling C. S., 1973. Resilience and stability of ecological systems. *Annual Review of Ecology and Systematics*, **4**: 1–23.

Hoogerhoud R. J. C., 1986. Taxonomic and ecological aspects of morphological plasticity in molluscivorous haplochromines (Pisces, Cichlidae). *Annales du Musée royal d'Afrique centrale (Sciences Zoologiques)*, **251**: 131–4.

Hopkins C. D., 1981. On the diversity of electric signals in a community of mormyrid electric fish in West Africa. *American Zoologist*, **21**: 211–22.

Hopson A. J., 1972. A study of the Nile perch (*Lates niloticus* L., Pisces: Centropomidae) in Lake Chad. Overseas Research Publications, London, 19.

Hopson A. J. (ed.), 1982. *Lake Turkana. A Report on the Findings of the Lake Turkana Project 1972–1975*. Overseas Development Administration, London.

Hopson A. J. & Ferguson J. D., 1982. The food of zooplanktivorous fishes. In Hopson A. J. (ed.), *Lake Turkana. A Report on the Findings of the Lake Turkana Project 1972–1975*, pp. 1505–61. Overseas Development Administration, London.

Hopson A. J. & Hopson J., 1982. The fishes of Lake Turkana. In Hopson A. J. (ed.), *Lake Turkana. A Report on the Findings of the Lake Turkana Project 1972–1975*, pp. 281–346. Overseas Development Administration, London.

Hopson A. J., Bayley P. B. & McLeod A. A. Q. R., 1982. The biology of *Hydrocynus forskalii* in Lake Turkana. In Hopson A. J. (ed.), *Lake Turkana. A Report on the Findings of the Lake Turkana Project 1972–1975*, pp. 753–65. Overseas Development Administration, London.

Hopson J., 1975. Preliminary observations on the biology of *Alestes baremoze* (Joannis) in Lake Rudolf. *Symposium on Hydrobiology and Fisheries of Lake Rudolf*, Molo, 25–29 May 1975.

Hora S. L. & Pillay T. V. R., 1962. Handbook of fish culture in the Indo-Pacific region. *FAO Fisheries Biological Technical Papers*, 14.

Hori M., 1983. Feeding ecology of thirteen species of *Lamprologus* (Teleostei, Cichlidae) coexisting on a rocky shore of lake Tanganyika. *Physiology and Ecology, Japan*, **20**: 129–49.

Hori M., 1987. Mutualism and commensalism in a fish community in Lake Tanganyika. In Kawano S., Connell J. H. & Hidaka T. (eds), *Evolution and Coadaptation in Biotic Communities*, pp. 219–39. University of Tokyo Press, Tokyo.

Hori M., Yamaoka K. & Takamura K., 1983. Abundance and microdistribution of cichlids on a rocky shore in Lake Tanganyika. *African Studies Monograph, Kyoto University*, **3**: 25–38.

Horwitz R. J., 1978. Temporal variability patterns and the distributional patterns of stream fishes. *Ecological Monographs*, **48**: 307–20.

Hougard J. M., Poudiougou P., Guiller P., Back C., Akpoboua L. K. B. & Quillévéré D., 1993. Criteria for the selection of larvicide by the Onchocerciasis Control Programme in West Africa. *Annals of Tropical Medicine and Parasitology*, **87**: 435–42.

Hough J. L., 1988. Obstacles to effective management of conflicts between national parks and surrounding human communities in developing countries. *Environmental Conservation*, **15**: 129–36.

Howel F. C. & Bourlière F. (eds.), 1964. *African Ecology and Human Evolution*. Methuen, London.

Howell P., Lock M. & Cobb S., 1988. *The Jonglei Canal: Impact and Opportunity*. Cambridge University Press, Cambridge.

Howells G., Calamari D., Gray J. & Wells P. G., 1990. An analytical approach to assessment of long-term effects of low levels of contaminants in the marine environment. *Marine Pollution Bulletin*, **21**: 371–5.

Howes G. J., 1980. The anatomy, phylogeny and classification of bariliine cyprinid fishes. *Bulletin British Museum (Natural History)* Zoology, **37**: 129–98.

Howes G. J., 1983. Additional notes on bariliine cyprinid fishes. *Bulletin British Museum (Natural History)* Zoology, **45**: 95–101.

Howes G. J., 1984. A review of the anatomy, taxonomy, phylogeny and biogeography of the African neoboline cyprinid fishes. *Bulletin British Museum (Natural History)* Zoology, **47**: 151–85.

Howes G. J., 1987. The phylogenetic position of the Yugoslavian cyprinid fish genus *Aulopyge* Heckel, 1841, with an appraisal of the genus *Barbus* Cuvier & Cloquet, 1861 and the subfamily Cyprininae. *Bulletin British Museum (Natural History)* Zoology, **52**: 165–96.

Howes G. J., 1991. Systematics and biogeography: an overview. In Winfield I. J. & Nelson J. S. (eds.), *Cyprinid Fishes: Systematics, Biology and Exploitation*, pp. 1–33. Fish and Fisheries Series 3 Chapman & Hall, London.

Howes G. J. & Teugels G. G., 1989. New bariliine species from West Africa, with a consideration of their biogeography. *Journal of Natural History*, **23**: 873–902.

Huang C. M., Chang S. L., Cheng H. J. & Liao I. C., 1988. Single gene inheritance of red body coloration in Taiwanese red tilapia. *Aquaculture*, **74**: 227–32.

Hubbell S. P. & Foster R. B., 1986. Biology, chance and history and the structure of tropical rain forest trees. In Diamond J. & Case T. J. (eds.), *Community Ecology*, pp. 314–29. Harper & Row, New York.

Huet M., 1949. Aperçu des relations entre la pente et la population des eaux courantes. *Schweizerische Zeitschrift für Hydrologie*, **11**: 333–51.

Hughes N. F., 1986. Changes in the feeding biology of the Nile perch, *Lates niloticus* (L.) (Pisces, Centropomidae) in Lake Victoria, East Africa since its introduction in 1960, and its impact on the native fish community of the Nyanza Gulf. *Journal of Fish Biology*, **29**: 541–8.

Hughes N. F., 1992. Nile perch, *Lates niloticus*, predation on the freshwater prawn, *Caridina nilotica*, in the Nyanza

Gulf, Lake Victoria, East Africa. *Environmental Biology of Fishes*, **33**: 307–9.

Hugueny B., 1989. West African rivers as biogeographic islands: species richness of fish communities. *Oecologia*, **79**: 236–43.

Hugueny B., 1990a. Geographic range of west African freshwater fishes: role of biological characteristics and stochastic processes. *Acta Oecologica*, **11**: 351–75.

Hugueny B., 1990b. Richesse des peuplements de poissons dans le Niandan (Haut Niger, Afrique) en fonction de la taille de la riviére et de la diversité du milieu. *Revue d'Hydrobiologie tropicale*, **23**: 351–64.

Hugueny B., 1990c. Biogéographie et structure des peuplements de poissons d'eau douce de l'Afrique de l'Ouest: approches quantitatives. Travaux et Documents Microédités, No. 65. ORSTOM, Paris.

Hugueny B. & Lévêque C., 1994. Freshwater fish zoogeography in West Africa: faunal similarities between river basins. *Environmental Biology of Fishes*, **39**: 365–80.

Hugueny B. & Paugy D., 1995. Unsaturated fish communities in African rivers. *American Naturalist*, **146**: 162–9.

Huisman E. A., 1986. The aquacultural potential of the African catfish (*Clarias gariepinus* Burchell, 1822). In Huisman E. A. (ed.), *Aquaculture Research in the Africa Region*, pp. 175–87. Pudoc, Wageningen.

Hulscher-Emeis T. M., 1986. Are colour markings in *Tilapia zillii* associated with motivational systems ? *Annales du Musée royal d'Afrique centrale (Sciences Zoologiques)*, **251**: 35–8.

Humphries C. J., Ladiges P. Y., Roos M. & Zandee M., 1988. Cladistic biogeography. In Myers A. A. & Giller P. S. (eds.), *Analytical Biogeography: An Integrated Approach to the Study of Animal and Plant Distributions*, pp. 371–404. Chapman & Hall, London.

Huntley B. J. (ed.), 1989. *Biotic Diversity in Southern Africa*. Oxford University Press, Cape Town.

Hutchinson G. E., 1959. Homage to Santa Rosalia, or why are there so many kinds of animals? *American Naturalist*, **93**: 145–59.

Hyder M., 1970. Gonadal and reproductive patterns in *Tilapia leucosticta* (Teleostei: Cichlidae) in an equatorial lake, Lake Naivasha (Kenya). *Journal of Zoology, London*, **162**: 179–95.

Hyslop E. J., 1986. The food habits of four small-sized species of Mormyridae from the floodplain pools of the Sokoto-Rima river basin, Nigeria. *Journal of Fish Biology*, **28**: 147–51.

Iles T. D., 1960. An opinion as to the advisability of introducing a non-indigenous zooplankton feeding fish into Lake Nyasa. In CCTA/CSA, *Third Symposium on*

Hydrology and Inland Fisheries. Problems of Major Lakes, Lusaka, August 1960, p. 165. Publications du Conseil Scientifique pour l'Afrique au Sud du Sahara, 63.

Iles T. D., 1973. Dwarfing or stunting in the genus *Tilapia* (Cichlidae); a possibly unique recruitment mechanism. In Parish B. B. (ed.), *Fish Stocks and Recruitment*, pp. 247–54. *Rapport et Procès-verbaux du Conseil permanent international d'exploration de la mer*, 164.

Illies J. & Botosaneanu L., 1963. Problèmes et méthodes de la classification de la zonation écologique des eaux courantes, considérées surtout du point de vue faunistique. *Mitteilungen-Internationale Vereinigung für Theoretische und Angewandte Limnologie*, 12: 1–57.

Im B. H., 1977. Etude de l'alimentation de quelques espèces de *Synodontis* (Poisons, Mochocidae) du Tchad. Thèse de Doctorat de Spécialité, Université Paul Sabatier, Toulouse.

Im B. H. & Lauzanne L., 1978. La sélection des proies chez *Hemisynodontis membranaceus* (Pisces, Mochocidae) du lac Tchad. *Cahiers ORSTOM, série Hydrobiologie*, 12: 237–44.

International Council of Scientific Unions (ICSU), 1986. *The International Geosphere-Biosphere Programme: A Study of Global Change*. Final Report of the *ad-hoc Planning Group*; ICSU, Paris.

Irvine F. R., 1947. *The Fish and Fisheries of the Gold Coast*. Crown Agents for the Colonies, London.

Ishwaran N., 1992. Biodiversity, protected areas and sustainable development. *Nature & Resources*, 28: 18–25.

Ita E. O., 1984. Kainji (Nigeria). In Kapetsky J. M. & Petr T. (eds.), *Status of African Reservoir Fisheries*, pp. 43–103. CIFA Technical Paper, 10.

Ita E. D. & Petr T., 1983. Selected Bibliography on Major African Reservoirs. CIFA Occasional Paper, 10. FAO, Rome.

IUCN, 1980. *World Conservation Strategy: Living Resource Conservation for Sustainable Development*. IUCN–UNEP–WWF, Gland, Switzerland.

Jackson P. B. N., 1961. The impact of predation, especially by the tiger fish (*Hydrocynus vittatus* Castelnau) on African freshwater fishes. *Proceedings of the Zoological Society of London*, 136: 603–22.

Jackson P. B. N., 1971. The African Great Lakes fisheries: past, present and future. *African Journal of Tropical Hydrobiology and Fisheries*, 1: 35–49.

Jackson P. B. N., 1986. Fish of the Zambezi system. In Davies B. R. & Walker K. F. (eds.), *The Ecology of Rivers Systems*, pp. 269–88. W. Junk Publishers, The Hague.

Jackson P. B. N., 1988. Aquaculture in Africa. In Lévêque C., Bruton M. N. & Ssentongo G. W. (eds.), *Biology and Ecology of African Freshwater Fishes*, pp. 459–80. ORSTOM, Paris.

Jackson P. B. N., 1989. Prediction of regulation effects on natural biological rhythms in south-central African freshwater fishes. *Regulated Rivers: Research and Management*, 3: 205–20.

Jackson P. N. B. & Ssentongo G. W., 1988. Fisheries science in Africa. In Lévêque C., Bruton M. N. & Ssentongo G. W. (eds.), *Biology and Ecology of African Freshwater Fishes*, pp. 427–48. ORSTOM, Paris.

Jackson P. B. N., Marshall B. E. & Paugy D., 1988. Fish communities in man-made lakes. In Lévêque C., Bruton M. N. & Ssentongo G. W. (eds.), *Biology and Ecology of African Freshwater Fishes*, pp. 325–50. ORSTOM, Paris.

Jaeger L., 1976. *Monatskarten des Niederschlag fur die ganze Erde*. Beritche Deutschen Wetterdienster, 139.

Jaenisch R., 1988. Transgenic animals. *Science*, 240: 1468–74.

Jalabert B. & Zohar Y., 1982. Reproductive physiology in cichlid fishes, with particular reference to *Tilapia* and *Sarotherodon*. In Pullin R. S. V. & Lowe-McConnell R. H. (eds.), *The Biology and Culture of Tilapias*, pp. 129–40. ICLARM Conference Proceedings 7.

Jalabert B., Kammacher P. & Lessent P., 1971. Déterminisme du sexe chez les hybrides entre *Tilapia macrochir* et *Tilapia nilotica* : étude de la sexratio dans les recroisements des hybrides de première génération par les espèces parentes. *Annales de Biologie animale, biochimie, biophysique*, 11: 155–65.

Jalabert B., Moreau J., Planquette P. & Billard R., 1974. Déterminisme du sexe chez *T. macrochir* et *T. nilotica*: action de la methyltesterone dans l'alimentation des alevins sur la différenciation sexuelle; proportion des sexes dans la descendance des mâles 'inversés'. *Annales de Biologie animale, biochimie, biophysique*, 14: 729–39.

James N. P. E. & Bruton M. N., 1992. Alternative life-history traits associated with reproduction in *Oreochromis mossambicus* (Pisces: Cichlidae) in small water bodies of the eastern Cape, South Africa. *Environmental Biology of Fishes*, 34: 379–92.

Janvier P., 1986. Les nouvelles conceptions de la phylogénie et de la classification des "Agnathes" et des Sarcoptérygiens. *Océanis*, 12: 123–38.

Jégu M. & Lévêque C., 1984a. Les espèces voisines et synonymes de *Labeo parvus* (Pisces, Cyprinidae) en Afrique de l'Ouest. *Cybium*, 8: 873–902.

Jégu M. & Lévêque C., 1984b. Le genre *Marcusenius* (Pisces, Mormyridae) en Afrique de l'Ouest. *Revue d'Hydrobiologie tropicale*, 17: 335–58.

Jobling M., 1995. *Environmental Biology of Fishes*. Fish and Fisheries Series 16. Chapman & Hall, London.

John D. M., 1986. The inland waters of tropical West Africa. An introduction and botanical review. *Archiv für Hydrobiologie*, 23: 1–244.

John D. M., Lévêque C. & Newton L. E., 1993. Western Africa. In Whigam D., Dykyjova D. & Hejny S. (eds.), *Wetlands of the World: Inventory, Ecology and Management*, Vol. 1, pp. 47–78. Kluwer Academic Publishers, Dordrecht.

Johnels A. G., 1954. Notes on fish from Gambia River. *Arkiv for Zoologi*, 6: 326–411.

Johnels A. G. & Svensson G. S. O., 1954. On the biology of *Protopterus annectens* (Owen). *Arkiv for Zoologi*, 7: 131–64.

Johnson G.D., 1992. Monophyly of the euteleostean clades – Neoteleostei, Eurypterygii and Ctenosquamata. *Copeia*, 1: 8–25.

Johnson G.D., 1993. Percomorph phylogeny: progress and problems. *Bulletin of Marine Sciences*, 52: 3–28.

Johnson G. D. & Patterson C., 1993. Percomorph phylogeny: a survey of acanthomorphs and a new proposal. *Bulletin of Marine Sciences*, 52: 554–626.

Johnson R. P., 1974. Synopsis of Biological Data on *Sarotherodon galilaeus*. FAO Fisheries Synopsis, 90. FAO, Rome.

Jubb R. A., 1965. Freshwater fishes of the Cape Province. *Annals of the Cape Provincial Museums, Natural History*, 4: 1–72.

Jubb R. A., 1967. *Freshwater fishes of Southern Africa*. A. A. Balkema, Cape Town.

Junk W. J., Bayley P. B. and Sparks R. E., 1989. The flood pulse concept in river-floodplain systems. In Dodge D. P. (ed.) *Proceedings of the International Large River Symposium*, pp. 110–27. Canadian Special Publication in Fisheries and Aquatic Sciences, 106.

Kalk M., McLachlan A. J. & Howard-Williams C. (eds.), 1979. *Lake Chilwa, Studies of Change in a Tropical Ecosystem*. Monographiae Biologicae 35. W. Junk Publishers, The Hague.

Källqvist T., Lien L. & Liti D., 1988. *Lake Turkana. Limnological Study 1985–1988*. Report, Norwegian Institute for Water Research.

Kapetsky J. M., 1974. The Kafue River floodplain: an example of preimpoundment potential for fish production. In Balon K. & Coche A. G. (eds.), *Lake Kariba: A Man-made Tropical Ecosystem in Central Africa*, pp. 497–523. W. Junk Publishers, The Hague.

Kapetsky J. M., 1981. Some considerations for the management of coastal lagoon and estuarine fisheries. *FAO, Fisheries Technical Papers*, 218: 48.

Kapetsky J. M. & Petr T., 1984. *Status of African Reservoir Fisheries*. CIFA Technical Paper, No. 10. FAO, Rome.

Kapoor B. G., Smit H. & Verighina I. A., 1975. The alimentary canal and digestion in teleosts. *Advances in Marine Biology*, 13: 109–239.

Kapuscinski A. R. & Hallerman E. M., 1990. Transgenic fish and public policy: anticipating environmental impacts of transgenic fishes. *Fisheries*, 15: 2–11.

Karr J. R., 1981. Assessment of biotic integrity using fish communities. *Fisheries*, 6: 21–7.

Karr J. R., 1991. Biological integrity: a long-neglected aspect of water resource management. *Ecological Applications*, 1: 66–84.

Karr J. R. & Dudley L. A., 1981. Ecological perspective in water quality goals. *Environmental Management*, 6: 21–7.

Karr J. R. & Freemark K. E., 1985. Disturbance and vertebrates: an integrative perspective. In Pickett S. T. A. & White P. S. (eds.), *The Ecology of Natural Disturbance and Patch Dynamics*, pp. 153–67. Academic Press, New York.

Karr J. R., Fausch K. D., Angermeier P. L., Yant P. R. & Schlosser I. J., 1986. *Assessing Biological Integrity of Running Waters: A Method and Its Rationale*. Special Publication 5. Illinois Natural History Survey, Champaign, IL.

Karr J. R., Yant P. R., Fausch K. D. & Schlosser I. J., 1987. Spatial and temporal variability of the index of biotic integrity in three midwestern streams. *Transactions of the American Fisheries Society*, 116: 1–11.

Kawanabe H., 1986. Cooperative study on the ecology of lake Tanganyika between Japanese and African scientists, with special reference to mutual interactions among fishes. *Physiological Ecology Japan*, 23: 119–128.

Kawanabe H., 1987. Niche problems in mutualism. *Physiological Ecology Japan*, 24 (special No.): 75–80.

Kawabata H. & Mihigo N. K., 1982. Littoral fish fauna near Uvira, north-western end of Lake Tanganyika. *African Study Monographs*, Kyoto, 2: 133–43.

Kawanabe H., Cohen J. E. & Iwasaki K. (eds.), 1993. *Mutualism and Community Organization: Behavioural, Theoretical and Food Web Approaches*. Oxford University Press, Oxford.

Keenleyside M. H. A., 1991. Parental care. In Keenleyside M. H. A. (ed.), *Cichlid Fishes: Behaviour, Ecology and Evolution*, pp. 191–208. Fish and Fisheries Series 2. Chapman & Hall, London.

Kendall R. L., 1969. An ecological history of the Lake Victoria basin. *Ecological Monographs*, 39: 121.

Kenmuir D. H. S., 1972. Report on a study of the ecology of the tigerfish *Hydrocynus vittatus* Castelnau in Lake Kariba. *Lake Kariba Fisheries Research Institute*, Project Report 6(1): 1–99.

Kenmuir D. H. S., 1975. The diet of fingerling tigerfish, *Hydrocynus vittatus* Cast., in Lake Kariba, Rhodesia. *Arnoldia, Rhodesia*, 9: 1–8.

Kennedy V. S., 1990. Anticipated effects of climate change on estuarine and coastal fisheries. *Fisheries*, 15: 16–24.

Khallaf E. A. & Authman M. N., 1992. Changes in diet, prey size and feeding habit in *Bagrus bajad* and possible interactions with *B. docmac* in a Nile canal. *Environmental Biology of Fishes*, 34: 425–31.

Kiener A., 1963. *Poissons, pêche et pisciculture à Madagascar.* Publication du Centre Technique Forestier tropicale, 24.

Kime D. E., 1982. The control of gonadal androgen biosynthesis in fish. In Richter C. J. J. & Goos H. J. T. (eds.), *Proceedings of the International Symposium on Reproductive Physiology of Fish.* Pudoc, Wageningen, The Netherlands.

Kime D. E. & Hyder M., 1983. The effects of temperature and gonadotrophin on testicular steroidogenesis in *Sarotherodon (Tilapia) mossambicus in vitro. General and Comparative Endocrinology*, 50: 105–115.

Kingsland S., 1985. *Modelling Nature.* University of Chicago Press, Chicago.

Kinzie R. A., 1988. Habitat utilization by Hawaiian stream fishes with reference to community structure in oceanic island streams. *Environmental Biology of Fishes*, 22: 179–92.

Kirk R., 1967. The fishes of Lake Chilwa. *The Malawi Journal Society*, 20: 1–14.

Kirkpatrick M., 1987. Sexual selection by females choice in polygynous animals. *Annual Review of Ecology and Systematics*, 18: 43–70.

Kirshbaum F., 1984. Reproduction of weakly electric teleosts: just another example of convergent development? *Environmental Biology of Fishes*, 10: 3–14.

Kirshbaum F., 1987. Reproduction and development of a weakly electric fish, *Pollimyrus isidori* (Mormyridae, Teleostei) in captivity. *Environmental Biology of Fishes*, 20: 11–31.

Klinkhardt M. B., 1991. A brief comparison of methods for preparing fish chromosomes: an overview. *Cytobios*, 67: 193–208.

Kohda M., 1991. Intra- and interspecific social organization among three herbivorous cichlid fishes in Lake Tanganyika. *Japanese Journal of Ichthyology*, 38: 147–63.

Kohler C. C. & Stanley J. G., 1984. A suggested protocol for evaluation proposed exotic fish introductions in the United States. In Courtenay W. R. & Stauffer J. R. (eds.), *Distribution, Biology and Management of Exotic Fishes*, pp. 387–406. John Hopkins University Press, Baltimore, MA.

Kolasa J., 1989. Ecological systems in hierarchical perspective: breaks in community structure and other consequences. *Ecology*, 70: 36–47.

Kolasa J. & Rollo C. D., 1989. Introduction: the heterogeneity of heterogeneity: a glossary. In Kolasa J. & Pickett S. T. A., *Ecological Heterogeneity*, pp. 1–23. Springer-Verlag, New York.

Kolding J., 1989. The fish resources of Lake Turkana and their environment. Cand. Scient. Thesis, University of Bergen and final report of project KEN-043 Trial Fishery 1986–1987, NORAD, Oslo.

Kolding J., 1992. A summary of Lake Turkana: an ever-changing mixed environment. *Mitteilungen-Internationale Vereinigung für Theoretische und Angewandte Limnologie*, 23: 25–35.

Kolding J., 1993. Population dynamics and life-history styles of Nile tilapia, *Oreochromis niloticus*, in Ferguson's Gulf, Lake Turkana, Kenya. *Environmental Biology of Fishes*, 37: 25–46.

Kolding J., 1994. On the ecology and exploitation of fish in fluctuating tropical freswater systems. Ph.D. thesis, Department of Fisheries and Marine Biology, University of Bergen, Norway.

Konings A., 1988. *Tanganyika Cichlids.* Verduijn Cichlids & Lake Fish Movies, Zevenhuizen, Holland.

Kornfield I., 1978. Evidence for rapid speciation in African cichlid fishes. *Experientia*, 34: 335–6.

Kornfield I., 1991. Genetics. In Keenleyside H. A. (ed.), *Cichlid Fishes: Behaviour, Ecology and Evolution*, pp. 103–28. Fish and Fisheries Series 2. Chapman & Hall, London.

Kornfield I. L., Ritte U., Richler C. & Wahrman J., 1979. Biochemical and cytological differentiation among cichlid fishes of the Sea of Galilee. *Evolution*, 33: 1–14.

Kotliar N. B. & Wiens J. A., 1990. Multiple scales of patchiness and patch structure. A hierarchical framework for the study of heterogeneity. *Oikos*, 59: 253–60.

Kouassi N., 1978. Données écologiques et biologiques sur les populations d'*Alestes baremoze* (Joannis), poisson characidae du lac de barrage de Kossou. Thèse de Doctorat, Abidjan.

Kramer B., 1990. *Electro-communication in Teleost Fishes.* Behavior and Experiments. Zoophysiology, Vol. 29. Springer-Verlag, Berlin.

Kramer D. L., 1983*a*. Aquatic surface respiration in the

fishes of Panama: distribution in relation to risk of hypoxia. *Environmental Biology of Fishes*, 8: 49–54.

Kramer D. L., 1983*b*. The evolutionary ecology of respiratory modes in fishes: an analysis based on the cost of breathing. *Environmental Biology of Fishes*, 9: 145–58.

Kramer D. L., 1987. Dissolved oxygen and fish behaviour. *Environmental Biology of Fishes*, 18: 81–92.

Kramer D. L. & McClure M., 1982. Aquatic surface respiration, a widespread adaptation to hypoxia in tropical freshwater fishes. *Environmental Biology of Fishes*, 7: 47–55.

Kramer D. L., Lindsey C. C., Moodie G. E. E. & Stevens E. D., 1978. The fishes and the aquatic environment of the central Amazon basin, with particular reference to respiratory patterns. *Canadian Journal of Zoology*, 56: 717–29.

Kramer D. L., Manley D. & Bourgeois R., 1983. The effects of respiratory mode on oxygen concentration on the risk of aerial predation in fishes. *Canadian Journal of Zoology*, 61: 653–65.

Kruger E. J. & Polling L., 1984. First attempts at artificial breeding and larval rearing of butter catfish *Eutropius depressirostris* (Pisces, Schilbeidae). *Water S.A.*, 10: 97–104.

Krysanov E. Yu. & Golubtsov A. S., 1993. Karyotypes of three *Garra* species from Ethiopia. *Journal of Fish Biology*, 42: 465–7.

Kühme W., 1964. Eine chemisch ausgelste Brutpflege- und Scwarmreaktion bei *Hemichromis bimaculatus* (Pisces). *Zietschrift für Tierpsychologie*, 20: 688–704.

Kuwamura T., 1986. Parental care and mating systems of cichlid fishes in Lake Tanganyika: a preliminary field survey. *Journal of Ethology*, 4: 129–46.

Kuwamura T., 1987. Distribution of fishes in relation to the depth and substrate at Myako, east-middle coast of Lake Tanganyika. *African Study Monographs, Kyoto University*, 7: 1–14.

Kwetuenda M. K., 1988. Seasonal changes of climatic conditions and water level in the north-western part of Lake Tanganyika. In Kawanabe H. & Kwetuenda M. K. (eds.), *Ecological and Limnological Study on Lake Tanganyika and Its Adjacent Regions*, pp. 66–7. Part V Report, Kyoto University.

Lae R., 1992. Influence de l'hydrologie sur l'évolution des pêcheries du delta central du Niger, de 1966 à 1989. *Aquatic Living Resources*, 5: 115–26.

Lae R., 1995. Climatic and anthropogenic effects on fish diversity and fish yields in the Central delta of the Niger River. *Aquatic Living Resources*, 8: 43–58.

Lae R., Maïga M., Raffray J. & Troubat J. J., 1994.

Evolution de la pêche. In Quensière J. (ed.), *La pêche dans le delta central du Niger*, pp. 143–64. ORSTOM-Khartala, Paris.

Lagler K. F., Kapetsky J. M. & Stewart D. J., 1971. *The Fisheries of the Kafue River Flats, in Zambia, in Relation to the Kafue George Dam*. Rep. Central Fisheries Research Institute, Chilanga, Zambia. Rep. FAO, UNDP/ORA Project 365020 FI: ST/ZAM 11. Technical Report 1.

Lam T. J., 1983. Environmental influences on gonadal activity in fish. In Hoar W. S., Donaldson E. M. & Randall D. J. (eds.), *Fish Physiology*, Vol. 9B. Academic Press, Orlando, FL.

Lamarck J. B., 1809. *Philosophie zoologique, ou exposition des considérations relatives à l'histoire naturelle des animaux*. Paris.

Lande R., 1981. Models of speciation by sexual selection on polygenic traits. *Proceedings of the National Academy of Sciences of the United States of America*, 78: 3721–5.

Lane R. P. & Marshall J. E., 1981. Geographical variation, races and subspecies. In Greenwood P. H. & Forey P. L. (eds.), *Chance, Change and Challenge: The Evolving Biosphere*, pp. 9–19. British Museum (Natural History), London and Cambridge University Press, Cambridge.

Langton R., 1993. In defence of captive breeding of endangered fish. *Aquatic Survival*, 2: 1–4.

Lanzing W., 1974. Sound production in the cichlid *Tilapia mossambica* Peters. *Journal of Fish Biology*, 6: 341–7.

Latif A. F. A., 1984. Lake Nasser-the new man-made lake in Egypt. In Taub F. B. (ed.), *Lakes and Rivers*, pp. 385–410. Elsevier, Amsterdam.

Lauder G. V. & Liem K. F., 1983. The evolution and interrelationships of the Actinopterygian fishes. *Bulletin of the Museum of Comparative Zoology*, 150: 95–196.

Lauder G.V., 1981. Form and function. Structural analysis in evolutionary biology. *Palaeobiology*, 7: 430–442.

Laurec A. & Le Guen J. C., 1981. Dynamique des populations marines exploitées, concepts et modèles. Tome I. Rapport Scientifique et Technique, CNEXO, 45.

Lauzanne L., 1969. Etude quantitative de la nutrition des *Alestes baremoze*. *Cahiers ORSTOM, série Hydrobiologie*, 3: 15–27.

Lauzanne L., 1970. La sélection des proies chez *Alestes baremoze* (Pisces, Characidae). *Cahiers ORSTOM, série Hydrobiologie*, 4: 71–6.

Lauzanne L., 1972. Régimes alimentaires des principales espèces de poissons de l'archipel oriental du lac Tchad. *Verhandlungen-Internationale Vereinigung für Theoretische und Angewandte Limnologie*, 18: 636–46.

Lauzanne L., 1975a. La sélection des proies chez trois poissons malacophages du lac Tchad. *Cahiers ORSTOM, série Hydrobiologie*, 9: 3–7.

Lauzanne L., 1975b. Régime alimentaire d'*Hydrocyon forskalii* (Pisces, Characidae) dans le lac Tchad. *Cahiers ORSTOM, série Hydrobiologie*, 9: 105–21.

Lauzanne L., 1976. Régimes alimentaires et relations trophiques des poissons du lac Tchad. *Cahiers ORSTOM, série Hydrobiologie*, 10: 267–310.

Lauzanne L., 1977. Aspects qualitatifs et quantitatifs de l'alimentation des poissons du Tchad. Thèse Université Paris VI.

Lauzanne L., 1978a. Croissance de *Sarotherodon galilaeus* (Pisces, Cichlidae) du lac Tchad. *Cybium*, 3: 5–14.

Lauzanne L., 1978b. Etude quantitative de l'alimentation de *Sarotherodon galilaeus* (Pisces, Cichlidae) du lac Tchad. *Cahiers ORSTOM, série Hydrobiologie*, 12: 71–81.

Lauzanne L., 1982. Les *Orestias* (Pisces, Cyprinodontidae) du petit lac Titicaca. *Revue d'Hydrobiologie tropicale*, 15: 39–70.

Lauzanne L., 1983. Trophic relationships of fishes in Lake Chad. In Carmouze J. P., Durand J. R. & Lévêque C. (eds.), *Lake Chad: Ecology and Productivity of a Shallow Tropical Ecosystem*, pp. 489–518. Monographiae Biologicae 53, W. Junk Publishers, The Hague.

Lauzanne L. & Iltis A., 1975. La sélection de la nourriture chez *Tilapia galilaea* (Pisces, Cichlidae) du lac Tchad. *Cahiers ORSTOM, série Hydrobiologie*, 9: 193–199.

Lawson R. & Robinson M. A., 1983. The needs and possibilities for the management of canoe fisheries in the CECAF Region. CECAF/Tech 83/47.

Lawton J. H., 1989. What is the relationship between population density and body size in animals? *Oikos*, 55: 429–34.

Lawton J. H. & Brown V. K., 1993. Redundancy in ecosystems. In Schulze E. D. & Mooney H. A. (eds.), *Biodiversity and Ecosystem Function*, pp. 255–70. Ecological Studies 99. Springer-Verlag, New York.

Lazard J., 1990. Transferts de poissons et développement de la production piscicole. Exemple de trois pays d'Afrique subsaharienne. *Revue d'Hydrobiologie tropicale*, 23: 251–65.

Lazard J., Weigel J. Y., Stomal B. & Lecomte Y., 1990. Bilan orientation de la pisciculture en Afrique francophone subsaharienne. Rapport CTFT-CIRAD, Nogent sur Marne, France.

Lazzaro X., 1987. A review of planktivorous fishes: their evolution, feeding behaviours, selectivities and impacts. *Hydrobiologia*, 146: 97–167.

Le Berre M., 1989. *Faune du Sahara. 1 – Poissons–Amphibiens–Reptiles*. Terres Africaines, Lechevalier éditeur, Paris.

Le Cren E. D. & Lowe-McConnell R. H. (eds.), 1980. *The Functioning of Freshwater Ecosystems*. International Biological Programme No. 22. Cambridge University Press, Cambridge.

Lecointre G., 1994. Aspects historiques et heuristiques de l'ichtyologie systématique. *Cybium*, 18: 339–430.

Legendre M. & Albaret J. J., 1991. Maximum observed length (MOL) as an indicator of growth rate in tropical fishes. *Aquaculture*, 94: 327–41.

Legendre M. & Ecoutin J. M., 1989. Suitability of brackish water tilapia species from the Ivory Coast for lagoon aquaculture. I- Reproduction. *Aquatic Living Resources*, 2: 71–79.

Legendre M. & Ecoutin J. M., 1996. Aspects of the reproductive strategy of *Sarotherodon melanotheron* (Ruppel, 1852): comparison between a natural population (lagoon Ebrié, Côte d'Ivoire) and different cultured populations. In Pullin R. V. S., Lazard J., Legendre M., Amon Kotias J. B. & Pauly D. (eds.), *Third International Symposium on Tilapia in Aquaculture* ICLARM Conference Proceedings 41.

Legendre M. & Teugels G. G., 1991. Développement et tolérance à la température des oeufs de *Heterobranchus longifilis*, et comparaison des développements larvaires de *H. longifilis* et de *Clarias gariepinus* (Teleostei, Clariidae). *Aquatic Living Resources*, 4: 227–40.

Legendre M. & Trebaol L., 1996. Efficacité de l'incubation buccale et fréquence de ponte de *Sarotherodon melanotheron* en milieu d'élevage (lagune Ebrié, Côte d'Ivoire). In Pullin R. V. S., Lazard J., Legendre M., Amon Kotias J. B. & Pauly D. (eds.), *Third International Symposium on Tilapia in Aquaculture*. ICLARM Conference Proceedings 41.

Lehman J. T. & Branstrator D. K., 1993. Effects of nutrients and grazing on the phytoplankton of Lake Victoria. *Verhandlungen-Internationale Vereinigung für Theoretische und Angewandte Limnologie*, 25: 850–5.

Leigh E. G. Jr, 1990. Community diversity and environmental stability: a re-examination. *Trends in Ecology and Evolution*, 5: 340–4.

Lek S., 1979. Biologie des petits Mormyridae du bassin tchadien. Thèse de 3ème cycle, Université Paul Sabatier, Toulouse.

Lek S. & Lek S., 1977. Ecologie et biologie de *Micralestes acutidens* (Peters, 1852) (Pisces, Characidae) du bassin du lac Tchad. *Cahiers ORSTOM, série Hydrobiologie*, 11: 255–68.

Lek S. & Lek S., 1978a. Régime alimentaire d'*Ichthyborus*

besse besse (Joannis, 1835) (Pisces, Citharinidae) du bassin du lac Tchad. *Cybium*, 3: 59–75.

Lek S. & Lek S., 1978*b*. Ecologie et biologie d'*Ichthyborus besse besse* (Joannis, 1835) (Pisces, Citharinidae) du bassin du lac Tchad. *Cybium*, 3: 65–86.

Lelek A., 1968. The vertical distribution of fishes in the Ebo stream and notes to the fish occurrence in the Lake Bosumtwi, Ashanti, Ghana. *Zoologicke Listy*, 17: 245–52.

Lelek A. & El Zarka S., 1973. Ecological comparison of the pre-impoundment and post-impoundment fish faunas of the River Niger and Kainji Lake, Nigeria. *Geophysical Monograph*, 17: 655–60.

Lemoalle J., 1974. Bilan des apports en fer au lac Tchad (1970–1973). *Cahiers ORSTOM, série Hydrobiologie*, 8: 35–40.

Lenfant C. & Johansen K., 1968. Respiration in the African lungfish *Protopterus eathiopicus*. I. Respiratory properties of blood and normal patterns of breathing and gas exchange. *Journal of Experimental Biology*, 49: 437–52.

Leopold L. B., Wolman M. B. & Miller J. P., 1964. *Fluvial processes in geomorphology*. W. H. Freeman, San Francisco.

LeRoy Poff N. & Ward J. V., 1989. Implications of streamflow variability and predictability for lotic community structure: a regional analysis of streamflow patterns. *Canadian Journal of Fisheries and Aquatic Sciences*, 46: 1805–18.

Levan A., Fredga K. & Sandberg A.A., 1964. Nomenclature for centromeric position of chromosomes. *Hereditas* (Lund), 52: 201–20.

Lévêque C., 1972. Mollusques benthiques du lac Tchad. Ecologie, étude des peuplements et estimation des biomasses. *Cahiers ORSTOM, série Hydrobiologie*, 6: 3–45.

Lévêque C., 1989*a*. Remarques taxinomiques sur quelques petits *Barbus* (Pisces, Cyprinidae) d'Afrique de l'Ouest (premiére partie). *Cybium*, 13: 165–80.

Lévêque C., 1989*b*. Remarques taxinomiques sur quelques petits *Barbus* (Pisces, Cyprinidae) d'Afrique de l'Ouest (deuxiéme partie). *Cybium*, 13: 197–212.

Lévêque C., 1989*c*. The use of insecticides in the Onchocerciasis Control Programme and aquatic monitoring in West Africa. In Bourdeau P., Haines J.A., Klein W. & Krishna Murti C. R. (eds.), *Ecotoxicology and Climate*, pp. 317–35. SCOPE 38. IPCS Joint Symposia 9. John Wiley & Sons, Chichester.

Lévêque C., 1990*a*. Impact de la lutte antivectorielle sur l'environnement aquatique. *Annales de Parasitologie Humaine et Comparée*, 65, Supplement 1: 119–24.

Lévêque C., 1990*b*. Relict tropical fish fauna in Central

Sahara. *Ichthyological Exploration of Freshwaters*, 1: 39–48.

Lévêque C., 1994*a*. Le concept de biodiversité: de nouveaux regards sur la nature. *Nature, Sciences, Sociétés*, 2: 243–54.

Lévêque C., 1994*b*. Introduction générale: biodiversité des poissons africains. In Teugels G. G., Guégan J. F. & Albaret J. J. (eds.), *Biological Diversity of African Fresh- and Brackish Water Fishes*, pp. 7–16. Annales Sciences zoologiques, Vol. 275. Musée royal d'Afrique centrale, Tervuren, Belgique.

Lévêque C., 1995*a*. Role and consequences of fish diversity in the functioning of African freshwater ecosystems: a review. *Aquatic Living Resources*, 8: 59–78.

Lévêque C., 1995*b*. L'habitat: être au bon endroit au bon moment? *Bulletin Français de Pêche et de pisciculture*, 337–338–339: 9–20.

Lévêque C. & Bigorne R., 1983. Révision des *Leptocypris* et *Raiamas* (Pisces, Cyprinidae) de l'Afrique de l'Ouest. *Revue d'Hydrobiologie tropicale*, 16: 373–93.

Lévêque C. & Bigorne R., 1985*a*. Le genre *Hippopotamyrus* (Pisces, Mormyridae) en Afrique de l'Ouest, avec la description de *Hippopotamyrus paugyi* n. sp. *Cybium*, 9: 175–92.

Lévêque C. & Bigorne R., 1985*b*. Répartition et variabilité des caractéres méristiques et métriques des espéces du genre *Mormyrus* (Pisces, Mormyridae) en Afrique de l'Ouest. *Cybium*, 9: 325–40.

Lévêque C. & Bigorne R., 1987. Caractères morphologiques et distribution de *Ichthyborus quadrilineatus* (Pellegrin, 1904) et *Phago loricatus* Günther, 1865 (Pisces, Distichodontidae) en Afrique de l'Ouest. *Revue d'Hydrobiologie tropicale*, 20: 49–56.

Lévêque C. & Daget J. 1984. Cyprinidae. In Daget J., Gosse J. P., Thys van den Audenaerde D. F. E. (eds.), CLOFFA 1 : *Check List of the Freshwater Fishes of Africa*, pp. 217–342. ORSTOM (Paris) – MRAC (Tervuren).

Lévêque C. & Guégan J. F., 1990. Les grands *Barbus* (Pisces, Cyprinidae) d'Afrique de l'Ouest: révision systématique et parasitofaune branchiale. *Revue d'Hydrobiologie tropicale*, 23: 41–65.

Lévêque C. & Herbinet P., 1982. Caractéres méristiques et biologie d'*Eutropius mentalis* dans les rivières de Côte d'Ivoire (Pisces, Schilbeidae). *Revue de Zoologie africaine*, 96: 366–92.

Lévêque C. & Quensière J., 1988. Les peuplements ichtyologiques des lacs peu profonds. In Lévêque C., Bruton M. N. & Ssentongo G. W. (eds.), *Biology and*

Ecology of African Freshwater Fishes. pp. 303–24, ORSTOM, Paris.

Lévêque C., Odei M. & Puch Thomas M., 1979. The Onchocerciasis Control Programme and the monitoring of its effects on the riverine biology of the Volta River Basin. In Perring F. H. and Mellanby K. (eds.), *Ecological Effects of Pesticides*, pp. 133–43. Linnean Society Symposium Series, No. 5.

Lévêque C., Dejoux C. & Iltis A., 1983. Limnologie du fleuve Bandama (Côte d'Ivoire): *Hydrobiologia*, **100**: 113–41.

Lévêque C., Thys van den Audenaerde D. F. E. & Traore K., 1987. Description de *Barbus parawaldroni* n. sp. (Pisces, Cyprinidae) d'Afrique occidentale. *Cybium*, **11**: 347–55.

Lévêque C., Bruton M. N. & Ssentongo G. W. (eds.), 1988a. *Biology and Ecology of African Freshwater Fishes/Biologie et écologie des poissons d'eau douce africains.* ORSTOM, Paris.

Lévêque C., Fairhurst C., Abban K., Paugy D., Curtis M. S. & Traore K., 1988b. Onchocerciasis Control Programme in West Africa: ten years monitoring of fish populations. *Chemosphere*, **17**: 421–40.

Lévêque C., Teugels G. G., Thys van den Audenaerde D. F. E., 1988c. Description de trois nouvelles espéces de *Barbus* d'Afrique de l'Ouest. *Cybium*, **12**: 179–87.

Lévêque C., Paugy D., Teugels G. G. & Romand R., 1989. Inventaire taxonomique et distribution des poissons d'eau douce des bassins côtiers de Guinée et de Guinée Bissau. *Revue d'Hydrobiologie tropicale*, **22**: 107–27.

Lévêque C., Paugy D. & Teugels G. G., 1990. *Faune des poissons d'eau douce et saumtre d'Afrique de l'Ouest*, Vol. 1. MRAC–ORSTOM, Tervuren and Paris.

Lévêque C., Paugy D. & Teugels G. G., 1991. Annotated check-list of the freshwater fishes of the Nilo-Sudan river basins, in Africa. *Revue d'Hydrobiologie tropicale*, **24**: 131–54.

Lévêque C., Paugy D. & Teugels G. G., 1992. *Faune des poissons d'eau douce et saumtre d'Afrique de l'Ouest*, Vol. 2. MRAC–ORSTOM, Tervuren and Paris.

Lewis D. S. C., 1974. The food and feeding habits of *Hydrocynus forskalii* Cuvier and *Hydrocynus brevis* Günther in Lake Kainji, Nigeria. *Journal of Fish Biology*, **6**: 349–63.

Lewis D. S. C., 1981. Preliminary comparisons between the ecology of haplochromine cichlid fishes of Lake Victoria and Lake Malawi. *Netherlands Journal of Zoology*, **31**: 746–61.

Lewis D. S.C., 1982a. A revision of the genus *Labidochromis*

(Teleostei, Cichlidae) from Lake Malawi. *Zoological Journal of the Linnean Society*, **75**: 189–265.

Lewis D. S. C., 1982b. *Problems of Species Definition in Lake Malawi Cichlid Species (Pisces, Cichlidae).* Special publication No. 23. J. L. B. Smith Institute of Ichthyology, Grahamstown.

Lewis W. M., 1970. Morphological adaptations of cyprinodontoids for inhabiting oxygen deficient waters. *Copeia*, **1970**: 319–26.

Lézine A. M., 1987. Paléoenvironnements végétaux d'Afrique nord-tropicale depuis 12 000 BP. Unpublished thesis, University of Aix-Marseille 2.

Lézine A. M. & Casanova J., 1989. Pollen and hydrological evidence for the interpretation of past climates in tropical West Africa during the Holocene. *Quaternary Science Reviews*, **8**: 45–55.

Liem K. F., 1973. Evolutionary strategies and morphological innovations: cichlid pharyngeal jaws. *Systematic Zoology*, **22**: 424–41.

Liem K. F., 1978. Modulatory multiplicity in the functional repertoire of the feeding mechanism in cichlid fishes. 1. Piscivores. *Journal of Morphology*, **158**: 323–60.

Liem K. F., 1980. Adaptative significance of intra- and interspecific differences in the feeding repertoire of cichlid fishes. *American Zoologist*, **20**: 295–314.

Liem K. F., 1991. Functional morphology. In Winfield I. J. & Nelson J. S. (eds.), *Cyprinid Fishes: Systematics, Biology and Exploitation*, pp. 129–50. Fish and Fisheries Series 3. Chapman & Hall, London.

Liem K. F. & Greenwood P. H., 1981. A functional approach to the phylogeny of the pharyngognath teleosts. *American Zoologist*, **21**: 83–101.

Liem K. F. & Stewart D. J., 1976. Evolution of the scale eating cichlid fishes of Lake Tanganyika: a generic revision with a description of a new species. *Bulletin of the Museum of Comparative Zoology*, **147**: 319–50.

Liese B., 1994. The challenge of success. Land settlement and environmental change in the Onchocerciasis Control Programme Area. Committee of Sponsoring Agencies, Onchocerciaisis Control Programme. World Bank.

Ligtvoet W., 1989. The Nile Perch in Lake Victoria: a blessing or a disaster? *Annales du Musée royal d'Afrique centrale (Sciences Zoologiques)*, **257**: 151–6.

Ligtvoet W. & Witte F., 1991. Perturbation through predator introduction: effects on the food web and fish yields in Lake Victoria (East Africa). In Ravera O. (ed.), *Terrestrial and Aquatic Ecosystems. Perturbation and Recovery*, pp. 263–8. Ellis Horwood, Chichester and New York.

Lindsey C. C., 1978. Form, function and locomotry habits in

fish. In Hoar W. S. & Randall D. J. (eds.), *Fish Physiology*, Vol. VII, pp. 1–100. Academic Press, New York.

Lindsey C. C. & Arnason A. N., 1981. A model for responses of vertebral numbers in fish to environmental influences during development. *Canadian Journal of Fisheries and Aquatic Sciences*, 38: 334–47.

Litterick M. R., Gaudet J. J., Kalff J. & Melak J. M., 1979. *The Limnology of an African Lake, Lake Naivasha, Kenya.* Workshop on African Limnology, Nairobi, Kenya.

Livingstone D. A., 1980. Environmental changes in the Nile headwaters. In Williams F. M. & Faure H. (eds.), *The Sahara and the Nile*, pp. 339–59. Balkema, Rotterdam.

Livingstone D. A., Rowland M. & Bailey P. E., 1982. On the size of African riverine fish faunas. *American Zoologist*, 22: 361–9.

Lock J. M., 1982. The biology of siluriform fishes in Lake Turkana. In Hopson A. J. (ed.), *Lake Turkana. A Report of the Findings of the Lake Turkana Project 1972–1975*, pp. 1021–281. Overseas Development Administration, London. Published by the University of Sterling, Sterling.

Loehle C., 1991. Managing and monitoring ecosystems in the face of heterogeneity. In Kolasa J. & Pickett S. T. A. (eds.), *Ecological Heterogeneity*, pp. 144–59. Ecological Studies 86. Springer-Verlag, New York.

Loiselle P. V., 1985. The care and breeding of *Sarotherodon occidentalis*. *Buntbarsche Bulletin*, 109: 14–24.

Loiselle P. V. & Barlow G. W., 1979. Do fishes lek like birds? In Reese E. & Lighter F. J. (eds.), *Contrasts in Behaviour: Adaptations in the Aquatic and Terrestrial Environments*, pp. 31–75. John Wiley & Sons, Chichester.

Losseau-Hoebeke M., 1992. The biology of four haplochromine species of Lake Kivu (Zaïre) with evolutionary implications. Thesis, Department of Ichthyology, Rhodes University, Grahamstown.

Loubens G., 1969. Etude de certains peuplements ichtyologiques par des pêches au poison. 1ère note. *Cahiers ORSTOM, série Hydrobiologie*, 3: 45–73.

Loubens G., 1970. Etude de certains peuplements ichtyologiques par des pêches au poison. 2éme note. *Cahiers ORSTOM, série Hydrobiologie*, 4: 45–61.

Loubens G., 1974. Quelques aspects de la biologie de *Lates niloticus* du Tchad. *Cahiers ORSTOM, série Hydrobiologie*, 8: 3–21.

Lovelock J. E., 1988. The earth as a living organism. In Wilson E. O. (ed.), *Biodiversity*, pp. 486–9. National Academic Press, Washington, DC.

Lovelock J. E., 1989. Geophysiology: the science of Gaia. *Reviews of geophysics*, 27: 215–22.

Lovelock J. E., 1992. Geophysiological aspects of biodiversity. In Solbrig O. T., van Emden H. M. & van Oordt P. G. W. J. (eds.), *Biodiversity and Global Change*, pp. 57–70. Monograph No. 8. IUBS, Paris.

Løvtrup S., 1979. The evolutionary species: fact or fiction. *Systematic Zoology*, 28: 386–92.

Løvtrup S., 1986. On progressive evolution and competitive extinction. *Environmental Biology of Fishes*, 17: 3–12.

Løvtrup S., 1988. Design, purpose and function in evolution: meditations on a classical problem. *Environmental Biology of Fishes*, 22: 241–7.

Lowe R. H., 1952. Report on the *Tilapia* and other fish and fisheries of Lake Nyasa, 1945–1947. *Fisheries Publication of the Colonial Office, London*, 1(2).

Lowe R. H., 1953. Notes on the ecology and evolution of Nyasa fishes of the genus *Tilapia* with a description of *T. saka* Lowe. *Proceedings of the Zoological Society of London*, 122: 1035–41.

Lowe R. H., 1958. Observations on the biology of *Tilapia nilotica* (Linne) in East African waters. *Revue de Zoologie et de Botanique africaine*, 57: 130–70.

Lowe-McConnell R. H., 1975. *Fish Communities in Tropical Freshwaters: Their Distribution, Ecology and Evolution.* Longman, London.

Lowe-McConnell R. H., 1979. Ecological aspects of seasonality in fishes in tropical waters. In Miller P. J. (ed.), *Fish Phenology*, pp. 219–41. Symposia of the Zoological Society, London, No. 44.

Lowe-McConnell R. H., 1982. Tilapia in fish communities. In Pullin R. V. S. & Lowe-McConnell R. H. (eds.), *The Biology and Culture of Tilapias*, pp. 83–113. ICLARM Conference Proceedings 7, Manila.

Lowe-McConnell R. H., 1985. The biology of the river systems with particular reference to the fishes. In Grove A. T. (ed.), *The Niger and Its Neighbours*, pp. 101–40. Balkema, Rotterdam.

Lowe-McConnell R. H., 1987. *Ecological Studies in Tropical Fish Communities.* Cambridge Tropical Biology Series. Cambridge University Press, Cambridge.

Lowe-McConnell R. H., 1988. Broad characteristics of the ichthyofauna. In Lévêque C., Bruton M. & Ssentongo G. W. (eds.), *Biology and Ecology of African Freshwater Fishes*, pp. 93–110. ORSTOM, Paris.

Lowe-McConnell R. H., 1991. Ecology of cichlids in South American and African waters, excluding the African Great Lakes. In Keenleyside M. H. A. (ed.), *Cichlid Fishes: Behaviour, Ecology and Evolution*, pp. 60–85. Fish and Fisheries Series 2. Chapman & Hall, London.

Lowe-McConnell R. H., 1993. Fish faunas of the Africa Great Lakes: origins, diversity, and vulnerability. *Conservation Biology*, 7: 634–43.

Lubchenco J., Olson A. M., Brubaker L. R., Carpenter S. R., Holland M. M., Hubbell S. P., Levin S. A., MacMahon J. A., Matson P. A., Melillo J. M., Mooney H. A., Peterson C. H., Pulliam H. R., Real L. A., Regal P. J. & Risser P. G., 1991. The Sustainable Biosphere Initiative: an ecological research agenda. *Ecology*, 72: 371–412.

Lundberg J. G., 1993. African–South American freshwater fish clades and continental drift: problems with a paradigm. In Goldblatt P. (ed.), *Biological Relationships Between Africa and South America*, pp. 156–99. Yale University Press, New Haven, CT.

Lykkoboe G., Johansen K. & Maloiy G. M. O., 1975. Functional properties of haemoglobins in the teleost *Tilapia grahami*. *Journal of Comparative Physiology*, 104: 1–11.

Lynch J. D., 1988. Refugia. In Myers A. A. & Giller P. S. (eds.), *Analytical Biogeography: An Integrated Approach to the Study of Animal and Plant Distributions*, pp. 311–42. Chapman & Hall, London.

MacArthur R. H., 1955. Fluctuations of animal populations, and a measure of community stability. *Ecology*, 36: 533–6.

MacArthur, R. H., 1969. Patterns of comunities in the tropics. *Biological Journal of the Linnean Society*, 1, 19–30.

MacArthur R. H., 1972. *Geographical Ecology*. Harper & Row, New York.

MacArthur R. H. & Wilson E. O., 1963. An equilibrium theory of insular zoogeography. *Evolution*, 17: 373–87.

MacArthur R. H. & Wilson E. O., 1967. *The Theory of Island Biogeography*. Princeton University Press, Princeton, NJ.

Maclean N., Penman D & Zhu Z., 1987. Introduction of novel genes into fish. *Biotechnology*, 5: 257–61.

Maetz J. & Bornancin M., 1975. Biochemical and biophysical aspects of salt excretion by chloride cells in teleosts. *Fortschritte der Zoologie*, 22: 322–62.

Magid A. & Babiker M. M., 1975. Oxygen consumption and respiratory behaviour of three Nile fishes. *Hydrobiologia*, 46: 359–67.

Magnuson J. J., 1990. The invisible present. *BioScience*, 40: 495–501.

Magnuson J. J., Bowser C. J. & Beckel A. L., 1983. The invisible present: long term ecological research on lakes. *L & S Magazine*, University of Wisconsin, Madison Fall, 1983: 3–6.

Magurran A. E., 1988. *Ecological Diversity and Its Measurement*. Cambridge University Press, Cambridge.

Mahboudi M., Ameur R., Crochet J. Y. & Jaeger J. J., 1984. Implications paléogéographiques de la découverte d'une nouvelle localité Eocène à Vertébrés continentaux en Afrique nord-occidentale: El Kohol (sud-oranais, Algérie). *Geobios*, 17: 625–9.

Mahdi M. A., 1973. Studies of factors affecting survival of Nile fish in the Sudan. III- The effect of oxygen. *Marine Biology*, 18: 96–8.

Mahé G., 1993. *Les écoulements fluviaux sur la façade atlantique de l'Afrique. Etude des éléments du bilan hydrique et variabilité interannuelle, analyse de situations hydroclimatiques moyennes et extrêmes*. Etudes et Thèses, ORSTOM, Paris.

Mahé G., Lerique J. & Olivry J. C., 1990. Le fleuve Ogoué au Gabon. Reconstitution des débits manquants et mise en évidence de variations climatiques à l'équateur. *Hydrologie continentale*, 5: 105–24.

Mahon R. & Portt C. B., 1985. Local size related segregation of fishes in streams. *Archiv für Hydrobiologie*, 103: 267–71.

Main A. R., 1982. Rare species; precious or dross? In Groves R. H. & Ride W. D. L. (eds.), *Species at Risk: Research in Australia*, pp. 163–74. Australians Academy of Science, Canberra.

Maitland P. S. & Lyle A. A., 1992. Conservation of freshwater fish in the British Isles: proposals for management. *Aquatic Conservation. Marine and Freshwater Ecosystems*, 2: 165–83.

Majumdar K. C. & McAndrew B. J., 1986. Relative DNA content of somatic nuclei and chromosome studies in three genera, *Tilapia*, *Sarotherodon* and *Oreochromis* of the tribe Tilapiini (Pisces, Cichlidae). *Genetica*, 68: 175–88.

Malaisse F., 1969. La péche collective par empoisonnement au "buba" (*Tephrosia vogeli* Hook). Son utilisation dans l'étude des populations de poissons. *Les Naturalistes belges*, 50: 481–500.

Malaisse F., 1976. *Ecologie de la rivière Luanga*. Cercle Hydrobiologique de Bruxelles.

Maley J., 1977. Paleoclimates of central Sahara during the early Holocene. *Nature*, 269: 573–7.

Maley J., 1980. Les changements climatiques de la fin du Tertiaire en Afrique: leurs conséquences sur l'apparition du Sahara et de sa végétation. In Williams M. A. J. & Faure H. (eds.), *The Sahara and the Nile*, pp. 63–86. Balkema, Rotterdam.

Maley J., 1981. *Etudes palynologiques dans le bassin du Tchad et paléoclimatologie de l'Afrique nord tropicale de 30 000 ans à l'époque actuelle*. Travaux et Documents de l'ORSTOM, 129. ORSTOM, Paris.

Maley J., 1983. Histoire de la végétation et du climat d'Afrique nord tropicale au Quaternaire récent. *Bothalia*, **14**: 377–89.

Maley J., 1986. Fragmentation et reconstitution de la forêt dense humide ouest-africaine au cours du Quaternaire récent: hypothèse sur le rôle des up-wellings. In Faure H., Faure L. & Diop E. S. (eds.), *Changements globaux durant le Quaternaire. Passé-présent-futur*, pp. 281–2. Tavaux et Documents, No. 97. ORSTOM, Paris.

Maley J., 1987. Fragmentation de la forêt dense humide africaine et extension des biotopes montagnards au quaternaire récent: nouvelles données polliniques et chronologiques. Implications paléoclimatiques et biogéographiques. *Paleoecology of Africa*, **18**: 307–34.

Maley J., 1989. Late quaternary climatic changes in the African rain forest: forest refugia and the major role of sea surface temperature variations. In Leinene M. & Sarnthein M. (eds.), *Paleoclimatology and Paleometeorology: Modern and Past Patterns of Global Atmospheric Transport*, pp. 585–616. Kluwer Academic Publishers, Boston, MA.

Maley J., 1991. The African rain forest vegetation and paleoenvironments during late quaternary. *Climatic Change*, **19**: 79–98.

Maley J., Livingstone D. A., Giresse P., Brenac P., Kling G., Stager C., Thouveny N., Kelts K., Haag M., Fournier M., Bandet Y., Williamson D. & Zogning A., 1991. West Cameroon Quaternary lacustrine deposits: preliminary results. *Journal of African Earth Sciences*, **12**: 147–57.

Mamonekene V. & Teugels G. G., 1993. *Faune des poissons d'eaux douces de la réserve de la biosphére de Dimonoka (Mayombe, Congo)*. Annales Sciences zoologiques, Vol. 272, Musée Royal de l'Afrique centrale, Tervuren, Belgique.

Mann R. H. K. & Mills C. A., 1979. Demographic aspects of fish fecundity. *Symposia of the Zoological Society of London*, **44**: 161–77.

Mann R. H. K., Mills C. A. & Crisp D. T., 1984. Geographical variation in the life-history tactics of some species of freshwater fish. In Potts G. W. & Wootton R. J. (eds.), *Fish Reproduction: Strategies and Tactics*, pp. 171–86. Academic Press, London.

Margalef R., 1969. Diversity and stability: a practical proposal and a model of interdependence. In Woodwell G. M. & Smith H. H. (eds.), *Diversity and Stability in Ecological Systems*, pp. 25–37. Brookhaven Symposium in Biology, 22.

Marsh A. C., 1981. A contribution to the ecology and systematics of the genus *Petrotilapia* (Pisces, Cichlidae) in Lake Malawi. M.Sc. thesis, Rhodes University, South Africa.

Marsh A. C. & Ribbink A. J., 1985. Feeding site utilization in three sympatric species of *Petrotilapia* (Pisces, Cichlidae) from Lake Malawi. *Biological Journal of the Linnean Society*, **25**: 331–8.

Marsh A. C. & Ribbink A. J., 1986. Feeding schools among Lake Malawi cichlid fishes. *Environmental Biology of Fishes*, **15**: 75–9.

Marsh B. A., Marsh A.C & Ribbink A. J., 1986. Reproductive seasonality in a group of rock-frequenting cichlid fishes in Lake Malawi. *Journal of Zoology, London (A)*, **209**: 9–20.

Marshall B. E., 1982. The fishes of Lake McIlwaine. In Thornton J. A. (ed.), *Lake McIlwaine. The Eutrophication and Recovery of a Tropical African Man-made Lake*, pp. 156–88. Monographiae Biologicae 49. W. Junk Publishers, The Hague.

Marshall B. E., 1984a. Kariba (Zimbabwe/Zambia). In Kapetsky J. M. & Petr T. (eds.), *Status of African Reservoir Fisheries*, pp. 106–53. CIFA Technical Paper, 10.

Marshall B. E., 1984b. *Small Pelagic Fishes in African Inland Waters*. FAO, CIFA Technical Paper, 14.

Marshall L.G., 1988. Extinction. In Myers A. A. & Giller P. S. (eds.), *Analytical Biogeography: An Integrated Approach to the Study of Animal and Plant Distributions*, pp. 219–54. Chapman & Hall, London.

Marten G. G., 1979a. Predator removal: effect on fisheries yields in Lake Victoria (East Africa). *Science*, **203**: 646–8.

Marten G. G., 1979b. The impact of fishing on the inshore fishery of Lake Victoria (East Africa). *Journal of the Fisheries Research Board of Canada*, **36**: 891–900.

Martens K., Coulter G. & Goodeeris B., 1994. Speciation in Ancient Lakes – 40 years after Brooks. *Archiv für Hydrobiologie. Ergebnisse der Limnologie*, **44**: 75–96.

Masters J. C., Rayner R. J., McKay I. J., Potts A. D., Nails D., Ferguson J. W., Weissenbacher B. K., Allsopp M. & Anderson M. L., 1987. The concept of species: recognition versus isolation. *South African Journal of Science*, **83**: 534–7.

Matthes H., 1963. A comparative study of the feeding mechanisms of some African Cyprinidae (Pisces, Cypriniformes). *Bijdragen tot de Dierkunde, Leiden*, **33**: 1–35.

Matthes H., 1964. Les poissons du lac Tumba et de la région d'Ikela. *Annales du Musée royal d'Afrique centrale (Sciences Zoologiques)*, **126**: 1–204.

Matthes H., 1968. The food and the feeding habits of the tiger fish, *Hydrocyon vittatus* (Cast., 1861) in Lake Kariba. *Beaufortia*, **15**: 145–53.

Matthes H., 1985. L'état des stocks et la situations des pêches au lac Itasy. Rapport préparé pour le projet Développement des pêches continentales et de l'aquaculture. FAO, DP/MAG/82/014.

Matthiessen P. & Johnson J. S., 1978. Accumulation of the organophosphate blackfly larvicide Abate (temephos) in *Sarotherodon mossambicus*, with reference to the larvicidal control of *Simulium damnosum*. *Journal of Fish Biology*, **13**: 575–86.

Mavuti K. M., 1990. Ecology and role of zooplankton in the fishery of Lake Naivasha. *Hydrobiologia*, **208**: 131–40.

Mavuti, K. M., 1983. Studies on the community structure, population dynamics and production of the limnetic zooplankton of a tropical lake, Lake Naivasha, Kenya. Ph.D. thesis, University of Nairobi.

May R., 1972. Will a large complex system be stable? *Nature*, **238**: 413–14.

May R., 1973. *Stability and Complexity in Model Ecosystems*. Princeton University Press, Princeton, NJ.

May R., 1981. The role of theory in ecology. *American Zoologist*, **21**: 903–10.

Mayden R. L., 1986. Speciose and depauperate phylads and tests of punctuated and gradual evolution: fact or artefact? *Systematic Zoology*, **35**: 591–602.

Maynard Smith J., 1986. *The Problems of Biology*. Oxford University Press, Oxford.

Maynard Smith J., 1989. *Evolutionary Genetics*. Oxford University Press, Oxford.

Mayr E., 1942. *Systematics and the Origin of Species*. Columbia University Press, New York.

Mayr E., 1963. *Animal Species and Evolution*. Harvard University Press, Cambridge, MA.

Mayr E., 1974. Behavior programs and evolutionary strategies. *American Scientist*, **62**: 650–9.

Mayr E., 1976. *Evolution and the Diversity of Life*. Harvard University Press, Cambridge, MA.

Mayr E., 1982. Processes of speciation in animals. In Barigozzi C. (ed.), *Mechanisms of Speciation*, pp. 1–19. Alan R. Liss, New York.

Mayr E., 1988. *Towards a New Philosophy of Biology: Observations of an Evolutionist*. Harvard University Press, Cambridge, MA.

Mayr E. & O'Hara R. J., 1986. The biogeographic evidence supporting the Pleistocene forest refuge hypothesis. *Evolution*, **40**: 55–67.

McAllister D. E., Platania S. P., Schueler F. W., Baldwin M. E. & Lee S. D., 1986. Ichthyofaunal patterns on a geographic grid. In Hocutt C. H. & Wiley E. O. (eds.), *The Zoogeography of North American Fishes*, pp. 17–51. John Wiley & Sons, New York.

McAndrew B. J. & Majumdar K.C., 1983. Tilapia stock identification using electrophoretic markers. *Aquaculture*, **30**: 249–61.

McAndrew B. J. & Majumdar K. C., 1984. Evolutionary relationships within three Tilapiine genera (Pisces, Cichlidae). *Zoological Journal of the Linnean Society*, **80**: 421–35.

McAndrew B. J., Roubal F. R., Roberts R. J., Bullock A. M. & McEwen I. M., 1988. The genetics and histology of red, blond, and associated colour variants in *Oreochromis niloticus*. *Genetica*, **76**: 127–37.

McCauley J. F., Breed C. S., Schaber G. G., McHugh W. P., Issawi B., Haynes C. V., Grolier M. J. & Ali el Kilani, 1986. Paleodrainages of the eastern Sahara – the radar rivers revisited (SIR-A/B implications for a mid-Tertiary trans-African drainage system). *IEEE Transactions on Geoscience and Remote Sensing*, GE-24 (4): 624–47.

McConnell R.B., 1972. Geological development of the rift system of eastern Africa. *Bulletin of the Geological Society of America*, **83**: 2549–72.

McCune A. R., Thompson K. S. & Olson P. E., 1984. Semiontid fishes from the Mesozoic Great lakes of North America. In Echelle A. A. & Kornfield I. (eds.), *Evolution of Fish Species Flocks*, pp. 27–44. University of Maine, Orono Press, Orono, ME.

McDowall R. M., 1987. The occurrence and distribution of diadromy among fishes. *American Fisheries Society Symposium*, **1**: 1–13.

McElroy D. M. & Kornfield I, 1990. Sexual selection, reproductive behavior, and speciation in the mbuma species flock of Lake Malawi. *Environmental Biology of Fishes*, **28**: 273–84.

McIntosh R. P., 1987. Pluralism in ecology. *Annual Review of Ecology and Systematics*, **18**: 231–41.

McKaye K. R., 1983. Ecology and feeding behaviour of a cichlid fish *Cyrtocara eucinostomus* on a large lek in Lake Malawi, Africa. *Environmental Biology of Fishes*, **8**: 81–6.

McKaye K. R., 1984. Behavioral aspects of cichlid reproductive strategies. Patterns of territoriality and brood defence in Central American substratum-spawners and African mouth-brooders. In Potts G. W. & Wootton R. J. (eds.), *Fish Reproduction: Strategies and Tactics*, pp. 245–73. Academic Press, London.

McKaye K. R., 1991. Sexual selection and the evolution of the cichlid fishes of Lake Malawi, Africa. In Keenleyside H. A. (ed.), *Cichlid Fishes: Behaviour, Ecology and*

Evolution, pp. 241–57. Fish and Fisheries Series 2. Chapman & Hall, London.

McKaye K. R. & Kocher T., 1983. Head ramming behaviour by three paedophagous cichlids in Lake Malawi, Africa. *Animal Behaviour*, 31: 206–10.

McKaye K. R. & Stauffer J. R. Jr, 1988. Seasonality, depth and habitat distribution of breeding males of *Oreochromis* spp. 'Chambo' in Lake Malawi National Park. *Journal of Fish Biology*, 33: 825–34.

McLeod A. A. Q. R., 1982. The biology of *Lates longispinis* Worthington in Lake Turkana. In Hopson A. J. (ed.), *Lake Turkana. A Report on the Findings of the Lake Turkana Project 1972–1975*, pp. 1305–29. Overseas Development Administration, London.

McNeely J. A., 1989. *The Economic Benefits of Conserving Biological Diversity*. IUCN, Gland.

McNeely J. A., 1992. The sinking ark: pollution and the worldwide loss of biodiversity. *Biodiversity and Conservation*, 1: 2–18.

McNeely J. A., Miller K. R., Reid W. V., Mittermeier R. A. & Werner T. B., 1990. *Conserving the World's Biological Diversity*. IUCN, Gland and WRI, World Bank, Conservation International and WWF-US, Washington, DC.

Meffe G. K., 1986. Conservation genetics and the management of endangered fishes. *Fisheries*, 11: 14–23.

Meffe G. K., 1987. Conserving fish genomes: philosophies and practices. *Environmental Biology of Fishes*, 18: 3–9.

Meglitsch P.A., 1954. On the nature of the species. *Systematic Zoology*, 3: 49–65.

Menge B. A. & Olson A. M., 1990. Role of scale and environmental factors in regulation of community structure. *Trends in Ecology and Evolution*, 5: 52–7.

Mérona B. de, 1980. Ecologie et biologie de *Petrocephalus bovei* (Pisces, Mormyridae) dans les rivières de Côte d'Ivoire. *Cahiers ORSTOM, série Hydrobiologie*, 13: 117–27.

Mérona B. de, 1981. Zonation ichthyologique du bassin du Bandama (Côte d'Ivoire). *Revue d'Hydrobiologie tropicale*, 14: 63–75.

Mérona B. de, 1983. Modéle d'estimation rapide de la croissance des poissons. Application aux poissons d'eau douce d'Afrique. *Revue d'Hydrobiologie tropicale*, 16: 103–13.

Mérona B. de, Hecht T. & Moreau J., 1988. Croissance des poissons d'eau douce africains. In Lévêque C., Bruton M. N. & Ssentongo G. W. (eds.), *Biology and Ecology of African Freshwater Fishes*, pp. 191–219. ORSTOM, Paris.

Merron G. S., 1993. Pack-hunting in two species of catfish,

Clarias gariepinus and *C. ngamensis*, in the Okavango Delta, Botswana. *Journal of Fish Biology*, 43: 575–84.

Merron G. S., Holden K. K. & Bruton M. N., 1990. The reproductive biology and early development of the African pike, *Hepsetus odoe*, in the Okavango delta, Botswana. *Environmental Biology of Fishes*, 28: 215–35.

Meybeck M., 1979. Concentration des eaux fluviales en éléments majeurs et apports en solution aux océans. *Revue de Géologie dynamique et de Géographie physique*, 21: 215–46.

Meyer A., 1993. Phylogenetic relationships and evolutionary processes in East African fishes. *Trends in Ecology and Evolution*, 8: 279–84.

Meyer A., 1994. Molecular phylogenetic studies of fish. In Beaumont A. R. (ed.), *Genetic and Evolution of Aquatic Organisms*, pp. 219–49. Chapman & Hall, London.

Meyer A., Kocher T. D., Basasibwaki P. & Wilson A. C., 1990. Monophyletic origin of Lake Victoria cichlid fishes suggested by mitochondrial DNA sequences. *Nature*, 347: 550–3.

Meyer A., Kocher T. D. & Wilson A. C., 1991. African fishes. *Nature*, 350: 467–8.

Meyer A., Montero C. & Spreinat A., 1994. Evolutionary history of the cichlid species flocks of the East African great lakes inferred from molecular phylogenetic data. In Martens K. F., Goddeeris B. & Coulter G. (eds.), *Speciation in Ancient Lakes*, pp. 407–23. *Archiv für Hydrobiologie*, 44.

Meyer A. & Wilson A. C., 1990. Origin of Tetrapods inferred from their mitochondrial DNA affiliation to lungfish. *Journal of Molecular Evolution*, 31: 359–64.

Michel P., 1973. *Les bassins des fleuves Sénégal et Gambie. Etude géomorphologique*, Vol. 2. Mémoires ORSTOM No. 63. ORSTOM, Paris.

Michele J. L. & Takahashi C. S., 1977. Comparative cytology of *Tilapia rendalli* and *Geophagus brasiliensis* (Cichlidae, Pisces). *Cytologia*, 42: 535–7.

Milinski M. & Bakker T. C. M., 1990. Female sticklebacks use male coloration in mate choice and hence avoid parasitized males. *Nature*, 344: 330–3.

Miller P. J., Wright J. & Wongrat P., 1989. An Indo-Pacific goby (Teleostei; Gobioidei) from West Africa, with systematic notes on *Butis* and related eleotridine genera. *Journal of Natural History*, 23: 311–24.

Mills L. S., Soulé M. E. & Doak D. F., 1993. The keystone-species concept in ecology and conservation. *BioScience*, 43: 219–24.

Mina M. V., 1991. *Microevolution of Fishes*. Academy of Sciences of the USSR (1986). Translated from Russian and published by Amerind Publishing, New Delhi.

Minshull J. L., 1978. Preliminary investigations of the ecology of juvenile *Sarotherodon macrochir* (Boulenger). M.Sc. Thesis, University of Rhodesia, Salisbury.

Mironowa, N. V., 1977. Energy expenditure on egg production in young *Tilapia mossambica*, and effect of maintenance conditions on their reproductive rate. *Journal of Ichthyology*, 17: 708–14.

Mitchell S. A., 1976. The marginal fish fauna of Lake Kariba. *Kariba Studies*, No. 8: 109–62.

Mitchell S. A., 1978. The diurnal activity patterns and depth zonation of marginal fish as shown by gill netting in Lake Kariba. *Kariba Studies*, No. 9: 163–73.

Mitson R. B., 1978. A review of biotelemetry techniques using acoustic tags. In Thorpe J. E. (ed.), *Rhythmic Activity of Fishes*, pp. 269–283. Academic Press, London.

Mittermeier R. A. & Bowles I. A., 1993. The GEF and biodiversity conservation: lessons to date and recommendations for future action. *Conservation International*, May 1993, pp. 1–21.

Mo T., 1991. *Anatomy, Relationships, and Systematics of the Bagridae (Teleostei: Siluroidei) with a Hypothesis of Siluroid Phylogeny*. Theses Zoologicae, 17. Koeltz Scientific Books, Champaign, IL.

Mok M., 1975. Biométrie et biologie des *Schilbe* (Pisces, Siluriformes) du bassin tchadien. II- Biologie comparée des deux espèces. *Cahiers ORSTOM, série Hydrobiologie*, 9: 33–60.

Moller P., 1980. Electroreception and the behaviour of mormyrid electric fishes. *Trends in Neurosciences*, 3: 105–9.

Moller P. & Bauer R., 1973. 'Communication' in weakly electric fish, *Gnathonemus petersii* (Mormyridae). II. Interaction of electric organ discharge activities of two fish. *Animal Behaviour*, 21: 501–12.

Moller P., Serrier J. & Belbenoit P., 1976. Electric organ discharges of a weakly electric fish *Gymnarchus niloticus* (Mormyriformes) in its natural habitat. *Experientia*, 32: 1007.

Moller P., Serrier J., Belbenoit P. & Push S., 1979. Notes on ethology and ecology of the Swashi River Mormyrids (Lake Kainji, Nigeria). *Behavioral Ecology and Sociobiology*, 4: 357–68.

Moller P., Serrier J., Squire A. & Boudinot M., 1982. Social spacing in the mormyrid fish *Gnathonemus petersii* (Pisces): a multisensory approach. *Animal Behaviour*, 30: 641–50.

Mondeguer A., Ravenne C., Masse P. & Tiercelin J. J., 1989. Sedimentary basins in an extension and stripe-slip background: the "south Tanganyika troughs complex" East Africa Rift. *Société Géologique de France*, Bulletin Spécial "Bassins en extension: structure et modélisation", (8), V, 3: 501–22.

Monod T., 1928. *L'industrie de pêches au Cameroun*. Société d'Etudes géographiques et maritimes coloniales, Paris.

Monod T., 1951. Contribution à l'étude des peuplements de la Mauritanie. Poissons d'eau douce. *Bulletin de l'Institut français d'Afrique noire (A)*, 13: 802–12.

Monod T., 1954. Contribution à l'étude des peuplements de la Mauritanie. Poissons d'eau douce (2e note). *Bulletin de l'Institut français d'Afrique noire (A)*, 16: 295–99.

Monteil C., 1932. *Djenné, métropole du delta central du Niger*. Société d'Etudes géographiques et maritimes coloniales, Paris, 1–304.

Moran P., Kornfield I. & Reinthal P., 1994. Molecular systematics and radiation of the haplochromine cichlids (Teleostei: Perciformes) of Lake Malawi. *Copeia*, 1994: 274–88.

Moreau J., 1979. Biologie et évolution des peuplements de Cichlidae (Pisces) introduits dans les lacs malgaches d'altitude. Thèse de Doctorat d'Etat, Institut National Polytechnique, Toulouse.

Moreau J., 1980. Le lac Alaotra à Madagascar, 50 ans d'aménagement des pêches. *Cahiers ORSTOM, série Hydrobiologie*, 13: 171–9.

Moreau J., 1982. Etude du cycle reproducteur de *Tilapia rendalli* et *Sarotherodon macrochir* dans un lac tropical d'altitude: le lac Alaotra (Madagascar). *Acta Oecologica*, 3: 3–22.

Moreau J., Arrignon J. & Jubb R. A., 1988. Les introductions d'espèces étrangères dans les eaux continentales africaines. Intérêt et limites. In Lévêque C., Bruton M. N. & Ssentongo G. W. (eds.), *Biology and Ecology of African Freshwater Fishes*, pp. 395–425. ORSTOM, Paris.

Moriarty D. J. W., 1973. The physiology of digestion of blue green algae in the cichlid fish *Tilapia nilotica*. *Journal of Zoology, London*, 171: 25–39.

Moriarty D. J. W. & Moriarty C. M., 1973. The physiology of digestion of blue green algae in the cichlid fish, *Tilapia nilotica*. *Journal of Zoology, London*, 171: 25–39.

Moses B., 1987. The influence of flood regime on fish catch and fish communities of the Cross River floodplain ecosystem, Nigeria. *Environmental Biology of Fishes*, 18: 51–65.

Motwani M. P. & Kanwai Y., 1970. Fish and fisheries of the coffer dammed right channel of the River Niger at Kainji. In Visser S.A (ed.), *Kainji: A Nigerian Man-made Lake*, pp. 27–48. Kainji Lake Studies, Vol. 1 – Ecology. Nigerian Institute of Social and Economic Research, Ibadan.

Moyle P. B. & Leidy R. A., 1992. Loss of biodiversity in

aquatic ecosystems: evidence from fish faunas. In Fiedler P. L. & Jain S. K. (eds.), *Conservation Biology: The Theory and Practice of Nature Conservation, Preservation and Management*, pp. 127–69. Chapman & Hall, London and New York.

Mraja A. H. S., 1982. The biology of *Barbus bynni* in Lake Turkana. In Hopson A. J. (ed.), *Lake Turkana. A Report on the Findings of the Lake Turkana Project 1972–1975*, pp. 817–27. Overseas Development Administration, London.

Mugidde R., 1993. The increase in phytoplankton primary productivity and biomass in Lake Victoria (Uganda). *Verhandlungen-Internationale Vereinigung für Theoretische und Angewandte Limnologie*, 25: 846–9.

Munro A. D., 1990. Tropical freshwater fish. In Munro A. D., Scott A. P. & Lam T. J. (eds.), *Reproduction and Seasonality in Teleosts: Environmental Influences*, pp. 145–239. CRC Press, Boca Raton, FL.

Myers G. S., 1949. Usage of anadromous, catadromous and allied terms for migratory fishes. *Copeia*, 1949: 89–97.

Myers G. S., 1951. Freshwater fishes and East Indian zoogeography. *Stanford Ichthyological Bulletin*, 4: 11–21.

Myers G. S., 1955. Notes on the classification and names of cyprinodont fishes. *Tropical Fish Magazine*, 4: 7.

Myers J. M., 1986. Tetraploid induction in *Oreochromis* spp. *Aquaculture*, 57: 281–7.

Myrberg A. A. Jr, 1965. A descriptive analysis of the behaviour of the African cichlid fish *Pelmatochromis guentheri* (Sauvage). *Animal Behaviour*, 13: 312–29.

Myrerberg A., Kramer E. & Heineke P., 1965. Sound production by cichlid fishes. *Science*, 149: 555–8.

Nagelkerke L. A. J., Sibbing F. A., Boogaart J. G. M. van den, Lammens E. H. R. R. & Osse J. W. M., 1994. The barbs (*Barbus* spp.) of Lake Tana: a forgotten species flock? *Environmental Biology of Fishes*, 39: 1–22.

Nagoshi M., 1985. Growth and survival in larval stage of the genus *Lamprologus* (Cichlidae) in Lake Tanganyika. *Verhandlungen-Internationale Vereinigung für Theoretische und Angewandte Limnologie*, 22: 2663–70.

Naiman R. J., Lonzarich D. G., Beechie T. J. & Ralph S. C., 1992. General principles of classification and the assessment of conservation potential in rivers. In Boon P. J., Calow P. & Petts G. E., *River Conservation and Management*, pp. 93–123. John Wiley & Sons, Chichester.

Nakai K., 1988. Breeding ecology of a cichlid fish, *Lamprologus elongatus*. In Kawanabe H. & Kwetuenda M. K. (eds.), *Ecological and Limnological Study on Lake Tanganyika and Its Adjacent Regions*, pp. 30–2. Kyoto University, Kyoto.

Nakai K., 1993. Foraging of brood predators restricted by territoriality of substrate-brooders in a cichlid fish assemblage. In Kawanabe H., Cohen J. & Iwasaki K. (eds.), *Mutualism and Community Organisation: Behavioural, Theoritical and Food Web Approaches*, pp. 85–108. Oxford University Press, Oxford.

Nakai K., Kawanabe H. & Gashagaza M. M., 1994. Ecological studies on the littoral cichlid communities of Lake Tanganyika: the coexistence of many endemic species. *Archiv für Hydrobiologie*, 44: 373–89.

Nayyar R. P., 1966. Karyotype studies in thirteen species of fishes. *Genetica*, 37: 78–92.

Nei M., 1972. Genetic distances between populations. *American Naturalist*, 106: 283–92.

Nei M., 1978. Estimation of the average heterozygosity and genetic distance from a small number of individuals. *Genetics*, 89: 583–90.

Nelissen M. H. J., 1977a. Sound production by *Haplochromis burtoni* (Gunther) and *Tropheus moori* Boulenger (Pisces, Cichlidae). *Annales de la Société royale Zoologique de Belgique*, 106: 155–66.

Nelissen M. H. J., 1977b. Rhythms of activity of some Lake Tanganyika cichlids. *Annales de la Société royale Zoologique de Belgique*, 107: 147–54.

Nelissen M. H. J., 1978. Sound production by some Tanganyikan cichlid fishes and a hypothesis for the evolution of the communication mechanisms. *Behaviour*, 64: 137–47.

Nelissen M. H. J., 1991. Communication. In Keenleyside H. A. (ed.), *Cichlid Fishes: Behaviour, Ecology and Evolution*, pp. 225–40. Fish and Fisheries Series 2. Chapman & Hall, London.

Nelson G. J., 1973. Relationships of clupeomorphs, with remarks on the structure of the lower jaw in fishes. *Zoological Journal of the Linnean Society*, 53: 333–49.

Nelson G. J., 1978. From Candolle to Croizat: comments on the history of biogeography. *Journal of the History of Biology*, 11: 269–305.

Nelson J. S., 1994. *Fishes of the World*, 3rd edn. John Wiley & Sons, Chichester.

Nevo E., Beiles A. & Ben-Shlomo R., 1984. The evolutionary significance of genetic diversity, ecological, demographic and life history correlates. Evolutionary dynamics of genetic diversity. *Lecture Notes in Biomathematics*, 53: 13–213.

Nijjhar B., Nateg C. K., Amedjo S. D., 1983. Chromosome studies on *Sarotherodon niloticus*, *S. multifasciatus* and *Tilapia busumana*. In Fishelson L. & Yaron Z. (eds.), *Proceedings: International Symposium on Tilapia in Aquaculture*, pp. 256–60. Tel Aviv.

Nishida M., 1991. Lake Tanganyika as evolutionary reservoir of old lineages of East African cichlid fishes: inferences from allozyme data. *Experientia*, **47**: 974–9.

Nixon K. C. & Wheeler Q. D., 1990. An amplification of the phylogenetic species concept. *Cladistics*, **6**: 211–23.

Noakes D. L. G., 1991. Ontogeny of behaviour in cichlids. In Keenleyside H. A. (ed.), *Cichlid Fishes: Behaviour, Ecology and Evolution*, pp. 209–24. Fish and Fisheries Series 2. Chapman & Hall, London.

Noakes D. L. G. & Balon E. K., 1982. Life histories of tilapias: an evolutionary perspective. In Pullin R. S. V. & Lowe-McConnell R. H. (eds.), *Biology and Culture of Tilapias*, pp. 61–82. ICLARM Conference Proceedings 7, Manila.

Norgaard R.B., 1988. Sustainable development: a coevolutionary view. *Futures*, **20**: 606–20.

Norris S. M. & Douglas M. E., 1992. Geographic variation, taxonomic status, and biogeography of two widely distributed African freshwater fishes: *Ctenopoma petherici* and *C. kingsleyae* (Teleostei: Anabantidae). *Copeia*, **1992**: 709–24.

Norris S. M. & Teugels G. G., 1990. A new species of *Ctenopoma* (Pisces, Anabantidae) from southeastern Nigeria. *Copeia*, **1990**: 492–9.

Northcote T. G., 1979. Migratory strategies and production in freshwater. In Gerking S. D. (ed.), *Ecology of Freshwater Fish Production*, pp. 326–59. Blackwell Scientific Publications, Oxford.

Northcote T. G., 1988. Fish in the structure and function of freshwater ecosystems: a "top-down" view. *Canadian Journal of Fisheries and Aquatic Sciences*, **45**: 361–79.

Norton J., 1993. Lake Victoria haplochromine cichlids. *Aquatic Survival*, **2**: 13–16.

Noss R. F., 1987. Saving species by saving ecosystems? *Conservation Biology*, **1**: 175–77.

Noss R. F., 1990. Indicators for monitoring biodiversity: a hierarchical approach. *Conservation Biology*, **4**: 355–64.

Novacek M. J. & Marshall L. G., 1976. Early biogeographic history of ostariophysan fishes. *Copeia*, **1976**: 1–12.

Nunney L. & Campbell K. A., 1993. Assessing minimum viable population size: demography meets population genetics. *Trends in Ecology and Evolution*, **8**: 234–9.

Nwadiaro C. & Okorie P., 1987. Feeding habits of the African bagrid *Chrysichthys filamentosus* in a Nigerian Lake. *Japanese Journal of Ichthyology*, **33**: 376–83.

O'Neill R. V., DeAngelis D. L., Waide J. B. & Allen T. F.H., 1986. *A hierarchical concept of the ecosystem*. Princeton University Press, Princeton, NJ.

Oberdorff T., Ozouf-Costaz C. & Agnese J. F., 1990. Chromosome banding in African catfishes: nucleolar organiser regions in five species of the genus *Synodontis* and some of the genus *Hemisynodontis* (Pisces, Mochokidae). *Caryologia*, **43**: 9–16.

Ochumba P. B. O., 1990. Massive fish kills within the Nyanza Gulf of Lake Victoria, Kenya. *Hydrobiologia*, **208**: 93–9.

Oellermann L. K. & Skelton P. H., 1990. Hexaploidy in yellowfish species (*Barbus*, Pisces, Cyprinidae) from southern Africa. *Journal of Fish Biology*, **37**: 105–15.

Ogari J. T. N., 1982. The biology of *Haplochromis macconneli* Greenwood in Lake Turkana. In Hopson A. J. (ed.), *Lake Turkana. A Report on the Findings of the Lake Turkana Project 1972–1975*, pp. 1369–75. Overseas Development Administration, London.

Ogutu-Ohwayo R., 1990*a*. Changes in the prey ingested and the variations in the Nile perch and other fish stocks of Lake Kyoga and the northern waters of Lake Victoria (Uganda). *Journal of Fish Biology*, **37**: 55–63.

Ogutu-Ohwayo R., 1990*b*. The decline of the native fishes of lakes Victoria and Kyoga (East Africa) and the impact of introduced species, especially the Nile perch, *Lates niloticus*, and the Nile tilapia, *Oreochromis niloticus*. *Environmental Biology of Fishes*, **27**: 81–96.

Ogutu-Ohwayo R., 1990*c*. The reduction in fish species diversity in lakes Victoria and Kyoga (East Africa) following human exploitation and introduction of non-native fishes. *Journal of Fish Biology*, **37**: 207–8.

Ogutu-Ohwayo R., 1993. The effects of predation by Nile perch, *Lates niloticus* L., on the fish of Lake Nabugabo, with suggestions for conservation of endangered endemic cichlids. *Conservation Biology*, **7**: 701–11.

Oijen M. J. P. van, 1982. Ecological differentiation among the haplochromine piscivorous species of Lake Victoria. *Netherlands Journal of Zoology*, **32**: 336–63.

Oijen M. J. P. van, 1989. Notes on the relationship between prey size and pharyngeal jaw size in piscivorous haplochromine cichlids from Lake Victoria. *Annales du Musée royal d'Afrique centrale (Sciences Zoologiques)*, **257**: 73–6.

Oijen M. J. P. van, Witte F. & Witte-Maas E. L. M., 1981. An introduction to ecological and taxonomic investigations on the haplochromine cichlids from Mwanza Gulf of Lake Victoria. *Netherlands Journal of Zoology*, **31**: 149–74.

Okach J. O. & Dadzie S., 1988. The food, feeding habits and distribution of a siluroid catfish, *Bagrus docmac* (Forskal) in the Kenya waters of Lake Victoria. *Journal of Fish Biology*, **32**: 85–94.

Okada T. & Nagahama Y. (eds.), 1993. Biotechnology of

aquatic animals. *Biology International*, special issue No. 28.

Okedi J., 1969. Observations on the breeding and growth of certain mormyrid fishes from Lake Victoria basin. *Revue de Zoologie et de Botanique africaine*, 79: 34–64.

Okedi J., 1970. A study of the fecundity of some mormyrid fishes from L. Victoria. *East African Agriculture and Forestry Journal*, 35: 436–42.

Okorie O., 1973. Lunar periodicity and the breeding of *Tilapia nilotica* in the northern part of Lake Victoria. *East African Freshwater Fisheries and Research Organisation, Annual Report*, Appendix E, pp. 50–8.

Oliver M. K., 1984. Systematics of African cichlid fishes: determination of the most primitive taxon, and studies of the haplochromines of Lake Malawi (Teleostei, Cichlidae). Ph.D. thesis, Yale University, New Haven, CT.

Olivry J. C., 1986. *Fleuves et rivières du Cameroun.* Monographies Hydrologiques, 9. ORSTOM, Paris.

Oppenheimer J. R., 1970. Mouthbreeding in fishes. *Animal Behaviour*, 18: 493–503.

Orians G. H., 1974. Diversity, stability and maturity in natural ecosystems. In van Dobben W. H. & Lowe-McConnell R. H. (eds.), *Unifying Concepts in Ecology*, pp. 139–50. W. Junk Publishers, The Hague.

Otobo F. O., 1974. The potential clupeid fishery in Lake Kainji, Nigeria. *African Journal of tropical Hydrobiology and Fisheries*, 3: 123–34.

Otobo F. O., 1978. The reproductive biology of *Pellonula afzeluisi* Johnels, and *Sierrathrissa leonensis* Thys van den Audenaerde in Lake Kainji, Nigeria. *Hydrobiologia*, 61: 99–112.

Owen H. G., 1983. Some principles of physical palaeogeography. In Sims R. W., Price J. H. & Whalley P. E. S. (eds.), *Evolution, Time and Space*, pp. 85–114. Academic Press, London.

Owen R. B., Crossley R., Johnson T. C. *et al.*, 1990. Major low levels of Lake Malawi and implications for speciation rates in cichlid fishes. *Proceedings of the Royal Society of London*, B, 240: 519–53.

Ozouf-Costaz C. & Foresti F., 1992. Fish cytogenetic research: advances, applications and perspectives. *Netherlands Journal of Zoology*, 42: 277–90.

Ozouf-Costaz C., Teugels G. G. & Legendre M., 1990. Karyological analysis of three strains of the African Clariidae catfish *Clarias gariepinus* used in aquaculture. *Aquaculture*, 87: 271–7.

Paine R. T., 1969. A note on trophic complexity and community stability. *American Naturalist*, 100: 65–75.

Panchen A. L., 1992. *Classification, Evolution and the Nature of Biology*, Cambridge University Press, Cambridge.

Pandare D. & Romand R., 1989. Feeding rates of *Aphyosemion geryi* (Cyprinodontidae) on mosquito larvae in the laboratory and in the field. *Revue d'Hydrobiologie tropicale*, 22: 251–58.

Pandian T. J. & Varadaraj K., 1987. Techniques to regulate sex-ratio and breeding in tilapia. *Current Science*, 56: 337–43.

Parenti L., 1981. A phylogenetic and biogeographic analysis of cyprinodontiform fishes (Teleostei, Atherinomorpha). *Bulletin of the American Museum of Natural History*, 168: 341–557.

Park E.H., 1974. A list of chromosome numbers of fishes. *College Review. College of Liberal Arts and Sciences. Seoul National University*, 20: 346–72.

Parsons P. A., 1988. Adaptation. In Myers A. A. & Giller P. S. (eds.), *Analytical Biogeography: An Integrated Approach to the Study of Animal and Plant Distributions*, pp. 165–84. Chapman & Hall, London.

Parzefall J., 1993. Behavioural ecology of cave-dwelling fishes. In Pitcher T. J. (ed.), *Behaviour of Teleost Fishes*, pp. 573–606. Chapman & Hall, London.

Pasteur N., Pasteur G., Bonhomme F., Catalan J. & Britton-Davidian J., 1988. *Practical Isozyme Genetics*. Ellis Horwood, Chichester.

Pastouret L. Chamley H., Delibrias G., Duplessy J. C. & Thiede J., 1978. Late Quaternary climatic changes in western tropical Africa deduced from deep-sea sedimentation off the Niger delta. *Oceanologica Acta*, 1: 217–32.

Pate J. S. & Hopper S. D., 1993. Rare and common plants in ecosystems with special reference to the South-West Australian flora. In Schulze E. D. & Mooney H. A. (eds.), *Biodiversity and Ecosystem Function*, pp. 193–325. Ecological Studies 99, Springer-Verlag, New York.

Paterson H. E., 1980. A comment on "mate recognition systems". *Evolution*, 34: 330–1.

Paterson H. E., 1982. Perspective on speciation by reinforcement. *South African Journal of Science*, 78: 53–7.

Paterson H. E., 1985. The Recognition Concept of species. In Vrba E. S. (ed.), *Species and Speciation*, pp. 21–9. Transvaal Museum Monographs 4. Transvaal Museum, Pretoria.

Patterson C., 1975. The distribution of Mesozoic freshwater fishes. *Mémoires du Muséum d'Histoire naturelle de Paris*, 88: 156–73.

Patterson C., 1981. The development of the North American fish fauna – a problem of historical biogeography. In Greenwood P. H. & Forey P. L. (eds.), *Chance, Change and Challenge: The Evolving Biosphere*, pp. 265–81. British Museum (Natural History), London and Cambridge University Press, Cambridge.

Patterson C., 1982. Morphology and interrelationships of primitive actinopterygian fishes. *American Zoologist*, 22: 241–59.

Patterson C., 1984. *Chanoides*, a marine Eocene otophysan fish (Teleostei; Ostariophysi). *Journal of Vertebrate Paleontology*, 4: 430–56.

Patterson C. & Rosen D.E., 1977. Review of ichthyodectiform and other Mesozoic teleost fishes and the theory and practice of classifying fossils. *Bulletin of the American Museum of Natural History*, 158: 81–172.

Paugy D., 1978. Ecologie et biologie des *Alestes baremoze* (Pisces, Characidae) des riviéres de Côte d'Ivoire. *Cahiers ORSTOM, série Hydrobiologie*, 12: 245–75.

Paugy D., 1980a. Ecologie et biologie des *Alestes imberi* (Pisces, Characidae) des riviéres de Côte d'Ivoire. Comparaison méristique avec *Alestes nigricauda*. *Cahiers ORSTOM, série Hydrobiologie*, 13: 129–41.

Paugy D., 1980b. Ecologie et biologie des *Alestes nurse* (Pisces, Characidae) des riviéres de Côte d'Ivoire. *Cahiers ORSTOM, série Hydrobiologie*, 13: 143–59.

Paugy D., 1982. Synonymie d'*Alestes rutilus* Boulenger, 1916 avec *A. macrolepidotus* (Valenciennes, 1849). Biologie et variabilité morphologique. *Revue de Zoologie africaine*, 96: 286–315.

Paugy D., 1986. *Révision systématique des* Alestes *et* Brycinus *africains (Pisces, Characidae)*. Collection Etudes et Thèses. ORSTOM, Paris.

Paugy D., 1990. Note à propos des Petersiini (Teleostei: Characidae) d'Afrique occidentale. *Ichthyological Exploration of Freshwaters*, 1: 75–84.

Paugy D. & Bénech V., 1989. Poissons d'eau douce des bassins côtiers du Togo (Afrique de l'Ouest). *Revue d'Hydrobiologie tropicale*, 22: 295–316.

Paugy D. & Guégan J. F., 1989. Note à propos de trois espéces d'*Hydrocynus* (Pisces, Characidae) du bassin du Niger, suivie de la réhabilitation de l'espéce *Hydrocynus vittatus* (Castelnau, 1861). *Revue d'Hydrobiologie tropicale*, 22: 63–9.

Paugy D., Lévêque C., Teugels G. G. & Romand R., 1989. Freshwater fishes of Sierra Leone and Liberia: annotated checklist and distribution. *Revue d'Hydrobiologie tropicale*, 23: 329–50.

Paugy D., Guégan J. F. & Agnèse J. F., 1990. Three simultaneous and independent approaches to the characterisation of a new species of *Labeo* (Teleostei, Cyprinidae) from West Africa. *Canadian Journal of Zoology*, 68: 1124–31.

Paugy D., Traoré K. & Diouf P. S., 1994. Faune ichthyologique des eaux douces d'Afrique de l'Ouest. In Teugels G. G., Guégan J. F. & Albaret J. J. (eds.), *Biological Diversity of African Fresh- and Brackish Water Fishes*, pp. 35–66. Annales Sciences zoologiques, Vol. 275. Musée royal d'Afrique centrale, Tervuren, Belgique.

Pauly D., 1984. A mechanism for the juvenile-to-adult transition in fishes. *Journal du Conseil internationale d'Exploration de la mer*, 41: 280–4.

Pearce D.W., 1988. The sustainable use of natural resources in developing countries. In Turner R. K. (ed.), *Sustainable Environmental Management: Principles and Practice*. Belhaven Press, London.

Pearce D. W. & Turner R. K., 1989. *Economics of Natural Resources and the Environment*. The Johns Hopkins University Press, Baltimore, MA.

Pélissier C., Leung Tack D. & Gras G., 1982. Action du téméphos sur l'activité acétylcholinestérasique du cerveau de *Tilapia guineensis*. 2° partie: étude expérimentale lors d'une exposition de 24 heures au toxique. *Toxicological European Research*, 4: 309–14.

Pellegrin J., 1904. Characinidés nouveaux de la Casamance. *Bulletin du Muséum national d'Histoire naturelle, Paris*, 10: 218–21.

Pellegrin J., 1909. Description d'un *Barbus* nouveau du Sahara. *Bulletin du Muséum national d'Histoire naturelle, Paris*, 15: 239–40.

Pellegrin J., 1911. La distribution des poissons d'eau douce en Afrique. *Comptes rendus hebdomadaires des Séances de l'Académie des Sciences, Paris*, 153: 287–99.

Pellegrin J., 1914. Les vertébrés des eaux douces du Sahara. *Comptes rendus de l'Association française pour l'Avancement des Sciences*, pp. 346–52.

Pellegrin J., 1919a. Poissons du Tibesti, du Borkou et de l'Ennedi récoltés par la mission Tilho. *Bulletin de la Société Zoologique de France*, 44: 148–53.

Pellegrin J., 1919b. Sur un Cyprinidé nouveau du Tibesti appartenant au genre *Labeo*. *Bulletin de la Société Zoologique de France*, 44: 325–7.

Pellegrin J., 1921. Les poissons des eaux douces de l'Afrique du Nord française et leur distribution géographique. *Comptes rendus de l'Association française pour l'Avancement des Sciences, Congrès de Strasbourg*, pp. 269–73.

Pellegrin J., 1923. Les poissons des eaux douces de l'Afrique occidentale (du Sénégal au Niger). *Publications du Comité d'Etudes Historiques et Scientifiques de l'Afrique occidentale française*, Larose Ed., Paris.

Pellegrin J., 1931. Reptiles, Batraciens et poissons du Sahara central recueillis par le Prof. Seurat. *Bulletin du Muséum national d'Histoire naturelle, Paris*, 23: 216–18.

Pellegrin J., 1934. Reptiles, Batraciens et Poissons du Sahara central. *Mémoires de la Société d'Histoire naturelle d'Afrique du nord*, **4**: 50–7.

Pellegrin J., 1936. Sur la présence de la tilapie de Zill dans le sud marocain. *Bulletin de la Société de Sciences naturelles du Maroc*, **16**: 1–2.

Penman D. J., Shah M. S., Beardmore J. A. & Skibinski D. O. F., 1987*a*. Sex ratios of gynogenetic and triploid tilapia. In *Selection, Hybridization and Genetic Engineering in Aquaculture*, pp. 267–76. Heeneman, Berlin.

Penman D. J., Skibinski D. O. F. & Berdmore J. A., 1987*b*. Survival, growth rate and maturity in triploid tilapia. In *Selection, Hybridization and Genetic Engineering in Aquaculture*, pp. 277–88. Heeneman, Berlin.

Perrot C. H., 1989. Le système de gestion de la pêche en lagune Aby au XIXème siècle (Côte d'Ivoire). *Cahiers des Sciences Humaines*, **25**: 177–88.

Perrot R. A. & Street-Perrot F. A., 1982. New evidence for a late Pleistocene wet phase in northern intertropical Africa. *Paleoecology of Africa*, **14**: 57–73.

Peters H. M. & Berns S., 1978. Uber die Vorgeschichte der maulbrütenden Cichliden. *Aquarien magazin, Stuttgart* **1978** (5): 211–17 and (7): 324–31.

Peters H. M. & Berns S., 1982. Larvophile und ovophile Maulbrüter. *Tatsachen und Informationen aus der Aquaristik*, **58**: 19–22.

Peters N., 1963. Embryonale Anpassungen oviparer Zahkarpfen aus periodische autrocknenden Gewassern. *Internationale Revue der gesamten Hydrobiologie*, **48**: 257–313.

Peters N., 1965. Diapause und embryonale Missbildung bei eierlegenden Zahnkarfen. *Wilhelm Roux' Archiv für Entwicklungsmechanik der Organisem*, **156**: 75–87.

Peters R. L., 1991. Consequences of global warming for biological diversity. In Wyman R. (ed.), *Global Climate Change and Life on Earth*, pp. 99–118. Chapman & Hall, London.

Petersen R. C., Madsen B. L., Wilzbach M. A., Magadza C. H. D., Paarlberg A., Kullberg A. & Cummins K. W., 1987. Stream management: emerging global similarities. *Ambio*, **16**: 166–79.

Petit-Maire N., 1989. Interglacial environments in presently hyperarid Sahara: palaeoclimatic implications. In Leinen M. & Sarnthein M. (eds.), *Paleoclimatology and Paleometeorology: Modern and Past Patterns of Global Atmospheric Transport*, pp. 637–61. Kluwer Academic Publishers, Dordrecht.

Petr T., 1968. Distribution, abundance and food of commercial fish in the Black Volta and the Volta man-made lake in Ghana, during its first period of filling (1964–1966). I. Mormyridae. *Hydrobiologia*, **32**: 417–48.

Petr T., 1970. The bottom fauna of the rapids of the Black Volta River. *Hydrobiologia*, **36**: 399–418.

Petr T., 1978. Tropical man made lakes. Their ecological impact. *Archiv für Hydrobiologie*, **81**: 368–85.

Petr T., 1986. The Volta river system. In Davies B. R. & Walker K. F. (eds.), *The Ecology of River Systems*, pp. 163–99. W. Junk Publishers, Dortrecht.

Pezzey J., 1989. "Definitions of sustainability". Discussion Paper 9, UK Centre for Economic and Environmental Development, London.

Philippart J. C. & Ruwet J. C., 1982. Ecology and distribution of Tilapias. In Pullin R. V. S. & Lowe-McConnell R. H. (eds.), *The Biology and Culture of Tilapias*, pp. 15–59. ICLARM Conference Proceedings 7, Manila.

Pianka E. R., 1983. *Evolutionary Ecology*, 3rd edn. Harper & Row, New York.

Pickett S. T. A., 1991. Long term studies: past experience and recommendations for the future. In Risser P. G. (ed.), *Long-term Ecological Research*, pp. 71–88. SCOPE 47. John Wiley & Sons, Chichester.

Pickett S. T. A. & White P. S., 1985. Patch dynamics: a synthesis. In Pickett S. T. A. & White P. S. (eds.), *The Ecology of Natural Disturbance and Patch Dynamics*, pp. 371–84. Academic Press, New York.

Pickett S. T. A., Parker V. T. & Fiedler P. L., 1992. The new paradigm in ecology: implications for conservation biology above the species level. In Fiedler P. L. & Jain S. K. (eds.), *Conservation Biology. The Theory and Practice of Nature Conservation, Preservation and Management*, pp. 65–88. Chapman & Hall, London and New York.

Pienaar U., de V., 1968. The freshwater fishes of the Krüger National Park. National Parks Board of South Africa. *Koedoe*, **11**.

Pietsch T. W., 1978. Evolutionary relationships of the sea moths (Teleostei, Pegasidae) with a classification of Gasterosteiform families. *Copeia*, **1978**: 517–29.

Pimm S. L., 1982. *Food Webs*. Chapman & Hall, London.

Pimm S. L., 1991. *The Balance of Nature? Ecological Issues in the Conservation of Species and Communities*. The University of Chicago Press, Chicago.

Pimm S. L., 1993. Biodiversity and the balance of nature. In Schulze E. D. & Mooney H. A. (eds.), *Biodiversity and Ecosystem Function*, pp. 347–59. Ecological Studies 99. Springer-Verlag, New York.

Pimm S. L. & Gilpin M. E., 1989. Theoretical issues in conservation biology. In Roughgarden J., May R. M. &

Levin S. A. (eds.), *Perspectives in Ecological Theory*, pp. 287–305. Princeton University Press, Princeton, NJ.

Pinchot G., 1947. *Breaking New Ground*. Harcourt, Brace, New York.

Place A. R. & Powers D. A., 1979. Genetic variation and relative catalytic efficiencies: lactate dehydrogenase B allozymes of *Fundulus heteroclitus*. *Proceedings of the National Academy of Sciences of the United States of America*, 76: 2354–8.

Plisnier P. D., 1990. Ecologie comparée et exploitation rationnelle de deux populations d'*Haplochromis* spp. (Teleostei, Cichlidae) des lacs Ihema et Muhazi (Rwanda). Dissertation pour le grade de Docteur en Sciences Agronomiques, Université Catholique de Louvain-la-Neuve, Belgique.

Plisnier P. D., Micha J. C. & Frank V., 1988. *Biologie et exploitation des poissons du lac Ihema (bassin Akagera, Rwanda)*. Presses Universitaires, Namur.

Pliya J., 1980. *La pêche dans le sud-ouest du Bénin*. Agence de coopération culturelle et technique, Paris.

Plotkin H. C., 1988. *The Role of Behavior in Evolution*. MIT Press, Cambridge, MA.

Polis A. G. & Holt R. D., 1992. Intraguild predation: the dynamics of complex trophic interactions. *Trends in Ecology and Evolution*, 7: 151–4.

Poll M., 1950. Histoire du peuplement et origine des espèces de la faune ichtyologique du lac Tanganika. *Annales de la Société royale zoologique de Belgique*, 81: 111–40.

Poll M., 1963. Zoogéographie ichtyologique du cours supérieur du Lualaba. *Publications de l'Université d'Elizabethville*, 6: 95–106.

Poll M., 1967. Contribution à la faune ichtyologique de l'Angola. Companhia de diamantes de Angola (Museo de Dundo). *Publicaçoes culturais*, 75: 1–381.

Poll M., 1971. Révision des *Synodontis* africains (Famille Mochokidae). *Annales du Musée royal d'Afrique centrale*, **191**.

Poll M., 1973. Nombre et distribution géographique des poissons d'eau douce africains. *Bulletin du Muséum national d'Histoire naturelle, Paris*, 150: 113–28.

Poll M., 1976. *Exploration du Parc National de l'Upemba*. Mission G. F. de Witte: Poissons. Fondation pour favoriser les Recherches Scientifiques en Afrique, **73**.

Poll M., 1986. Classification des cichlidae du lac Tanganyika: tribus, genres, et espèces. *Académie royale de Belgique, Mémoires de la classe des sciences*. Coll. in 8°-27me série, tome **XLV**, fascicule 2: 1–163.

Poll M., 1987. Un genre inédit pour une espèce nouvelle du lac Tanganyika: *Trematochromis schreyeni* gen. n. sp.

Statut de *Tilapia trematocephala* Boulenger, 1901. *Cybium*, 11: 167–72.

Poll M. & Gosse J. P., 1963. Contribution à l'étude systématique de la faune ichtyologique du Congo central. *Annales du Musée royal d'Afrique centrale*, in-8° Zool., 116: 41–101.

Pomerol C. & Renard M., 1989. *Éléménts de géologie*. Armand Collin, Paris.

Popper D. & Lichatowich T., 1975. Preliminary success in predator contact of *Tilapia mossambica*. *Aquaculture*, 5: 213–14.

Popper K. R., 1962. *Conjectures and Refutation: The Growth of Scientific Knowledge*. Routledge and Kegan Paul, London.

Post A. (von), 1965. Vergleichende Untersuchungen der Chromozomenzahlen bei Süsswasser Teleosteen. *Zeitschrift für Zoologische Systematik und Evolutionforschung*, 3: 47–93.

Potts W. T. W., Foster M. A., Rudy P. P. & Howells G. P., 1967. Sodium and water balance in the cichlid teleost, *Tilapia mossambica*. *Journal of Experimental Biology*, 47: 461–70.

Pouyaud L., 1994. Génétique des populations de tilapias d'intérêt aquacole en Afrique de l'Ouest. Relations phylogénétiques et structurations populationnelles. Diplôme de Doctorat, Université de Montpellier II, Sciences et Techniques du Languedoc.

Powers D. A., 1989. Fish as model systems. *Science*, **246**: 352–8.

Price M. F., 1992. The evolution of global environmental change. Issues and research programmes. *Impact of Science on Society*, 166: 171–82.

Priem R., 1914. Sur des poissons fossiles et en particulier des siluridés du Tertiaire supérieur et des couches récentes d'Afrique. *Mémoires de la Société Géologique de France*, 21(3)49: 1–13.

Priem R., 1920. Poissons fossiles du Miocéne d'Egypte (Burdigalien de Moghara, désert lybique). In Fourtau R., *Contribution à l'étude des Vertébrés Miocènes de l'Egypte*. Government Press, Cairo.

Pruginin Y., 1967. Report to the Government of Uganda on the experimental fish culture project in Uganda, 1965–1966. Report on Fisheries TA Reports 2446. FAO, Rome.

Pruginin Y., Rothbard S., Wohlfarth G., Halevey A., Moav R. & Hulata G., 1975. All-male broods of *Tilapia nilotica* × *T. aurea* hybrids. *Aquaculture*, 6: 11–21.

Prunet P. & Bornancin M., 1989. Physiology and salinity tolerance in tilapia: an update of basic and applied aspects. *Aquatic Living Resources*, 2: 91–7.

Pulliam H. R., 1989. Individual behavior and the procurement of essential resources. In Roughgarden J., May R. M. & Levin S. A. (eds.), *Perspectives in Ecological Theory*, pp. 25–38. Princeton University Press, Princeton, NJ.

Pullin R. V. S. & Capili J. B., 1988. Genetic improvement in Tilapias: problems and prospects. In Pullin R. V. S. (ed.), *Tilapia Genetic Resources for Aquaculture*, pp. 87–94. ICLARM, Manila.

Pullin R. V. S. & Lowe-McConnell R. H.(eds.), 1982. *The Biology and Culture of Tilapias*. ICLARM Conference Proceedings 7, Manila.

Pursel V. G., Pinkerts C. A., Miller K. F., Bolt D. J., Cambell R. G., Palmiter R. D., Brinster R. L. & Hammer R. E., 1989. Genetic engineering of livestock. *Science*, **244**: 1281–8.

Pyke G. H., 1984. Optimum foraging theory: a critical review. *Annual Review of Ecology and Systematics*, **15**: 523–75.

Quelennec G., Miles J. W., Dejoux C. & Merona B. de, 1977. Chemical monitoring for temephos in mud, oysters and fish from a river within the Onchocerciasis Control Programme in the Volta basin area. Report WHO/VBC/683.

Queller D. C., Strassmann J. E. & Hughes C. R., 1993. Microsatellites and kinship. *Trends in Ecology and Evolution*, **8**: 285–8.

Quensière J., 1976. Influence de la sécheresse sur les pêcheries du delta du Chari (1971/1973). *Cahiers ORSTOM, série Hydrobiologie*, **10**: 3–18.

Quensière J. (ed.), 1994. *La pêche dans le Delta central du Niger. Approche pluridisciplinaire d'un système de production halieutique*. ORSTOM-Karthala-IER, Paris.

Quézel P., 1965. *La végétation du Sahara, du Tchad à la Mauritanie*. Fischer-Verlag, Stuttgart.

Rab P., 1981. Karyotypes of two African barbels, *Barbus bariloides* and *Barbus holotaenia*. *Folia Zoologica*, **30**: 181–94.

Rab P., Machordom A., Percides A. & Guégan J. F., 1996. Karyotypes of three small *Barbus* species (Cyprinidae) from Guinea (West Africa) followed by a review on karyology of African small *Barbus. Caryologia*. (In press)

Rabelahatra A., 1988. Etudes nationales pour le développement de l'aquaculture en Afrique. 22. Madagascar. FAO Circ. Pêches, 770/22. FAO, Rome.

Rahman M. A. & Maclean N., 1992. Production of transgenic Tilapia (*Oreochromis niloticus*) by one-cell-stage microinjection. *Aquaculture*, **105**: 219–32.

Raminasoa N., 1987. Ecologie et biologie d'un poisson téléostéen: *Ophicephalus striatus* (Bloch, 1793), introduit à Madagascar. Thèse de 3ème cycle, Université d'Antananarivo, Madagascar.

Rana K. J. & McAndrew B. J., 1989. The viability of cryopreserved tilapia spermatozoa. *Aquaculture*, **76**: 335–45.

Rana K. J., Muiruri R. M., McAndrew B. J. & Glimour A., 1990. The influence of diluents, equilibration time and prefreezing storage time on the viability of cryopreserved *Oreochromis niloticus* (L.) spermatozoa. *Aquaculture and Fisheries Management*, **21**: 25–30.

Rankin C. & Moller P., 1986. Social behaviour of the African catfish *Malapterurus electricus*, during intra and interspecific encounters. *Ethology*, **73**: 177–90.

Rapport D. J., Regier H. A. & Hutchinson T. C., 1985. Ecosystem behavior under stress. *American Naturalist*, **125**: 617–40.

Reebs S. G. & Colgan P. W., 1991. Nocturnal care of eggs and circadian rhythms of fanning activity in two normally diurnal cichlid fish, *Cichlasoma nigrofasciatum* and *Heterotilapia multispinosa*. *Animal Behaviour*, **41**: 303–11.

Reed W., 1967. *Fish and Fisheries of Northern Nigeria*. Ministry of Agriculture and Fisheries, Northern Nigeria.

Regan C. T., 1911. Classification of the Teleostean fishes of the order Ostariophysi. 1. Cyprinoidea. *Annals and Magazine of Natural History*, (8)**8**(43): 13–32.

Regan C. T., 1922. The classification of the fishes of the family Cichlidae. II – On African and Syrian genera not restricted to the Great Lakes. *Annals and Magazine of Natural History*, (9)**10**: 249–64.

Regier H. A. & Meisner J. D., 1990. Anticipated effects of climate change on freshwater fishes and their habitats. *Fisheries*, **15**: 10–15.

Regier H. A., Holmes J. A. & Payly D., 1990. Influence of temperature changes on aquatic ecosystems: an interpretation of empirical data. *Transactions of the American Fisheries Society*, **119**: 374–89.

Regier H. A., Welcomme R. L., Steedman R. J. & Henderson H. F., 1989. Rehabilitation of degraded river ecosystems. In Dodge D. P. (ed.), *Proceedings of the International Large River Symposium*, pp. 86–97. Canadian Special Publication in Fisheries and Aquatic Sciences 106.

Reid G. McG., 1985. *A Revision of African Species of* Labeo *(Pisces, Cyprinidae)*. Theses Zoologicae, 6. Cramer Publications.

Reid G. McG., 1990. Captive breeding for the conservation of cichlid fishes. *Journal of Fish Biology*, 37, Supplement A: 157–66.

Reid G. McG., 1991. Threatened rainforest cichlids of Lower

Guinea, West Africa. A case for conservation. *Annales du Musée royal d'Afrique centrale (Sciences Zoologiques)*, **262**: 109–19.

Reid G. McG., 1993. Workshop on Conservation of Diversity *in situ* and *ex situ* (Dakar, Senegal). Conclusions and recommendations. *Aquatic Survival*, **2** (4): 7–10.

Reid W. V. & Miller K. R., 1989. *Keeping the Options Alive. The Scientific Basis for Conserving Biodiversity*. World Resource Institute, Washington, DC.

Reinthal P. N., 1990. The feeding habits of a group of herbivorous rock-dwelling cichlid fishes (Cichlidae, Perciformes) from Lake Malawi, Africa. *Environmental Biology of Fishes*, **27**: 215–33.

Reinthal P. N., 1993. Evaluating biodiversity and conserving Lake Malawi's Cichlid fish fauna. *Conservation Biology*, **7**: 712–18.

Reinthal P. N. & Stiassny M. L. J., 1991. The freshwater fishes of Madagascar: a study of an endangered fauna with recommendations for a conservation strategy. *Conservation Biology*, **5**: 231–43.

Reite O. B., Maloiy G. M. O. & Aasehaug B., 1974. pH, salinity and temperature tolerance of Lake Magadi *Tilapia*. *Nature*, **247**: 315.

Reizer C., 1974. Définition d'une politique d'aménagement des ressources halieutiques d'un écosystème aquatique complexe par l'étude de son environnement abiotique, biotique et anthropique. Le fleuve Sénégal Moyen et Inférieur. Doctorat en Sciences de l'Environnement. Dissertation Arlon, Fondation Universitaire Luxembourgeoise. 4 vols.

Resh V. H., Brown A. V., Covich A. P., Gurtz M. E., Li H. W., Minshall G. W., Reice S. R., Sheldon A. L., Wallace J. B. & Wissmar R. C., 1988. The role of disturbance in stream ecology. *Journal of North American Benthological Society*, **7**: 433–455.

Reynolds J. D., 1971. Biology of the small pelagic fishes in the new Volta Lake in Ghana. Part II – Schooling and migrations. *Hydrobiologia*, **38**: 79–91.

Reynolds J. E. & Greboval D. F., 1988. *Socio-economic Effects of the Evolution of Nile Perch Fisheries in Lake Victoria: A Review*. CIFA Technical Paper No. 17. FAO, ROME.

Ribbink A. J., 1977. Cuckoo among Lake Malawi cichlid fish. *Nature*, **267**: 243–4.

Ribbink A. J., 1984. The feeding behaviour of a cleaner and scale, skin and fish eater of Lake Malawi (*Docimus evelynae*, Pisces, Cichlidae). *Netherlands Journal of Zoology*, **34**: 182–96.

Ribbink A. J., 1986. The species concept, sibling species and speciation. *Annales du Musée royal d'Afrique centrale (Sciences Zoologiques)*, **251**: 109–16.

Ribbink A. J., 1987. African lakes and their fishes: conservation scenarios and suggestions. *Environmental Biology of Fishes*, **19**: 3–26.

Ribbink A. J., 1988. Evolution and speciation of African cichlids. In Lévêque C., Bruton M. N. & Ssentongo G. W. (eds.), *Biology and Ecology of African Freshwater Fishes*, pp. 35–51. ORSTOM, Paris.

Ribbink A. J., 1990. Alternative life-history styles of some African cichlid fishes. *Environmental Biology of Fishes*, **28**: 87–100.

Ribbink A. J., 1991. Distribution and ecology of the cichlids of the African Great Lakes. In Keenleyside H. A. (ed.), *Cichlid Fishes: Behaviour, Ecology and Evolution*, pp. 36–59. Fish and Fisheries Series 2. Chapman & Hall, London.

Ribbink A. J., 1994. Lake Malawi. In Martens K. F., Goddeeris B. & Coulter G. (eds.), *Speciation in Ancient Lakes*, pp. 27–33. *Archiv für Hydrobiologie*, **44**.

Ribbink A. J. & Eccles D., 1988. Fish communities in the East African Great Lakes. In Lévêque C., Bruton M. N. & Ssentongo G. W. (eds.), *Biology and Ecology of African Freshwater Fishes*, pp. 277–301. ORSTOM, Paris.

Ribbink A. J. & Twentyman-Jones V., 1989. Captive propagation as a conservation tool. In Proceedings Workshop Biological Conservation Cichlids, 1988. *Annales du Musée royal d'Afrique centrale (Sciences Zoologiques)*, **257**: 145–50.

Ribbink A. J., Marsh A. C., Marsh B. A. & Sharp B. J., 1980. Parental behaviour and mixed broods among cichlid fish of Lake Malawi. *South African Journal of Zoology*, **15**: 1–6.

Ribbink A. J., Marsh A. C. & Marsh B. A., 1981. Nest building and communal care of young by *Tilapia rendalli* Duméril (Pisces, Cichlidae) in Lake Malawi. *Environmental Biology of Fishes*, **6**: 219–22.

Ribbink A. J., Marsh A. C., Marsh B. A. & Sharp B. J., 1983*a*. The zoogeography, ecology and taxonomy of the genus *Labeotropheus* Ahl, 1927, of Lake Malawi (Pisces, Cichlidae). *Zoological Journal of the Linnean Society*, **79**: 223–43.

Ribbink A. J., Marsh B. A., Marsh A. C., Ribbink A. C. & Sharp B. J., 1983*b*. A preliminary survey of the cichlid fishes of rocky habitats in Lake Malawi. *South African Journal of Zoology*, **18**: 149–310.

Richter C. J. J., Viveen W. J. A. R., Eding E. H., Sukkel M., Van Hoof M. F. P. M., van der Berg F. G. J. & Van Oordt P. G. W. J., 1987. The significance of photoperiodicity, water temperature and an inherent endogenous rhythm for the production of viable eggs by the African catfish, *Clarias gariepinus*, kept in subtropical ponds in Israel and under Israel and Dutch hatchery conditions. *Aquaculture*, **63**: 169–85.

Ricklefs R. E., 1987. Structure in ecology. *Science*, **236**: 206–7.

Ridley M., 1978. Paternal care. *Animal Behaviour*, **26**: 904–32.

Ridley M., 1989. The cladistic solution to the species problem. *Biology and Philosophy*, **4**: 1–16.

Riedel D., 1962. Der Margaritensee (Sudabessinien) Zugleich ein Beitrag zur Kenntnis der Abessinischen Graben-See. *Archiv für Hydrobiologie*, **58**: 435–66.

Rind D., 1988. The doubled CO_2 climate and the sensitivity of the modelled hydrologic cycle. *Journal of Geophysical Research*, **93**: 5385–412.

Rinne J. N. & Wanjala A. B., 1983. Maturity, fecundity and breeding seasons of the major catfishes in Lake Victoria, E. Africa. *Journal of Fish Biology*, **23**: 357–63.

Risch L., 1986. Het gemnus *Chrysichthys* Bleeker 1858, en aanverwante genera (Pisces, Siluriformes, Bagridae). Een systtematishe, morfologishe, anatomische en zoogeografishce studie. Thèse de l'Université Catholique de Leuven.

Riser J. & Petit-Maire N., 1986. Paleohydrographie du bassin d'Araouane à l'Holocène. *Revue de Géologie dynamique et de Géographie physique*, **27**: 205–12.

Ritchie J. C., Eyles C. H. & Haynes C.V., 1985. Sediment and pollen evidence for an early to mid-Holocene humid period in the eastern Sahara. *Nature*, **314**: 352–5.

Robben J. & Thys van den Audenaerde D. K. F., 1984. A preliminary study of age and growth of a cyprinid fish *Barilius moori* (Blgr) in Lake Kivu. *Hydrobiologia*, **108**: 153–62.

Roberts T. R., 1969. Osteology and relationship of characoid fishes, particularly the genera *Hepsetus*, *Salminus*, *Hoplias*, *Ctenolucius* and *Acestrorhynchus*. *Proceedings of the California Academy of Sciences*, **36**: 391–500.

Roberts T. R., 1975. Geographical distribution of African freshwater fishes. *Zoological Journal of the Linnean Society*, **57**: 249–319.

Roberts T. R. & Stewart D. J., 1976. An ecological and systematic survey of fishes in the rapids of the lower Zaïre or Congo River. *Bulletin of the Museum of Comparative Zoology*, **147**: 239–317.

Robinson A. H. & Robinson P. K., 1969. A comparative study of the food habits of *Micralestes acutidens* and *Alestes dageti* (Pisces, Characidae) from the northern basin of Lake Chad. *Bulletin de l'Institut Fondamental d'Afrique noire (A)*, **31**: 951–64.

Roche M. A., 1970. Evaluation des pertes du lac Tchad par abandon superficiel et infiltrations marginales. *Cahiers ORSTOM, série Géologie*, **2**: 67–80.

Roest F. C., 1988. Predator–prey relations in Northern Lake Tanganyika and fluctuations in the pelagic fish stocks. In Lewis D. (ed.), *Predator–prey Relationships, Population Dynamics and Fisheries Productivities of Large African Lakes*, pp. 104–29. CIFA (Committee of Inland Fisheries of Africa) Occasional papers, No. 15. FAO, Rome.

Roest F. C., 1992. The pelagic fisheries resources of Lake Tanganyika. *Mitteilungen-Internationale Vereinigung für Theoretische und Angewandte Limnologie*, **23**: 11–15.

Roff D. A., 1992. *The Evolution of Life Histories. Theories and Analysis*. Chapman & Hall, New York and London.

Rogers J., 1972. *Measures of Genetic Similarity and Genetic Distances*. Studies in Genetics VII, University of Texas, Austin.

Rohwer S., 1978. Parent cannibalism of offspring and egg raiding as a courtship strategy. *American Naturalist*, **112**: 429–40.

Roman B., 1966. Les poissons des hauts bassins de la Volta. *Annales du Musée royal d'Afrique centrale*, in-8°, **150**.

Roman B., 1971. *Peces de Rio Muni, Guinea ecuatorial (Aguas dulces y salobres)*. Fondacion La Salle de Ciencias naturales, Barcelona.

Romand R., 1992. Cyprinodontidae. In Lévêque C., Paugy D. & Teugels G. G. (eds.), *Faune des poissons d'eaux douces et saumâtres d'Afrique de l'Ouest*, tome 2, pp. 586–654. Collection Faune tropicale No. 28. ORSTOM Paris.

Rompey J. van, Ahmadyar Z. & Verheyen E., 1988. General protein patterns of muscle homogenate of some Lake Victoria haplochromines (Pisces, Cichlidae). *Journal of Zoology, London*, **214**: 505–18.

Rosen B. R., 1988. Biogeographical patterns: a perceptual overview. In Myers A. A. & Giller P. S. (eds.), *Analytical Biogeography: An Integrated Approach to the Study of Animal and Plant Distributions*, pp. 23–55. Chapman & Hall, London.

Rosen B. R., 1988. From fossils to earth history: applied historical biogeography. In Myers A. A. & Giller P. S. (eds.), *Analytical Biogeography: An Integrated Approach to the Study of Animal and Plant Distributions*, pp. 437–81. Chapman & Hall, London.

Rosen D. E., 1978. Vicariant patterns and historical explanation in biogeography. *Systematic Zoology*, **27**: 159–88.

Rosen D. E., 1979. Fishes from the uplands and intermontane basins of Guatemala: revisionary studies and comparative biogeography. *Bulletin of the American Museum of Natural History*, **162**: 267–376.

Rosen D. E., 1982. Teleostean interrelationships, morphological function and evolutionary inference. *American Zoologist*, **22**: 261–73.

Rosen D. E., 1985. An essay on euteleostean classification. *American Museum Novitates*, 2827: 1–57.

Rosen D. E., Forey P. L., Gardiner B. G. & Patterson C., 1981. Lungfishes, tetrapods, palaeontology, and plesiomorphy. *Bulletin of the American Museum of Natural History*, 167: 159–276.

Rosen D. E. & Greenwood P. H., 1970. Origin of the Weberian apparatus and the relationships of the ostariophysan and gonorynchiform fishes. *American Museum Novitates*, 2428: 1–25.

Ross S. T., 1986. Resource partitioning in fish assemblages: a review of field studies. *Copeia*, 1986: 352–88.

Roughgarden J., 1976. Resource partitioning among competing species – a coevolutionary approach. *Theoretical Population Biology*, 9: 388–424.

Roughgarden J., 1989. The structure and assembly of communities. In Roughgarden R., May M. & Levin S. A. (eds.), *Perspectives in Ecological Theory*, pp. 203–26. Princeton University Press, Princeton, NJ.

Rowland W. J., 1978. Sound production and associated behavior in the jewel fish, *Hemichromis bimaculatus*. *Behaviour*, 64: 125–36.

Ryder R. A., 1978. Fish yield assessment of large lakes and reservoir – a prelude to management. In Gerking S. A. (ed.), *Ecology of Freshwater Fish Production*, pp. 402–23. Blackwell Scientific, Oxford.

Rzoska J. (ed.), 1976. *The Nile: Biology of an Ancient River*. W. Junk Publishers, The Hague.

Saad M. A. H., Amuzu A. T., Biney C., Calamari D., Imevbore A. M., Naeve H. & Ochumba P. B. O., 1994. Domestic and industrial organic loads. In Calamari D. & Naeve H. (eds.), *Review of Pollution in the African Aquatic Environment*, pp. 23–32. CIFA Technical Paper, No. 25, FAO, Rome.

Sacca R. & Burggren W., 1982. Oxygen uptake in air and water in the air breathing reedfish *Calamoichthys calabaricus*: role of skin, gills and lungs. *Journal of Experimental Biology*, 97: 179–86.

Sage R. D., Loiselle P. V., Basasibwaki P. & Wilson A.C., 1984. Molecular versus morphological change among cichlid fishes of Lake Victoria. In Echelle A. A. & Kornfield I. (eds.), *Evolution of Fish Species Flock*, pp. 185–201. University of Maine at Orono Press, Orono, ME.

Saglio P., 1992. La communication chimique chez les poissons. *La Recherche*, 248: 1282–93.

Sagua V. O., 1979. Observations on the food and feeding habits of the African electric fish *Malapterurus electricus* (Gmelin). *Journal of Fish Biology*, 15: 61–9.

Salard-Cheboldaeff M., 1981. Palynologie maestrichienne et tertiaire du Cameroun. Résultats botaniques. *Review of Palaeobotany and Palynology*, Amsterdam, 32: 401–39.

Sale P. F., 1978. Coexistence of coral reef fishes – a lottery for living space. *Environmental Biology of Fishes*, 3: 85–102.

Salthe S.N., 1985. *Evolving Hierarchical Systems: Their Structure and Representation*. Columbia University Press, New York.

Sandon M. & Amin al Tayib, 1953. The food of some common Nile fish. *Sudan Notes and Records*, 34: 205–29.

Santiago C. B., Albada M. B., Abuan E. F. & Laron M. A., 1985. The effects of artificial diets on fry production and growth of *Oreochromis niloticus* breeders. *Aquaculture*, 47: 193.

Sargent R. C. & Gross M. R., 1986. Williams principle: an explanation of parental care in teleost fishes. In Pitcher T. J. (ed.), *The Behaviour of Teleost Fishes*, pp. 275–93. Croom Helm, London.

Sargent R. C., Taylor P. D. & Gross M. R., 1987. Parental care and the evolution of egg sizes in fishes. *American Naturalist*, 129: 32–46.

Sato J., 1986. A brood parasitic catfish of mouthbrooding cichlid fishes in Lake Tanganyika. *Nature*, 323: 58–9.

Sauvage H. E., 1891. Histoire naturelle des Poissons. In Grandidier A., *Histoire physique, naturelle, et politique de Madagascar*, 16.

Schaffer W. M., 1979. The theory of life-history evolution and its application to the Atlantic salmon. *Symposia of the Zoological Society of London*, 44: 307–26.

Scheel J. J., 1966. Notes on the phenotypy, distribution and systematics of *Aphyosemion bivittatum* (Lönnberg), with remarks on the chromosome number in the Rivulinae. *Ichthyologia*: 261–78.

Scheel J. J., 1968. *Rivulins of the Old World*. T.H.F. Publications, Neptune City, New Jersey.

Scheel J. J., 1972. Cytotaxonomic studies: The *Aphyosemion elegans* group. *Zeitschrift für Zoologische Systematik und Evolutionforschung*, 10: 122–7.

Scheel J. J., 1975a. The karyotypic evolution in *Aphyosemion*. *Journal of the American Killifish Association*, 8: 191–3.

Scheel J. J., 1975b. The karyotypes of some *Fundulopanchax* forms. *Journal of the American Killifish Association*, 8: 194–9.

Scheel J. J., 1981. Notes on a certain group of Killifish. 1. *British Killifish Association, Kill News*, 185: 65–8.

Scheringa E. J. F., Strick J. J. T. & Antwi L. A. K., 1981. Fish brain acetylcholinesterase activity after Abate applications against *Simulium damnosum* in the Volta river basin area. Unpublished report to OCP.

Schlesinger W. H., 1989. Ecosystem structure and function.

In Roughgarden J., May R. M. & Levin S. A. (eds.), *Perspectives in Ecological Theory*, pp. 268–74. Princeton University Press, Princeton, NJ.

Schliewen U. K., Tautz D. & Pääbo S., 1994. Sympatric speciation suggested by monophyly of crater lake cichlids. *Nature*, **368**: 629–32.

Schlosser I. J., 1982. Fish community structure and function along two habitat gradients in a headwater stream. *Ecological Monographs*, **52**: 395–414.

Schlosser I. J., 1985. Flow regime, juvenile abundance, and the assemblage structure of stream fishes. *Ecology*, **66**: 1484–90.

Schlosser I. J., 1987. The role of predation in age and size-related habitat use by stream fishes. *Ecology*, **68**: 651–9.

Schoener T. W., 1986. Overview: kinds of ecological communities. Ecology becomes pluralistic. In Diamond J. M. & Case T. J. (eds.), *Community Ecology*, pp. 467–79. Harper & Row, New York.

Scholz C. A. & Rosendahl B. R., 1988. Low lake stands in Lake Malawi and Tanganyika, delineated with multifold seismic data. *Science*, **240**: 1645–8.

Scholz C. A. & Rosendahl B. R., Versfelt J. W. & Rach N., 1991. Results of high-resolution echo sounding of Lake Victoria. *Journal of African Earth Sciences*, **11**: 25–32.

Schültz L. P., 1942. The freshwater fishes of Liberia. *Proceedings of the U.S. National Museum*, **92**(3152): 301–48.

Schwanck E., 1986. Filial cannibalism in *Tilapia mariae*. *Journal of Applied Ichthyology*, **2**: 65–74.

Schwanck E., 1987. Lunar periodicity in the spawning of *Tilapia mariae* in the Ethiop River, Nigeria. *Journal of Fish Biology*, **30**: 533–7.

Schwanck E., 1989. Parental care of *Tilapia mariae* in the field and in aquaria. *Environmental Biology of Fishes*, **24**: 251–65.

Schwarz A. L., 1985. The behavior of fishes in their acoustic environment. *Environmental Biology of Fishes*, **13**: 3–15.

Schweitzer J., 1992. Conserving biodiversity in developing countries. *Fisheries*, **17**: 35–8.

Scott D. B. C., 1979. Environmental timing and the control of reproduction in teleost fish. *Symposia of the Zoological Society of London*, **44**: 105–32.

Scudder T., 1989. River basin projects in Africa. *Environment*, **31**: 4–32.

Scudder T. & Conelly T., 1985. Management systems for riverine fisheries. *FAO, Fisheries Technical Papers*, **263**: 85.

Seaman M. T., Scott W. E., Walmsley R. D., van der Waal B. C. W. & Toerien D. F., 1978. A limnological investigation of Lake Liambezi, Caprivi. *Journal of the Limnological Society of Southern Africa*, **4**: 129–44.

Serruya C. & Pollingher U., 1983. *Lakes of the Warm Belt*. Cambridge University Press, Cambridge.

Servant M., 1983. *Séquences continentales et variations climatiques: évolution du bassin du Tchad au Cénozoïque supérieur*. Travaux et Documents 159. ORSTOM, Paris.

Servant, M. & Servant-Vildary S., 1980. L'environnement quaternaire du bassin du Tchad. In Williams F. M. & Faure H. (eds.), *The Sahara and the Nile*, pp. 133–62. Balkema, Rotterdam.

Seyoum S., 1989. Stock identification and the evolutionary relationships of the genera *Oreochromis*, *Sarotherodon* and *Tilapia* (Pisces, Cichlidae) using allozyme analysis and restriction endonuclease analysis of mitochondrial DNA. Ph.D. thesis, University of Waterloo, Waterloo, Ontario.

Seyoum S., 1990. Allozyme variations in subspecies of *Oreochromis niloticus*. *Isozymes Bulletin*, **23**: 97.

Seyoum S. & Kornfield I., 1992. Taxonomic note on the *Oreochromis niloticus* subspecies complex (Pisces, Cichlidae) with a description of a new subspecies. *Canadian Journal of Zoology*, **70**: 2161–5.

Sharp B. J., 1981. An ecological survey of territoriality in four cichlid species resident on rocky shores near Monkey Bay, Lake Malawi. M.Sc. thesis, Rhodes University, South Africa.

Sheldon A. L., 1988. Conservation of stream fishes: patterns of diversity, rarity and risk. *Conservation Biology*, **2**: 149–56.

Sibly R. M. & Calow P., 1983., An integrated approach to life-cycle evolution using selective landscapes. *Journal of Theoretical Biology*, **102**: 527–47.

Sibly R. M. & Calow P., 1986. *Physiological Ecology of Animals: an Evolutionary Approach*. Blackwell, Oxford.

Siddiqui A. Q., 1977. Reproductive biology, length-weight relationship and relative condition of *Tilapia leucosticta* (Trewavas) in lake Naivasha, Kenya. *Journal of Fish Biology*, **10**: 251–60.

Siddiqui A. Q., 1979. Changes in fish species composition in Lake Naivasha, Kenya. *Hydrobiologia*, **64**: 131–8.

Siegfried W. R. & Davies B. R., 1982. *Conservation of Ecosystems: Theory and Practice*. South African National Scientific Programmes, Report No. 61.

Signor P. W., 1990. The geological history of diversity. *Annual Review of Ecology and Systematics*, **21**: 509–39.

Simberloff D., 1988. The contribution of population and community biology to conservation science. *Annual Review of Ecology and Systematics*, **19**: 473–511.

Simpson B. R. C., 1979. The phenology of annual killifishes. *Symposia of the Zoological Society of London*, **44**: 243–61.

Simpson G. G., 1944. *Tempo and Mode in Evolution.* Columbia University Press, New York.

Simpson G. G., 1961. *Principles of Animal Taxonomy.* Columbia University Press, New York.

Sindayigaya E., 1991. Quelques données concernant la surveillance de la contamination des poissons du lac Tanganyika. In Cohen A. (ed.), *Report of the First International Conference on the Conservation and Biodiversity of Lake Tanganyika*, p. 110.

Söjlander S., 1972. Feldbeobachtungen an einigen westafrikanischen Cichliden. *Aquarien und Terrarien*, **19**: 42–5.

Skelton P. H., 1986. Fish of the Orange-Vaal system. In Davies B. R. & Walker K. F. (eds.), *The Ecology of Rivers Systems*, pp. 143–61. W. Junk Publishers, The Hague.

Skelton P. H., 1987. *South African Red Data Book – Fishes.* South African National Scientific Programmes Report 137.

Skelton P. H., 1988. The distribution of African freshwater fishes. In Lévêque C., Bruton M. & Ssentongo G. W. (eds.), *Biology and Ecology of African Freshwater Fishes*, pp. 65–91. ORSTOM, Paris.

Skelton P. H., 1994. Diversity and distribution of freshwater fishes in East and Southern Africa. In Teugels G. G., Guégan J. F. & Albaret J. J. (eds.), *Biological Diversity of African Fresh- and Brackish Water Fishes*, pp. 95–131. Annales Sciences zoologiques, Vol. 275. Musée royal d'Afrique centrale, Tervuren, Belgique.

Skelton P. H. & Cambray J. A., 1981. The freshwater fishes of the middle and lower Orange River. *Koedoe*, **24**: 51–66.

Skelton P. H., Bruton M. N., Merron G. S. & van der Waal B. C. W., 1985. The fishes of the Okavango drainage system in Angola, South West Africa and Botswana: taxonomy and distribution. *Ichthyological Bulletin of the J. L. B. Smith Institute of Ichthyology*, **50**: 1–21.

Skelton P. H., Tweddle D. & Jackson P. B. N., 1991. Cyprinids of Africa. In Winfield I. J. & Nelson J. S. (eds.), *Cyprinid Fishes: Systematics, Biology and Exploitation*, pp. 211–39. Fish and Fisheries Series 3. Chapman & Hall, London.

Slobodkin L. B. & Rapoport A., 1974. An optimal strategy of evolution. *Quarterly Review of Biology*, **49**: 181–200.

Slooteweg R., 1989. Proposed introduction of *Astatoreochromis alluaudi*, an East African mollusc-crushing cichlid, as a means of snail control. *Annales du Musée royal d'Afrique centrale (Sciences Zoologiques)*, **257**: 61–4.

Smith C., 1989. An investigation into the problem of conspecific predation among the fry of the Nile tilapia, *Oreochromis niloticus* (Linnaeus, 1757) in an intensive culture system. M.Sc. thesis, Plymouth Polytechnic.

Smith C. J. & Haley S. R., 1988. Steroid profiles of the female tilapia *Oreochromis mossambicus*, and correlation with oocyte growth and mouth-brooding behavior. *General and Comparative Endocrinology*, **69**: 88–98.

Smith C. & Reay P., 1991. Cannibalism in teleost fish. *Reviews in Fish Biology and Fisheries*, **1**: 41–64.

Sodsuk P. & McAndrew B. J., 1991. Molecular systematics of three tilapiine genera *Tilapia*, *Sarotherodon* and *Oreochromis* using allozyme data. *Journal of Fish Biology*, Supplement A, 39.

Sola L., Catandella S. & Capanna E., 1981. New developments in vertebrate cytotaxonomy. III — Karyology of bony fishes: a review. *Genetica*, **54**: 285–328.

Solbrig O.T., 1991a. *Biodiversity. Scientific Issues and Collaborative Research Proposals.* MAB Digest 9. UNESCO, Paris.

Solbrig O.T., 1991b. *From Genes to Ecosystems: A Research Agenda for Biodiversity.* IUBS, Paris.

Solbrig O. T. & Nicolis G. (eds.), 1991. *Perspectives on Biological Complexity.* IUBS, Monograph Series No. 6, Paris.

Sombroek W. G. & Zonneveld I. S., 1971. *Ancient Dune Fields and Fluviatile Deposits in the Rima-Sokoto Basin (N. W. Nigeria).* Netherlands Soil Survey Inst. Wageningen, Soil Survey Paper 5.

Sorbini L., 1973. Evoluzione e distribuzione del genera fossile Eleolates e susi rapporti con il genere attuale *Lates* (Pieces, Centropomidae). *Studi e ricerchi si giacimenti terziari di Bolca*, **2**: 1–43.

Soriano M., Moreau J., Hoenig J. M. & Pauly D., 1990. New functions for the analysis of two-phase growth of juvenile and adult fishes, with application to Nile perch. ICES C. M. 1990/D: 16, Statistics Committee.

Soulé M. E., 1980. Thresholds for survival: maintaining fitness and evolutionary potential. In Soulé M. E. & Wilcox B. A. (eds.), *Conservation Biology: An Evolutionary–Ecological Perspective*, pp. 151–69. Sinauer Associates, Sunderland, MA.

Soulé M. E. (ed.), 1986. *Conservation Biology. The Science of Scarcity and Diversity.* Sinauer Associates, Sunderland, MA.

Soulé M. E., 1987. Where do we go from here? In Soulé M. E. (ed.), *Viable Populations for Conservation*, pp. 175–83. Cambridge University Press, Cambridge.

Soulé M. E., 1991. Conservation: tactics for a constant crisis. *Science*, **253**: 744–50.

Sousa W., 1984. The role of disturbance in natural

communities. *Annual Review of Ecology and Systematics*, **15**: 353–65.

Southwood T. R. E., 1977. Habitat, the templet for ecological strategies? *Journal of Animal Ecology*, **46**: 337–65.

Southwood T. R. E., 1987. The concept and nature of the community. In Gee J. H. R. & Giller P. S. (eds.), *Organizations of Communities: Past and Present*, pp. 3–27. Blackwell Scientific Publications, Oxford.

Southwood T. R. E., 1988. Tactics, strategies and templets. *Oikos*, **52**: 3–18.

Spliethoff P. C., de Iongh H. H. & Frank V. C., 1983. Success of the introduction of the freshwater clupeid *Limnothrissa miodon* (Boulenger) in Lake Kivu. *Fisheries Management*, **14**: 17–31.

Srinn K. Y. 1976. Biologie *d'Hydrocynus forskalii* (Pisces, Characidae) du bassin tchadien. Thèse de doctorat de Spécialité, Université Paul Sabatier, Toulouse.

Ssentongo G., 1988. Population structure and dynamics. In Lévêque C., Bruton M. N. & Ssentongo G. W. (eds.), *Biology and Ecology of African Freshwater Fishes*, pp. 363–77. ORSTOM, Paris.

Stanley S. M., 1979. *Macroevolution: Patterns and Process*. Freeman, San Francisco, CA.

Stanley S. M., 1989. Fossils, macroevolution and theoretical ecology. In Roughgarden J., May R. M. & Levin S. A. (eds.), *Perspectives in Ecological Theory*, pp. 125–34. Princeton University Press, Princeton, NJ.

Statzner B., 1987. Characteristics of lotic ecosystems and consequences for future research directions. *Ecological Studies*, **61**: 365–90.

Statzner B. & Higler B., 1986. Stream hydraulics as a major determinant of benthic invertebrate zonation patterns. *Freshwater Biology*, **16**: 127–39.

Statzner B., Dejoux C. & Elouard J. M., 1984. Field experiments on the relationship between drift and benthic densities of aquatic insects in tropical streams (Ivory Coast). I – Introduction: review of drift literature, methods and experimental conditions. *Revue d'Hydrobiologie tropicale*, **17**: 319–34.

Statzner B., Gore J. A. & Resh V. H., 1988. Hydraulic stream ecology: observed patterns and potential applications. *Journal of the North American Benthological Society*, **7**: 307–60.

Stauch A., 1966. *Le bassin camerounais de la Bénoué et sa péche*. Mémoires ORSTOM No. 15. ORSTOM, Paris.

Stearns S. C., 1983. A natural experiment in life-history evolution: field data on the introduction of mosquito fish (*Gambusia affinis*) to Hawaii. *Evolution*, **37**: 601–17.

Stearns S. C., 1989. The evolutionary significance of phenotypic plasticity. *Bio-Science*, **39**: 436–45.

Stearns S. C., 1992. *The Evolution of Life Histories*. Oxford University Press, Oxford.

Stearns S. C. & Crandall R. E., 1984. Plasticity for age and size at sexual maturity: a life-history response to unavoidable stress. In Potts G. W. & Wootton R. J. (eds.), *Fish Reproduction*, pp. 13–33. Academic Press, London.

Steedman R. J. & Regier H. A., 1990. Ecological bases for an understanding of ecosystem integrity in the Great Lakes basin. In Edwards C. J. & Regier H. A. (eds.), *An Ecosystem Approach to the Integrity of the Great Lakes in Turbulent Times*, pp. 257–70. Great Lakes Fishery Commission Special Publication 90–4.

Stephens D. W. & Krebs J. R., 1987. *Foraging Theory*. Princeton University Press, Princeton, NJ.

Stevens G. C., 1989. The latitudinal gradient in geographical range: how so many species coexist in the tropics. *American Naturalist*, **133**: 240–56.

Stewart D. J., 1977. Geographic variation of *Barbus radiatus* Peters, a widely distributed African Cyprinid. *Environmental Biology of Fishes*, **1**: 113–25.

Stiassny M. L. J., 1981. Phylogenetic versus convergent relationships between piscivorous cichlid fishes from Lakes Malawi and Tanganyika. *Bulletin British Museum (Natural History)* Zoology, **40**: 67–101.

Stiassny M. L. J., 1987. Cichlid familial intrarelationships and the placement of the Neotropical genus *Cichla* (Perciformes, Labroidei). *Journal of Natural History*, **21**: 1311–31.

Stiassny M. L. J., 1989. A taxonomic revision of the African genus *Tylochromis* (Labroidei, Cichlidae); with notes on the anatomy and relationships of the group. *Annales du Musée royal d'Afrique centrale (Sciences Zoologiques)*, **258**.

Stiassny M. L. J., 1990. Notes on the anatomy and relationships of the betodiid fishes of Madagascar, with a taxonomic revision of the genus *Rheocles*. *American Museum Novitates*, **2979**: 1–33.

Stiassny M. L. J., 1991. Phylogenetic interrelationships of the family Cichlidae: an overview. In Keenleyside H. A. (ed.), *Cichlid Fishes: Behaviour, Ecology and Evolution*, pp. 1–35. Fish and Fisheries Series 2. Chapman & Hall, London.

Stiassny M. L. J., 1992. Phylogenetic analysis and the role of systematics in the biodiversity crisis. In Eldredge N. (ed.), *Systematics, Ecology and the Biodiversity Crisis*, pp. 109–20. Columbia University Press, New York.

Stiassny M. L. J. & Gerstner C. L., 1992. The parental behaviour of *Paratilapia polleni* (Perciformes, Labroidei), a phylogenetically primitive cichlid from Madagascar, with a discussion of the evolution of maternal care in the family Cichlidae. *Environmental Biology of Fishes*, **34**: 219–33.

Stiassny M. L. J. & Jensen J. S., 1987. Labroid intrarelationships revisited: morphological complexity, "key innovations", and the study of comparative diversity. *Bulletin of the Museum of Comparative Zoology*, **151**: 269–319.

Stiassny M. L. J. & Moore J. A., 1992. A review of the pelvic girdle of acanthomorph fishes, with comments on hypotheses of acanthomorph interrelationships. *Zoological Journal of the Linnean Society*, **104**: 209–42.

Stiassny M. L. J. & Raminosoa N., 1994. The fishes of the inland waters of Madagascar. In Teugels G. G., Guégan J. F. & Albaret J. J. (eds.), *Biological Diversity of African Fresh- and Brackish Water Fishes*, pp. 133–49. Annales Sciences zoologiques, Vol. 275. Musée royal d'Afrique centrale, Tervuren, Belgique.

Strahler A. N., 1957. Quantitative analysis of watershed geomorphology. *Transactions American Geophysical Union*, **38**: 913–20.

Street F. A. & Grove A. T., 1976. Environment and climatic implications of late Quaternary lake-level fluctuations in Africa. *Nature*, **261**: 385–90.

Stromer E., 1916. Die Entdeckung und die bedeutung des Land-und Süsswasser bewohneneden Wirbeitiere im Tertir iund der Keride Ägyptens. *Zeischrift der Deutschen Geologischen Gesellschaft*, **68**: 397–425.

Stuart S. N., Adams R. J. & Jenkins M. D., 1990. *Biodiversity in Sub-Saharan Africa and its Islands. Conservation, Management and Sustainable Use.* Occasional Papers of the IUCN Species Survival Commission No. 6.

Sturmbauer C. & Meyer A., 1992. Genetic divergence, speciation and morphological status in a lineage of African cichlid fishes. *Nature*, **358**: 578–81.

Sturmbauer C. & Meyer A., 1993. Mitochondrial phylogeny of the endemic mouthbrooding lineages of cichlid fishes from lake Tanganyika, East Africa. *Molecular Biology and Evolution*, **10**: 751–68.

Sutherland W. J., 1992. Genes map the migratory route. *Nature*, **360**: 625–6.

Svensson G. S. O., 1933. Freshwater fishes of the Gambia River (British West Africa). Results of the Swedish Expedition 1931. *Kungliga Svenska Vetenskapsakademiens Handlingar*, **12**(3): 1–102.

Sydenham D. H. J., 1977. The qualitative composition and longitudinal zonation of the fish fauna of the River Ogun. *Revue de Zoologie africaine*, **91**: 974–96.

Sydenham D. H. J., 1978. The qualitative composition and longitudinal zonation of the fish fauna of the river Ogun,

Western Nigeria. Errata. *Revue de Zoologie africaine*, **92**: 1022.

Taborsky M., 1984. Broodcare helpers in the cichlid fish *Lamprologus brichardi*: their costs and benefits. *Animal Behaviour*, **32**: 1236–52.

Taborsky M., Hudde B. & Wirtz P., 1987. Reproductive behavior and ecology of *Symphodus ocellatus*, a European wrasse with four type of male. *Behaviour*, **102**: 82–118.

Taborsky M. & Limberger D., 1981. Helpers in fish. *Behavioural Ecology and Sociobiology*, **8**: 143–5.

Takamura K., 1983. Interspecific relationship between two Aufwuchs eaters *Petrochromis polyodon* and *Tropheus moorei* (Pisces, Cichlidae) of Lake Tanganyika with a discussion on the evolution and functions of a symbiotic relationship. *Physiology and Ecology, Japan*, **20**: 59–69.

Takamura K., 1984. Interspecific relationships of aufwuchs-eating fishes in Lake Tanganyika. *Environmental Biology of Fishes*, **10**: 225–41.

Talbot M. R., 1980. Environmental responses to climatic change in the West African Sahel over the past 20 000 years. In Williams M. A. J. & Faure H. (eds.), *The Sahara and the Nile*, pp. 37–62. Balkema, Rotterdam.

Talbot M. R. & Delibrias G., 1977. Holocene variations in the level of Lake Bosumtwi, Ghana. *Nature*, **268**: 722–4.

Talling J. F., 1965. The photosynthetic activity of phytoplankton in Lake Victoria (East Africa). *Internationale Revue der gesamten Hydrobiologie*, **51**: 1–32.

Talling J. F., 1966. The annual cycle of stratification and phytoplankton growth in Lake Victoria (East Africa). *Internationale Revue der gesamten Hydrobiologie*, **51**: 545–621.

Talling J. F., 1992. Environmental regulation in African shallow lakes and wetlands. *Revue d'Hydrobiologie tropicale*, **25**: 87–144.

Tave D., Rezk M. & Smitherman R. O., 1989. Gold-colored tilapia now possible. *Highlights of Agricultural Research*, **36**: 4.

Taverne L., 1974. *Parachanos* Arambourg et Schneegans (Pisces, Gonorhynchiformes) du Crétacé inférieur du Gabon et de Guinée Equatoriale et l'origine des Téléostéens Ostariophysi. *Revue de Zoologie africaine*, **88**: 683–7.

Taverne L., 1975. Sur l'existence d'un poisson Ostéoglossoide fossile proche parent de l'actuel genre *Heterotis* dans le Crétacé moyen du Kwango (Zaïre). *Revue de Zoologie africaine*, **89**: 964–8.

Taverne L., 1976. A propos de trois téléostéens Salmoniformes fossiles du Crétacé inférieur (Wealdien) du Zare, précédemment décrits dans les genres *Leptolepis* et *Clupavus* (Pisces, Teleostei). *Revue de Zoologie africaine*, **89**: 481–504.

Tavolga W. N., 1971. Sound production and detection. In Hoar V. & Randall D. J. (eds.), *Fish Physiology*, pp. 135–205. Academic Press, New York.

Temple S. A., 1981. Applied island biogeography and the conservation of endangered island birds in the Indian Ocean. *Biological Conservation*, **20**: 147–61.

Terborgh J., 1974. Preservation of natural diversity: the problem of extinction-prone species. *BioScience*, **24**: 715–22.

Terkatin-Shimony A., Ilan Z., Yaron Z. & Johnson D. W., 1980. Relationship between temperature, ovarian recrudescence and plasma cortisol levels in *Tilapia aurea* (Cichlidae, Teleostei). *General and Comparative Endocrinology*, **40**: 143–8.

Teugels G. G., 1982. A systematic outline of the African species of the genus *Clarias* (Pisces, Clariidae) with an annoted bibliography. *Annales du Musée royal d'Afrique centrale (Sciences Zoologiques)*, **236**.

Teugels G. G., 1986. A systematic revision of the African species of the genus *Clarias* (Pisces, Clariidae). *Annales du Musée royal d'Afrique centrale (Sciences Zoologiques)*, **247**.

Teugels G. G. & Thys van den Audenaerde D. F. E., 1992. Cichlidae. In Lévêque C., Paugy D. & Teugels G. G. (eds.), *Faune des poissons d'eaux douces et saumâtres d'Afrique de l'Ouest*, tome 2, pp. 714–79. Collection Faune tropicale No. 28. ORSTOM, Paris.

Teugels G. G., Lévêque C., Paugy D. & Traoré K., 1988. Etat des connaissances sur la faune ichtyologique des bassins côtiers de Côte d'Ivoire et de l'Ouest du Ghana. *Revue d'Hydrobiologie tropicale*, **21**: 221–37.

Teugels G. G., Denayer B. & Legendre M., 1990. A systematic revision of the African catfish genus *Heterobranchus* (Pisces, Clariidae). *Zoological Journal of the Linnean Society*, **98**: 237–57.

Teugels G. G., Risch L., de Vos L. & Thys van den Audenaerde D. F. E., 1991. Generic review of the African bagrid catfish genera *Auchenoglanis* and *Parauchenoglanis* with description of a new genus. *Journal of Natural History*, **25**: 499–517.

Teugels G. G., Ozouf-Costaz C., Legendre M. & Parrent M., 1992*a*. A karyological analysis of the artificial hybridization between *Clarias gariepinus* (Burchell, 1822) and *Heterobranchus longifilis* Valenciennes, 1840 (Pisces, Clariidae). *Journal of Fish Biology*, **40**: 81–6.

Teugels G. G., Guyomard R. & Legendre M., 1992*b*. Enzymatic variation in African clariid catfishes. *Journal of Fish Biology*, **40**: 87–96.

Teugels G. G., Reid G. McG. & King R. P., 1992*c*. Fishes of the Cross River Basin (Cameroon-Nigeria). Taxonomy, zoogeography, ecology and conservation. *Annales du Musée royal d'Afrique centrale (Sciences Zoologiques)*, **266**.

Teyssedre C. & Moller P., 1982. The optomotor response in weak-electric fish mormyrid: can they see? *Zietschrift für Tierpsychologie*, **60**: 306–312.

Thiero Yatabary N., 1983. Contribution à l'étude du régime alimentaire de *Synodontis schall* (Bloch-Schneider, 1801) dans le delta central du fleuve Niger. *Revue d'Hydrobiologie tropicale*, **16**: 277–86.

Thines G., 1969. *L'évolution régressive des poissons cavernicoles et abyssaux*. Masson, Paris.

Thomas D. H. L., Jimoh M. A. & Matthes H., 1993. Fishing. In Hollis G. E., Adams W. M. & Aminu-Kano M. (eds.), *The Hadejia-Nguru Wetlands: Environment, Economy and Sustainable Development*, pp. 97–115. IUCN, Gland, Switzerland.

Thomas D. S. G. & Shaw P. A., 1988. Late cenozoic drainage evolution in the Zambezi basin: geomorphological evidence from the Kalahari rim. *Journal of African Earth Sciences*, **7**: 611–618.

Thompson K. W., 1981. Karyotypes of six species of African Cichlidae (Pisces, Perciformes). *Experientia*, **37**: 351–2.

Thornton J.A. (ed.), 1982. *Lake McIlwaine. The Eutrophication and Recovery of a Tropical African Man-made Lake*. Monographiae Biologicae 49. W. Junk Publishers, The Hague.

Thys van den Audenaerde D. F. E., 1963. La distribution géographique des *Tilapia* au Congo. *Bulletin des Séances de l'Académie royale des Sciences d'Outre-Mer*, **9**: 570–605.

Thys van den Audenaerde D. F. E., 1966. Les *Tilapia* (Pisces, Cichlidae) du Sud Cameroun et du Gabon. Etude systématique. *Annales du Musée royal d'Afrique centrale (Sciences Zoologiques)*, **153**.

Thys van den Audenaerde D. F. E., 1967. The freshwater fishes of Fernando Poo. *Verhandelingen van de Koninklijke Vlaamse Academie voor Wetenschappen, Letteren en Schone Kunsten van België*, **29** (100).

Thys van den Audenaerde D. F. E., 1968. An annotated bibliography of *Tilapia* (Pisces, Cichlidae). *Document zoologique du Musée royal d'Afrique centrale*, **14**.

Thys van den Audenaerde D. F. E., 1971. Some new data concerning the *Tilapia* species of the subgenus *Sarotherodon* (Pisces, Cichlidae). *Revue de Zoologie et de Botanique africaine*, **84**: 203–16.

Thys van den Audenaerde D. F. E., 1986. The circadian activity patterns of *Haplochromis* in Lake Kivu. *Annales* du Musée royal d'Afrique centrale (Sciences Zoologiques), 251: 61–5.

Thys van den Audenaerde D. F. E., Coenen E., Robben J. & Vervoot D., 1982. Fisheries research on Lake Kivu. MRAC, Tervueren.

Tiercelin J. J. & Mondeguer A., 1991. The geology of the Tanganyika trough. In Coulter G. W (ed.), Lake Tanganyika and its Life, pp. 7–48. Natural History Museum Publications and Oxford University Press, Oxford.

Tiercelin J. J., Scholz C. A., Mondeguer A., Rosendahl B. R. & Ravenne C., 1989. Discontinuités séismiques et sédimentaires dans la série de remplissage du fossé du Tanganyka, Rift Est Africain. Comptes rendus hebdomadaires des Séances de l'Académie des Sciences, Paris, 309, série II: 1599–606.

Tilman D., 1982. Resource Competition and Community Structure. Princeton University Press, Princeton, NJ.

Tobor J. G., 1972. The food and the feeding habits of some lake Chad commercial fishes. Bulletin de l'Institut Fondamental d'Afrique noire (A), 34: 179–211.

Toerring M. J. & Serrier J., 1978. Influence of water temperature on the electric organ discharge (EOD) of the weakly electric fish Marcusenius cyprinoides. Journal of Experimental Biology, 74: 133–50.

Toews D., 1975. Limnology of Lake Bangweulu. UN/FAO Report FR: DP/SAM/681511/7/REV:1. FAO, Rome.

Tonn W. M., 1990. Climate change and fish communities: a conceptual framework. Transactions of the American Fisheries Society, 119: 337–52.

Townsend C. R. & Winfield I. J., 1985. The application of foraging theory to feeding behaviour in fish. In Tyler P. & Calow P. (eds.), Fish Energetics New Perspectives, pp. 67–98. Croom Helm, London.

Townsend C. R., 1989. The patch dynamics concept of stream community ecology. Journal of the North American Benthological Society, 8: 36–50.

Travers R.A., 1984. A review of the Mastacembeloidei, a suborder of synbranchiform teleost fishes. Part II: Phylogenetic analysis. Bulletin British Museum (Natural History) Zoology, 47: 83–150.

Travis J. & Mueller L. D., 1989. Blending ecology and genetics: progress toward a unified population biology. In Roughgarden J., May R. M. & Levin S. A. (eds.), Perspectives in Ecological Theory, pp. 101–24. Princeton University Press, Princeton, NJ.

Trebaol L., 1991. Biologie et potentialités aquacoles du Carangidae Trachinotus teraia (Cuvier & Valenciennes, 1832) en milieu lagunaire ivoirien. Collection études et thèses. ORSTOM, Paris.

Trendall J., 1988a. Recruitment of juvenile mbuna (Pisces, Cichlidae) to experimental rock shelters in Lake Malawi, Africa. Environmental Biology of Fishes, 22: 117–31.

Trendall J., 1988b. The distribution and dispersal of introduced fish at Thumbi West Island in Lake Malawi, Africa. Journal of Fish Biology, 33: 357–69.

Trewavas E. 1966. A preliminary review of fishes of the genus Tilapia in the eastward-flowing rivers of Africa, with proposals of two new specific names. Revue de Zoologie et de Botanique africaine, 74: 394–424.

Trewavas E., 1973. On the cichlid fishes of the genus Pelmatochromis with proposal of a new genus for P. congicus; on the relationship between Pelmatochromis and Tilapia and the recognition of Sarotherodon as a distinct genus. Bulletin British Museum (Natural History) Zoology, 25: 1–26.

Trewavas E., 1974. The freshwater fishes of Rivers Mungo and Meme and Lakes Kotto, Mboandong and Soden, west Cameroon. Bulletin British Museum (Natural History) Zoology, 26: 331–419.

Trewavas E., 1982. Generic groupings of Tilapiini used in aquaculture. Aquaculture, 27: 79–81.

Trewavas E., 1983. Tilapiine Fishes of the Genera Sarotherodon, Oreochromis and Danakilia. British Museum (Natural History), London.

Tricart J., 1965. Rapport de la mission de reconnaissance géomorphologique de la vallée moyenne du Niger. Mémoire de l'Institut français d'Afrique noire, Dakar, 72.

Tudorancea C., Fernando C. H. & Paggi J. C., 1988. Food and feeding ecology of Oreochromis niloticus (Linnaeus, 1758) juveniles in Lake Awassa (Ethiopia). Archiv für Hydrobiologie, Supplement 79: 267–89.

Turner B. J., 1964. An introduction to the fishes of the genus Notobranchius. Part I. African Wildlife, 18: 117–24.

Turner B. L. et al., 1990. Two types of global environmental change: definition and spatial-scale issues in their human dimensions. Global Environmental Change: Human and Policy Dimensions, 1: 14–22.

Turner G. F., 1986. Teleost mating systems. In Pitcher T. J. (ed.), The Behaviour of Teleost Fishes, pp. 253–74. Croom Helm, London.

Turner G. F., 1994. Fishing and the conservation of the endemic fishes of Lake Malawi. In Martens K. F., Goddeeris B. & Coulter G. (eds.), Speciation in Ancient Lakes, pp. 481–94. Archiv für Hydrobiologie, 44.

Turner J. L., 1977. Some effects of demersal trawling in Lake Malawi (Lake Nyasa) from 1968 to 1974. Journal of Fish Biology, 10: 262–71.

Turner J. L., 1982. Lake flies, water fleas and sardines. FAO

Fishery Expansion Project, Malawi FI: DP/MLW/75/O19, *Technical Report* I: 165–82.

Turner W. R., 1970. The fish population of newly impounded Kainji Lake. *Kainji Lake Research Project, Nigeria*. FAO: FI/SF/NIR, Technical Report 24.

Tweddle D., 1975. Age and growth of the catfish *Bagrus meridionalis* Günther in southern lake Malawi. *Journal of Fish Biology*, 7: 677–85.

Tweddle D., 1982. Fish breeding migrations in the North Rukuru area of Lake Malawi with a note on gillnet colour selectivity. *Luso: Malawi Journal of Science & Technology*, 3: 67–74.

Tweddle D., 1992. Conservation and threats to the resources of Lake Malawi. *Mitteilungen-Internationale Vereinigung für Theoretische und Angewandte Limnologie*, 23: 17–24.

Tweddle D. & Magasa J. H., 1989. Assessment of multispecies cichlid fisheries of the Southern Arm of Lake Malawi, Africa. *Journal du Conseil international d'Exploration de la Mer*, 45: 209–22.

Tweddle D. & Turner J. L., 1977. Age, growth and natural mortality rates of some cichlids fishes of Lake Malawi. *Journal of Fish Biology*, 10: 385–98.

Udoidiong O. M., 1988. A comparative study of the fish communities of two Nigerian headwater streams in relation to man-made perturbation. *Biological Conservation*, 4: 93–108.

Ulyel A. P., 1991. Ecologie alimentaire des *Haplochromis* spp. (Teleostei; Cichlidae) du lac Kivu en Afrique centrale. Ph.D. thesis, University of Leuven, Belgique.

Ulyel A. P., Ollevier F., Ceusters R. & Thys van den Audenaerde D. F. E., 1990. Régime alimentaire des *Haplochromis* (Teleostei: Cichlidae) du lac Kivu en Afrique. 1 – Relations trophiques interspécifiques. *Belgian Journal of Zoology*, 120: 143–55.

Underwood A. J., 1989. The analysis of stress in natural populations. *Biological Journal of the Linnean Society*, 37: 51–78.

University of Idaho *et al.*, 1971. *Ecology of Fishes in the Kafue River*. Report prepared for FAO/UN acting as executing agency for UNDP. University of Idaho, Moscow, Idaho: FI:SF/ZAM 11. Technical Report 2.

Ursin E., 1979. Principles of growth in fishes. *Symposium of the Zoological Society of London*, 44: 63–87.

Urushido T., Takahashi E. & Taki Y., 1975. Karyotypes of three species of fishes in the order Osteoglossiformes. *Chromosome Information Service*, 18: 20–2.

Urushido T., Takahashi E., Taki Y. & Kondo N., 1977. A karyotype study of polypterid fishes, with notes on their phyletic relationships. *Proceedings of the Japan Academy*, 53: 95–8.

Uyeno T., 1973. A comparative study of chromosomes in the teleostean fish order Osteoglossiformes. *Japanese Journal of Ichthyology*, 20: 211–17.

Van Couvering J. A. H., 1977. Early records of freshwater fishes in Africa. *Copeia*, 1: 163–6.

Van Couvering J. A. H., 1982. *Fossil Cichlid Fish of Africa*. Special papers in Paleontology, No. 29, The Paleontological Association, London.

Van Damme D., 1984. *The Freshwater Mollusca of Northern Africa*. Developments in Hydrobiologia 25. W. Junk Publishers, Dordrecht.

Van den Sande P., 1991. Some considerations with regard to the necessity of starting breeding programs for endangered fishes in collaboration with existing organisations and institutions. *Annales du Musée royal d'Afrique centrale (Sciences Zoologiques)*, 262: 121–6.

Van der Bank F. H., 1993. Allozyme variation in *Synodontis leopardinus* (Pisces, Siluriformes). *Comparative Biochemistry and Physiology*, 105B: 333–6.

Van Neer W., 1987. A study on the variability of the skeleton of *Lates niloticus* (Linnaeus, 1758) in view of the validity of *Lates maliensis* Gayet, 1983. *Cybium*, 11: 411–425.

Van Neer W., 1989. Recent and fossil fish from the Sahara and their palaeohydrological meaning. *Palaeoecology of Africa*, 20: 1–18.

Van Neer W. & Gayet M., 1988. Etude des poissons en provenance des sites holocènes du bassin de Taoudeni, Araouane (Mali). *Bulletin du Muséum national d'Histoire naturelle, Paris*, 10, section C, No. 4: 343–84.

Van Valen L., 1972. Body size and number of plants and animals. *Evolution*, 27: 27–35.

Van Valen L., 1973. A new evolutionary law. *Evolution Theory*, 1: 1–33.

Van Zidderen Bakker E. M., 1982. African paleoenvironments 18 000 yrs BP. *Paleoecology of Africa*, 15: 77–99.

van der Waal B. C. W., 1976. 'n Visekologiese studie van die Liambezimeer in die Oos-Caprivo met verwysing na visontginning deur die Bantoe bevolking. Ph.D. Thesis, Rand University, Johannesburg, South Africa.

van der Waal B. C. W., 1985. Aspects of the biology of larger fish species of Lake Liambezi, Caprivi, South West Africa. *Madoqua*, 14: 101–44.

van der Waal B. C. W. & Skelton P. H., 1984. Check list of fishes of Caprivi. *Madoqua*, 13: 303–20.

Vanden Bossche J. P. & Bernacsek G. M., 1990. *Source Book for the Inland Fishery Resources of Africa*, Vols. 1 and 2. CIFA Technical Paper, No. 18/1 and 18/2. FAO, Rome.

Vanden Bossche J. P. & Bernacsek G. M., 1991. *Source Book*

for the Inland Fishery Resources of Africa Vol. 3. CIFA Technical Paper, No. 18/3. FAO, Rome.

Vannote R. L., Minshall G. W., Cummins K. W., Sedell J. R. & Cushing C. E., 1980. The river continuum concept. *Canadian Journal of Fisheries and Aquatic Sciences*, 37: 130–7.

Vareschi E., 1978. The ecology of Lake Nakuru (Kenya). I. Abundance and feeding of the lesser flamingo. *Oecologia (Berlin)*, 32: 11–35.

Vareschi E., 1979. The ecology of Lake Nakuru (Kenya). II. Biomass and spatial distribution of fish (*Tilapia grahami* Boulenger = *Sarotherodon alcalinum grahami* Boulenger). *Oecologia (Berlin)*, 37: 321–35.

Vareschi E. & Jacobs J., 1984. The ecology of Lake Nakuru (Kenya). V. Production and consumption of consumer organisms. *Oecologia (Berlin)*, 61: 83–98.

Vari R. P., 1979. Anatomy, relationships and classification of the families Citharinidae and Distichodontidae (Pisces, Characoidea). *Bulletin British Museum (Natural History)* Zoology, 36: 261–344.

Verbeke J., 1959. Le régime alimentaire des poissons des lacs Edouard et Albert. Exploration scientifique et hydrobiologique des Lacs Kivu, Edouard et Albert (1952–1954). IRSNB, 3.

Verdeaux F., 1986. Du pouvoir des génies au savoir scientifique: les métamorphoses de la lagune Ebrié (Côte d'Ivoire). *Cahiers d'Etudes africaines*, No. 101–102: 145–71.

Verdeaux F., 1989. Généalogie d'un phénoméne de surexploitation: lagune Aby (Côte d'Ivoire) 1935–1982. *Cahiers de Science humaines*, 25: 191–211.

Verheyen E., 1989. Adaptative radiation and speciation in African lacustrine Cichlidae: genetical and ecophysiological aspects. Ph.D. thesis, Biology Department, Antwerp University, Belgium.

Verheyen E. & Rompaey J. van, 1986. Genetics and speciation in African lacustrine cichlids. *Annales du Musée royal d'Afrique centrale (Sciences Zoologiques)*, 251: 95–101.

Verheyen E., Blust R. & Doumen C., 1985a. The oxygen uptake of *Sarotherodon niloticus* L. and the oxygen binding properties of its blood and hemolysate (Pisces, Cichlidae). *Comparative Biochemistry and Physiology*, 81A: 423–6.

Verheyen E., Rompaey J. van & Selens M., 1985b. Enzyme variations in haplochromine cichlid fishes from Lake Victoria. *Netherlands Journal of Zoology*, 35: 469–78.

Vervoort A., 1979. Karyotype and DNA content of *Phractolaemus ansorgei* Blgr (Teleostei, Gonorynchiformes). *Experientia*, 35: 479–80.

Vervoort A., 1980a. Tetraploidy in *Protopterus* (Dipnoi). *Experientia*, 36: 294–6.

Vervoort A., 1980b. Karyotypes and nuclear DNA contents of Polypteridae (Osteichthyes). *Experientia*, 36: 646–7.

Vervoort A., 1980c. The caryotypes of seven species of *Tilapia* (Teleostei; Cichlidae). *Cytologia*, 45: 651–6.

Vidy G., 1976. Etude du régime alimentaire de quelques poissons insectivores dans les rivières de Côte d'Ivoire. Recherche de l'influence des traitements insecticides effectués dans le cadre de la lutte conre l'onchocercose. Rapport du centre ORSTOM de Bouaké, No. 2.

Vidy G., 1983. Pêche traditionnelle en bordure du Grand Yaéré nord-camerounais: le Logomatia. *Revue d'Hydrobiologie tropicale*, 16: 353–72.

Viner A. B. & Smith I. R., 1973. Geographical, historical and physical aspects of Lake George. *Proceedings of the Royal Society of London, B*, 184: 235–70.

Volckaert F. A. & Agnèse J. F., 1996. Evolutionary and population genetics of Siluroidei, *Aquatic Living Resources*, 9, Supplement 1. (In press)

Voss J., 1977. Les livrées ou patrons de coloration chez les poissons cichlidés africains. Leur utilisation en éthologie et en systématique. *Revue française d'Aquariologie*, 4(2): 33–82.

Voss J., 1980. *Color Patterns of African Cichlids*. T.H.F. Publications, Neptune City, NJ.

Vrijenhoeck R. C., Douglas M. E. & Meffe G. K., 1985. Conservative genetics of endangered fish populations in Arizona. *Science*, 229: 400–12.

Wallace A. R., 1878. *Tropical Nature and Other Essays*. Macmillan, New York and London.

Walter H., Harnickell E. & Mueller-Dombois D., 1975. *Climate-diagram Maps of the Individual Continents and the Ecological Climatic Regions of the Earth*. Springer-Verlag, Berlin.

Wangead C., Geater A. & Tansakul R., 1988. Effects of acid water on survival and growth rate of Nile tilapia (*Oreochromis niloticus*). In Pullin R. S. V., Bhukasswan T., Tonguthai K. & Maclean J. L. (eds.), *The Second International Symposium on Tilapia in Aquaculture*, pp. 43–37. ICLARM Conference Proceedings 15.

Wanink J. H., 1991. Survival in a perturbed environment: the effects of Nile perch introduction on the zooplanktivorous fish community of Lake Victoria. In Ravera O. (ed.), *Terrestrial and Aquatic Ecosystems, Perturbation and Recovery*, pp. 269–75. Ellis Horwood, Chichester and New York.

Wanink J. H. & Goudswaard P. C., 1994. Effects of Nile perch (*Lates niloticus*) introduction in Lake Victoria, East

Africa, on the diet of the pied kingfisher (*Ceryle rudis*). *Hydrobiologia*, **279–280**, 367–73.

Ward J. V. & Stanford J. A., 1989. Riverine ecosystems: the influence of man on catchment dynamics and fish ecology. In Dodge D. P. (ed.), *Proceedings of the International Large River Symposium*, pp. 56–64. Canadian Special Publication in Fisheries and Aquatic Sciences, 106.

Wasson J. G., 1989. Eléments pour une typologie fonctionnelle des eaux courantes: I. Revue critique de quelques approches existantes. *Bulletin d'Ecologie*, **20**: 109–27.

Watanabe W. O. & Kuo C. M., 1985. Observations on the reproductive performance of Nile tilapia (*Oreochromis niloticus*) in laboratory aquaria at various salinities. *Aquaculture*, **49**: 315–23.

Watanabe W. O., Kuo C. M. & Huang M-C., 1985. Salinity tolerance of Nile tilapia fry (*Oreochromis niloticus*) spawned and hatched at various salinities. *Aquaculture*, **48**: 159–76.

WCED, 1987. *Our Common Future*. Oxford University Press, Oxford.

Wcislo W.T., 1989. Behavioral environments and evolutionary change. *Annual Review of Ecology and Systematics*, **20**: 137–69.

Webb P. W., 1982. Locomotor patterns in the evolution of Actinopterygian fishes. *American Zoologist*, **22**: 329–42.

Webb P. W., 1984*a*. Form and function in fish swimming. *Scientific American*, **251**: 58–68.

Webb P. W., 1984*b*. Body form, locomotion and foraging in aquatic vertebrates. *American Zoologist*, **24**: 107–20.

Webb P. W. & de Buffrénil V., 1990. Locomotion in the biology of large aquatic vertebrates. *Transactions of the American Fisheries Society*, **119**: 629–41.

Weiler W., 1926. Mitteilungen über die Wirbeltierreste aus dem Mittelpliozn des Natronales (Ägypten). 7. Selachii und Acanthopterygii. *Sitzungsberichte der Bayerischen Akademie der Wissenchaften*, pp. 317–40.

Welcomme R. L., 1967. Observations on the biology of the introduced species of *Tilapia* in Lake Victoria. *Revue de Zoologie et de Botanique africaine*, **76**: 249–79.

Welcomme R. L., 1972. An evaluation of the acadja method of fishing as practised in the coastal lagoons of Dahomey (West Africa). *Journal of Fish Biology*, **4**: 39–55.

Welcomme R. L., 1979. *Fisheries Ecology of Floodplain Rivers*. Longman, London.

Welcomme R. L., 1985. *River Fisheries*. FAO Fisheries Technical Paper No. 262. FAO, Rome.

Welcomme R. L., 1986. The effects of the Sahelian drought on the fishery of the central delta of the Niger River. *Aquaculture and Fisheries Management*, **17**: 147–54.

Welcomme R. L., 1988. *International Introductions of Inland Aquatic Species*. FAO Fisheries Technical Paper No. 294, FAO, Rome.

Welcomme R. L., 1989. Review of the present state of knowledge of fish stocks and fisheries of African rivers. In Dodge D. P. (ed.), *Proceedings of the International Large River Symposium*, pp. 515–32. Canadian Special Publication of Fisheries and Aquatic Science, 106.

Welcomme R. L., 1992. The conservation and environmental management of fisheries in inland and coastal waters. *Netherlands Journal of Zoology*, **42**: 176–89.

Welcomme R. L. & Hagborg D., 1977. Towards a model of a floodplain fish population and its fishery. *Environmental Biology of Fishes*, **2**: 7–22.

Welcomme R. L. & Merona B de, 1988. Fish communities of rivers. In Lévêque C., Bruton M. N. & Ssentongo G. W. (eds), *Biology and Ecology of African Freshwater Fishes*, pp. 251–76. ORSTOM, Paris.

Welcomme R. L., Ryder R. A. & Sedell J. A., 1989. Dynamics of fish assemblages in river systems – a synthesis. In Dodge D. P. (ed.), *Proceedings of the International Large River Symposium*, pp. 569–77. Canadian Special Publication of Fisheries and Aquatic Sciences, 106.

Werner E. E., 1986. Species interactions in freshwater fish communities. In Diamond J. & Case T. (eds.), *Community Ecology*, pp. 344–58. Harper & Row, New York.

Werner E. E. & Gilliam J. F., 1984. The ontogenic niche and species interaction in size-structured populations. *Annual Review of Ecology and Systematics*, **15**: 393–425.

West-Eberhard M. J., 1983. Sexual selection, social competition, and evolution. *Quarterly Review of Biology*, **58**: 155–83.

West-Eberhard M. J., 1989. Phenotypic plasticity and the origins of diversity. *Annual Review of Ecology and Systematics*, **20**: 249–78.

Wetherald R. T., 1991. Changes of temperature and hydrology caused by an increase of atmospheric carbon dioxide as predicted by general circulation models. In Wyman R. (ed.), *Global Climate Change and Life on Earth*, pp. 1–17. Chapman & Hall, London.

Whigam D., Dykyjova D. & Hejny S., 1993. *Wetlands of the World: Inventory, Ecology and Management*, Vol. 1. Kluwer Academic Publishers, Dordrecht.

White E., 1983. *The Vegetation of Africa*. A descriptive memoir to accompany the UNESCO/AETFAT/UNSO vegetation map of Africa. UNESCO, Paris.

White G. F., 1988. The environmental effects of the high dam at Aswan. *Environment*, 30: 5–40.

White P. S. & Pickett S. T. A., 1985. Natural disturbance and patch dynamics: an introduction. In Pickett S. T. A. & White P. S. (eds.), *The Ecology of Natural Disturbance and Patch Dynamics*, pp. 3–13. Academic Press, San Diego, CA.

Whitehead P. J. P., 1959. Notes on a collection of fishes from the Tana River below Garissa Kenya. *Journal of the East Africa Natural History Society*, 23: 167–71.

Whitfield A. K. & Blaber S. J. M., 1979. The distribution of the freshwater cichlid *Sarotherodon mossambicus* in estuarine systems. *Environmental Biology of Fishes*, 4: 77–81.

Whittaker, R. H., 1970. *Communities and Ecosystems*. Macmillan, London.

Wickbom T., 1945. Cytological Studies on Dipnoi, Urodela, Anura and *Emys*. *Hereditas*, 31: 241.

Wickler W., 1962. "Egg-dummies" as natural releasers in mouth-breeding cichlids. *Nature*, 194: 1092–4.

Wiens J. A., 1984. On understanding a non-equilibrium world: myth and reality in community patterns and processes. In Strong *et al.* (eds.), *Ecological Communities: Conceptual Issues and the Evidence*, pp. 439–57. Princeton University Press, Princeton, NJ.

Wiens J. A., Addicott J. F., Case T. J. & Diamond J., 1986. Overview: the importance of spatial and temporal scale in ecological investigations. In Diamond J. & Case T. J. (eds.), *Community Ecology*, pp. 145–53. Harper & Row, New York.

Wiley E. O., 1981. *Phylogenetics: The Theory and Practice of Phylogenetics Systematics*. Wiley Interscience, New York.

Wiley E. O., 1988. Vicariance biogeography. *Annual Review of Ecology and Systematics*, 19: 513–42.

Wilhelm W., 1980. The disputed feeding behaviour of a paedophagous haplochromine cichlid (Pisces) observed and discussed. *Behaviour*, 74: 310–23.

Williams C. B., 1964. *Patterns in the Balance of Nature and Related Problems in Quantitative Ecology*. Academic Press, New York.

Williams G. C., 1966. Natural selection, the costs of reproduction, and a refinement of the Lack's principle. *American Naturalist*, 100: 687–90.

Williams M. A. J., Clark J. D., Adamson D. A. & Gillespie R., 1975. Recent Quaternary research in central Sudan. *Bulletin de l'Association française pour l'étude du Quaternaire, Paris*, 46: 75–86.

Williams, M. A. J., Adamson D. A., Williams F. M., Morton W. H. & Parry D. E., 1980. Jebel Marra volcano: a link between the Nile valley, the Sahara and Central Africa. In Williams F. M. & Faure H. (eds.), *The Sahara and the Nile*, pp. 305–37. Balkema, Rotterdam.

Williams R., 1971. Fish ecology of the Kafue River and floodplain environment. *Fisheries Research Bulletin of Zambia*, 5: 305–30.

Williamson M., 1988. Relationship of species number to area, distance and other variables. In Myers A. A. & Giller P. S. (eds.), *Analytical Biogeography: An Integrated Approach to the Study of Animal and Plant Distributions*, pp. 91–115. Chapman & Hall, London.

Willoughby N. G., 1974. The ecology of the genus *Synodontis* (Pisces, Siluroidei) in Lake Kainji, Nigeria. Ph.D. thesis, University of Southampton, Southampton.

Wilson E. O. & Peters F. M. (eds.), 1988. *Biodiversity*. National Academy of Sciences Press, Washington, DC.

Wimpenny R. S., 1934. The fisheries of Egypt. *Scientific Progress in the Twentieth Century, London*, 29: 210–27.

Winemiller K. O., 1991. Comparative ecology of *Serranochromis* species (Teleostei, Cichlidae) in the upper Zambezi River floodplain. *Journal of Fish Biology*, 39: 617–39.

Winemiller K. O., 1992. Life history strategies and the effectiveness of sexual selection. *Oikos*, 62: 318–27.

Winemiller K. O. & Kelso-Winemiller L. C., 1994. Comparative ecology of the African pike, *Hepsetus odoe*, and tiger fish, *Hydrocynus forskhalii*, in the Zambezi River floodplain. *Journal of Fish Biology*, 45: 211–25.

Winter J. D., 1983. Underwater biotelemetry. In Nielsen L. A. & Johnson D. L. (eds.), *Fisheries Techniques*, pp. 371–95. American Fisheries Society, Bethesda, MD.

Witte F., 1981. Initial results of the ecological survey of the haplochromine cichlid fishes from the Mwanza Gulf of Lake Victoria (Tanzania): breeding patterns, trophic and species distribution, with recommendations for commercial trawl-fisher. *Netherlands Journal of Zoology*, 31: 175–202.

Witte F., 1984a. Consistency and functional significance of morphological differences between wild-caught and domestic *Haplochromis squamipinnis* (Pisces, Cichlidae). *Netherlands Journal of Zoology*, 34: 596–612.

Witte F., 1984b. Ecological differentiation in Lake Victoria haplochromines: comparison of cichlid species flocks in African lakes. In Echelle A. A. & Kornfield I. L. (eds.), *Evolution of Fish Species Flocks*, pp. 155–67. Oklahoma State University Press, Stillwater.

Witte F. & Goudswaard P. C., 1985. Aspects of haplochromine fishery in southern Lake Victoria. In CIFA report of the 3rd session of the subcommittee for the development and the management of the fisheries of Lake Victoria, pp. 81–8. *FAO Fisheries Report*, 335.

Witte F. & van Oijen M. J. P., 1990. Taxonomy, ecology and fishery of Lake Victoria haplochromine trophic groups. *Zoologische Verhandelingen, Leiden,* **262**.

Witte F., Barel C. D. N. & Hoogerhoud J. C., 1990. Phenotypic plasticity of anatomical structures and its ecomorphological significance. *Netherlands Journal of Zoology,* **40**: 278–98.

Witte F., Goldschmidt T., Goudswaard P. C., Ligtvoet W., van Oijen M. P. J. & Wanink J. H., 1992*a*. Species extinction and concomitant ecological changes in Lake Victoria. *Netherlands Journal of Zoology,* **42**: 214–32.

Witte F., Goldschmidt T., Wanink J., van Oijen M., Goudswaard K., Witte-Maas E. & Bouton N., 1992*b*. The destruction of an endemic species flock: quantitative data on the decline of the haplochromine cichlids of Lake Victoria. *Environmental Biology of Fishes,* **34**: 1–28.

Witte-Maas E. L. M., 1981. Egg snatching: an observation on the feeding behaviour of *Haplochromis barbarae* Greenwood, 1967 (Pisces, Cichlidae). *Netherlands Journal of Zoology,* **31**: 786–9.

Wohlfarth G. W. & Wedekind H., 1991. The heredity of sex determination in tilapias. *Aquaculture,* **92**: 143–56.

Wolf N. G., 1985. Air breathing and risk of aquatic predation in the dwarf gourami *Colisa lalia. American Zoologist,* **25**: 89A (abstract).

Wootton R. J., 1984. Introduction: strategies and tactics in fish reproduction. In Potts G. W. & Wootton R. J. (eds.), *Fish Reproduction: Strategies and Tactics,* pp. 1–12. Academic Press, London.

Wootton R. J., 1990. *Ecology of Teleost Fishes.* Fish and Fisheries Series 1. Chapman & Hall, London.

Worthington E. B., 1932. Scientific results of the Cambridge expedition to the east African lakes, 1930–1932. 2. Fishes other than Cichlidae. *Journal of the Linnean Society, Zoology,* **38**: 121–34.

Worthington E.B., 1937. On the evolution of fish in the Great Lakes of Africa. *Internationale Revue der gesamten Hydrobiologie,* **35**: 304–17.

Worthington E.B., 1940. Geographical differentiation in freshwaters with especial reference to fish. In J. Huxley (ed.), *The New Systematics.* Oxford, Clarenton Press.

Worthington E. B. & Ricardo C. K., 1936. Scientific results of the Cambridge expedition to east African lakes, 1930–31. 15 – The fish of Lake Rudolf and Lake Baringo. *Journal of the Linnean Society, Zoology,* **39**: 353–89.

Worthington S. & Worthington E. B., 1933. *Inland Waters of Africa. The results of two expeditions to the great lakes of Kenya and Uganda, with accounts of their biology, native tribes and development.* Macmillan, London.

Wourms J. P., 1972. The developmental biology of annual fishes. III- Pre-embryonic and embryonic diapause of variable duration in the eggs of annual fishes. *Journal of Experimental Zoology,* **182**: 389–414.

Wright S. J., 1983. Species-energy theory: an extension of species-area theory. *Oikos,* **41**: 496–506.

Wyman R. (ed.), 1991. *Global Climate Change and Life on Earth.* Chapman & Hall, London.

Yamaoka K., 1983. Feeding behaviour and dental morphology of algae scraping cichlids (Pisces, Teleostei) in Lake Tanganyika. *African Study Monographs,* **4**: 77–89.

Yamaoka K., 1991. Feeding relationships. In Keenleyside M. H. A. (ed.), *Cichlid Fishes. Behaviour, Ecology and Evolution,* pp. 150–72. Chapman & Hall, London.

Yamaoka K., Hori M. & Kuratani S., 1986. Eco-morphology of feeding in 'goby-like' cichlid fish in Lake Tanganyika. *Physiology and Ecology, Japan,* **23**: 17–29.

Yaméogo L., 1994. Impact sur les entomocénoses aquatiques des insecticides utilisés pour contrôler les stades larvaires de *Simulium damnosum* Theobald (Diptera: Simuliidae), vecteur de l'onchocercose humaine en Afrique de l'Ouest. *Travaux et Documents Microfichés,* No. 123. ORSTROM, Paris.

Yaméogo L., Lévêque C., Traore K. & Fairhurst C., 1988. Dix ans de surveillance de la faune aquatique des rivières d'Afrique de l'Ouest. *Naturaliste Canadien,* **115** (3/4): 287–98.

Yaméogo L., Tapsoba J. M., & Calamari D., 1991. Laboratory toxicity of potential blackfly larvicides on some African fish species in the Onchocerciasis Control Programme area. *Ecotoxicology and Environmental Safety,* **21**: 248–56.

Yan H. Y., 1987. Size at maturity in male *Gambusia heterochir. Journal of Fish Biology,* **30**: 731–41.

Yanagisawa Y., 1985. Parental care strategy of the cichlid fish *Peridossus microlepis,* with particular reference to intraspecific brood "farming out". *Environmental Biology of Fishes,* **12**: 241–9.

Yanagisawa Y., 1986. Parental care in a monogamous mouthbrooding cichlid *Xenotilapia flavipinnis* in Lake Tanganyika. *Japanese Journal of Ichthyology,* **33**: 249–61.

Yanagisawa Y., 1987. Social organization of a polygynous cichlid *Lamprologus furcifer* in Lake Tanganyika. *Japanese Journal of Ichthyology,* **34**: 82–90.

Yanagisawa Y. & Nshombo M., 1983. Reproduction and parental care of the scale-eating cichlid fish *Peridossus microlepis* in Lake Tanganyika. *Physiology and Ecology, Japan,* **20**: 23–31.

Yodzis P., 1986. Competition, mortality, and community structure. In Diamond J. & Case T. J. (eds.), *Community Ecology,* pp. 480–91. Harper & Row, New York.

Yuma M., 1993. Competitive and cooperative interactions in Tanganyikan fish communities. In Kawanabe H., Cohen J. & Iwasaki K. (eds.), *Mutualism and Community Organisation: Behavioural, Theoretical and Food Web Approaches*, pp. 213–27. Oxford University Press, Oxford.

Zahner E., 1977. A karyotype analysis of fifteen species of the family Cichlidae. Ph.D. dissertation, St John's University, New York.

Zalewski M. & Naiman R. J., 1985. The regulation of fish communities by a continuum of abiotic–biotic factors. In Alabaster J. S. (ed.), *Habitat Modification and Freshwater Fisheries*, pp. 3–9. Butterworths, London.

Zan R., Song Z. & Liu W., 1986. Studies on karyotypes and nuclear DNA contents of some cyprinoid fishes, with notes on fish polyploids in China. In Uyeno T. *et al.* (eds.), *Indo-Pacific Fish Biology*, pp. 877–85. Proceedings of the Second International Conference on Indo-Pacific Fishes. Ichthyological Society of Japan, Tokyo.

Zhang T., Rawson D. M. & Morris G. J., 1993. Cryopreservation of pre-hatch embryos of zebrafish (*Brachydanio rerio*). *Aquatic Living Resources*, 6: 145–53.

Zhu Z., Liu G., He L. & Chen S., 1985. Novel gene transfer into fertilized eggs of goldfish (*Carassius auratus* L., 1785). *Zeitschrift für Angewandte Ichthyologie*, 1: 31–4.

Zhu Z., Xu K., Li G., Xie Y. & He L., 1986. Biological effects of human growth hormone gene microinjected into the fertilized eggs of loach, *Misgurnus anguillicaudatus*. *Kexue Tongbao Academia Sinica*, 31: 988–90.

Species Index

Subject Index